THE CHEMISTRY OF ENZYME ACTION

New Comprehensive Biochemistry

Volume 6

General Editors

A. NEUBERGER
London

L.L.M. van DEENEN
Utrecht

ELSEVIER
AMSTERDAM · NEW YORK · OXFORD

The Chemistry of Enzyme Action

Editor

Michael I. PAGE

Department of Chemical Sciences, The Polytechnic, Huddersfield (Great Britain)

1984

ELSEVIER
AMSTERDAM · NEW YORK · OXFORD

ISBN for the series: 0444 80303 3
ISBN for the volume: 0444 80504 4

Published by:
Elsevier Science Publishers B.V.
P.O. Box 1527
1000 BM Amsterdam, The Netherlands

Sole distributors for the U.S.A. and Canada
Elsevier Science Publishing Company Inc.
52 Vanderbilt Avenue
New York, NY 10017, U.S.A.

Printed in The Netherlands

Preface

Recognition is of fundamental importance to living systems. How *do* proteins and other macromolecules distinguish between molecules of similar shape or ions of similar size? Recognition is controlled by the intermolecular forces between the 'host' and 'guest'. The binding energy resulting from the mutual satisfying of these forces are ultimately responsible for the catalysis and specificity of enzyme-catalysed reactions. Understanding how enzymes efficiently transform their substrates is not only a question of reaction mechanisms, describing the routes of bond making and breaking processes, but also one of recognising that the interactions between the 'non-reacting' parts of the substrate and enzyme play a crucial role in the activation step. The forces responsible for the chemical mechanism adopted by the enzyme are closely related to those which account for recognition of the 'non-reacting' parts. The interplay of these forces is fundamental to an appreciation of enzymic catalysis.

This volume describes the physical and organic basis of enzyme action. The background knowledge required to understand the chemistry of enzyme action is presented by major scientists in their own field. The borderline area between disciplines are stimulating and rewarding and this is reflected by the high calibre of the contributors to this volume. The level of understanding enzyme-catalysed reactions is dependent upon the techniques employed. Determining reaction mechanisms requires a detailed knowledge of kinetic techniques and discussions of these topics are followed by examples of their applications. The level of understanding enzyme catalysis that has been reached by using physical organic methods is illustrated by some biologically important examples. Finally the important contribution that biomimetic studies have made to understanding the recognition and catalysis exhibited by enzymes is emphasised by leading exponents in the field.

Michael I. Page
Huddersfield, October 1983

Contents

Preface v

Chapter 1
The energetics and specificity of enzyme – substrate interactions
Michael I. Page (Huddersfield) 1

1. Introduction 1
2. Enzyme structure 2
3. Michaelis – Menten kinetics 6
4. Intra- and extra-cellular enzymes 7
5. Regulation and thermodynamics 9
6. Specificity and k_{cat}/K_m 11
7. Rate enhancement and specificity 12
8. Specificity, induced fit and non-productive binding 14
9. Approximation, entropy and intramolecular reactions 16
10. Decreasing the activation energy 22
11. Utilisation of binding energy 27
12. Intramolecular force fields 34
Bond stretching, 35 – Bond angle bending, 35 – Torsion, 35 – Disulphide links, 36 – Non-bonded interaction, 37 –
13. Intermolecular force fields 38
Hydrogen bonding, 39 – Electrostatic interactions, 40 – Hydrophobicity, 44 – Dispersion forces, 45 –
14. Stress and strain 47
15. Estimation of binding energies 50

References 53

Chapter 2
Non-covalent forces of importance in biochemistry
Peter Kollman (San Francisco) 55

1. Introduction 55
   1.1. The basis of the non-covalent forces 55
2. The thermodynamics of non-covalent interactions 57
   2.1. Gas phase interactions 57
   2.2. Solution phase association 59
3. Examples of biologically important non-covalent interactions 62
   3.1. Electrostatic forces 62
   3.2. Dispersion forces 65
   3.3. Hydrophobic interactions 66
4. Summary 69

Acknowledgement 70
References 70

Chapter 3
Enzyme kinetics
Paul C. Engel (Sheffield) 73

1. Introduction: aims and approaches in enzyme kinetics 73
2. Steady-state kinetics 76
2.1. Michaelis–Menten equation 76
2.2. Relationship between K_m and dissociation constant 78
2.3. Experimental determination of kinetic constants 78
(a) Experimental design, 78 – (b) Lineweaver–Burk plot, 79 – (c) Eadie–Hofstee plot, 79 – (d) Hanes plot, 79 – (e) Eisenthal and Cornish–Bowden plot, 82 – (f) Does the choice of plotting method matter? 82 –
2.4. Non-linearity 82
2.5. Inhibition 86
(a) Definition, 86 – (b) Competitive inhibition, 86 – (c) Uncompetitive inhibition, 87 – (d) Non-competitive inhibition, 89 – (e) Mixed inhibition, 90 –
2.6. Multi-substrate kinetics 91
(a) Types of mechanism, 91 – (b) Overall strategy, 93 – (c) Deriving a rate equation, 93 – (d) Experimental determination of the rate equation for an individual enzyme, 97 – (e) Drawing conclusions from the experimentally determined rate equation, 99 – (f) Inhibition experiments, 104 – (g) Isotope exchange at equilibrium, 106 – (h) Whole time-course studies, 106 –
3. Rapid reaction kinetics 107
References 109

Chapter 4
Aspects of kinetic techniques in enzymology
Kenneth T. Douglas and Michael T. Wilson (Colchester) 111

1. Introduction 111
2. Use of steady-state techniques 113
2.1. Note on measurement of initial velocities 114
3. Experimental treatment of transients 115
3.1. Determination of k_{obs} 116
4. Stopped-flow methods 119
4.1. Binding reactions 119
4.2. Burst kinetics 121
5. Relaxation methods (temperature jump) 123
5.1. Ligand binding 123
5.2. Coupled reactions, linked redox reactions and structural rearrangements 124
6. Conclusion 125
References 125

Chapter 5
Free-energy correlations and reaction mechanisms
Andrew Williams (Canterbury) 127

1. Introduction 127
2. Brønsted relationships 128

2.1. Simple proton transfer 128
2.2 Molecular basis of the Brønsted relationship and interpretation of the exponents 129
2.3. Statistical treatments 130
2.4. The extended Brønsted relationship 132
2.5. Meaning of the Brønsted exponents 133
(a) Effective charges in transition states, 134 – (b) Additivity of 'effective' charge, 135 – (c) Transition state index (α), 136 –
2.6. Curvature in Brønsted correlations 137
(a) Eigen curvature, 137 – (b) Marcus curvature, 139 –
2.7. Anomalies 141
3. Indices of electronic and steric effects 142
3.1. The Hammett equation 143
(a) Additivity of sigma values, 147 – (b) Resonance and inductive effects, 148 – (c) Sigma-minus parameters, 148 – (d) Sigma-plus parameters, 149 – (e) More than one transmission path for σ, 151 – (f) The Yukawa–Tsuno equation, 155 –
3.2. Separation of inductive, steric and resonance effects 156
(a) Taft's polar (σ *) and steric (E_s) parameters, 158 – (b) Taft's steric parameter (E_s), 161 – (c) Relationship between σ * and σ_I, 162 – (d) Meaning and use of E_s and δ, 163 – (e) Other steric parameters, 164 – (f) Values of σ * for alkyl groups, 166 –
4. Hydrophobic interactions 166
4.1. Other hydrophobic parameters 168
4.2. Non-linear hydrophobic relationships 170
4.3. Molar refractivity 170
4.4. Additivity 171
4.5. Ambiguities arising from interrelationships between parameters 172
4.6. Application to non-biochemical reactions 172
5. Solvent effects 173
5.1. Reporter groups 175
6. General equations of reactivity 177
6.1. Swain–Scott and Edwards relationship 177
7. Cross-correlation and selectivity–reactivity 179
7.1. Cross-correlation 179
7.2. Reactivity–selectivity 183
8. Estimation of ionisation constants 183
9. Elucidation of mechanism 186
9.1. Mechanistic identity 186
(a) Correlations with model reactions, 187 –
9.2. Changes in mechanism 189
9.3. Change in rate-limiting step 191
9.4. Dependece on concentration as a free energy correlation 192
9.5. Distinction between kinetic ambiguities 195
9.6. Proton transfer 195
References 197
General references 200

Chapter 6
Isotopes in the diagnosis of mechanism
Andrew Williams (Canterbury) 203

1. Theoretical background 203
2. Measurement of isotope effects 205

3. Equilibrium isotope effects 207
4. Primary isotope effects 209
4.1. Variation in primary isotope effects 211
4.2. Solvent isotope effects 213
(a) Nucleophilic versus general base catalysis, 213 – (b) Fractionation factors, 214 –
4.3. Heavy atom isotope effects 218
5. Secondary isotope effects 219
6. Labelling techniques 220
6.1. Position of bond cleavage 220
6.2. Proton transfer 222
6.3. Detection of intermediates by isotope exchange 222
6.4. Isoracemisation 223
6.5. Double-labelling experiments 225
6.6. Isotopic enrichment 226
References 226
General references 227

Chapter 7
The mechanisms of chemical catalysis used by enzymes
Michael I. Page (Huddersfield) 229

1. Introduction 229
2. General acid base catalysis 230
2.1. Catalysis by stepwise proton transfer (trapping) 231
2.3. Catalysis by preassociation 235
2.3. Catalysis by hydrogen bonding 236
2.4. Concerted catalysis 236
2.5. Intramolecular general acid–base catalysis 237
3. Covalent catalysis 240
4. Metal-ion catalysis 243
5. Catalysis by coenzymes 246
5.1. Pyridoxal phosphate coenzyme 246
5.2. Thiamine pyrophosphate coenzyme 249
5.3. Adenosine triphosphate (ATP) 252
5.4. Coenzyme A 253
6. Oxidation and reduction 256
6.1. Hydride transfer 256
6.2. Nicotinamide coenzymes 257
6.3. Flavin coenzymes 259
(a) Oxidation of amino and hydroxy acids and carboxylic acids, 260 – (b) Oxidation of thiols, 261 – (c) Reductive activation of oxygen by dihydroflavins, 261 – (d) Flavomonooxygenase, 262 –
6.4. Electron transfer with metals 262
(a) Thermodynamic stability of metal complexes, 263 – (b) Kinetic effects, 263 – (c) Redox metalloproteins and oxygen, 264 – (d) Iron-containing proteins and enzymes, 265 – (e) Copper-containing oxidases and monooxygenases, 267 –
References 269

Chapter 8
Enzyme reactions involving imine formation
Donald J. Hupe (Rahway) 271

1. Introduction 271
2. Iminium ion formation 272
3. Activation of carbonyl groups by iminium ion formation 276
4. Aldolases 279
5. Transaldolase 286
6. Acetoacetate decarboxylase 288
7. Pyruvate-containing enzymes 291
8. Dehydratases 294
9. Conclusions 298
Acknowledgement 298
References 299

Chapter 9
Pyridoxal phosphate-dependent enzymic reactions: mechanism and stereochemistry
Muhammad Akhtar, Vincent C. Emery and John A. Robinson (Southampton) 303

1. Historic background: Braunstein–Snell hypothesis 303
2. Pyridoxal phosphate-dependent reactions involving C_α–CO_2H bond cleavage 306
3. Pyridoxal phosphate-dependent enzymic reactions involving C_α–H bond cleavage 314
   3.1. Aminotransferases 314
   (a) Metabolic background, 314 – (b) Transaminations at the C_α of amino acid, 315 – (c) Mechanistic studies on miscellaneous transaminations, 315 –
   3.2. Racemases 319
   3.3. Serine hydroxymethyltransferase (SHMT) 320
   3.4. 5-Aminolevulinate synthetase (ALA synthetase) 327
4. Pyridoxal phosphate-dependent reactions involving modifications at C_β 331
   4.1. General introduction 331
   4.2. β-Replacement reactions 331
   (a) Tryptophan synthetase, 331 –
   4.3. β-Elimination–deamination reactions 336
   (a) Tryptophan synthetase–β_2 protein, 337 – (b) Tryptophanase, 338 – (c) Miscellaneous enzymes, 339 –
5. Pyridoxal phosphate-dependent reactions occurring at C_γ 343
   5.1. Enzymic aspects 343
   5.2. Stereochemical aspects 344
6. Structure and molecular dynamics within the binary and ternary complexes 349
   6.1. Electronic spectrum of the coenzyme chromophore 349
   6.2. Chemical studies on binary (coenzyme–enzyme) complexes 353
   6.3. Stereochemical aspects of the reduction of Schiff base at C-4′ with $NaBH_4$ 354
   6.4. Structure and stereochemistry of the substrate–coenzyme bond in ternary complexes 355
   6.5. Stereochemical and mechanistic events at C_α of the substrates and at C-4′ of the coenzyme during catalysis 359
   6.6. Orientation of the pyridinium ring of the coenzyme in the binary and ternary complexes 366
References 367

Chapter 10
Transformations involving folate and biopterin cofactors
S.J. Benkovic and R.A. Lazarus (University Park) 373

1. Introduction 373
2. Structure 373
3. Reduction 375
4. Methylene transfer 376
5. Formyl transfer 379
6. Methyl transfer 381
7. Hydroxylation 381
References 385

Chapter 11
Glycosyl transfer – The Physicochemical Background
Michael L. Sinnott (Bristol) 389

1. Introduction 389
2. Effects of the direction of the lone pairs of oxygen in space 390
   2.1. The anomeric effect 391
   2.2. Geometrical changes in oxocarbonium ion formation 395
   2.3. Stereoelectronic control of reactions of acetals? 396
3. The chemistry of processes occurring with electrophiles or acids 398
   3.1. Lifetimes of oxocarbonium ions 398
   3.2. Preassociation mechanisms 399
   3.3. Chemical synthesis of glycosides 402
   3.4. Effect of oxocarbonium ion structure 403
   3.5. Intramolecular nucleophilic assistance 406
   3.6. Electrostatic stabilisation? 407
4. Processes occurring via the application of acidic or electrophilic assistance to the departure of oxygen-leaving groups 408
   4.1. Specific acid catalysis of acetals, metals and glycosides 408
   4.2. Intramolecular nucleophilic assistance in specific acid-catalysed processes 413
   4.3. Intermolecular general acid catalysis of the hydrolysis of acetals and ketals 413
   4.4. Intramolecular general acid catalysis of the hydrolysis of acetals, ketals and glycosides . . . 417
   4.5. Intramolecular general acid catalysis concerted with intramolecular nucleophilic (or electrostatic) assistance 419
   4.6. Electrophilic catalysis 420
5. Acid- and electrophile-assisted departure of nitrogen, sulphur and fluorine from oxocarbonium ion centres 421
   5.1. Hydrolysis of glycosylamines 422
   5.2. Hydrolysis of nucleosides and deoxy nucleosides 423
   5.3. Hydrolysis of hemithioacetals, hemithioketals, and thio- and thia-glycosides 425
   5.4. Hydrolysis of glycosyl fluorides 426
6. Envoi 427
References 427

Chapter 12
Vitamin B_{12}
Kenneth L. Brown (Arlington) 433

1. Introduction and scope of this chapter 433
2. Structure 433
3. Oxidation states 435
4. Enzymology of the B_{12} coenzymes 437
5. Chemical reactivity of organocobalt complexes 438
5.1. Reactions in which carbon–cobalt bonds are formed 439
(a) Synthesis of organocobalt complexes via cobalt(I) reagents, 439 – (b) Synthesis of organocobalt complexes via cobalt(II) reagents, 441 – (c) Synthesis of organocobalt complexes via cobalt(III) reagents, 443 –
5.2. Reactions in which carbon–cobalt bonds are cleaved 444
(a) Mode I cleavage of carbon–cobalt bonds, 445 – (b) Mode II cleavages of carbon–cobalt bonds, 447 – (c) Mode III cleavages of carbon–cobalt bonds, 450 –
5.3. Axial ligand substitution reactions 452
5.4. Reactions of cobalt-bound organic ligands 455
6. Concluding remarks 457
References 457

Chapter 13
Reactions in micelles and similar self-organized aggregates
Clifford A. Bunton (Santa Barbara) 461

1. Introduction 461
2. Formation of normal micelles 462
3. Micellar structure in water 464
4. Kinetic and thermodynamic effects 468
4.1. Micellar effects upon reaction rates and equilibria 468
4.2. Quantitative treatments of micellar effects in aqueous solution 471
4.3. Quantitative treatment of bimolecular reactions 472
4.4. Second-order rate constants in the micellar pseudophase 475
5. Reactive counterion micelles 479
6. Reactions in functional micelles 482
7. Stereochemical recognition 487
8. Submicellar and non-micellar aggregates 487
9. Micelles in non-aqueous systems 490
9.1. Normal micelles in non-aqueous media 490
9.2. Reverse micelles in aprotic solvents 491
10 Related systems 493
10.1. Reactions in microemulsions 493
10.2. Reactions in vesicles 495
11. Photochemical reactions 496
12. Isotopic enrichment 497
13. Preparative and practical aspects 498
Note added in proof 499
Acknowledgements 500
References 500

Chapter 14
Cyclodextrins as enzyme models
Makoto Komiyama and Myron L. Bender (Tokyo and Evanston) 505

1. Introduction 505
2. Formation of an inclusion complex 506
3. Catalysis by cyclodextrins 511
   3.1. Catalysis by the hydroxyl groups 511
   3.2. Effect of reaction field 513
4. Specificity in cyclodextrin catalysis 514
   4.1. Substrate specificity 514
   4.2. Product specificity 517
   4.3. D,L Selectivity 519
5. Catalysis by modified cyclodextrins 520
   5.1. Models of hydrolytic enzymes 520
   5.2. Model of carbonic anhydrase 524
   5.3. Model of metalloenzymes 525
   5.4. Introduction of a coenzyme moiety 525
6. Conclusion 526

Acknowledgements 526
References 526

Chapter 15
Crown ethers as enzyme models
J. Fraser Stoddart (Sheffield) 529

1. Introduction 529
2. Ground-state binding and recognition 530
   2.1. Binding forces 530
   2.2. Complexation 538
   2.3. Enantiomeric differentiation 540
   2.4. Substrate recognition 542
   2.5. Allosteric effects 544
3. Binding and recognition at the transition state 546
   3.1. Enzyme mimics: hydrogen-transfer reactions 546
   3.2. Enzyme mimics: acyl transfer reactions 550
   3.3. Enzyme analogues: Michael addition reactions 556
4. Conclusion 558

References 559

Subject Index 563

The energetics and specificity of enzyme–substrate interactions

MICHAEL I. PAGE

Department of Chemical and Physical Sciences, The Polytechnic, Queensgate, Huddersfield HD1 3HD, Great Britain

1. Introduction

Life is a dynamic process which depends upon the recognition (interaction) between inanimate molecules. Living organisms have the capacity to utilize external energy and matter to maintain and propagate themselves. The interactions between substrate and enzyme that account for the catalysis and specificity of the reactions of intermediary metabolism also account for the energy-coupling processes which enable the exchange of chemical energy and allow the system to do work.

Because of their relative masses the electron density distribution controls the movement of nuclei. Electronic interactions and distortions of electron density distributions are responsible not only for the formation of structures and complexes but also for chemical reactions themselves. It is thus logical to examine the forces that hold enzymes in their unique 3-dimensional structure, then the energetics of enzyme–substrate interactions and finally the forces or mechanism by which the bond-making and -breaking processes occur during the reaction catalysed by the enzyme.

Enzymes are usually globular proteins and are distinguished from fibrous proteins by the ability of the pieces of secondary structure to associate and give a stable 3-dimensional structure. The forces responsible for protein folding are similar to those used in the formation of antibody–antigen or hormone–receptor complexes and to those giving rise to enzymic catalysis. The problem with protein folding is how does the system gain enough energy from hydrophobicity, hydrogen bonds and all the various electrostatic interactions to overcome the loss of conformational entropy and steric strain that occurs in the folded state? The problem with complex formation is how does the system gain enough energy to overcome the loss of translational, rotational and vibrational entropy that occurs upon complexation? The problem with enzymic catalysis is how are these interactions expressed in the transition state but not in the ground state of the enzyme–substrate complex?

It is necessary to understand the structures of folded proteins so that we are aware of the environment or force field generated by the enzyme which in turn controls the interaction between substrate and enzyme. This chapter will briefly

Michael I. Page (Ed.), The Chemistry of Enzyme Action

review the problems associated with understanding the energetics of enzymic catalysis, the strength and geometrical requirements of the various forces available to molecular systems (which is dealt with in detail in Chapter 2) and how these forces stabilise transition states.

2. *Enzyme structure*

Proteins, with a specific function and isolated from a single source, usually have a homogeneous population of molecules all with the same unique amino acid sequence. Yet with 20 different amino acids possible at each position in a polypeptide chain of n residues, 20^n different primary structures are theoretically possible. Furthermore, the great majority of all molecules of a natural protein may exist in a unique conformation despite the degrees of freedom formally permitted by rotation about the peptide backbone (motility) and side chains (mobility). For example, with only 3 conformations defined per residue, a polypeptide chain of 210 residues would have a theoretical possibility of existing in 10^{100} different conformations.

Reversible unfolding of proteins has been known for some time. The folding of several proteins occurs spontaneously showing that the required information for folding is present in the protein's primary structure [1].

The forces stabilising the folded state are presumably similar to those that bind the substrate to enzymes. The conventional stress with protein structures is on (i) the primary structure of the peptide sequence, i.e. the linear covalent linkage, (ii) the secondary structure of the hydrogen bonds between the peptide links, (iii) the tertiary structure formed by hydrophobic interaction which is largely responsible for many protein folds, together with charge–charge interactions and disulphide linkages.

The production of a disordered polypeptide chain by removing just a few of the interactions which normally contribute to the stability of the folded state is well illustrated by breaking the 4 disulphide bonds in ribonuclease A – even in the absence of a denaturant, the reduced protein is fully unfolded despite all other favourable interactions, such as hydrogen bonds and hydrophobic forces, still being possible [2].

Detailed models of the folded states of proteins depend almost entirely on X-ray diffraction analysis of the protein crystal. Although side chains and flexible loops on the surface may be mobile in solution, protein conformation in solution is essentially that determined in the solid crystal. The atoms of folded proteins are generally well fixed in space.

The overall structure of a folded protein is remarkably compact. The extended chain of carboxypeptidase, with 307 residues, would be 1114 Å long but the *maximum* dimension of the folded molecule is only about 50 Å. However, the polypeptide topology in roughly spherical, globular proteins has never been observed to form knots [3].

The non-polar side chains tend to be on the inside, shielded from the solvent, and generate globular structures. The ionised hydrophilic side chains are nearly always on the outside and help maintain solubility. Many polar groups are buried in the hydrophobic interior but these tend to be "neutralised" by hydrogen bonding [3]. However, internal hydrogen bonds probably provide no substantial net stabilisation energy to the folded state because similar hydrogen bonds would be formed with water in the unfolded state. Hydrogen bonds are probably important in limiting conformational fluctuations in the folded protein. The major difference between folded and unfolded molecules is probably the decreased exposure of the non-polar groups to water.

Upon folding, there is a large decrease, amounting to several thousand $Å^2$, in the surface area of the protein exposed to water. Even choosing the transfer of amino acids from water to a non-polar solvent as an estimation of these changes (which, as we shall see later, underestimates the change) yields hundreds of kJ/mole. The free energy of folding amounts to only 15–60 kJ/mole which shows the large unfavourable entropic contribution which accompanies folding as a result of the loss of internal rotations.

The polar external surface of a protein which is in contact with water is usually much more mobile than the interior. Generally, the higher the proportion of charged amino acids in a protein the greater is its flexibility. Most enzymes have a high proportion of hydrophobic amino acids relative to charged ones. Together with disulphide bridges this makes proteins have low motility and mobility. An exception is the kinases which are not globular, have a high content of charged amino acids, are not cross-linked, and are relatively flexible [6].

Charge–charge interactions between two opposite charges in a buried hydrophobic region of low dielectric constant could be important in strengthening the fold.

Proteins found in membranes can experience very different solvation environments. In the membrane the environment may be similar to a hydrocarbon solvent while at its surface the medium slowly changes to an aqueous region. Proteins or parts of proteins located in the non-polar region of the membrane are expected to have very few exposed charged groups.

The atoms *within* a protein molecule are very closely packed with few holes – about 75% of the volume of the interior is filled with atoms [4]. However, the packing density does appear to vary somewhat which may be important for the flexibility of the molecule. Any holes present within the molecule tend to be occupied by solvent molecules as in chymotrypsin and carboxypeptidase. In micelles, by contrast, the packing density is low with the volume occupied by the surfactant atoms being large. Enzymes are not like micelles and the "oil drop" model of protein structure is incorrect. The low packing density of micelles accounts for the much lower efficiency of micellar catalysis (Chapter 13) compared with enzymic catalysis; the surface area of contact between micelle and substrate is much less than that between enzyme and substrate. The close packing of protein interiors prevents water molecules being trapped in non-polar cavities and maximises the packing energy.

Most of the larger protein structures are composed of 2 or 3 globular units, each of 40–150 residues. However, disulphide bonds always link cysteine residues within the same domain. Each domain is closely packed with flexibility between the domains.

Proteins from different species are often functionally indistinguishable; even though their primary sequences may be different their folded conformations are often very similar. Even more surprising is the observation that functionally unrelated proteins sometimes have similar folded conformations, e.g. superoxide dismutase and immunoglobulin domain [5]. Structural homology between proteins of different amino acid sequence may be a packing phenomenon.

Although the high degree of time-averaged order of the individual atoms of folded proteins in the crystalline state permits their location in space many protein molecules exhibit varying degrees of flexibility. The motions have very different activation energies ranging from 5 to over 100 kJ/mole, but their importance, if any, is often unknown. For example, in most proteins aromatic rings of tyrosine and phenylalanine residues flip by 180° rotations at a rate greater than 10^4/s. In general, experimental techniques such as fluorescence quenching and relaxation, phosphorescence and NMR indicate a rather fluid, dynamic structure for globular proteins [6].

It is of interest to note that the smaller the substrate, apparently, the greater is the requirement for a rigid enzyme. Cytochrome *c* transfers electrons and is very immobile, the solution structure is almost identical to that found in the solid state [6]. The flavodoxins, transferring hydride ion; catalase, of molecular weight 284 000 with hydrogen peroxide as substrate and carbonic anhydrase, of molecular weight 31 000 with carbon dioxide as substrate show a slight increase in mobility. Lysozyme, of molecular weight 14 000 with a polysaccharide as substrate has the same outline fold in solution and in the solid state but many aromatic rings and aliphatic groups are mobile. Kinases, molecular weight 45 000, transfer phosphate from ATP and show even greater mobility but like all enzymes still have well defined structures.

NMR relaxation parameters, by virtue of their sensitivity to both the frequencies and the amplitudes of motions, are potentially the best source of experimental information on macromolecular dynamics, largely for motions in the frequency range 10^6–10^{12}/sec. However, it is difficult to find a unique model and often several models, differing largely in the formulation of amplitude factors, account equally well for the experimental data. Physically non-existent motions may erroneously be assumed to exist or motions which actually do contribute to relaxation may be suppressed in the analysis [7]. Succinctly, some techniques indicate a rigid inflexible system whereas others suggest that enzymes are fluid and flexible. The apparent dichotomy may be attributable to the application of concepts derived from macroscopic observations (density, elasticity, heat capacity, etc.) of an assembly of a large number of molecules or atoms to *one* molecule of an enzyme. There is no clear-cut or sharp boundary between statistical mechanics and thermodynamics. A macroscopic system undergoes incessant and rapid transitions among its microstates. Extensive parameters, such as energy and volume, have *average* values equal to the sum of the values in each of the subsystems and undergo macroscopic fluctuations. Statistical distribution functions give the probability that any specified value of a fluctuating extensive parameter will be realised at a given moment. The distribution function for macroscopic systems is so sharply peaked that average values and most probable values are nearly identical. The most probable values are those that maximise the distribution function. The average value of the deviation from the average parameter is clearly zero, but the average square of the deviation is non-zero and is called the mean square deviation, a convenient measure of the magnitude of the fluctuations, although it is only a partial specification of the distribution. The distribution becomes increasingly sharp as the size of the system increases [8].

Familiar macroscopic systems consist of many discrete particles, e.g. 6×10^{23} molecules per mole, and their thermodynamic properties are very sharp. In contrast the individual molecules of proteins are very small systems consisting of relatively few particles, a few thousand atoms, and statistical fluctuations will cause the thermodynamic parameters to be blurred (Fig. 1).

A typical globular protein may have a molecular weight of 25 000 and each molecule a mass of 4.2×10^{-20} g and a volume of 3.2×10^{-20} cm^3. The root mean square volume fluctuation is about 0.2% per molecule. A system with similar gross thermal properties but one hundredth of the size of the protein molecule, i.e. 3.2×10^{-22} cm^3 would show a volume fluctuation of about 2%, whereas a system of volume 3.2 cm^3 would show a volume fluctuation of only 2×10^{-11}%.

Although globular proteins appear to be only marginally stable in the folded state this does not imply conformational flexibility. Indeed environmental changes do not easily induce changes in the folded state. Crystallisation under different conditions does *not* produce large changes in conformation and even the binding of ligands tends only to alter the relative orientations of sub-units or domains. In no instance has a substantial change of conformation of a globular protein been observed even

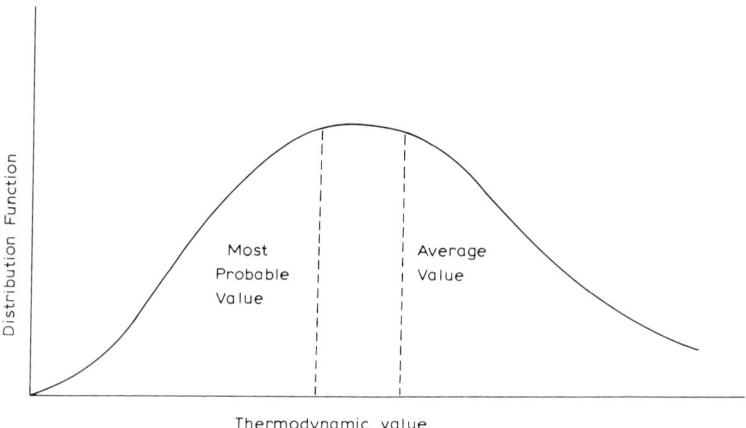

Fig. 1. The distribution function for thermodynamic values showing the difference between the most probable and average value of the thermodynamic function.

when the binding energies of the ligands are comparable to the net stability of the folded conformation [9]. Substantial energy barriers must therefore separate alternative conformations of globular proteins. Of course, proteins have considerable thermal energy and consequently their structures, like all molecules, fluctuate but the atoms within the molecule do not appear to deviate far from their positions determined crystallographically. Of course this is not true of all proteins, surfaces and structural units, for example, may show high mobility and motility [6].

3. Michaelis–Menten kinetics

Experimentally, the initial rate, v, of enzyme catalysed reactions is found to show saturation kinetics with respect to the concentration of the substrate, S. At low concentrations of substrate the initial rate increases with increasing concentration of S but becomes independent of [S] at high or saturating concentrations of S (Fig. 2). This observation was interpreted by Michaelis and Menten in terms of the rapid and reversible formation of a non-covalent complex (ES), from the substrate (S) and enzyme (E), which then decomposes into products (P) (Eqn. 1).

$$\mathrm{E} + \mathrm{S} \underset{K_\mathrm{S}}{\rightleftharpoons} \mathrm{ES} \xrightarrow{k_\mathrm{cat}} \mathrm{EP} \rightarrow \mathrm{E} + \mathrm{P} \quad (1)$$

This scheme led to the familiar Michaelis–Menten equation.

$$v = \frac{k_\mathrm{cat}[\mathrm{E}_0][\mathrm{S}]}{K_\mathrm{m} + [\mathrm{S}]} = \frac{V_\mathrm{max}[\mathrm{S}]}{K_\mathrm{m} + [\mathrm{S}]} \quad (2)$$

where K_m is the Michaelis constant which is the concentration of substrate at which the initial rate is *half* the maximal rate at saturation, V_max. The first-order rate constant for the decomposition of ES is k_cat (the turnover number). At low concentration of substrate, where $[\mathrm{S}] \ll K_\mathrm{m}$, v is given by Eqn. 3, with $k_\mathrm{cat}/K_\mathrm{m}$ being an apparent second-order rate constant.

$$v = \frac{k_\mathrm{cat}}{K_\mathrm{m}}[\mathrm{E}_0][\mathrm{S}] \quad (3)$$

At high concentrations of S, where $[\mathrm{S}] \gg K_\mathrm{m}$, v is given by Eqn. 4 and becomes independent of S.

$$v = k_\mathrm{cat}[\mathrm{E}_0] = V_\mathrm{max} \quad (4)$$

Although the Michaelis–Menten equation (Eqn. 2) is valid for several enzyme-catalysed reactions the mechanism (Eqn. 1) is not always followed. The measured K_m and k_cat values are not always equal to the dissociation constant, K_s, for the

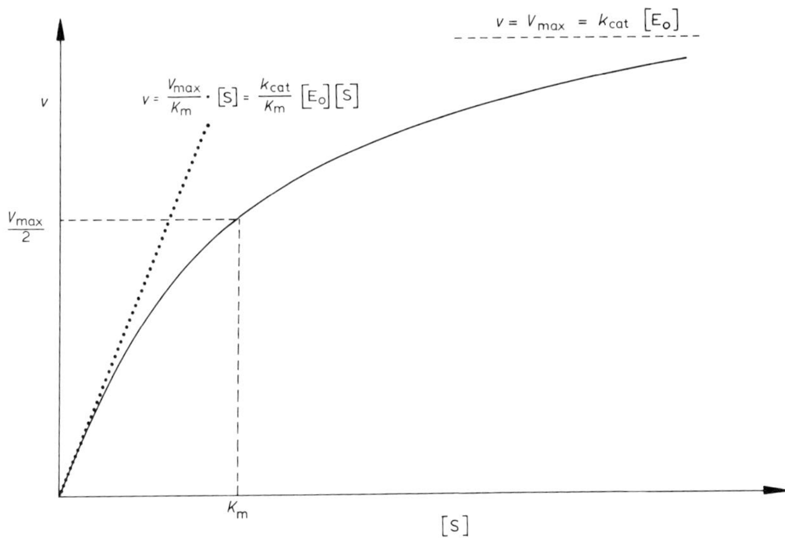

Fig. 2. Reaction rate–substrate concentration profile for a reaction obeying Michaelis–Menten (or saturation) kinetics.

enzyme–substrate complex and the rate constant for decomposition of ES, respectively. The apparent dissociation constant K_m can be less than K_s, i.e. apparent tighter binding of substrate, if additional intermediates, covalently or non-covalently bound, are formed during the reaction pathway and the rate-limiting step is the reaction of one of these intermediates. Similarly, $K_m > K_s$ if the rate of dissociation of ES to E and S is comparable or slower than the forward rate of reaction of ES (Briggs–Haldane kinetics). The measured value of k_{cat} may also be a function of various and several microscopic rate constants (see Chapter 3).

When the activation energies for the catalytic steps have been sufficiently lowered, the binding of the substrate or the desorption of the product may become at least partially rate-limiting.

4. Intra- and extra-cellular enzymes

Extracellular enzymes show little motility and are not very sensitive to the ionic strength of the solutions in which they are dissolved. They are synthesised in one environment but are then placed in another outside the cell. In particular, salt concentrations vary between the two environments; for example, calcium concentration outside the cell is 10^4 times greater than in the cell. Furthermore many extracellular enzymes are produced as zymogens and are activated by removing a section of the protein. Insensitivity to the environment is an essential requirement of these enzymes if activity is to be maintained in extracellular fluids of uncertain composition. Extracellular enzymes are usually cross-linked several times by disulphide bridges. The more oxidising environments generally encountered outside cells make the disulphide bonds correspondingly more stable. Stability depends upon the thiol redox potential of the environment but, of course, the disulphide bond could be formed inside the cell.

Intracellular enzymes exist in a more controlled medium and as the requirement for rigidity is less very few intracellular enzymes have disulphide cross-links. Some enzymes, particularly those involved in electron transfer, have a very fixed fold. Intracellular enzymes, in distinction to most extracellular ones, have a quaternary structure.

Individual enzymes in vivo have different constraints and requirements. Generally, intracellular enzymes are required to maintain a constant concentration of the various metabolites and this may be achieved by having a wide variation in the reaction flux of the material through the various metabolic pathways. The reaction rate will vary with [S] if the enzyme is working below saturation, $K_m \gg [S]$, and the rate is given by Eqn. 3. However, extracellular enzymes are often faced with dramatic changes in the concentration of their substrates and yet are required to maintain a steady flow of material for absorption and use by the cell. The reaction rate will be independent of [S] if the enzyme is working under saturation conditions, $[S] \gg K_m$, and the rate is given by Eqn. 4.

Given the obedience to Michaelis–Menten kinetics one may obtain information

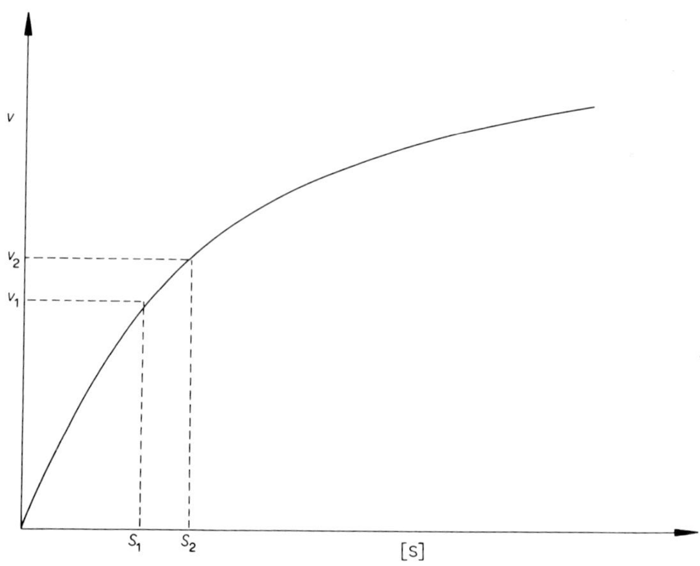

Fig. 3. A reaction rate showing saturation kinetics has different rates, v_1, and v_2, at different substrate concentrations, S_1 and S_2, respectively.

about the optimal physiological concentration for the substrate [10]. Obviously, the maximal change in rate for a change in substrate concentration is when S and v are minimal (dv/dS is maximal at $S = 0$, $v = 0$). If there is an optimal physiological concentration for the substrate, enzymes could have evolved to bind the substrate more or less tightly to maximise the changes in rate with respect to changes in substrate concentration. The problem is illustrated in Fig. 3, with two substrate concentrations S_1 and S_2 with respective rates v_1 and v_2. The minimal fractional change in substrate concentration (S_1/S_2) to obtain a given change in velocity x, occurs at the K_m for the reaction.

From Eqn. 2 the substrate concentration is given by

$$[S_1] = \frac{v_1 K_m}{V_{max} - v_1} \tag{5}$$

If $v_2 = v_1 + x$, the concentration of S_2 is given by

$$[S_2] = \frac{(v_1 + x) K_m}{V_{max} - (v_1 + x)} \tag{6}$$

The derivative of S_2/S_1 is zero when $v_1 + v_2 = V_{max}$ (Eqns. 7–9), i.e. at the K_m of the reaction.

$$\frac{[S_2]}{[S_1]} = \frac{V_{max} v_1 + V_{max} x - x v_1 - v_1^2}{V_{max} v_1 - v_1^2 - x v_1} \tag{7}$$

$$\frac{d[S_2]/[S_1]}{dv_1} = \frac{V_{max} x (x + 2v_1 - V_{max})}{(V_{max} v_1 - v_1^2 - x v_1)^2} = 0 \tag{8}$$

$$V_{max} = x + 2v_1 = v_1 + v_2 \tag{9}$$

The optimal physiological concentration of the substrate is at its K_m [10,11].

5. *Regulation and thermodynamics*

Biochemical reactions exist to bring about the net formation of a compound which may be required either of itself or as the starting material of a further process. A metabolic pathway could involve the conversion of a large, or constantly maintained, concentration of substrate S into an equally well maintained pool of product P (Eqn. 10).

$$S \underset{}{\overset{w \curvearrowleft \curvearrowright x}{\rightleftharpoons}} B \rightleftharpoons C \underset{}{\overset{y \curvearrowleft \curvearrowright z}{\rightleftharpoons}} D \rightleftharpoons P \tag{10}$$

The sequence may involve the formation of several intermediates, B, C, D and require the conversion of coenzymes w and y into x and z, respectively. The chemical flux out of S and P may be controlled by: (i) changing the concentrations of S or P, (ii) changing the concentrations of the coenzymes, (iii) changing the activities of the enzymes involved in the pathway.

Although the thermodynamics of the step $S \rightarrow B$ may be unfavourable, net synthesis can occur from left to right if the *overall* equilibrium constant is favourable. Thermodynamically, Eqn. 10 may be described by Eqn. 11 where ΔG is the Gibbs free energy change for the reaction under the given conditions and ΔG^0 is the change in Gibbs energy for the reaction with reactants and products at the standard state concentrations.

$$\Delta G = \Delta G^0 + RT \ln \frac{(P)(x)(z)}{(S)(w)(y)} \tag{11}$$

At thermodynamic equilibrium there can be no net flow. Enzymes do not alter the position of equilibrium between *unbound* substrates and products. If B, C or D are removed rapidly their concentration will not correspond to their equilibrium value but rather to their steady state concentrations. The total chemical flux *out of* C is given by the sum of the fluxes out of C into B and D. The total chemical flux *into* C is given by the sum of the flux of B into C and that of D into C. The rate of appearance of C is then given by the difference of these two sums (Eqn. 12).

$$\frac{d[C]}{dt} = \Sigma\phi_c - \Sigma\phi_{-c} \tag{12}$$

The flux through a sequence of reactions cannot generally be controlled by suppressing the activity of the enzyme catalysing *any* one of the reactions. The positions of the reaction with respect to its equilibrium value and with respect to the degree of enzyme saturation are important. If any of the steps in Eqn. 10 are near equilibrium the rate of the reverse reaction will be similar to that of the forward reaction and this step cannot, therefore, limit the rate of production of P. It is often suggested that control of metabolism through alteration of enzyme activity should be exerted at reactions which are far from thermodynamic equilibrium. If the substrate concentration is not in excess and if the dissociation constants for substrate and product, K_s and K_p, respectively, are very different it is conceivable that the overall equilibrium constant K (Eqn. 14) is not a good guide to the value of the equilibrium constant, K_E (Eqn. 13).

$$E + S \underset{K_s}{\rightleftharpoons} ES \overset{K_E}{\rightleftharpoons} EP \overset{K_p}{\rightleftharpoons} E + P \tag{13}$$

$$K = \frac{K_p \cdot K_E}{K_s} \tag{14}$$

In this case, it is possible for an enzyme to apparently "alter" the position of equilibrium between substrate and product.

6. *Specificity and k_{cat}/K_m*

Obedience to Michaelis–Menten kinetics yields interesting conclusions about the specificity of enzyme-catalysed reactions. In vitro "non-specific" substrates are sometimes described as "poor" because they show a low value of k_{cat} or a high value of K_{m}. However, in vivo specificity results from a *competition* of substrates for the active site of the enzyme. If two substrates, S and S′, compete for the same enzyme different conclusions could be reached about their relative specificity if rates of reaction or the K_{m}s of the individual substrates are compared instead of their relative values of $k_{\mathrm{cat}}/K_{\mathrm{m}}$. If the enzyme catalyses the reaction of both S and S′ (Eqn. 15) the relevant equations may be obtained by the usual procedures (Eqns. 16–18).

$$\begin{array}{ccccc} \mathrm{E}+\mathrm{S}+\mathrm{S'} & \underset{K_m}{\rightleftharpoons} & \mathrm{ES} & \xrightarrow{k_{cat}} & \mathrm{P} \\ K'_{\mathrm{m}} \Updownarrow & & & & \\ \mathrm{ES'} & \xrightarrow{k'_{cat}} & \mathrm{P'} & & \end{array} \tag{15}$$

$$(\mathrm{ES}) = \frac{(\mathrm{E}_0)(\mathrm{S})K'_{\mathrm{m}}}{K_{\mathrm{m}}K'_{\mathrm{m}} + K_{\mathrm{m}}(\mathrm{S'}) + K'_{\mathrm{m}}(\mathrm{S})} \tag{16}$$

$$(\mathrm{ES'}) = \frac{(\mathrm{E}_0)(\mathrm{S'})K_{\mathrm{m}}}{K_{\mathrm{m}}K'_{\mathrm{m}} + K_{\mathrm{m}}(\mathrm{S'}) + K'_{\mathrm{m}}(\mathrm{S})} \tag{17}$$

$$\frac{\mathrm{d(P)}}{\mathrm{d(P')}} = \frac{k_{\mathrm{cat}}(\mathrm{S})[K'_{\mathrm{m}} + (\mathrm{S'}) + K'_{\mathrm{m}}(\mathrm{S})/K_{\mathrm{m}}]}{k'_{\mathrm{cat}}(\mathrm{S'})[K_{\mathrm{m}} + (\mathrm{S}) + K_{\mathrm{m}}(\mathrm{S'})/K'_{\mathrm{m}}]} \tag{18}$$

The relative rate of the two reactions is given by Eqn. 19 whether the enzyme is working below or above saturation for both substrates. Furthermore Eqn. 18 reduces to Eqn. 19 even if the individual K_{m}s or substrate concentrations are such that the enzyme is working below saturation for one substrate but above saturation for the other.

$$\frac{v}{v'} = \frac{k_{\mathrm{cat}}/K_{\mathrm{m}}(\mathrm{S})}{k'_{\mathrm{cat}}/K'_{\mathrm{m}}(\mathrm{S'})} \tag{19}$$

Specificity between competing substrates is therefore given simply by relative values of k_{cat}/K_m and not by the individual values of k_{cat} or K_m [11]. Specificity can apparently be reflected in poor binding (high K_s) and/or slow catalytic steps (low k_{cat}) but specificity between competing substrates is controlled only by their relative values of k_{cat}/k_m.

According to transition-state theory the second-order rate constant k_{cat}/K_m is directly related to the free-energy difference (ΔG^{+}) between the enzyme–substrate transition state (ES^{+}) and the *free* unbound substrate and enzyme (Eqn. 20) by Eqn. 21.

$$E + S \overset{\Delta G^{+}}{\rightleftharpoons} ES^{+} \tag{20}$$

$$k_{cat}/K_m = \frac{kT}{h} \exp(-\Delta G^{+}/RT) \tag{21}$$

It follows, therefore, that the maximum relative rate of the enzyme-catalysed reactions of two substrates S_1 and S_2, of the *same chemical reactivity,* is given by the difference in the free energy of binding the *non-reacting* parts of the substrates to the enzyme in the *transition state* minus the difference between their free energies in the ground state. The latter difference may be insignificant in some cases but in others, e.g. replacing a hydrophobic by a hydrophilic group, it may make the major contribution. Ground state differences could be reflected in the solubilities of the two substrates.

7. Rate enhancement and specificity – the two striking phenomena associated with enzyme-catalysed reactions are the rate enhancement and specificity

Enormous rate enhancements can be achieved by enzymes compared with non-enzyme-catalysed reactions, but the quantitative evaluation of the rate difference is not always straightforward. For example, hydrolytic enzymes often exhibit rate enhancements of 10^8–10^{12} compared with the spontaneous water-catalysed – or the acid- or base-catalysed – reaction at around neutral pH. However, the *mechanism* by which the enzyme-catalysed reaction occurs is different from these relatively simple non-enzyme-catalysed reactions. Enzymes generally utilize their functional groups to act as nucleophilic, electrophilic, general acid or base catalysts (Chapter 5). It has been tempting, therefore, to speculate that it is the "*chemical catalysis*" or "*mechanism*" which is responsible for the large rate enhancement brought about by enzymes. There are thus two aspects of enzymic catalysis which should be distinguished: (i) the rate enhancement brought about by "chemical catalysis" relative to the "uncatalysed" or "solvent-catalysed" reaction; (ii) the rate enhancement broughht about by the reaction occurring within the enzyme–substrate complex compared with the *same* chemical reaction in the absence of enzyme. There appears to be nothing unusual in the pathways by which covalent bonds are formed and broken in enzyme-catalysed reactions; the mechanisms used to account for ordinary chemical reactions are also applicable to nature's reactions. Chemical catalysis alone cannot explain the rate enhancement brought about by enzymes. The forces of interaction between the *non-reacting* parts of the substrate and enzyme are used to lower the activation energy of the reaction. The enzyme succinyl-CoA-acetoacetate transferase catalyses reaction 22, ($R_1CO_2^-$ = acetoacetate and $R_2CO_2^-$ = succinate) and proceeds by the initial formation of an enzyme–CoA intermediate in which the coenzyme A is bound to the enzyme as a thiol ester of the γ-carboxyl group of glutamate (Eqn. 23).

$$R_1CO_2^- + R_2COS\text{-}CoA \rightleftharpoons R_1COS\text{-}CoA + R_2CO_2^- \quad (22)$$

$$R_2COS\text{-}CoA + Enz\text{-}CO_2^- \rightleftharpoons R_2CO_2^- + Enz\text{-}COS\text{-}CoA \quad (23)$$

In turn this intermediate is generated by nucleophilic attack of the glutamate carboxylate on succinyl-CoA to give an anhydride intermediate (I). The second-order rate constant for this reaction I is 3×10^{13}-fold greater than the analogous reaction of acetate with succinyl-CoA (II) [12]. It seems unlikely that the chemical reactivity of acetate and the enzyme's glutamate will be vastly different. Similar chemical reactions are therefore being compared and yet the *non-reacting* part of the enzyme lowers the activation energy by 78 kJ/mole ($RT \ln 3 \times 10^{13}$).

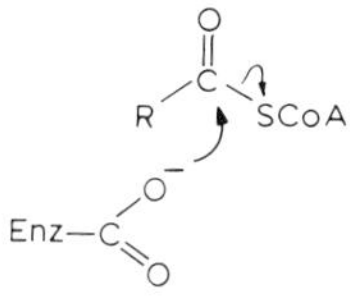

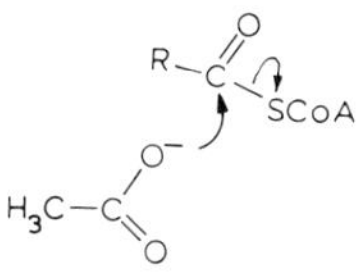

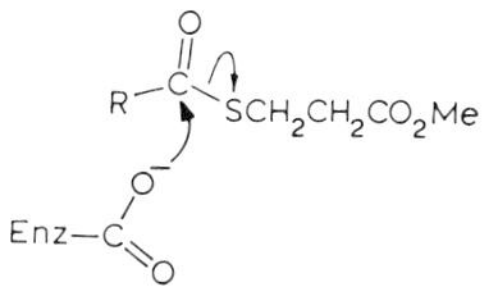

The same example may be used to illustrate specificity. The enzyme also forms an anhydride from the "non-specific" thiol ester, succinyl methyl mercaptopropionate (III). However, the enzyme reacts with succinyl-CoA (I) 3×10^{12}-fold faster than with III. The chemical reactivities of the two substrates are similar, for example, towards alkaline hydrolysis [12]. The small substrate, succinyl methyl mercaptopropionate, should be able to fit into the active site. Therefore the *non-reacting* part of succinyl-CoA of molecular weight ca. 770 lowers the activation energy by 72 kJ/mole ($RT \ln 3 \times 10^{12}$).

It is relatively easy to rationalise how enzymes discriminate against substrates which are larger than the specific substrate — the substrate is simply too large to fit into the active site. It is more difficult to explain why some small non-specific substrates that can bind to the active site react very slowly. In the example cited above, it is not too difficult to imagine that the non-covalent interactions between the *non-reacting* part of succinyl-CoA and the enzyme can give rise to binding energies on the order of 70–80 kJ/mole which in turn can be utilised to increase the reaction rate. However, it is not so easy to understand how, for example, an enzyme which has iso-leucine (IV) as a specific substrate can discriminate against valine (V).

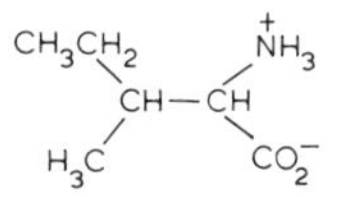

$$(CH_3CH_2)(H_3C)CH{-}CH(\overset{+}{N}H_3)CO_2^- \quad \text{IV} \qquad\qquad (H_3C)(H_3C)CH{-}CH(\overset{+}{N}H_3)CO_2^- \quad \text{V}$$

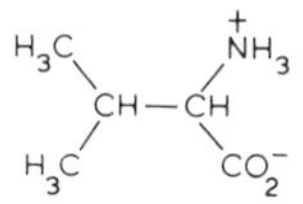

It is of interest to note that the efficient enzymes, catalase, carbonic anhydrase and the nitrogenases have very small substrates (H_2O_2, CO_2 and N_2, respectively) for which it is not easy to distinguish between "reacting" and "non-reacting" parts. It is a great challenge to quantitatively understand the forces of interaction between these substrates and their enzymes.

8. Specificity, induced fit and non-productive binding

According to the Induced Fit model of specificity the active site of the free enzyme (E) is in the "wrong" conformation and is catalytically inactive. The binding of a "good" substrate induces a conformational change in the enzyme making it catalytically active (E′) (Eqn. 24).

$$\begin{array}{ccccc} E & \overset{K}{\rightleftharpoons} & E' & & \\ \updownarrow & & \updownarrow & & \\ ES & \rightleftharpoons & E'S & \xrightarrow{k_{cat}} & ES^{\ddagger} \end{array} \tag{24}$$

A "poor" substrate does not have enough favourable binding energy to compensate for the unfavourable conformational change in the enzyme. Whereas some of the binding energy between the "good" substrate and the enzyme is used to "pay for" the conversion to the unfavourable, but *active* conformation of the enzyme (E′). The free energy of binding good substrates is greater than the free energy of distortion of the enzyme. Induced fit can be defined to explain specificity between very good and very poor substrates but not between substrates that all have sufficient binding energy to compensate for the unfavourable conformational change. It is an energetically expensive, but sometimes necessary, mechanism of specificity. It mediates against catalysis in the sense that the enzyme is not as catalytically effective as it would be if the enzyme existed in the catalytically active conformation in the free state [11,12].

From Eqn. 24, where $K = (E')/(E)$, it can be seen that the observed k_{cat}/K_m is reduced by a factor K, equivalent to the free energy required to distort the enzyme, compared with the situation where the enzyme is initially present in the active conformation.

$$(k_{cat}/K_m)_{obs} = K\,k_{cat}/K_m \tag{25}$$

Provided that the substrates being compared have sufficient binding energy to compensate for the unfavourable conformational change of the enzyme, then induced fit cannot explain specificity between competing substrates. Compared with the situation where the enzyme is initially in the active conformation, the induced fit mechanism reduces the value of k_{cat}/K_m for *all* substrates by the *same* fraction and therefore does not affect their relative rates.

If the conformational change occurs upon initial binding to form the Michaelis complex then the observed binding constant, K_m, will be increased by a factor $1/K$ but k_{cat} will be the same as it would be if all the enzyme were in the active conformation.

Induced fit can therefore only be used to explain the rates of reactions of very poor substrates compared with very good ones, e.g. the rate of phosphoryl transfer to water compared with that to glucose catalysed by hexokinase. Although water can almost certainly bind to the active site it must have insufficient binding energy to induce the necessary conformational change in the enzyme.

Non-productive binding is sometimes suggested to account for the low reactivity of "poor" substrates by suggesting that they bind at sites on the enzyme where the catalytic reaction cannot take place.

$$\begin{array}{ccccc} E + S & \underset{K_s}{\rightleftharpoons} & ES & \xrightarrow{k_{cat}} & ES^{\ddagger} \\ K'_S \Big\updownarrow & & & & \\ ES' & & & & \end{array} \tag{26}$$

At saturation, only a fraction of the substrate is productively bound and k_{cat}(ES) is lowered. The observed binding constant, K_m, is lower than K_s because the additional binding site leads to apparently tighter binding. The in vitro conclusion could therefore be that the binding energy between the "poor", but smaller, substrate and the "wrong" sites of the enzyme are used to prevent rapid reaction at the active site; the larger specific substrate can only bind at the active site and is sterically prevented from binding to the "wrong" site. Although this is an appealing idea it cannot explain the discrimination between competing substrates in vivo. The free-energy difference between the unbound substrate and enzyme (E + S) and the transition state (ES^{+}) is unaffected by alternative modes of binding, e.g. to ES′ (Eqn. 26), i.e. the free energy of activation, represented by k_{cat}/K_m, is not changed by non-productive binding.

9. Approximation, entropy and intramolecular reactions

It has long been thought that enzymes are effective by bringing reactants together at the active site of the enzyme which is referred to as approximation or proximity. However, it was not until 10 years ago that this effect was fully understood [14]. The importance of this effect may be illustrated by considering a reaction between 2 molecules A and B to give a transition state AB^{+} and comparing this with the same reaction occurring at the surface of an enzyme but without involving chemical catalysis by the enzyme (Eqn. 27) i.e. the enzyme is used to bring the reactants together but none of the functional groups on the enzyme are used to facilitate the reaction (see also p. 32).

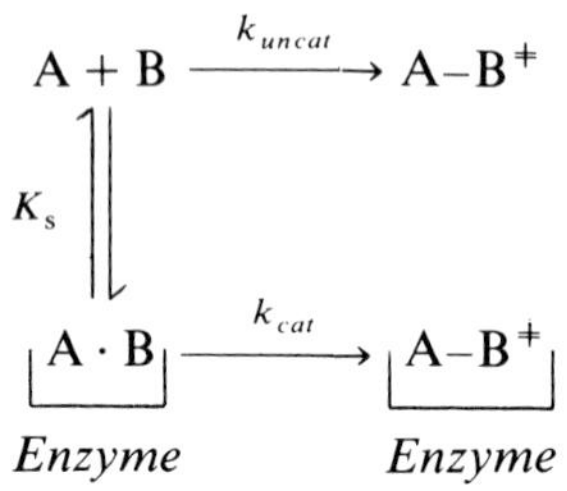
A + B
k uncat
A–B+
Ks
A · B
k cat
A–B+
Enzyme
Enzyme

(27)

Bringing 2 molecules together to form a transition state in a bimolecular reaction is generally accompanied by a large loss of entropy as the transition state is a more ordered system. The formation of the transition state from the independently moving and rotating reactants is accompanied by a large loss of translational and rotational entropy.

Historically, there was a delay in appreciating the magnitude of the entropic contribution to intramolecular and enzyme-catalysed reactions because of the wide variation observed in the rate enhancements of intramolecular reactions. It is well known that intramolecular reactions (Eqn. 28) in which the reactants are covalently bonded to one another, often proceed at very much faster rates than those of the analogous intermolecular reactions (Eqn. 29). These intramolecular reactions are frequently taken as models for enzyme-catalysed reactions where the reactants are held close together in the enzyme–substrate complex (Eqn. 30).It is therefore often

A
B
C

(28)

$$A + B \longrightarrow C \qquad (29)$$

$$A + B + \underset{\text{Enzyme}}{\llcorner\quad\lrcorner} \rightleftharpoons \underset{\text{Enzyme}}{\llcorner A.B \lrcorner} \longrightarrow C \qquad (30)$$

suggested that the enzyme also facilitates the reaction by this approximation of reactants.

The problem, then, is to estimate the maximum rate difference between k_{cat} (for the enzyme-catalysed reaction) and k_{uncat} (for the uncatalysed reaction) (Eqn. 27) simply on the basis of approximation and in the absence of strain or solvation effects and without involving chemical catalysis by the enzyme.

Typical rate enhancements and favourable equilibria of intramolecular reactions are illustrated by some reactions of succinic acid and its derivatives [15]. The equilibrium constant for succinic anhydride formation from succinic acid (Eqn. 31) is 3×10^5 moles/l more favourable than that for acetic anhydride formation from acetic acid (Eqn. 32), an analogous intermolecular reaction. A similar situation exists

$$\begin{matrix} CH_2CO_2H \\ | \\ CH_2CO_2H \end{matrix} \rightleftharpoons \begin{matrix} CH_2CO \\ | \quad\; O \\ CH_2CO \end{matrix} + H_2O \qquad (31)$$

$$\begin{matrix} CH_3CO_2H \\ + \\ CH_3CO_2H \end{matrix} \rightleftharpoons \begin{matrix} CH_3CO \\ \quad\; O \\ CH_3CO \end{matrix} + H_2O \qquad (32)$$

for the rates of reactions. For example, the rate of succinic anhydride formation from the succinate monoester, by intramolecular nucleophilic attack of the carboxylate group on the ester group (VI), is also about 10^5 moles/l greater than that of the analogous intermolecular formation of acetic anhydride (VII). A special explanation

VI

VII

for the rate differences which is peculiar to the activated complex is not required since the favourable reactions of intramolecular systems over their intermolecular counterparts are manifested in both rates and equilibria [16].

The rate enhancements and favourable equilibria have units of concentration because a unimolecular reaction is being compared with a bimolecular one. For this reason the rate enhancement is sometimes called the "effective concentration" or "effective molarity", which is the hypothetical concentration of one of the reactants in the intermolecular reaction required to make the intermolecular reaction proceed at the same rate as that of the intramolecular one [15].

In general, however, the effective concentrations or rate enhancements of intramolecular reactions do not show a constant value but cover a very wide range from about 1 to 10^{15} moles/l [15]. For example, the intramolecular general base catalysed enolization reaction of 4-oxopentanoate (VIII) in which the neighbouring carboxylate group removes a proton in a 5-membered ring transition state, shows an effective concentration of 0.5 mole/l compared with the analogous intermolecular reaction of (IX). However, the equilibrium constant for the dehydration of dimethylamaleic acid (X) to give the 5-membered ring acid anhydride (XI) is about 10^{12} moles/l times more favourable than that for acetic anhydride formation from acetic acid (Eqn. 32).

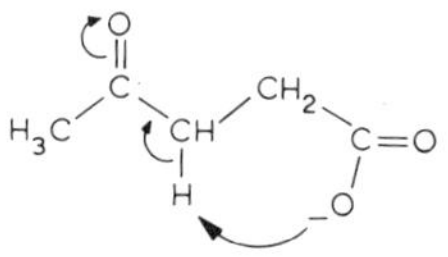

VIII

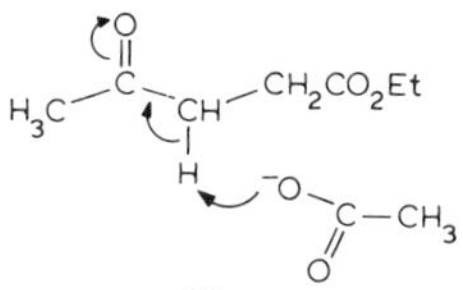

IX

X

XI

$+ H_2O$

It is quite easy to set an upper limit to the rate enhancement, in the absence of strain, that may be brought about by covalently binding the reactants together, as in an intramolecular reaction, or by binding them to the active site of an enzyme. This may be done by considering the different entropy changes that occur in bimolecular and unimolecular reactions [14,16].

There are different degrees of freedom lost in intramolecular and analogous intermolecular reactions which gives rise to large differences in the entropy change between the two systems. For a non-linear molecule containing n atoms there are 3 degrees of translational freedom, i.e. its freedom to move along the 3 axes in space; there are also 3 degrees of rotational freedom representing the freedom of the whole molecule to rotate about its centre of gravity. This leaves $3n - 6$ degrees of freedom associated with internal motions in the molecule, i.e. vibrations and internal rotations. On forming the product of a bimolecular association reaction (Eqn. 33) there is a loss of 3 degrees of translational freedom and a loss of 3 degrees of rotational

	A	+	B	$\rightleftharpoons$	A − B	
Translation	3		3		3	
Rotation	3		3		3	(33)
Vibration	$3n - 6$		$3n' - 6$		$3n + 3n' - 6$	

freedom. There is a gain of 6 new vibrational modes in the product. However, in the uni- and intramolecular reaction (Eqn. 34) there is no net change in the number of degrees of freedom of translation, rotation, and vibration upon forming the product. The entropy associated with these motions may easily be calculated from the

	A B	⇌	A—B
Translation	3		3
Rotation	3		3
Vibration	$3n-6$		$3n-6$

(34)

partition functions, which are simply an indication of the number of quantum states that the molecule may occupy for each type of movement [16].

In Table 1 are shown some typical values of these entropy terms. For a standard state of 1 mole/l the translational entropy is normally about 120–150 J/K/mole; this value is not very dependent upon the molecular weight. For the majority of medium-sized molecules the entropy of rotation is about 85–115 J/K/mole and not normally very dependent upon the size and structure of the molecule. The exceptionally small value for water, 44 J/K/mole, is because in this molecule nearly all the mass is concentrated at the centre of gravity and hence the moment of inertia, and thus the entropy, is small. The entropy of internal rotation makes a comparatively small contribution to the total entropy of the molecule and the vibrational contribution is even smaller except for the very low vibrational frequencies.

In the gas phase, for a standard state of 1 mole/l and at 298 K the total loss of

TABLE 1

Typical entropy contributions from translational, rotational, and vibrational motions at 298 K [14]

Motion	S^0 (J/K/mole)
Three degrees of translational freedom; molecular weights 20–200; standard state 1 mole/l	120–148
Three degrees of rotational freedom	
water	44
n-propane	90
endo-dicyclopentadiene $C_{10}H_{12}$	114
Internal rotation	13–21
Vibrations $\omega = 1000$ (cm^{-1})	0.4
800	0.8
400	4.2
200	9.2
100	14.2

translational and rotational entropy for a bimolecular association reaction is about −220 J/K/mole and this value has only a small dependence upon the masses, sizes and structures of the molecules involved. However, this loss is often compensated to varying extents by low frequency vibrations in the product or transition state. For several reactions having "tight" transition states or covalently bonded products the change in internal entropy is about +50 J/K/mole so the total entropy change is predicted to be about −170 J/K/mole, which is the value observed experimentally for many bimolecular gas phase reactions [14,16]. In solution, the entropy change for a bimolecular reaction is estimated to be only about 20 J/K/mole less than that in the gas phase, at the same standard state of 1 mole/l.

Since this large loss of entropy for a bimolecular reaction is avoided if the reactants are bound to an enzyme active site (Eqn. 30) or converted to an intramolecular reaction (Eqn. 28), the maximum entropic advantage from approximation may now be estimated (Eqn. 35).

$$A + B \rightleftharpoons A \cdots B \rightleftharpoons AB \qquad (35)$$

	"Loose" transition state or product	"Tight" transition state or product	Maximum
ΔS (J/K/mole)	−40	−150	−200
Effective concentration (mole/l)	10^2	10^8	10^{11}

The theoretical most negative entropy change for a bimolecular reaction is about −200 J/K/mole which is equivalent to 60 kJ/mole at 298 K and makes such a reaction unfavourable by antilog (60/2.303 RT), a factor of 10^{11} moles/l. However, this is very rare and a more general situation is that this large loss of translational and rotational entropy is compensated for by an entropy change of about +50 J/K/mole, resulting from changes in internal motions on going from the reactants to the product or transition state and from differences in entropies of vaporisation of reactants and products giving bimolecular reactions a total entropy change of about −150 J/K/mole. Since this loss of entropy does not take place for a reaction in which the reactants are bound to the active site of an enzyme (Eqn. 30) or for an intramolecular reaction (Eqn. 28), the approximate maximum effective concentration or rate enhancement for these reactions from entropic factors alone is about 10^8 moles/l. In the comparatively rare situation of a bimolecular reaction having a very "loose" transition state or product then association will be even less entropically unfavourable since the entropy of the low frequency vibrations in the "loose" complex will counterbalance the large loss of translational and rotational entropy [14–16]. The rate enhancement for the analogous intramolecular reaction will, therefore, be smaller, perhaps in the range of 100 moles/l or less.

In summary, rate enhancements of about 10^8 moles/l may occur for intramolecular and enzymic reactions simply on the basis of the difference in entropy changes between bimolecular and unimolecular reactions. Reactions which show rate enhancements greater than this are probably the result of additional contributions from potential energy differences. Smaller rate enhancements may result either from a "loose" transition state or product or from unfavourable entropy changes, such as the loss of internal rotations, and/or potential energy changes in the intramolecular reaction [15,16].

When allowance is made for the changes in strain energy and loss of entropy of internal rotations that occur upon cyclisation in intramolecular reactions the variations in "effective concentrations" previously mentioned may be nicely accounted for [15,16].

The formation of the lactone (XIII) from the hydroxy-amide (XII) is free of strain and solvation effects and is, incidentally, a good model for the mechanism of hydrolysis of amides catalysed by chymotrypsin. The rate enhancement for this intramolecular reaction is 10^8 M and therefore agrees well with the theoretical prediction [40].

HO CONHR

XII

O — C =O

XIII

If the reactants A and B are bound to the enzyme tightly and in close proximity (Eqn. 27) there will be little loss of entropy upon forming the transition state. Reaction within the enzyme–substrate complex therefore has an entropic advantage over the uncatalysed reaction and k_{cat} may be 10^8 times greater than k_{uncat} even though there is no chemical catalysis by the enzyme. However, initial binding of the reactants A and B to the enzyme is, of course, entropically unfavourable and any increase in the rate of reaction brought about by the enzyme is given by the energy of binding the enzyme and substrates, A and B, less the entropy of association of the transition state and the enzyme [13,17] (see p. 32). The binding energy may not appear as enthalpy as it is likely to be hydrophobic or electrostatic or have compensating enthalpy/entropy effects (see Section 13). Any excess of the intrinsic binding energy between the enzyme and substrates over this entropy loss will appear as catalysis of the reaction.

The idea of the binding energy being directly responsible for the rate enhancement brought about the enzyme has been substantiated by the observation that β-galactosidase catalyses the S_N1 hydrolysis of β-galactopyranosyl pyridinium salts. The enzyme increases the rate of reaction by a factor of 10^{10} and yet there is no chemical catalysis by the enzyme [18]. Similarly, enzymes which catalyse [3,3]-sigmatropic rearrangements, such as the Claisen type found in the conversion of chorismate (XIV) to prephenate (XV), are unlikely to involve chemical catalysis by the enzyme. The binding energy between the substrate and enzyme in the transition state must be responsible for catalysis.

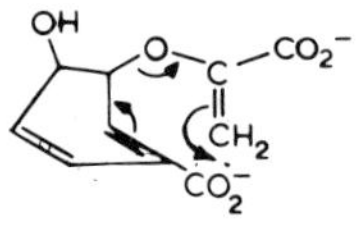

XIV 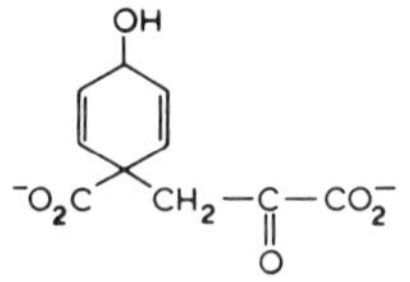 XV

10. Decreasing the activation energy

In principle, the activation energy for any chemical reaction may be decreased by raising the energy of the ground state or lowering that of the transition state. The important contribution of enzyme–substrate binding energy in decreasing the activation energy of an enzyme-catalysed compared with a non-enzyme-catalysed reaction has been recognised for many years. However, there has been much discussion as to whether the binding energy between the enzyme and substrate, between the enzyme and product or between the enzyme and some intermediate is the most important. It is now generally considered that maximum binding energy, i.e. stabilisation, occurs between the substrate and enzyme in the transition state of the reaction. It has also become common practice to use energy diagrams to rationalise aspects of enzymic catalysis [10].

A simple enzyme-catalysed reaction (Eqn. 36)

$$\mathrm{E} + \mathrm{S} \underset{K_s}{\rightleftharpoons} \mathrm{ES} \xrightarrow{k_{cat}} \mathrm{ES}^{\ddagger} \qquad (36)$$

may be depicted as in Fig. 4. The "pit" into which the enzyme–substrate complex falls reflects the favourable exergonic process of binding, the free-energy change associated with this is dependent upon the choice of standard states. For a standard state of 1 mole/l, the free-energy change reflects the difference in free-energy between *1 mole* of enzyme–substrate complex, *at a concentration of 1 M, and 1 mole each* of enzyme and substrate, also each *at a concentration of 1M.* Fig. 4 does *not* exemplify the actual free energy of the species in the reaction mixture as carried out under the conditions of an experiment and as computed from their chemical potentials. If the system is at equilibrium then E, S and $\mathrm{ES}^{\ddagger}$ are at the same energy level as there is zero free-energy difference between the states at equilibrium. For example, if the equilibrium constant for *binding,* $1/K_s$, is 10^3/M then the free energy change at 298 K when the standard state is chosen as *1* M is -17.3 kJ/mole and Fig. 3 could be used to illustrate this negative free-energy change. However, if the standard state is chosen as 10^{-5} M then the corresponding free-energy change would be $+11.5$ kJ/mole, which would give the free-energy difference between *1 mole* of enzyme–substrate complex, *at a concentration of* 10^{-5} *M,* and *1 mole each* of enzyme and substrate, *also each at a concentration of* 10^{-5} *M.* This could be illustrated by Fig. 5. Altering the standard state changes the entropy difference between reactants and products in a bimolecular reaction.

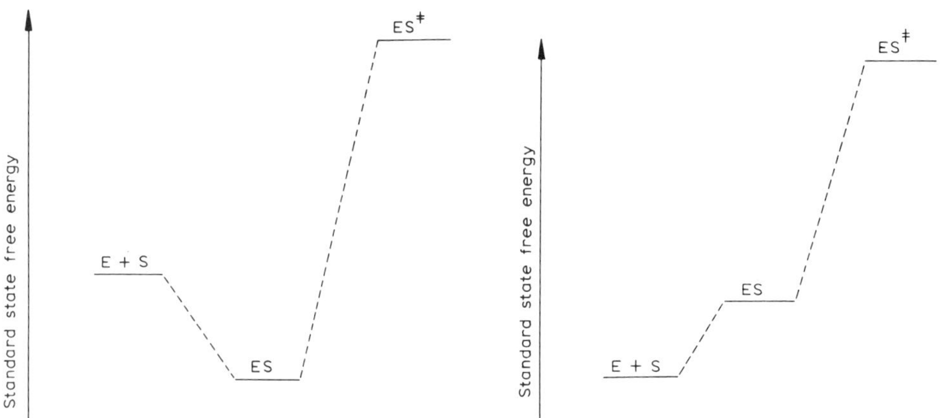

Fig. 4. Standard state free-energy changes for an enzyme-catalysed reaction showing saturation kinetics, (Eqn. 36), with the standard state chosen so that the reaction occurs above saturation.

Fig. 5. Standard state free-energy changes for an enzyme-catalysed reaction showing saturation kinetics, (Eqn. 36), with the standard state chosen so that the reaction occurs below saturation.

The free-energy change accompanying the conversion of the enzyme–substrate complex to the enzyme-bound transition state is *independent* of the choice of standard states. The standard free energy of activation for this process represents the difference in free energy between *1 mole* of $ES^{\ddagger}$ and *1 mole* of ES and, apart from Henry's law effects, it does not matter whether the concentration of these species is say, 10^{-7} M or 10^{-3} M. If the reaction is plotted on a free-energy diagram such as those illustrated in Figs. 4 and 5 the ordinate does *not* indicate the actual free energy of the species in the reaction mixture. According to the Transition State Theory, reactants, either E and S or ES, are in *equilibrium* with the transition state, $ES^{\ddagger}$, and therefore there is *zero* free-energy difference between them.

If energy diagrams are used to answer questions about enzymic catalysis sometimes it is necessary to specify the standard states chosen or the working concentrations and sometimes it is not. There are two cases commonly used, one is the comparison of enzyme-catalysed with non-enzyme-catalysed reactions and the other is a series of enzyme-catalysed reactions in which the enzyme or substrate is assumed to be modified in a particular manner [10,13,19,20].

Consider the hypothetical case where an enzyme binds equally well to the ground state and the transition state.

Let us first compare a reaction where the transition state is the same in the absence and presence of enzyme (Eqn. 37).

$$k_S = k_{ES} \qquad K_{ES} = K_{ES^{‡}}$$

It is necessary to state whether the comparison is made above or below saturation. Above saturation, two first-order processes are being compared, k_S and k_{ES}, and it is not necessary to specify the standard state. The rate of reaction will be the same in the enzyme substrate complex as it is in the absence of enzyme. There can be *no* catalysis or rate enhancement brought about by the enzyme (Fig. 6a). Below saturation the difference in the free energies of activation to give $S^{‡}$ and $ES^{‡}$ will depend upon the choice of standard state because the uncatalysed reaction is a first-order process and the enzyme-catalysed reaction is a second-order one (Fig. 6b). However, there can be *no* catalysis by the enzyme as the free energy of activation to give $ES^{‡}$ will always be greater than that to give $S^{‡}$.

Let us now compare a series of enzyme-catalysed reactions in which the enzyme

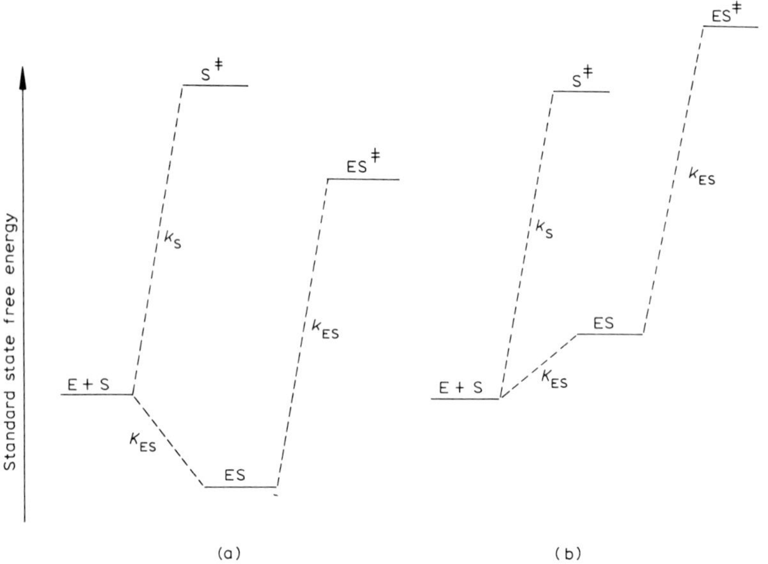

Fig. 6. Standard state free-energy changes for a reaction with the same transition state in the absence and presence of enzyme. Hypothetical case where the enzyme–substrate (in the ground state) and enzyme–transition state complexes are equally stabilised. There can be no catalysis either above saturation (a) or below saturation (b).

or substrate is modified so that the ground state and transition state are bound more strongly by *equal* amounts. If the rates of enzyme-catalysed reactions are to be compared it is necessary to specify the working concentrations of enzyme and substrate to determine whether the reaction occurs above or below saturation. This is independent of the arbitrary standard state chosen, the enzyme and substrate do not know the standard state which we have prescribed!

Above saturation, there can be no difference in the rates of reaction upon binding the ground state and transition state more tightly by the same amount (Fig. 7a). However, *below saturation* there would be an increased rate, (Fig. 7b), but there is a limit to this type of catalysis because if the enzyme binds the transition state and ground state very tightly, although the free energy of $ES^{\neq}$ will be reduced, the concentrations of enzyme and substrate required to maintain non-saturation conditions will be decreased. These concentrations may be so low that catalysis would not be observed.

Specificity, therefore, could be reflected by the enzyme stabilising *both* the transition state and the ground state for the better substrate, but this is true for only small changes in stabilisation energy. If a non-reacting substituent of a specific substrate contributes a large amount of binding energy it is essential that this is not expressed in the ground state or intermediate states in order to avoid saturation

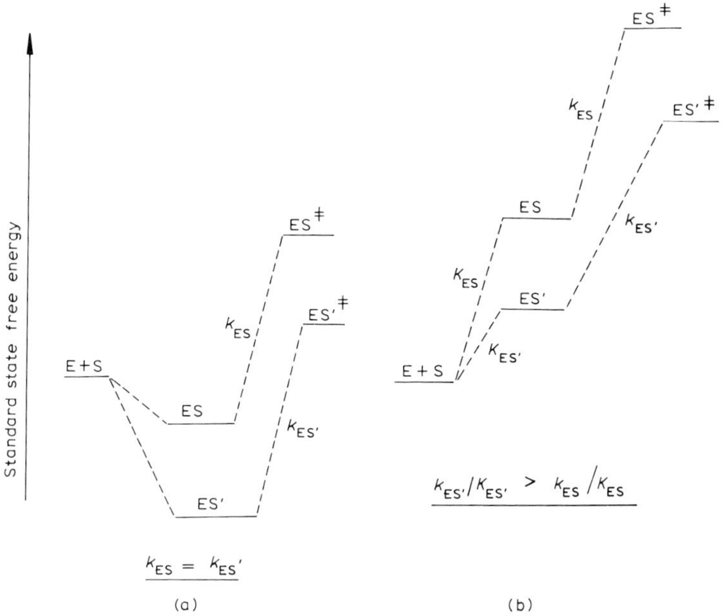

Fig. 7. Standard state free-energy changes for a series of enzyme-catalysed reactions where the substrate (or enzyme) is modified so that the ground states and transition states are equally stabilised by the modification. There can be no catalysis above saturation (a). There is a limited amount of catalysis (see text) below saturation (b).

conditions and the low concentrations of E and S required to observe k_{cat}/K_m [19].

For a given amount of binding energy between the substrate and enzyme the most effective catalysis will be obtained if this is used to stabilise the transition state, which maximises the value of k_{cat}/K_m. For a given free energy of activation $\Delta G^{\ddagger}$ for the process $E + ES \rightarrow ES^{\ddagger}$, i.e. for a fixed value of k_{cat}/K_m, and for a given substrate concentration the maximum rate is obtained if the substrate is bound *weakly*, i.e. a high K_m (Fig. 8). A low value of K_m, i.e. the tight binding of the substrate or intermediate state mediates against catalysis [21]. In fact, the physiological concentrations of most substrates are below their K_m values [21].

If the enzyme stabilises the transition state *more* than the ground state and if the transition state is the same in the absence and presence of enzyme the rate constant within the enzyme–substrate complex, k_{ES}, will be greater than that in the absence of enzyme, k_s (Eqn. 38).

$$\begin{array}{ccc} S & \xrightarrow{k_s} & S^{\ddagger} \\ K_{ES}\, \Big\Updownarrow & & \Big\Updownarrow\, K_{ES^{\ddagger}} \\ ES & \xrightarrow{k_{ES}} & ES^{\ddagger} \end{array} \qquad (38)$$

$$K_{ES^{\ddagger}} > K_{ES};\ k_{ES} > k_s$$

At saturation, the amount of catalysis would be given by $K_{ES}/K_{ES^{\ddagger}}$ i.e. the increased energy of binding the transition state compared with the ground state to the enzyme. This conclusion is independent of the standard state chosen – the rate constants k_s and k_{ES} have the same units.

Below saturation, the rate would be given by Eqn. 39

$$\text{Rate} = k_s[S] + k_{ES}K_{ES} \cdot [E][S] \qquad (39)$$

and $K_{ES} \cdot [S]$ would be less than unity. With the substrate concentration greater than that of the enzyme the amount of catalysis depends upon the "strength" of the binding of the substrate to the enzyme. If binding is weak, non-saturation conditions can hold with relatively high concentrations of enzyme and substrate and catalysis can be effective. If binding is strong very low concentrations of substrate and enzyme will be required to obtain second-order kinetics and catalysis may not be observed.

Catalysis by the enzyme can also be effective if binding of the transition state is better than that of the ground state even when the transition state structure is different within the enzyme and when the enzyme changes conformation when it binds the transition state [20].

If a series of enzyme-catalysed reactions are compared in which the enzyme or

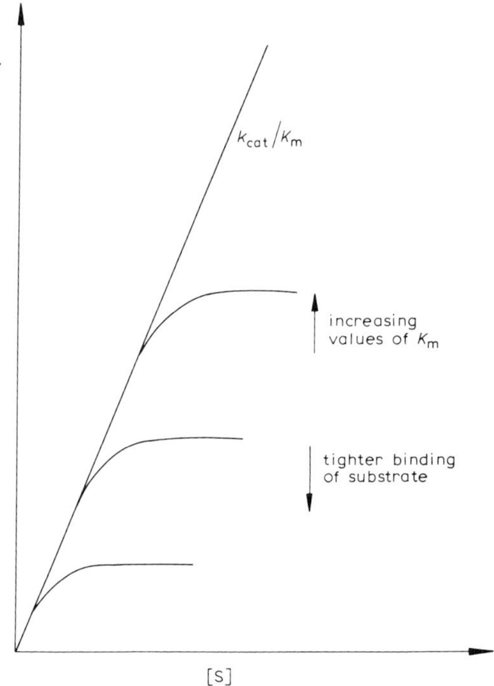

Fig. 8. Reaction rate–substrate concentration profiles for a given free energy of activation (fixed k_{cat}/K_m) but with a varying degree of binding of the substrate and enzyme. A higher maximal rate is achieved with weak binding of the substrate.

substrate is modified so that the transition state is bound more strongly than the ground state, catalysis will be more effective both above and below saturation.

11. Utilisation of binding energy

The binding energy between substrate and enzyme may conceivably be used in a variety of mechanisms to lower the activation energy of the reaction. The main mechanisms are charge, geometrical, solvation and entropic effects. The electron density distribution in a molecule determines the nuclear configuration and the first two mechanisms are not always easily separable, but as molecules may have the same shape and yet different charge distributions there are important differences, particularly in the design of inhibitors such as transition state analogues. The idea of the enzyme being complementary in shape and charge to the transition state is not separable from the problem of "rigid" or "flexible" enzymes discussed in the first section.

If the enzyme is rigid and *remains undistorted* during the catalytic process the enzyme can be complementary in structure to only one particular molecular geometry and charge distribution during the conversion of the substrate to product.

The concept of the maximum binding energy occurring between the enzyme and the initial substrate structure was introduced 85 years ago by Fischer with his "lock and key" analogy [22]. However, if the enzyme is rigid and remains undistorted during the reaction it is catalytically advantageous for the enzyme to be complementary to the structure of the transition state rather than to that of the ground state of the substrate.

Although the kinetics of binding may control the reaction if enzyme complementarity is too good and the mobility of the enzyme is too restricted, this is unlikely to be a general situation. Even if, because of weak binding, the rate of dissociation of an initially formed enzyme–substrate complex was fast, say 10^8/sec, there is still plenty of time for conformational changes which may take place in less than 10^{-10} sec. This is compatible with those enzyme-catalysed reactions where the rate-limiting step is diffusion-controlled encounter of the enzyme and substrate.

Because of the way that we make models and "draw" reaction mechanisms we often exaggerate the differences in geometry of ground-state and transition state structures whereas the actual distance that atoms move to reach a transition state is not *tremendously* more than that experienced by normal vibrations. For example, at 298 K mean vibrational amplitudes commonly range from about 0.05 to 0.10 Å and bending amplitudes of $\pm 10°$ are not rare. The important factor contributing to the increased binding energy of the transition state to the enzyme around the reactant site is that formation of the transition state is often accompanied by large changes in the electron density surrounding the reacting atoms which provides the increased favourable interaction between the substrate and enzyme.

The idea that enzymes should not stabilise the substrate too much upon binding is sometimes interpreted to mean that the substrate should be *destabilised* in the ES complex. This is illustrated in Fig. 9 where the energy of the ES complex is raised by some unspecified mechanism which has the effect of increasing k_{cat} but has no effect upon k_{cat}/K_m. Such an enzyme would be more effective above saturation but would have the same efficiency below saturation. Specificity, in the sense of discrimination between competing substrates and which depends upon the value of k_{cat}/K_m, would not be affected by having the enzyme–substrate complex destabilised. The introduction of strain or any other mechanism of destabilising the substrate cannot therefore directly affect specificity.

The ground state electron distribution of an isolated molecule invariably determines the type of charge with which it favourably interacts. For example, a carbonyl group would be stabilised by an electron acceptor adjacent to the carbonyl oxygen while it would be energetically favourable to have an electron donor near the carbonyl carbon (XVI). The ground state charge distribution, of course, also

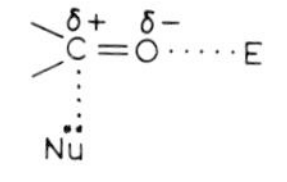

$$\overset{\delta+}{C}=\overset{\delta-}{O}\cdots\cdots E$$
$$\vdots$$
$$\ddot{Nu}$$

XVI

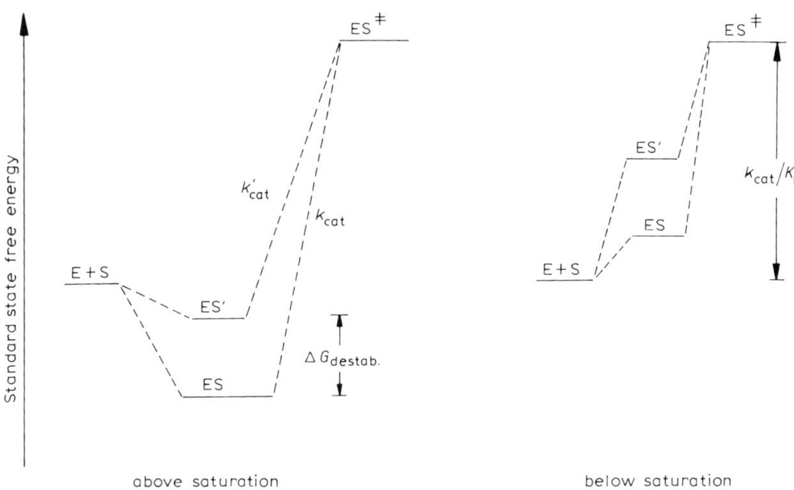

Fig. 9. Standard state free-energy changes for an enzyme-catalysed reaction showing saturation kinetics in which the enzyme–substrate complex is destabilised by a modification. Destabilisation increases the rate above saturation (a) but has no effect on the rate below saturation (b). The second-order rate constant, k_{cat}/K_m, and hence specificity is unaffected by such destabilisation.

determines the type of chemical reactions the molecule readily undergoes initially. The stabilisation of a molecule brought about by solvation or binding to a complementary charged surface, (e.g. XVI), is simply the first stage of perturbation which can lead to bond making and breaking. If the perturbation of electron density brought about by the initial interaction between substrate and enzyme changes the geometry of the substrate this would rightly be interpreted as using the binding energy to distort or strain the substrate.

The addition of electrons to molecules changes their shape e.g. NO_2^+ (trigonal planar) to NO_2^- (V-shaped) SO_3 (trigonal planar) to SO_3^{2-} (tetrahedral). The addition of electron donors (nucleophiles) to electron-deficient centres increases the coordination number of the acceptor and also changes its shape. In general, electron donation into antibonding orbitals will perturb the geometry of the molecule (increase bond lengths). Exceptions to this generalisation are electron donation into vacant bonding π orbitals and a reduction in the oxidation state of some inorganic complexes. What is not immediately apparent is the extent to which electron transfer can take place without perturbing the molecular geometry. The shape of the electron donor atom changes little upon coordination, although, of course, bond lengths and angles change slightly as the non-bonded pair of electrons is replaced by an atom upon coordination.

Alkenes and carbonyls are normally planar. However, even a distant interaction of a nucleophile or electrophile (including substituents within the molecule) produces a small but discernible pyramidalisation of the trigonal carbon atom [23]. The direction of this pyramidalisation is the same as the direction in which alkene or carbonyl carbon pyramidalises preferentially upon attack by a nucleophile or electrophile suggesting that there is a relationship between this molecular distortion and the stereoselectivities of addition reactions. It should be emphasised that these distortions are small, corresponding to only a few per cent of pyramidalisation to a tetrahedral geometry.

Enzymes are not completely inflexible and steric complementarity does not seem to make a major contribution to the efficiency of enzymic catalysis.

Enzyme substrates usually consist of several atoms undergoing chemical reaction, the reaction centre, and many other atoms which do not undergo any electron density or bonding changes, the non-reacting part. Let us compare a 'non-specific' and a 'specific' substrate in which the reaction centres are identical and the intrinsic binding energy between the reacting groups undergoing electron density changes and the enzyme are the same. How is the binding energy of the non-reacting part of the specific substrate used to stabilise the transition state but not used to over-stabilise the enzyme–substrate or enzyme–intermediate complex? Can this binding energy be used to 'destabilise' the substrate? That this is a real problem is reflected by the fact

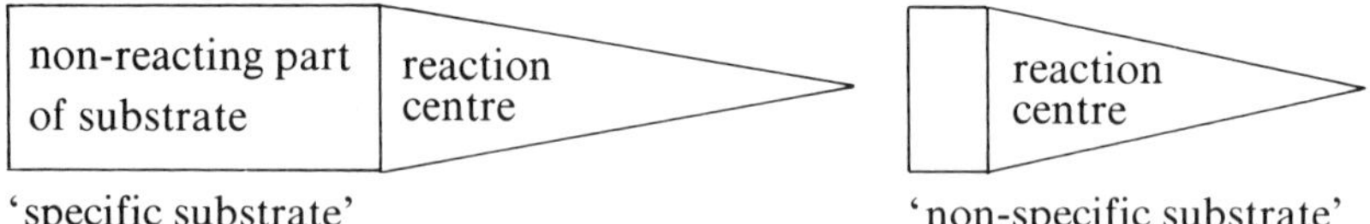

that the *observed* binding constant of a specific substrate is often apparently 'weaker' or no 'stronger' (in a thermodynamic sense) than that for a non-specific substrate.

The molecular mechanisms for stabilisation of the transition state relative to the ground state at the active site are available for specific and small non-specific substrates. The geometry, 'solvation' and general charge neutralisation provided for the transition state by the active site of the enzyme are presumably available for the reaction site of both "specific" and small "non-specific" substrates. The free-energy difference between E + S and $ES^{\ddagger}$, i.e. k_{cat}/K_m, is reduced for the specific substrate by the free energy of binding the non-reacting part of the substrate. The binding energy of the non-reacting part of the substrate may be used either to stabilise the transition state but not the ground state or any intermediate, or to destabilise the substrate by entropy, geometrical, solvation or electrostatic changes in the ground state but not in the transition state, i.e. the binding energy of the non-reacting parts may be used to compensate for unfavourable processes which are apparent in the ground state but not in the transition state.

At the reaction centre electron density and geometrical changes in the substrate on going to the transition state may be stabilised by complementary charges and shape of the active site in the enzyme. These electron-density changes could be accompanied by a conformational change so that a large non-reacting group not bound to the enzyme in the ground state becomes bound in the transition state. In general, it is less easy to see how the binding energy of the non-reactive part of the substrate is prevented from being fully expressed in the ES complex unless it is used to compensate for unfavourable processes. If the environment of the enzyme is unfavourable to the ground state structure of the substrate, the binding energy from the non-reacting part of the substrate could be used to force the substrate into this 'unwelcoming hole'. Examples are (1) a 'rigid' enzyme that has an active site complementary in shape to the transition state but not the ground state, (2) an active site which is non-polar and conducive to stabilising a neutral transition state but destabilising a charged substrate, (3) desolvation or solvation changes of groups on the substrate or enzyme, (4) an active site where electrostatic charges on substrate and enzyme are similar. If the binding energy is used to compensate for the induction of 'strain' in the substrate it is *essential* that the 'strain' is *relieved* in the transition state in order to increase the rate of reaction. This would be the case if the changes in geometry of the substrate were in the direction which accompanies the reaction mechanism. The observed binding energy would be what is 'left over' after the strain has been 'paid for' and may thus appear to be weak. It is also essential that the non-reacting group only exhibits its binding energy when the *reactive centre* of the substrate is bound to the active site. Alternative binding modes would otherwise result leading to non-productive binding which does not affect k_{cat}/K_m and specificity, but does decrease K_m leading to saturation conditions at lower concentrations of substrate. The binding energy of the substituent of the specific substrate may be used to prevent non-productive binding. Amino acid side chains, not necessarily those near the active site, may have their exposure to water changed during catalysis. For example, a polar group may become exposed or protected from water on going from the ground state to transition state which would make a favourable and unfavourable contribution to the free energy of activation, respectively.

Probably the most important way that the binding energy is 'used' is to compensate for the unfavourable entropy change that accompanies formation of the ES and $ES^{\ddagger}$ complexes. The entropy loss that is required to reach the transition state may be partially or completely lost in the ES complex. The binding energy of a non-specific substrate may be insufficient to restrict the degree of freedom of the substrate when bound to the enzyme. Consequently, more entropy has to be lost to reach the transition state, which reduces k_{cat}, compared with a specific substrate where the binding energy of the additional substituent may compensate for the required loss of entropy which may occur in ES or $ES^{\ddagger}$. Thus, although the free energy change accompanying formation of the ES complex may be similar for specific and non-specific substrates the former may be more "tightly" bound (in an entropic sense).

At 25°C for a standard state of 1 M, the complete restriction of medium-sized substrates requires a decrease in entropy and an increase in energy of about 150 kJ/mole, an unfavourable factor of 10^8 in rates or equilibria. This entropy change is that typically required to form a covalent bond in which the atoms are necessarily confined to a relative small volume. If the enzyme-catalysed reaction requires the formation of a covalent bond then the binding energy of the non-reacting part of the substrate may compensate for the unfavourable entropy change. A non-specific substrate may have insufficient binding energy to compensate for the required entropy loss resulting in a reduced value of k_{cat}/K_m. If the chemical mechanism of the catalysis requires the involvement of other functional groups such as general acids or bases, metal ions or a change in solvation then a further entropic advantage can result from enzymic catalysis. However, the contribution is smaller than that from covalent catalysis because these secondary effects require less restriction of degrees of freedom because the 'flexibility' of hydrogen bonds and metal ion co-ordination is greater than that for covalent bonds.

Suppose two compounds A and B react to form a transition state $AB^{\ddagger}$ in an uncatalyzed reaction. The free energy of activation may be arbitrarily separated into the entropy and enthalpy of activation (Eqn. 40).

$$
\begin{array}{ccccc}
 & & \mathrm{A+B} \xrightleftharpoons{-T\Delta S^{\ddagger}} & \mathrm{A.B} & \xrightleftharpoons{\Delta H^{\ddagger}} & \mathrm{AB}^{\ddagger} \\
 & & & \Big\Updownarrow \Delta G_x & & \Big\Updownarrow \Delta G_{cat} \\
\mathrm{A+B+Enzyme} & \xrightleftharpoons{\Delta G_s} & & \underset{\text{Enzyme}}{\mathrm{A.B}} & \xrightarrow[\text{or } \Delta H^{\ddagger}]{k_{cat}} & \underset{\text{Enzyme}}{\mathrm{AB}^{\ddagger}}
\end{array}
\qquad (40)
$$

This *same* transition state may be reached with the reactants bound to the active site of an enzyme, i.e. we are considering a process in which there is no chemical catalysis by the enzyme, the function of which is simply to bring the reactants together. If the reactants A and B are bound to the enzyme tightly and in close proximity (Eqn. 40) there will be little loss of entropy upon forming the transition state. Reaction within the enzyme–substrate complex therefore has an entropic advantage over the uncatalysed reaction and k_{cat} may be 10^8 times greater than k_{uncat} even though there is no chemical catalysis by the enzyme. However, initial binding of the reactants A and B to the enzyme is, of course, entropically unfavourable and any increase in the rate of reaction brought about by the enzyme is given by the energy of binding the enzyme and the transition state, $AB^{\ddagger}$, less the entropy of association of the transition state and the enzyme [17] (Eqn. 41).

$$\Delta G_{cat} = \Delta H_s - T\Delta S_x \qquad (41)$$

The term $T\Delta S_x$ in Eqn. 41 represents the unfavourable loss of entropy in having the enzyme act as a catalyst. Any excess of the intrinsic binding energy between the enzyme and substrates over this entropy loss will thus appear as catalysis of the reaction. Pictorially, the entropy effect may be regarded as "entropic strain" that is compensated by the binding energy [17]. The binding energy may not appear as enthalpy as it is likely to be hydrophobic or electrostatic or have compensating enthalpy/entropy effects (see Section 13).

The advantage of covalent catalysis, where an electrophilic or nucleophilic group on the peptide chain of the enzyme forms a covalent bond with the substrate, is immediately apparent by considering the difference in entropy changes between the uncatalysed (Eqn. 42) and enzyme-catalysed reaction (Eqn. 43).

$$A + B \longrightarrow AB^{+} \tag{42}$$

$$A + B\text{–Enzyme} \underset{}{\overset{\Delta G_s}{\rightleftharpoons}} \underset{\text{Enzyme}}{\llcorner AB \urcorner} \xrightarrow{k_{cat}} \underset{\text{Enzyme}}{\llcorner A\text{–}B^{+} \urcorner} \tag{43}$$

In Eqn. 43 one of the reactants, B, is covalently bonded to the enzyme and a comparison of this reaction with that of Eqn. 42 illustrates the advantage of binding the substrate to the enzyme even if the chemical reactivity of B in the enzyme may be similar to that of B in intermolecular reaction (Eqn. 42). Any free-energy advantage of the enzymic reaction is then given by Eqn. 44 which may be simplified to Eqn. 45 if it is assumed that the entropy of activation $T\Delta S^{+}$ is equal to the entropy change of binding the substrate A to the enzyme, $T\Delta S_s$.

$$\Delta G_{cat} = \Delta H_s + T\Delta S^{+} - T\Delta S_s \tag{44}$$

$$\Delta G_{cat} = \Delta H_s \tag{45}$$

In this case the amount of catalysis is given directly by the binding energy between the enzyme and substrate and shows the advantage of covalent catalysis in enzymic reactions.

It should be emphasised that the quantities referred to here will not be reflected in measured enthalpies and entropies of reaction which are dominated by solvent effects.

The classical separation of specificity and catalysis may not be valid. Often the specificity of the substrate binding process, which may be described by some sort of "lock and key" model is treated as a separate process from the catalytic rate acceleration which is often erroneously explained in solely chemical terms. In fact, many enzymes apparently exhibit their specificity in their maximal rates of catalysis rather than in substrate binding.

The assumption involved in the derivation of Eqn. 45 is that the entropy of activation in the intermolecular reaction (Eqn. 42) is approximately equal to the entropy change of binding the substrate A to the enzyme (Eqn. 43). This may be true for good substrates that bind tightly. However, poor substrates which have a smaller enthalpy of binding will bind more loosely to the enzyme and the formation of the transition state will consequently be accompanied by a negative entropy change. Since enough entropy will then not have been lost to reach the transition state k_{cat} would be smaller but the apparent free energy of binding would be unchanged for the poorer substrate. Much or all of the more favourable intrinsic binding energy of good substrates may therefore appear as a rate acceleration, increased k_{cat} (Eqn. 43), rather than as an increase in the binding constant.

Bimolecular reactions which proceed through "loose" transition states are entropically less unfavourable than those involving "tight" transition states [15,16]. This is because the entropy of low frequency motions in the transition state compensates for the large loss of translational and rotational entropy (see Eqn. 35). Such a situation exists in proton transfer reactions and all reactions involving intramolecular general acid or general base catalysis have low effective concentrations. This appears to be true for proton transfer to or from both electronegative atoms and carbon atoms [16]. Presumably this is because such reactions either have a transition state which is a very loose hydrogen-bonded complex or have a rate-limiting step which is diffusion controlled. The rate enhancement that can be brought about by having a general acid or general base catalyst as part of the protein structure (Eqn. 46) as opposed to the chemically equivalent intermolecular mechanism (Eqn. 47) thus appears to be minimal.

$$\mathrm{S + H{-}A{-}Enzyme} \rightleftharpoons \underset{\mathrm{Enzyme}}{\underline{\mathrm{S.HA}}} \longrightarrow \underset{\mathrm{Enzyme}}{\underline{\mathrm{S \cdots H \cdots A^{+}}}} \quad (46)$$

$$\mathrm{S + H{-}A} \longrightarrow \mathrm{S \cdots H \cdots A^{+}} \quad (47)$$

Of course, it may be necessary for the proton acceptor or proton donor to be at the active site as the equivalent intermolecular catalyst may be sterically prevented from reaching the enzyme-bound substrate and therefore in the absence of the general acid or general base on the enzyme a slower rate of reaction would result. However, the fact that the general acid or base is *part* of the enzyme apparently makes little contribution to the enormous rate enhancement brought about by the enzyme.

We have seen that the binding energy may appear in two ways. First, it gives the observed binding (which is useful at concentrations below the K_m). Second, it may be used to cause entropy loss of the bound molecule which gives a rate acceleration.

12. Intramolecular force fields [16]

To understand the energetics and specificity of enzyme–substrate interactions it is necessary to know the magnitude and geometry of intermolecular and intramolecular forces. It is of interest to know whether the intermolecular force field generated by the environment (solvent, crystal, enzyme etc.) is strong enough to overcome the conformational and geometrical dictates of the intramolecular force field of the substrate. Can the enzyme distort the substrate or vice versa? We must first examine the forces holding atoms together in the ground state configuration and conformation of a molecule of substrate or enzyme.

The forces controlling the distortion of a molecule may be partitioned into the following contributions: (i) bond stretching; (ii) bond angle bending; (ii) torsional effects; (iv) attractive and repulsive non-bonded interactions; (v) electrostatic interactions such a dipole–dipole and polar effects. Unfortunately, there is by no means universal agreement upon the values of the parameters to be used in the quantitative estimation of these effects, and although they all have a physical reality (probably containing some areas of overlap), there is a tendency to treat them as adjustable parameters.

Bond stretching
Small deformations in bond lengths are usually assumed to have harmonic restoring forces and thus obey Hooke's law. The energy is proportional to the square of the deformation (Eqn. 48)

$$E_r = k_r(r - r_0)^2 \tag{48}$$

where k_r is the force constant and r_0 is the 'strain-free' or normal bond length. Using the standard bond lengths found in *n*-alkanes and the corresponding force constants, typical equations, with r in Å, are:

$$E_{c-c} = 1369(r - 1.53)^2 \text{ kJ/mole}$$

$$E_{c-H} = 1333(r - 1.09)^2 \text{ kJ/mole}$$

Deformation of bond lengths is thus very difficult and rarely occurs in 'normal' molecules, and is unlikely to occur as a result of intermolecular interactions.

Bond angle bending
The deformation of bond angles from their 'normal' value is also usually assumed to be controlled by a harmonic potential (Eqn. 49).

$$E_\theta = k_\theta(\theta - \theta_0)^2 \tag{49}$$

The necessary force constants, k_θ, are usually obtained from spectroscopic measurements and the values of the normal bond angle, θ_0, are those found experimentally in supposedly 'strain-free' molecules. For the *n*-alkanes, there is a separate k_θ and θ_0 for the $C\hat{C}C$, $H\hat{C}H$, and $H\hat{C}C$ angles but it is often assumed that changes in these angles are linearly related to one another, and hence only one effective methylene group force constant is required to give the total angle strain at a given carbon atom (Eqn. 50).

$$E_\theta = 0.113(\theta - 111)^2 \text{ kJ/mole} \tag{50}$$

Bond angle deformation is fairly easy: e.g. from Eqn. 50, a 10° change costs 11.3 kJ/mole, and this is the pathway commonly used to relieve unfavourable non-bonded interaction within a molecule and also it is a conceivable distortion to be caused by intermolecular interactions.

Torsion
A clear understanding of the origin of the barrier to internal rotation in ethane is a

long-standing problem in quantum chemistry. Except when non-bonded interactions are chosen simply to fit this barrier, it is generally agreed that van der Waals forces alone cannot account for the effect. Therefore the torsional potential is considered to be a separate contribution to the total energy and is usually assumed to be represented by a cosine function:

$$E_{(\phi)} = E_{\phi}(1 \pm \cos n\phi)$$

E_{ϕ} is half of the barrier height, n depends on the symmetry of rotation. ϕ is the dihedral angle between bonds, and the plus and the minus signs are taken depending on whether $E_{(\phi)}$ has a maximum or a minimum, respectively, at $\phi = 0°$.

A typical equation for rotation about C–C single bonds is given by

$$E_{(\phi)} = 5.65(1 + \cos 3\phi) \text{ kJ/mole} \tag{51}$$

For dihedral angles up to about 20°, Eqn. 51 may be replaced by a quadratic function (Eqn. 52). Separate values are, of course, needed when rotation occurs around a bond which is adjacent to a C=C or a C=O bond and some other environments.

$$E_{(\phi)} = 11.30 - 0.00769\ (\phi)^2 \tag{52}$$

Torsional energy is the 'softest' of all the potential energy terms, and hence distortion of dihedral angles is relatively easy. The hydrolysis of saccharides usually proceeds through a carbonium ion-like transition state (Chapter 11). In the case of glucopyranosides this involves the conversion of 6-membered ring structure from the chair conformation to the half-chair conformation. It is often suggested that a factor contributing to the efficiency of the lysozyme-catalysed hydrolysis of polysaccharides is the distortion of the ring from the normal chair to the half-chair conformation upon binding to the enzyme. However, the increase in energy on introducing two sp^2 centres into a 6-membered ring is less than 10 kJ/mole. Even if such a conformational change was induced by the enzyme, it is therefore unlikely that this is a significant factor contributing to catalysis. We shall see later that such geometrical distortion does not, in any case, take place.

Disulphide links

Disulphide bridges are usually buried in the interior of proteins or deeply embedded in grooves. Recently, it has been suggested that these bonds provide elasticity and allow the enzyme to deform under stress. However, S–S bonds are strong covalent bonds which are not easily stretched or distorted. Although the stretching frequency of an S–S bond (~ 500/cm) is lower than that for C–C stretching (~ 1200/cm), the reduced mass is greater. The force constant for S–S stretching is about 1300 kJ/mole/Å^2 similar to that for C–C stretching. When both atoms attached by a single bond possess non-bonding electron pairs the barriers to rotation are increased.

The hindered rotation about the S–S bond in disulphides is about 30 kJ/mole. It seems that the classic view of disulphides providing rigidity is correct.

Non-bonded interaction

Unfortunately, probably the most important but the least understood energy function is that describing non-bonded interaction. As a consequence not only is there a variety of functional forms used to describe this interaction, but there is a range of values reported for the parameters of the same functions. Most of the functions in the literature are evaluated empirically from data on gas viscosity, molecular scattering, and other data relating to *intermolecular* forces. By analogy with intermolecular concepts, the intramolecular interaction energy is assumed to be the sum of *short-range repulsive* forces and *long-range attractive* dispersion or London forces. Most treatments have made use of the Buckingham exp-6 (Eqn. 53) or the Lennard–Jones 12-6 (Eqn. 54) potential functions.

$$E_{nb} = \mathrm{A}\exp(-\mathrm{B}r) - \mathrm{C}r^{-6} \tag{53}$$

$$E_{nb} = \mathrm{D}r^{-12} - \mathrm{E}r^{-6} \tag{54}$$

The attractive potential is taken as the inverse 6th power of the internuclear distance, r, and values of C or E may be derived from atomic polarisabilities.

By the use of an empirical relationship the three parameters in Eqn. 53 may be reduced to two. The variation in the parameters used for CC and HH non-bonded interaction is exemplified by Eqns. 55 and 56, with E in kJ/mole and r in Å.

$$\begin{aligned} E_{\mathrm{C...C}} &= 4.018 \times 10^5 \exp(-4.53r) - 795.8r^{-6} \\ E_{\mathrm{H...H}} &= 2.079 \times 10^5 \exp(-4.53r) - 411.8r^{-6} \end{aligned} \tag{55}$$

$$\begin{aligned} E_{\mathrm{C...C}} &= 6.263 \times 10^4 \exp(-3.15r) - 2690r^{-6} \\ E_{\mathrm{H...H}} &= 1.108 \times 10^4 \exp(-3.74r) - 114.4r^{-6} \end{aligned} \tag{56}$$

The calculation of meaningful non-bonded interaction energies is beset by a number of complications. Unlike the free atoms, those in molecules do not possess spherical symmetry. To allow for this anisotropic character of non-bonded interactions it has been suggested that, say, the centre of a hydrogen atom should be shifted along the C–H bond, but still treated as spherical. Another problem is that the effective dielectric constant of the molecule may influence the transmission of the forces involved. Finally, the calculations apply to the gas phase and in solution the attractive part of the non-bonded interaction would be decreased by the solvent.

The relative ease of distortion of a molecule by the above parameters is illustrated in Fig. 10 (p. 39).

13. Intermolecular force fields

Unlike covalent bonds, the strengths and geometrical properties of other interactions between atoms are immensely variable. These non-covalent interactions are dealt with in more detail in Chapter 2, but here we are concerned whether intermolecular forces between the substrate and enzyme can overcome the intramolecular force field controlling the position of the atoms in the substrate. At one extreme are electrostatic forces which are long range with low barriers to distortion but as multipole interactions increase the range of the force decreases with Van der Waals forces being the shortest range.

The origin of attractive intermolecular interactions and the factors determining the geometry of molecular complexes is still being sought after decades of numerous experimental and theoretical techniques. The interaction energy, ΔE, is the difference between the energy of the species AB and the two isolated molecules A and B. ΔE may be partitioned as in Eqn. 57

$$\Delta E = E_{es} + E_{pol} + E_{ex} + E_{ct} + E_{corr} \tag{57}$$

where the individual components have the following physical significance:

E_{es} – the electrostatic interaction i.e. the interaction between the undistorted electron distributions of A and B. This contribution includes the interactions of all permanent charges and multipoles such as dipole–dipole, dipole–quadrupole etc. and may be either attractive or repulsive.

E_{pol} – the polarisation interaction i.e. the effect of the distortion of the electron distribution of A by B, and vice versa, and the higher-order coupling resulting from such distortions. This component includes the interactions between all permanent charges and induced multipoles and is always an attractive interaction.

E_{ex} – the exchange repulsion i.e. the interaction caused by exchange of electrons between A and B. Physically, this is the short-range repulsion due to overlap of the electron distribution of A with that of B.

E_{ct} – charge transfer of electron delocalisation interaction i.e. the interaction caused by charge transfer from occupied molecular orbitals of B to vacant MOs of A.

E_{corr} – correlation energy. A part of the intermolecular correlation term is the dispersion energy which is the second-order attraction between fluctuating charges on A and B and is relatively unimportant for interactions between polar molecules.

The factors determining the equilibrium geometry depend upon the nature of the complex but the electrostatic interaction usually makes the major contribution. For example, in hydrogen bonding, with a small distance separating the proton donor and acceptor, E_{es}, E_{ct} and E_{pol} can all be important attractive components competing against a large E_{ex} repulsion. At longer distances, for the *same complex,* the short range attractions E_{ct} and E_{pol} are usually unimportant and E_{es} is the only important attraction. However, the importance of an individual component depends on the type of hydrogen bonding, but the electrostatic term is dominant in "normal" hydrogen bonds between neutral electronegative atoms.

Hydrogen bonding

Two aspects of the geometry of hydrogen bonding are important in determining enzyme–substrate interactions. (1) Can the intermolecular force field generated by hydrogen bonding override the intramolecular force field determining the geometry of the substrate? (2) How does the energy of intermolecular hydrogen bonding change with perturbations from optimal geometry? The distance between the electronegative atoms in hydrogen bonding is important, but not critical. The O...O and the O...N distance in hydrogen-bonded systems varies between 2.4 and 3.0 Å and 2.65 and 3.15 Å, respectively. The force constant for O...O stretching in ice is 121 kJ/mole/Å and, if distortion from the equilibrium O...O distance obeyed a quadratic expression such as Eqn. 48, the energy change as a function of distance would be as illustrated in Fig. 10. It is easier to stretch a hydrogen bond than it is to

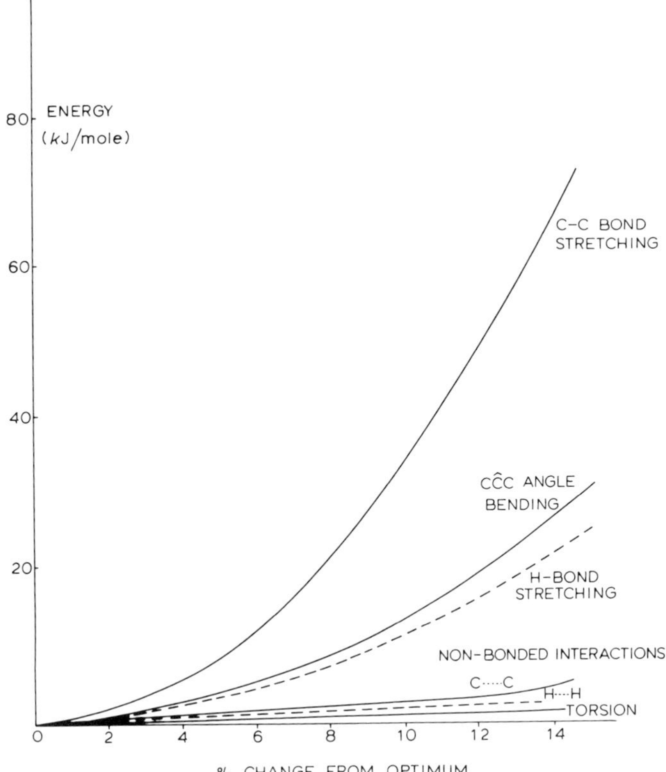

Fig. 10. The increase in energy brought about a change in the optimum parameter of a molecule as indicated.

distort a $C\hat{C}C$ bond angle. A quadratic probably overemphasises the energy dependence upon distance and theoretical calculations indicate that, for example, O...O distance can vary ±0.3 Å for less than 5 kJ/mole.

Although at one time it was thought that "normal" hydrogen bonds were linear deviations from linearity are now accepted as the normal situation, especially in solids. An examination of the distribution of X–$\hat{H}$... Y angles shows that deviations by 15° from linearity occur as frequently as strictly linear hydrogen bonds. The O–$\hat{H}$...O and N–$\hat{H}$...O angles in intermolecular hydrogen-bonded systems vary from 140 to 180° and from 130 to 180°, respectively. Theoretical calculations show that large changes in angle correspond to only 1 or 2 kJ/mole; for example ±30° for O$\hat{H}$...O=C costs ca. 3 kJ/mole. Incidentally, experimental and computational data do not support the hypothesis that the H bond lies along the maximum of the electron donor lone pair density.

In summary, distortion from the optimum hydrogen bond geometry is easy for changes in orientation and slightly less facile for changes in distance. Conversely, the intermolecular force field generated by a hydrogen bond is unlikely to cause major distortions, except for conformational changes, in the geometry of the substrate. Furthermore, the strength of a hydrogen bond between the substrate and enzyme will be very similar to that between substrate and water.

Electrostatic interactions

One difference between an enzyme and, say, water in providing solvation shell for the substrate and transition state is that individual water molecules orientate themselves towards the substrate at the expense of their interaction with other water molecules. The dipoles within the one molecule of a rigid enzyme could be preorientated to create an electric field complementary to the charge distribution of the transition state. The electrostatic energy dominates the medium to long-range interaction between two molecules and it seems reasonable therefore to emphasise its importance.

An extreme example of the importance of electrostatic interactions is the solubility of salts which also highlights the role of metal ions in enzymes. The solubility product, K_s, of a salt is related to the free energy of solution ΔG_s^0 by Eqn. 58

$$-\Delta G_s^0 = RT \ln K_s \qquad (58)$$

A change of a factor of 10 in the solubility product corresponds to a change of only 5.7 kJ/mole in the free energy of solution, so that the energetic difference between 'soluble' and 'insoluble' salts is a very small one. For example, the standard free energy of solution of potassium nitrate is less than 12 kJ/mole more negative than that of potassium perchlorate, yet the first is classified as soluble and the second as insoluble.

The thermodynamic cycle shown in Eqn. 59 allows a discussion of the solubilities of ionic salts, where ΔG_l^0 and ΔG_h^0 are the free energy changes associated with the sublimation of the lattice into the separate gaseous ions at a standard state and the immersion of these ions in water to form an infinitely dilute solution, respectively. The free energy of solution is the sum of the lattice energy and hydration energy (Eqn. 60) both of which are of the order of hundreds of kJ/mole, but of opposite sign.

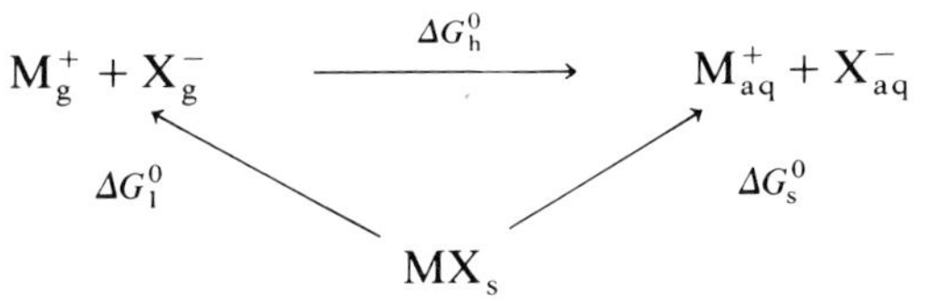

$$M_g^+ + X_g^- \xrightarrow{\Delta G_h^0} M_{aq}^+ + X_{aq}^-$$

$$\Delta G_l^0 \qquad MX_s \qquad \Delta G_s^0 \tag{59}$$

$$\Delta G_s^0 = \Delta G_l^0 + \Delta H_h^0 \tag{60}$$

Typical free energies of solution are between +80 and −80 kJ/mole and it is not easy to rationalise trends in these values as the result of the difference between two large quantities.

The major component of ΔG_l^0 is the lattice energy of the solid which is nearly always greater than 400 kJ/mole, so if the free energy of solution is to be negative, powerful forces must hold the ions and solvent molecules together. The free energy of sublimation of the lattice is given by Eqn. 61 and, for uni-univalent salts by Eqn. 62.

$$\Delta G_l^0 = \Delta H_l^0 - T\Delta S_l^0 \tag{61}$$

$$= \frac{1075}{r^+ + r^-} - 31 \text{ kJ/mole} \tag{62}$$

An interpretation of the consistency and differences in the free energies of hydration of ions is based on the suggestion of Born that if a sphere of radius r has a charge of z units placed upon it in a vacuo or solvent of dielectric constant D then the amount of electrostatic work done is $z^2e^2/2r$ and $z^2e^2/2rD$, respectively. The free energy of solvation is then given by Eqns. 63 and 64 for water.

$$\Delta G_h^0 = \frac{-N_o z^2 e^2}{2r}\left(1 - \frac{1}{D}\right) \tag{63}$$

$$= -689z^2/r \text{ kJ/mole} \tag{64}$$

Of course, the radius of an ion is unlikely to be the same in the gas and solvent phases, but in the case of water the dielectric constant is so large that the second term in the bracket is negligible compared with the first. Therefore the gas phase radius is the value required in Eqn. 63, which is about 0.7 Å bigger than the crystal radius for cations and 0.3–0.6 Å bigger for anions. It is also probably inappropriate to use the macroscopic value of the dielectric constant as the high electric fields operational close to the ions would considerably reduce the value required.

ΔG_s^0 is thus given by the difference between an energy dependent upon the reciprocal of the *sum* of ionic radii (Eqn. 62), and another dependent upon the sum of two reciprocals of the separate radii (Eqn. 63).

The general conclusion is that a salt should be very soluble when there is a severe mismatch in cation and anion size. The presence of large ions means that lattice energies are low, especially if anion–anion or cation–cation contact occurs, and the free energy of hydration of the small ion is almost, by itself, sufficient to ensure dissolution.

Selectivity in charge–charge interactions depends on the magnitude of the charge, the size of the group or atom carrying the charge, the degree of hydration and the hydrogen-bonding potential.

There are no strong directional or distance dependencies on the electrostatic interactions between Na^+, K^+, Ca^{2+} and NH_4^+ cations with anions. Calcium ions have high coordination numbers but little stereochemical demand and fast exchange of ligands and it seems unlikely that these ions will restrict internal mobility of the enzyme or the mobility of the substrate.

For estimating the forces between dipoles of the enzyme and that of the substrate a simplified approach is to use partial charges on the individual atoms, and to calculate the electrostatic interaction by Coulomb's law (Eqn. 65) as a function of the distance, r Å, between the partial charges, q expressed in terms of the electronic charge, in a medium of dielectric constant D.

$$E_{el} = \frac{q_1 q_2}{Dr} \times 1389 \text{ kJ/mole} \tag{65}$$

The use of Coulomb's law should be regarded as a purely empirical procedure because, when two partial charges are not well separated, the solvent molecules and the rest of the solute between and around the two charges do not behave like a continuous medium of constant dielectric constant, and it is also difficult to know where the point dipoles should be located. For two partial charges separated by greater than one width of water layer it has been suggested that the effective dielectric constant approaches that of bulk water, 80; hence electrostatic interactions would be negligible at these distances.

If the pK_a of a group in an enzyme is similar to the corresponding pK_a in water there must be significant stabilisation of the charged groups by bound water molecules, permanent and induced dipoles of the enzyme.

The reaction field of the highly polarisable solvent water decreases the field near the molecular surface of the solute. The earliest attempts to take into account solute–solvent interactions focussed on varying the dielectric constant in electrostatic energy terms used to describe pairwise interactions between species in the solute molecule, but, of course, there will be preferred hydration sites. Because of the intrinsic dielectric properties of the solute and the voids produced by unfavourable solute group–solvent interactions the "local dielectric" behaviour about the solute is heterogeneous and can be significantly different in magnitude from that of the bulk solvent. No model has been developed in which the local dielectric varies with spatial position. Attempts have been made to treat the dielectric effect as due to the polarisation of individual atoms leading to induced dipoles. It is concluded that an enzyme can stabilise charge better than water because, in contrast to water, the enzyme dipoles are kept oriented toward the charge even when the field from the charge is small. In water the dipoles of the first few solvation shells cannot be strongly polarised towards the solute charges because they interact with the surrounding randomly oriented bulk dipoles.

A promising lead in the estimation of electrostatic interactions is the use of electrostatic molecular potentials [24]. This method simulates the coulombic contribution to the intermolecular interaction by substituting appropriately chosen point charges, which may be fractional, to replace actual neighbouring molecules. The potential is a function of the electronic distribution and the position of the nuclei in the molecule and is computed from molecular wave functions. It is the basic premise of the use of electrostatic potentials that E_{pol}, E_{ct} and E_{disp} are much less important than electrostatic effects in determining structural features of complex formation. The deformations of the special charge distribution of a set of atoms bound together to form a molecule are quite complex and the electrostatic potential, like all other 1-electron observables, is correct to the second order in the Hartree–Fock approximation. Formally, the electrostatic potential is the value of the interaction between a molecule and a point charge placed at the point k (Eqn. 66).

$$E_{es}(AB) = \sum_{k} V_A(k)_{qkB} \tag{66}$$

Similar chemical groups have portions of space where the shape of V is reasonably similar. An interpretation of V can be obtained by considering V as the sum of local contributions, transferable from molecule to molecule. This transferability may be exploited to get a rationale of the shape of V in different molecules and to shed light on the effect of chemical substitution on the values assumed by V near a given group in a set of related molecules. In the next section we will see that it is difficult to quantitative explain the large binding energies observed between small substituents, particularly alkyl, and proteins. Alkanes show impressive features in the electrostatic potential, notably a negative region (i.e. attracts an incoming positive charge) associated with CH_3 and CH_2 with the minima at the axis of local symmetry. For example, the minima for CH_3 and CH_2 in propane are 17.2 and 13.0 kJ/mole, respectively (Fig. 11) [24]. Incidentally, V remains essentially unaltered if one changes to a larger basis set calculation. The contributions of the neighbouring groups are sometimes important in determining the shape of V; distant groups can even reverse the sign of V near the CH_3 or CH_2 groups e.g. in glycine. A general property of electrostatic potentials is that different atoms with different local geometries can yield the same electrostatic potential and thus, presumably, the same binding energy. This could be important in the design of enzyme inhibitors and of drugs in general. The combined action of several atoms bound together is not at all

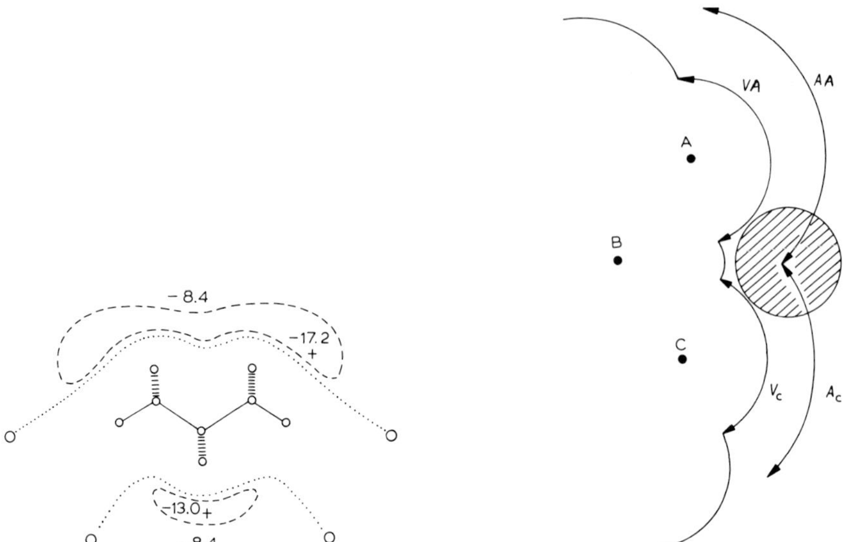

Fig. 11. The electrostatic potential of propane showing the negative regions. The data is from ref. 24.

Fig. 12. The concept of accessible surface area. The three atoms A, B and C have Van der Waals envelopes, V_A, V_B and V_C respectively which define the surface of the protein. Atoms A and C have accessible surface represented by the arcs A_A and A_C. A and C sterically prevent any contact between the water molecule (shaded) and atom B which, therefore, has no accessible surface. Reproduced from ref. 27 with permission from the Copyright holder.

apparent from a "ball and stick" model but is an automatic result of electrostatic potentials.

Considerable electrostatic interaction can arise from the dipole moment of 3.5 *D* for the peptide unit. In an α-helix the peptide dipole moments are aligned nearly parallel to the helix axis giving rise to a significant electrical field [25]. It's dipole runs from the C-terminal to the positive end of the N-terminal. For example, the potential at 5 Å from the N-terminus of an α-helix of 10 Å length is ca. 0.5 V so that, ignoring the effect of solvent, a negatively charged group at that position involves an attractive energy of ca. 50 kJ/mole. It has been suggested [26] that the electrostatic field from the N-terminal α-helix favours proton transfer from CysH-25 to His-159 by stabilising charge separation in $CysS^- \ldots HIm^+$.

Hydrophobicity

The factors so far examined have ignored the contribution of the solvent entropy to the thermodynamic balance of association. If water molecules that surround the enzyme and substrate are released upon binding this may make a favourable contribution to the free energy of binding if the water molecules released are less restricted in the bulk solvent. The hydrophobic energy of enzyme–substrate complexes is proportional to the loss of accessible surface area that occurs upon binding.

For an atom in the enzyme or the substrate to interact with the solvent it must be able to form Van der Waals contact with water molecules. The accessible surface area of an atom is defined as the area on the surface of a sphere, radius R on each point of which the centre of a solvent molecule can be placed in contact with the atom without penetrating any other atoms of the molecule (Fig. 12). R is the sum of the Van der Waals radii of the atom and solvent molecule [27]. There is a linear relationship between the solubility of hydrocarbons and the surface area of the cavity they form in water [28]. It has been estimated that the hydrophobicity of residues in proteins is 100 J/mole/Å^2 of accessible surface area [29]. The surface tension of water is 72 dynes/cm^2 so to form a free surface area of water of 1 Å^2 costs 435 J/mole/Å^2. The implication is that the free energy of cavity formation in water to receive the hydrophobic group is offset by favourable interactions (dispersion forces) between the solute and water.

To achieve the same interaction with each other as they do with water the surfaces of the enzyme and the substrate must be complementary to give a closely packed complex.

Dispersion forces

The energy of attraction between two *isolated* atoms or molecules was first evaluated by London who obtained the approximate expression (Eqn. 67),

$$U_{(r)} = -\frac{C_1}{r^6} - \frac{C_2}{r^8} - \frac{C_3}{r^{10}} \tag{67}$$

where r is the distance between the two molecules and the constants C_i depend on the physical properties of the interacting partners. The three terms represent, in order, dipole–dipole, dipole–quadrupole and quadrupole–quadrupole interactions. Although the last two terms are often neglected in numerical calculations this is only justified for $r > 8$ Å. Molecules such as CO_2, N_2 and O_2 do not possess dipole moments but can be characterised by quadrupole moments, although these are often not known with certainty.

Physical adsorption of molecules onto solid surfaces possesses certain similarities to enzyme–substrate binding. The atoms in a solid present a closely packed surface and in some cases, for example graphite, the interatomic distance is similar to that found in covalently bonded molecules. Instead of the normal r^{-6} dependence of the London attractive potential between a pair of molecules there is a r^{-3} dependence for the interaction between a molecule and a solid. This longer range interaction is the basic reason why gases are adsorbed at pressures lower than those at which they condense to liquids or solids. The differential heat of adsorption is often of the order of twice the heat of condensation of the adsorbed vapour.

Dispersion energy is additive i.e. the total energy for a collection of molecules can be expressed as a sum of pair-wise interactions over all pairs of the configuration. So although the potential energy curve for a molecule approaching a surface is similar to the standard Lennard–Jones type curve commonly shown in textbooks for the interaction between a *pair* of molecules, the distance effective interaction is more extended in the case of the surface. Summation procedures are tedious but qualitative information can be obtained by integration.

The enzyme surface may be replaced by a semi-infinite continuum with a mathematically plane bounding surface. The density of the continuum, ρ, represents the number density of enzyme atoms and if z is the shortest distance between the bound atom and the surface the total interaction energy is given by Eqn. 68

$$W_{(z)} = \int_{z}^{\infty} \int_{-\infty}^{\infty} \int_{-\infty}^{\infty} U_{(r)}\, \mathrm{d}x\, \mathrm{d}y\, \mathrm{d}z \qquad (68)$$

where x and y are orthogonal coordinates in the plane of the surface [30]. Solving Eqn. 68 shows that the attractive potential falls off as z^{-3} in contrast to the interaction potential between *isolated* atoms which diminishes a r^{-6}.

Because dispersion and repulsion potentials between two partners fall off so rapidly with distance the binding energy depends very much upon the shape of the surface. For example, a molecule bound at the centre of a hemispherical cavity of radius z_0 (Fig. 13a) is subject to a dispersion potential 4 times as great as for a molecule located on a smooth surface [31]. Inside a long, narrow capillary, terminated by a hemisphere at whose centre the molecule is located (Fig. 13b) the attractive potential is roughly 7 times that of a plane surface [31]. Inside a topographic depression a bound molecule is in close contact with a larger number of atoms than on a smooth surface.

It is a consequence of the additivity law that the adsorption potential will be greatest if a maximum number of atoms in a molecule are in close contact with the surface. If other forces constrain the molecule to be in a different orientation the dispersion potential will be considerably reduced. For example, electrostatic forces tend to localise atoms at positions and orientations favourable for maximising electrostatic energies which may not be those which maximise the dispersion potential.

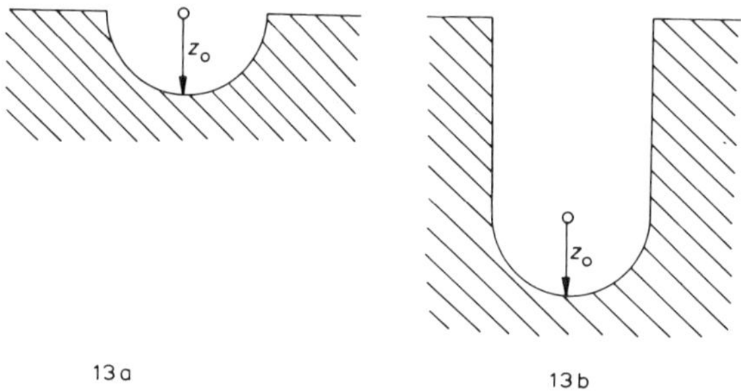

Fig. 13. Molecule bound at surfaces with a hemispherical cavity of radius Z_0 (a) and with a long capillary terminated by a hemisphere also of radius Z_0 (b).

14. Stress and strain

Stress is a force tending to change the dimensions of an object and may be tensile (stretching) compressive, shear or torsional. Strain is dimensionless and is given by the distortion divided by the original dimension. Measuring the stress necessary to cause a given strain in a solid gives a curve of the type shown in Fig. 15. The slope of the curve is known as the elastic modulus (stiffness) and is a measure of rigidity. The stress–strain curves of typical elastomers show marked deviations from the straight line required by Hooke's law (Fig. 14).

When an elastomer is stretched work is done on it and its free energy is changed. The elastic work W_{el} is given by $f\mathrm{d}l$ where f is the retractive force (Eqn. 69), and $\mathrm{d}l$ is the change in length.

$$f = \left(\frac{\mathrm{d}G}{\mathrm{d}l}\right) = \left(\frac{\mathrm{d}H}{\mathrm{d}l}\right) - T\left(\frac{\mathrm{d}S}{\mathrm{d}l}\right) \tag{69}$$

An ideal elastomer has zero enthalpy change upon stretching and the retractive force

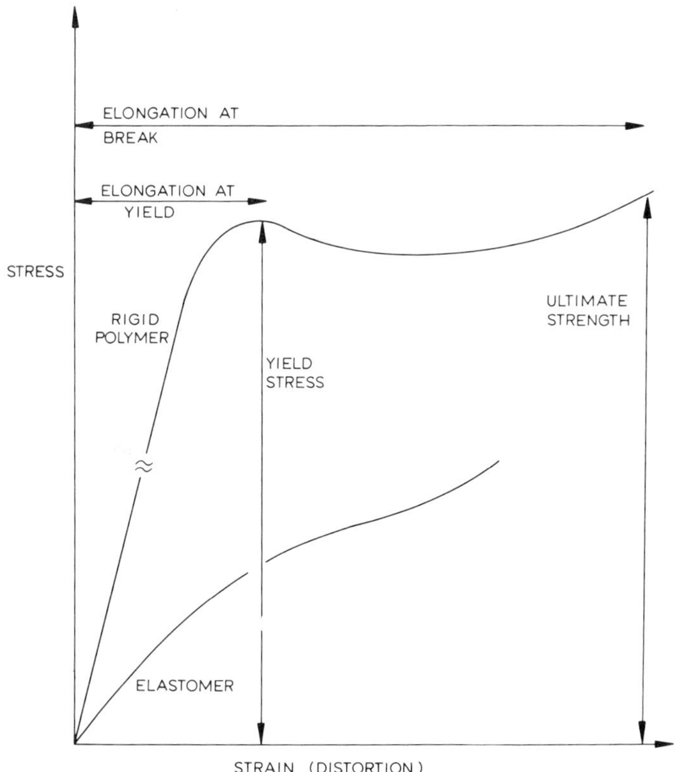

Fig. 14. Stress–strain curves of elastomers and rigid polymers.

is given by its decrease in entropy on extension. The molecular origin of this entropy elasticity is the distortion of the polymer chains from their most probable conformations in the unstretched sample i.e. a stretched elastomer is in a more ordered state. Unlike the restoring force of a spring obeying Hooke's law, the retractive force of an elastomer increases with increasing temperature, i.e. elastomers heat on adiabatic stretching and contract on heating. Mobile polymers store energy on impact of force – they deform.

The polypeptide chain of a globular protein that folds upon itself to form local regions of secondary structure, α-helices and β-sheets, gives a highly crystalline, ordered, closely packed system which is highly rigid. The regions of no specified secondary structure are much more mobile and elastic. If a force is applied to such a system then deformation and energy storage, largely $\delta\Delta S$, would occur in the regions of the polymer which are elastic.

One possible mechanism which would allow the binding energy to be expressed in the transition state but not in the ground state is if the energy is stored by making flexible regions of the enzyme more ordered in the ground state but not in the transition state. Diagrammatically this is illustrated in Fig. 15 – the free enzyme comprises of predominantly rigid regions with a few flexible areas. Binding of the substrate could decrease the entropy of the more elastic regions so that the observed free energy of binding is less favourable than it would be in the absence of this change. If, in the transition state, the flexible regions of the enzyme are in the higher

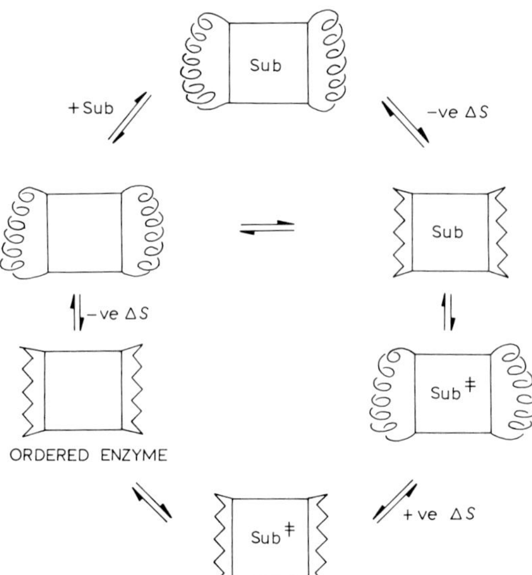

Fig. 15. Hypothetical enzyme with flexible regions which are disordered in the free enzyme. Binding of the substrate causes the enzyme to decrease its flexibility and entropy. Formation of the transition state returns the enzyme to its disordered state.

entropy state found in the free enzyme the full binding energy could be realised. This mechanism requires the enzyme to become "strained" upon initial binding of the substrate, by decreasing its entropy, but to increase its entropy upon reaching the transition state by the temporarily ordered regions returning to their more flexible state. The beauty of this mechanism is that strain is induced in the ground state by decreasing the entropy of the system and not by geometrical deformation of the substrate or enzyme, which, as we have seen, is very difficult. It also has a mechanical analogy in that if a force (binding energy) is applied to a system consisting of a rigid body (the substrate) held to other rigid bodies connected by flexible regions (the enzyme) deformation will occur in the elastic regions.

Geometrical distortion of the substrate by the enzyme is unlikely to be a significant factor in catalysis. The intermolecular forces of binding are generally too weak and flexible to distort the substrate. The intermolecular force field cannot overcome the intramolecular force field of the substrate.

Quantum mechanical and semi-empirical calculations support this suggestion. For example, it has been shown that, contrary to earlier suggestions, lysozyme does not distort the chair conformation of polysaccharide pyranoside units [32]. Similarly, it has been suggested that the serine proteases distort the trigonal planar arrangement around the carbonyl carbon of amides and peptides (XVII) to a non-planar conformation which resembles the tetrahedral intermediate (XVIII). Support for this hypothesis came from X-ray studies of the trypsin–trypsin inhibitor complex [33].

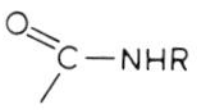

O=C—NHR　　O=C—NHR

XVII 　　XVIII

However, the resonance energy of amides is about 70 kJ/mole and distortion to a non-planar arrangement is energetically expensive. It has recently been shown that the X-ray data are not unambiguous and the trypsin inhibitor itself exhibits distorted structures when not bound to the enzyme [34].

So-called transition state analogues which bind to the enzyme more tightly than natural substrates probably do so not so much because of a geometrical complementarity to the enzyme but more as a result of compatible electrostatic interactions. For example, it seems more likely that lactones (XIX) bind to lysozyme better

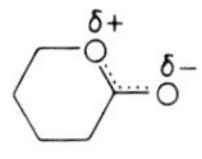

δ+ O　δ− O

XIX

than sugar substrates because of the attractive force between the electron-deficient carbonyl carbon (resembling that of the carbonium ion intermediate) and the carboxylate of Asp-52 rather than because of a geometrical analogy between the lactone and the intermediate carbonium ion.

It seems that a major contribution to the rate enhancement and specificity exhibited by enzymes is complementarity between the electrostatic potentials of the substrate and the enzyme in the transition state.

15. Estimation of binding energies

Observed free energies of binding substrates to enzymes are often less than the intrinsic binding energy because much of the binding energy is "used up" in bringing about the required loss of entropy and in inducing any destabilisation of the substrate or enzyme. Observed enthalpy and entropy changes for binding or activation cannot be used to deduce binding energies because of the dominant role that the solvent water plays in determining the measured values. The intrinsic binding energy of a small group A may be obtained from a comparison of the free energies of binding the larger molecule B and the molecule A–B, where A is a covalently bound substituent. The tight binding of the small substituent or "molecule" A to an enzyme requires the loss of its translational and rotational entropy which makes an unfavourable contribution to the overall free-energy change. However, when A is a substituent in the molecule A–B most of this entropy loss is accounted for in the binding of B because the loss of entropy upon binding A–B will be much the same as that upon binding B because the total translational and rotational entropy of the large molecule B will be much the same as that for A–B. The difference in the free energies of binding A–B and B therefore gives an estimation of the true binding energy of A (Eqns. 70–72).

$$\Delta G_{AB} - \Delta G_B = \Delta H_{AB} - T\Delta S_{AB} - \Delta H_B + T\Delta S_B \tag{70}$$

$$T\Delta S_{AB} = T\Delta S_B \tag{71}$$

$$\Delta G_{AB} - \Delta G_B = \Delta H_{AB} - \Delta H_B = \Delta H_A \tag{72}$$

The difference, $\Delta G_{AB} - \Delta G_B$, is very much greater than ΔG_A, the observed free energy of binding A, because it is free of the unfavourable entropy term accompanying the binding of A. Consequently, if $(\Delta G_{AB} - \Delta G_B)$ gives a very favourable negative free-energy change for the intrinsic binding energy of A it would be *incorrect* to conclude that the favourable binding of A–B was mainly due to the favourable binding of A. The binding energy of B in A–B has been 'used' to compensate for the large loss of translational and rotational entropy upon binding B and A–B to the enzyme. The maximum increase in free energy from the entropy loss of binding is ca. 150 kJ/mole. If the anchor molecule B is only loosely bound then the intrinsic binding energy of A in A–B may cause a greater restriction of motion in A–B so that $T\Delta S_{AB} > T\Delta S_B$ and there will be no change, or even a decrease, in the observed binding constant. Intrinsic binding energies estimated in this way are, therefore, likely to be lower limits.

A comparison of ΔG_A with $(\Delta G_{AB} - \Delta G_B)$ provides a lower limit to the intrinsic loss of entropy upon binding A (Eqn. 73).

$$\Delta G_{AB} - \Delta G_B - \Delta G_A = \Delta H_{AB} - \Delta H_B - \Delta H_A + T\Delta S_A = T\Delta S_A = K_A K_B / K_{AB} \tag{73}$$

At a standard state of 1 M the maximum loss of entropy upon binding is about -150 J/K/mole and therefore a comparison of $T\Delta S$, obtained as in Eqn. 73, with this value allows an estimation of the tightness of binding for A. ΔG_{AB} may be up to 45 kJ/mole (a factor of 10^8 M in binding constants) more favourable than the sum of ΔG_A and ΔG_B.

The most interesting problem to arise from the estimation of intrinsic binding energies is the apparently large values associated with small substituents. Examples are 9–16 kJ/mole for CH_2 [35]; 21 kJ/mole for SCH_3 [36]; 34 kJ/mole for OH [37] and 23–38 kJ/mole for SH [38]. These intermolecular interactions are much greater than those generally observed between solute and solvent or between molecules within a crystal.

If the interior of a protein is analogous to a liquid the transfer of a group from water to a non-polar liquid may be taken as a model for the transfer of the group from water to the enzyme (Eqns. 74–75).

$$\underset{\text{water}}{\text{-CH}_2\text{-}} \xrightarrow[\text{kJ/mole}]{\Delta G = -13} \underset{\text{enzyme}}{\text{-CH}_2\text{-}} \qquad (74)$$

$$\underset{\text{water}}{\text{-CH}_2\text{-}} \xrightarrow[\text{kJ/mole}]{\Delta G = -4.2} \underset{\text{non-polar liquid}}{\text{-CH}_2\text{-}} \qquad (75)$$

However, the change in energy for a methylene group, about -4.2 kJ/mole for transfer to a non-polar liquid (it is about -2.1 kJ/mole for transfer to dioxane or ethanol) is at least 3 times *less* favourable than the change upon transfer to an enzyme. There are many other examples which show that these estimations of the interaction between a substituent and a non-polar liquid underestimate the forces between the same substituent and an enzyme.

The comparison between the two processes (Eqns. 74 and 75) involves several changes in interaction energies. As the initial state is the same in the two systems, changes in energy when the solute is removed from water will be the same in each case. When a solute is transferred to a non-polar liquid (Eqn. 75) there is a gain of solute–solvent interactions and a loss of solvent–solvent interactions, which is equivalent to the free energy of formation of a cavity in the solvent of a suitable size to accommodate the solute molecule.

Transfer to an enzyme (Eqn. 74) involves a gain of solute–enzyme interactions if water is initially absent from the binding site. If water is initially bound to the site of binding and is displaced by the substituent there is a gain of water–water interactions and a loss of water–enzyme interactions. However, the sum of the free-energy changes of these two interactions would be thermodynamically unfavourable, otherwise water would not initially be bound to the enzyme, and hence the observed free energy of transfer of a substituent from water to an enzyme provides a *lower* limit to the magnitude of enzyme–substrate interactions.

Because protein crystals are usually highly hydrated (27–65%) it is usually assumed that protein structures determined in the solid state are identical to those in solution. A protein is surrounded by a "hydration shell' composed of several types of water. The similarity of crystal and solution structures are dependent on the force fields generated by the crystal and the "hydration shell" respectively. There are a few water molecules (ca. 10–20 per protein) that are held very strongly in specific locations such as the 16 in the interior of chymotrypsin, the 9 in the active site of carbonic anhydrase C and the 8 within subtilisin. The presence of these water molecules within the protein is inconsistent with the oil drop model of globular proteins. Another class of water in aqueous protein solutions is bound water which is much more abundant than tightly bound water and forms roughly a monolayer about the protein with rotation times near 10^{-9} sec. The enthalpy of water–peptide hydrogen bonds (H_2O as the proton donor), is smaller than the bulk water–water interactions, as can be seen from IR shifts which are related to hydrogen-bonding enthalpy. The free energy of binding water to the amide NH is 1.3 kJ/mole and to the amide CO 7.9 kJ/mole compared with 10.1 kJ/mole in liquid bulk water. The free energy of solvation of amides and peptides is ca. 35–40 kJ/mole [39].

If water is *not* initially bound to the enzyme at the site where the substituent will be bound then the transfer in Eqn. 74 could be thermodynamically more favourable than that in Eqn. 75 because the latter process involves a loss of solvent–solvent interactions, i.e. the free energy of cavity formation, and thus the free-energy change for Eqn. 75 provides an underestimation of solute–solvent interactions. According to the scaled particle theory, and other calculations, the free energy of cavity formation in non-polar liquids is of similar magnitude, but, of course, of opposite sign, to the free energy of interaction of the solute with the solvent. This, therefore, may explain, in some cases, part of the discrepancy between Eqns. 74 and 75 because the enzyme may have a ready-made "cavity" and the binding of a substituent to this site would not result in the unfavourable changes which accompany cavity formation in liquids.

An important factor is that the *number* of the favourable dispersion or Van der Waals interactions of the substituent with the enzyme are greater than with a liquid solvent. The atoms in an enzyme molecule are very closely packed as they are linked together by covalent bonds and, in some cases, the polypeptide chains are cross-linked by covalent disulphide bonds. The fractions of space occupied by the atoms in a molecule of protein is 0.75 which may be compared with 0.74, the value for the most efficient known way of packing spheres, cubic close-packing or hexagonal close-packing [22]. In a liquid the fraction of space occupied is very much less, for example upper limits for water, cyclohexane and carbon tetrachloride are 0.36, 0.44, and 0.44, respectively. The inside of a protein thus contains little space and may be more analogous to a solid than to a liquid. If Eqn. 74 is compared with the free energy of transfer of a methylene group from water to a solid Eqn. 76,

$$\underset{\text{water}}{\overline{\underline{-CH_2-}}} \longrightarrow \underset{\text{pure liquid}}{\overline{\underline{-CH_2-}}} \longrightarrow \underset{\text{solid}}{\overline{\underline{-CH_2-}}} \qquad (76)$$

of about −8.3 kJ/mole, it is still less favourable than transfer to its binding site on the protein [35].

The atoms in an enzyme surrounding the substrate are very compact and most are separated only by their covalent radii. In a liquid or a solid nearest neighbour interactions are controlled by their Van der Waals radii. The effective coordination number is thus much greater in the enzyme than it is in a liquid or solid. For example, the Van der Waals radii for carbon, nitrogen, and oxygen are about 2.2 times greater than their respective covalent radii and a very simple treatment of the packing of spheres shows that the coordination number of a sphere is about 2.5 times greater when surrounded by other spheres separated by such covalent radii than when they are separated by Van der Waals radii [15]. In a liquid, and even in a solid, a large fraction of the surface of each molecule is surrounded by empty space between other molecules.

References

1 Anfinsen, C.B. (1973) Science 181, 223.
2 Tanford, C. (1968) Adv. Protein Chem. 23, 121.
3 Creighton, T.E. (1978) Progr. Biophys. Mol. Biol. 33, 231.
4 Richards, F.M. (1977) Annu. Rev. Biophys. Bioengng. 6, 151.
5 Richardson, J.S. et al. (1976) J. Mol. Biol. 102, 221.
6 Williams, R.J.P. (1979) Biol. Rev. 54, 389.
7 Jardetzky, O. (1981) Acc. Chem. Res. 14, 291.
8 Callen, H.B. (1960) Thermodynamics, Wiley and Sons, New York, Ch. 15.
9 Weber, G. (1975) Adv. Protein Chem. 29, 1.
10 Page, M.I. (1983) in: D.J. Galas (Ed.), Accuracy in Molecular Biology, Marcel Dekker, New York, in press.
11 Fersht, A.R. (1977) Enzyme Structure and Mechanism, Freeman, San Francisco, CA.
12 Jencks, W.P. (1983) in: From Cyclotrons to Cytochromes – A Symposium in Honor of Martin Kamen's 65th Birthday, August 27–31, 1978, La Jolla, CA, in press.
13 Jencks, W.P. (1975) Adv. Enzymol. 43, 220.
14 Page, M.I. and Jencks, W.P. (1971) Proc. Natl. Acad. Sci. (U.S.A.) 68, 1678.
15 Page, M.I. (1977) Angew. Chem. Int. Ed. Engl. 16, 449.
16 Page, M.I. (1973) Chem. Soc. Rev. 2, 295.
17 Jencks, W.P. and Page, M.I. (1972) Proc. Eighth FEBS Meeting, Amsterdam 29, 45.
18 Jones, C.C., Sinnott, M. and Souchard, I. (1977) J. Chem. Soc. Perkin II, 1191.
19 Jencks, W.P. (1980) in: F. Chapeville and A.L. Haenai (Eds.), Chemical Recognition in Biology, Springer, New York.
20 Page, M.I. (1980) Int. J. Biochem. 11, 331.
21 Fersht, A.R. (1974) Proc. Roy. Soc. London Ser. B 187, 397.
22 Fischer, E. (1894) Ber. Dtsch. Chem. Ges. 27, 2985.
23 Schweizer, W.B., Procter, G., Kaftory, M. and Dunitz, J.D. (1978) Helv. Chim. Acta 61, 2783; Rondan, N.G., Paddon-Row, M.N., Caramella, P. and Houk, K.N. (1981) J. Amer. Chem. Soc. 103, 2436; Cieplak, A.S. (1981) J. Am. Chem. Soc. 103, 4540.
24 Scrocco, E. and Tomasi, J. (1978) Adv. Quantum Chem. 11, 115; Pullman, A. and Berthod, H. (1978) Theoret. Chim. Acta 48, 269.
25 Hol, W.G.T., van Duijnen, P.T. and Berendsen, H.J.C. (1978) Nature (London) 273, 443.
26 Van Duijnen, P.T., Thole, B.T., Broer, R. and Nieuwpoort, W.C. (1980) Int. J. Quant. Chem. 17, 651.
27 Lee, B. and Richards, F.M. (1971) J. Mol. Biol. 55, 379.
28 Hermann, R.B. (1972) J. Phys. Chem. 76, 2754.
29 Chothia, C. (1974) Nature (London) 248, 338.
30 Honig, J.M. (1954) Ann. N.Y. Acad. Sci. 58, 741.

31 De Boer, J.H. and Custers, J.F.H. (1934) Z. Physik. Chem. B 25, 225; De Boer, J.H. (1936) Trans. Faraday Soc. 32, 10.
32 Levitt, M. (1974) in: E.R. Blout, F.A. Bovey, M. Goodman and N. Lotan (Eds.), Peptides, Polypeptides and Proteins, Wiley, New York, pp. 99–113; Warshel, A. and Levitt, M. (1976) J. Mol. Biol. 103, 227.
33 Huber, R. and Bode, W. (1978) Acc. Chem. Res. 11, 114.
34 Fersht, A.R., personal communication.
35 Page, M.I. (1976) Biochem. Biophys. Res. Commun. 72, 456.
36 Currier, S.F. and Mautner, H.G. (1976) Biochem. Biophys. Res. Commun. 69, 431.
37 Abeles, R.H. and Stubbe, J.A. (1980) Biochemistry 19, 5505.
38 Fersht, A.R. and Dingwall, C. (1979) Biochemistry 18, 1245, 1254.
39 Wolfenden, R. (1978) Biochemistry 17, 201.
40 Morris, J.J. and Page, M.I. (1980) J. Chem. Soc. Perkin II, 679.

CHAPTER 2

Non-covalent forces of importance in biochemistry

PETER KOLLMAN

School of Pharmacy, Department of Pharmaceutical Chemistry, University of California, San Francisco, CA 94143, U.S.A.

1. Introduction

The nature of non-covalent forces makes them important in biological function because they are specific without conferring as much "rigidity" as covalent forces. We define covalent forces as those quantum mechanical forces which determine the nature of electron pair chemical bonding. Non-covalent forces, on the other hand, do not involve electron pairing effects. Examples of covalent forces are those that make the reaction of two hydrogen atoms to form a hydrogen molecule an energetically favorable process by ~ 400 kJ/mole: the hydrogen bond between two water molecules in the complex HO–H... OH_2 is a typical example of a non-covalent bond, which has a strength of ~ 20 kJ/mole.

These non-covalent forces play an essential role in determining the shape of all molecules and the macromolecules which are important in biology are no exception. The base stacking and hydrogen bonding in the DNA double helix, the hydrogen bonding in the peptide α helix and the U shape of prostaglandin molecules are examples where non-covalent forces determine shape. The specificity in intermolecular association between proteins and nucleic acids and between enzymes and their inhibitor is also a result of the nature of these non-covalent forces.

1.1. The basis of the non-covalent forces

The non-covalent forces have a quantum mechanical basis, but much of their nature can be understood using a classical or semi-classical picture. Studies of intermolecular interactions in the gas phase [1] suggest that the intermolecular (non-covalent) interaction energy between two molecules can be broken down into 5 main components: (1) electrostatic, (2) exchange repulsion, (3) dispersion, (4) polarization and (5) charge transfer.

We now proceed to give a physical explanation for each of these energy components. Examples of such interactions, their magnitude, and their angular and distance dependence are presented in Table 1.

The electrostatic energy (1) is the energy from the interaction between the charge distribution on one molecule with that of the other. Examples of this are ion–ion interactions, such as between Na^+ and Cl^-, ion–dipole interactions such as between Na^+ and H_2O, and dipole–dipole interactions, such as between two H_2O molecules.

Michael I. Page (Ed.), The Chemistry of Enzyme Action

TABLE 1
Summary of important contributions to noncovalent association

Energetic (enthalpic) contributions	Functional form	Magnitude of ΔH(gas phase)	Magnitude of ΔH(soln)
1. Electrostatic			
ion–ion	$\frac{q_A q_B}{r_{AB}}$	− − −	~ 0
ion–dipole	$\frac{\mu_A \mu_B \cos\theta}{r_{AB}}$	− −	~ 0
dipole–dipole	$\frac{\mu_A \mu_B}{r_{AB}} \cdot f(\theta_A, \theta_B)$	−	~ 0
2. Exchange repulsion	$A_e^{-\alpha r_{AB}}$	+	+
3. Dispersion	$\frac{-B}{r_{AB}^6}$	−	−
4. Polarization	$\frac{-1/2 q_A^2 \alpha_B}{r_{AB}^4}$	−	−
5. Charge transfer	$Ce^{-Br_{AB}}$	−	−

Entropic contributions	ΔS(gas phase)	ΔS(aqueous soln)
1. Trans/rotation/vib. freq. changes	− −	−
2. Hydrophobic effects	0	+

Although the electrostatic interaction arises from the interaction of fixed nuclei and spread out electron density distribution of the molecules, its nature can be adequately represented with simple models which locate net charges (or higher moments) on each atom and classically evaluate the interactions between such charges [2].

The exchange repulsion energy (2) is an energy component which is always repulsive, i.e., which is responsible for the finite density of molecules. It arises from the Pauli principle repulsion between the electron clouds of the molecules. It should be emphasized that this component is fundamentally quantum mechanical, i.e., unlike the electrostatic component, there is no classical "model" for this phenomenon. Although classically, the repulsion between the positively charged nuclei would be expected to keep atoms and molecules apart, such a classical repulsion would occur at much shorter distances than that due to the Pauli principle repulsion. The Pauli principle repulsion is much shorter range than typical electrostatic components and is represented in analytical models with functions that depend on the inverse 9th, inverse 12th or exponential power of the atom–atom distances [3].

The dispersion attraction (3) is a universal attraction that comes from "instantaneous" dipole–dipole attractions of induced charge distributions on one molecule due to the other. This attraction, unlike the electrostatic component, even exists between rare gas atoms, such as helium and comes about because, when these atoms approach each other, the fluctuations in the electron distributions on one atom resulting in an "instantaneous" dipole $\delta^+ He \delta^-$ are felt by the nearby molecule and cause its electrons to polarize in the following manner $\delta^+ He \delta^-$, such that the electronic configuration of the dimer $\delta^+ He \delta^- \ldots \delta^+ He \delta^-$ is net attractive. Such an attraction is relatively short range, having an inverse 6th power dependence on the intermolecular separation [4].

The polarization energy (4) is somewhat analogous to dispersion in that it is always *attractive* and arises because of electron redistribution of one molecule due to the presence of the second. It is different than dispersion, however, because it comes from the permanent charge asymmetry on one molecule causing a charge redistribution on the other. The distance dependence of such an energy term depends on the charge distribution of the molecules, being inverse 4th power for ion-induced dipole and inverse 6th power for dipole-induced dipole.

The final energy component is charge transfer (5), which is a quantum mechanical term arising from electron delocalization from the occupied orbitals of one molecule into unoccupied molecules of the other. This energy component depends on the overlap between these orbitals, which has an approximate exponential dependence on the atom–atom distance and has a directional dependence which depends on how the orientation of the orbitals effects their overlap.

Table 1 summarizes the pertinent facts about these energy components. In quantum mechanical studies of the interactions of small molecules [1,2], all the components have chemically significant values, although the largest terms are typically electrostatic and exchange repulsion. In fact, we have argued [2] that one can reasonably model the relative strength and directionality of hydrogen bonding interactions with only the electrostatic term, provided that one constrains the distance between molecules to physically reasonable values. This success in modeling such interactions is often due to the somewhat fortuitous cancellation of the attractive charge transfer and repulsive exchange repulsion terms, the relatively small size of the polarization energy and the relative lack of directionality of the dispersion energy. Nonetheless, it suggests a basis for the success of the many models [5–7] which simulate biologically interesting molecules using only the electrostatic, exchange repulsion and dispersion components. By using "effective" electrostatic charges, one can often simulate polarization effects; and by reducing the exchange repulsion terms involving hydrogen bonding hydrogens, one can often compensate for the lack of inclusion of the charge transfer component. It should be emphasized that such simple models are much less appropriate in modeling molecules containing atoms beyond the second row of the periodic table.

2. *The thermodynamics of non-covalent interactions*

2.1. *Gas phase interactions*

An accurate calculation of the energetics of a non-covalent interaction at 0^0K

A + B → AB (ΔE_0) tells us nothing about whether one can observe such a complex experimentally. At temperature T, one needs to know the change in free energy, ΔG_T, for the interaction. The free energy change can be determined from the equations:

$$\Delta G_T = \Delta H_T - T\Delta S \qquad (1)$$

$$\Delta G_T = \Delta E_T + \Delta(PV) - T\Delta S_T, \qquad (2)$$

which for ideal gases reduces to

$$\Delta G_T = \Delta E_T + (\Delta n)RT - T\Delta S_T \qquad (3)$$

where T is the absolute temperature, ΔS_T is the entropy change for the process at temperature T, ΔE_T the energy change for the process at temperature T, R the universal gas constant and Δn the charge in the number of molecules in the "reaction", i.e., complex formation. To calculate ΔS_T and determine ΔE_T from ΔE_0, we need to use statistical mechanics. This "textbook" [8] calculation is simple if we assume the molecules are rigid rotors and harmonic oscillators, and requires a knowledge of the molecules' masses, their moments of inertia and their vibrational frequencies. These can be quite accurately determined or estimated for small molecules. We can illustrate with a sample calculation of the dimerization free energy of water $2H_2O \rightarrow (H_2O)$ at 300 K. Accurate quantum mechanical calculations suggest that $\Delta E_0 = -5.5$ kcal/mole, discussed in refs. 9–11, which is in reasonable agreement with the most direct experimental determination of this quantity. To determine ΔE_{300} from ΔE_0, we estimate the energy in each vibrational, rotational and translation mode in the monomer and dimer and subtract twice this average monomer "motional" energy from that in the dimer. Because the two monomers have a lower average energy than the dimer, ΔE_{300} is about 2 kcal/mole less negative (a less favorable interaction) than ΔE_0. ΔH_{300} is about 0.6 kcal/mole more negative than ΔE_{300} and thus ΔH_{300} (calculated) is about -4.1 kcal/mole; the experimentally determined value [12] is -3.6 ± 0.5 kcal/mole.

The entropy for the process A + B → AB is very negative. This is because 3 translational and 3 rotational degrees of freedom in the monomers are converted to 6 low frequency vibrations in the dimer. Translations and rotations have more freedom than vibrations and, thus, entropy is lost upon dimerization. The lower the frequency of vibrational modes, the higher the contribution to the entropy of the mode. Thus, in non-covalent interactions involving the "rigidification" of the active site to bind ligand or the "freezing" of conformations to obtain an energetically favorable fit, the entropy change is unfavorable for the binding process. For $(H_2O)_2$, the ΔS_{298} is estimated [10] to be $+9$ kcal/mole (standard state = 1 atm) and thus ΔG_{298} is $+5$ kcal/mole. We thus see how, despite an energetically favorable E_0 for water dimerization, the free energy for such dimerization is not very favorable; thus there is little tendency for water dimerization in the gas phase at equilibrium; only under nozzle beam conditions [13] has a water dimer been observed.

In summary, there are two main factors one must consider in gas phase non-covalent interactions of molecules; the first is the energy of the interaction, which, provided the molecules are oriented favorably and are not like charged, is attractive (favors associations). On the other hand, the entropy of association is quite negative and disfavors association. The two terms are often correlated, i.e., the stronger and more favorable the ΔE, the more rigid the complex and the larger and more negative the ΔS.

2.2. Solution phase association

The difference between gas phase and solution association can be best understood with the help of a thermodynamic cycle:

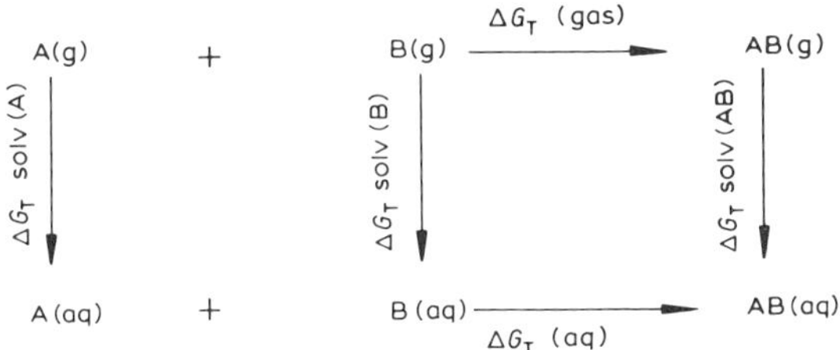

The free energy of association of A and B in aqueous solution at temperature T ΔG_{T}(aq) is related to the solvation free energies of A(ΔG_{T}solv(A)), B(ΔG_{T}solv(B)) and AB(ΔG_{T}solv(AB)) and the gas phase free energy of association ΔG_{T}(gas) by

$$\Delta G_{\mathrm{T}}(\mathrm{aq}) = \Delta G_{\mathrm{T}}(\mathrm{gas}) + \Delta G_{\mathrm{T}}\mathrm{solv}(\mathrm{AB}) - \Delta G_{\mathrm{T}}\mathrm{solv}(\mathrm{A}) - \Delta G_{\mathrm{T}}\mathrm{solv}(\mathrm{B}) \quad (4)$$

These solvation free energies are characterized by much more complicated processes than the gas phase energies, since they involve interactions between A, B and AB and many water molecules. However, there is experimental data on the free energy of transfer of molecules from the gas phase to water, as well as from water to other solvents. If A, B and AB are ions, the free energies of transfer for these molecules from the gas phase to water are very favorable (ΔG is large and negative) because of the strength and large number of favorable ion–water dipolar interactions. If A, B and AB contain mainly polar and hydrogen bonding groups, the free energy of transfer for such molecules from the gas phase to water is quite favorable because of dipole–dipole (hydrogen bonding) interactions. The same energy components which dominate the gas phase associations are operative here. On the other hand, the free energy of transfer from the gas phase to water of molecules that contain mainly non-polar groups is relatively unfavorable, despite the attractive dispersion and polarization interactions between the non-polar groups and water. This lack of solubility of hydrocarbons in water is well known and is due to the "hydrophobic effect", explained in more detail below.

It is clear from Eqn. 4 how the solvent water dramatically changes the free energy of association of A with B from its value in the gas phase. For example, for A = Na^+ and B = Cl^-, ΔG_{T}(gas) is of the order of -400 to -600 kJ/mole. ΔG_{T}(solv, A) and

ΔG_T(solv, B) are much more negative than ΔG_T(solv, AB) because AB is neutral rather than charged and the sum of the last 3 terms on the right side of Eqn. 4 is nearly identical in magnitude to the first term on the right side and of the opposite sign. Thus, because of the much greater interactions of water with A and B than AB, there is relative little tendency for ion pairing in water, in contrast to the situation in the gas phase. Water has a great leveling effect, in that the stronger the non-covalent interactions between A and B in the gas phase, the stronger they interact with water. Thus, electrostatic dominated interactions such as ion pairing, ion–dipole and dipole–dipole, which are the intrinsically strongest gas phase interactions, often contribute little to aqueous solution free energies of association, because the ion–water and dipole–water interactions which are lost when A and B associate are approximately equal to those gained in the association.

On the other hand, if A and B are non-polar molecules, their association is likely to be significantly more favorable in aqueous solution than in the gas phase, for reasons that are rather subtle. This hydrophobically driven association [14,15], which has stimulated much research, is due to the "release" of water molecules from A and B when the molecules associate and shows up mainly in the change in entropy (ΔS_T) upon association. This effect must be initially analyzed by examining thermodynamic data for the transfer of a hydrocarbon from a non-polar solvent or the gas phase to water. The ΔG(soln, A) for such a process is positive (unfavorable), despite the fact that ΔH_{soln} for the process is either near zero or slightly negative (favorable). On the other hand ΔS(soln, A) for such a process is very negative. Initially, one might expect ΔH(soln, A) to be positive because, by placing a hydrocarbon in water, one is replacing a strong attractive electrostatic water–water interaction with a less attractive (dispersion and polarization only) hydrocarbon–water interaction. They very negative ΔS(soln, A) gives a clue. Computer simulations of such processes [16,17] indicate that the water molecules in the vicinity of the hydrocarbon compensate for the weak hydrocarbon–water interactions by forming stronger, more favorable water–water interactions and thus become more ordered because of stronger water–water hydrogen bonding. Thus little net H-bonding energy is lost, but the motion of the waters are restricted and the ΔS(soln, A) is very unfavorable (negative). On the other hand, when A and B associate and cause less of their non-polar groups to be exposed to water, this *release* of water has a favorable entropy and free energy and the association occurs. This hydrophobic effect is extremely important in many biological processes.

Table 2 contrasts the free energy cycle for association of a proton with NH_3 (a very strong gas phase electrostatic dominated interaction) with that of two methane molecules (a very weak gas phase association). As one can see, the aqueous solution free energies are very different from those in the gas phase.

One can see (Table 2) that the greater tendency for hydrocarbons to associate in water than found for association in the gas phase comes from the very unfavorable free energy of solution of the methanes. A methane dimer, by being associated and disturbing much less of the water structure than two monomers, has a much *less unfavorable* solvation free energy than two monomers. Because of this solvation

TABLE 2
Estimate of free energies in Eqn. 4 for protonation of NH_3 and dimerization of methane

Reaction	kcal/mole				
	ΔG_T(aq)	ΔG_Tsolv(AB)	ΔG_Tsolv(A)	ΔG_Tsolv(B)	ΔG_T(gas)
$H_3N+H^+ \rightarrow NH_4^+$ [a]	−4	−78	−2	−270	−198
$2CH_4 \rightarrow (CH_4)_2$	(−0.8)	(6.1) [b]	+6.3 [c]	+6.3 [c]	(+5.7) [d]

[a] See ref. 18 and discussion therein.
[b] Using the solvation free energy of ethane as a model for $(CH_4)_2$ solvation. The solvation free energy of butane is also very similar to this value (see ref. 19).
[c] Ref. 19.
[d] Estimated assuming a $\Delta H = -1$ kcal/mole in the gas phase and a ΔS identical to water dimer (−22.6 eu for a mole/l standard state) [10].

effect, the solution free energy of ~ −1 kcal/mole is much more favorable than that in the gas phase. This "hydrophobic" association due to the solvation effects compensates for the unfavorable translational/rotational entropy loss, which makes gas phase association so unfavorable.

The situation is completely different for the strong ion/polar association of NH_3 with H^+. Here the gas phase association is very favorable, but because of the much greater solution free energy of H^+ than NH_4^+, the solution free energy of association is only slightly favorable, and, at pH values above the pK_a, ΔG is no longer negative.

Let us now summarize the important energetic aspects of non-covalent interactions of A and B in solution, stressing the differences between gas phase and solution association. Polar and ionic molecules interact strongly with each other, but, because A–B interactions are compensated by A–H_2O and B–H_2O interactions, the net ΔH(soln) may be near zero even for strong A... B electrostatic interactions. On the other hand, these strong A... B interactions are important for biological specificity, since any electrostatically unfavorable A... B interactions would extract a heavy energy penalty.

One might expect that dispersion attractions would contribute little to AB solution association, because any A... B dispersion attraction would be compensated by loss of H_2O... A dispersion and H_2O... B dispersion upon association. However, the density of atoms in proteins is greater than that of water and so that ligand–protein dispersion attraction may be larger than the sum of ligand–H_2O and protein–H_2O and thus dispersion attraction may play an important energetic role in A... B interactions in aqueous solution.

The loss of translational and rotational entropy upon AB association is likely to be less important in solution than in the gas phase, but is still a significant thermodynamic contributor to a negative ΔS for AB association. On the other hand, for the association of non-polar molecules A and B, the hydrophobic effects provide a significant *positive* ΔS for AB association which is often larger than the negative ΔS due to translational and rotational entropy loss.

Two general papers on the thermodynamics of protein ligand association should be mentioned here. Sturdevant [20] has reviewed the contributors to ΔH and ΔS of association and has pointed out that, whereas these thermodynamic variables are very temperature dependent, the change in heat capacity (ΔC_p) is not and can give insight into the nature of the interaction. For example, the association of NAD^+ to LADH has a favorable ΔS at 0°C and is thus hydrophobically driven, whereas at 40°C the association is strongly enthalpically driven and has a negative ΔS of association. This comes about because the hydrophobic part of ΔS becomes less important as temperature increases (less ordered H_2O structure), and the translational/rotational part of ΔS becomes more important as temperature increases. There is no simple physical picture for the change in ΔH, except to note that this change depends on ΔC_p (which is relatively constant over the temperature range) which is large and negative, characteristic of a hydrophobic association. Throughout the temperature range, ΔG association stays relatively constant.

Ross and Subramanian [21] have suggested, on the basis of analysis of ΔH and ΔS of protein–ligand interactions, that a negative ΔH for association could come from enhanced polar interactions in the relatively lower dielectric medium of the protein (compared to water). Thus, even though such interactions are clearly less important than for gas phase association, they may contribute along with dispersion to non-covalent macromolecule ligand association.

3. Examples of biologically important non-covalent interactions

Although we can precisely understand and predict the major factors determining non-covalent interactions of small molecules in the gas phase, we must stress how difficult it is to find an unambiguous interpretation of aqueous solution association thermodynamics in terms of specific atomic interactions [20,21]. However, one has some circumstantial evidence that the most important energy terms which determine the strength and specificity of non-covalent aqueous solution associations are electrostatic, dispersion and hydrophobic. We now give examples of cases where one of these terms is the key determinant of the strength and/or the specificity of the association.

3.1. Electrostatic forces

Fersht [22] has nicely summarized the relative binding energies (or relative catalytic rates k_{cat}/K_m) for different amino acids and des-NH_3^+ amino acids interacting with aminoacyl tRNA synthetases (Table 3). The absence of the NH_3^+ group in the des-NH_3^+ amino acids causes a loss in binding free energy of 3.4–4.5 kcal/mole relative to the amino acids. Although this is far smaller than a "gas phase" loss of an ion pair, it is larger than one might expect, in view of our earlier discussion that ionic effects are larger damped out in solution. However, if, in the enzyme active site there is a Glu or Asp-COO^- near where the NH_3^+ from the amino acids binds, then the absence of the NH_3^+ group will leave the COO^- with at best a single water near

TABLE 3
Binding energy estimates from aminoacyl tRNA synthetases [a]

Substrate	Aminoacyl tRNA synthetase	k_{cat}/K_m (relative) [b]	ΔG
Isoleucine	Isoleucyl	1	
Valine	Isoleucyl	0.0006	3.0
Tyrosine	Tyrosyl	1	
Phenylalanine	Tyrosyl	3.6×10^{-5}	6.1
Valine	Valyl	1	
des-NH_3^+ valine	Valyl	–	4.5

[a] See ref. 22.

[b] Estimated relative *binding* energies (not k_{cat}/K_m) for the comparison of des-NH_3^+ valine with valine. A comparison of K_m rather than k_{cat}/K_m for the isoleucyl and tyrosyl tRNA synthetase examples gives similar qualitative results.

it. In the absence of substrate, this Glu or Asp likely had a more complete solvation shell, which is largely displaced on binding the amino acid substrate.

An even larger loss in binding free energy (6.1 kcal/mole) occurs upon binding phenylalanine rather than tyrosine to tyrosyl-tRNA synthetase. Again, the most logical explanation of this result is that the enzyme contains hydrogen bond donors or acceptors for the tyrosine OH, which are unable to form hydrogen bonds (or fewer, more distorted ones) when phenylalanine binds because it lacks the aromatic OH group.

Andrea et al. [23] have determined the binding affinities of thyroxine analogs to the plasma protein prealbumin and have noted the order of binding affinities is des-NH_3^+ analog > L-amino acid > des-COO^- analog, consistent with the preference of the protein for more anionic ligands. Although both Lys 15 and Glu 54 are in the binding channel, Lys 15 is apparently closer to the amino acid side chain.

The specificity of trypsin for peptides which contain Lys or Arg next to the catalytic site can be clearly understood from the X-ray structure, which shows Glu 189 in the position to bind the side chain near the to be cleaved peptide bond [24]. In the absence of substrate, Glu 189 is likely heavily solvated by waters, with little tendency to ion pair. The binding of substrate displaces the water, and, without a cationic side chain nearby, there is a large loss in $COO^- \ldots H_2O$ interactions without compensating $COO^- \ldots {}^+H_3N$ interactions.

In Fig. 1 we show an electrostatic potential molecular surface view of the active site of trypsin and of trypsin inhibitor [25]. The complementarity of the electrostatic potentials (blue = positive; red = negative) is apparent for the orientation shown here.

In summary, there are many examples of *specificity* in non-covalent interactions of biological interest, but little evidence to suggest that the presence of such interactions (ion pairing, H bonding) provides a thermodynamic driving force for that association.

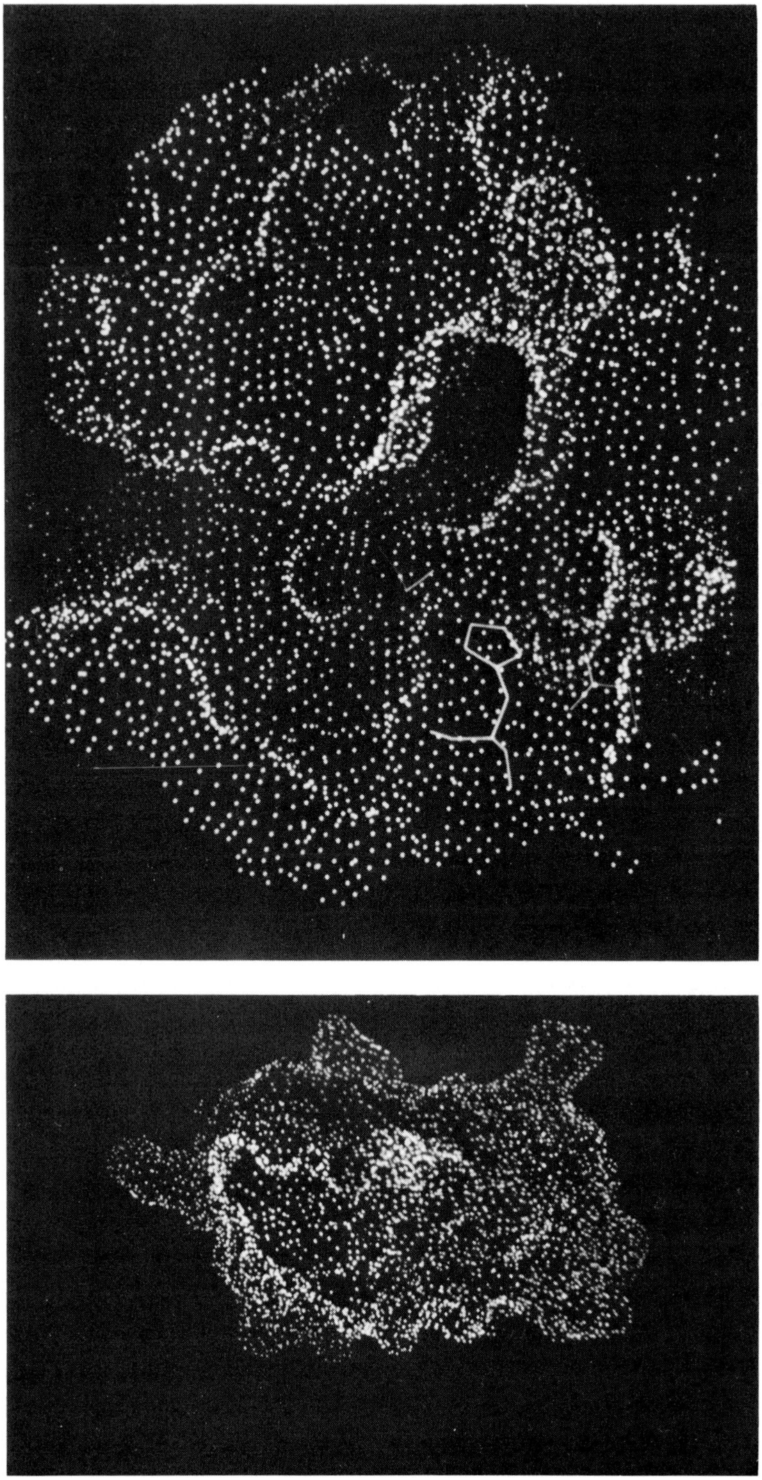

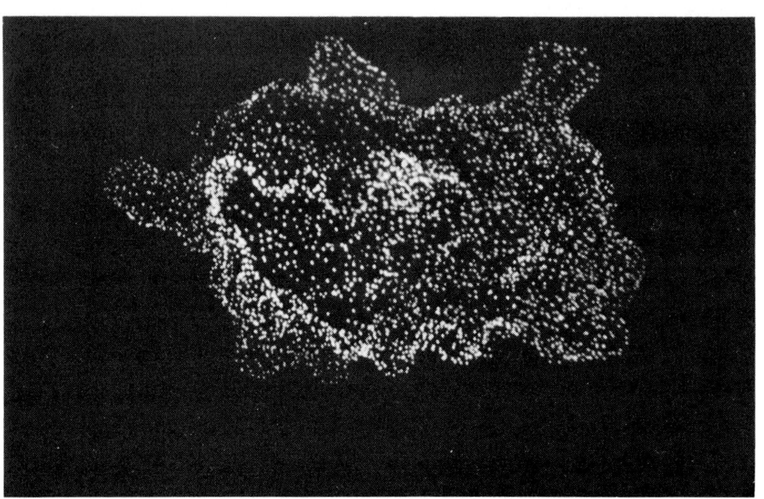

Fig. 1. An electrostatic potential molecular surface for the trypsin active site and for trypsin inhibitor [25].

3.2. Dispersion forces

The unequivocal demonstration of the importance of dispersion forces in protein–ligand association is difficult, since individual atom–atom energies are rather small and it is the sum of many atom–atom interactions which determines whether an association is given by the dispersion interactions or not.

Again, the comparison of the binding of isoleucine and valine with isoleucyl-tRNA synthetase is illuminating [22]. The difference in binding energy is 3 kcal/mole, whereas the difference in octanol/water partition coefficient (thought to be a reasonable model for the hydrophobic contribution) only amounts to 0.7 kcal/mole of free energy difference. Placing valine in isoleucyl-tRNA synthetase apparently leaves a hole the size of a $-CH_2$ group in the complex; since the dispersion force dominated sublimation energy/CH_2 group of crystalline hydrocarbons [22] is ~ 2 kcal/mole, this interpretation is reasonable.

Wolff et al. [26] have suggested, on the basis of the thermodynamics of steroid binding to glucocorticoid receptor protein, that both hydrophobic and dispersion contributions are required to explain the binding energies observed.

The intercalation of flat planar drugs between DNA base pairs is an example of a biologically important interaction where both electrostatics and dispersion are apparently important [27].

Finally, Grieco et al. [28] in their analysis of protein ligand binding data with linear free energy (QSAR) methods have noted many examples where polar group contributions to binding free energy correlate with the molar refractivity of the group. This has been interpreted as being due to dispersion attraction, since both the dispersion attraction coefficient and molar refractivity increase as the polarizability and number of electrons of the chemical group does.

How might dispersion attractions provide a driving force for ligand association with the protein? Low and Richards [29] have noted that proteins are more dense than water. Since dispersion attractions are relatively non-specific and depend mainly on the number of close contracts, it is sensible that a ligand can gain more dispersion attraction by binding to the protein than remaining in the water. Wolff et al. [26] estimated that such a density difference could result in a 4–5 kcal/mole contribution to steroid–receptor binding energies. The correlation observed by Grieco et al. [28] between molar refractivity and binding for polar groups on ligands suggest that, although both water and the protein can satisfy the ligand's polar interactions, the protein provides more dispersion attraction for the ligand and, thus, this enhances binding.

The computer graphics observation by Blaney et al. [31] that there was a "hole" in the thyroxine prealbumin complex and that the replacement of the outer phenyl ring by a naphthyl ring was sterically acceptable led to the testing of naphthyl analogs (Fig. 2). The fact that these analogs bound more tightly to prealbumin is suggestive of a dispersion contribution to the relative binding free energy when one replaces an analog which leaves a hole in the binding channel with one that fits into it more snugly.

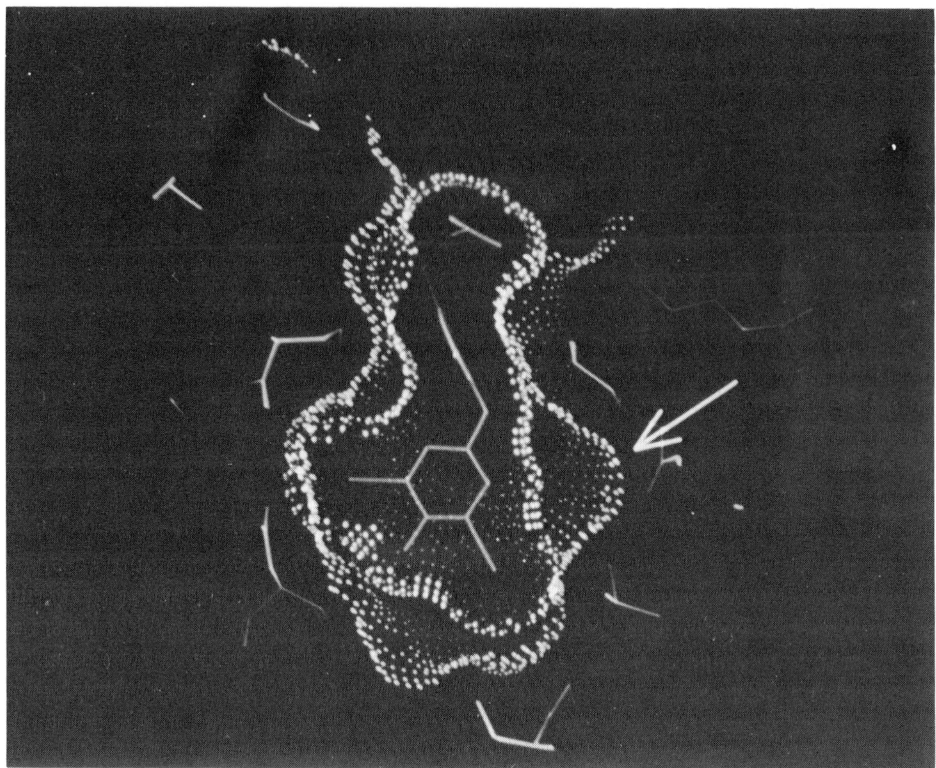

Fig. 2. A view of the thyroxine–prealbumin complex, with the arrow designating the "naphthyl hole" [31].

3.3. Hydrophobic interactions

It is clear that hydrophobic interactions are essential to the formation of the "native" structure of globular proteins. Again, thermodynamic studies [20] on such systems plus the analysis of protein crystal structures [30] are strong evidence for the essential role of these hydrophobic interactions. Globular proteins in their native conformation do have their hydrophobic atoms on the "inside", away from the water, and the addition of non-aqueous solvents to the water tends to destabilize this structure, since the exposure of non-polar groups is not so energetically costly in e.g., mixed alcohol/water solvents as it is in pure water.

How might one distinguish "hydrophobic interactions" from dispersion force driven interactions? From the analysis above, one might expect that hydrophobic interactions would be characterized by a positive ΔS of association, whereas dispersion interactions would be characterized by a negative ΔS of association (and a compensatingly more negative ΔH). For simple reactions over a small temperature range ΔS and ΔH are relatively temperature independent and such a separation of ΔG into ΔH and ΔS terms by studying the temperature dependence of ΔG is meaningful. However, for many biologically important reactions, ΔH and ΔS are very temperature dependent, as in the example of NAD^+ association to LADH mentioned above [20]. In such cases, a large negative ΔC_p characterizes a hydrophobically driven process, since studies of octanol–water partition coefficients of simple model compounds have found a significant and negative ΔC_p for the water $\rightarrow$ octanol partition process of a non-polar group. A plausible molecular interpretation is as follows: by ordering the water molecules around it, a non-polar group stabilizes their hydrogen bonded network and provides more "bonds" to be broken in absorbing heat into the system. Thus, the heat capacity is raised by placing a non-polar solute in water. When these non-polar solutes get transferred to octanol or associate in water, the heat capacity change, ΔC_p is negative because the water structuring by the non-polar groups is decreased. In summary, a large negative and relatively temperature independent ΔC_p is characteristic of a hydrophobically driven process.

Recent examples of steroid–receptor binding also appear to be consistent with this analysis [26]. These are characterized by large positive ΔS association at temperatures near 0°C and slightly negative ΔS values near 40°C. Again, the ΔC_p for these associations is large and negative, suggestive of an important hydrophobic contribution to association. The decrease in ΔS is also characteristic of the greater contribution of translation and rotational entropy loss at the higher temperatures.

It was recognized some time ago that the binding and catalysis of α-chymotrypsin substrates correlated with their hydrophobicity [32] and subsequent X-ray studies [33] clearly showed a large cavity, lined by hydrophobic residues, that supported this interpretation. More recently, more "subtle" evidence for the hydrophobic contribution to ligand–protein association comes from the pioneering work of Hansch and co-workers on the combined QSAR/computer graphics studies of binding of ligands to papain [34] and dihydrofolate reductase [35]. In both of the above cases, the linear regression analysis of binding preceded the computer graphics study and indicated a correlation of relative binding affinities of substituents at particular locations with the octanol–water partition coefficient of the group. For example, in the case of the papain substrates the binding correlated with the partition coefficient of the more

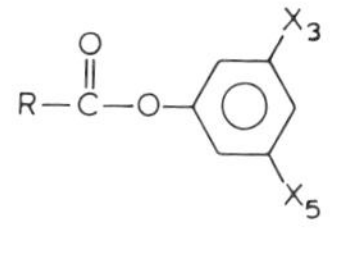

R—C(=O)—O—C_6H_3(X_3)(X_5)

hydrophobic of the two substituents, not the sum of the two. Computer graphics visualization showed the non-equivalence of the two substituents when in contact with the enzyme, with one group in contact with hydrophobic atoms on the protein and the other sticking into solution [Fig. 3).

A second very interesting example was the difference in the QSAR for identical triazine analogs binding to two different (*L. casei* and chicken liver) dihydrofolate reductases. In one case, the dependence of binding on octanol partition coefficient π for long *O*-alkyl chains O–$(CH_2)_n$ was ~ 1, indicating an active site similar in

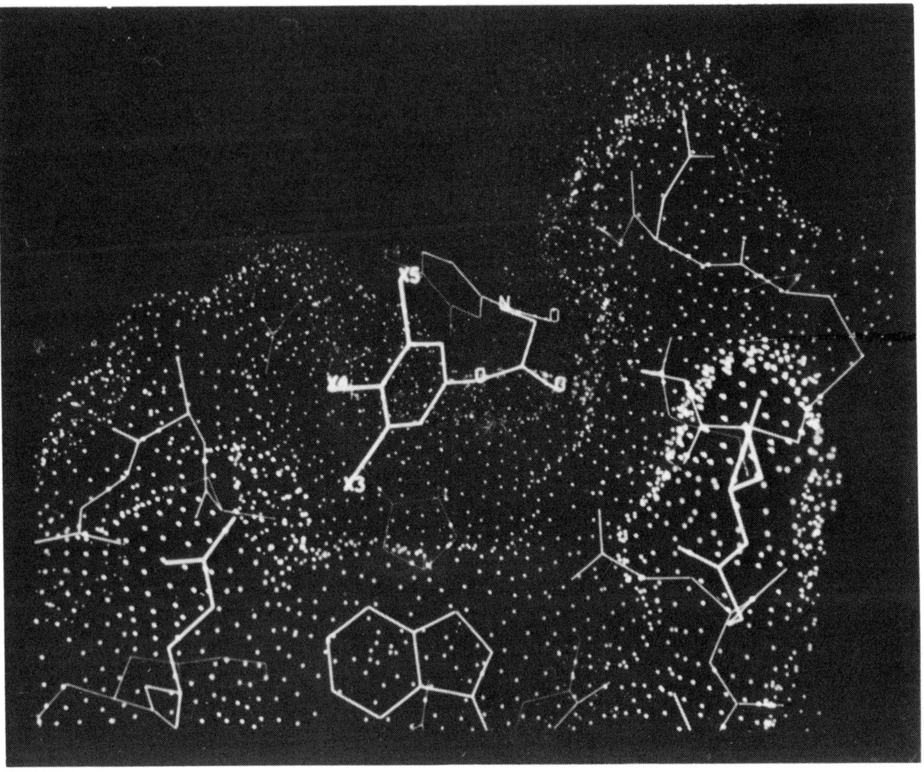

Fig. 3. A view of the binding of hippurate substrates to papain [34].

hydrophobicity to octanol, but beyond $n = 6$, binding affinity leveled off and there was no dependence on hydrophobicity. In the other isozyme, the dependence of binding on π was much less steep, but extended to $n = 11$. The reasons for the differences in the QSAR became very clear upon observation of the binding geometries (one from X-ray crystallography, the other model built) (Fig. 4). In the first case, the alkyl group was in a tight hydrophobic channel, but for chains longer than $n = 6$, the hydrocarbon stuck out into solution. In the second case, the alkyl chain sat on a long hydrophobic "floor", long enough to accommodate chains up to $n = 11$, but with only one side of the chain in contact with the protein and the other side solvated.

The role of the hydrophobic concept in drug design is quite clear. Ariens [36] has noted that receptors have hydrophobic and hydrophilic regions, and that one could design pharmacologically interesting molecules by adding hydrophobic groups to an existing active molecule and varying the chain length between the initial structure and the hydrophobic group. Muscarinic acetylcholine antagonists [37] and histamine H_1 antagonists [38] are examples of molecules which contain the basic features of the

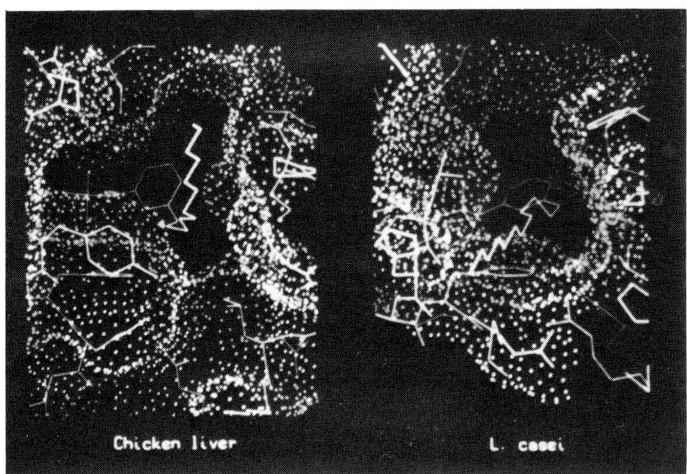

Fig. 4. A comparison of the postulated binding geometries of alkyl triazines to *E. coli* and chicken liver dihydrofolate reductase [35].

parent molecule (prolonged amine and ester in the case of muscarinic antagonists and protonated amine in the case of histamine H_1 antagonists), with large hydrophobic groups on the molecule providing a substantial binding affinity.

4. *Summary*

It appears that the most important attractive non-covalent forces for biological association in aqueous solution are electrostatic, dispersion and hydrophobic. Electrostatic interactions are probably more important in providing specificity than in contributing to the overall thermodynamic driving force for association. It is likely that dispersion is important and hydrophobic terms essential in most protein–ligand interactions.

It is disappointing, but must be emphasized, that the connection between the experimental thermodynamic data and an unambiguous molecular interpretation of these in complex systems in solution is circumstantial at best. Nonetheless, thermodynamic studies on model system and QSAR/computer graphics studies of ligand protein structures provide strong evidence to support the role of hydrophobic interactions in protein–ligand association. The relative binding affinity studies on aminoacyl + RNA synthetases with slightly wrong substrates provide beautiful examples of the role of dispersion forces, ion pairing and hydrogen bonding in determining biological specificity. These results facilitate the (hopefully intelligent) extrapolation and interpretation of binding energies and specificity in systems where precise structural data is not yet available.

More extensive discussions on the nature of intermolecular interactions and thermodynamics in biological systems are given by Jencks [39] and Fersht [22].

Acknowledgement

The author would like to acknowledge research support from the NSF (CHE-80-26560) and helpful discussions with Jeff Blaney.

References

1 Morokuma, K. (1977) Acc. Chem. Res. 10, 294.
2 Kollman, P. (1977) Acc. Chem. Res. 10, 377; (1978) J. Am. Chem. Soc. 100, 2974.
3 Hager, A., Huler, E. and Lifson, S. (1974) J. Am. Chem. Soc. 96, 5319.
4 Margenau, H. and Kestner, N. (1971) Theory of Intermolecular Forces, 2nd Edn., Pergamon Press, Oxford; Hirshfelder, J., Curtis, C. and Bird, R. (1954) Molecular Theory of Gases and Liquids, Wiley, New York.
5 Shipman, L. and Scheraga, H. (1979) Proc. Natl. Acad. Sci. (U.S.A.) 76, 3585.
6 Gelin, B. and Karplus, M. (1979) Proc. Natl. Acad. Sci. (U.S.A.) 76, 3585.
7 Levitt, M. (1974) J. Mol. Biol. 82, 393.
8 Davidson, N. (1962) Statistical Mechanics, McGraw-Hill, New York.
9 Joesten, M. and Schaad, L. (1974) Hydrogen Bonding, Marcel Dekker, New York.
10 Kollman, P. (1979) in: M. Wolff (Ed.), Burger's Medicinal Chemistry, Part 1; The Basis of Medicinal Chemistry, Wiley, New York.
11 Kollman, P. (1977) in: H.F. Schaefer (Ed.), Modern Theoretical Chemistry: Applications of Electronic Structure Theory, Vol. 4, Plenum, New York.
12 Curtiss, L., Frurip, M. and Blander, M. (1978) Chem. Phys. Lett. 55, 469.
13 Dyke, T.R. and Muenter, J.S. (1974) J. Chem. Phys. 60, 2929.
14 Kauzmann, W. (1959) Adv. Prot. Chem. 14, 1.
15 Tanford, C. (1973) The Hydrophobic Effect, Wiley, New York.
16 Owicki, J. and Scheraga, H. (1977) J. Am. Chem. Soc. 99, 7413.
17 Swaminathan, S., Harrison, S. and Beveridge, D. (1978) J. Am. Chem. Soc. 100, 5705.
18 Aue, D., Webb, H. and Bowers, M. (1976) J. Am. Chem. Soc. 98, 311, 318.
19 Battino, R. (1977) Chem. Rev. 77, 219.
20 Sturdevant, J. (1977) Proc. Natl. Acad. Sci. (U.S.A.) 74, 2236.
21 Ross, P. and Subramanian, S. (1981) Biochemistry 20, 3096.
22 Fersht, A. (1977) Enzyme Structure and Mechanism, W. Freeman, Reading and San Francisco.
23 Andrea, T., Cavalieri, R., Goldfine, I. and Jorgensen, E. (1980) Biochemistry 19, 55.
24 Huber, R., Kukla, D., Bode, W., Schwager, P., Bartels, K., Deisenhofer, J. and Steigemann, W. (1974) J. Mol. Biol. 89, 73.
25 Weiner, P., Langridge, R., Blaney, J., Schaefer, R. and Kollman, P. (1983) Electrostatic Potential Molecular Surfaces, Proc. Natl. Acad. Sci. (U.S.A.), 79, 3754.
26 Wolff, M., Baxter, J., Kollman, P., Lee, D., Kuntz, I., Bloom, F., Matulich, D. and Morris, J. (1978) Biochemistry 17, 3201.
27 Kollman, P., Weiner, P. and Dearing, A. (1981) Ann. N.Y. Acad. Sci. 357, 250–268.
28 Grieco, C., Silipo, C., Vittoria, A. and Hansch, C. (1977) J. Med. Chem. 20, 586; Silipo, C. and Hansch, C. (1975) J. Am. Chem. Soc. 97, 6849.
29 Low, B. and Richards, F. (1954) J. Am. Chem. Soc. 76, 2511.
30 Kuntz, I.D. (1979) in: M. Wolff (Ed.), Burger's Medicinal Chemistry, Vol. I, John Wiley, New York, p. 285.

31 Blaney, J., Jorgensen, E., Connolly, M., Ferrin, T., Langridge, R., Oatley, S., Burridge, J. and Blake, C. (1982) J. Med. Chem., 25, 785.
32 Knowles, J. (1965) J. Theor. Biol. 9, 213.
33 Blow, D. (1976) Acc. Chem. Res. 9, 145.
34 Smith, R., Hansch, C., Kim, K., Omiya, B., Fukumara, G., Selassie, C., Jow, P., Blaney, J. and Langridge, R. (1982) Arch. Biochem. Biophys., 215, 319.
35 Blaney, J., Hansch, C. and Matthews, D. (1983) J. Am. Chem. Soc., to be submitted.
36 Ariens, E. (1971) in: E. Ariens (Ed.), Drug Design, Academic Press, New York.
37 Pauling, P. and Datta, N. (1980) Proc. Natl. Acad. Sci. (U.S.A.) 77, 708.
38 Ganellin, C. (1980) in: M.E. Wolff (Ed.), Burger's Medicinal Chemistry, Vol. 2, Ch. 48, John Wiley, New York.
39 Jencks, W. (1969) Catalysis in Chemistry and Enzymology, McGraw-Hill, New York.

CHAPTER 3

Enzyme kinetics

PAUL C. ENGEL

Department of Biochemistry, University of Sheffield, Sheffield S10 2TN, Great Britain

1. Introduction: aims and approaches in enzyme kinetics

Enzyme-catalysed reactions occur as a result of the intimate interaction of individual molecules of substrate(s) with individual enzyme molecules. In the case of a single-substrate reaction:

$$\mathrm{A} \overset{\mathrm{E}}{\rightarrow} \mathrm{Product(s)}$$

where E represents the enzyme and A the substrate, this fundamental assumption is represented as shown:

$$\mathrm{E} + \mathrm{A} \rightleftharpoons \mathrm{EA} \rightarrow \mathrm{Product(s)} + \mathrm{E} \qquad \text{Scheme 1}$$

One may postulate isomerisations of EA preceding the release of products, but the basic point is that a specific, stoichiometric enzyme–substrate complex is formed. This fact is now so totally taken for granted by biochemists that it is easy to forget that its establishment was the first major milestone in enzyme kinetics. The analysis of Scheme 1 by Brown [1] and Henri [2] and subsequently by Michaelis and Menten [3] and Briggs and Haldane [4] provided an explanation of the previously puzzling observation that the rate of a typical enzyme reaction plotted as a function of substrate concentration increases asymptotically to a maximum (Fig. 1). In Scheme 1 the overall rate of the catalysed reaction, i.e. of product formation, is proportional to $[\mathrm{EA}]/([\mathrm{E}]+[\mathrm{EA}])$, the fraction of the total enzyme present as the productive complex EA; at low substrate concentration this fraction is proportional to [A], whereas at high substrate concentration the fraction approaches 1, and the rate is then limited only by the rate constant for conversion of EA to E + products.

In fact the majority of reactions in biochemistry involve more than one substrate, and therefore a more complex scheme than Scheme 1 is inevitably required. Choices immediately present themselves. Thus, if we consider the reaction

$$\mathrm{A} + \mathrm{B} \overset{\mathrm{E}}{\rightleftharpoons} \mathrm{P} + \mathrm{Q}$$

the enzyme *may* work by bringing A and B close together at its active site in a

Michael I. Page (Ed.), The Chemistry of Enzyme Action

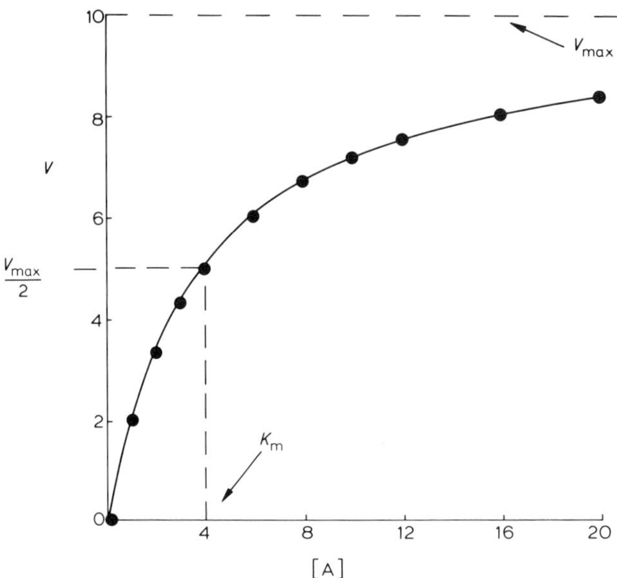

Fig. 1. Saturation kinetics for an enzymatic reaction. At low substrate concentrations, [A], the rate of reaction is almost proportional to [A], but an upper limit is approached with high concentrations. The points plotted here are obtained by substituting in Eqn. 4 $V_{max} = 10$ and $K_m = 4$. Notice how in looking at such a direct plot of v against [A] the eye tends to underestimate the value of V_{max}.

ternary complex EAB. Alternatively each substrate may conclude its transaction with the enzyme separately in an enzyme substitution mechanism. Each of these options then entails further alternatives. There are thus a number of possible courses that a 2-substrate reaction could follow. Ultimately one hopes, through a consideration of an enzyme's structure and the general principles of chemical catalysis, to be able to give a reasonably full account of the binding processes and the bond-making and -breaking events that occur within enzyme–substrate complexes. Understanding at this level has now been approached for a few enzymes, e.g. hen egg white lysozyme [5], chymotrypsin [6] and pancreatic ribonuclease [7]. An essential step in this process, however (and, as we have seen, by no means a trivial one if more than one substrate is involved), is to establish the sequence and composition of the complexes that actually occur in the mechanism, and identify potentially rate-determining steps. These more limited goals are among the principle objectives of enzyme kinetics.

In addition to this perspective of the enzyme mechanism as a subject for study in its own right, there is also the view of the enzyme as a component of an intricate biological system. The metabolic biochemist wishes to know what rates of throughput can be achieved by a given amount of enzyme protein under physiological conditions of substrate concentration, pH, ionic strength and temperature, and also how the throughput may be regulated. Understanding an enzyme's function thus depends principally upon the information provided by enzyme kinetics. This is clearly true not only in the physiological context but also for pharmacological applications and for the increasingly widespread use of enzymes for assay procedures, chemical syntheses etc.

Enzyme kinetics, then, is the study of the rates of reactions involving enzymes. This includes the study not only of overall reactions catalysed by enzymes but also of the individual component steps if they can be measured separately. The topic can be divided into steady-state kinetics and rapid-reaction kinetics. Roughly speaking, steady-state measurements are usually made over a time scale of minutes, and involve small molar concentrations of the enzyme under study and relatively much higher concentrations of substrates. What is measured is the gradual disappearance of a substrate or accumulation of a product as enzyme molecules undergo multiple catalytic cycles. Rapid-reaction studies, on the other hand, typically involve much higher molar concentrations of enzyme, similar to the concentrations of substrates. They are usually on a time scale of 1–1000 msec and concentrate on the anatomy of a single turnover. In rapid-reaction studies it is often possible to observe the changing concentrations of individual enzyme–substrate complexes as well as substrates and products.

These two approaches require different apparatus. Steady-state kinetics usually requires only the apparatus of the routine biochemical laboratory. Most steady-state studies are spectrophotometric. If the reaction does not involve a convenient absorbance change, it may be necessary to use other methods of measurement, but often it is possible to obtain an absorbance change by using coupling enzymes or secondary reactants. Rapid-reaction studies require more sophisticated equipment for precise and rapid mixing of enzymes with reactant solutions and for synchronous rapid recording of the appropriate signal.

On the whole, since the main biological function of enzymes is the catalysis of net chemical changes, steady-state kinetics must be more relevant to a consideration of enzyme function (as opposed to mechanism) than the kinetics of the transient state. The requirement for only very small amounts of enzyme in order to obtain useful information also provides a major practical argument in favour of the steady-state approach. However, in the context of mechanistic studies, conclusions drawn from steady-state studies are inferred or rejected because they are or are not compatible with the mathematical behaviour of the system. Rapid reaction studies, by contrast, involve less guesswork because the postulated complex can often be directly observed by virtue of distinctive physical properties (e.g. absorbance or fluorescence).

The differing intellectual and aesthetic appeal of the two approaches has perhaps encouraged kineticists to divide into two doctrinaire camps. For instance, in prefatory remarks to an excellent introduction to rapid-reaction kinetics [8], Halford dismisses the steady-state approach by remarking that, "The contents of a letter cannot be revealed by looking at the envelope". The analogy is hardly fair! A better postal analogy would be to consider the attempt to find out how the postal system works (a) by posting many letters of different sorts at different times and seeing whether and how soon they reached the intended destination or (b) by going to watch the sorting at the post office. The investigator denied an on-the-spot inspection would certainly be handicapped, but so would the colleague who spent the whole time looking at the machinery without trying out the service! In truth the approaches are complementary: rapid-reaction kinetics cannot get very far without a solid steady-state framework to support it; equally the tentative inferences of steady-state kinetics are greatly strengthened by the direct demonstrations offered by rapid-reaction kinetics.

In the almost 20 years since this volume's predecessor appeared enzyme kinetics has come of age. Extended theory and the availability of much more sensitive measuring equipment have made possible incisive kinetic analysis of multi-substrate enzymes. One must also add, however, that the full potential of the method has been achieved in rather few cases. Much of the published information has been collected by investigators primarily interested in function rather than mechanism, and is therefore of descriptive value only. Even when a more thorough-going analysis is attempted, it is often difficult and tedious to obtain enough data to remove all ambiguity. Hence doubt and controversy regarding the mechanisms of many important enzymes remain. In the space available here it is not possible to go into much detail about individual cases. The intention, therefore, is to sketch out current approaches and problems.

2. *Steady-state kinetics*

2.1. *Michaelis–Menten equation*

If we consider the reaction shown in Scheme 1, it is clear that, at the instant of mixing solutions of E and A (A being in a very large molar excess over E), the concentrations of EA and of products are zero. It is assumed, however, that over a short period of time after mixing, the concentration of EA builds up to a value *which then remains fairly constant* for a longer period of time. As the net reaction proceeds towards equilibrium, accumulating products gradually contribute to a significant flux in the reverse reaction, but it is assumed that this is negligible in the early stages of reaction. The usual aim experimentally, therefore, is to measure a so-called *initial rate* – i.e. the rate of product formation once the concentration of enzyme–substrate complex has reached a steady level and before the reverse reaction has become significant. The assumption that [EA] remains constant long enough to allow measurement of a true initial rate is obviously an approximation. It is justified in many cases by the success and internal consistency of the deductions based upon it, but it should not be forgotten that it is an assumption. The assumption may take one of two forms. The more extreme assumption, adopted both by Henri [2] and by Michaelis and Menten [3], in their analysis of Scheme 1, is that EA actually attains equilibrium with E and A very rapidly, so that formation of products is essentially a leak out of the equilibrium mixture. If so:

$$[\mathrm{E}]k_1[\mathrm{A}] = [\mathrm{EA}]k_{-1}$$

$$\therefore [\mathrm{E}] = [\mathrm{EA}]\frac{k_{-1}}{k_1[\mathrm{A}]} \tag{1a}$$

Note that [A] is assumed to be so large relative to enzyme concentration that it is negligibly decreased by formation of EA. Briggs and Haldane [4] introduced the less restrictive *steady-state assumption*. Here it is postulated only that the rates of formation and removal of EA *through whatever route* are equal. This avoids arbitrary assumptions about the relative values of k_2 and k_{-1}. Thus now:

$$[\mathrm{E}]k_1[\mathrm{A}] = [\mathrm{EA}](k_{-1} + k_2)$$

$$\therefore [\mathrm{E}] = [\mathrm{EA}]\frac{(k_{-1} + k_2)}{k_1[\mathrm{A}]} \tag{1b}$$

Regardless of the assumption adopted in order to obtain the unknown concentrations [E] and [EA] in terms of known quantities, we need a second equation relating [E] and [EA]. In fact we know that:

$$e = [\mathrm{E}] + [\mathrm{EA}] \tag{2}$$

This, the *"enzyme conservation equation"*, simply indicates that the total enzyme concentration in all forms, e, remains constant. If we substitute for E from Eqn. 1a, we obtain

$$e = [\mathrm{EA}]\left(1 + \frac{k_{-1}}{k_1[\mathrm{A}]}\right)$$

We also know that the overall rate of product formation, v, is given by

$$v = [\mathrm{EA}]k_2 \tag{3}$$

$$\therefore \frac{v}{e} = \frac{k_2}{1 + \dfrac{k_{-1}}{k_1[\mathrm{A}]}}$$

$$\therefore v = \frac{k_2 e[\mathrm{A}]}{[\mathrm{A}] + \dfrac{k_{-1}}{k_1}}$$

If [A] becomes very large, v will approach $k_2 e$, which is therefore defined as V_{max}, the *maximum rate.* k_{-1}/k_1 can also be replaced by a single constant, K_m, the

Michaelis constant, to give

$$v = \frac{V_{max}[A]}{[A] + K_m} \tag{4}$$

K_m may be operationally defined as that substrate concentration which gives half the maximum rate, as may readily be verified by substituting $[A] = K_m$. Eqn. 4 is the Michaelis–Menten equation.

If we adopt the Briggs–Haldane steady-state assumption and substitute from Eqn. 1b rather than Eqn. 1a, we obtain an equation of exactly the same form except that K_m now is not k_{-1}/k_1 but $(k_{-1} + k_2)/k_1$.

K_m and V_{max}, then, are the kinetic parameters which define the rate behaviour of an enzyme-catalysed reaction as substrate concentration is varied – provided that the reaction obeys Eqn. 4. These parameters are useful as descriptors of the enzyme (e.g. for the purpose of differentiating isoenzymes), although it must be emphasised that they are valid only for a single set of conditions of temperature, pH etc.

2.2. *Relationship between K_m and dissociation constant*

As shown above, the 'equilibrium assumption' leads to a version of the Michaelis–Menten equation in which the Michaelis constant is equivalent to the dissociation constant for the enzyme–substrate complex EA. This unfortunately encourages many biochemists to assume that Michaelis constants can always be so equated. The misleading statement that K_m reflects "an enzyme's affinity for its substrate" is often encountered. Even if we consider only 1-substrate enzymes, the Briggs–Haldane version of K_m will only approximate to k_{-1}/k_1 if $k_2 \ll k_{-1}$. However, if, for instance, $k_2 = 10k_1$, then K_m is 11 times greater than the dissociation constant for EA. If $k_2 \gg k_1$, then $K_m = k_2/k_1$ rather than k_{-1}/k_1.

In the 1-substrate case one can at least say that the dissociation constant may not exceed K_m; there is therefore a formal mathematical relationship between the two constants. For some mechanisms involving more than one reactant, not even this limited degree of linkage exists. Michaelis constants are empirical kinetic parameters. They have an entirely adequate definition in kinetic terms and should not be equated with thermodynamic constants without sound theoretical or experimental justification.

2.3. *Experimental determination of kinetic constants*

(a) Experimental design

Experimentally K_m and V_{max} are obtained by measuring v for a series of different values of [A]. Plotting v against [A] usually involves an uncertain extrapolation to V_{max}, and so several linear plots have been introduced. The greater ease of extrapolation does not remove the need for well-designed experiments, however; it is essential to obtain points sufficiently numerous and reliable to establish (a) whether they really do fit a straight line – i.e. whether Eqn. 4 is obeyed and (b) what the slope and intercepts are. This demands a range of substrate concentrations extending well above and well below K_m. To establish linearity, ideally a range of at least 1000-fold should be explored. If linearity is assumed, then a more restricted range, say 0.25 K_m up to 5 K_m, should be sufficient to allow reasonably precise estimation of the kinetic parameters. Often experimental constraints limit the accessible range, however.

(b) Lineweaver–Burk plot

This, the most widely used of the linearisations, is obtained [9] by inverting Eqn. 4:

$$\frac{1}{v} = \frac{1}{V_{max}} + \frac{K_m}{V_{max}[A]} \tag{5}$$

Thus a plot of $1/v$ against $1/[A]$ should give a straight line with an ordinate intercept of $1/V_{max}$, a slope of K_m/V_{max} and an abscissa intercept of $-1/K_m$ (Fig. 2a). The plot has been much criticised [10–12] because it compresses the display of the higher values of v in any set of data. It is often assumed that the highest values of v are necessarily the most accurate ones in the data set, but this depends on the method of measurement and in particular on whether the rate is measured through appearance of product or disappearance of substrate. There can be no doubt, however, that the Lineweaver–Burk plot distorts the relative weighting of the points. Moreover, if the desirable wide range of substrate concentration is used, it becomes necessary to construct two or more plots on different scales in order to display the results adequately.

(c) Eadie–Hofstee plot

Multiplying Eqn. 5 by v, we obtain

$$1 = \frac{v}{V_{max}} + \frac{K_m v}{V_{max}[A]}$$

$$\therefore v = V_{max} - K_m \frac{v}{[A]} \tag{6}$$

A plot [10,13] of v against $v/[A]$ therefore gives a straight line with ordinate intercept V_{max} and slope $-K_m$ (Fig. 2b). The abscissa intercept is V_{max}/K_m. This plot is particularly useful from the point of view of display if a wide range of substrate concentration (say 100–1000-fold) is covered, because, whatever the range of v and [A], the span of the plot is limited by the positive intercepts on both axes.

(d) Hanes plot [14,15]

Multiplying Eqn. 5 by [A] gives

$$\frac{[A]}{v} = \frac{[A]}{V_{max}} + \frac{K_m}{V_{max}} \tag{7}$$

A plot of $[A]/v$ against [A] gives a slope of $1/V_{max}$, an ordinate intercept of K_m/V_{max} and an abscissa intercept of K_m (Fig. 2c). An investigation by Dowd and Riggs [11] showed that both this plot and the Eadie–Hofstee plot were markedly superior to the Lineweaver–Burk plot from a statistical standpoint – i.e. in the

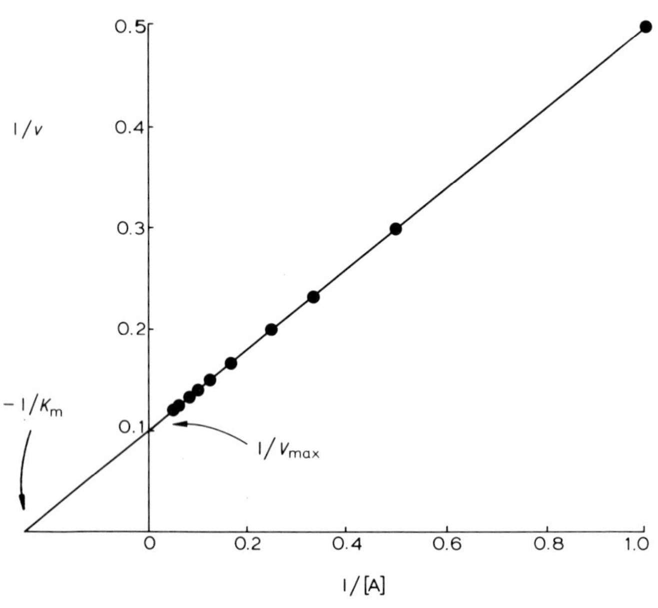

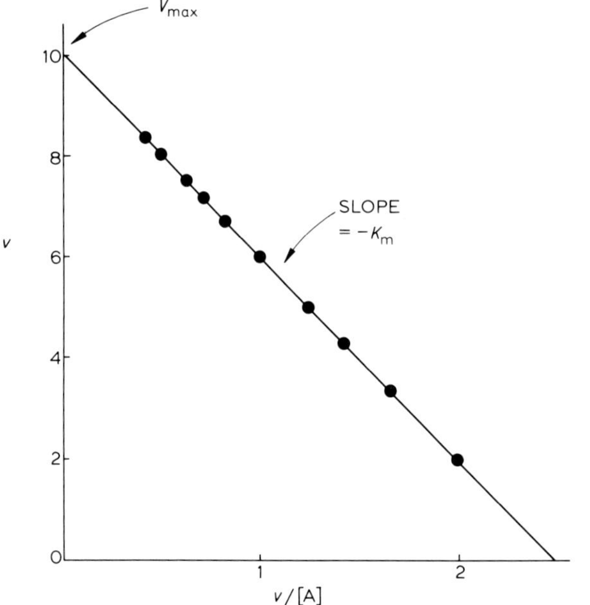

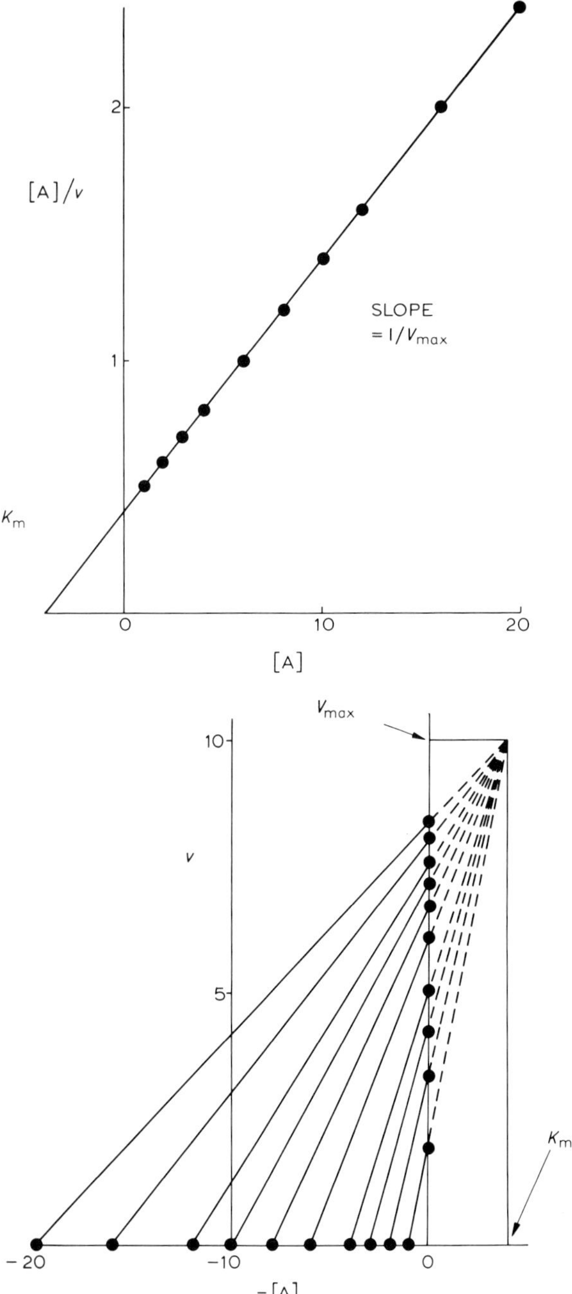

Fig. 2. Different linear plotting methods for initial rate measurements which obey Michaelis–Menten kinetics. The imaginary, mathematically perfect data are the same as those plotted in Fig. 1, i.e. $V_{max} = 10$, $K_m = 4$, and [A] = 1, 2, 3, 4, 6, 8, 10, 12, 16, 20, ranging from 0.25 K_m to 5 K_m. The plots are (a) the Lineweaver–Burk plot, (b) The Eadie–Hofstee plot, (c) the Hanes plot, and (d) the Eisenthal–Cornish-Bowden plot.

presence of experimental error they were likely to provide truer estimates of the kinetic parameters.

(e) Eisenthal and Cornish-Bowden plot [12,16,17]

The direct linear plot of Eisenthal and Cornish-Bowden is a more recent introduction. It is not an attractive method from the point of view of display since each measurement is represented as a line rather than a point (Fig. 2d). Its claim, however, is to statistical superiority for the purposes of (non-manual) computation. The estimation of kinetic parameters by this method is claimed to be less sensitive to undue distortion by 'outliers' – i.e. a few exceptionally inaccurate rate measurements.

The axes of the graph are v and [A] as ordinate and abscissa respectively. For each measurement of v and [A] a line is drawn joining the value of v on the ordinate to $-$[A] on the abscissa. The extrapolation of this line, which is given by Eqn. 8,

$$V_{max} = \frac{v}{[A]} K_m + v \tag{8}$$

into the first quadrant (positive v and [A]) gives all the possible values of V_{max} and K_m consistent with the experimental value of v and [A]. Hence the intersection of 2 such lines gives, in theory, the true K_m and V_{max}. Real data are subject to experimental error and the method relies on taking the median of all the $n(n-1)/2$ points of intersection for n combinations of v and [A].

(f) Does the choice of plotting method matter?

Some of the relative strengths and weaknesses of the various plotting procedures are indicated above, and a thorough consideration of these matters is to be found in the review paper of Markus et al. [18]. It is as well to remember, however, that the importance of the choice of the plotting procedure increases as the quality of the data decreases! Despite its evident deficiencies, the Lineweaver–Burk plot still enjoys the widest currency, perhaps because v and [A] are kept separate in the plotted variables, so that the graph is easier to understand and use than the other representations. The whole argument is in any case becoming increasingly irrelevant in view of the widespread adoption of computer methods relying on the fitting of a rectangular hyperbola to the plot of v against [A] without any linear transformation [19–22]. The linear plot thus becomes merely a convenient method of display rather than the means of obtaining the constants.

2.4. Non-linearity

To what extent does Eqn. 4 in fact account adequately for the initial-rate behaviour of known enzymes? Inspection of any textbook of biochemistry reveals large numbers of quoted K_m values, indicating that the enzymes in question are regarded as conforming to Eqn. 4. On the other hand the view has been forcefully expressed by Bardsley and his colleagues in numerous publications (e.g. [23–26]) that 'Michaelis–Menten behaviour' is the exception rather than the rule. Experimenters are indeed often too ready, on the basis of very few measurements over a narrow range of substrate concentration, to state that an enzyme conforms to Eqn. 4, dismissing obvious deviations as reasonable experimental error. In that sense the warning is timely, and Bardsley et al. rightly advocate the use of much wider ranges of substrate concentration than are customarily employed. However, if the wide range includes high molar concentrations, this may well entail changes in the bulk properties of the solvent, giving rise to deviations from Eqn. 4 which no longer reflect the properties of unaltered enzyme in the initially defined solvent. Although such disturbances may legitimately be regarded as part of the overall profile of an enzyme's behaviour, it is nevertheless doubtful whether they in any way invalidate mechanistic conclusions based on Michaelis–Menten behaviour observed with lower concentrations of substrate. A case recently considered is that of fumarase from pig heart: Bardsley et al. showed very marked departures from Michaelis–Menten behaviour with high concentrations of fumarate [25,26] apparently requiring a rate equation containing terms with substrate concentration raised to the power 3, but in fact the anomalous behaviour appears to be due to perturbation of the pH [27]. Clearly not all cases of departure from the classical pattern are attributable to trivial causes, but this example does emphasise that non-linearity in Lineweaver–Burk plots needs to be documented with at least as much care as linearity.

Bardsley [28] and Childs and Bardsley [29] have provided a substantial body of mathematical theory to facilitate the categorisation of detailed curve shapes in cases where the data do not fit the linear transformations of Eqn. 4. This approach may be seen as essentially inductive. It is an attempt to set up rigorous procedures for empirical mathematical description of enzymes' kinetic behaviour. Such description should in theory define minimum levels of complexity for physical models. Application of this approach will severely test the precision of rate measurements in real cases, and there is a risk that valid mechanisms may be ruled out on the basis of apparent subtleties of curve shape that are no more than experimental error. This, however, is certainly no excuse for ignoring genuine non-linearity.

Deductive approaches to non-linearity have on the whole been more popular. On the basis of available information on enzymes' structure and biological function, "plausible" physical models have been formulated. These then generate mathematical predictions and experimental data are then tested against the predictions. The risk here, of course, is that information may be overlooked or discarded in the attempt to substantiate a favoured model.

Irrespective of the interpretative approach, it is now widely recognised that many enzymes do show marked deviations from Michaelis–Menten behaviour, and the deviation is often interpretable in terms of regulatory function in vivo. Thus, for example, a number of enzymes, including threonine deaminase [30] and aspartate transcarbamylase [31] as 'textbook' cases, show a sigmoid, rather than hyperbolic dependence of rate upon substrate concentration. This, like the oxygen saturation curve of haemoglobin, permits a response to changes in substrate concentration (within a restricted range) greater than that predicted by simple proportionality (Fig. 3). Many of the enzymes that display such behaviour also respond to "heterotropic effectors" [32] i.e. substances that are not obviously related to the substrates; an example is the regulation of aspartate transcarbamylase [31] by CTP (inhibitor) and ATP (activator), neither nucleotide being a substrate for this enzyme.

Kinetic and structural studies of these enzymes led to the formulation of the concept of *allosteric control* [33]. The fundamental idea is that the flexibility of protein structure allows a regulator substance, by binding at one site on an enzyme molecule, to influence binding and/or catalysis at another site viz. the active site. It was noted both by Monod et al. [32] and Koshland et al. [34], who formulated two of the main models for allosteric regulation (Figs. 4 and 5), that most allosteric enzymes have oligomeric molecules, and these models rely on subunit interaction to account for the transmission of information between binding sites. There is no doubt that the scope for co-operativity is increased by the oligomeric structure of many regulatory proteins; in haemoglobin the highly co-operative binding of oxygen is made possible by the fact that the molecule has 4 oxygen binding sites, one contributed by each subunit. On the other hand there are documented cases of monomeric enzymes displaying allosteric behaviour [35,36], and there are also allosteric enzymes which, although they are oligomeric, possess within each subunit both a catalytic site and one or more regulatory sites [37]. There is no space within the present chapter to go further into the details of the allosteric models. Suffice it to say:

(i) That allosteric interaction between sites is not the only possible explanation of 'non-classical' kinetic behaviour (see below).

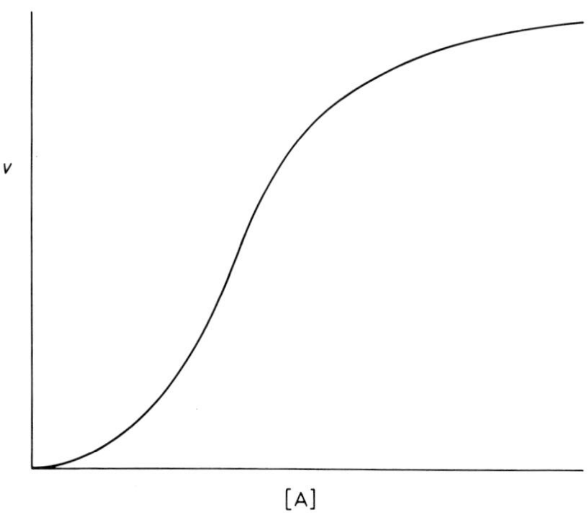

Fig. 3. A sigmoidal dependence of rate upon substrate concentration. As in the case of a hyperbolic dependence, the curve is asymptotic to an upper limit. There is a point of inflexion in the curve, however, and in the middle region of the curve a tangent would cross the abscissa axis well to the right of the origin. This means that over that region a given fractional increase in [A] results in a greater fractional increase in V. Such 'amplification' over a critical concentration range may often be of physiological value.

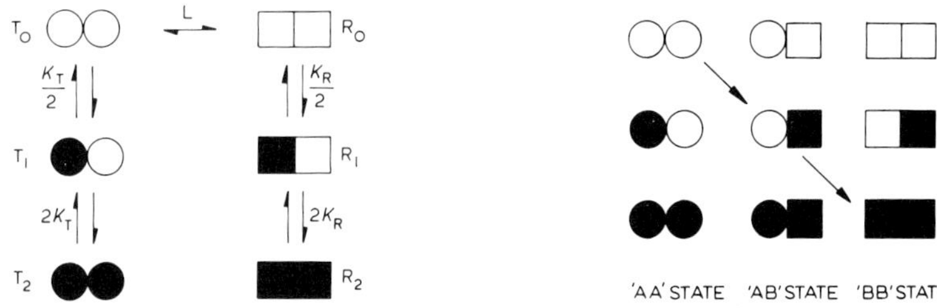

Fig. 4. The Monod–Wyman–Changeux allosteric model. In this representation the circles and squares denote 2 conformational states of the protein, the 'T state' and the 'R state' respectively. The simplest case, that of a dimeric protein, is illustrated. The 'allosteric constant', L, governs the conformational transition ($L = [\mathrm{To}]/[\mathrm{Ro}]$). Clearly this umbrella constant must take account, not only of the intrinsic stability of T and R monomers, but also of the strength of subunit interactions in the two states. These contributions are not differentiated however. There are no hybrid dimers; it is assumed that if one subunit changes conformation, both (all) change. The model is often therefore referred to as the 'symmetry model'. The filled symbols represent subunits with the substrate or other ligand bound, and hence the subscripts in T_1, R_2 etc. denote the number of occupied subunits in the molecule. K_R and K_T represent the dissociation constants governing attachment of the ligand *to a single R or T subunit,* and accordingly are multiplied by statistical factors for each net conversion step – e.g. since T_2 has 2 subunits able to release ligand whereas T_1 has only 1 vacant subunit able to bind ligand the dissociation constant for the $T_1 \rightleftharpoons T_2$ step is 2 K_T. In this model, positive homotropic co-operativity is accounted for by assuming that in the absence of the ligand the protein is predominantly in the form (T) that binds it less tightly. Inevitably, therefore, addition of the ligand swings the conformational equilibrium over in favour of the tight binding form. The symmetry principle then ensures that a ligand molecule binding at one site creates new tight binding sites (one in the case of a dimer) on other subunits. A heterotropic effector acts by disturbing the conformational equilibrium, an inhibitor by favouring the T form, an activator by favouring the R form.

Fig. 5. The Koshland–Némethy–Filmer allosteric model. The symbols are essentially as defined for Fig. 4, but this model has one crucial difference in the inclusion of hybrid conformational states – i.e. the middle column here. The overall allosteric constant L of the Monod–Wyman–Changeux model is broken down into the separate contributory factors. Thus there is a 'transformation constant' reflecting the intrinsic stability of each conformational state *for the monomer*, and 3 stability constants defining the strength of the subunit interactions in the 3 possible modes of pairing. It follows that this model may give not only positive homotropic co-operativity but also 'negative co-operativity', with appropriate values of the constants. In one extreme version of the model, known as the 'simple sequential model', the saturation process follows the path indicated by the diagonal arrows – i.e. the B state of a subunit is exclusively and compulsorily promoted by the binding of the ligand. This does not necessarily imply any judgement regarding the nature of transitory intermediates – i.e. whether binding precedes the conformational change or vice versa.

(ii) Allosteric interaction can occur in monomeric enzymes.

(iii) the models of Monod et al. and the simple sequential model of Koshland et al. are by no means the only possibilities within the framework of inter-subunit interaction. The Monod–Wyman–Changeux model and the simple-sequential model are extremes within a more general scheme which Wyman calls the 'parent model' [38].

(iv) Most consideration of allosteric enzymes has focussed on interactions affecting the tightness of binding of substrate to the enzyme rather than the rate of catalytic conversion of bound substrate.

(v) Although kinetic analysis is the usual means of detecting allosteric behaviour in an enzyme, it is difficult to achieve a convincing description of an allosteric mechanism without a considerable amount of independent information about structure, binding constants etc. Without such information there are usually so many different permutations of the various possible elements of a model that, even with excellent data and penetrating mathematical analysis, it is difficult or impossible to distinguish between alternatives. Even in the relatively simple case of haemoglobin it is unlikely that the detailed mechanism of subunit interaction would be accessible without the structural analysis achieved by X-ray crystallography [39].

2.5. *Inhibition*

(a) Definition

Many compounds other than the true substrate are capable of binding to an enzyme molecule and some of these may influence the rate of the enzyme-catalysed reaction either by blocking the active site or by bringing about a change in the structure affecting the efficiency of the active site. If the rate is decreased, the compound is termed an *inhibitor*. Some inhibitors act irreversibly, usually by modifying the covalent structure of the enzyme and these can be very useful in identifying functionally important amino acid residues. Of greater interest within the narrower context of enzyme kinetics, however, is the effect of non-covalent *reversible* inhibitors. These can be removed from an enzyme by dialysis, chromatography etc. with complete restoration of the uninhibited rate. The pattern of such inhibition can be quite revealing in the analysis of mechanism, especially for multi-substrate enzymes.

(b) Competitive inhibition

The basic inhibition patterns are readily derived in the case of the simple 1-substrate mechanism illustrated in Scheme 1. Let us consider the most easily understood pattern, that of competitive inhibition. The assumption here is that binding of the inhibitor and binding of the substrate to the enzyme are mutually exclusive. Thus the inhibitor, I, can bind to E to form EI but cannot bind to EA to form EIA (Scheme 2). The usual basis for this is that the inhibitor has structural features sufficiently

$$\begin{array}{l} \mathrm{E} + \mathrm{A} \rightleftharpoons \mathrm{EA} \rightarrow \text{Product(s)} + \mathrm{E} \\ +\mathrm{I} \\ \upharpoonleft\downharpoonright \\ \mathrm{EI} \end{array} \qquad \text{Scheme 2}$$

similar to those of the substrate to enable it to occupy all or part of the substrate binding site. In theory, though, an allosteric inhibitor bearing no structural resemblance to the substrate, could also prevent substrate binding by bringing about a conformational change.

If we now want to work out the initial-rate equation for this inhibition scheme, the equation (1a or 1b) relating [E] to [EA] still stands unaltered, as does Eqn. 3 relating v to [EA]. What has to expand, however, is Eqn. 2, the enzyme conservation equation, since we now have a third enzyme species, EI, to consider. We thus need an equation relating [EI] to the concentrations of the other enzyme forms. Since the route from E to EI is a 'dead end', we may assume that equilibrium is established between E, I and EI. This will be governed by a dissociation constant K_i, so that we may write

$$[\mathrm{EI}] = [\mathrm{E}]\frac{[\mathrm{I}]}{K_i} \tag{9}$$

Thus

$$[\mathrm{E}] + [\mathrm{EI}] = [\mathrm{E}]\left(1 + \frac{[\mathrm{I}]}{K_i}\right)$$

The rate equation follows very simply if we now replace [E] in Eqn. 2 by $[\mathrm{E}](1 + [\mathrm{I}]/K_i)$. Since the term representing [E] in the denominator gives rise to K_m in Eqn. 4, we now arrive at an equation in which K_m is multiplied by $(1 + [\mathrm{I}]/K_i)$

$$v = \frac{V_{max}[\mathrm{A}]}{[\mathrm{A}] + K_m\left(1 + \frac{[\mathrm{I}]}{K_i}\right)} \tag{10}$$

Qualitatively this mean that the maximum rate is unaltered by a fixed concentration of inhibitor. This is to be expected since substrate at infinite concentration must be able to compete successfully with inhibitor at finite concentration. The effect of the inhibitor is seen most clearly at lower substrate concentrations; the higher K_m means that a higher substrate concentration is required in order to attain half-saturation in the presence of the inhibitor.

This type of inhibition gives rise to a characteristic pattern in the Lineweaver–Burk plot (Fig. 6). Since V_{max} is unaltered, the lines for various inhibitor concentrations intersect at a common point on the ordinate. Since K_m is increased, the slopes of the lines increase with increasing inhibitor concentrations.

(c) Uncompetitive inhibition

We may now consider the case of uncompetitive inhibition (Scheme 3). This may be thought of almost as the opposite of competitive inhibition, since the assumption is that the inhibitor, I, can bind *only* to the enzyme–substrate complex EA, to form

$$\begin{array}{l} \mathrm{E} + \mathrm{A} \rightleftharpoons \mathrm{EA} \rightarrow \mathrm{Product(s)} + \mathrm{E} \\ \qquad\qquad\; +\mathrm{I} \\ \qquad\qquad\;\; \upharpoonleft\downharpoonright \\ \qquad\qquad\; \mathrm{EIA} \end{array}$$

Scheme 3

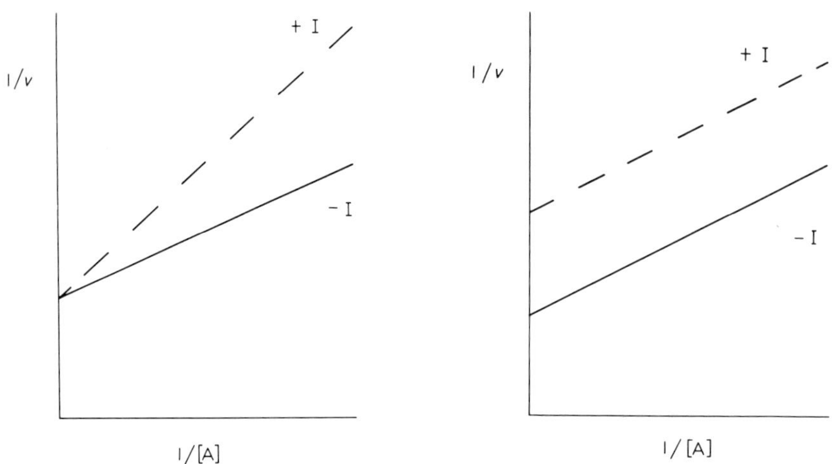

Fig. 6. Competitive inhibition. The diagram shows the characteristic pattern of Lineweaver–Burk plots intersecting on the ordinate for data obtained with and without an added competitive inhibitor, I.

Fig. 7. Uncompetitive inhibition.

EIA. This clearly implies that the enzyme sites for I and A must be different. At first sight the assumption may seem a far-fetched abstraction included for mathematical and logical completeness. Indeed this pattern of inhibition is extremely uncommon for 1-substrate enzymes. However, for multi-substrate enzymes it is quite common. One can readily envisage, for example, that if we have a sequential mechanism $E \rightleftharpoons EA \rightleftharpoons EAB \rightarrow \ldots$, an inhibitory analogue of B would very likely be unable to bind to the enzyme in the absence of A (just as B itself is also unable to bind to E).

In Scheme 3 we have an equilibrium between EA, I and EIA so that

$$[\mathrm{EIA}] = [\mathrm{EA}]\frac{[\mathrm{I}]}{K_\mathrm{i}}$$

Proceeding as above, we can see that this time it is the term representing [EA], which gives rise to [A] in the denominator in Eqn. 4, that is multiplied by $(1 + [\mathrm{I}]/K_\mathrm{i})$. Thus

$$v = \frac{V_\mathrm{max}[\mathrm{A}]}{[\mathrm{A}]\left(1 + \dfrac{[\mathrm{I}]}{K_\mathrm{i}}\right) + K_\mathrm{m}}$$

The equation is most obviously meaningful, however, when written in a form in which the denominator is [A] + constant. Therefore, dividing both top and bottom by $1 + [\mathrm{I}]/K_i$ we obtain

$$\frac{\left[V_{max}/\left(1+\frac{[\mathrm{I}]}{K_i}\right)\right][\mathrm{A}]}{[\mathrm{A}]+K_m/\left(1+\frac{[\mathrm{I}]}{K_i}\right)} \tag{11}$$

From this form of the equation we can see that in the presence of the inhibitor the maximum rate is decreased by the factor $1 + [\mathrm{I}]/K_i$, but also the Michaelis constant is *decreased* by the same factor. This contrasts with the increased K_m in the case of competitive inhibition. Since I binds only to EA, it is to be expected that I will promote the binding of A, but, once A has bound, the enzyme is divided between the productive complex EA and the unproductive complex EAI, so that the maximum rate, inevitably, is decreased.

Fig. 7 shows that since K_m and V_{max} are decreased by the same factor, the effect on the Lineweaver–Burk plot is to produce a family of parallel lines for increasing inhibitor concentrations.

(d) Non-competitive inhibition

This label includes the more general case of so-called mixed inhibition and the special case that is sometimes designated *simple* non-competitive inhibition. Starting first with the simpler case, let us assume that our inhibitor can bind with equal ease

$$\begin{array}{ccl} \mathrm{E} + \mathrm{A} & \rightleftharpoons & \mathrm{EA} \rightarrow \text{Product(s)} + \mathrm{E} \\ +\mathrm{I} & & +\mathrm{I} \\ \upharpoonleft\downharpoonright & & \upharpoonleft\downharpoonright \\ \mathrm{EI} & & \mathrm{ETA} \end{array}$$

Scheme 4

to either E or EI, but that the enzyme–substrate–inhibitor complex is inactive (Scheme 4). We now have *two* extra terms to add to the enzyme conservation equation. Since $[\mathrm{EI}] = [\mathrm{E}](1 + [\mathrm{I}]/K_i)$ and also $[\mathrm{EIA}] = [\mathrm{EA}](1 + [\mathrm{I}]/K_i)$, the appropriate equation has the contributions of both [E] and [EA] to the denominator multiplied by $1 + [\mathrm{I}]/K_i$. Thus

$$v = \frac{V_{max}[\mathrm{A}]}{([\mathrm{A}]+K_m)\left(1+\frac{[\mathrm{I}]}{K_i}\right)}$$

Since, again, we need to have the denominator in the form [A] + constant, we divide through by $1 + [\mathrm{I}]/K_i$ in both numerator and denominator to obtain

$$v = \frac{V_{max}[\mathrm{A}]/\left(1+\frac{[\mathrm{I}]}{K_i}\right)}{[\mathrm{A}]+K_m} \tag{12}$$

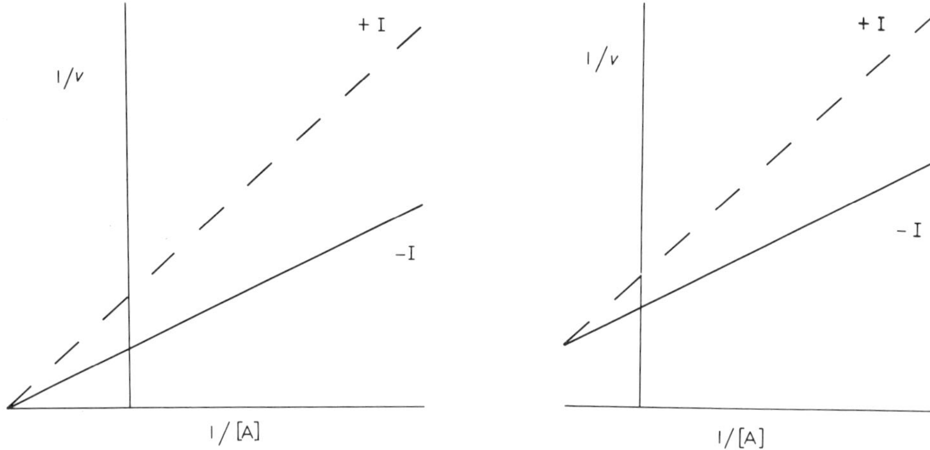

Fig. 8. Non-competitive inhibition.

Fig. 9. Mixed inhibition. This plot illustrates the case where $K_i < K_{i(A)}$ and the inhibitor affects K_m more than it affects V_{max}.

From Eqn. 12 it is clear that simple non-competitive inhibition leaves the Michaelis constant unaltered, but decreases the maximum rate by the factor $1 + [I]/K_i$. This will give in the Lineweaver–Burk plot a set of lines all intersecting on the abscissa and with the slope increasing with inhibitor concentration (Fig. 8).

(e) Mixed inhibition

Turning to the more general case now, we can see that the assumption that I binds equally well to both E and EA is arbitrary. It might seem likely that, if the interaction is sufficient to abolish catalysis in the presence of I, there would also be some effect on binding of A. In the case of *mixed inhibition* we still assume that I binds to both E and EA but now with two distinct dissociation constants K_i and $K_{i(A)}$. Once again both denominator terms will be affected but not by the same factor. When we divide through by $1 + I/K_{i(A)}$, therefore, in order to get [A] on the left-hand side of the denominator, we arrive at an equation in which both V_{max} and K_m are altered. V_{max} is simply decreased by the factor $1 + I/K_{i(A)}$, but the Michaelis constant is multiplied by $(1 + I/K_i)/(1 + I/K_{i(A)})$. This factor could be either greater or less than 1 depending on whether $K_i < K_{i(A)}$ or $K_{i(A)} < K_i$. If $K_i < K_{i(A)}$ the K_m will be increased and the Lineweaver–Burk plots will intersect in the second quadrant (i.e. to the left of the ordinate but above the abscissa in the usual representation) (Fig. 9); if $K_i > K_{i(A)}$, the K_m will be decreased and the lines will intersect in the third quadrant, i.e. below the abscissa (Fig. 10). Table 1 summarises these inhibition patterns.

TABLE 1

Type of inhibition	Effect on V_{max}	Effect on K_m	Effect on slope of L-B plot	Position of intersection of L-B plots
Competitive	No change	Increase	Increase	Ordinate axis
Uncompetitive	Decrease	Decrease	No change	None
Non-competitive				
(a) Simple	Decrease	No change	Increase	Abscissa axis
(b) Mixed: $K_i < K_{i(A)}$	Decrease	Increase	Increase	Second quadrant
(c) Mixed: $K_i > K_{i(A)}$	Decrease	Decrease	Increase	Third quadrant

2.6. *Multi-substrate kinetics*

(a) Types of mechanism

As we have mentioned, once we start considering enzyme-catalysed reactions involving more than one substrate in each direction, possible mechanisms proliferate. As an example let us consider a 2-substrate, 2-product reaction:

$$A + B \overset{E}{\rightleftharpoons} P + Q$$

The possible mechanisms may be categorised first of all as "*sequential*" or "*non-sequential*" according to whether the substrates A and B follow one another onto the enzyme surface before any product is released. If they do not, then, in the

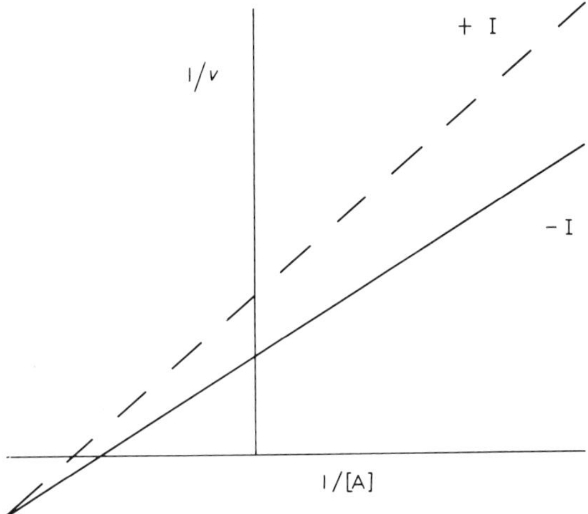

Fig. 10. Mixed inhibition. This plot illustrates the case where $K_i > K_{i(A)}$ so that the inhibitor affects V_{max} more than K_m.

non-sequential (or *enzyme-substitution* or *ping pong*) mechanism, one or other substrate must react with the enzyme, leaving behind a chemical function (Scheme 5). After the departure of the first product, the second substrate arrives to receive the donation. The enzyme thus serves as a drop-off point so that the reacting substrates do not actually have to meet.

$$E + A \underset{k_2}{\overset{k_1}{\rightleftharpoons}} EA \underset{k_4}{\overset{k_3}{\rightleftharpoons}} E'P \underset{k_6}{\overset{k_5}{\rightleftharpoons}} E' + P \qquad \text{(Scheme 5)}$$

$$E' + B \underset{k_8}{\overset{k_7}{\rightleftharpoons}} E'B \underset{k_{10}}{\overset{k_9}{\rightleftharpoons}} EQ \underset{k_{12}}{\overset{k_{11}}{\rightleftharpoons}} E + Q$$

The mechanism comprises two distinct halves, each able to occur without the other. This mechanism is followed for example by pyridoxal phosphate-dependent transaminases, the amino group being transferred from the donor substrate to the enzyme's prosthetic group and so, ultimately, to the acceptor substrate [40,41].

In a sequential mechanism, on the other hand, such as is followed by most $NAD(P)^+$-linked dehydrogenases [42], the enzyme forms a *ternary complex* i.e. a complex containing the enzyme itself and both substrates. This allows for several further possibilities. There may be either *random-order* or *compulsory-order* binding. If there is a compulsory order of substrate addition and product release, there are 4 possible sequences:

A on, B on, Q off, P off
A on, B on, P off, Q off
B on, A on, Q off, P off
B on, A on, P off, Q off

The random-order mechanism is illustrated below

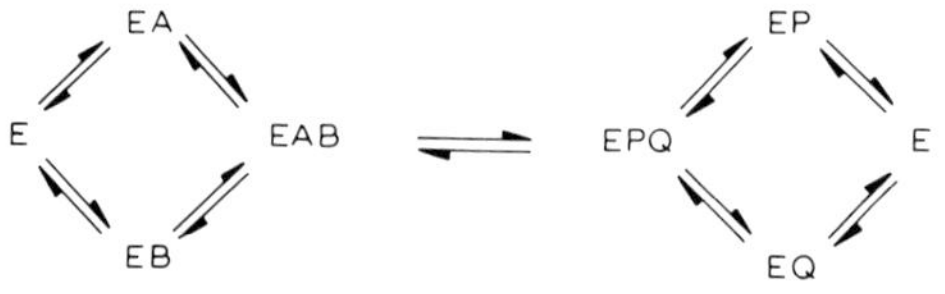

$$E \rightleftharpoons EA \rightleftharpoons EAB;\; E \rightleftharpoons EB \rightleftharpoons EAB \quad EAB \rightleftharpoons EPQ \quad EPQ \rightleftharpoons EP \rightleftharpoons E;\; EPQ \rightleftharpoons EQ \rightleftharpoons E \qquad \text{Scheme 6}$$

This is shown with two distinct ternary complexes, EAB and EPQ, the intervening step representing the chemical conversion. The definition of two such complexes is arbitrary. It is possible that there might be more than two well-defined chemical states of the ternary complex along the reaction path. Equally there might only be one. The same is true for the compulsory-order mechanisms. The choice between one, or two, or more ternary complexes cannot be made through steady-state kinetic studies, however, although it might be made through rapid-reaction studies. One situation which can be distinguished is the so-called Theorell–Chance mechanism [43]. This is at first sight a paradoxical mechanism which is sequential but has *no* ternary complex (Scheme 7).

$$E + A \underset{k_2}{\overset{k_1}{\rightleftharpoons}} EA \underset{B \quad k_8 \quad P}{\overset{B \quad k_3 \quad P}{\rightleftharpoons}} EQ \underset{k_{10}}{\overset{k_9}{\rightleftharpoons}} E + Q$$

Scheme 7

What this really means is that the ternary complex has such a transitory existence that it never makes up a significant fraction of the total amount of enzyme. Steady-state kinetics concerns itself only with those complexes which, by their existence, detectably alter the pattern of dependence of reaction rate on substrate concentration. The Theorell–Chance mechanism may be seen perhaps as a manifestation of highly effective catalysis. Certainly, in the case of the enzyme for which it was first described, horse liver alcohol dehydrogenase, the mechanism is obeyed for 'good' substrates i.e. short-chain primary alcohols; with secondary alcohols, which are poor substrates, the ternary complex becomes kinetically significant – because it works less well [44].

(b) Overall strategy

Fortunately, as the possible mechanisms proliferate, the analytical power of enzyme kinetics increases, since the kinetic patterns lead to diagnostic tests for various mechanisms. The task of unravelling 2- and 3-substrate mechanisms began in earnest in the 1950's with the theoretical framework laid by Alberty [45,46], Dalziel [47,48] and Cleland [49–51]. The detailed experimental investigations of horse liver alcohol dehydrogenase by Theorell, Dalziel and their collaborators [42,52] and of beef heart lactate dehydrogenase by Schwert's group [53] were important milestones.

The fundamental approach can be summarised a below:

(i) Write out all possible reasonable mechanisms.

(ii) Set out the rate equation for each mechanism.

(iii) Carry out experiments with systematic variation of the concentrations of substrates over appropriately chosen ranges. Dead-end and/or product inhibitors may also be used.

(iv) Obtain the kinetic parameters from the experimental data by computer analysis or graphical methods.

(v) Compare the relations between the kinetic parameters with the predictions for the various mechanisms. Comparisons may also be made with independently measurable parameters (equilibrium constants, binding constants etc.).

The whole process is iterative in nature since steps (iv) and (v) invariably lead to new experiments.

(c) Deriving a rate equation

Let us consider this process for a 2-substrate, 2-product reaction. We have already examined the likely range of possible patterns of substrate addition and product release. The second stage is in theory the most tedious, but in practice is now a trivial task since (a) there are tabulations of rate equations and derivations available for most of the mechanisms likely to come into consideration [54,55] and (b) there are now computer programs to provide a rate equation given the mechanism [56,57]. Nevertheless, it is as well to be aware what the process involves and how it relates to the simple derivation that we have already seen for a 1-substrate mechanism in Section 2.1. We shall illustrate this with the relatively simple case of a sequential, compulsory-order mechanism with 2 ternary complexes. This may be represented in various ways:

$$\mathrm{E} \underset{k_2}{\overset{k_1}{\rightleftharpoons}} \mathrm{EA} \underset{k_4}{\overset{k_3}{\rightleftharpoons}} \mathrm{EAB} \underset{k_6}{\overset{k_5}{\rightleftharpoons}} \mathrm{EPQ} \underset{k_8}{\overset{k_7}{\rightleftharpoons}} \mathrm{EP} \underset{k_{10}}{\overset{k_9}{\rightleftharpoons}} \mathrm{E} \qquad \text{Scheme 8}$$

or in the Cleland representation [49]

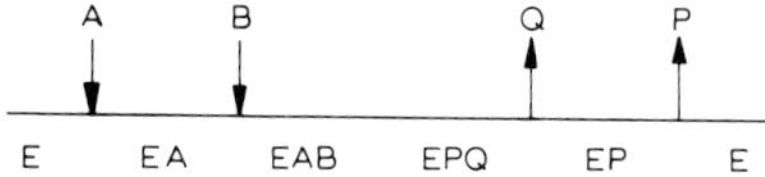

or in the more recent "bracket notation" of Ainsworth [58]

$$(\mathrm{A}\ 1), (\mathrm{B}\ 1), (1\ 1), (1\ \mathrm{Q}), (1\ \mathrm{P})^{=}$$

In this latter notation 1 denotes a reaction step that does not involve the binding of a substrate. The two symbols within each bracket denote the two directions of the reaction interconverting a pair of enzyme complexes. This mechanism contains 5 distinct enzyme species, E, EA, EAB, EPQ, EP, and, in solving the equations for this mechanism, the concentrations of these 5 species are the unknowns. The known variables are the total concentration of enzyme and the initial concentrations of all the reactants, since these are experimentally fixed. Just as in the 1-substrate case we can solve for the unknowns and determine how the enzyme is distributed among the various complexes in the mechanism by using the enzyme conservation equation and a set of equations relating the concentration of each complex to those of its neighbours in the mechanism. These equations are expressions of the steady-state assumption (Section 2.1). In the present case we may simplify the task by assuming in advance that we are, for the moment, only interested in the initial rate in the absence of products. By setting [P] = [Q] = 0 we eliminate the reaction vectors leading from E to EP and from EP to EPQ [59].

$$\mathrm{E} \underset{k_2}{\overset{k_1}{\rightleftharpoons}} \mathrm{EA} \underset{k_4}{\overset{k_3}{\rightleftharpoons}} \mathrm{EAB} \underset{k_6}{\overset{k_5}{\rightleftharpoons}} \mathrm{EPQ} \overset{k_7}{\rightharpoonup} \mathrm{EP} \overset{k_9}{\rightharpoonup} \mathrm{E}$$

We can solve the distribution in terms of the concentration of any one of the complexes, but now it is convenient to use [EP] or [EPQ] because the overall net rate of reaction will be given simply as k_7[EPQ] or k_9[EP]. We shall use [EP] here.

Steady state for EP. The steady-state assumption tells us that the rate of formation of EP equals the rate of its removal. Thus:

$$[\text{EP}]k_9 = [\text{EPQ}]k_7$$

$$\therefore [\text{EPQ}] = [\text{EP}]\frac{k_9}{k_7}$$

Steady state for EPQ

$$[\text{EPQ}](k_6 + k_7) = [\text{EAB}]k_5$$

$$\therefore [\text{EAB}] = [\text{EPQ}]\frac{k_6 + k_7}{k_5} = [\text{EP}]\frac{(k_6 + k_7)k_9}{k_5 k_7}$$

Steady state for EAB

$$[\text{EAB}](k_4 + k_5) = [\text{EPQ}]k_6 + [\text{EA}]k_3[\text{B}]$$

$\therefore$ Substituting for [EAB] from above,

$$[\text{EA}]k_3[\text{B}] = [\text{EPQ}]\frac{(k_4 + k_5)(k_6 + k_7)}{k_5} - k_6$$

$$= [\text{EPQ}]\frac{(k_4 k_6 + k_4 k_7 + k_5 k_7)}{k_5}$$

Hence, substituting again

$$[\text{EA}] = [\text{EP}]\frac{k_9(k_4 k_6 + k_4 k_7 + k_5 k_7)}{k_3 k_5 k_7\,[\text{B}]}$$

Steady state for E

$$[\text{E}]k_1[\text{A}] = [\text{EP}]k_9 + [\text{EA}]k_2$$

$$\therefore [\text{E}] = [\text{EP}]\frac{k_9}{k_1[\text{A}]} + \frac{k_2 k_9(k_4 k_6 + k_4 k_7 + k_5 k_7)}{k_1 k_3 k_5 k_7[\text{A}][\text{B}]}$$

At this stage we have expressions for all the unknowns in terms of a single unknown, [EP], and can solve by introducing the conservation equation:

$$e = [\text{EP}] + [\text{EPQ}] + [\text{EAB}] + [\text{EA}] + [\text{E}]$$

$$= [\text{EP}]\left[1 + \frac{k_9}{k_7} + \frac{k_9(k_6 + k_7)}{k_5 k_7} + \frac{k_9(k_4 k_6 + k_4 k_7 + k_5 k_7)}{k_3 k_5 k_7\,[\text{B}]} + \frac{k_9}{k_1}\right.$$

$$+\frac{k_2k_9(k_4k_6+k_4k_7+k_5k_7)}{k_1k_3k_5k_7[\mathrm{A}][\mathrm{B}]}\Bigg] \tag{13}$$

Within the large bracketed expression in Eqn. 13 the 3 constant terms represent the concentrations of EP, EPQ, EAB, the forms among which the enzyme will be distributed when it is saturated with both A and B; the term in 1/[B] represents the contribution of [EA] and the terms in 1/[A] and 1/[A][B] both represent the free enzyme, E.

The initial rate of product formation is given by

$$v = k_9\,[\mathrm{EP}] \tag{14}$$

Dividing Eqn. 13 by Eqn. 14 we get the initial-rate equation in its reciprocal form.

$$\frac{e}{v}=\frac{1}{k_9}+\frac{1}{k_7}+\frac{k_6+k_7}{k_5k_7}+\frac{k_4k_6+k_4k_7+k_5k_7}{k_3k_5k_7[\mathrm{B}]}+\frac{1}{k_1[\mathrm{A}]}$$
$$+\frac{k_2(k_4k_6+k_4k_7+k_5k_7)}{k_1k_3k_5k_7[\mathrm{A}][\mathrm{B}]} \tag{15}$$

Casting the equation in this form shows us clearly that the expression for e/v is linear in both 1/[B] and 1/[A]. It follows that with [B] fixed the enzyme should give a linear Lineweaver–Burk plot for varied [A] and vice versa. If we group the terms according to which variables they contain, Eqn. 15 may be rewritten in a simpler form [47]:

$$\frac{e}{v}=\Phi_0+\frac{\Phi_\mathrm{A}}{[\mathrm{A}]}+\frac{\Phi_\mathrm{B}}{[\mathrm{B}]}+\frac{\Phi_\mathrm{AB}}{[\mathrm{A}][\mathrm{B}]} \tag{16}$$

Φ_0, Φ_B and Φ_AB are composite constants, functions of the rate constants in the mechanism, and Φ_A is simply $1/k_1$.

If [B] is to be fixed and [A] varied, Eqn. 16 can be rearranged:

$$\frac{e}{v}=\left[\Phi_0+\frac{\Phi_\mathrm{B}}{[\mathrm{B}]}\right]+\left[\Phi_\mathrm{A}+\frac{\Phi_\mathrm{AB}}{[\mathrm{B}]}\right]\frac{1}{[\mathrm{A}]} \tag{17}$$

Apart from the fact that the rate is expressed here as a specific rate (i.e. divided by enzyme concentration), Eqn. 17 is of the same form as Eqn. 5, the equation for the linear Lineweaver–Burk plot for a 1-substrate reaction. Thus if we plot e/v against 1/[A] at fixed [B], a straight line should result (Fig. 11). Note however, that the $V_{\max}$ and K_m values it gives will be only *apparent* $V_{\max}$ and K_m, because the two composite 'constants' in Eqn. 17 both in fact depend on [B]. Only by saturating with B can one get a plot giving a true $V_{\max}$, e/Φ_0, and a true K_m for A, Φ_A/Φ_0.

Since the form of Eqn. 16 is symmetrical with respect to [A] and [B], a similar process can be carried by fixing [A] and varying [B]. The derivation of Eqn. 15, though obviously more lengthy than that for a single-substrate mechanism is still not difficult. Although the problem remains simple in principle even for more complex mechanisms, the algebra becomes rapidly more cumbersome as one includes products, introduces branch points in the mechanism, and so on. To remove some of the drudgery and reduce the risk of error, King and Altman [60] introduced a routine procedure, applicable to any mechanism, for writing out the rate equation. It is based on the analysis of simultaneous equations by the use of determinants. Even this procedure, which involves drawing dozens of arrow diagrams is somewhat tedious and the availability of computer programmes to carry out this task is a boon [56,57].

(d) Experimental determination of the rate equation for an individual enzyme

Once the theoretical rate equations are available one must compare them with the experimentally obtained rate equation. The appropriate strategy is suggested by Eqn. 17. If an enzyme obeys Eqn. 16, i.e. if it shows "linear kinetics" with respect to both substrates, then by varying the concentration of one substrate with several fixed concentrations of the other, and measuring the enzymatic rate with each combination of concentrations, one may evaluate all 4 Φ constants. Thus if [B] is fixed at several values, judiciously chosen after a preliminary experiment to establish the approximate range of values of [B] required to give rate variation, one would obtain a family of lines as shown in Fig. 11.

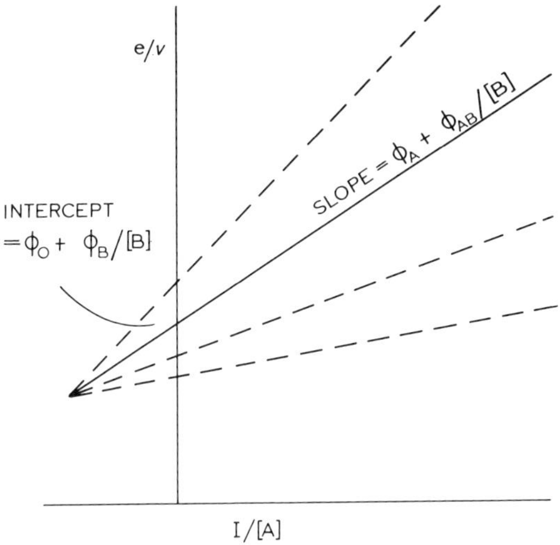

Fig. 11. Schematic double reciprocal plot for a 2-substrate reaction obeying Eqn. 16. The primary plot here is analogous to the Lineweaver–Burk plot for a 1-substrate reaction, but the V_{max} and K_m obtained from such a plot as the solid line here are *apparent* V_{max} and K_m, true only for the fixed value of [B] at which they were determined. As shown, both the slope and the ordinate intercept may depend upon [B]. The dashed lines indicate plots for other fixed values of [B].

The ordinate intercepts would be given by

$$I = \Phi_0 + \frac{\Phi_B}{[B]}$$

and the slopes by

$$S = \Phi_A + \frac{\Phi_{AB}}{[B]}$$

The intercepts and slopes could therefore be replotted against 1/[B]. The plot of S would then give another straight line with a slope of Φ_{AB} and an intercept of Φ_A on the ordinate (Fig. 12a). The plot of I likewise would yield values of Φ_A and Φ_0 (Fig. 12b). This procedure, described for one specific mechanism by Florini and Vestling [61] and in a more generally applicable form by Dalziel [47], can also be extended to 3-substrate enzymes [62,63].

If a 2-substrate enzyme does *not* obey Eqn. 16 then this may show up either in the 'primary' plots of e/v against 1/[substrate 1] or in the 'secondary' replots of primary slopes and intercepts against 1/[substrate 2], or both. Detection of the non-linearity usually depends on using a wide range of substrate concentrations. Curvature is usually sufficiently gradual that it may be mistakenly dismissed as experimental error if only a narrow range is employed. As has been discussed earlier non-linear kinetics may be due to allosteric interaction or to alteration of bulk properties of the solvent at very high substrate concentrations. Non-linearity may,

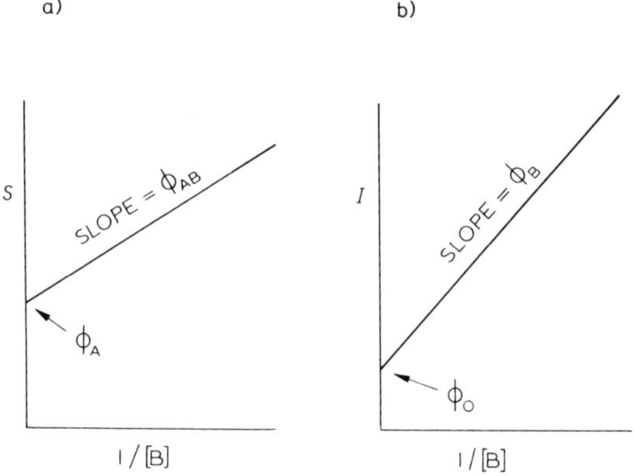

Fig. 12. Secondary plots of 2-substrate initial-rate data. From a set of primary plots such as is schematically illustrated in Fig. 11, one obtains a set of slopes and intercepts. The replot (a) of slopes, S, against 1/[B] yields estimates of Φ_A and Φ_{AB}, and the replot (b) of intercepts, I, gives Φ_0 and Φ_B. If Eqn. 16 is a valid description of the kinetic behaviour over the substrate range investigated and if the experimental data are adequate, these 4 constants then allow calculation of the initial rate for *any* combination of substrate concentrations over the range for which Eqn. 16 is obeyed.

however, also be an intrinsic feature of fairly simple mechanisms. A steady-state mechanism in which a given substrate can combine with the enzyme at more than one step will give non-linearity provided that these different points of combination are connected by reversible steps [50]. The most obvious example of this situation is a steady-state random-order mechanism [48,64]. In the one illustrated earlier, A may combine with either E or EB, and B may combine with either E or EA, so that this mechanism will give an initial rate non-linear in both [A] and [B] unless special restrictions are imposed. Another such source of non-linearity is the formation of so-called *abortive complexes*. If, for example, in Scheme 4, substrate B is able to combine not only productively with EA to form a reactive ternary complex, but also with EP to form a dead-end ternary complex EPB, then one may expect non-linearity in plots of e/v against 1/[B].

(e) Drawing conclusions from the experimentally determined rate equation

If, however, the primary plots are linear, what may we hope to discover from them? First of all, if the enzyme follows a ping-pong mechanism, the initial rate equation in its reciprocal form has no term in 1/[A][B] [47] – i.e. the final term in Eqn. 16 is missing, $\Phi_{AB} = 0$. This is a consequence of the fact that the points at which A and B combine with the enzyme are separated in the mechanism by irreversible steps – viz. product release at zero concentration [50]. This means that in Eqn. 17 the slope term will be invariant as [B] is changed. Thus a set of primary plots (against either 1/[A] or 1/[B] will be a family of parallel lines rather than converging to the left of the ordinate as in Fig. 11. If the slopes vary systematically, one can immediately rule out a ping-pong mechanism. If they do not vary, i.e. if the lines are parallel, it is always possible that the range of substrate concentrations that will give accurately measurable rates does not allow the Φ_{AB} term to become prominent even though it is present [47]. Competitive inhibitors have been used in such cases to enhance a masked slope variation [65]. If the primary plots point to a ping-pong mechanism then further evidence may be sought through the demonstration of partial reactions – i.e. one substrate should be able to form its corresponding product in the presence of the enzyme even if the other substrate is absent. Reference to Scheme 5 will show that such net conversion (of A to P for example) cannot exceed the molar concentration of the enzyme itself. If enough enzyme is available, the change may nevertheless be measurable and also, especially if the enzyme has a conveniently measurable physical property (e.g. the spectral characteristics of pyridoxal phosphate-dependent transaminases or flavin-dependent dehydrogenases or oxidases), it may be possible to demonstrate the conversion of the enzyme into its substituted form ($E \rightarrow E'$ in Scheme 5). Even if the enzyme is not plentiful in molar terms, the partial reactions may be demonstrated by isotope exchange: in Scheme 5 the presence of a catalytic amount of enzyme will cause radioactive label initially present in A to be transferred to the pool of P, or vice versa, in the absence of B and Q. This would not be so in a sequential mechanism.

Further evidence for a ping-pong mechanism may be obtained, however, from the initial rate parameters. Derivation of the rate equation for the mechanism in Scheme

5 gives the following:

$$\Phi_0 = \frac{1}{k_3} + \frac{1}{k_5} + \frac{1}{k_9} + \frac{1}{k_{11}} + \frac{k_{10}}{k_9 k_{11}} + \frac{k_4}{k_3 k_5}$$

$$\Phi_A = \frac{k_2 k_4 + k_2 k_5 + k_3 k_5}{k_1 k_3 k_5}$$

$$\Phi_B = \frac{k_8 k_{10} + k_8 k_{11} + k_9 k_{11}}{k_7 k_9 k_{11}}$$

As might be expected, Φ_0 is a function of all those rate constants which are still liable to govern the rate even when A and B are both saturating. However, if we examine Φ_A we note that the rate constants in the expression all have subscripts from 1 to 5 – i.e. they are in the first half of the mechanism. The symmetrically related expression for Φ_B is a function of rate constants entirely from the second half of the mechanism. Now, since B and Q are not involved in any way in the first half of the mechanism, Φ_A must be independent of the chemical nature of B, and likewise Φ_B must be independent of the nature of A. If, therefore, the enzyme has a broad enough specificity to allow one to use several different substrates A with a single substrate B and vice versa, this presents a very useful test of mechanism. If the prediction is clearly violated by the experimental results, this mechanism may be discarded. If the prediction is satisfactorily met, how strong a piece of evidence is this in the mechanism's favour? To answer this we must know whether the prediction is unique or whether it is a common feature of two or more mechanisms. If, for example, we inspect Eqn. 15, we find that, for the compulsory-order mechanism shown in Scheme 8, $\Phi_A = 1/k_1$. This is indeed independent of the nature of B since k_1 is the rate constant for combination of A with the free enzyme E. However, $\Phi_B = (k_4 k_6 + k_4 k_7 + k_5 k_7)/k_3 k_5 k_7$. This involves 5 rate constants k_3, k_4, k_5, k_6 and k_7 all of which involve both A and B or, in the case of k_6 and k_7, both corresponding products. Thus Φ_B is unlikely to be independent of the nature of A for this mechanism. Clearly this criterion will distinguish between two mechanisms shown in Schemes 5 and 8, and in fact the prediction *is* unique to the ping-pong mechanism. Another distinctive test of the ping-pong mechanism is the *Haldane relationship*. Haldane pointed out that from the initial rate kinetic parameters for the forward and reverse directions of a reversible enzyme-catalysed-reaction it was possible to obtain an expression for the overall equilibrium constant [66].

The equilibrium constant may be determined independently of the rate measurements. For the ping-pong mechanism of Scheme 5 the equilibrium constant is:

$$K_{eq} = \frac{k_2 k_4 k_6 k_8 k_{10} k_{12}}{k_1 k_3 k_5 k_7 k_9 k_{11}} \tag{18}$$

Now the denominators of Φ_A and Φ_B contain the 6 rate constants with odd-number subscripts which make up the denominator of the expression for K_{eq} in Eqn. 18. The initial rate parameters for the reverse reaction, P + Q → A + B, may be written down by inspection from considerations of symmetry. One finds, on doing this, that the expressions for Φ_P and Φ_Q contain all the even-number rate constants and that Φ_P has the same numerator as Φ_B, and Φ_Q the same as Φ_A. In consequence

$$\frac{\Phi_A \Phi_B}{\Phi_P \Phi_Q} = K_{eq} \tag{19}$$

This Haldane relationship for the ping-pong mechanism is not shared by any of the other 2-substrate mechanisms.

If we turn now to focus our attention on sequential mechanisms, first of all the initial rate equation will contain the Φ_{AB} term, so that primary plots will intersect at a point to the left of the ordinate. In the case of the 2-substrate compulsory-order mechanism, as we have seen, $\Phi_A = 1/k_1$. Not only does this mean that Φ_A should be independent of B, but also it means that Φ_A is defined by the value of a single rate constant which may well be directly measurable by rapid reaction techniques if the combination of A with E leads to a measurable physical change in either A or E. This relationship could in fact also be true for the ping-pong mechanism: if we take the expression for Φ_A for the ping-pong mechanism and impose the restrictions that $k_2 k_4 \ll k_3 k_5$ and $k_2 \ll k_3$ then it reduces to $k_3 k_5 / k_1 k_3 k_5 = 1/k_1$. However, we already have plenty of other criteria with which to decide for or against that mechanism.

Another relationship which emerges from the Φ parameters for the compulsory-order mechanism is the following:

$$\frac{\Phi_{AB}}{\Phi_B} = \frac{k_2}{k_1} \tag{20}$$

This is the dissociation constant for the leading substrate A. This should both be independent of the nature of B and be measurable – e.g. by spectrophotometric titration or by equilibrium dialysis. Taken together, with our expression for Φ_A, Eqn. 20 gives

$$\frac{\Phi_{AB}}{\Phi_A \Phi_B} = k_2 \tag{21}$$

This clearly is not an independent relationship when one has the other two. However, as Dalziel first pointed out [47], it is the basis of a further test if one also has the kinetic parameters for the reverse reaction. The rate constant k_2 governs a compulsory product release step in the reverse reaction and this imposes an upper limit on the rate of that reaction. Thus if this mechanism is followed, we have the following two maximum-rate relationships:

$$\frac{\Phi_{AB}}{\Phi_A \Phi_B} \geqslant \frac{1}{\Phi_0'} \quad \frac{\Phi_{PQ}}{\Phi_P \Phi_Q} \geqslant \frac{1}{\Phi_0}$$

(the 'prime' in Φ_0' denotes the reverse reaction). The availability of Φ parameters for both directions of reaction again allows us to test a Haldane relationship. In this case

$$\frac{\Phi_{AB}}{\Phi_{PQ}} = K_{eq} \tag{22}$$

Reverting to the Theorell–Chance mechanism (Scheme 7), one can see that this is a special case of the mechanism we have just considered, in which EAB and EPQ have effectively disappeared because of the high values of some of the rate constants relative to others in the mechanism. Accordingly, in Scheme 7 the apparently idiosyncratic numbering of rate constants is to retain comparability with Scheme 8. The rate equation is very easily derived, and the parameters are:

$$\Phi_0 = \frac{1}{k_9} \qquad \Phi_A = \frac{1}{k_1} \qquad \Phi_B = \frac{1}{k_3} \qquad \Phi_{AB} = \frac{k_2}{k_1 k_3}$$

From these and the corresponding, symmetrically related parameters for the reverse direction of reaction it may be seen that, as one would expect, all the relationships listed above for the more general case also apply to the Theorell–Chance mechanism. It does, however, have some diagnostic tests that are uniquely its own. First, not only, as before, is Φ_A independent of the nature of B, but now also Φ_0, and therefore the maximum rate, is independent of B. This is because in the Theorell–Chance mechanism the maximum rate is determined solely by the rate of dissociation of the 'outer' product, P. It has been shown, for instance, that horse liver alcohol dehydrogenase gives essentially the same maximum rate for a series of primary alcohols [44] and this may be equated with the rate of dissociation of NADH. Conversely, of course, Φ_0' is independent of the nature of Q. From Eqn. 21, which is still obeyed, one may now see that the Dalziel maximum-rate relationships become equalities for the Theorell–Chance mechanism:

$$\frac{\Phi_{AB}}{\Phi_A \Phi_B} = \frac{1}{\Phi_0'} \qquad \frac{\Phi_{PQ}}{\Phi_P \Phi_Q} = \frac{1}{\Phi_0}$$

This mechanism, moreover, has its own unique Haldane relationship in addition to Eqn. 22:

$$\frac{\Phi_A \Phi_B \Phi_0}{\Phi_P \Phi_Q \Phi_0'} = K_{eq}$$

It should be re-emphasised that for a given enzyme the compulsory-order mechanism (with or without kinetically significant ternary complexes) is not just a single mechanism, because for a real enzyme the symbols denote real substrates so that there is the question of which is the leading substrate in each direction. Some of the relationships devised above, therefore, help not only to distinguish the compulsory-order mechanism from other mechanisms, but also to distinguish among the different possible compulsory-order mechanisms.

This brings us to the final mechanism we need to consider for a 2-substrate reaction, namely a random-order mechanism. We have assumed that we would be alerted to the possibility of a steady-state random-order mechanism by non-linear primary or secondary plots, but it is possible to get linear kinetics with a random-order mechanism. If we make the assumption that the further reaction of the ternary complex EAB is much slower than the network of reactions connecting E to EAB via EA and EB, then there are only 4 kinetically significant complexes and their concentrations are related to one another by substrate concentrations and dissociation constants. This is the *rapid-equilibrium random-order mechanism,* and the assumption made is analogous to the Michaelis–Menten equilibrium assumption for a 1-substrate mechanism.

EA
K_A
$K_{B(A)}$
E
EAB
$\xrightarrow{k}$ E + Products
K_B
$K_{A(B)}$
EB

Each complex contributes one of the 4 Φ parameters

$$\Phi_0 = \frac{1}{k} \qquad \text{(EAB)}$$

$$\Phi_A = \frac{K_{A(B)}}{k} = \frac{K_{B(A)}K_A}{K_B k} \qquad \text{(EB)}$$

$$\Phi_B = \frac{K_{B(A)}}{k} \qquad \text{(EA)}$$

$$\Phi_{AB} = \frac{K_{B(A)}K_A}{k} \qquad \text{(E)}$$

In the list above, the bracketed complex in each case is the one that contributes that particular term to the rate equation – e.g. $\Phi_A/[A]$ is proportional to the relative concentration of EB. From the ratios of the 4 Φ parameters we may obtain all 4 dissociation constants:

$$K_A = \Phi_{AB}/\Phi_B \qquad K_B = \Phi_{AB}/\Phi_A$$

$$K_{A(B)} = \Phi_A/\Phi_0 \qquad K_{B(A)} = \Phi_B/\Phi_0$$

For this mechanism, therefore, we no longer expect Φ_A to be independent of B or Φ_B of A, but we still have Eqn. 20, and therefore Φ_{AB}/Φ_B *will* be independent of the nature of B. For this mechanism, however, we also have the prediction that Φ_{AB}/Φ_A

should be independent of the nature of A. The two relationships would not both be satisfied for a compulsory-order mechanism.

It is also easily shown that this mechanism also obeys Eqn. 22 for its Haldane relationship.

Perhaps the most obvious prediction, however, for such a mechanism is that it should be possible to demonstrate the binding of each substrate to the enzyme in the absence of the other substrate. If the dissociation constants can be measured they may be compared with the predictions from ratios of Φ parameters.

(f) Inhibition experiments

The previous sections may give the impression that it is an easy matter to establish the mechanism of a multi-substrate enzyme. In fact, more often than not uncertainty and controversy surround such mechanisms for many years despite an abundance of experimental work. We have assumed an ideal situation whereas there are a number of possible obstacles in practice. For example the reaction may be effectively irreversible so that it is only possible to measure the kinetic parameters for one direction of reaction; the substrate specificity may be so stringent that it is impossible to apply tests which rely on using a range of alternative substrates; the available methods of rate measurement may not be sufficiently sensitive to allow all the kinetic parameters to be determined reliably. The last problem at least is one that allows some hope: the kinetic study of NAD-dependent dehydrogenases became much more incisive once fluorescence measurement took over from absorbance measurement as the method of choice [52,57]. Nevertheless there is clearly a need for as many criteria of mechanism as may be mustered; and the study of inhibition patterns is a valuable adjunct to the methods already discussed.

We have seen in Section 2.5 how the pattern of inhibition is determined by the relative positions along the reaction sequence at which the inhibitor and the varied substrate combine: if they combine with the same enzyme form, and only with that form, they are competitive, and so on. This approach is useful in the dissection of multi-substrate mechanisms because one now has the possibility of taking each substrate in turn as the varied substrate and attempting to arrive at an internally consistent interpretation of the different inhibition patterns. One can no longer simply state that substance I is a competitive inhibitor of enzyme E; it may be competitive with respect to substrate A and non-competitive with respect to substrate B, for example, and this must be explicable in terms of a valid mechanism.

In a multi-substrate mechanism the inhibitors may be of two types. We still have, as for the 1-substrate case, the possibility of using *dead-end inhibitors.* These are not normal participants in the catalytic reaction, although often they may be structural analogues of one of the substrates. Malonate as an inhibitor of succinate dehydrogenase or ADP-ribose as an inhibitor of alcohol dehydrogenase would be typical examples. Such inhibitors do not alter the distribution ratios among the pre-existing complexes [59, pp. 70–71]; what they do, however is draw some of the enzyme off the main reaction pathway into a *dead-end*, so that the amount of enzyme to be distributed among the productive complexes is now less.

$$\dashrightarrow E_p \longrightarrow E_q \dashrightarrow \text{main reaction pathway}$$
$$E_p \rightleftharpoons E_pI$$

A second type of inhibition is possible, because one may add one of the products in the absence of the other(s) without initiating the reverse reaction. In the case of a reaction $A + B \rightarrow P + Q$ one might use a starting reaction mixture containing A, B and Q, for example. A *product inhibitor*, such as Q in this case, inhibits not by siphoning enzyme off from the reaction pathway, but by reversing or opposing one of the forward steps along that pathway [46,51]. In this case, therefore, the inhibitor *does* affect the distribution ratio among the complexes in the mechanism. An added factor in some cases is that a product inhibitor may cause dead-end inhibition as well as simple product inhibition by binding at a second point in the mechanism to form an abortive complex (as discussed in Section 2.6(d)).

$$\dashrightarrow E_xP \underset{P}{\overset{P}{\rightleftharpoons}} E_x \dashrightarrow \text{main reaction pathway}$$

Although the basis of product inhibition is somewhat different from that of dead-end inhibition, the end results may nevertheless be categorised in the same way, depending upon whether the inhibitor affects the slopes, the intercepts or both in plots of e/v against $1/[S]$. Thus the terms competitive, non-competitive, uncompetitive and mixed still apply. If we consider Eqn. 17, a slope effect will be seen, with A as the varied substrate, if the inhibitor affects the Φ_A term, the Φ_{AB} term or both; an intercept effect will be seen if it affects the Φ_0 term, the Φ_B term or both. To find out the overall pattern of inhibition, therefore, one needs to know with which complex the inhibitor combines and which term(s) this will affect. The appropriate rate equation in the case of a product inhibitor can be obtained either by deriving the equation with that single product added back and all others held at zero concentration, or by deriving the general equation with all substrates and products present and then setting all products other than the chosen inhibitor to zero. At a more qualitative level, the pattern of inhibition may be predicted by applying the rules given by Cleland [51]. An intercept effect, i.e. a decreased V_{max}, means that saturation with the variable substrate does not displace the inhibitor. Therefore intercepts are altered when the inhibitor combines with any enzyme form other than that with which the variable substrate combines. A dead-end inhibitor affects the slope of the Lineweaver–Burk plot if it combines with the *same* enzyme form as the variable substrate *or* if it combines with other enzyme complexes upstream from that with which the variable substrate combines and connected to it through *reversible* steps. Thus any inhibitor which opposes the saturation of the enzyme with the variable substrate gives a slope effect. In applying the same rules to a product inhibitor it must be borne in mind that the very presence of a product inhibitor re-establishes the reversibility of one of the reaction steps in a way that the presence of a dead-end inhibitor combining at the same point in the mechanism does not.

The main practical problem with product inhibition studies is that the progress trace, from which an initial rate must be estimated, tends to show more curvature. There is, as always, also the problem of choosing suitable ranges of concentrations of substrates and inhibitor so that any slope and intercept effects will be seen.

(g) Isotope exchange at equilibrium

A method that is potentially very powerful in distinguishing between various candidate mechanisms for an enzyme is that of isotope exchange at equilibrium [68–70]. It has been relatively little used, although the principle is elegant, because it is experimentally tedious; each piece of information represents many laborious measurements. The principle may be easily illustrated by considering Schemes 8 and 6. Let us suppose that we believe our enzyme to follow the mechanism shown in Scheme 8, a compulsory-order sequence with A and P as the leading substrates in either direction. If we set up a reaction mixture and allow it to come to equilibrium, net reaction ceases. We could now add a tiny amount of radioactively labelled A with a balancing amount of unlabelled product without perturbing the equilibrium at all. Nonetheless, since equilibrium is merely a balance between opposing reactions which do not cease, a transfer of label would occur from A to the corresponding product P, and by sampling the reaction mixture at intervals, inactivating the enzyme and analysing for radioactive A and P, the rate of this exchange reaction could be measured. The exchange depends upon there being a significant dissociation of EA to E and A, and of EP to E and P. Suppose that we now steadily increase the concentrations of B and Q, maintaining their ratio so that the equilibrium position is unaltered. In Scheme 8 the steady-state concentrations of EA and EP would gradually dwindle as the enzyme was forced more and more into the ternary complex forms. As the concentrations of the outer, binary complexes became vanishingly small the exchange rate would drop effectively to zero. On the other hand exchange between B and Q would not be eliminated by raising the concentrations of A and P. For the opposite compulsory sequence the patterns would of course be reversed. In the random-order mechanism (Scheme 6), on the other hand, increasing [B] and [Q] cannot abolish exchange between A and P, because the A and P can still dissociate from the central complexes to form EB and EQ which can then recombine with the A and P from the free pools. Thus for the random mechanism each reactant pair should give an exchange rate that tends to a constant finite level as the concentrations of the other pair are increased.

(h) Whole time-course studies

By integrating the rate equation it should be possible to obtain a mathematical description of the entire time course of a reaction. Conversely it is evident that in measuring only the initial rate of a reaction one is discarding a large amount of information. Various attempts have been made to use integrated rate equations in order to obtain the kinetic parameters for enzymatic reactions [e.g. 71–75] but only with enzymes which had previously been studied by the initial-rate method. The integrated-rate equation has not found favour as a primary analytical tool, despite the obvious attraction of economies in time and material, and is mentioned here only for completeness. Probably the foremost objection is that a declining rate might reflect enzyme instability rather than approach to equilibrium, whereas a good initial-rate estimate from a linear trace can presumably be used with greater confidence. On the other hand the need to use multiple reaction mixtures is in itself a major source of experimental error.

3. Rapid reaction kinetics

The analysis discussed up to this point relies on the establishment of a state in which the concentrations of the various enzyme complexes remain constant for long enough to allow the measurement of a steady net rate. The transient phase during which the steady-state levels become established is not normally observed in these measurements. However, if it can be observed, information may be extracted about the nature of the intermediates in the mechanism, the rates at which they are formed and the values of some of the rate constants. It is no coincidence that this experimental approach was first adopted not with an enzyme but with the oxygen-binding protein, haemoglobin [76]. In that case a steady-state approach was impossible since haemoglobin does not catalyse a net reaction. Rapid-reaction studies may be carried out either by following the course of reaction after mixing the reacting components or [77] by taking a pre-established equilibrium and then perturbing it and observing the process of 'relaxation' to a new equilibrium. The success of the first approach depends on the design of efficient mixing chambers which will allow the solutions to settle down to non-turbulent flow after mixing within an acceptably small dead time. The second approach requires the means of rapidly delivering a perturbation, say by temperature or pressure jump. The original studies of haemoglobin were not seriously limited by the availability of either the protein or the ligands, and a *continuous flow* apparatus was used, in which the time course of reaction was studied simply by moving the observation point along a tube containing a rapidly flowing solution of the mixed reactants. On the other hand, in studies of enzymes, the protein and often also the substrate tend to be in short supply, and this led to the development of the *stopped-flow* method [78,79] in which, after a short period of flow, the mixed solutions are brought to rest and observed at fixed point as they age. This requires rapid recording facilities to allow 'real-time' acquisition and storage of a changing signal. Stopped flow is now the most widely used approach to rapid reaction studies.

The scope for profitable rapid-reaction studies depends upon the characteristics of each system. Light absorbance and fluorescence are the physical properties most used for monitoring rapid reactions and the measurements depend therefore upon selection of suitable wavelengths for characterising intermediates or products. For example, if a reaction produces NADH, the absorbance at 340 nm and fluorescence at 430 nm provide valuable signals. Now, not only the final product will give these signals, but also any NADH-containing complex along the reaction path will also give measurable signals in these absorbance and emission regions. It is likely though, that the spectral maxima and fluorescence yields will differ somewhat and therefore time courses of reaction monitored at different wavelengths by both absorbance and fluorescence may allow one to sort out the buildup and decay of various NADH-containing species.

Protein fluorescence is also a useful signal in such studies. In some cases, in addition to its aromatic amino acid residues, an enzyme may possess other useful chromophores. Thus rapid reaction studies of haemoproteins and flavoproteins, for example, have relied heavily on the spectral properties of the prosthetic groups of these enzymes.

Even if the enzyme and its natural substrate do not offer very useful spectral changes for rapid-reaction studies, it is often possible to substitute artificial substrates which do not give a large change in absorbance and/or fluorescence upon reaction. Such substrates have played a large part in rapid-reaction studies of proteases and alkaline phosphatase, for example.

Enzyme mechanisms are made up of a series or sometimes a network of individual steps which are all either first- or second-order reactions. When such reactions can be isolated, the analysis and solution for individual rate constants pose no problem [80], but the analysis of the time course of a series of reactions is potentially much more difficult. In enzyme kinetics there are several strategies for overcoming these difficulties.

First of all it is possible to study segments of a mechanism separately. In a ping-pong mechanism one may study the half-reactions separately. In a sequential mechanism one may study the formation of binary complexes in the absence of the other substrate(s). One may preform the binary complex and mix it with the next substrate and so on. Also it may be possible to simplify the mechanism by saturating with individual substrates. Very often, however, the recorded transient is clearly multiphasic. If the rate constants governing the successive phases differ sufficiently, it is possible to work backwards, subtracting out the contribution of the slowest reaction at each stage and deriving rate constants for each phase. Clearly, as for steady-state kinetics, there exists the possibility of varying substrate concentrations, in this case to identify second-order reaction steps and evaluate their rate constants. A further possibility is to use alternative substrates to establish whether the appropriate rate constants are either dependent upon or independent of the nature of the substrates. A special aspect of this approach is the use of isotopically substituted substrates: if a reaction step is thought to involve the breaking of a C–H bond, for example, then a substantial isotope effect on the rate constant is predicted if deuterium replaces hydrogen in that position.

Sometimes the differential equations describing a transient reaction sequence are not solvable by the standard procedures of algebra and calculus. In this case it becomes necessary to seek iterative numerical solutions to fit the observed results. Here, naturally, the power of modern computers is indispensible.

In summary, rapid-reaction kinetics, like steady-state kinetics, is not without its pitfalls and formidable assumptions. One of the best aspects of the technique is the direct qualitative demonstration of the existence of some of the complexes that are postulated. The two approaches are mutually interdependent, however, and even their combined application leads to conclusions which should only be finally trusted when they have been corroborated by a variety of independent methods.

References

1 Brown, A.J. (1902) J. Chem. Soc. 81, 373–388.
2 Henri, V. (1903) Lois Générales de l'Action des Diastases, Herman, Paris.
3 Michaelis, L. and Menten, M.L. (1913) Biochem. Z. 49, 333–369.
4 Briggs, G.E. and Haldane, J.B.S. (1925) Biochem. J. 19, 338–339.
5 Imoto, T., Johnson, L.N., North, A.C.T., Phillips, D.C. and Rupley, J.A. (1972) in: P.D. Boyer (Ed.), The Enzymes, 3rd Edn., Vol. 7, Academic Press, New York, pp. 665–868.
6 Fersht, A.R., Blow, D.M. and Fastrez, J. (1973) Biochemistry 12, 2035–2041.
7 Richards, F.M. and Wyckoff, H.W., in:P.D. Boyer (Ed.), The Enzymes, 3rd Edn., Vol. 4, Academic Press New York, pp. 647–806.
8 Halford, S.E. (1974) in: A.T. Bull, J.R. Lagnado, J.O. Thomas and K.F. Tipton (Eds.), Companion to Biochemistry, Longman, London, pp. 197–226.
9 Lineweaver, H. and Burk, D. (1934) J. Am. Chem. Soc. 56, 658–666.
10 Hofstee, B.H.J. (1959) Nature (London) 184, 1296–1298.
11 Dowd, J.E. and Riggs, D.S. (1965) J. Biol. Chem. 240, 863–869.
12 Eisenthal, R. and Cornish-Bowden, A. (1974) Biochem. J. 139, 715–720.
13 Eadie, G.A. (1942) J. Biol. Chem. 146, 85–93.
14 Hanes, C.S. (1932) Biochem. J. 26, 1406–1421.
15 Endrenyi, L. and Kwong, F.H.F. (1972) In: Analysis and Simulation, North-Holland, Amsterdam, p. 219.
16 Cornish-Bowden, A. and Eisenthal, R. (1974) Biochem. J. 139, 721–730.
17 de Miguel Merino, F. (1974) Biochem. J. 143, 93–95.
18 Markus, M., Hess, B., Ottaway, J.H. and Cornish-Bowden, A. (1976) FEBS Lett. 63, 225–230.
19 Wilkinson, G.N. (1961) Biochem. J. 80, 324–332.
20 Cleland, W.W. (1963) Nature (London) 198, 324–332.
21 Cleland, W.W. (1967) Adv. Enzymol. 29, 1–32.
22 Hoy, T.G. and Goldberg, D.M. (1971) Int. J. Bio-Med. Computing 2, 71–77.
23 Bardsley, W.G. and Childs, R.E. (1975) Biochem. J. 149, 313–328.
24 Hill, C.M., Waight, R.D. and Bardsley, W.G. (1977) Mol. Cell. Biochem. 15, 173–178.
25 Crabbe, M.J.C. and Bardsley, W.G. (1976) Biochem. J. 157, 333–337.
26 Bardsley, W.G., Leff, P. Kavanagh, J. and Waight, R.D. (1980) Biochem. J. 187, 739–765.
27 Andersen, B. (1980) Biochem. J. 189, 653–654.
28 Bardsley, W.G. (1976) Biochem. J. 153, 101–117.
29 Childs, R.E. and Bardsley, W.G. (1976) J. Theor. Biol. 63, 1–18.
30 Changeux, J.-P. (1962) J. Mol. Biol. 4, 220–225.
31 Gerhart, J.C. and Pardee, A.B. (1962) J. Biol. Chem. 237, 891–896.
32 Monod, J., Wyman, J. and Changeux, J.-P. (1965) J. Mol. Biol. 12, 88–118.
33 Monod, J., Changeux, J.-P. and Jacob, F. (1963) J. Mol. Biol. 6, 306–329.
34 Koshland, D.E.Jr., Némethy, G. and Filmer, D. (1966) Biochemistry 5, 365–385.
35 Panagou, D., Orr, M.D., Dunstone, J.R. and Blakley, R.L. (1972) Biochemistry 11, 2378–2388.
36 Anderson, P.M. (1977) Biochemistry 16, 583–586.
37 Goldin, B.R. and Frieden, C. (1971) Curr. Top. Cell. Regul. 4, 77–117.
38 Wyman, J. (1972) Curr. Top. Cell. Regul. 6, 209–226.
39 Perutz, M.F. (1970) Nature (London) 228, 726–739.

40 Velick, S.F. and Vavra, J. (1962) J. Biol. Chem. 237, 2109–2122.
41 Henson, C.P. and Cleland, W.W. (1964) Biochemistry 3, 338–345.
42 Dalziel, K. (1975) in: P.D. Boyer (Ed.), The Enzymes, 3rd Edn. Vol. 11, Academic Press, New York, pp. 1–60.
43 Theorell, H. and Chance, B. (1951) Acta Chem. Scand. 5, 1127–1144.
44 Dalziel, K. and Dickinson, F.M. (1965) Biochem. J. 100, 34–46.
45 Alberty, R.A. (1953) J. Am. Chem. Soc. 75, 1928–1932.
46 Alberty, R.A. (1958) J. Am. Chem. Soc. 80, 1777–1782.
47 Dalziel, K. (1957) Acta Chem. Scand. 11, 1706–1723.
48 Dalziel, K. (1958) Trans. Farad. Soc. 54, 1247–1253.
49 Cleland, W.W. (1963) Biochim. Biophys. Acta 67, 104–137.
50 Cleland, W.W. (1963) Biochim. Biophys. Acta 67, 173–187.
51 Cleland, W.W. (1963) Biochim. Biophys. Acta 67, 188–196.
52 Theorell, H. (1958) Adv. Enzymol. 20, 31–49.
53 Schwert, G.W. and Winer, A.D., in: P.D. Boyer, H. Lardy and K. Myrbäck (Eds.), The Enzymes, 2nd Edn. Vol. 7, Academic Press, New York, pp. 127–148.
54 Fromm, H.J. (1975) Initial Rate Enzyme Kinetics, Springer, Berlin.
55 Segal, I.H. (1975) Enzyme Kinetics, Wiley, New York.
56 Fisher, D.D. and Schulz, A.R. (1970) Int. J. Bio-Med. Computing 1, 221–235.
57 Kinderlerer, J. and Ainsworth, S. (1976) Int. J. Bio-Med. Computing 7, 1–20.
58 Ainsworth, S. (1975) J. Theor. Biol. 50, 129–151.
59 Engel, P.C. (1981) Enzyme Kinetics: The Steady-State Approach, Chapman and Hall, London.
60 King, E.L. and Altman, C. (1956) J. Phys. Chem. 60, 1375–1378.
61 Florini, J.R. and Vestling, C.S. (1957) Biochim. Biophys. Acta 25, 575–578.
62 Dalziel, K. (1969) Biochem. J. 114, 547–556.
63 Engel, P.C. and Dalziel, K. (1970) Biochem. J. 118, 409–419.
64 Ferdinand, W. (1966) Biochem. J. 98, 278–283.
65 Koster, J.F. and Veeger, C. (1968) Biochim. Biophys. Acta 151, 11–19.
66 Haldane, J.B.S. (1930) Enzymes, Longman, London.
67 Dalziel, K. (1962) Biochem. J. 84, 244–254.
68 Boyer, P.D. (1959) Arch. Biochem. Biophys. 82, 387–410.
69 Boyer, P.D. and Silverstein, E. (1963) Acta Chem. Scand. 17, S195–S202.
70 Silverstein, E. and Sulebele, G. (1969) Biochemistry 8, 2543–2550.
71 Schwert, G.W. (1969) J. Biol. Chem. 244, 1278–1284.
72 Schwert, G.W. (1969) J. Biol. Chem. 244, 1285–1290.
73 Bates, D.J. and Frieden, C. (1973) J. Biol. Chem. 248, 7878–7884.
74 Bates, D.J. and Frieden, C. (1973) J. Biol. Chem. 248, 7885–7890.
75 Wharton, C.W. and Szawelski, R.J. (1982) Biochem. J. 103, 351–360.
76 Hartridge, H. and Roughton, F.J.W. (1925) Proc. Roy. Soc. A 107, 654–683.
77 Eigen, M. and de Maeyer, L. (1963) in: S.L. Friess, E.S. Lewis and A. Weissberger, (Eds.), Techniques of Organic Chemistry, Vol. 8, Interscience, New York, pp. 895–1054.
78 Chance, B. (1963) in: S.L. Friess, E.S. Lewis and A. Weissberger (Eds.), Techniques of Organic Chemistry, Vol. 8, Interscience, New York.
79 Gibson, Q.H. and Milnes, L. (1964) Biochem. J. 91, 161–171.
80 Frost, A.A. and Pearson, R.G. (1961) Kinetics and Mechanism, Interscience, New York.

CHAPTER 4

Aspects of kinetic techniques in enzymology

KENNETH T. DOUGLAS and MICHAEL T. WILSON

Department of Chemistry, University of Essex, Wivenhoe Park, Colchester CO4 3SQ, Great Britain

1. Introduction

A simple formulation of a typical enzyme-catalysed reaction is given in Eqn. 1 where E is enzyme, S is substrate, P is product and ES is the physical Michaelis complex (see Ch. 3 for details).

$$E + S \underset{k_b}{\overset{k_f}{\rightleftharpoons}} ES \overset{k}{\rightarrow} E + P \tag{1}$$

This can be divided broadly into two phases, (a) the formation (and dissociation back to starting materials S and E) of the enzyme–substrate complex ES and (b) chemical reaction in this complex to form products. Generally, processes involved with (a) are often, but not always, more rapid than those in (b). Stated another way, section (a) ($E + S \rightleftharpoons ES$) is often at or close to equilibrium and this ligand-binding "equilibrium" is perturbed by chemical leakage to E + P in the processes of (b).

For 2-substrate reactions, we can write schematically Eqn. 2:

$$E + S_1 + S_2 \rightleftharpoons ES_1S_2 \rightarrow E + P \tag{2}$$

Such a reaction can occur by many routes, but one simple pathway might be the ordered process of Eqn. 3, in which substrate S_1 must add before substrate S_2 can complete the reactive enzyme–substrate complex.

$$E + S_1 \rightleftharpoons ES_1 \overset{+S_2}{\rightleftharpoons} ES_1S_2 \rightarrow E + P \tag{3}$$

This type of reaction lends itself to a straightforward study by provision of only one substrate S_1 to the enzyme, i.e. only the process corresponding to Eqn. 4 is possible and this is a simple ligand-binding equilibrium, albeit one whose rate processes are rapid.

$$E + S_1 \rightleftharpoons ES_1 \tag{4}$$

Michael I. Page (Ed.), The Chemistry of Enzyme Action

Considering the above from an experimental point of view, one can study enzyme reactions in two broad rate categories. The traditional steady-state methods, dealt with elsewhere in this volume [1], refer largely to section (b) of Eqn. 1, modified by the effects of section (a). Transient or presteady-state kinetics deal with the rapid processes of (a) which lead to ES (Eqn. 1) or ES_1 (Eqn. 3), etc.

The presteady-state region requires knowledge of the forward (k_f) and backward (k_b) rates in the formation and reversal of enzyme–substrate complexation. Most commonly such studies involve fast-mixing techniques (e.g. stopped-flow). Of course, for very fast k_f and k_b processes, the ratio of k_f and k_b is effectively an equilibrium constant. This simplification allows the use of special relaxation techniques (e.g. temperature jump) and gives access to some very rapid steps which could not otherwise be probed. In addition, the number of component rate processes in k_f or k_b can be assessed using such procedures. There are several examples of enzymes which use a string of enzyme substrate complexes, as depicted in Eqn. 5.

$$E + S \rightleftharpoons ES_1 \rightleftharpoons ES_2 \text{ etc.} \tag{5}$$

In the case of reaction of aspartate aminotransferase with *erythro-β*-hydroxyaspartate, a total of 8 relaxation processes were resolved by Hammes and Haslam [2]. Steady-state kinetics could not have detected such isomerisations of an enzyme–substrate complex as they do not involve reaction with any additional ligand.

The rate coefficients obtained by any of the above approaches (viz. steady-state, rapid-mixing or relaxation procedures) are fortunately not constants in any real sense of the word, in spite of often being loosely called rate constants. Indeed, it is this lack of constancy which makes kinetics of great value as a tool for mechanistic study. Rate coefficients of enzyme reactions are particularly sensitive to the usual changeable parameters of chemistry although some such parameters are considerably less useful for enzymic reactions than they would be in the study of their non-enzymatic counterparts. For example, temperature effects are of limited value as frequently enzymes are thermosensitive, restricting the range of possible measurement. Perhaps more important is the highly complex nature of most experimentally observed rate parameters in enzyme kinetics (e.g. k_{cat}, K_m – see Ch. 3).

A glance at Eqns. 2–5 will indicate that the interpretation of activation parameters, for example, for any enzyme-catalysed reaction demands that knowledge of that enzyme reaction has reached a level of sophistication deep enough that we are truly studying an elementary process (a single rate coefficient). However, the use of cryoenzymology [3], the study of enzymes at extremely low temperatures, has led to the recognition and indeed isolation, of a number of transient intermediates, whose existence at room temperatures is ephemeral.

The sensitivity of rate coefficients to structural changes in substrate or inhibitor, to covalent modifications of the enzyme either at sites proximate to, or distal to, the active site, to pH changes etc. does provide a powerful probe. The pH dependence of individual steps in a mechanism for an enzyme can sometimes now be isolated, the sites of action in a reaction pathway of complex (e.g. mixed type) inhibitors can sometimes be defined using combinations of steady-state and presteady-state approaches. We shall look in turn at these two general approaches.

2. Use of steady-state techniques

Steady-state techniques remain a major mode of enzyme investigation forming the basis of most enzyme assays, inhibition studies, pH profiles, etc. The detailed kinetics of this area are discussed well elsewhere in this volume [1] and our coverage here is selective of the common *uses* of steady-state kinetics.

In spite of their structural complexity and, indeed of the often chemically highly complicated processes which they catalyse, enzymes acting in the steady-state frequently follow a very simple hyperbolic rate equation, the Michaelis–Menten equation (Eqn. 6), where $S_0 \gg E_0$.

$$v_0 = \frac{k_{cat} \cdot E_0 \cdot S_0}{(K_m + S_0)} \tag{6}$$

in which v_0 is the observed initial velocity of the enzyme-catalysed reaction, k_{cat} the turnover number ($k_{cat}E_0 = V_{max}$, where V_{max} is the maximal value of v_0 when $S_0 \gg K_m$). E_0 and S_0 are the initial enzyme and substrate concentrations respectively and K_m is an operationally defined coefficient, the Michaelis constant, which simply provides the substrate concentration required to attain half-maximal velocity ($V_{max}/2$).

For the purposes of our discussion we shall look in turn at the uses of each term in Eqn. 6. The v_0 term should be noted to be a velocity (units are concentration or amount/time) and not a rate coefficient. The terms k_{cat} and K_m (whose components can sometimes be dissected by means of fast reaction techniques, vide infra) are used to describe the sensitivity of the enzyme reaction to a variety of changes, as described above. Their ratio k_{cat}/K_m, with dimensions of a second-order rate coefficient (M^{-1} sec^{-1}) is also useful and has been called the "specificity constant" by Brot and Bender [4]. For a single, covalent intermediate pathway (e.g. Eqn. 7, e.g. an acyl-enzyme pathway, this composite constant is insensitive to problems such as non-productive substrate binding, whilst its components are complicated by such problems.

$$E + S \rightleftharpoons ES \rightarrow ES' + P_1 \rightarrow E + P_2 \tag{7}$$

It is clear that *whatever* the relative values of S_0 and K_m in Eqn. 6, as long as $S_0 \gg E_0$, the observed initial velocity is proportional to the concentration of functioning enzyme (E_0) present in solution. This is the basis of rate assays for enzymes. By working at $S_0 \gg K_m$, i.e. $v_0 = V_{max}$, the observed velocity is not complicated by errors made in values of S_0. To obtain a value of E_0 itself from such a velocity measurement requires at least an accurate knowledge of k_{cat} (for $S_0 \gg K_m$

conditions) or even of k_{cat} and K_m (if $S_0 \gg K_m$ cannot be achieved). Such determinations of enzyme concentration are dangerous to use as they rely on the degree of purity of the enzyme for their calculation. With new purification techniques, values of k_{cat} may be reassessed. Obviously preferable would be some titration procedure capable of registering the absolute molarity of functioning active sites. Such active-site titrations are possible in suitable cases and will be discussed later.

However, not all enzymes under steady-state conditions exhibit hyperbolic v_0 versus S_0 profiles (see Ch. 3). Some enzymes contain a number of subunits (often identical or closely similar) each with a substrate binding site. Of these, a sub-class called allosteric enzymes, exhibits steady-state kinetics which deviate from the Michaelis–Menten formulation (Eqn. 6). The phenomenological interpretation of this behaviour is framed generally in terms of interactions between substrate binding sites leading to the enhancement or depression of substrate-binding constants. The former, called positive cooperativity, is more common and gives rise to sigmoidal v_0 versus S_0 curves. Negative cooperativity is also well established [5]. All the major control enzymes exhibit such site–site interactions as this confers both altered sensitivity of reaction rate towards substrate concentration changes and the possibility of controlling rate by ligands other than substrate. The mechanism underlying such cooperativity has been expressed in terms of two major models, both of which involve interconversions between two conformational states [6,7] available to the enzyme (the *R*, or relaxed, state, which is distinguished from the *T*, or tense state, by its tighter substrate binding). In favourable circumstances fast reaction techniques have allowed studies of such interconversions [7a].

2.1. Note on measurement of initial velocities

The easiest case is when $S_0 \gg K_m$ and $v_0 = V_{max}$. In this case the progress curve is truly zero order in substrate for a considerable time and it is easy to obtain an accurate initial slope from such a linear plot. As S_0 becomes smaller (e.g. $S_0 \simeq K_m$ or $S_0 \ll K_m$) the progress curve becomes markedly non-linear and there is considerable practical difficulty in obtaining a true initial tangent. It is most common to underestimate the true critical tangent to the progress curve unless some special procedure is adopted. One approach is to use a mirror across the progress curve at the first point on the progress curve and, by obtaining a symmetrical reflection in this, construct a normal, and hence a tangent, to the initial point on the curve. Even this procedure produces errors, not only the obvious subjective visual problem of deciding on a truly symmetrical image, but also the problem of how to estimate the amount of reaction which has been missed in the time between mixing and starting of recording. A procedure to circumvent this is to accurately note the time of mixing and fit the observed progress curve to a polynomial such as Eqn. 8, where $[P]_t$ is the amount of product formed at *real* time *t*, or some experimentally measured parameter proportional to it (frequently absorbance).

$$[P]_t = a + bt + ct^2 + dt^3 \ldots qt^n \tag{8}$$

From Eqn. 8, differentiation gives

$$\frac{dP}{dt} = b + 2ct + 3dt^2 \ldots nqt^{(n-1)} \tag{9}$$

Thus, if one obtains a best fit polynomial in real time (i.e. allowing for mixing time which must be kept short) by regression analysis, the second coefficient (b) is equal to dP/dt at $t = 0$ (from Eqn. 9). On-line minicomputers for spectrophotometers make such procedures straightforward.

3. Experimental treatment of transients

As a general rule many of the complexities of analysing transient kinetic measurements may be circumvented by careful choice of experimental design. The objective is to employ conditions, where possible, under which the transient follows an exponential time course (or the sum of a small number of exponentials). In this circumstance the kinetic process can be described by a single parameter, a rate coefficient, which is independent of the amplitude of the transient.

Perturbation methods, in which systems are displaced slightly from a pre-existing equilibrium position (say by a rapid temperature rise of a few degrees) and monitored as they relax to the new equilibrium position, fortunately always generate exponential (or sum of exponential) progress curves (see below).

The more conventional kinetic method in which reagents (e.g. enzymes and substrates) are mixed and the subsequent approach to equilibrium followed by monitoring the concentration of reactants or products may or may not yield exponential curves. If the reaction being considered is first-order, or rate-limited by a first-order step, the reaction proceeds in conformity with the equations $A = A_0 e^{-kt}$ and $B = B_0(1 - e^{-kt})$: A and B are the concentrations of reactant and product, respectively, and A_0 and B_0 their initial concentrations. In either case an exponential decay or build-up is expected.

Often, however, the reaction under study proceeds by a second-order mechanism of the form of Eqns. 10, 11 or 12.

$$\text{(a) } \mathrm{A} + \mathrm{B} \xrightarrow{k_1} \mathrm{C}(+\mathrm{D}\ldots) \tag{10}$$

or

$$\text{(b) } \mathrm{A} + \mathrm{B} \underset{k_{-1}}{\overset{k_1}{\rightleftharpoons}} \mathrm{C} \tag{11}$$

or

$$\text{(c) } \mathrm{A} + \mathrm{B} \underset{k_{-1}}{\overset{k_1}{\rightleftharpoons}} \mathrm{C} + \mathrm{D} \text{ etc.} \tag{12}$$

The formation of product(s) and depletion of reactants on mixing A and B in general follow time courses which are complicated by both the second-order nature of the reaction and the fact that for reversible processes the reaction approaches equilibrium and not completion. There are a number of ways of dealing with these complexities but the simplest is to arrange the initial concentrations of A and B such that the rate equations describing the kinetics collapse to forms which closely conform to exponentials. This is best achieved by working with one reactant in large excess over the other such that the reactant is driven to near completion. In practice an order of magnitude difference in the concentrations of A and B will often suffice providing the equilibrium constant for the reaction is not adverse.

For the mechanisms depicted above (Eqns. 10–12) and where B is the component in excess, the concentration of component A follows Eqn. 13.

$$A = A_0 \exp(-k_{\text{obs}} \cdot t) \tag{13}$$

The parameter k_{obs} is a function of the fundamental rate constants and concentrations but, with excess of B_0, will be given by Eqn. 14.

$$k_{\text{obs}} = k_1[B_0] \tag{14}$$

and measurement of k_{obs} will yield directly the value of k_1. For perturbation methods k_{obs} retains its more complex form (see below).

3.1. Determination of k_{obs}

Plotting the logarithm of the concentration of A (or of a quantity proportional to concentration such as absorbance) versus time transforms Eqn. 13 into a linear form, the slope of which is k_{obs} (or $k_{\text{obs}}/2.303$ if logarithms to base 10 are used).

Fig. 1 shows such a plot and illustrates that the error in the value of log A increases with time, i.e. as the concentration approaches its limiting value at $t = \infty$, and is asymmetrically distributed. As the values of log A at long times are less reliable than those determined at short times, it is not formally correct to use a simple linear regression analysis to fit such a logarithmic plot and to determine k_{obs}. Rather a non-linear regression with the data weighted to account for the increase in error with time and its asymmetric distributions should be employed. Nevertheless, it is often adequate to use a simple linear regression provided that only data collected over, say, the first 90% of the reaction (i.e. ~ 3 half-lives) are used (see Fig. 2).

With the advent of cheap minicomputers much of the analysis may be carried out at the time of the experiment. This requires that the output from the detector used to follow the reaction, say a photomultiplier, is first converted to a digital form and then transferred to the computer. This process of coupling the detection system to the computer is termed "interfacing" and although easily stated may, in practice, present a considerable problem unless the compatibility of the detector output, analogue to digital converter, minicomputer and the "soft-ware" (ensuring rapid

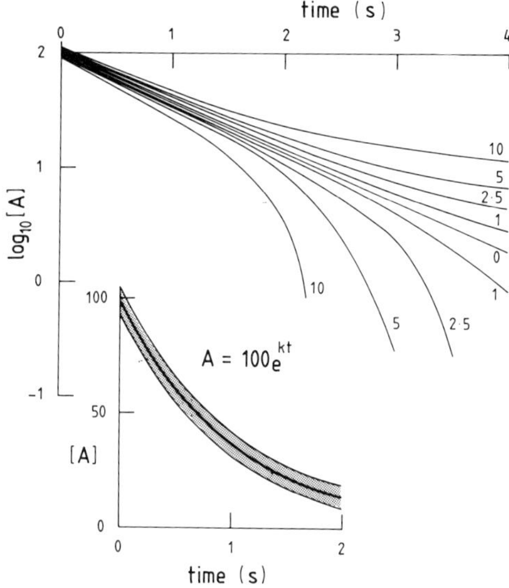

Fig. 1. Errors in simple exponential transients. The lower panel shows a simple exponential decay. The shaded area indicates a consistent error of $\pm 5\%$ of the initial value A_0. Such errors are typical of slightly displaced baselines (A_∞ values). The upper panel displays the logarithmic plot of the exponential. The family of curves indicate how the errors are distributed as a function of time. The numbers give the systematic percentage errors.

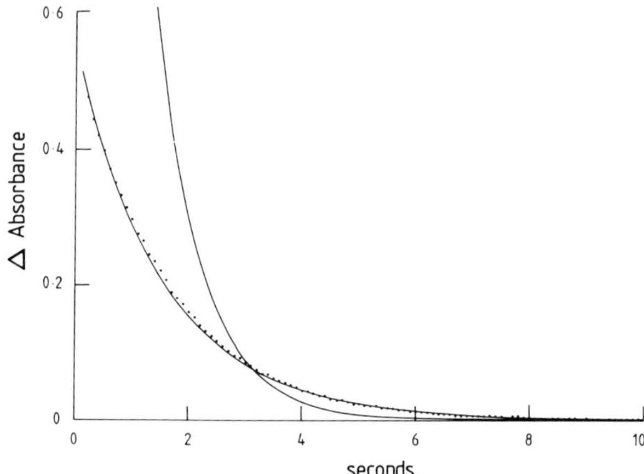

Fig. 2. Exponential fitting using linear regression analysis. The data points were obtained from an experiment in which 25 μM ferrocytochrome *c* was mixed in the stopped-flow apparatus with 3 μM cytochrome oxidase in the presence of oxygen. The temperature was 20°C and the monitoring wavelength 550 nm. The solid line through the points represents the 'best fit' exponential as judged by a linear regression obtained as a logarithmic transformation of the data collected over the first 6.5 sec. The second line is the 'best fit' based on the same rationale using all the data points. The obvious discrepancy is due to errors at long times being strongly weighted in this procedure and points below the 'true' exponential gaining undue weight, see Fig. 1.

data transfer and the speed of the transformation of the analogue data to digital form) have been carefully considered. There are numerous ways these problems may be handled but for those not familiar with these problems we give a block diagram of a system suitable for on-line use with spectrophotometric devices, such as stopped-flow spectrophotometers (Fig. 3).

Fig. 3 indicates that a typical experiment has two parts – data collection and data analysis. Although we have cited a fast-reaction spectrometer, the analogue signal could arise from any device used to follow reaction progress. For most practical purposes it is convenient to collect the data in a transient recorder, which serves both as an analogue-to-digital convertor taking typically 1024 data points over the reaction time and as a memory, storing these data. The advantage of this is that the memory may be examined repeatedly in a number of different ways, without going through a microprocessor analysis each time. Thus, the transient may be displayed directly on a simple (non-storage) oscilloscope for visual assessment. This step is extremely useful as it allows the operator to decide on the quality of the data and any complexities in it such as instrumental artefacts or polyphasic kinetics. Decisions on such matters are best taken before computer analysis. In the case of stopped-flow studies, one does not wish to waste time analysing the kinetics of

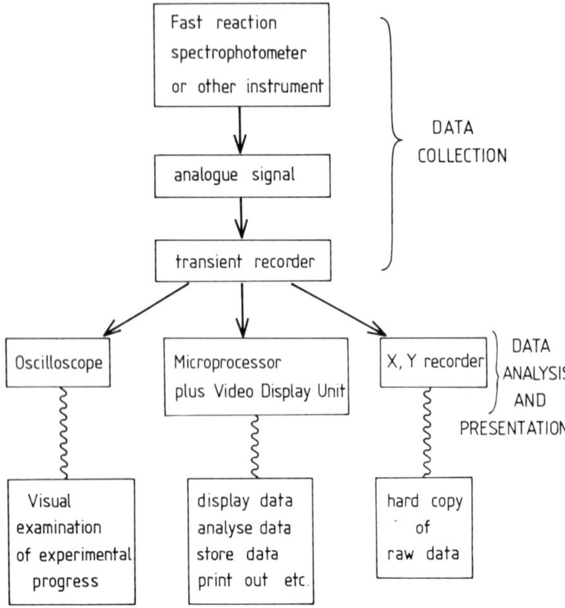

Fig. 3. Block diagram of data acquisition and analysis system. In our laboratories two systems based on the above have been built. One is a Durrum–Gibson stopped-flow spectrophotometer coupled through a Datalab DL901 transient recorder to a Commadore PET 32K minicomputer (used for data in Fig. 2). Another system consists of a Canterbury SF-3A stopped-flow instrument interfaced, via a Datalab DL901 transient recorder to an Exidy Sorcerer minicomputer. Software is available.

bubble formation in the chamber, for example, Once an acceptable data set has been collected, the data can be treated in a number of ways. For a new reaction system for which the background kinetic details have not been fully evaluated, it is best to collect the data as an expanded-scale version on a hard-copy chart recorder. Careful analysis to choose appropriate mathematical models can then be carried out. Once the mathematical behaviour is known, appropriate analytical software can be written or chosen for the microprocessor (see Fig. 2, for example). It is at this stage that the microprocessor comes into its own, facilitating rapid on-line data treatment. Not only can analyses such as regression treatments of data be effected, but also transmittance–absorbance conversions, amplitude–wavelength measurements etc.

4. *Stopped-flow methods* [8]

In general, these procedures involve rapid-mixing devices incorporated into rather conventional instruments for following reactions (e.g. spectrophotometers, spectrofluorimeters). A number of such devices are available commercially and have been described extensively [8]. Most recent improvements have been in data capture/analysis and in the development of rapid spectral scanning instruments capable of recording time dependence of whole spectral changes, as opposed to recording absorbance changes at a single (or two) monitoring wavelength(s). We discuss some of the applications of stopped-flow techniques in the following examples.

4.1. Binding reactions

Reactions of the type depicted in Eqn. 4, i.e. simple ligand binding, have been studied extensively by stopped-flow techniques. The method depends on some measurable change in a property of the system on complexation, e.g. absorbance changes. As an example we may consider the reaction between the metalloprotein cytochrome *c* and imidazole as this illustrates both simple binding and some more complex features [9]. This binding reaction involves the displacement of an intrinsic sulphur ligand by a nitrogen atom of imidazole and leads to a change in the haem spectrum around 410 nm.

If experiments are performed at low protein concentrations ($\sim \mu$M) over a range of imidazole concentrations, the reaction conforms to a simple formulation (Eqn. 15).

$$\text{P (Protein)} + \text{L (Ligand)} \underset{k_{-1}}{\overset{k_1}{\rightleftharpoons}} \text{PL (Complex)} \tag{15}$$

$$K = k_1/k_{-1}$$

The general solution of the differential equations describing this process is complex, and often of limited value in practice, so that it is very difficult to obtain the values of the fundamental rate constants k_1 and k_{-1}. However, one may simplify the situation by arranging [L] to be both $\gg$ [P] and $1/K$ with the result that the second-order rate equation collapses to the simpler pseudo-first-order form and the reaction is driven almost completely to the right. Under these circumstances complex formation follows an exponential time course characterised by a single rate constant, k_{obs}, which may be determined as described above by following, for example, optical absorption at a given wavelength with time. The relationship between k_{obs}, k_1 and k_{-1} is then given by Eqn. 16.

$$k_{obs} = k_{-1} + k_1 [L] \tag{16}$$

A plot of k_{obs} versus [L] thus enables both k_{-1} and k_1 to be determined and thus the binding equilibrium constant. For tightly binding ligands k_{-1} is often too small to measure by this method, i.e. the line appears to pass through the origin and only k_1 the second-order rate coefficient is obtainable. But, as Fig. 4 shows, when the affinity is relatively low both values may be determined. The initial (approximately) linear portion of the plot yields $k_1 = 22/\text{M}/\text{sec}$, $k_{-1} = 1.2/\text{sec}$.

At higher imidazole concentrations, and thus at higher binding rates, the second-order process gives way to first-order behaviour Fig. 4. The binding becomes rate limited at 8/sec by a first-order process which, in this case, may be interpreted as the dissociation (k_{off}) of methionine from the central iron atom of cytochrome *c* prior to imidazole binding.

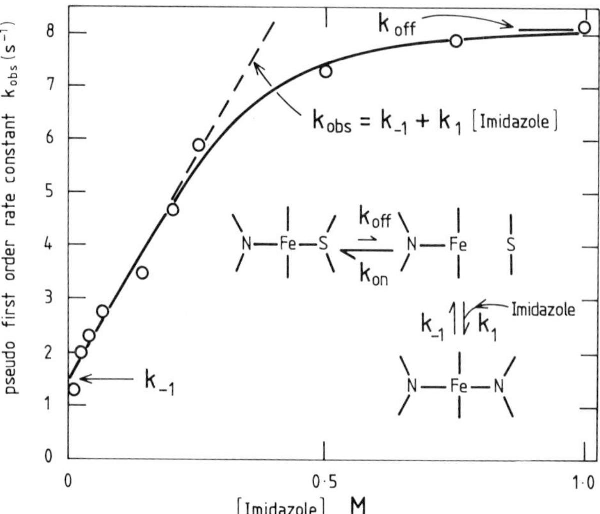

Fig. 4. Imidazole concentration dependence of the rate of binding of imidazole to ferricytochrome *c*. A solution of 7.5 μM horse heart cytochrome *c* was mixed with solutions containing known concentrations of imidazole in a stopped-flow apparatus. The buffer was 0.1 M sodium borate pH 9.0. The reaction was followed at 410 nm, temperature 21°C. The inset depicts the mechanism by which the sulphur of methionine (80) dissociates from the central iron and is subsequently replaced by a nitrogen atom of imidazole.

4.2. Burst kinetics

If the α-chymotrypsin-catalysed hydrolysis of 4-nitrophenyl acetate [10] is monitored at 400 nm (to detect 4-nitrophenolate ion product) using relatively high concentrations of enzyme, the absorbance time trace is characterised by an initial burst (Fig. 5a). Obviously the initial burst cannot be instantaneous and if one uses a rapid-mixing stopped-flow spectrophotometer to study this reaction, the absorbance time trace appears as in Fig. 5b. Such observations have been reported for a number of enzymes (e.g. α-chymotrypsin [11], elastase [12], carboxypeptidase Y [13]) and interpreted in terms of an acyl-enzyme mechanism (Eqn. 7) in which the physical Michaelis complex, ES, reacts to give a covalent complex, ES′ (the acyl-enzyme) and one of the products (monitored here at 400 nm). This acyl-enzyme then breaks down to regenerate free enzyme and produce the other products. The dissociation constant of ES is K_s; k_2 is the rate coefficient of acylation of the enzyme and k_3 is the deacylation rate coefficient. Detailed kinetic analysis of this system [11] has shown

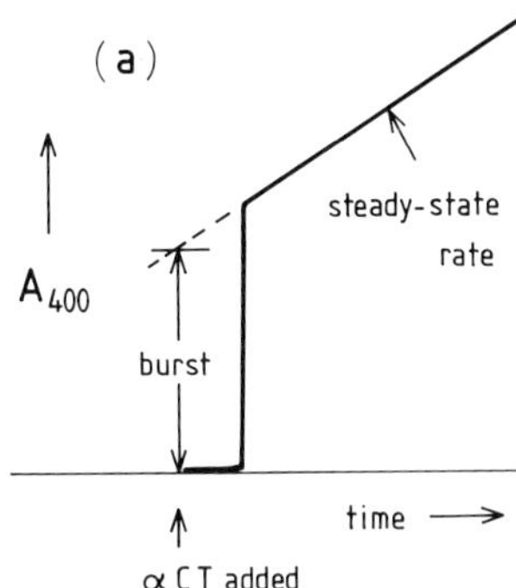

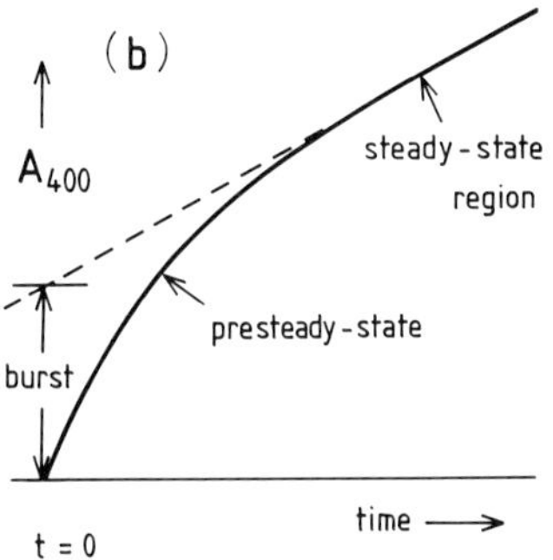

Fig. 5. (a) Schematic diagram of the absorbance–time trace recorded spectrophotometrically for the burst kinetics exhibited in the α-chymotrypsin (αCT)-catalysed cleavage of 4-nitrophenyl acetate followed at 400 nm (A_{400}). (b) Schematic diagram similar to that shown in (a) but on a shorter time scale.

that the steady-state region (Fig. 2a) follows Eqn. 6 with

$$k_{cat} = \frac{k_2 k_3}{k_2 + k_3} \text{ and } (K_m) = K_s \frac{k_3}{k_2 + k_3}$$

The time course of the reaction is described by Eqn. 17.

$$[P_1] = At + B(1 - e^{-bt}) \tag{17}$$

By means of Eqn. 17 and stopped-flow studies at various values of S_0, k_2, and k_3, K_s can be separately determined and studied. For carboxypeptidase Y, which also shows such burst kinetics with 4-nitrophenyl trimethylacetate (I), enzyme preparations with differing amounts of attached carbohydrate, reacted with closely similar steady-state parameters (k_{cat}, K_m, k_3) but differences were apparent for presteady-state parameters (k_2, K_s) [14].

Not only can one obtain rate/mechanistic information from such studies but burst kinetics have provided a method for active-site titration, which determines the concentration of *functioning* active sites in a suitable enzyme preparation.

The size of the burst (π) is given by Eqn. 18.

$$\pi = E_0 \cdot \frac{\left(\frac{k_2}{k_2 + k_3}\right)^2}{\left(1 + \frac{K_m}{S_0}\right)^2} \tag{18}$$

The simplest case is when $k_2 \gg k_3$ and $K_m \ll S_0$ for then $\pi = E_0$, but Eqn. 18 can be applied even when such useful inequalities do not hold. In the studies with carboxypeptidase Y mentioned above [14], the difference in molecular weights of enzyme forms, because of different amounts of carbohydrate attached, was first noticed in burst titration studies and then confirmed using ultracentrifugation, polyacrylamide gels, etc.

Obviously the sensitivity of the active-site titration procedure is not high if carried out spectrophotometrically (the burst being stoichiometric, 1 : 1, with the active-site concentration) even if strongly absorbing products are used. Using spectrofluorimetry for burst studies is one way of improving sensitivity by orders of magnitude. Using I with carboxypeptidase Y allows detection of ~ 3 μM enzyme.

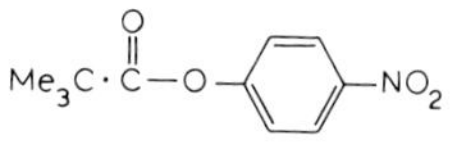

$Me_3C \cdot C(=O){-}O{-}C_6H_4{-}NO_2$

(I)

$Me_3C \cdot C(=O){-}O{-}$(4-methylcoumarin-7-yl)

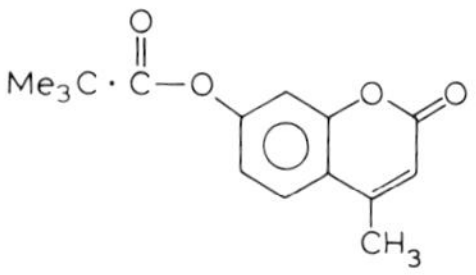

(II)

In preliminary studies using II, which is cleaved by this enzyme to give a fluorescent product, one can easily detect at 0.2 μM enzyme [15]. Active-site titration of isoleucyl-tRNA synthetase, using only 0.1 ml of 1 μM enzyme has been carried out [16] by means of an aliquot-sampling procedure of radioactive pyrophosphate released as P_i in a mechanism formally analogous to that of Eqn. 7. Active-site titration has been reviewed [17].

5. Relaxation methods (temperature jump)

In all relaxation methods a system at equilibrium is perturbed by changing one of the thermodynamic variables which govern the equilibrium. Provided the perturbation is rapid and leads to a significant change in the concentrations of the reactants, the system may be observed to relax to a new equilibrium position with a rate determined by the fundamental rate constants of the steps and by the equilibrium concentrations of the components.

Provided also that the concentration changes are small compared with the equilibrium concentrations the differential equations governing the approach to the new equilibrium can always be linearised such that the kinetic progress curves are exponential or the sums of exponentials. In principle, the number of relaxations which may be observed, equals the number of *independent* equilibria involved in the mechanism of the reaction. Each relaxation is characterised by a relaxation time (τ) with ($1/\tau$) being equivalent to k_{obs} (see above). The theoretical basis of chemical relaxation has been extensively discussed by Eigen and de Maeyer [18] and the temperature jump method by Brunori [19]. In this section we will briefly illustrate some applications of temperature-jump methods to proteins and enzymes.

5.1. Ligand binding

Consider a reaction of the type in Eqn. 19

$$A + B \underset{k_{21}}{\overset{k_{12}}{\rightleftharpoons}} C \tag{19}$$

For example,

$$\text{Deoxymyoglobin} + O_2 \rightleftharpoons \text{Oxymyoglobin}$$

or

$$\text{Enzyme} + \text{Inhibitor} \rightleftharpoons \text{Enzyme–inhibitor complex}$$

(Rate constants are generally assigned 2-number subscripts to indicate direction and step in mechanism.)

The reciprocal relaxation time ($1/\tau$) is given by Eqn. 20

$$\frac{1}{\tau} = k_{12}(\bar{A} + \bar{B}) + k_{21} \tag{20}$$

The barred quantities indicate equilibrium concentrations. We draw attention to this simple case as it is of such widespread importance (e.g. O_2 binding to respiratory proteins; binding of substrate or inhibitor to an enzyme) and to illustrate the equivalence for this mechanism between perturbation and flow methods (see above) when one component is in excess. Indeed, flow methods may be considered as a "concentration-jump" relaxation method.

5.2. Coupled reactions, linked redox reactions and structural rearrangements

The copper protein, azurin and cytochrome c_{551} form part of the soluble respiratory chain of the bacterium *Pseudomonas aeruginosa* and their reaction together [20] conforms at a given pH value to the following scheme (Eqn. 21):

$$\mathrm{Cr} + \mathrm{Ao} \underset{k_{21}}{\overset{k_{12}}{\rightleftharpoons}} \mathrm{Co} + \begin{array}{c}\mathrm{Ar}\\ k_{23}\downarrow\uparrow k_{32}\\ \mathrm{A'r}\end{array} \tag{21}$$

$$K_{12} = \frac{k_{12}}{k_{21}}; \; K_{23} = \frac{k_{23}}{k_{32}}$$

The suffixes o and r refer to the oxidised and reduced protein respectively. Thus, the electron transfer reaction between the redox proteins is coupled to a transition between two forms of reduced azurin, only one of which participates directly in the redox reaction. As the mechanism involves two independent equilibria, two relaxation processes are observed following an increase in temperature. Fig. 6 shows a typical progress curve monitored at a wavelength characteristic of reduced cytochrome c_{551}. This transient is comprised of a rapid increase in absorbance followed by a slower decrease. The rapid relaxation reflects the initial re-equilibration of the second-order redox reaction, an increase in temperature favouring ferrocytochrome C^{2+}. The slower relaxation reflects re-equilibration between the forms of reduced azurin and favour A′r. This step is spectrally silent but is reported through the coupled redox reaction which is displaced towards ferricytochrome c_{551} as Ar is converted to A′r.

The exponential components in Fig. 6 may be separated and their characteristic reciprocal relaxation times determined, as described above. The model predicts the concentration dependences of these relaxations as

$$\frac{1}{\tau_f} = k_{12}(\overline{\mathrm{Cr}} + \overline{\mathrm{Ao}}) + k_{21}(\overline{\mathrm{Ar}} + \overline{\mathrm{Co}}) + k_{23} + k_{32}$$

$$\frac{1}{\tau_s} = \frac{\{[k_{12}(\overline{\mathrm{Cr}} + \overline{\mathrm{Ao}}) + k_{21}(\overline{\mathrm{Ar}} + \overline{\mathrm{Co}})](k_{23} + k_{32})\} - k_{23}k_{21}\overline{\mathrm{Co}}}{k_{12}(\overline{\mathrm{Cr}} + \overline{\mathrm{Ao}}) + k_{21}(\overline{\mathrm{Ar}} + \overline{\mathrm{Co}}) + k_{23} + k_{32}}$$

where τ_f and τ_s are the relaxation times for the fast and slow processes respectively.

Fig. 6 illustrates the results of an experiment in which oxidised azurin was added progressively to reduced cytochrome c_{551}. Under these circumstances the equations

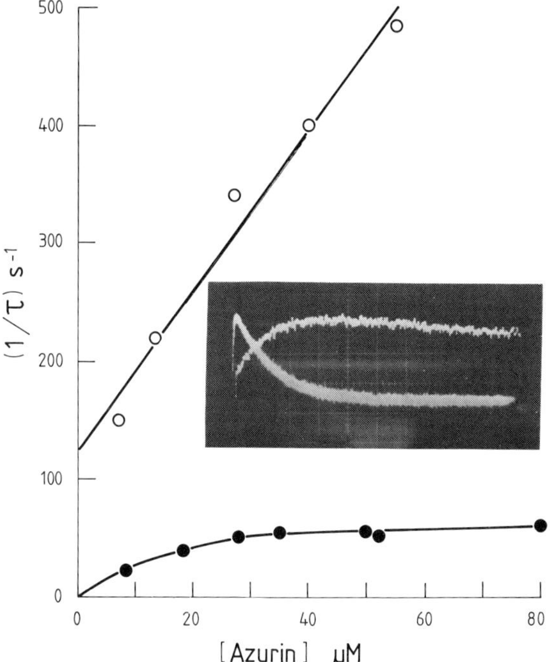

Fig. 6. Reaction between cytochrome c_{551} and azurin. Dependence on total azurin concentration of the reciprocal relaxation time for the fast (○) and slow (●) processes. Condition 0.1 M phosphate buffer pH 7.0 and 25°C (after jump). Solid lines are calculated (see text). The insert shows a temperature-jump relaxation of the reaction between 38 μM cytochrome c_{551} and 76 μM azurin. The monitoring wavelength was 550 nm and ΔE (the total absorbance change) = 0.0017. Absorbance scale 1 msec/cm (ascending) and 20 msec/cm (descending).

for the relaxation times may be simplified and the relaxation data fitted quantitatively with $k_{12} = 6 \times 10^6$/M/sec, $k_{21} = 3.4 \times 10^6$/M/sec, $k_{23} = k_{32} = 40$/sec.

6. Conclusion

Clearly many of the operations of modern enzymology rely on kinetic measurements in some form. Most frequently rather simple analysis (e.g. zero or first-order rate equations) is applicable and we have therefore emphasised the practical aspects of this, with concomitant problems and limitations.

References

1 Chapter 3.
2 Hammes, G.G. and Haslam, J.L. (1969) Biochemistry 8, 1591–1597.

3 Douzou, P. (1971) Biochimie 53, 1135–1145; Fink, A.L. (1976) J. Theor. Biol. 61, 419–445; Makinen, M.W. and Fink, A.L. (1977) Ann. Rev. Biophys. Bioeng. 6, 301–343.
4 Brot, F.E. and Bender, M.L. (1969) J. Am. Chem. Soc. 91, 7187–7191.
5 Engel, P.C. and Dalziel, K. (1969) Biochem. J. 115, 621.
6 Monod, J., Wyman, J. and Changeux, J.-P. (1965) J. Mol. Biol. 12, 88–128.
7 Koshland Jr., D.E., Nemethy, G. and Filmer, D. (1966) Biochemistry 5, 365–385.
7a Kirschner, K., Eigen, M., Bittman, R. and Voight, B. (1966) Proc. Natl. Acad. Sci. (U.S.A.) 54, 1661.
8 Hague, D.N. (1971) Fast Reactions, Wiley-Interscience, London; Chance, B., Eisenhardt, R.H., Gibson, Q.H. and Lonberg-Holm, K.K. (1964) Rapid Mixing and Sampling Techniques in Biochemistry, Academic Press, New York.
9 Al-Ayash, A.I. and Wilson, M.T. (1979) Biochem. J. 177, 641–648.
10 Hartley, B.S. and Kilby, B.A. (1954) Biochem. J. 56, 288–297.
11 Bender, M.L., Kezdy, F.J. and Wedler, F.L. (1966) J. Chem. Educ. 44, 84–88.
12 Bender, M.L. and Marshall, T.H. (1968) J. Am. Chem. Soc. 90, 201–207.
13 Douglas, K.T., Nakagawa, Y. and Kaiser, E.T. (1976) J. Am. Chem. Soc. 98, 8231–8236.
14 Margolis, H.C., Nakagawa, Y., Douglas, K.T. and Kaiser, E.T. (1978) J. Biol. Chem. 253, 7891–7897.
15 Douglas, K.T. and Birchall, N.F. (1979) Unpublished observations.
16 Fersht, A.R. and Kaethner, M.M. (1976) Biochemistry 15, 818–823.
17 Kezdy, F.J. and Kaiser, E.T. (1970) Methods Enzymol. 19, 3–20.
18 Eigen, M. and de Maeyer, K. (1974), Tech. Chem. (N.Y.) 6 (II), 63–89.
19 Brunori, M. (1978) Methods Enzymol. 54, 64–84.
20 Wilson, M.T., Greenwood, C., Brunori, M. and Antonini, E. (1975) Biochem. J. 145, 449–457.

CHAPTER 5

Free-energy correlations and reaction mechanisms

ANDREW WILLIAMS

University Chemical Laboratory, Canterbury, Kent, Great Britain

1. Introduction

A reaction mechanism is essentially a sequence of structures taken by a molecule in its passage from reactant to product. The study of reaction mechanisms involves the delineation of the "average structures" along the path taken during the reaction. In principle it is possible to determine structures at fixed points along the path corresponding to intermediates; these structures exist in energy "wells". It is conceptually impossible to determine structures at other points on a reaction path which do not correspond to discrete molecules. Energy maxima on the reaction path namely transition states may be examined if they are defined as discrete structures; structures determined for transition states must be regarded with caution as they are estimated within this definition. Fortunately the structures which can be estimated for transition states may be used predictively and extensive and detailed considerations indicate that transition states behave as if they are discrete molecules [1,2]. A discrete molecule has a lifetime sufficient to allow energy equilibration by redistribution of vibrations within the structure; structures at points other than energy "wells" on the reaction path do not survive long enough for energy redistribution.

Transition-state structures must be estimated indirectly and the most general method involves comparison of the *effect* of substituents on the free energy with the effect on a model equilibrium (free energy) involving known structures. The difference in substituent effects gives structural information concerning difference in ground and transition states *relative* to the known structural differences between ground and product states in the standard equilibrium. Correlations of free energy are essentially a manifestation of a basic scientific approach namely the comparison of unknown with known.

The comparison of unknown reactions with a parameter derived from a "known" reaction becomes more successful the closer the model is to the reaction in question. People have sought universal parameters (σ_I, σ, σ^*, E_s, n etc. – see later) but these are all related to particular models and unfortunately their origin is often forgotten in the interpretation of correlations.

In this article we aim to indicate how free-energy correlations can be used in mechanistic studies; the audience will be graduate students or workers not actively engaged in studies of free-energy correlations. We include tables of parameters in order to make the article relatively self-contained so that it may be used as a

Michael I. Page (Ed.), The Chemistry of Enzyme Action

handbook. We stress the application of free-energy correlations and problems which may arise; we refer the reader to more advanced texts when detailed considerations are required.

2. *Brønsted relationships*

2.1. *Simple proton transfer*

Let us consider the proton transfer from acetone to a series of bases [3] and compare the logarithm of the rate constant with the pK of the conjugate acid of the base. A linear relationship is observed (Fig. 1) which is an example of a *Brønsted* correlation. If we had plotted log k_f (a measure of the free-energy difference between ground and transition state) versus log k_r/k_f (a measure of the free energy of the reaction) it is clear intuitively that the transition state would be "product-like" for a slope of unity and "reactant-like" for zero slope. It is difficult to measure equilibrium constants such as in Eqn. 1 but ionisation constants are easily estimated (using pH-titration equipment for example) so that the majority of comparisons are with these. Inspection of Eqns. 1 and 2 shows that the only identities are the base and the acid; comparison of oxonium ion with acetone and water with the conjugate base of acetone is doubtful.

$$CH_3COCH_3 + B \underset{k_r}{\overset{k_f}{\rightleftharpoons}} \overset{+}{B}H + \overset{-}{C}H_2{\cdots}C(=O)CH_3 \tag{1}$$

$$H_3\overset{+}{O} + B \rightleftharpoons B\overset{+}{H} + H_2O \tag{2}$$

In the event, the correlation of Fig. 1 is tantamount to a correlation between rate and equilibrium constant for Eqn. 1 because the two equilibria are related by Eqn. 3 and the value K_{SH} is constant throughout the series.

$$K_{SH} = [\overline{C}H_2(CH_3)CO][H^+]/[CH_3COCH_3]$$

$$K_{BH} = [H^+][B]/[B\overset{+}{H}]$$

$$K = k_f/k_r = K_{SH}/K_{BH} \tag{3}$$

The Brønsted relationship for proton *abstraction* is usually expressed in the linear form (Eqn. 4)

$$\log k_f = \beta pK_{HB} + C \tag{4}$$

and for proton *donation* in Eqn. 5

$$\log k_r = \alpha pK_{HB} + C' \tag{5}$$

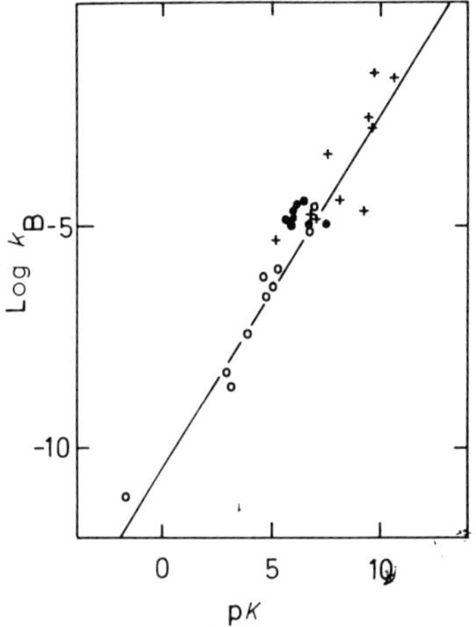

Fig. 1. Brønsted relationship for the base-catalysed halogenation of acetone [3]; +, amines; ●, pyridines; ○, carboxylate ions and water.

The sign of α is defined as positive when it is derived from a *negative* slope; we may then calculate from Eqns. 3, 4 and 5 the following relationship:

$$\log k_f/k_r = \log K_{SH} - \log K_{BH}$$

$$\equiv \beta pK_{HB} + \alpha pK_{HB} + C''$$

Thus

$$1 \equiv \alpha + \beta \qquad (6)$$

2.2. Molecular basis of the Brønsted relationship and interpretation of the exponents

A reasonably good qualitative picture of the origin of the Brønsted relationship can be obtained from a consideration of the energy profile for the proton transfer (Fig. 2). The profile is essentially a superimposition of two Morse curves – one for the proton dissociation ($AH \rightarrow A^- + H^+$) and the other for association ($B^- + H^+ \rightarrow BH$). If the acid (BH) is varied the right-hand Morse curve will alter and the change in energy of activation (ΔE_a) will vary according to the change in overall energy by the relationship (Eqn. 7) where m_1 and m_2 are the slopes of the two curves at the point of intersection.

$$\Delta\Delta E_a = \{m_1/(m_1 + m_2)\}\Delta\Delta E_0$$

$$\equiv \beta\Delta\Delta E_0 \qquad (7)$$

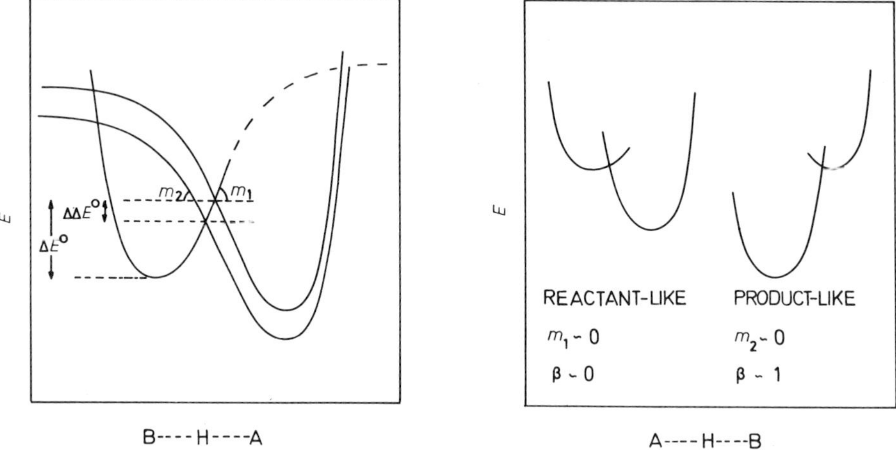

Fig. 2. A derivation of the Brønsted relationship from interesting Morse-type energy profiles [4].

Fig. 3. A qualitative derivation of the relationship between β and the position of the transition state relative to reactant and product states.

Eqn. 7 has the same form as the Brønsted equation. Of course the requirements for the derivation are that the curves are linear at the intersection and that the variation in ΔE_0 does not alter linearity over the whole range of acids. We also assume that the variation in acid gives a vertical change. The curves are of course *potential*-energy curves and not *free* energies.

Qualitatively we may use β as an index of the transition state along the reaction path. If the transition state is reactant-like $m_1 \approx 0$ (Fig. 3) and β is thus close to zero; when the transition state is product-like $m_2 \approx 0$ (Fig. 3) and β is thus unity. The usually linear Brønsted relationships occurring in nature presumably result from a relatively small change in equilibrium constant; as the reaction changes from endothermic to exothermic there should be a smooth change from $\beta = 1$ to $\beta = 0$.

2.3. *Statistical treatments*

The linear correlation between free energies may be represented by Eqn. 8

$$y = mx + c \tag{8}$$

where x is the independent variable. Slope and intercept (m and c) should be estimated by the method of least squares which calculates m and c for a correlation line where $\Sigma(y - y_{calc})^2$ is a minimum. The procedure is analogous to drawing the "best" straight line with a ruler but is less subjective. The goodness of fit may be estimated as the "correlation coefficient" (r); the various equations are given below.

$$m = \{\Sigma xiyi - \Sigma xi \Sigma yi/n\}/\{\Sigma xi^2 - (\Sigma xi)^2/n\} \tag{9}$$

$$c = (\Sigma yi - m\Sigma xi)/n \tag{10}$$

$$r = m\frac{\left(\Sigma xi^2 - (\Sigma xi)^2/n\right)^{1/2}}{\left(\Sigma yi^2 - (\Sigma yi)^2/n\right)^{1/2}} \tag{11}$$

These parameters are easily computed on cheap hand held calculators with integral programmes for statistics but it is useful to understand the rationale and suitable texts should be consulted [5,6].

It is important at this stage to realise that "scatter" or poorness of fit in a correlation can arise from two sources. Error in measurement of rate or equilibrium constant is a source of scatter; model error is a natural effect caused by small differences in the interaction between medium and reaction centre and that of the model reaction. These differences could be enhanced if, for example, a rate constant for a reaction in one medium is compared with an equilibrium constant measured in another. As far as possible the conditions for comparison should be identical.

A further cause of scatter is the neglect of statistical factors for acids and bases with multiple donor and acceptor sites; thus the reactivity of a base with q identical acceptor sites should not simply be halved in a Brønsted correlation because the ionisation constant of the conjugate acid will also be affected. Sophisticated treatments including the use of symmetry numbers are available [7] but the original one

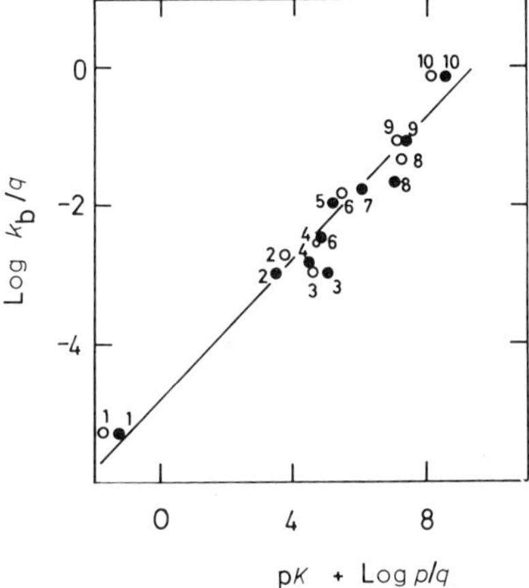

Fig. 4. General base-catalysed hydrolysis of ethyl dichloroacetate [8]. The figure illustrates uncorrected (○) and corrected (●) points: 1, water; 2, HCO_2^-; 3, $PhNH_2$; 4, acetate; 5, pyridine; 6, succinate dianion; 7, 4-picoline; 8, phosphate dianion; 9, imidazole; 10, trishydroxymethylaminomethane.

of Brønsted is widely used (Eqns. 12 and 13).

$$\log k_B/q = \beta \log p/q + \beta \, pK + C \tag{12}$$

$$\log k_A/p = -\alpha \log q/p + \alpha pK + C' \tag{13}$$

The values of p (the number of equivalent protons) and q are reasonably straightforward for an acid–base pair such as triethylamine ($p = 1, q = 1$), diaminoethane dication → monocation ($\overset{+}{N}H_3CH_2CH_2\overset{+}{N}H_3 \rightarrow \overset{+}{N}H_3CH_2CH_2NH_2$; $p = 6, q = 1$), dihydrogen phosphate → monohydrogen phosphate ($p = 2, q = 3$) but there is uncertainty about the oxonium ion (H_3O^+) which is surrounded by three additional waters and could be regarded as $H_9O_4^+$. The species RO^- which has *three* electron pairs which could accept a proton is regarded as having $q = 1$. There has been argument relative to the ammonium ion where several protons are lost from the *same* atom but it is now generally agreed that a correction should be made. Statistical corrections of the type mentioned above are not usually large but their inclusion improves the Brønsted correlation. Corrected and uncorrected points are illustrated in the Brønsted plot of Fig. 4 [8].

2.4. The extended Brønsted relationship

Brønsted's original correlation was for proton-transfer reactions but many workers now use the ionisation of an acid as a general measure of polarity for reactions involving substituent groupings similar to the model acid. For example nucleophilic substitution by aryloxide ions into 2,4-dinitrochlorobenze (Eqn. 14) [9] may be

Cl, NO_2, O^-, X, NO_2, NO_2, NO_2, X, H^+, OH (14)

correlated with the ionisation of phenols and to distinguish this "extended" Brønsted correlation the exponent is given the symbol β_N or β_{nuc} because substituent variation is in the nucleophile [10]. Another type of reaction involves substituent change in the electrophile as in the alkaline hydrolysis of aryl esters (Eqn. 15) [11].

$CH_3CO{-}O{-}C_6H_4X \xrightarrow{k_{OH}} CH_3CO_2^- + {}^-O{-}C_6H_4X$ (15)

The Brønsted exponent is identified as β_L or $\beta_{l.g.}$ to denote variation in the leaving group.

2.5. Meaning of the Brønsted exponents

Together with other workers in the field Jencks [10] has extended the Brønsted correlation to include a meaning for the exponents provided one is prepared to work within a given framework of assumptions. A model for the equilibrium between acyl esters and their hydrolysis products (Eqn. 16) might be the ionisation of the phenol where the charge on phenol and phenolate oxygen is defined as zero and −1 respectively.

$$\underset{\text{Ar–O–COR}}{(+0.7)} + \text{H}_2\text{O} \xrightleftharpoons{\beta_{\text{EQ}} = -1.7} \underset{\text{Ar–O}}{(-1)} + 2\,\text{H}^+ + \text{RCO}_2^- \tag{16}$$

$$\underset{\text{Ar–O–H}}{(0)} \xrightleftharpoons{\beta_{\text{EQ}} = -1} \underset{\text{Ar–O}}{(-1)} + \text{H}^+$$

We would emphasise here that the defined charge on the oxygen in the phenol ionisation is not the same as the *actual* charge. Reference to the book of Pople and Beveridge [12] reveals that absolute charge on atoms in molecules is not integral. The β_{EQ} for phenol ionisation is by definition unity thus β_{EQ} for the acyl group transfer reaction will indicate that the substituents are more or less effective in stabilising the transition from ester to phenolate than from phenol to phenolate. The effect of the substituents on the acyl group transfer equilibrium is "as if the ether oxygen in the ester possesses a charge of +0.7" relative to zero and −1 unit charge on phenol and phenolate oxygen. It must be appreciated that the ionisation is the standard and since the defined charges are not *actual* we must distinguish the estimated values as *effective* charges.

Measurements of β_{EQ} may be made directly from a series of equilibrium constants; the latter parameters are, however, quite rare owing to experimental difficulties but β_{EQ} can be estimated without explicitly measuring equilibrium constants. The equilibrium constant for the esterification reaction (Eqn. 17) is related to that between phenyl acetates, phenol and acetic anhydride (Eqn. 18) by a simple equation (Eqn. 19).

$$\text{CH}_3\text{CO}_2\text{H} + \text{ArOH} \xrightleftharpoons{K} \text{CH}_3\text{COOAr} + \text{H}_2\text{O} \tag{17}$$

$$\text{CH}_3\text{COOAr} + \text{CH}_3\text{CO}_2\text{H} \xrightleftharpoons{K_1} \text{CH}_3\text{CO–O–COCH}_3 + \text{ArOH} \tag{18}$$

$$\text{CH}_3\text{CO–O–COCH}_3 + \text{H}_2\text{O} \xrightleftharpoons{K_2} 2\,\text{CH}_3\text{CO}_2\text{H}$$

$$K = [\text{CH}_3\text{COOAr}][\text{H}_2\text{O}]/[\text{CH}_3\text{CO}_2\text{H}][\text{ArOH}]$$

$$\equiv \frac{[\text{CH}_3\text{COOAr}][\text{CH}_3\text{CO}_2\text{H}]}{[\text{CH}_3\text{COOCOCH}_3][\text{ArOH}]} \cdot \frac{[\text{CH}_3\text{COOCOCH}_3][\text{H}_2\text{O}]}{[\text{CH}_3\text{CO}_2\text{H}]^2}$$

$$\equiv 1/K_2K_3 \tag{19}$$

Since K_2 is substituent independent β_{EQ} for K is identical with that for K_1. Ths approach is clearly very useful as the equilibrium of Eqn. 17 is not easily accessible whereas that of Eqn. 18 is. The approach is general and enables β_{EQ} to be estimated for *any* constant donor acyl function.

An example of the above method is the estimation of β_{EQ} for the formation of ketene from aryl esters of 9-fluorene carboxylic acid; these esters ionise and the conjugate base decomposes to ketene with the liberation of phenolate ion [13]. The overall equilibria are illustrated in Eqn. 20 and we may assume that β_{EQ} for the overall reaction is 1.7; the decomposition of ketene to acid is of course substituent-

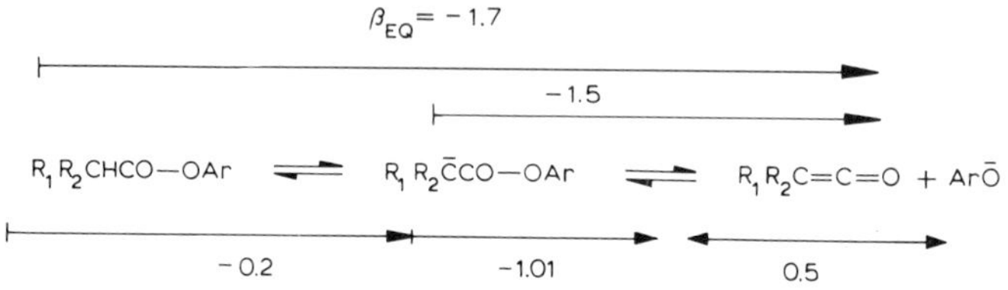

$\beta_{EQ} = -1.7$

-1.5

$$R_1R_2CHCO{-}OAr \rightleftharpoons R_1R_2\bar{C}CO{-}OAr \rightleftharpoons R_1R_2C{=}C{=}O + Ar\bar{O} \qquad (20)$$

-0.2 $\quad -1.01 \quad$ 0.5

independent. The β_{EQ} value for the ionisation of the fluorene esters was measured (-0.2) [13] leading to a β_{EG} of -1.5 for the decomposition of the anion; this indicates a lower "effective" charge on the ether oxygen than in neutral esters in agreement with expectation. It is interesting that we may also estimate the β_N for attack of phenolate anions on ketenes since β_L for the decomposition of the conjugate base is known.

(a) Effective charges in transition states

We can extend the effective charge concept to the transition state provided we know the overall β_{EQ} and either β_L or β_N. If we assume that there is a monotonic change in charge on the atom in question from ground to product state then the effective charge calculated as in Eqn. 21 is a good measure of the charge relative to the two states [14].

$$E_{TS} = E_{GS} + \beta_L \qquad (21)$$

$$E_{GS} = E_{PS} - \beta_{EQ}$$

A useful example of this calculation is that for the attack of aryl oxide ions on cyclic sulphonate esters (Eqn. 22) [15].

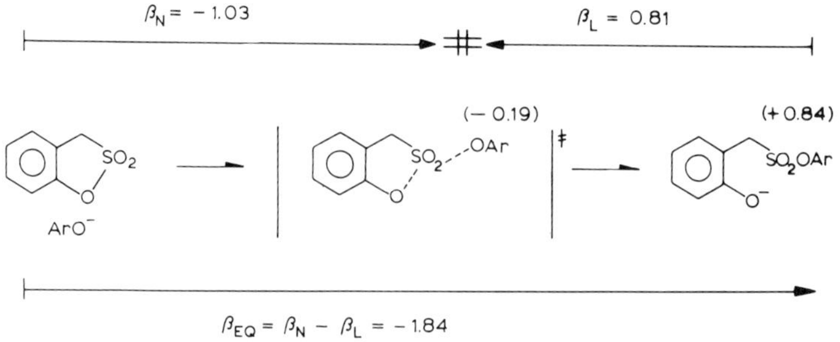

$\beta_N = -1.03$ $\quad \ddagger \quad$ $\beta_L = 0.81$

(22): cyclic sulphonate ester (SO_2) + ArO^- → transition state [SO_2---OAr, (−0.19)]‡ → ring-opened SO_2OAr, O^- (+0.84)

$\beta_{EQ} = \beta_N - \beta_L = -1.84$

The total change in effective charge on the aryl oxygen is $+1.84$ from phenolate to ester; the transition state lies $-1.03/(-1.03-0.81)$ i.e. $\beta_N/(\beta_N-\beta_L)$ along from ground to product state thus the change in effective charge (left to right) is $\beta_N/(\beta_N-\beta_L)$ of the total change (1.84) and is $+0.81$ units (β_L); the "effective" charge on the ether oxygen in the transition state is thus -0.19.

(b) Additivity of "effective" charge

Naively we might expect that effective charges in a molecule might be additive; thus esters possess positive effective charge on the ether oxygen which should be balanced by negative charge elsewhere in the molecule (for example Eqn. 23).

+0.31 ArO–C(=O)(NH) −1.31[16]   +0.4 R–O–C(=O)(O) −1.4[17]   +0.7 R–O–C(=O)CH$_3$ −0.7   (23)

This assumption is not valid because the effective charge on the ether in these examples has been defined in a particular way which does not apply to other atoms in the molecule. In the case of the cyclic sulphonates [15] we are able to use the additivity principle because effective charges on two phenolic oxygens are in question. The effective charge on the exocyclic ether is determined by substituent effects on the attacking or leaving phenolate ion while the endocyclic ether effective charge comes from varying the ring substituents (Eqn. 24).

β_{EQ} = 1.69

X-substituted cyclic sulphonate (SO$_2$, O +0.69) → transition state (S–OPh, O (−0.16)) → X-substituted phenolate (SO$_2$Ph, O$^-$)   (24)

β_L = −0.85 β_N = +0.84

The change in effective charge from ground to transition state is -0.85 leaving an effective charge of -0.16 on the endocyclic oxygen. We can deduce from this that the effective charge lost from the attacking phenolate ion ($+0.81$ units) is almost exactly balanced by that gained on the endocyclic ether (-0.85 units) in the transition state. These results are consistent with a concerted displacement of the sulphonate ester but not with a stepwise process. The effective charges on the ether oxygen in open and closed sulphonate are of similar magnitude as might be expected.

Table 1 illustrates effective charges determined for ether, thio ether and amine adjacent to acyl functions. It is interesting to note that acyl functions, even those bearing double negative charge are more electropositive than the proton. Where ionisation of protons on the acyl function is possible the charge on the adjacent atom varies in accord with expectation thus carbamate ether oxygens are more positive than those in the anionic species [16a].

TABLE 1
Effective charges on atoms adjacent to acyl functions [15]

+0.3 ArO–C(O)(NH)–	+0.4 ArS(R)–C(=O)CH_3	+0.2 ArO–PO$(OR)_2$
+0.8 ArO–C(=O)NH_2	+0.7 ArNH–C(=O)CH_3	X-pyridine, +1.07 N–$PO_3^=$
+0.4 RO–C(O)(O)–	+0.5 RNH–C(=O)H	X-pyridine, +1.6 N–C(=O)CH_3
+0.7 ArO(R)–C(=O)CH_3	+0.3 RNH–C(O)(O)–	X-pyridine, +1.25 N–SO_3^-
+0.8 ArO–SO_2R	X-benzo ring with SO_2–O, +0.7	+0.7 ArO–SO_3^-
+1.4 ArO–C(=S)NHAr	+0.7 ArO–C(S)(NAr)–	

(c) Transition state index (α)

Leffler [16b] originally proposed that the ratio β_f/β_{EQ} measures the extent of the structure change at the transition state as a function of the total change from ground to product state. The assignment of effective charges to particular atoms in the transition state is an alternative method of using the Leffler assumption. Where suitable model reactions are not available to estimate "effective" charges it is profitable to use the Leffler ratio and we propose that this is called the *transition state index* (α).

Let us consider the reaction of aryl isothiocyanates with aryl oxide anions (Eqn. 25) [18];

$$ArNCS + Ar'O^- \rightleftharpoons ArN{-}\bar{C}SOAr' \rightleftharpoons ArNHCSOAr' \qquad (25)$$

variation of the phenolate substituent presents no problems with regard to effective charge estimation as the relevant standard equilibrium is the ionisation of phenols. Variation of the substituent in the isothiocyanate does not possess an obvious model; the ionisation of anilines is a *convenient* standard but will not yield meaningful effective charges. Treatment of both substituent effects by the Leffler approach will yield two transition state indices. Since only one bond is being formed or broken in the reaction the α values should be identical and the observation of similar values gives us an experimental test of the method.

The reaction of aryl oxide anions with 2-aryloxazolinones (Eqn. 26) [19] is another example where the application of the effective charge concept is not appropriate.

$$\text{Ar-oxazolinone} + \text{Ar'O}^- \rightleftharpoons \text{Ar}\cdots\text{COOAr'} \rightleftharpoons \text{ArNHCOCH}_2\text{COOAr'} \tag{26}$$

Variation on the 2-acyl group (Ar) can be modelled against Hammett sigma and the Leffler transition state index evaluated for the cleavage of the endocyclic CO bond. It is possible to measure an effective charge for the ether oxygen in the exocyclic CO bond formation; the overall charge (+0.86) is in agreement with that normally expected for esters. The Leffler transition state indices calculated for endocyclic and exocyclic CO bond cleavage and formation sum to approximately unity consistent with a concerted reaction (Eqn. 27).

0.49 —OAr' ‡

0.37 (27)

2.6. *Curvature in Brønsted correlations*

(a) Eigen curvature

Brønsted and Pedersen [20] indicated that the rate constant for proton transfer from acid to a base cannot continue to increase in accord with a linear Brønsted law but must be limited by an encounter rate. This prediction was confirmed by Eigen's school [21] who showed that β changed from 1 to zero as the pK of the donor acid fell below that of the acceptor base (Fig. 5). Eigen [21] considered the following scheme (sometimes called the Eigen mechanism) for proton transfer from HX to Y where reactions in brackets occur in the encounter complex (Eqn. 28). The overall rate constants are given in Eqns. 29 and 30.

$$\underset{\text{a}}{\text{HX}+\text{Y}} \rightleftharpoons \left[\underset{\text{b}}{\text{XH}\cdots\text{Y}} \rightleftharpoons \underset{\text{b}'}{\text{X}\cdots\text{HY}}\right] \rightleftharpoons \underset{\text{a}'}{\text{X}+\text{HY}} \tag{28}$$

$$\vec{k} = k_{ab}k_{bb'} \cdot k_{b'a'}/N \tag{29}$$

$$\overleftarrow{k} = k_{a'b'} \cdot k_{b'b} \cdot k_{ba}/N \tag{30}$$

$$N = (k_{ba} + k_{bb'})k_{b'a'} + k_{ba} \cdot k_{b'b}$$

$$= (k_{b'a'} + k_{b'b}) + k_{b'a'} \cdot k_{bb'}$$

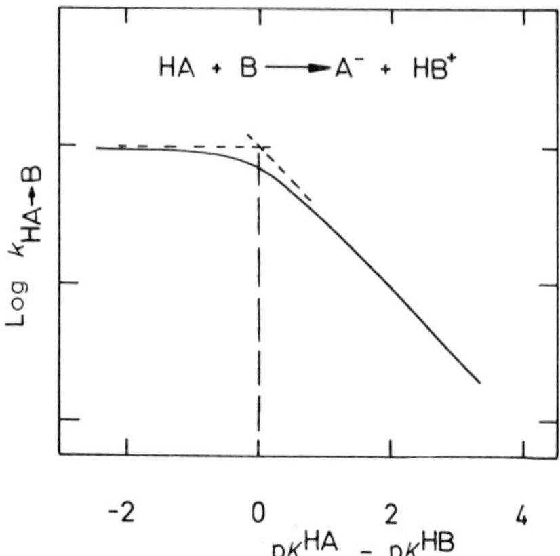

Fig. 5. Brønsted plot for proton transfer showing the limiting rate constant due to diffusion becoming rate determining at low $\Delta\mathrm{p}K$ [21].

Diffusion control occurs if the reactions inside the encounter complex are faster than the rate of encounter complex formation or decomposition (i.e. $k_{ba} \ll k_{bb'}$ or $k_{b'a'} \ll k_{b'b}$). Under the latter conditions Eqns. 29 and 30 collapse to Eqns. 31 and 32;

$$\vec{k} = k_{ab}/(1 + K_b \cdot k_{ba}/k_{b'a'}) \tag{31}$$

$$\overleftarrow{k} = k_{a'b'}/(1 + k_{b'a'}/k_{ba} \cdot K_b) \tag{32}$$

these predict the non-linear Brønsted plot (Fig. 5) with a break point at K_b = unity. The qualitative mechanistic rule (see later) that a correlation with a break with curvature "convex upwards" indicates a change in rate-limiting step is correct for this result; at $\Delta\mathrm{p}K = 0$ the rate-limiting step changes from formation to decay of the complex. It should be seen at this point that the intra-encounter complex proton transfer reaction has to be " very fast" for the classical Brønsted curvature ($\beta = 1 - 0$) to occur. In terms of *observation* these plots are seen for proton transfers between heteroatoms; the correlation is strictly according to Eqns. 31 and 32 where at $\Delta\mathrm{p}K = 0$ the rate constant is $0.5 \times$ its maximal value, The change from $\beta = 1$ to $\beta = 0$ should be very sharp occurring over about 0.5 pK unit either side of $\Delta\mathrm{p}K = 0$. There is thus only a short range of $\Delta\mathrm{p}K$ where the slope β could be between 0 and 1 but for many reactions involving proton transfer at *carbon* the β is mid way between 0 and 1 over a *wide* range of $\Delta\mathrm{p}K$ values. The explanation for this is that the proton transfer in the encounter complex of the Eigen mechanism is slower than the diffusion rate constant. Despite much current thought this problem has not had a satisfactory explanation; there are carbon acids where the proton transfer is fast in the encounter complex and it is believed that these involve the minimal of electronic structure rearrangement on ionisation. Examples of such acids are sulphones and nitriles ($RSO_2-CH_2SO_2R$, $CH_2(CN)_2$); ketones and nitroalkanes are examples of slow proton-transfer acids [22].

The meaning of β at this stage might appear to have little relevance; this is not so – the unit value for proton transfer between heteroatoms at $\Delta pK < 0$ indicates that the transition state of the rate-limiting step is close to products as indeed it might be for an endothermic reaction. At $\Delta pK > 0$, $\beta = 0$ and the transition state would be predicted to be close to reactants in agreement with fact for this *exothermic* reaction.

(b) Marcus curvature

Brønsted correlations for proton transfer to carbon are often strictly linear over a very wide range of ΔpK; many examples show a slight curvature which is not consistent with the "classical" Eigen mechanism as exemplified in the proton transfer from acetylacetone to a series of bases (Fig. 6) [21]. The origin of this curvature must reside in the step in the encounter complex namely the chemical activation barrier; linearity will only be maintained over the range of ΔpK where the two intersecting curves corresponding to the potential energy/reaction coordinate are linear.

Marcus was the first to investigate how the energy profile might vary with ΔG^0 (equivalent to ΔpK); the assumption was that the intersecting curves are two parabolae and the effect of changing ΔG^0 is to shift the vertical relationship [23,24]. Using this model the relationship (Eqn. 33) was obtained.

$$\Delta G^{\pm} = \left(1 + \frac{\Delta G^0}{4\Delta G_0^{\pm}}\right)^2 \Delta G_0^{\pm}$$

$$\beta = \frac{d\Delta G_0^{\pm}}{d\Delta G^0} = \tfrac{1}{2}\left(1 + \frac{\Delta G^0}{4\Delta G_0^{\pm}}\right) \tag{33}$$

$$\frac{d\beta}{d\Delta G^0} = \tfrac{1}{8} \cdot \Delta G_0^{\pm}$$

The term $\Delta G_0^{\pm}$ is the "intrinsic barrier" namely the activation energy for $\Delta G^0 = 0$.

The problem lies in the model representing the forces on the proton in its transfer and two other models give results qualitatively similar to the Marcus theory. One of these involves intersecting Morse curves and the other the proton moving in an electrostatic force field between two negative charges a fixed distance apart [25].

In principle it should be possible to investigate the role of the solvent in proton-transfer reactions by the use of the "Marcus curvature". Let us treat the encounter and separation of products as discrete steps (Eqn. 34) with the associated energies w^r and w^p respectively.

$$XH + Y \overset{w^r}{\rightleftharpoons} XH \cdot Y \overset{\Delta G^0}{\rightleftharpoons} X \cdot HY \overset{w^p}{\rightleftharpoons} X + HY \tag{34}$$

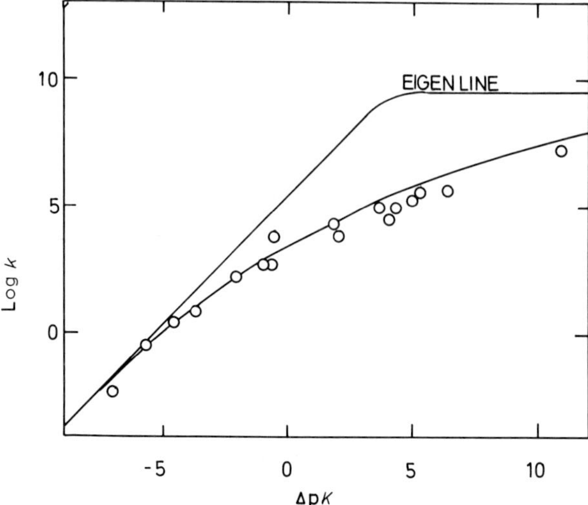

Fig. 6. Deprotonation of acetylacetone by bases [21]. The points fall below the line predicted by the classical "Eigen mechanism" [21] and are correlated by a "Marcus" type equation.

We may rewrite Eqn. 34 in terms of these quantities (Eqn. 35); assuming w^r, w^p and ΔG_0^I are constant in a series they may be determined from the experimental data from the coefficients of the quadratic equation in ΔG^0.

$$\Delta G^{\pm} = w^r + 1 + \left\{ \frac{\left[(\Delta G^0) - w^r + w^p \right]}{4\Delta G_0^{\pm}} \right\}^2 \Delta G_0^{\pm} \tag{35}$$

Table 2 collects some data on $\Delta G_0^{\pm}$ and w^r values [22a,26]. The values of w^r are too large for simple encounter of reactants and it is possible that this term includes the energy to orient an already juxtaposed reactant pair for proton transfer [27].

TABLE 2

"Marcus" parameters for some proton-transfer reactions [22a,26]

Reaction			$\Delta G_0^{\neq}$	W^r
Aromatic protonation			10	10
Ionisation—$\overset{\vert}{C}HCO$			8	6
$CH_3CH(OH)_2 + AH$	⟶	CH_3CHO	5	13
$-CHN_2 \quad + AH$	⟶	$-CH_2OH$	1 -5	8 -14
$B^a + AH^b$	⟶		2	3

[a] B is a "normal" nitrogen or oxygen base.

[b] AH is a phenol or carboxylic acid.

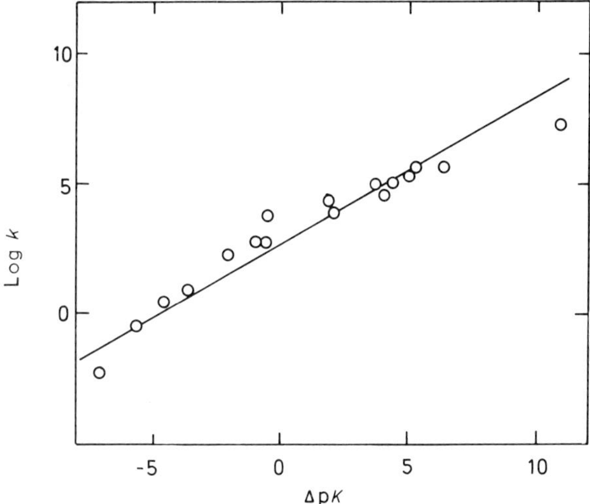

Fig. 7. Data from Fig. 6 correlate reasonably well with a linear function.

Table 2 includes data for proton transfer between heteroatoms; low values of $\Delta G_0^{\pm}$ and w^r for these systems are consistent with classical Eigen curvature. As a general rule it should be noted that intrinsically "fast" reactions (low $\Delta G_0^{\pm}$) will exhibit "sharp" curvature.

Interpretation of the slightly curved Brønsted correlations in terms of Marcus type theory is clouded by the curvature not being accurately defined. This is quite apart from any reservations about the validity of the energy-profile equations. The data can often be force fitted to a linear plot with little difference in correlation coefficient from that to the quadratic (Fig. 7).

Marcus-type considerations do not alter the simple interpretation of β; equations for β derived from various models [28] predict a variation in accord with the Hammond postulate. Thus the Marcus equation gives $\beta = 0.5$ when there is zero driving force for the reaction ($\Delta G^0 = 0$); for an increasingly unfavourable driving force the value of β increases while with a favourable one it decreases (see Eqn. 33).

2.7. Anomalies

The numerical values of the forward and reverse Brønsted coefficients in a simple proton transfer should sum to unity. Neither α nor β should exceed unity or be less than zero. Bordwell and his co-workers [29] discovered that in the nitroalkane acid–base system (Eqn. 36) the introduction of electron-donating substituents into R lowered the rate constant for reaction but *increased* the overall equilibrium constant.

$$R_2CHNO_2 + OH^- \rightleftharpoons R_2C{=}NO_2^- + H_2O \qquad (36)$$

The resultant β (+1.5) gives rise to a value of α of -0.5 since $\alpha+\beta=1$. Originally this anomalous behaviour was used as evidence that Brønsted parameters are not valid measures of transition-state structures. In effect the Brønsted parameter indicates that the transition state is behaving *as if* it had a structure "beyond" that of the product rather than "between" ground and product states.

Current thought is that the anomalous parameters reflect an unusual transition state where there is no monotonic change from ground to product states. Bordwell considers that the anomalies arise from the rehybridisation required as well as the proton transfer in passing from nitroalkane to aci-species (Eqn. 37) [29c].

$$\bar{B} + H-C-NO_2 \rightleftharpoons \bar{B}\text{------}H\cdots C-NO_2 \quad \text{(slow step)} \quad BH + C{=}\overset{+}{N}(O^-)_2 \rightleftharpoons BH\text{-----}\bar{C}-\overset{+}{N}(O^-){=}O \tag{37}$$

Qualitatively the transition state structure for the rate-limiting step will bear little relationship to the product aci-compound to which the equilibrium is measured. If the equilibrium were measured to the pyramidal *carbanion* then normality should exist.

There are few examples of other anomalous Brønsted parameters; Jones and Patel found that detritiation of fluorine-substituted acetyl acetones had anomalous values of β and these were attributed to covalent hydration of the carbonyl group which will affect the measured acidities [30] but not the transition state (Eqn. 38).

$$CR_3-C(OH)_2-CHR-COCR_3 \rightleftharpoons CR_3COCHRCOCR_3 \rightleftharpoons CR_3\overline{CO\bar{C}RCO}CR_3 \tag{38}$$

It is likely that the factors discussed above will distort the β values of proton transfers from ketones and similar molecules but not sufficiently to produce marked effects.

The observation of anomalous β values is not confined to proton-transfer reactions. Brønsted parameters larger than unity are quite common in reactions involving no proton transfer; clearly $\beta_{EQ} \neq 1$ in these cases as it is in proton transfer.

3. Indices of electronic and steric effects

The quest for predictive methods has led physical organic chemists into a search for parameters which are apparently independent of standard models. Throughout this section involving Hammett, Taft and Charton type correlations it should be borne in mind that the electronic and steric parameters such as σ or E_s, while seemingly independent of reactions, are strictly derived from models. The danger in using these correlations is that the derivation of the parameters is easily forgotten by the user and the limitations imposed by the models are hidden.

3.1. The Hammett equation

Linear correlations between the logarithms of rate constants and pK values of benzoic acids were discovered independently by Hammett and Burkhardt in the 1930's [31,32]. This approach has since been used extensively as a means of estimating polarity of substituents with the aim of eliminating structural aspects of the model. The measure of polarity is simply the effect of substituents on the dissociation constant of benzoic acid. For the simple hydrolysis reaction (Eqn. 40) the ionisation is clearly a good model and a plot of log k_{OH} versus σ (Eqn. 41) gives a good straight line (Fig. 8).

$$X\text{-}C_6H_4\text{-}CO_2H \overset{K}{\rightleftharpoons} X\text{-}C_6H_4\text{-}CO_2^- + H^+ \tag{39}$$

$$X\text{-}C_6H_4\text{-}CO_2Et + OH^- \xrightarrow{k_{OH}} X\text{-}C_6H_4\text{-}CO_2^- + EtOH \tag{40}$$

$$\sigma = pK^H - pK^X \equiv \log K^X/K^H \tag{41}$$

Hammett relationships are possible for equilibria and Fig. 9 illustrates the linear dependence of pK for substituted phenylacetic acids (Eqn. 42).

$$X\text{-}C_6H_4\text{-}CH_2CO_2H \rightleftharpoons X\text{-}C_6H_4\text{-}CH_2CO_2^- + H^+ \tag{42}$$

The slope of the Hammett plot (ρ) does *not* bear a simple relationship with the transition-state structure (relative to ground state) because the model and the reaction or equilibrium in question are usually entirely dissimilar. Even the phenylacetic acid dissociation constants are not strictly comparable with those of benzoic acids because the polar effect is transmitted via different routes.

Comparison of the Hammett ρ values for phenylacetic acid (+0.47) and benzoic acid (+1.00) dissociation constants provides us with a measure of the transmission of the methylene group (0.47) compared with direct coupling. The ratio (0.47) is remarkably constant for attenuation of the polar effect down a methylene side chain and essentially the same value is obtained for the reaction of phenylacetic and

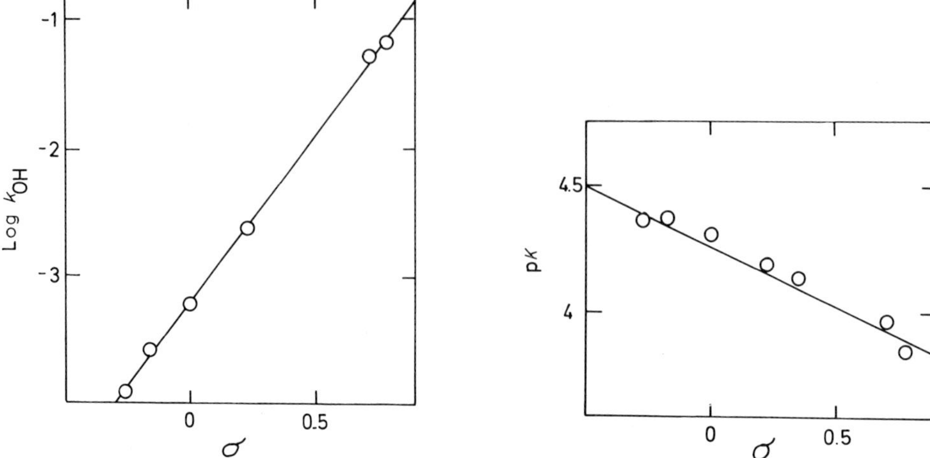

Fig. 8. Hammett relationship for the alkaline hydrolysis of ethyl benzoates.

Fig. 9. Correlation of the dissociation constants of phenylacetic acids with sigma.

benzoic acids with diphenyldiazomethane (0.45). It is interesting that alternation by the vinyl group measured by comparing benzoates and cinnamates is 0.54 and the ethylene has a factor of 0.22 which is close to the square of the attenuation factor for methylene. Table 3 collects some typical attenuation factors for groups attached to benzene.

The only information that we may obtain *a priori* from ρ the value for hydrolysis of ethyl benzoates (Eqn. 43) is that electron-withdrawing substituents (X) increase

$$\log k_{OH}^{X}/k_{OH}^{H} = \rho \cdot \sigma \tag{43}$$

the rate of reaction relative to (H) and thus the transition state has excess negative charge compared with the ground state. Values of ρ, which is dependent on the reaction and not on the substituent, have been measured for a large number of reactions (see for examples Table 4) and these lie within fairly close limits for similar reactions. It is only within this framework of knowledge that one may use ρ values to diagnose mechanism. For example the alkaline hydrolysis of substituted phenyl carbamates [33] has a very high ρ value (+2.87) compared with what is normally expected for the alkaline hydrolysis of aryl esters (~ 1) [11,34,35]. This is interpreted as due to a different mechanism; normally a tetrahedral intermediate is involved in the hydrolysis of esters such as phenyl acetates where addition is the rate-limiting step. Cleavage of the ArO–C bond in the transition state of the eliminative mechanism (Eqn. 44) gives rise to the high ρ value.

$$C_6H_5\text{-NHCO-OAr} \rightleftharpoons C_6H_5\text{-}\overset{\ominus}{N}\text{CO-OAr} \longrightarrow \text{PhNCO} + \text{Ar}\bar{\text{O}} \tag{44}$$

TABLE 3
Alternation factors for aromatic side chains

Interposed group	Factor
$-CH_2-$	0.47
$-(CH_2)_2-$	0.22
$-CH{=}CH-$	0.54
$-C{\equiv}C-$	0.39
Phenylene	0.24
 H H	0.39
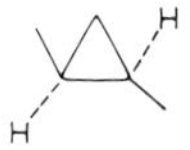 H H	0.28

TABLE 4
Some representative Hammett σ correlations

Reaction	σ	Substituent constant
$ArH + Cl_2 \rightarrow ArCl + HCl$	−8.06	σ^+
$ArPhCHCl + EtOH \rightarrow ArPhCHOEt$	−4.1	σ^+
$ArCHO + HCN \rightarrow ArCH(OH)CN$	2.33	σ
$ArNMe_2 + MeI \rightarrow ArNMe_3^+ I^-$	−3.3	σ^+
$ArCH_2Cl + H_2O \rightarrow ArCH_2OH$	−2.18	σ^+
$ArO^- + EtI \rightarrow ArOEt$	−0.99	σ
$ArMe + Cl_2 \rightarrow ArCH_2Cl$	−1.25	σ
$ArCO_2Me + OH^- \rightarrow ArCO_2^-$	2.23	σ
$ArCO_2H \rightarrow ArCO_2^- + H^+(H_2O)$	1.00	σ
$ArCO_2H \rightarrow ArCO_2^- + H^+(MeOH)$	1.54	σ
$ArCO_2H \rightarrow ArCO_2^- + H^+(EtOH)$	1.96	σ
$ArCH_2CO_2H \rightarrow ArCH_2CO_2^- + H^+(H_2O)$	0.47	σ
$ArCH_2CH_2CO_2H \rightarrow ArCH_2CH_2CO_2^- + H^+(H_2O)$	0.22	σ
$ArNH_2 + HCO_2H \rightarrow ArNHCHO$	−1.43	σ^-
$ArSO_2OCH_3 + OH^-$	1.25	σ
$ArCOCl + H_2O$	1.78	σ
$ArCOCl + PhNH_2$	1.22	σ
$ArCONH_2 + H^+ \rightarrow ArCO_2H + NH_4^+$	−0.48	σ
$ArCONH_2 + OH^- \rightarrow ArCO_2^- + NH_3$	1.36	σ
$ArO^- +$ 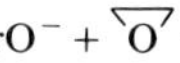 O $\rightarrow ArOCH_2CH_2OH$	−0.95	σ^-
2,4-Dinitrophenylarylether + $MeO^- \rightarrow ArOMe$	1.45	σ
$PhCOCl + ArNH_2$	−2.78	σ
2,4-Dinitrofluorobenzene + $ArNH_2$	−4.24	σ^-
N,*N*-Dimethylanilines + MeI	−3.3	σ^-
ArNCO + MeOH	2.46	σ
$ArCOCH_3 + Br_2 \rightarrow ArCOCH_2Br$	−0.46	σ
$2ArCHO \rightarrow ArCH_2OH + ArCO_2H$	3.63	σ

Fuller tables of Hammett correlations are given by H.H. Jaffe (1953) Chem. Rev. 53, 191.

TABLE 5
Sigma values as defined in Eqn. 41 [a]

Substituent	σ_m	σ_p	Substituent	σ_m	σ_p
H	0	0	COPh	0.34	0.46
Me	−0.07	−0.17	—CN	0.61	0.66
Et	−0.07	−0.15	$COCF_3$	0.65	
Pr^n	−0.05	−0.13	NH_2	−0.04	−0.66
Pr^i	−0.07	−0.15	NHMe	−0.3	−0.84
Bu^n	−0.07	−0.16	NHEt	−0.24	−0.61
Bu^i		−0.12	$NHBu^n$	−0.34	−0.51
CHMeEt		−0.12	OH	0.1	−0.37
Bu^t	−0.1	−0.2	O^-	−0.71	−0.52
CH_2Bu^i		−0.23	OMe	0.08	−0.27
CMe_2Et		−0.19	OEt	0.07	−0.24
CH_2OH	0.08	0.08	OPr^n	0.1	−0.25
CH_2Cl		0.18	OPr^i	0.1	−0.45
CH_2CN		0.01	OBu^n	0.1	−0.32
$CH_2CH_2CO_2H$	−0.03	−0.07	OC_5H_{11}	0.1	−0.34
CH=CHPh	0.14		$O(CH_2)_5CHMe_2$		−0.27
3,4-$[CH_2]_3$ fused		−0.26	OCH_2Ph		−0.42
3,4-$[CH_2]_4$ fused		−0.48	OPh	0.25	−0.32
Ph	0.06	0.01	3,4-O-CH_2-O—		−0.16
CH_2SiMe_3	−0.16	−0.21	OCF_3	0.36	0.32
CHO	0.36	0.22	OAc	0.39	0.31
CF_3	0.47	0.54	SH	0.25	0.15
CO_2H	0.37	0.41	SMe	0.15	0.00
CO_2^-	−0.1	0	SEt		0.03
CO_2Me	0.32	0.39	Spr^i		0.07
CO_2Et	0.37	0.45	SMe_2^+	1.00	0.9
$CONH_2$	0.28	0.36	SOMe	0.52	0.49
$COCH_3$	0.38	0.50	SO_2Me	0.68	0.72
SO_2NH_2	0.46	0.57	AsO_3H^-		−0.02
SO_3^-	0.05	0.09	SeMe	0.1	0
—SCN		0.52	SeCN		0.66
SAc	0.39	0.44	NMe_2	−0.05	−0.83
SCF_3	0.35	0.38	NH_3^+	1.13	1.70
$B(OH)_2$	0.01	0.45	NH_2Me^+	0.96	
$SiMe_3$	−0.04	−0.07	NMe_3^+	0.86	0.82
$GeMe_3$		0	NH_2Et^+	0.96	
$SnMe_3$		0	NHAc	0.21	
PO_3H^-	0.25	0.17	NHCOPh	0.22	0.08
F	0.34	0.06	$NHNH_2$	−0.02	−0.55
Cl	0.37	0.23	NHOH	−0.04	−0.34
Br	0.39	0.27	N=NPh		0.64
I	0.35	0.30	NO_2	0.71	0.78
IO_2	0.7	0.76	NO		0.12

[a] This table is essentially from G.B. Barlin and D.D. Perrin (1960) Quart. Rev. 20, 1.

Ortho substituents are omitted from the table of representative σ values (Table 5) because the steric effect which may be felt by the reaction in question will be largely absent from a σ value (the change in size and in solvation on ionisation of the carboxyl group is not very large). Nevertheless it is quite possible to have a linear correlation for a reaction with *constant ortho* substituents.

An example of the use of the *value* of ρ in diagnosing a mechanism comes from studies of phosphoramido ester hydrolyses (Eqn. 45) [36].

$(PhNH)_2P(O)OAr \overset{K}{\rightleftharpoons} Ph\bar{N}(PhNH)P(O)OAr$

$(PhNH)_2P(O)OAr \xrightarrow{k_2[OH^-]}$ Product

$Ph\bar{N}(PhNH)P(O)OAr \xrightarrow{k_1} PhN{=}P(O)NHPh \xrightarrow{\text{Fast}}$ Product (45)

The observed rate constant has the rate law (Eqn. 46) and k' has a Hammett ρ value of 1.58.

$$k_{obs} = k'/[1 + K_w/K[OH]] \tag{46}$$

Assuming k_1 is negligible $k' = k_2K_w/K$; since we know the ρ value for K (−1.27) and may calculate the ρ for k_2 (1.2) k' has an estimated ρ of $1.2 - 1.27 = -0.07$. The value of ρ for k_2 may be obtained from model studies on the hydrolysis of aryl phosphate esters where the eliminative mechanism does not hold. Disagreement between observed and calculated values of ρ for k' indicate that the simple attack of hydroxide on the neutral ester is not operative and that another, presumably the E1cB, is active.

(a) Additivity of sigma values

If the aromatic species has more than one substituent the Hammett relationship will hold if the σ value employed is the sum of all the sigma values for the substituents; we can write the general Hammett equation.

$$\log k/k_0 = \rho\Sigma\sigma \tag{47}$$

Provided the substituents are not *ortho* to the reactive site this relationship holds quite well.

(b) Resonance and inductive effects

Let us consider the σ value for 4-methoxybenzoic acid; the transmission of charge to the carboxyl group is by inductive and resonance pathways (Eqn. 48).

$$\text{MeO-C}_6\text{H}_4\text{-CO}_2\text{H} \longleftrightarrow \text{Me}\overset{+}{\text{O}}\text{=C}_6\text{H}_4\text{=C(O}^-\text{)OH} \quad (48)$$

We may make the assumption that σ_p is the sum of inductive and resonance effects ($\sigma_I + \sigma_R = \sigma_P$). To a near approximation $\sigma_m = \sigma_I$ because with the *meta* substituent there can be no "through" resonance. Special σ values have been defined (σ^0, σ^n, σ^g) where through resonance is absent; σ^0 values are usually derived from side-chain reactions insulated from the aromatic ring by one or two methylene groups. Values of σ^n are from ordinary σ values where a ρ value is obtained for a reaction using substituents free of resonance interaction. Values of σ may then be calculated from the basic Hammett regression line. Values of σ^g were defined from the rates of alkaline hydrolysis of ethyl arylacetates.

(c) Sigma-minus parameters

Substituents which can withdraw electrons mesomerically from a benzene ring often deviate from an otherwise perfect Hammett plot. The course of this deviation is the extra resonance pathway for transmitting the substituent effect not felt in the ionisation of benzoic acid. For example the ionisation of phenols does not correlate well with σ owing to interactions of the type in Eqn. 49.

$$^-\text{O-C}_6\text{H}_4\text{-}\overset{+}{\text{N}}\text{(=O)O}^- \longleftrightarrow \text{O=C}_6\text{H}_4\text{=}\overset{+}{\text{N}}\text{(O}^-\text{)}_2 \quad (49)$$

Perusal of the Hammett correlation indicates that *meta* substituents form a perfect linear relationship but the *para* ones deviate. A new parameter σ^- may be defined which will restore linearity to the plot; this is the value of sigma corresponding to the ionisation constant for the *para* substituents predicted from the *meta* regression line (Eqn. 50).

$$\sigma_x^- = \log(k_x/k_H)_{\text{standard reaction}}/\rho\sigma \quad (50)$$

Table 6 collects some useful σ^- values.

A simple mechanistic use of the σ^- correlation is the diagnosis that addition of hydroxide to the ester is the rate-limiting step in the alkaline hydrolysis of phenyl acetates. The rate constant k_{OH} correlates much better with σ than with σ^-

(providing we take sufficient substituents where a large difference between σ and σ^- exists). This indicates that the transition state does not "look" like the phenolate anion; there is little through resonance between substituent and reaction centre. Of the three possible transition states (Eqn. 51) the one where addition is rate-limiting is the most consistent. The concerted mechanism is excluded.

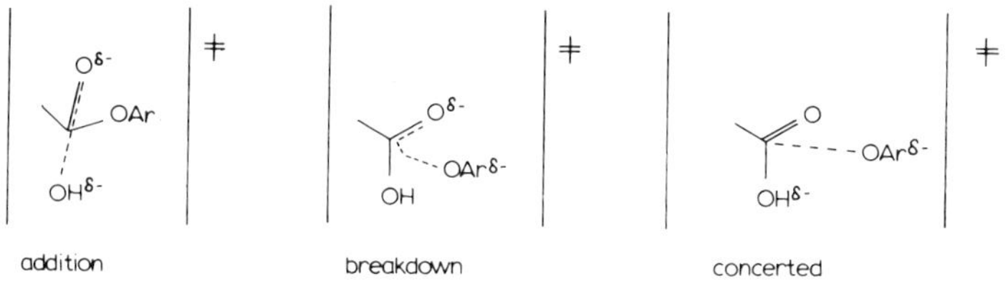

(51)

Examples of good σ^- correlations are for the alkaline hydrolysis of aminosulphonate esters ($MeNHSO_2$–OAr) [37], acyloxysilanes (Et_3SiOAr) [38] and acyl benzene sulphonates ($PhSO_2$–OAr) [39]. These results indicate that the ArO– bond is being cleaved in the transition state of the rate-limiting step. A typical example is illustrated (Fig. 10) for the alkaline hydrolysis of substituted phenyldimethylphosphinates [40] where the σ^- dependence is consistent with ArO–P cleavage in the transition state of the rate-limiting step (Eqn. 52).

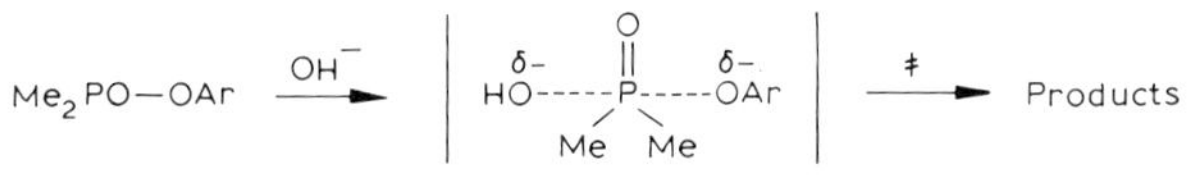

(52)

(d) Sigma-plus parameters

Reactions where the reaction centre becomes positively charged and may attract electrons from a substituent via a resonance pathway will give poor correlations with σ. The solvolysis of 2-phenyl-2-propylchlorides (Eqn. 53) gives a transition state where the almost fully formed carbenium ion can stabilise by resonance with a para

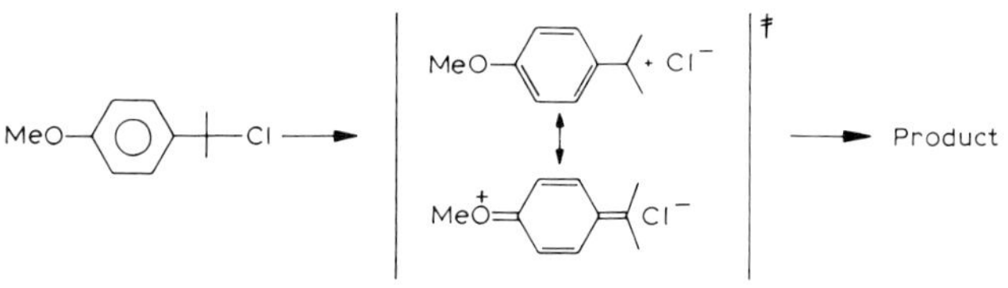

(53)

TABLE 6
Hammett σ^- values for *para* substituents

Substituent	σ^-
OCF_3	0.26
SCF_3	0.57
F	0.02
$COCH_3$	0.85
Br	0.26
CN	0.89
NO_2	1.25
SO_2CF_3	1.36
CH=CH—Ph	0.619
$—CONH_2$	0.627
$—CO_2CH_3$	0.636
$—CO_2C_2H_5$	0.678
$—CO_2H$	0.728
$—CF_3$	0.74
—CHO	1.03
$—SMe_2^+$	1.16
$—SOCH_3$	0.73
$—NO_2CH_3$	1.049
$—N_2^+$	3.2
$—N_3$	1.1

methoxyl group. A set of σ^+ values is defined using *meta* substituents to determine a *non-resonating* regression line (ρ_0) as in Eqn. 54 and a table of representative values is given (Table 7).

$$\sigma_x^+ = \log(k_x/k_H)_{\text{standard reaction}}/\rho\sigma \tag{54}$$

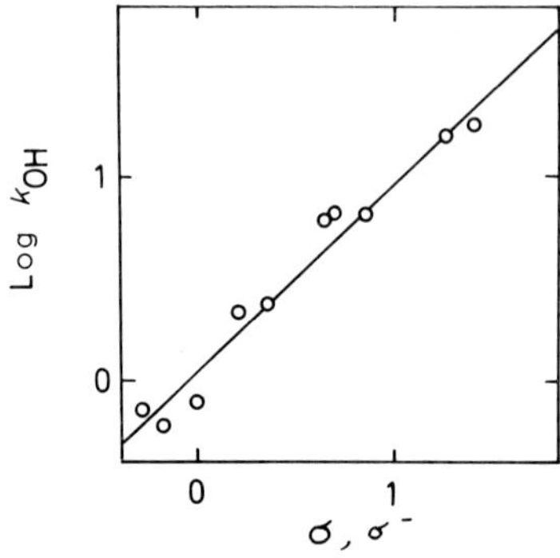

Fig. 10. Hammett sigma-minus correlation for the hydroxide ion catalysed hydrolysis of substituted phenyl dimethylphosphinates [40].

An excellent example of the use of the σ^+ parameter is in the distinction between ether or carbonyl protonation of substituted benzoic acids [41]. The excellent correlation with σ^+ (Fig. 11) indicates that resonance interactions are very important and that carbonyl protonation is therefore the major contributor (Scheme 1).

Scheme 1.

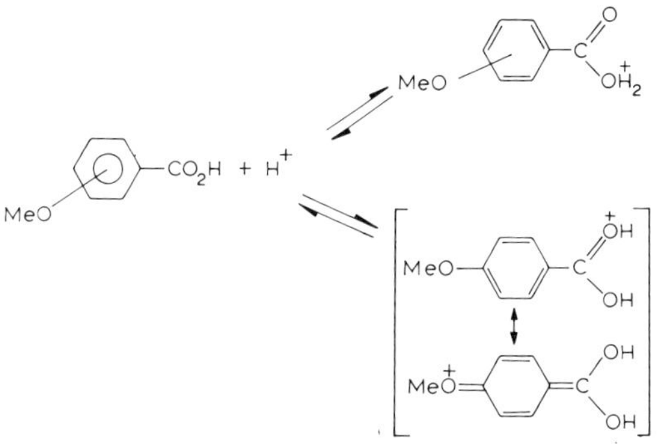

(e) More than one transmission path for σ

The transmission of the effect from substituent to reactive site can be considered to go through essentially one path. In the case of *meta* or *ortho* substituents there are two formal paths respectively through C_2 or C_4, C_5, C_6 and through C_3, C_4, C_5, C_6 or direct (Eqn. 55).

(55)

Because σ is defined by a similar model the action of *meta* or *ortho* substituents is tantamount to a single interaction pathway. When the transmission path in the ionisation of benzoic acids differs from that in the unknown reaction special treatment is required [42]. In the case of substituted aspirins the carboxyl group is involved in catalysis. Thus two pathways exist for substituent transmission not comparable with the simple Hammett definition of σ; an alternative approach is to look at the substituents and realise that a problem exists in assigning σ values (Eqn. 56).

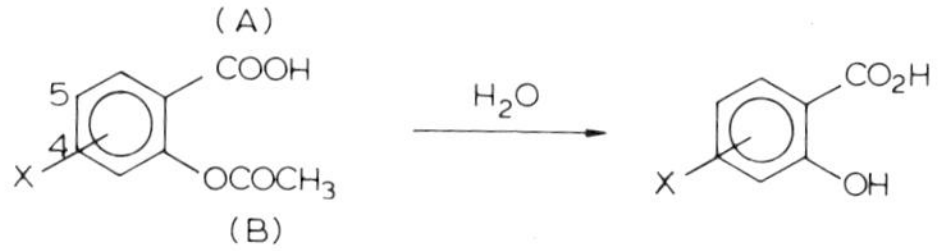

(56)

A substituent in the 4 position is *para* to A but *meta* to B. Two transmission routes exist, one to A and one to B, and the linear free energy correlation will be

TABLE 7
Hammett σ^+ values for *para* substituents

Substituent	σ^+
Dimethylamino	−1.7
Anilino	−1.4
Amino	−1.3
Hydroxy	−0.92
Acetylamino	−0.6
Benzoylamino	−0.6
Methoxy	−0.778
Phenoxy	−0.5
Methylthio	−0.604
Methyl	−0.311
Ethyl	−0.295
Isopropyl	−0.280
t-Butyl	−0.256
Phenyl	−0.179
3,4-C_4H_4 (β-naphthyl)	−0.135
$CH_2COOC_2H_5$	−0.164
Chloromethyl	−0.01
Hydrogen	0
Trimethylsilyl	0.021
Fluoro	−0.073
Chloro	0.114
Bromo	0.150
Iodo	0.135
Carboxy	0.421
Carbomethoxy	0.489
Carboethoxy	0.482
Trifluoromethyl	0.612
Cyano	0.659
Nitro	0.790
Carboxylate	−0.023
$[(CH_3)_3N—]^+$	0.408

given by Eqn. 57

$$\log k_X/k_H = \rho_A \cdot \sigma_A^X + \rho_B\sigma_B^X \tag{57}$$

where σ_A^X is the appropriate parameter for substituent X relative to A and σ_B^X that for transmission to B. Simply plotting $\log k_X/k_H$ versus σ for only one transmission route is not valid and gives considerable scatter (Fig. 12). Rearrangement of Eqn. 57

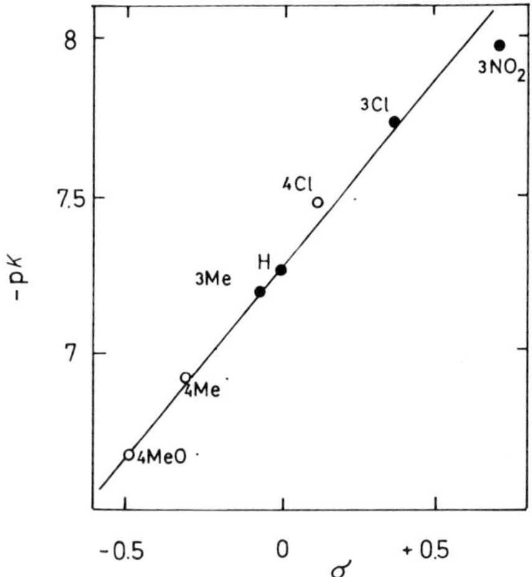

Fig. 11. Plot of pK for aromatic acids ($ArCO_2H_2^+ \rightleftharpoons ArCO_2H + H^+$) versus sigma-plus. Data from K. Yates (1960) J. Am. Chem. Soc. 82, 4059. Open and filled circles represent Hammett sigma values, open circles sigma-plus parameters.

leads to Eqn. 58 which correlates the data (Fig. 13).

$$\frac{\log k_X/k_H}{\sigma_A^X} = \rho_A + \rho_B \cdot \frac{\sigma_B^X}{\sigma_A^X} \tag{58}$$

The data lead to ρ_A and ρ_B values corresponding to interaction of the substituents with carboxyl and ester respectively of -0.52 and -0.96.

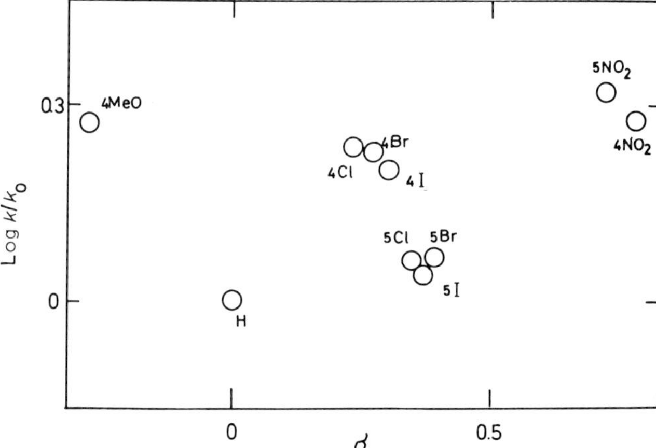

Fig. 12. Simple Hammett plot of the hydrolysis of the acidic form of substituted aspirins [43] versus sigma. The 5-position is taken as *meta* and the 4-position as *para*.

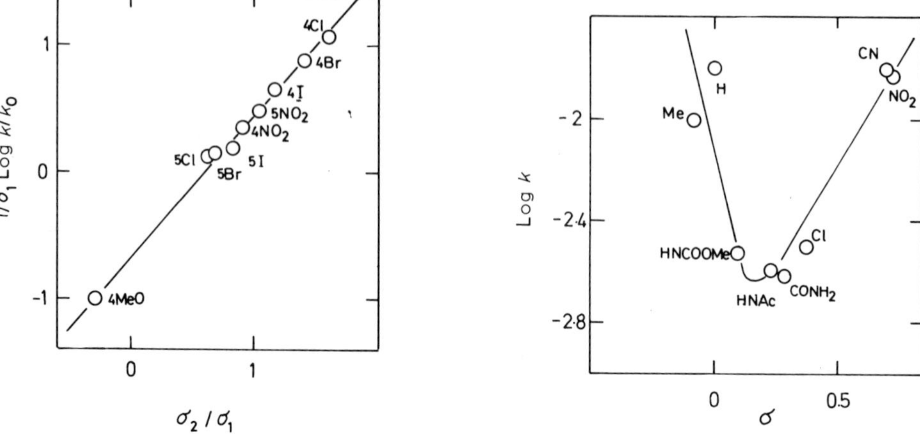

Fig. 13. Jaffé plot for the data in Fig. 12; notice that the point for hydrogen is not expressed.

Fig. 14. Simple Hammett plot for the ring closure of ω-hydroxythioesters (Eqn. 59) [44] falsely indicating a change in mechanism.

The 4-parameter Hammett plot will in any case give a better correlation than a 2-parameter one; we need some means of testing whether the correlation of Eqn. 57 is real. A good correlation between σ_m and σ_p for the chosen substituents will give a spurious correlation according to Eqn. 57 even when transmission of charge in reaction and model are identical. Jaffe [42] suggests that a correlation coefficient between σ_A and σ_B of less than 0.9 should be used as a criterion of the significance of a good correlation according to Eqn. 57. Jaffe [42b] cites a considerable number of examples of the dual transmission pathway.

The neglect of Jaffe's treatment can give misleading results if there is no random scattering in the simple approach as in Fig. 12. For example neglect of the dual transmission in a study of macrolide ring closure through 2-pyridyl thiol esters (Eqn. 59) led to the conclusion that a change in mechanism had occurred (Fig. 14) since the nitrogen *and* the ester group are affected by the substituent R through different transmission routes.

$$\mathrm{HO{-}(CH_2)_{13}{-}\overset{O}{\overset{\|}{C}}{-}S{-}C_5H_3N{-}R} \longrightarrow \mathrm{\overset{O}{\overset{\|}{C}}{-}O{-}(CH_2)_{13}}\ \text{(ring)} \qquad (59)$$

Application of the Jaffe 2-parameter equation yields a good correlation with no evidence for a change in mechanism (Fig. 15) [45]. Other recent examples of the use of the 2-parameter equation come from phosphate [46] and sulphonate chemistry [47].

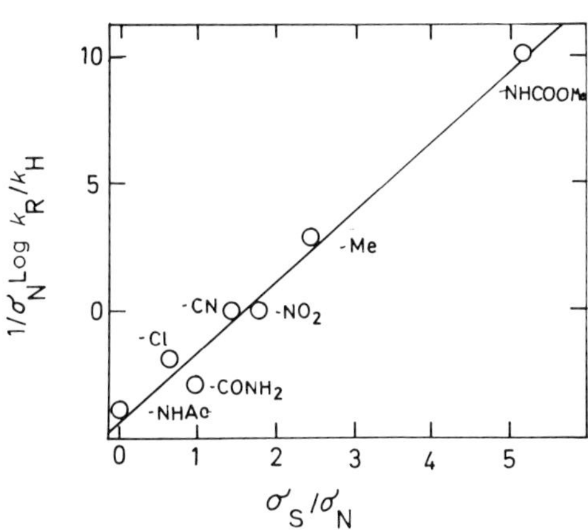

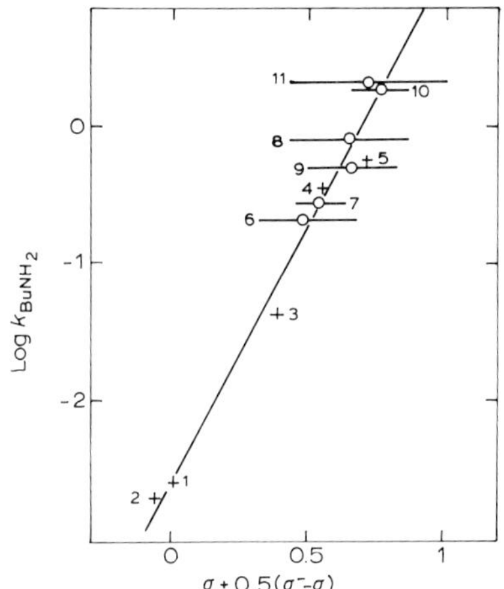

Fig. 15. Jaffé treatment of the data in Fig. 14 [45].

Fig. 16. Plot of butylamine-catalysed hydrolysis of triethylphenoxysilanes [38]. 1, H; 2, *m*-Me; 3, *m*-Ac; 4, *m*-CN; 5, *m*-nitro-; 6, *p*-NNPh; 7, *p*-ethoxycarbonyl-; 8, *p*-COPh; 9, *p*-Ac; 10, *p*-CN; 11, *p*-formyl. +, *meta* substituents; ○, *para* substituents. Extremities of horizontal lines represent: left, no resonance interaction and right, full resonance interaction.

(f) The Yukawa–Tsuno equation

The addition of extra terms will naturally improve a correlation and the σ^- and σ^+ correlations essentially do this for reactions with a resonance component. The simple approach neglects the possibility that the resonance and inductive transmissions may differ in model and reaction in question. Yukawa and Tsuno [48] proposed a correlation where ρ and ρ' (Eqn. 60) are essentially measures of transmission of substituent effects by induction and resonance.

$$\log k/k_{\mathrm{H}} = \rho\sigma + \rho'\sigma^+ \tag{60}$$

The form of equation used by Yukawa and Tsuno expressed the ρ' value as a linear function of the inductive one (Eqn. 61). Humffray and Ryan [38] extended Eqn. 61 to the electron-releasing substituents where (σ^-) replaces (σ^+) in Eqns. 60 and 61.

$$\log k/k_{\mathrm{H}} = \rho\left[\sigma + \mathrm{r}(\sigma^+ - \sigma)\right]$$

$$\equiv (\rho - \mathrm{r})\sigma + \rho\mathrm{r}\sigma^+ \tag{61}$$

When r is zero the correlation is a simple Hammett function and if r is unity the correlation is with σ^+ or σ^-. Fig. 16 illustrates the application of Eqn. 61 to the alkaline hydrolysis of triethylphenoxysilanes where the simple Hammett σ^- values give a poor correlation; σ^- correlation is better but the optimal correlation is with Eqn. 61 when $r = 0.5$ and $\rho = 3.52$. Table 8 collects some typical r values for the Yukawa–Tsuno equation.

3.2. Separation of inductive, steric and resonance effects

Hammett σ values are adequate for reactions and equilibria involving separation of substituent and reaction centre by a phenylene grouping. Examination of aliphatic species necessitates a more sophisticated set of σ values as transmission effects are now markedly different from the model. Resonance and inductive effects may be separated if we can use as models reactions with no resonance interactions so that σ^0 type values may be defined. Entirely new approaches have led to the definition of an inductive scale of σ parameters (σ_I). Although some of these parameters are now only of historic value we shall include a brief discussion of early attempts to define the σ_I scale.

Roberts and Moreland [49] recommend the dissociation constants of a series of bicyclo[222]octane-1-carboxylic acids to define a σ' scale of parameters (Eqn. 62).

$$\text{X–C}_8\text{H}_{12}\text{–CO}_2\text{H} \rightleftharpoons \text{X–C}_8\text{H}_{12}\text{–CO}_2^- + \text{H}^+$$

$$\sigma'_x = \frac{1}{1.46} \log K_x/K_H \qquad (62)$$

The ionisation was in 50% ethanol at 25° and the Hammett plot for the ionisation

TABLE 8
Some applications of the Yukawa–Tsuno equation

Reaction	p	r
Hydrolysis of $XC_6H_4C(CH_3)_2Cl$	−4.54	1.00
Brominolysis of $XC_6H_4B(OH)_2$	−3.84	2.29 [a]
Methanolysis of $(XC_6H_4)_2CHCl$	−4.02	1.23
Bromination of XC_6H_5	−5.28	1.15
Nitration of XC_6H_5	−6.38	0.90
Ethanolysis of $(XC_6H_4)_3CCl$	−2.52	0.88
Basicity of $XC_6H_4N{=}NC_6H_5$	−2.29	0.85
Decomposition of $XC_6H_4COCHN_2$	−0.82	0.56
Beckmann rearrangement of $XC_6H_4C(CH_3){:}NOH$	−1.98	0.43
Semicarbazone formation of XC_6H_4CHO	1.35	0.40
Rearrangement of $XC_6H_4CH(OH)CH{=}CHCH_3$ to $XC_6H_4CH{=}CHCH(OH)CH_3$	−4.06	0.40
Decomposition of $(XC_6H_4)_2CN_2$ with benzoic acid	−1.57	0.19
$H_2O + ArOSiEt_3$	2.84	0.95
$OH^- + ArOSiEt_3$	3.52	0.46
$OH^- + ⌀COOAr$	1.2	0.1

of substituted benzoic acids in the same solvent has $\rho = 1.46$. The divisor 1.46 facilitates comparison of σ' and σ values; the close comparison of σ' and σ_m (Table 9) indicates a lack of resonance in σ_m and confirms the use of this species to define σ^0 values. Transmission of substituent effect from the 4-position to the ionisation centre cannot include resonance or steric effects. A similar approach by Grob [50] has recently produced a much more extensive set of σ' (σ_I^Q) values from the protonation of 4-substituted quinuclidine derivatives. This modern set of data is derived from a model which is very suitable for the derivation of polar effects in that the quinuclidine is highly symmetrical and water soluble (Eqn. 63).

X–(quinuclidine)N + H$^+$ ⇌ X–(quinuclidine)N$^+$—H (63)

$$\sigma_I^Q = \log K^X/K^H$$

The net polar effect is directed along the axis of the molecule in line with the nitrogen lone pair. Table 10 collects some σ_I^Q values. The Roberts and Moreland approach, while giving excellent values absolutely independently of resonance or steric effects requires the synthesis of some very challenging molecules and the same might be said of Grob's method. Taft and Lewis [51] defined a σ_I value which is very accessible; substituted acetic acids are readily available as are their ionisation

$$X\text{–}CH_2CO_2H \overset{K}{\rightleftharpoons} XCH_2CO_2^- + H^+ \tag{64}$$

$$\sigma_I = \frac{1}{3.95} \log K_X/K_H$$

constants in water. In order to place these σ_I values on the same scale as the Roberts and Moreland σ' value the divisor 3.95 is employed. Charton [52] has extended this parameter and the table of σ_I values (Table 11) is taken from his work. Table 9 indicates how σ', σ_I and σ_m are related. The substituted acetic acid model clearly involves no resonance effect; we are justified in neglecting steric effects as the ionisation is not likely to have any large difference between solvation in ground or product states. Movement of the substituent closer to the carboxyl as in the model using substituted *formic* acids does not allow a useful σ_I scale to be generated because of the closeness of the substituent to the reaction centre (see below).

TABLE 9
Comparison of some polar constants

Substituent	σ'	σ_m	σ_I	σ_p
H	0	0	0	0
OH	0.283	0.121	0.25	−0.37
CO_2Et	0.297	0.37	0.34	0.45
Br	0.454	0.391	0.45	0.232
CN	0.579	0.56	0.58	0.660

TABLE 10
Some σ_I^Q values

R	σ_I^Q
H	–
CH_3	0.11
C_2H_5	0.03
i-C_3H_7	−0.08
t-C_4H_9	−0.15
CH_2OH	0.66
CH_2OCH_3	0.66
CH_2OCOCH_3	0.88
$CH_2OSO_2C_6H_4CH_3$-p	1.28
CH_2Cl	0.97
CH_2Br	1.02
CH_2I	1.04
$CH(OH)_2$	1.23
$CH{=}CH_2$	0.56
$C(CH_3){=}CH_2$	0.60
$C{\equiv}CH$	1.64
C_6H_5	0.94
COO^-	0.58
$COOCH_3$	1.70
$COOC_2H_5$	1.70
$COCH_3$	1.69
$CONH_2$	1.78
CN	3.04
NH_2	0.98
$NHCH_3$	0.80
$N(CH_3)_2$	0.97
$NHCOCH_3$	1.58
$NHCOOC_2H_5$	1.56
NO_2	3.48
OH	1.68
OCH_3	1.81
$OCOCH_3$	2.12
SCH_3	1.66
SO_2CH_3	3.23
F	2.57
Cl	2.51
Br	2.65
I	2.34

(a) Taft's polar (σ^*) and steric (E_s) parameters

Taft [53] utilised as a model of steric and polar inductive effects the hydrolysis of

TABLE 11
Some values of Charton's σ_I parameters

Substituent	σ_I	Substituent	σ_I
CH_2Cl	0.15	CH_2Br	0.18
CH_2I	0.16	EtO	0.27
Me_2COHCH_2	−0.04	$MeCHOHCH_2$	−0.01
$PhCH_2SCH_2$	0.06	$PhCH_2CH_2S$	0.23
PhCHOH	0.08		
i-Pr	−0.02	Pr	−0.03
Bu	−0.04	*t*-$BuCH_2$	−0.02
s-Bu	−0.03	HCONH	0.32
i-Pr	−0.04	$CCl_3CH_2CH_2$	0.04
EtO_2CNH	0.26	Cl_2CHCH_2	0.03
CCl_3CH_2	0.12	EtCONH	0.25
$Cl_2C{=}CH$	0.16	$AcNHCH_2$	0.07
CO_2^-	−0.17	H_2NCO	0.27
N_2NCONH	0.21	$C_2H_3CH_2$	0.01
C_2H_3	0.09	1-$C_{10}H_7$	0.12
H_2NCOCH_2	0.05	HC_2	0.35
PhO	0.39	SO_3^-	0.13
2-$C_{10}H_7$	0.12	NO_2	0.76
HC_2CH_2	0.13	PhS	0.30
SH	0.26	EtS	0.25
$MeSO_2$	0.59	PrS	0.24
MeS	0.25	Ph_3CS	0.11
i-PrS	0.25	Bu	−0.03
BuS	0.23	*t*-Bu	−0.07
$PhCH_2S$	0.25	EtCH=CH	0.05
i-Bu	−0.03	$Me_2C{=}CH$	0.03
MeCH=CH-	0.05	$BuCH_2CH_2$	−0.04
$MeCH{=}CHCH_2$	0.00	cy-C_6H_{11}	−0.02
$BuCH_2$	−0.04	cy-C_5H_9O	0.26
$Bu(CH_2)_3$	−0.06	cy-$C_6H_{11}S$	0.31
CF_3	0.42	Me_3Si	−0.13
cy-$C_6H_{11}O$	0.30	$PhMe_2Si$	−0.14
cy-$C_6H_{11}Se$	0.38	ONO_2	0.62
Me_3SiCH_2	−0.05	EtO_2C	0.34
AsO_3H^-	0.01	$PhSO_2$	0.57
Me_3N^+	0.73	SCN	0.58
MeO_2C	0.34	PrO	0.27
PhSO	0.52	s-BuO	0.26
N_2	0.42	CO_2H	0.39
BuO	0.27	CH_2CF_2	0.14
i-PrO	0.26	$SiMe_2OSiMe_3$	−0.13
CH_2CN	0.18	2-Thienyl	0.21
$CH_2C_3F_7$	0.14	Ac	0.29
3-Indolyl	−0.01	$PhNHCOCH_2$	0.00
BzNH	0.27	CH_2OMe	0.07

TABLE 11 (continued)

Substituent	σ_I	Substituent	σ_I
PhNHCO	0.25	$Me_2C=NO$	0.29
CH_2OH	0.05	$PhSO_2NH$	0.32
$PhCH_2CH_2$	−0.01	$Me_3Si(CH_2)_2$	−0.04
CH_2CO_2Me	0.17	$H_2NCO(CH_2)_2$	0.03
CH_2SH	0.10	cy-$C_6H_{11}CH_2O$	0.21
$MeSi(CH_2)_3$	−0.06	cy-$C_6H_{11}(CH_2)_4$	−0.06
$H_2NCO(CH_2)_3$	0.02	$MeNH_2^+$	0.60
cy-$C_6H_{11}CH_2$	−0.05	$EtNH_2^+$	0.60
Me_2NH^+	0.70	$BuNH_2^+$	0.60
NH_3^+	0.60	H_3NO^+	0.47
$PrNH_2^+$	0.60	Pr	−0.04
i-$BuNH_2^+$	0.60	MeCHOH	0.02
PhNAc	0.22	SCN	0.61
i-Pr	−0.03	SO_2Me	0.61
SeCN	0.58	SO_2Pr	0.59
$SCONH_2$	0.33	cy-$C_6H_{11}O$	0.29
SO_2Et	0.60	2-$C_{10}H_7NAc$	0.27
SO_2-*i*-Pr	0.59	2-Thienyl	0.15
1-$C_{10}H_7NAc$	0.26	3-Indolyl	0.00
2-Furyl	0.04	1-$C_{10}H_7CH_2$	0.07
2-Thienylmethyl	0.05	$Ph(CH_2)_2$	0.01
CF_3H	0.32	F	0.52
Et	−0.05	Cl	0.47
Me	−0.05	Br	0.45
H	0	I	0.39
$PhCH_2$	0.04	S-Bu	−0.07
Ph	0.10	OH	0.25
NHAc	0.28	CN	0.58
$CH_2NH_3^+$	0.36	$CH_2CO_2^-$	0.01
$Ph(CH_2)_2$	−0.01	CH_2NH_2	0.08
cy-C_6H_{11}	−0.02	$Ph(CH_2)_4$	−0.02
PhCHMe	0.06	MeO	0.25
t-Bu	0.01	$H(CH_2)_7$	0.03
CO_2Me	0.30	CH_2OH	0.09
		CH_2CO_2Et	0.13

substituted ethyl formate esters (XCOOEt). Two basic assumptions are made that steric and electronic effects are mutually exclusive; thus the fundamental equation defining the polar (P) and steric (S) effects is Eqn. 65.

$$\log k_R/k_{Me} = P + S$$

$$= \rho^*\sigma^* + \delta E_s \qquad (65)$$

The standard substituent is by convention taken as *methyl* whereas for σ_I it is hydrogen. Ingold [54] indicated that steric effects in acid and base hydrolysis of a given ester should be identical; this is supported by modern ideas on ester hydrolysis where both mechanisms involve similar transition states in their rate-limiting steps (Eqns. 66 and 67).

$$RCO_2Et \underset{}{\overset{H^+}{\rightleftharpoons}} R{-}C(\overset{+}{O}H)(OEt) \xrightarrow{\text{slow}} R{-}C(OH)_2(OEt)\ (\mathrm{A}) \xrightarrow{\text{fast}} \text{Product} \qquad (66)$$

$$RCO_2Et \xrightarrow[\text{slow}]{OH^-} R{-}C(O^-)(OH)(OEt)\ (\mathrm{B}) \xrightarrow{\text{fast}} \text{Product} \qquad (67)$$

The transition states, which will resemble A and B in Eqns. 66 and 67 are solvated and thus quite close in steric requirements. Eqns. 68 and 69 may be defined and lead to Eqn. 70 because δ_H and δ_{OH}, the steric requirements, are assumed identical.

$$\log k_R^{OH}/k_{Me}^{OH} = \rho_{OH}^*\sigma^* + \delta_{OH}E_s \qquad (68)$$

$$\log k_R^H/k_{Me}^H = \rho_H^*\sigma^* + \delta_H E_s \qquad (69)$$

$$\log k_R^{OH}/k_{Me}^{OH} - \log k_R^H/k_{Me}^H = (\rho_{OH}^* - \rho_H^*)\sigma^* \qquad (70)$$

In order to define σ^* on essentially the same scale as σ we define $(\rho_{OH}^* - \rho_H^*)$ as 2.48 the difference between ρ_{OH} and ρ_H for the hydrolysis of ethyl benzoates.

(b) Taft's steric parameter (E_s)

Many reactions may be correlated by a simple "Taft" relationship (Eqn. 71) and the success is due to the minimal effect of steric interactions in these reactions. When steric effects compete with electronic the extended "Pavelich–Taft" equation (Eqn. 65) is successful [53b,55]. Steric parameters may be defined by Eqn. 72 for the hydrolysis of ethyl esters.

$$\log k^R/k^{Me} = \rho^*\sigma^* \qquad (71)$$

$$\log k_R^H/k_{Me}^H = \delta E_s \qquad (72)$$

The electronic component of the equation ρ_H^* is expected to be negligible since ρ_H for the acid hydrolysis of ethyl benzoates is very close to zero. If we define $\delta = 1$ for acid-catalysed hydrolysis of ethyl formates we now have a standard reaction to measure E_s, the steric effect. Tables 12 and 13 collect data on E_s and σ^*.

TABLE 12
Some Taft σ^* values

Substituent	σ^*	Substituent	σ^*
Cl_3C	2.65	H	0.490
F_2CH	2.05	$C_6H_5CH{=}CH$	0.410
CH_3OOC	2.00	$(C_6H_5)_2CH$	0.405
Cl_2CH	1.940	$ClCH_2CH_2$	0.385
$(CH_3)_3\overset{+}{N}CH_2$	1.90	$CH_3CH{=}CH$	0.360
trans-$O_2NCH{=}CH$	1.70	$CF_3CH_2CH_2$	0.32
CH_3CO	1.65	$C_6H_5CH_2$	0.215
$C_6H_5C{\equiv}C$	1.35	$CH_3CH{=}CHCH_2$	0.13
$CH_3SO_2CH_2$	1.32	$CF_3CH_2CH_2CH_2$	0.12
$N{\equiv}CCH_2$	1.30	$C_6H_5(CH_3)CH$	0.11
trans-$Cl_3CCH{=}CH$	1.188	$C_6H_5CH_2CH_2$	0.080
FCH_2	1.10	$C_6H_5(C_2H_5)CH$	0.04
$HOOCCH_2$	1.05	$C_6H_5CH_2CH_2CH_2$	0.02
$ClCH_2$	1.050	CH_3	(0.000)
trans-$HOOCCH{=}CH$	1.012	*cyclo*-$C_6H_{11}CH_2$	−0.06
$BrCH_2$	1.000	C_2H_5	−0.100
CF_3CH_2	0.92	*n*-C_3H_7	−0.115
trans-$ClCH{=}CH$	0.900	*i*-C_4H_9	−0.125
trans-$Cl_2CHCH{=}CH$	0.882	*n*-C_4H_9	−0.130
$C_6H_5OCH_2$	0.850	*cyclo*-C_6H_{11}	−0.15
ICH_2	0.85	*t*-$C_4H_9CH_2$	−0.165
$C_6H_5(OH)CH$	0.765	*i*-C_3H_7	−0.190
$CH_2{=}CH$	0.653	*cyclo*-C_5H_9	−0.20
CH_3COCH_2	0.60	*sec*-C_4H_9	−0.210
C_6H_5	0.60	$(C_2H_5)_2CH$	−0.225
$HOCH_2$	0.555	$(CH_3)_3SiCH_2$	−0.26
CH_3OCH_2	0.520	(*t*-C_4H_9)(CH_3)CH	−0.28
$O_2NCH_2CH_2$	0.50	*t*-C_4H_9	−0.300

(c) Relationship between σ^* and σ_I

Due to the historical generation of σ^* and σ_I values from formate and acetate series two inductive parameters exist with different zeros namely methyl and hydrogen respectively. This duality is confusing as both types of inductive parameter are used extensively (the use of most others having died out). The σ_I value is the most utilised inductive parameter at present and is the constant of choice when steric effects are absent. The Pavelich–Taft approach is necessary when there are steric effects. The use of σ^* is somewhat confusing as this value can be defined using RCH_2CO_2Et as the reference [56]; under this definition hydrogen is taken as the zero. If σ^* for RCH_2– is taken then it is possible to compare σ_I with σ^* and Eqn. 73 holds.

$$\sigma_I^R = 0.45\ \sigma^*_{RCH_2} \qquad (73)$$

TABLE 13
Some Taft–Pavelich–steric parameters

Substituent	E_s	Substituent	E_s	Substituent	E_s
H	1.24	ϕCH_2	−0.38	ϕ(Et)CH	−1.50
CH_3	(0.00)	ϕCH_2CH_2	−0.38	Cl_2CH	−1.54
C_2H_5	−0.07	$\phi CH_2CH_2CH_2$	−0.45	$(CH_3)_3CCH_2$	−1.74
Cyclobutyl	−0.06	*i*-Propyl	−0.47	ϕ_2CH	−1.76
CH_3OCH_2	−0.19	Cyclopentyl	−0.51	Me(neopentyl)CH	−1.85
$ClCH_2$, FCH_2	−0.24	F_2CH	−0.67	Br_2CH	−1.86
$BrCH_2$	−0.27	Cyclohexyl	−0.79	Et_2CH	−1.98
CH_3SCH_2	−0.34	$MeOCH_2CH_2$	−0.77	Cl_3C	−2.06
ICH_2	−0.37	$ClCH_2CH_2$	−0.90	$(n\text{-Propyl})_2CH$	−2.11
n-Propyl	−0.36	*i*-Butyl	−0.93	$(i\text{-Butyl})_2CH$	−2.47
n-Butyl	−0.39	*cyclo*-$C_6H_{11}CH_2$	−0.98	Br_3C	−2.43
n-Amyl	−0.40	Me(Et)CH	−1.13	Me_2(neopentyl)C	−2.57
i-Amyl	−0.35	F_3C	−1.16	$(\text{neopentyl})_2CH$	−3.18
n-Octyl	−0.33	Cycloheptyl	−1.10	Me(*t*-Bu)CH	−3.33
$(CH_3)_3CCH_2CH_2$	−0.34	ϕ(Me)CH	−1.19	Me_2(*t*-Bu)C	−3.9
ϕOCH_2	−0.33	*t*-Butyl	−1.54	Et_3C	−3.8
				Me(*t*-Bu) (neopentyl) C	−4.0

The values of σ_I for R are less than σ^* simply because the methylene used in defining σ_I attenuates the inductive effect. Grunwald and Leffler [57] show that the attenuation for methylene may be estimated for σ^* values for pairs of substituents such as $ClCH_2CH_2$ and $ClCH_2$. The value appears to lie between 0.3 and 0.5 [58] and 0.4 is usually taken as a good average [56]; the attenuation per methylene group is assumed to be constant and for an alkylene chain is the *n*th power of the attenuation per methylene. Additivity in σ^* should be valid provided the reaction is not modified by steric requirements. Thus the *t*-butyl group should have a σ^* resulting from the sum of three ethyl groups and the success of this is indicated in Table 14.

(d) Meaning and use of E_s and δ

The δ value measures the steric requirements of a reaction relative to the nucleophilic

TABLE 14
The additivity of σ^*

Group C *XYZ*	σ^*	$\Sigma\sigma^*_{cx,cy,cz}$	
—CMe_3	−0.3	$3\times\sigma^*_{Et}$	= −0.30
—CHPhMe	+0.11	$\sigma^*_{PhCH_2}+\sigma^*_{Me}+\sigma^*_{Et}$	= +0.115
—CHPhEt	+0.04	$\sigma^*_{PhCH_2}+\sigma^*_{pr}+\sigma^*_{Me}$	= +0.10
—CHPhOH	+0.765	$\sigma^*_{PhCH_2}+\sigma^*_{CH_2OH}+\sigma^*_{Me}$	= +0.77
—$CHCl_2$	+1.940	$2\sigma^*_{CH_2Cl}+\sigma^*_{Me}$	= +2.10

attack by water or hydroxide ion on a trigonal ester molecule. Thus for a reaction where the transition state is more crowded relative to ground than in the model a δ value in excess of unity is expected because E_s values of bulky groups are negative (Table 15). It is not expected that E_s values will be additive and inspection of Table 13 confirms this.

As we have pointed out previously the addition of a further parameter is bound to improve the correlation and Hancock et al. [59] introduced a parameter h to modify the number of α-hydrogen atoms in the system. Thus E_s is considered as the sum of a corrected value (E_s^c) modified by a term which contributes to E_s from a "hyperconjugative" effect of the α-hydrogen atoms (Eqn. 74).

$$E_s = E_s^c + h(n-3) \tag{74}$$

Chapman and Shorter [60] considered C—C as well as C—H hyperconjugation and introduced yet a further parameter. In our opinion the addition of further parameters naturally makes for a better correlation but the physical meaning of the underlying models is rapidly becoming obscure. At present we would advocate going no further than a 2-parameter equation if meaning is not to be lost.

(e) Other steric parameters

Charton [61] has postulated a steric scale ν_x (Eqn. 75) which is the difference between the Van der Waal's radius for a symmetrical substituent and hydrogen

$$\nu_x = r_x - 1.20 \tag{75}$$

The values ν_x for primary (symmetrical) substituents were extended to unsymmet-

TABLE 15

Some reactions correlated by the δE_s equation

Reaction	δ
Acid-catalysed hydrolysis of *ortho*-substituted benzamides	0.812
Acid-catalysed methanolysis of RCOO-β-$C_{10}H_7$	1.376
Acid-catalysed 1-propanolysis of RCOO-β-$C_{10}H_7$	1.704
Acid-catalysed 2-propanolysis of RCOO-β-$C_{10}H_7$	1.882
Reaction of methyl iodide with 2-alkylpyridines	2.065
ΔH^0 for reaction of 2-alkylpyridines with diborane	3.322
ΔH^0 for reaction of 2-alkylpyridines with BF_3	5.49
ΔH^0 for reaction of 2-alkylpyridines with trimethylboron	6.36
Ionisation of arylaliphatic acids	0.25
Br_2 + olefins with branched alkyl substituent	0.96
RCMe=CH_2 + boranes	0.71
$RCOCH_3 + BH_4^-$	0.76

rical substituents empirically and the values (Table 16) correlate quite well with E_s. Incorporation of Charton's steric parameter into the free energy relationship is shown in Eqn. 76 where inductive, resonance and steric interactions are mutually exclusive [61,62].

$$\log k_x/k_H = \alpha\sigma_I + \beta\sigma_R + \psi\nu \tag{76}$$

TABLE 16
Some values of Charton's steric constant ν

R	ν	R	ν
H	0	Pr_2CH	1.54
Me	0.52	Bu_2CH	1.56
t-Bu	1.24	$(i\text{-}PrCH_2)_2CH$	1.70
CCl_3	1.38	$(t\text{-}BuCH_2)_2CH$	2.03
CBr_3	1.56	*i*-PrCHEt	2.11
CI_3	1.79	*t*-BuCHMe	2.11
CF_3	0.91	$t\text{-}BuCH_2CHMe$	1.41
Me_3Si	1.40	$c\text{-}C_6H_{11}$	0.87
F	0.27	$t\text{-}BuCMe_2$	2.43
Cl	0.55	$t\text{-}BuCH_2CMe_2$	1.74
Br	0.65	Et_3C	2.38
I	0.78	$PhCH_2$	0.70
Ph	1.66	$PhCH_2CH_2$	0.70
		$Ph(CH_2)_3$	0.70
Et	0.56	$Ph(CH_2)_4$	0.70
Pr	0.68	PhMeCH	0.99
Bu	0.68	PhEtCH	1.18
$BuCH_2$	0.68	Ph_2CH	1.25
$BuCH_2CH_2$	0.73	$ClCH_2$	0.60
$Bu(CH_2)_3$	0.73	$BrCH_2$	0.64
$Bu(CH_2)_4$	0.68	Cl_2CH	0.81
$i\text{-}PrCH_2$	0.98		
$t\text{-}BuCH_2$	1.34	Br_2CH	0.89
$s\text{-}BuCH_2$	1.00		
$i\text{-}PrCH_2CH_2$	0.68	Br_2CMe	1.46
$t\text{-}BuCH_2CH_2$	0.70	Me_2CBr	1.39
$t\text{-}BuCHEtCH_2CH_2$	1.01	CH_2F	0.62
$c\text{-}C_6H_{11}CH_2$	0.97	CH_2I	0.67
$c\text{-}C_6H_{11}CH_2CH_2$	0.70	CH_2CN	0.89
$c\text{-}C_6H_{11}(CH_2)_3$	0.71	CHI_2	0.97
i-Pr	0.76	CHF_2	0.68
s-Bu	1.02		
Et_2CH	1.51		

(f) Values of σ for alkyl groups*
A note of caution must be sounded with regard to σ^* values for alkyl groups. When σ^* is substantial there is little doubt that it represents the inductive effect of the group; small values might easily result from neglect of the very small steric effect difference in the Taft analysis [60,63,64]. In the event of the use of σ^* in correlations the small but different values for alkyl substituents probably do not make a significant difference.

4. Hydrophobic interactions

One of the main interactions available to enzymes for binding substrates is the so-called "hydrophobic bond". Hydrocarbon groups dissolved in water cause an "ordering" of the surrounding sheath of water molecules. The aggregation of hydrocarbons in water is a result of the minimisation of the ordering and a hydrophobic solute may be driven into a hydrophobic region of a protein by the regaining of entropy by water. A model of the ordering process is the partitioning of organic solute from water to a non-aqueous phase [65,66]; *n*-octanol is the conventional phase employed.

The partition coefficient (P) between water and *n*-octanol is easily measured and is related to π in Eqn. 77.

$$\pi = \log P_X - \log P_H \tag{77}$$

The substituent X has a π value defined from the partitioning of substituted phenoxyacetic acids between *n*-octanol and water. Thus π for the parent ($x = H$) is defined as zero. Both π and log P (Table 17) may be used as parameters, the former in a restricted series of substituted phenyl groups and the latter in more general correlations.

α-Chymotrypsin is inhibited by hydrocarbons (anthracene, benzene, toluene, azulene, etc.) [67] and the equilibrium constant (Eqn. 78) has a surprisingly good

$$E + I \rightleftharpoons EI$$

$$K_I = [E][I]/[EI] \tag{78}$$

correlation coefficient with log P [68]. The reaction under study and the model thus have very similar characteristics consistent with the hypothesis that α-chymotrypsin possesses a "hydrophobic" region associated with the active site. The hydrophobic region is sometimes called the "hydrophobic pocket" [69] or ρ_2 site [70]; it is thought that this pocket binds hydrophobic side chains of aromatic amino acid substrates [70].

The application of P or π correlations to general biological processes allows some degree of prediction in drug action studies (Table 18). The efficiency of drugs depends not only on their interaction with a receptor site but in their ability to

TABLE 17
Collection of π values [65]

Substituent	π	Substituent	π
H	0	3-CF_3	1.07
2-F	0.01	3-CH_2OH	−0.65
3-F	0.13	4-CH_2OH	−0.60
4-F	0.15	3-CH_2CO_2H	−0.40
2-Cl	0.59	4-CH_2CO_2H	−0.47
3-Cl	0.76	3-CO_2H	−0.15
4-Cl	0.70	4-CO_2H	0.12
2-Br	0.75		
3-Br	0.94	3-COOMe	−0.05
4-Br	1.02	2-Ac	0.01
2-I	0.92	3-Ac	−0.28
3-I	1.15	4-Ac	−0.37
4-I	1.26	3-CN	−0.3
2-Me	0.68	4-CN	−0.32
3-Me	0.51	2-OH	−0.54
4-Me	0.52	3-OH	−0.49
2-Et	1.22	4-OH	−0.61
3-Et	0.97	2-OMe	−0.33
3-*n*-pr	1.43	3-OMe	0.12
3-*i*-pr	1.30	4-OMe	−0.04
4-*i*-pr	1.40	3-OCF_3	1.21
3-*n*-Bu	1.90	3-OCH_2CO_2H	−0.70
4-*sec*-Bu	1.82	4-OCH_2CO_2H	−0.81
3-*t*-Bu	1.68	3-NH_2	−1.15
4-Cyclopentyl	2.14	4-NH_2	−1.03
4-Cyclohexyl	2.51	3-NMe_2	0.10
3-Ph	1.89	2-NO_2	−0.23
3,4-$(CH_2)_3$	1.04	3-NO_2	0.11
3,4-$(CH_2)_4$	1.39	4-NO_2	0.24
3-NHAc	−0.79	3-SCF_3	1.58
3-NHCOPh	0.72	3-SO_2Me	−1.26
4-N=NPh	1.71	3-SO_2CF_3	0.93
3-$NHCONH_2$	−1.01		
3-SMe	0.62	SO_2NH_2	−1.82

penetrate hydrophobic membranes and to be involved in other transport processes. The inability to assign effects to particular processes makes the unravelling of drug action correlations very complex. Provided the effect of changing the substituent or substrate can be confined to a particular reaction the slope of the correlation (a, in

TABLE 18
Correlation of activity with π or P and other parameters

Reaction	Equation
Hydrolysis of 4-nitrophenyl esters by serum esterase	$\log k = 0.95 \log P + 3.5E_s - 0.47$
Inhibition of chymotrypsin:	
with aromatic acids	$\log K_i = -0.94 \log P - 0.96\, pK + 3.66$
with PhCOR	$\log K_i = -0.31\pi - 2.53$
with ArOH	$\log K_i = -0.95 \log P + 1.88$
with hydrocarbons	$\log K_i = -1.47 \log P + 1.2$
Reaction of chymotrypsin:	
with $PhCONHCH_2CO_2R$	$\log K_m = -0.41\pi - 0.4E_s + 0.71$
with $PhCH_2CH_2CO_2R$	$\log K_m = -0.21\pi - 3.16$
Binding of barbiturates to bovine serum albumin	$\log 1/C = 0.58P + 2.39$
Binding of phenols to bovine serum albumin	$\log 1/C = 0.68\pi + 3.48$

Eqn. 79) should indicate more or less whether the environment binding the substrate is similar to that in *n*-octanol.

$$\log K = a \log P + b \tag{79}$$

The success of a correlation with π or $\log P$ implies that orientation requirements at the active or receptor site are not stringent. An interesting example of this is the hydrolysis of 4-nitrophenyl esters by serum esterase where Eqn. 80 holds [71].

$$\log \text{rate} = 0.95 \log P + 3.5E_s - 0.47 \; (r = 0.976) \tag{80}$$

While the hydrophobic site behaves in a manner similar to that of *n*-octanol the intrusion of a large δ term modifying E_s points to relative stringent steric requirements which could be associated with either the catalytic step or with a "hydrophobic cavity" of limited size.

4.1. Other hydrophobic parameters

Solubility in water has been proposed as a measure of hydrophobicity and has been shown to be related to the free energy of transfer from water to ethanol [72]. Fig. 17 illustrates a plot of $\log K_m$ versus $\log N_w$ (N_w = mole fraction of the substance at saturation in water) for substrates of α-chymotrypsin. A plot of $\log K_I$ versus $\log N_w$ for inhibitors of α-chymotrypsin is linear (Fig. 18).

Good correlations are also obtained between $\log P$ and $\log R_f$. The $\log K_m$ values for the hydrolysis of *N*-acetyl-L-amino acid methyl esters catalysed by α-chymotrypsin show a good correlation with the effluent volumes ($\log v$) of amino acids from cation exchange resins (Fig. 19).

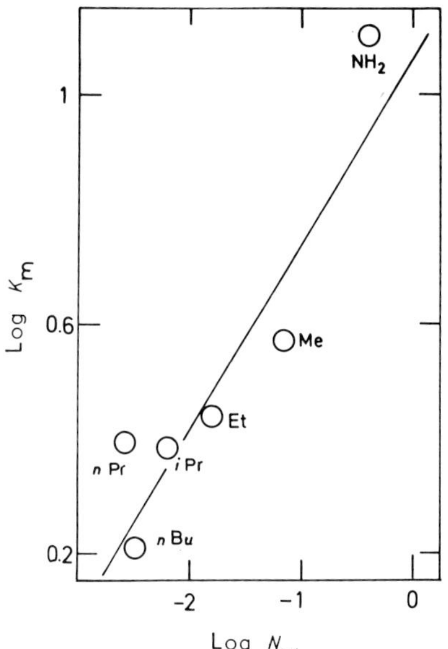

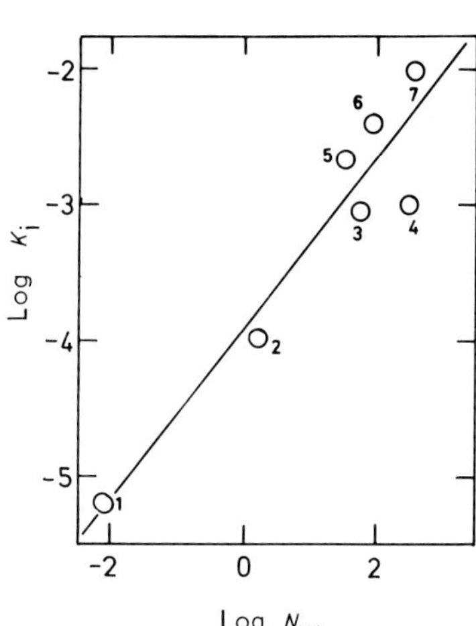

Fig. 17. Hippuric acid derivatives as substrates of α-chymotrypsin correlated with the solubility (N_w) of corresponding acetic acid derivatives in water.

Fig. 18. Inhibition of α-chymotrypsin (K_i) versus solubility of inhibitors (N_w) in water [72]. 1, anthracene; 2, naphthalene; 3, chlorobenzene; 4, nitrobenzene; 5, ethylbenzene; 6, toluene; 7, benzene.

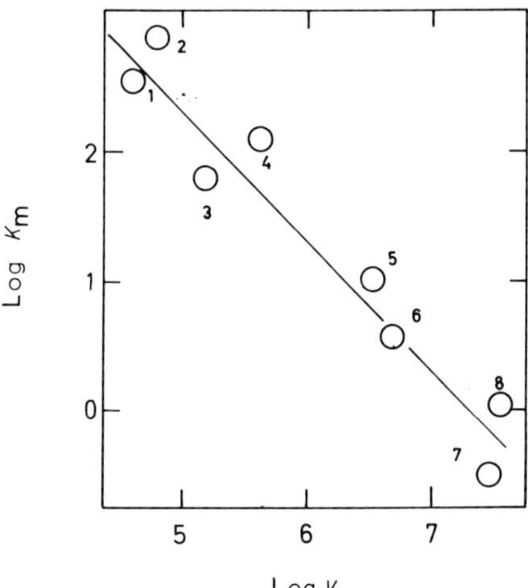

Fig. 19. Michaelis–Menten constant (K_m) for N-acetyl-L-amino acid methyl esters with α-chymotrypsin versus the effluent volume of the amino acid from a cation-exchange resin [72]. 1, glycine; 2, alanine; 3, aminoisobutyric acid; 4, valine; 5, methionine; 6, leucine; 7, tyrosine; 8, phenylalanine.

4.2. *Non-linear hydrophobic relationships*

Hansch [66] and his co-workers soon found that curved free-energy correlation plots were frequent and equations of the type shown in Eqn. 81 were found.

$$\log(1/C) = -k_1(\log P)^2 + k_2(\log P) + k_3\sigma + k_4 \tag{81}$$

These correlations arise from the movement of species through a series of membranes to their site of action. The problem was discussed analytically by Hansch [66] but a qualitative understanding may easily be obtained. Hydrophilic substrates will not penetrate lipophilic barriers easily and will have a low propensity for reaching the active site. Highly lipophilic molecules on the other hand will be strongly held in the first lipophilic membrane and will also not easily reach the active site. Reference to our discussions of free energy correlations and changes in rate-determining step reveal a remarkable similarity if the set of membranes leading to an active site may be regarded as a consecutive series of reactions. Changing the hydrophobicity may cause a particular step to be rate-determining. Examples of non-linear hydrophobic correlations are given in Table 19.

4.3. *Molar refractivity*

Pauling and Pressman [73] attempted to correlate binding constants of haptens and antibodies with molar refractivity of substrates (Eqn. 82)

$$MR = \frac{(n^2 - 1)MW}{(n^2 + 2)d} \tag{82}$$

where n = refractive index, d = density and MW = molecular weight. The applica-

TABLE 19
Non-linear biological response [66] ($\log 1/C = k_1(\log P)^2 + k_2 \log P + k_3$)

Response system	Compound	k_1	k_2	k_3
Mice hypnosis	Barbiturates	−0.44	1.58	1.93
Mice hypnosis	Acetylenic alcohols	−0.51	2.13	0.86
Mice hypnosis	Diacylureas	−0.18	0.6	1.89
Inhibition of *B. diphtheriae*	Hydrocupreines	−0.12	1.43	1.16
Inhibition of *Strep. hemolyticus*	Phenols	−0.17	2.14	−3.57
Inhibition of *S. aureus*	4-Hydroxyphenylsulphides	−0.15	1.73	−2.21
Inhibition of Aphids	RSCN	−0.1	1.14	1.37
I_{50}, frog ventricle	Miscellaneous	−0.13	1.49	0.36
Chymotrypsin (V_0)	3-(*n*-Alkanoyl)-*O*-benzoates	−0.66	0.58	−5.37
Elastase (k_{cat})	4-Nitrophenylalkanoates	−1.13	6.46	−5.79

tion of molar refractivity can be ambiguous because n does not vary significantly leading to $MR \alpha$ molecular volume. Thus a negative coefficient in the MR term of a correlation suggests a steric effect; a positive coefficient suggests a role for dispersion forces resulting from the polarisability identity of MR.

An interesting example of the application of molecular polarisability comes from data of this laboratory on papain analysed by Hansch and Calef [75]. The papain-catalysed hydrolysis of substituted phenyl methanesulphonylglycinates gave relatively poor correlations between k_0/K_m and σ, π, MR and π and σ. The correlation of k_0/K_m with MR and σ is significant (Eqn. 83) indicating that inductive and polarisable effects are involved so far as substitution on the phenyl-leaving group is concerned.

$$\log[k_0/K_m]/[k_0/K_m]_0 = 0.53\ MR + 0.37\ \sigma \tag{83}$$

4.4. *Additivity*

Like the inductive effect and unlike the steric effect (E_s) hydrophobic and molecular polarisability parameters are additive; thus it is possible to calculate partition

TABLE 20
Additive π values and their uncertainty units

Calculation step	π	UU
CH_2	0.5	0.02
Branching in C chain	−0.2	0.02
Branching in functional group	−0.2	0.05
Ring closure	−0.09	0.02
Double bond	−0.3	0.03
"Folding" [a]	−0.6	0.05
Intramolecular H bond	0.65	0.1
Equivalence of OH and NH_2	0	0.05
Aliphatic groups		
CO_2H	−0.65	0.03
OH	−1.16	0.03
NH_2	−1.16	0.03
CO	−1.21	0.03
CN	−0.84	0.04
O—	−0.98	0.05
$CONH_2$	−1.71	0.05
F	−0.17	0.03
Cl	0.39	0.04
Br	0.60	0.04
I	1.00	0.05

[a] For details of this step see A. Leo, C. Hansch and D. Elkins (1971) Chem. Rev. 71, 525.

coefficients of compounds from suitable reference molecules with considerable confidence. In concert with the calculation of pK values (see later) several routes may be used depending on the parent model for calculating $\log P$ values. An example is given for lactic acid ($\log P = -0.62$) where the parent is taken as glycollic acid ($\log P = -1.11$); glycollic acid requires the insertion of a methylene (+0.5, see Table 20) [77] group to yield lactic acid. The presence of a branched functional group requires an extra -0.2 units to yield a calculated $\log P = -0.81$. Summation of the uncertainty units (UU) gives an "error" of 0.12 indicating that the derived $\log P$ agrees quite well with the observed.

4.5. *Ambiguities arising from interrelationships between parameters*

We have already pointed out that over a small range of substituents MR may be proportional to molecular volume; thus it is not possible to assign an interaction unambiguously if a small set of substituents is used. MR and π may also be closely related so that over a small range of substituents a polarisability effect could be mistaken for a hydrophobic effect. Charge-transfer complexation between tetracyanoethylene and various molecules has been used as a model to derive a parameter (the charge-transfer constant, C_T) [78a]. The correlation of charge-transfer complex formation with σ [78b] indicates that some care is needed in the interpretation of the substituent effects as being different from polar effects.

The best course to take when investigating correlations such as the Taft–Pavelich or the hydrophobic correlations is to examine closely the correspondence between model and reaction in hand. For example a strict dependence on E_s in biological interaction could mean that the process mirrors the simple ester hydrolysis reaction in its steric requirements or that an identical steric requirement exists but which arises from a different mechanism such as entry of the substrate through a finite shaped cleft.

4.6. *Application to non-biochemical reactions*

The major application of the partition-coefficient parameters has been to biological processes and the simplest major application is to enzyme reactions. Murakami et al.'s [79b] work on host–guest interactions is a good example of a non-biological application. Hansch [79a] showed that Murakami et al.'s data for the reaction of an imidazolyl group attached to a paracyclophane with 4-nitrophenyl esters (Eqn. 84) correlated with π (Eqn. 85).

R–(imidazole, N–H) + R'CO—O4NP ⟶ R–(imidazole, N–COR') ⟶ R–(imidazole, N–H) + R'CO$_2$H (84)

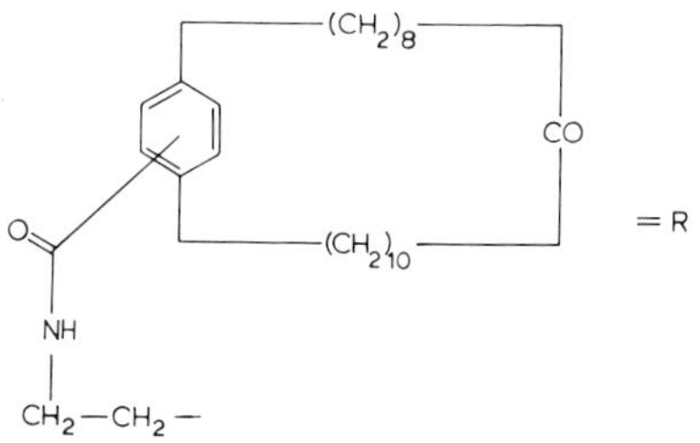

$$(CH_2)_8 \ldots CO \ldots (CH_2)_{10} \ldots NH-CH_2-CH_2- \;=\; R \qquad (85)$$

The possibility that micelle formation was giving rise to the correlation was excluded by carrying out the reaction with ester concentrations below the critical micelle concentration. Probably the "hydrophobic affinity" of the acyl groups of the ester to the cyclophane grouping assists reaction with the imidazolyl group.

5. *Solvent effects*

Grunwald and Winstein [80] developed an index for solvents based on the model reaction for the solvolysis of *t*-butylchloride. It is assumed that the solvolysis involves full carbenium ion development so that differences in reactivity from solvent to solvent will mirror solvating power relative to a standard solvent (80% ethanol/water). A parameter "Y" dependent on solvent is defined by Eqn. 86 and the reaction in question will obey Eqn. 87 when carried out in a series of solvents.

$$Y_{\text{solvent}} = \log\left[k^{\text{Bu}^t\text{Cl}}_{\text{solvent}}/k^{\text{Bu}^t\text{Cl}}_{\text{standard}}\right] \qquad (86)$$

$$\left[\log k_{\text{solvent}}/k_{\text{standard}}\right] = m \cdot Y \qquad (87)$$

The solvolysis of *t*-butylchloride is defined to have $m = 1$ so that unit m will point to an S_N1 mechanism by analogy. Difficulties involved in the Grunwald–Winstein approach are that the solvent may *change* the mechanism of the standard reaction, and that determination of Y values requires the elimination of the "internal return" problem. In the latter problem decomposition of the carbenium ion becomes rate-limiting. A further experimental difficulty is that the relationship only seems to work well when the solvents used are variable composition binary mixtures (such as ethanol/water); this presumably arises from the smooth change available for Y with these systems, and is analogous to the breakdown of substituent correlations when the structural type is varied over a wide range. Examples of some m values and a set of Y parameters are given in Tables 21 and 22.

Scott et al. [81] found that the m value for cyclisation of *N*-benzoyl 2-aminoethyl bromide to a dihydro oxazole (Eqn. 88) has a value (0.13) which is less than that for a bona-fide SN1 reaction (Bu^tBr has $m = 0.941$).

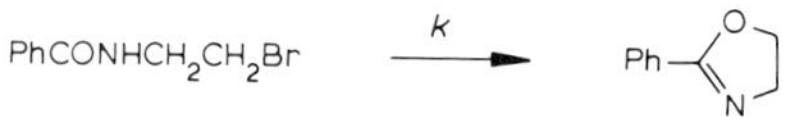

$$\text{PhCONHCH}_2\text{CH}_2\text{Br} \xrightarrow{k} \text{Ph-(oxazoline)} \qquad (88)$$

TABLE 21
Values of *m* for some representative reactions

Substrate	Solvent	*m*
EtBr	Aqueous ethanol	0.343
tBuCl	$EtOH/H_2O$	1.000
tBuBr	$EtOH/H_2O$ (0°)	1.02
tBuBr	$EtOH/H_2O$ (25°)	0.94
tBuBr	$AcOH/HCO_2H$	0.95
$HOCH_2CH_2Br$	$EtOH/H_2O$	0.23
$PhCH_2OTS$	$Acetone/H_2O$	0.65
Ac_2O	$Acetone/H_2O$	0.58
CH_3SO_2Cl	$Dioxane/H_2O$	0.47
$PhCHClCH_3$	$EtOH/H_2O$	1.00
$PhCHClCH_3$	$AcOH/H_2O$	1.14
$PhCHClCH_3$	$Dioxane/H_2O$	1.14
$PhC(CH_3)_2CH_2Cl$	$EtOH/H_2O$	0.83
Ph_2CHCl	$EtOH/H_2O$	0.74
Ph_2CHCl	$AcOH/H_2O$	1.56
Ph_3CF	$EtOH/H_2O$	0.89
Ph_3CF	$Acetone/H_2O$	1.58

An S_N2 reaction such as ethylbromide solvolysis has an *m* value of 0.343. It is therefore suggested [81] that an S_N2 intramolecular reaction is occurring (Eqn. 89) with a transition state with very little charge expression.

TABLE 22
Some *Y* values for various solvents and solvent mixtures

Solvent	*Y*	Solvent	*Y*
Ethanol/water		iPrOH	−2.73
100% Ethanol	−2.03	iBuOH	−3.26
95%	−1.29	100% HCO_2H	2.05
90%	−0.75	50% HCO_2H/H_2O	2.64
80%	0	100% AcOH	−1.64
70%	0.60	50% $AcOH/H_2O$	1.94
50%	1.66	50% $HCO_2H/AcOH$	0.76
30%	2.72	90% Dioxane	−2.03
Water	3.49	50% Dioxane	1.36
Methanol/water		90% Acetone	−1.86
100% Methanol	−1.09	50% Acetone	1.40
90%	−0.30	100% $HCONH_2$	0.60
70%	0.96	$nC_3F_7CO_2H$	1.70
50%	1.97		
30%	2.75		

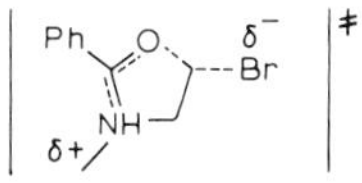

(89)

The solvolysis of $PhCH_2O–SO_2–C_6H_4 4Me$ has $m = 0.65$ indicating that the mechanism is close to the S_N1/S_N2 borderline; we shall see different evidence for this later on.

Probably a more convenient solvent parameter is Kosower's "Z" parameter [82]; this is derived from the energy of the charge-transfer spectrum of the pyridinium iodide (Eqn. 90) in the solvent in question.

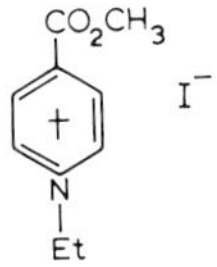

$$Z = h\nu \tag{90}$$

The electronic transition involves the transformation of an initial state consisting of a negative ion close to a positive ion to an excited state in which the two ions have partially neutralised each other. Since the process is similar to the reverse of the transformation of a polar C–Y bond into a "charge separated" bond in the conversion of reactant to transition state in an S_N1 reaction solvent effects on the two processes are expected to be inversely proportional to one another. "Z" values are collected in Table 23.

Reichert [83] proposed an alternative model for the absorption process involving transition from a zwitterionic ground state to the excited state (Eqn. 91).

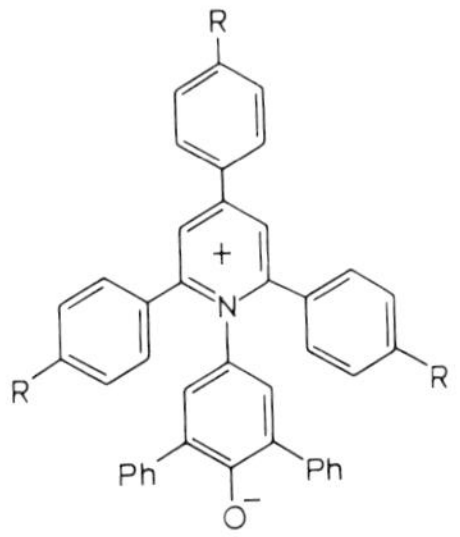

(91)

The parameter E_T ($= h\nu$) will increase with polar solvents as the ground state will be more stable than in non-polar solvents.

Correlations between Z, E_T and Y (Figs. 20 and 21) indicate that these parameters are useful solvent probes and it is probably better to use Z or E_T as these are more easily obtained.

5.1. Reporter groups

Some time ago Hille and Koshland [84] coined the phrase reporter group for a group which has medium-sensitive properties and which may be placed in defined locations

TABLE 23
Z values for some solvents

Solvent	*Z*/kcal/mole	Solvent	*Z*/kcal/mole
H_2O	94.6	$CHCl_3$	63.2
MeOH	83.6	Acetone	65.7
EtOH	79.6	DMF	68.5
Pr^nOH	78.3	CH_3CN	71.3
Bu^nOH	77.7	Pyridine	64.0
Pr^iOH	76.3	$Me_3CCH_2CHMe_2$	60.1
Bu^tOH	71.3	DMSO	71.1
Glycol	85.1	Formamide	83.3
$CHF_2CF_2CH_2OH$	86.3	*cyclo*-C_3H_2-$COCH_3$	65.4
$CHF_2(CF_2)_3CH_2OH$	84.4	Acetic acid	79.2
		CH_2Cl	64.2

in a biological molecule. The effect of the microscopic medium on the property when compared with a standard gives an indication of the nature of the surroundings of the reporter group. Hille and Koshland [84] reacted chymotrypsin with 4-nitro-2-bromoacetamidophenol (alkylation occurs at a methionine group (Eqn. 92)).

$$CT{-}CH_2{-}SMe \longrightarrow CT{-}CH_2{-}\overset{+}{S}(Me){-}CH_2CONH{-}C_6H_3(NO_2)(OH) \qquad (92)$$

NO2 / OH / NHCOCH2Br (4-nitro-2-bromoacetamidophenol)

The phenolic ionisation of the reporter group is modified by the presence of the positively charged centre; the absorption spectrum of the derivative indicates an environment more polar than water. Sigman and Blout [85] reacted α-bromo-4-nitroacetophenone with chymotrypsin and found an intense absorption at 350 nm not shown by enzyme or alkylating agent; the absorption was attributed to interaction of the nitrobenzene group with a neighbouring indolyl group close to the alkylation site. An interesting example from the laboratory of Kallos [86] involves measuring the λ_{max} of the nitrobenzene chromophore in different solvents including the active site of chymotrypsin where it is held as the sulphonate (Eqn. 93).

$$NO_2{-}C_6H_4{-}SO_2F + CTOH \longrightarrow CTO{-}SO_2{-}C_6H_4{-}NO_2 + HF \qquad (93)$$

The ultraviolet spectrum of methyl nitrobenzene sulphonate in various solvents

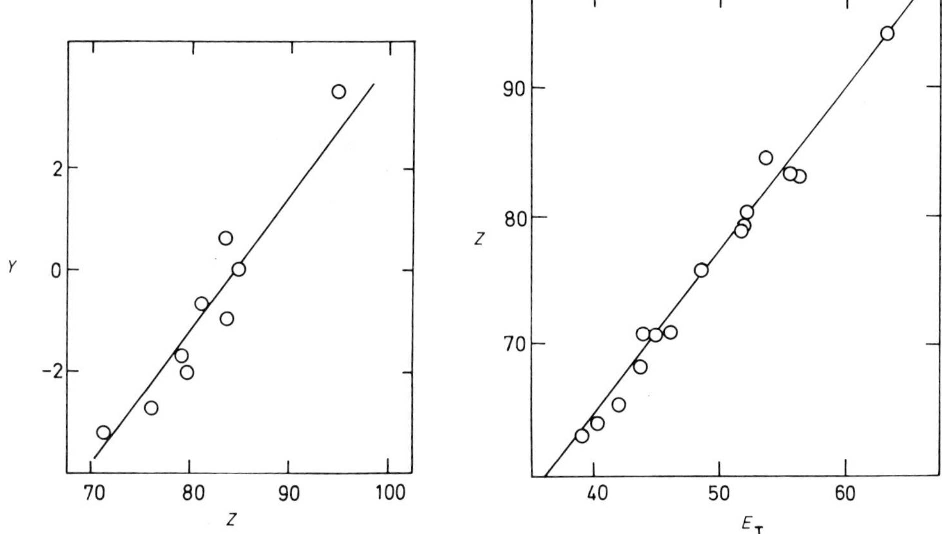

Fig. 20. Correlation between Y and Z parameters.

Fig. 21. Correlation between Z and E parameters.

has λ_{max}/nm: water (260), ethanol (255), ether (252), cyclohexane (250.8). The chymotrypsin derivative when the sulphonyl group is attached as an ester has λ_{max} 251 nm consistent with the nitrobenzene group being located in a hydrophobic region. Reference to the section on hydrophobic parameters indicates a similar result from a different approach.

It is of course clear that properties other than UV spectra are valuable as reporting properties; such properties as electron spin and nuclear magnetic resonance spectra and fluorescence yield valuable information about the environment in the protein.

6. *General equations of reactivity*

6.1. *Swain–Scott and Edwards relationship*

Swain and Scott [87] investigated a nucleophilicity scale using as the standard reaction attack of nucleophile on methyl bromide (Eqn. 94);

$$\mathrm{MeBr} + \mathrm{N}^- \rightarrow \mathrm{MeN} + \mathrm{Br}^-$$

$$\log k^{\mathrm{N}}/k^{\mathrm{H_2O}} = n \tag{94}$$

water is used as the standard nucleophile and several reactions have been shown to conform with Eqn. 95 where s is characteristic of the reaction.

$$\log k^{\mathrm{N}}/k^{\mathrm{H_2O}} = s \cdot n \tag{95}$$

Representative s values are collected in Table 24 and n values in Table 25. The order of nucleophlicity (n) in the Swain–Scott approach bears no relationship to the pK_a of the conjugate acid of the nucleophile. Since in a given constant series of nucleophiles basicity and nucleophilicity are related there have been attempts to correlate nucleophilicity with one or more parameters. If the relative importance of these parameters is the same as in the methyl bromide reaction then the Swain–Scott equation will hold.

Edwards and Pearson [88] indicated that a highly polarisable reagent is a better nucleophile in relation to its base strength than a less polarisable reagent. Edwards [89] formulated a reactivity Eqn. 96

$$\log k_{\mathrm{N}}/k_{\mathrm{H_2O}} = aE_{\mathrm{N}} + bH_{\mathrm{N}} \tag{96}$$

including polarisability (judged by E_{N}) and basicity in a linear combination. The value H_{N} is related to pK of the conjugate acid of the nucleophile ($H_{\mathrm{N}} = \mathrm{p}K + 1.74$; 1.74 is the "p$K$" of H_3O^+). The value E_{N} is the standard electrode potential of the reagent X^- (Eqn. 97) and is empirically related to molecular polarisability.

$$2\,X^- \rightleftharpoons X_2 + 2\,e^-$$

$$2\,\mathrm{H_2O} \rightleftharpoons \mathrm{H_4O_2^{2+}} + 2\,e^- \qquad E^0 = -2.60v \tag{97}$$

$$E_{\mathrm{N}} \equiv E^0 + 2.60$$

TABLE 24
Some representative Swain s values for reaction with nucleophiles

Reactant	s
C_2H_5OTs	0.66
$C_6H_5CH_2Cl$	0.87
β-Propiolactone	0.77
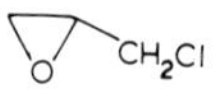 (epoxide)–CH_2Cl	0.93
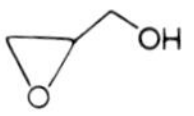 (epoxide)–CH_2OH	1.00
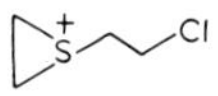 (episulfonium, S^+)–CH_2CH_2Cl	0.95
CH_3Br	1.00
$C_6H_5SO_2Cl$	1.25
C_6H_5COCl	1.43

TABLE 25
Some n values for representative nucleophiles

Nucleophile	n	Nucleophile	n
ClO_3^-, ClO_4^-, BrO_3^-, IO_3^-	< 0	Br^-	3.89
H_2O	0.00	N_3^-	4.00
TsO^-	< 1.0	Thiourea(S attack)	4.1
NO_3^-	1.03	OH^-	4.20
Picrate anion	1.9	Aniline	4.49
F^-	2.0	SCN^-	4.77
$SO_4^=$	2.5	I^-	5.04
Acetate ion	2.72	CN^-	5.1
Cl^-	3.04	SH^-	5.1
Pyridine	3.6	$SO_3^=$	5.1
HCO_3^-	3.8	$S_2O_3^=$	6.36
$HPO_4^=$	3.8	$HPSO_3^=$	6.6

The influence of two parameters is necessary to explain such problems as the differential attack of methoxide and thiophenoxide ion on neopentyl 4-toluene sulphonate (Eqn. 98).

$$Me_3CCH_2OTs \xrightarrow[(S \neq O)\ (C \neq O)]{MeO^-} Me_3CCH_2O^- + MeOTs$$
$$Me_3CCH_2OTs \xrightarrow{PhS^-} Me_3C{-}CH_2SPh + TsO^- \qquad (98)$$

Hudson [90] has discussed the factors affecting nucleophilic reactivity in terms of the fundamental thermodynamics.

7. *Cross-correlation and selectivity–reactivity*

Chemists have intuitively assumed for some time that selectivity is inversely related to reactivity. Thus a very fast reaction is likely to be less affected by substituents than is a slower one. There is at present considerable discussion as to the general validity of the reactivity–selectivity postulate (RSP) and Johnson [91] has indicated in his book that the linearity of the Hammett correlation is directly contrary to the postulate. The validity of the reactivity–selectivity postulate is directly related to that of Hammond (see later) which states that the structure of the transition state is closest to the state (ground or product) which has the highest potential energy.

7.1. Cross-correlation

If we can assume that nucleophile reactivity is governed by the Swain–Scott equation and that general acid catalysis is governed by Brønsted's law then the following equations (Eqns. 99 and 100) hold.

$$\log k_{ij} = \log k_{0j} + s_j \cdot n_i \tag{99}$$

$$\log k_{ij} = \log G_i^A - \alpha_i \mathrm{p}K_j \tag{100}$$

The individual acids in the acid series are denoted as *i* and the individual nucleophiles as *j* thus in the Brønsted equation a series of acids *i* is varied for a constant nucleophile *j*. Combining Eqns. 99 and 100 yields Eqn. 101;

$$\log k_{0j} + s_j n_i = \log G_i^A - \alpha_i \mathrm{p}K_j \tag{101}$$

rewriting Eqn. 101 for 2 values of *j* (1 and 2) gives Eqns. 102 and 103.

$$\log k_{0,2} + s_2 \cdot n_i = \log G_i^A - \alpha_i \mathrm{p}K_2 \tag{102}$$

$$\log k_{0,1} + s_1 \cdot n_i = \log G_i^A - \alpha_i \mathrm{p}K_1 \tag{103}$$

$$\log(k_{0,2}/k_{0,1}) + (s_2 - s_1)n_i = \alpha_i(\mathrm{p}K_1 - \mathrm{p}K_2) \tag{104}$$

$$\log(k_{0,2}/k_{0,1}) = \alpha_0(\mathrm{p}K_1 - \mathrm{p}K_2) \tag{105}$$

$$\therefore n_i(s_2 - s_1) = (\alpha_i - \alpha_0)(\mathrm{p}K_1 - \mathrm{p}K_2) \tag{106}$$

$$\therefore \frac{n_i}{(\alpha_i - \alpha_0)} \equiv \frac{(\mathrm{p}K_1 - \mathrm{p}K_2)}{(s_2 - s_1)} \equiv C \tag{107}$$

Eqn. 100 yields Eqn. 105 and since the left side of Eqn. 107 depends only on the nature of the nucleophile and the right side of the acid catalyst the two sides are *independently variable* and individually equal to a constant (C). A similar cross-correlation can be shown to exist between the Hammett and Brønsted equations (Eqn. 108) and Hammett and Swain relationships.

$$\frac{\mathrm{p}K_2 - \mathrm{p}K_1}{\rho_2 - \rho_1} = \frac{\sigma_i}{\alpha_0 - \alpha_i} = C^1 \tag{108}$$

These equations are essentially quantitative statements of the considerations of Hammond [92] and Leffler [16b]; consideration of the Hammond postulate indicates that as acid strength increases ($\mathrm{p}K_2 \downarrow$) the selectivity (s_2) in Eqn. 107 should decrease; thus C is negative [93]. The above relationships have been derived by a number of authors [94].

Cross-correlations are directly related to the three-dimensional potential-energy surface description of a reaction involving two coordinates. Most reactions involve major changes which are not located in a simple bond cleavage. For example the simple displacement reaction at an ester involves two bond cleavages and geometry variations; associated with these variations are solvation changes. The simple Hammond and the related reactivity–selectivity postulates refer to a single-bond cleavage and it is natural that they tend to fail as reactions predominantly are more complicated.

Recent discussions have been on reactions with two major changes in a single step and these may be described by a three-dimensional potential-energy surface. A qualitative picture of the surface for a reaction (Eqn. 109) is all that is necessary to

$$A + BC \rightarrow AB + C \tag{109}$$

predict the change in reaction parameters (β, ρ, s or m) as a function of reactivity caused by altering energies of the 4 corners (Fig. 22). Let us imagine that the substituent effect in A (β_A) and that in C (β_C) relate to horizontal and vertical axes; these axes may be calibrated in terms of β_A and β_C. The reaction path at the transition state ("saddle point") is a vector; the effect of the surface on the transition state can be "parallel" or "perpendicular" to the vector. Effects of the energies of product and reactant will be "parallel" and conform to the Hammond postulate while "perpendicular" effects (of A + B + C and ABC) will move the transition state in the direction of the *lower* energy ('anti-Hammond'). It is thus possible for the change in β_A or β_C to be in an 'anti-Hammond' direction if the perpendicular effect is dominant as the direction of movement is the sum of perpendicular and parallel vectors.

Thornton [95], Harris and Kurz [96], Jencks and Jencks [97], Critchlow [98] and Dunn [99] have discussed the above approach analytically using assumed equations to define the energy surface. The equations are not necessarily very reliable at the "edges" of the diagram but should describe the surface reasonably well at the "central" region. The Thornton [95], Kurz [96] and Jencks [97] equations are essentially parabolae along the edges and clearly do not give a double-energy well (Eqn. 110).

$$E = a(\beta_A)^2 + b(\beta_C)^2 + c\beta_A\beta_C + d\beta_A + e\beta_C + f \tag{110}$$

Along an edge (at $\beta_A = 0$) the equation is of the form of Eqn. 111 and has only 1 turning point – a maximum.

$$E = b(\beta_C)^2 + e\beta_C + f \tag{111}$$

Dunn [99] proposed a more sophisticated equation which gives an equation along an edge possessing 3 turning points comprising 2 minima and 1 maximum (Eqn. 112)

$$E = (\beta_C)^4/4 - (\alpha + 10)(\beta_C)^3/3 + 5\alpha\beta_C^2 + C \tag{112}$$

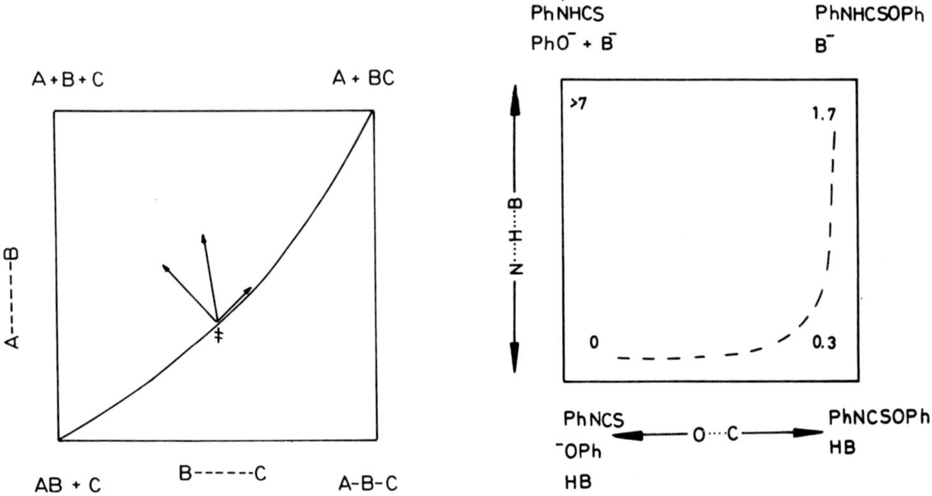

Fig. 22. Energy surface for the reaction A + BC → AB + C (Eqn. 109).

Fig. 23. Free-energy diagram for reaction of phenylisothiocyanate with phenolate anion in the presence of acid (HB) pK 7. The figures refer to $-\log K$ for the equilibrium constants for formation from isothiocyanate [18]. Reaction path is indicated by the dashed line.

at $\beta_C = 0$, α and 10. These equations are guesses which will reproduce a curve close to what might be expected for an energy surface in the region of interest. It is stressed that the surface will be smooth in the important regions and discontinuities will be absent except at the boundaries. The surfaces do not account explicitly for any changes in interatomic distances and angles not involved in the reaction and it is assumed that these will adjust to their most stable values on the surface [100]. The approach using three-dimensional surfaces was first applied by More O'Ferrall [101] to elimination reactions and the diagrams are usually referred to as *More O'Ferrall* diagrams.

We can use More O'Ferrall diagrams to account for the absence of general acid catalysis of phenolate ion addition to isothiocyanates in the presence of acid of pK 7. The numbers at the 4 corners of the diagram (Fig. 23) are logarithms of equilibrium constants related to the ground state and indicate that the surface is "skewed" by the high energy of the top left corner. The reaction path is forced along the bottom and right coordinates and thus involves no general acid catalysis.

Reaction of nucleophiles with carbonyl groups catalysed by general acid illustrates the vectorial combination of parallel and perpendicular effects (Eqn. 113).

$$\mathrm{N} \quad \mathrm{{>}C{=}O} \quad \mathrm{H{-}A} \longrightarrow \mathrm{N{-}\overset{|}{\underset{|}{C}}{-}OH} + \mathrm{A^-} \qquad (113)$$

Increasing the nucleophile reactivity (p$K^{NH}\uparrow$) effectively decreases α [93] in accord with expectation; the alternative mechanism involving general base catalysis (Eqn. 114) yields a More O'Ferrall diagram

$$A^- \curvearrowright H{-}N \curvearrowright \; >C{=}O \curvearrowright \;\rightarrow\; N{-}C{-}OH \;+\; A^- \qquad (114)$$

where increasing nucleophilicity raises the top left corner (AH $\bar{N}$ $>$C=O) and essentially only affects the transition state perpendicularly (to reaction coordinate); this causes a decrease in α. The observation of the decrease in α (which is essentially an *increase* in β) gives us a useful tool for diagnosing the ambiguous kinetic rate law for Eqns. 113 and 114 [93].

7.2. Reactivity–selectivity

It is the view of some authors that the "failure" of the Hammond and reactivity–selectivity postulates invalidates these and the use of linear free-energy relationships in mechanistic studies. We would be left with very little indeed to probe transition-state structures especially in solution if we excluded free-energy correlations. Indeed it is our view that free-energy correlations are derived from the transition-state/ground-state structure differences and the difficulty lies in our interpretation. Our view is therefore that free-energy relationships must indicate transition-state structure and that any deviations from the key postulates are particularly significant. From the above discussions it is apparent that we can apply Hammond-type arguments to segments of a reaction mechanism broken down from the whole.

It should be stressed here that the key postulates depend for their successful simple application on a single-bond cleavage associated with minor alterations in bond angles and bond lengths of bonds not involved in the major change. Potential energies should be considered rather than free energy and enthalpy is not necessarily equivalent to the former. Solvation should also be considered in the structures [102]. A classical example of an apparent failure of the reactivity–selectivity postulate is the correlation of nucleophilic reactivity against carbenium ions [103] where Eqn. 115 is valid over a reactivity range of about 10^6-fold in the rate constant for water reaction.

$$\log k_n/k_{H_2O} = N_+ \qquad (115)$$

Recently Pross [104] advanced a theory based on solvation effects which accommodates this apparent breakdown. More advanced texts should be consulted for detailed discussion of this problem which is the focus of much current thought [102–106].

8. Estimation of ionisation constants

The value of linear free-energy relationships in estimating pK values is beyond question; the method is especially useful in mechanistic studies where it is often

TABLE 26
Some Hammett relationships for ionisation constants

Acid	ρ	pK_0
$ArCO_2H$ (water)	1.00	4.20
$ArCO_2H$ (40% EtOH)	1.67	4.87
$ArCO_2H$ (70% EtOH)	1.74	6.17
$ArCO_2H$ (EtOH)	1.96	7.21
$ArCO_2H$ (MeOH)	1.54	6.51
o-$HOArCO_2H$	1.10	4.00
$ArCH_2CO_2H$	0.47	4.30
$ArCH_2CH_2CO_2H$	0.22	4.55
trans-$ArCH{=}CHCO_2H$	0.54	4.45
$ArCO_2H$ (50% butylcellosolve)	1.42	5.63
p-$ArC_6H_4CO_2H$ (50% butylcellosolve)	0.48	5.64
$ArC{\equiv}CCO_2H$ (35% dioxane)	0.80	3.26
$ArOCH_2CO_2H$	0.30	3.17
$ArSCH_2CO_2H$	0.32	3.38
$ArSeCH_2CO_2H$	0.35	3.75
$ArSOCH_2CO_2H$	0.17	2.73
$ArSO_2CH_2CO_2H$	0.25	2.51
$ArB(OH)_2$ (25% EtOH)	2.15	9.70
$ArPO(OH)_2$	0.76	1.84
$ArPO_2OH^-$	0.95	6.96
$ArAsO(OH)_2$	1.05	3.54
$ArSeO_2H$	0.90	4.74
α-$ArCH{=}NOH$	0.86	10.70
ArOH	2.23	9.92
$ArCOCH_2PPh_3^+$ (80% EtOH)	2.4	6.0
$ArCOCH_2CONHPh$ (water/dioxan)	0.79	9.4
$ArCOCH_2COMe$	1.72	8.53
ArSH (48.9% EtOH)	2.24	7.67
$ArNH_3^+$	2.89	4.58
$ArCH_2NH_3^+$	1.06	9.39
$ArNHNH_3^+$	1.17	5.19
ArNHPh	4.07	22.4
$ArNH_2$ (ammonia)	5.3	–
$ArC(OH){=}CHOMe$	1.10	8.24
$ArCOCH_2\overset{+}{S}(Me)Ph$	2.0	7.32
$PhCOCH_2\overset{+}{S}(Me)Ar$	1.4	7.32
$ArCH_2NO_2$	0.83	6.88
$ArCH_2NO_2$ (50% MeOH)	1.22	7.93
$ArCH(Me)NO_2$	1.03	7.39
$ArCH(Me)NO_2$ (50% dioxan)	1.62	10.30
$ArCH(NO_2)_2$	1.47	3.89
$ArCH_2CH(Me)NO_2$ (50% MeOH)	0.40	9.13
Fluorenes (water/DMSO)	6.3	22.1

TABLE 26 (continued)

Acid	ρ	pK_0
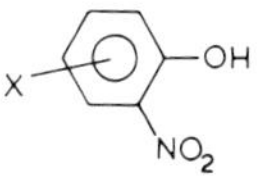 X —OH, NO$_2$	2.16	6.89
$ArSO_2NH_2$	1.06	10.0
$ArSO_2NHPh$	1.16	8.31
$PhSO_2NHAr$	1.74	8.31
$Ar\overset{+}{C}(OH)CH_3$	2.6	−6.0
$ArC(OH)_2^+$	1.2	−7.26
$ArNMe_2$	3.46	5.13
Pyridines	5.9	5.25
Quinolines [b]	5.9	4.90
Iso quinoline [c]	5.9	5.4
α-Naphthylamines [b]	2.81	3.9
β-Naphthylamines [d]	2.81	4.35

[a] Except where stated solvent is water.
[b] 2, 3 and 4 substituted.
[c] 1, 3 and 4 substituted.
[d] 3 and 4 substituted.

necessary to have an accurate idea of the pK of an intermediate which is too unstable to study explicitly. Brønsted, Taft, Taft–Pavelich, Charton and simple Hammett approaches have been used to correlate ionisation constants and Tables 26 and 27 summarise the data for some representative compounds.

The pK of the hydrate of formaldehyde is a useful example; this species is clearly unstable in water and simple titration techniques do not allow a measurement to be made. An indirect method [107] yields a value of 13.7. Reference to Table 27 indicates that the pK of RCH_2OH is governed by the equation $pK = 15.9 - 1.42\sigma^*$; allowing for the change in σ^* from H to OH and a statistical factor (0.3) as there are two hydroxyls leads to $pK = 14.4 - 1.42\sigma^*$. Substitution of σ^*_H gives $pK = 13.29$ in good agreement with experimental findings.

Fox and Jencks [108a] provide a useful set of examples for estimating the pK of the ammonium ion $R_1R_2C(OH)NH\overset{+}{R}_3R_4$ found as an intermediate in many acyl group transfer reactions. These authors use a standard acid of known pK and progress to the unknown using changes in σ_I and the ρ_I value. This approach usually gives reliable results as the standard can be as close to the acid in question as is possible. Table 27 indicates that primary, secondary and tertiary ammonium ions have a ρ^* of -3.2 ($\rho_I = -20$) for $X\overset{+}{N}HR_1R_2$. Allowing a factor of 2.5 on the transmission coefficient for the carbon leads to a value of $\rho_I = -8$ for $XCR_3R_3\overset{+}{N}HR_1R_2$. Starting with standard $Me\overset{+}{N}H_2OMe$ (pK 4.75) and correcting for OH ($-8.0 \times 0.25 = -2$) gives 2.75 for the pK of $HO—CH_2–\overset{+}{N}H_2OMe$. A start

TABLE 27
Some Taft relationships for ionisation constants [a]

Acid	ρ^*	pK_0
RNH_3^+	−3.14	13.23
$RR_1NH_2^+$	−3.23	12.13
$RR'R''NH^+$	−3.3	9.61
RPH_3^+	−2.64	2.46
$RR_1PH_2^+$	−2.61	5.13
$RR'R''PH^+$	−2.67	7.85
RCO_2H	−1.7	4.66
RCH_2CO_2H	−0.73	5.16
RCH_2OH	−1.42	15.9
RSH	−3.5	10.22
RCH_2SH	−1.47	10.54
RO^+H_2	−2.36	−2.18

[a] In the case of multi-substitution $\Sigma\sigma^*$ is employed.

with $Me_2\overset{+}{N}H_2$ (pK 10.7) and substitution of CH_3O for CH_3 $((-0.05+0.25)\times 20)$ and OH for H (8.0×0.25) gives a pK of 2.7. Substituting CH_2OH for H in $MeO\overset{+}{N}H_3$ (pK 4.6) alters the pK by (20×0.09) to give 2.8. These three independent calculations yield essentially identical values giving confidence in the resulting pK. It is clearly good practice especially when calculating inaccessible pKs to use as many routes as possible. Clark, Barlin and Perrin [108b,109] give some useful examples of pK estimation of both acids and bases.

9. *Elucidation of mechanism*

9.1. *Mechanistic identity*

A prerequisite that the mechanism in a series should be the same is a linear free-energy correlation; if a member of the series takes a different pathway then the alternative mechanism will be more favourable, will predominate over the normal and will be diagnosed as a *positive* deviant point. An example of this behaviour is the demonstration that general acid catalysis by primary amines of the iodination of acetone differs from that by carboxylic acids (Fig. 24) [3a]; simple proton transfer is identified with the carboxylic acid catalysis and it is thought that amines utilise Schiff's base formation (Eqn. 116) leading to a more favourable path.

$$H_2O\ \ H{-}CH_2{-}C(CH_3)O\ \ HOCOR \longrightarrow CH_2{=}C(CH_3)OH \xrightarrow{I_2} \text{Product}$$

$$\text{Acetone} + RNH_2 \rightleftharpoons (CH_3)_2C{=}\overset{+}{N}HR \longrightarrow CH_2{=}C(CH_3)NHR \xrightarrow{I_2} \text{Product} \qquad (116)$$

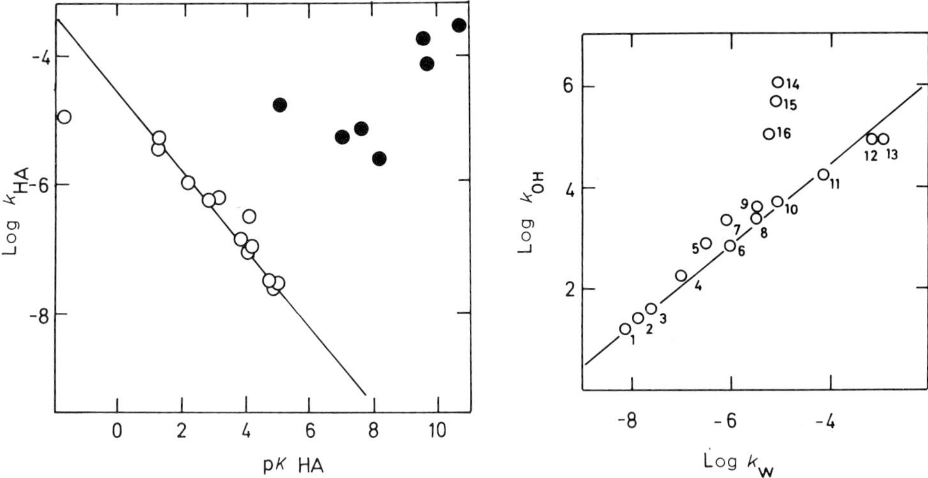

Fig. 24. Acid-catalysed enolisation of acetone: ○, carboxylic acids, pyridinium ions and oxonium ion; ●, primary and secondary amines [3].

Fig. 25. Correlation between hydroxide and water catalysis of hydrolysis of 2-nitrophenyl esters ($RCO\text{-}O_2NP$) R. 1, Et; 2, Me; 3, $PhCH_2$; 4, $EtSCH_2$; 5, CO_2^-; 6, $PhOCH_2$; 7, $Me_3NCH_2^+$; 8, $Me_2C(CN)$; 9, $BrCH_2$; 10, $ClCH_2$; 11, $C_5H_5NCH_2^+$; 12, EtOCO; 13, Cl_2CH; 14, $EtOCOCH_2$; 15, $NCCH_2$; 16, $Me_2SCH_2^+$. Data from Holmquist and Bruice [110].

The existence of a linear free energy correlation is *prima facie* evidence for a single mechanism within the range of substituents. It should be emphasised that the worker should be on the lookout for an *apparent* linearity due to the choice of a short range of substituents close to a break point.

(a) Correlations with model reactions

Holmquist and Bruice [110] correlated the alkaline hydrolysis of 2-nitrophenyl substituted acetates versus the water rate constant for the hydrolysis (Fig. 25). The slope of the excellent correlation is just less than unity indicating that the mechanisms have similar electronic and steric requirements. However, 2-nitrophenyl ethyl malonate, cyanoacetate and dimethylsulphonioacetate esters react faster with hydroxide ion than is expected from the relationship and are therefore judged to pass through different mechanisms (Eqns. 117 and 118).

Normal

$$RCO{-}ONP \xrightarrow{HO^-} R{-}\underset{OH}{\overset{O^-}{C}}{-}ONP \longrightarrow \text{Product} \qquad (117)$$

Accelerated

$$NC{-}CH_2CO{-}ONP \overset{HO^-}{\rightleftharpoons} NC{-}\overset{-}{C}HCOONP \longrightarrow NCCH{=}C{=}O \longrightarrow \text{Product} \qquad (118)$$

The deacylation of aryl-α-chymotrypsin provides a further example of this approach [111a]. A plot of log k_3, for the deacylation versus log k_{OH} for the hydrolysis of RCOOEt is essentially linear indicating electronic and steric requirements for the two mechanisms as might be expected. Three acyl functions deviate

$C_6H_5-CH_2-CH_2CO-$, $HO-C_6H_4-CH_2CH_2-CO-$ and (3-indolyl)$-CH_2CH_2CO-$

markedly in a positive direction (Fig. 26). This deviation is not explained in terms of mechanistic deviation but by an interaction of the acyl group side chain with a site on the enzyme architecture (ρ_2 site) effectively holding the carbonyl function in an orientation where it is most easily attacked by the catalytic groups of the active site. This "freezing" of the motion in a favourable orientation is not available to the smaller side chains and does not occur in the model. As studies are extended from simple models such effects will be apparent in non-enzymatic reactions. The above approach to the chymotrypsin deacylation reaction is probably the most effective way. Other workers have fitted the deacylation rate constants to Pavelich–Taft parameters and Fife and Milstien [111b] obtained Eqn. 119 for a series of aliphatic acyl groups (RCO—). Hexanoyl deviates presumably due to the reasons mentioned above.

$$\log k_3/k_3^0 = 1.90\sigma^* + 0.96\,E_s \qquad (119)$$

Acylation of chymotrypsin with 4-nitrophenyl esters (RCOONP) has a Pavelich–Taft

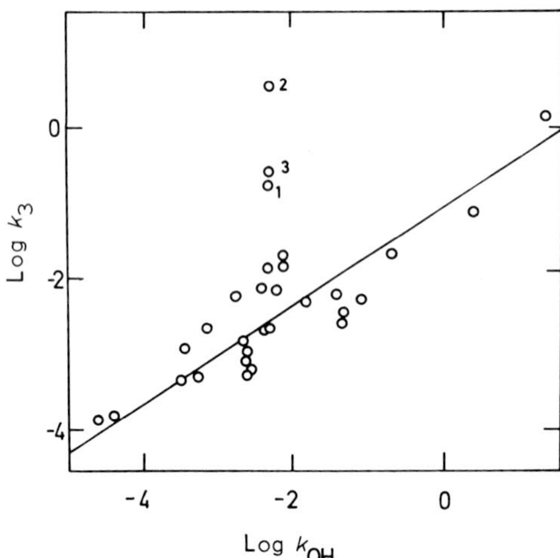

Fig. 26. Dependence of hydrolysis of acyl-α-chymotrypsins (k_3) on the reaction of hydroxide ion with the corresponding ethyl esters [111a]. Deviant points: 1, 3-phenylpropionyl; 2, 3-(4′-hydroxyphenyl)propionyl; 3, 3-(3′-indolyl)propionyl.

δ value of 0.95; there is little electronic effect due to the acyl function for aliphatic R groups and again hexanoate deviates positively [112,113].

9.2. Change in mechanism

If the equation for two mechanisms (Eqn. 120) accounts for the reaction from A to B the rate constant for the overall reaction is given by Eqn. 121 where k_I and k_{II} are linear functions of say σ.

$$A \underset{II}{\overset{I}{\rightleftarrows}} B \qquad (120)$$

$$\log k/k_0 = \log(k_I + k_{II})/(k_I^0 + k_{II}^0) \qquad (121)$$

The form of this equation is illustrated in Fig. 27; the predominant term of Eqn. 121 will depend on σ and ρ values for I or II. The change in correlation will always be *concave* upwards for a change in mechanism because the reaction flux will go through the most efficient path available and this *requires* an increase in rate constant over that predicted for no change in mechanism.

An interesting example of a change in mechanism diagnosed by non-linear free-energy correlations is from E1cB to BAc2 (Fig. 28) in the alkaline hydrolysis of hippurate esters [114]. Good leaving groups such as phenols stimulate the former mechanism (Eqn. 122) whereas strongly basic leaving groups lead to the latter (Eqn. 123).

$$PhCONHCH_2CO{-}OAr \overset{OH^-}{\rightleftharpoons} PhCO\overset{\ominus}{N}CH_2COOAr \longrightarrow \text{(oxazolone: Ph, N, O, =O)} \longrightarrow \text{Products} \qquad (122)$$

$$PhCONHCH_2COOR \longrightarrow PhCONHCH_2{-}\overset{O^-}{\underset{OH}{C}}{-}OR \longrightarrow \text{Products} \qquad (123)$$

A trap for the unwary is not to investigate fully the cause of the "U-shaped" non-linear free energy relationship. Young and Jencks [105c] studied the reaction of hydroxide with acetophenone bisulphites (Eqn. 124)

$$Ar{-}\overset{OH}{\underset{Me}{C}}{-}SO_3^- \xrightarrow{OH^-} ArCOCH_3 + SO_3^{2-} + H_2O \qquad (124)$$

and observed a non-linear free energy correlation (Fig. 29). Recognition that resonance interactions will also play a part in the Hammett correlation (Yukawa–Tsuno) leads to an equation (Eqn. 125) giving a *linear* correlation.

$$\log k/k_0 = \rho\sigma^n + \rho^r(\sigma^+ - \sigma^n) \qquad (125)$$

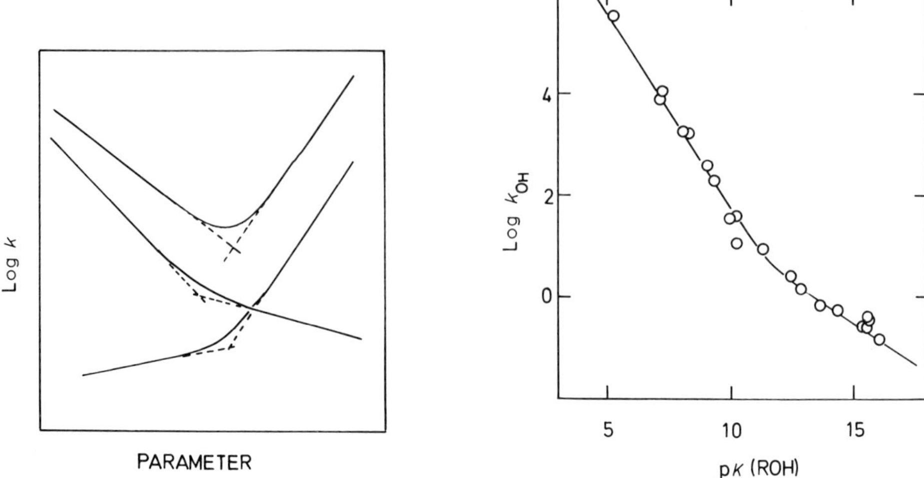

Fig. 27. Examples of free-energy correlations involving changes in mechanism ("U-shaped" or concave upwards).

Fig. 28. Brønsted plot of log k_{OH} versus pK of the leaving hydroxyl group in the hydrolysis of hippuric acid esters ($PhCONHCH_2COOR$). Data from A. Williams (1975) J. Chem. Soc. (Perkin 2) 947.

Application of a similar treatment to the well known cases of benzyl halide solvolyses where a change in mechanism is presumed to account for a "U-shaped" correlation indicates that these reactions may in fact have a single, common, mechanism [114].

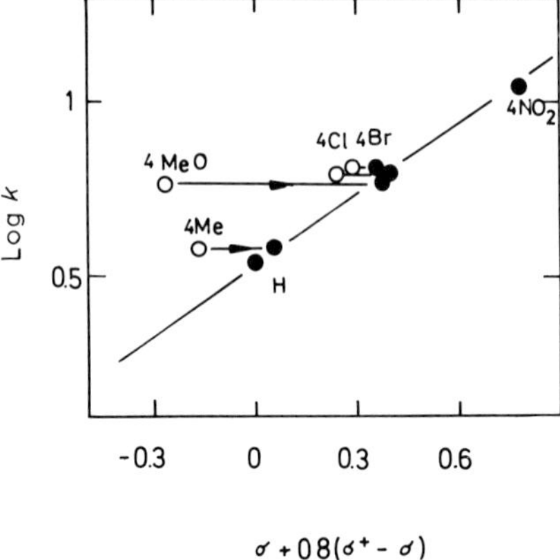

Fig. 29. Decomposition of acetophenone–bisulphite complexes [105c]. Inequality of inductive and resonance interactions causes a "U-shaped" free energy correlation in the simple Hammett approach. Yukawa–Tsuno type correction of the mesomerically donating groups is illustrated by the horizontal lines leading to a linear plot.

9.3. Change in rate limiting step

Eqn. 126 is that for the simplest conservative reaction; the overall rate constant is given by Eqn. 127.

$$\mathrm{A} \underset{k_{-1}}{\overset{k_1}{\rightleftarrows}} \mathrm{B} \overset{k_2}{\rightarrow} \mathrm{C} \tag{126}$$

$$k = k_1 k_2/(k_{-1} + k_2) \tag{127}$$

If a change in rate-limiting step occurs during variation of a substituent then a non-linear free-energy relationship will occur; the separate limbs of the correlation will obey equations of the type (Eqn. 128)

$$\log k/k_0 = \rho_1 \sigma \qquad (k_1 = \text{r.d.s.}) \tag{128}$$

$$\log k/k_0 = (\rho_1 + \rho_2 - \rho_{-1})\sigma \qquad (k_2 = \text{r.d.s.})$$

The non-linear correlation will always be "convex upwards" or "hump shaped" as illustrated in Fig. 30 because the change in rate-determining step causes a "slowing" of the reaction below that predicted in the absence of a change. An excellent example of substituent effect causing a change in rate-determining step is the cyclodehydration of 2-phenyl triarylcarbinols (Fig. 30) [115]. A possible mechanism involves pre-equilibrium protonation of the alcohol followed by elimination to yield carbenium ion (R^+, Eqn. 129). At low values of σ k_3 is rate limiting ($\rho = 2.67$) and at high σ k_2 is rate-limiting ($\rho = -2.51$).

$$\mathrm{ROH} + \mathrm{H}^+ \rightleftharpoons \mathrm{R\overset{+}{O}H_2} \underset{k_{-2}}{\overset{k_2}{\rightleftharpoons}} \mathrm{R}^+ \overset{k_3}{\rightarrow} \text{product} \tag{129}$$

The reaction of isocyanic acid with amines exhibits a change in rate-limiting step as diagnosed by a "hill-shaped" free-energy correlation (Fig. 31). The overall mechanism for this reaction (Eqn. 130) [116] involves rate-limiting proton transfer for the weakly basic amines and addition for the strongly basic amines.

$$\mathrm{RNH_2} + \mathrm{HNCO} \underset{k_{-1}}{\overset{k_1}{\rightleftharpoons}} \mathrm{R\overset{+}{N}H_2{-}C(=O)(\ominus)NH} \rightleftharpoons \mathrm{RNH{-}C(=O)(\ominus)NH} \text{ or } \mathrm{R\overset{+}{N}H_2{-}C(=O)NH_2} \rightleftharpoons \mathrm{RNHCONH_2} \tag{130}$$

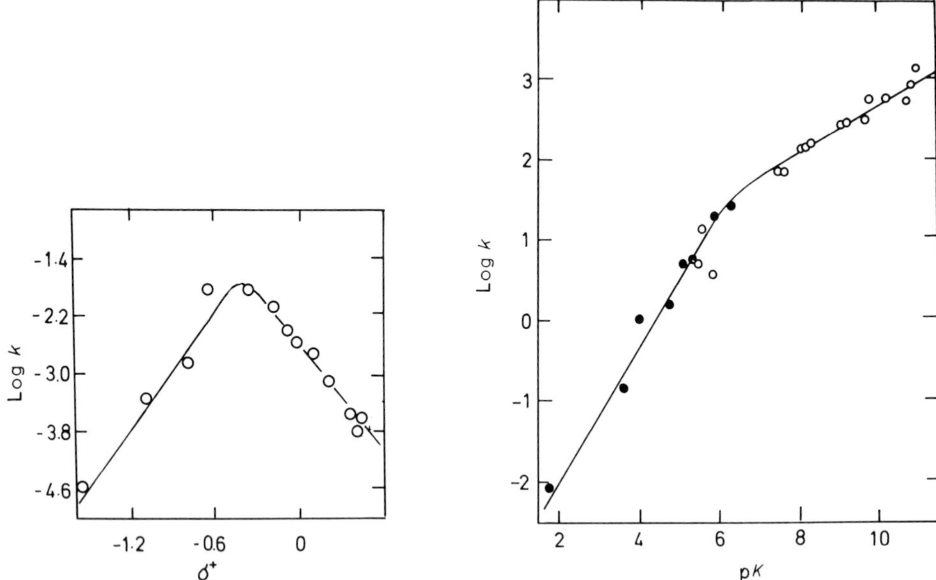

Fig. 30. Change in rate-limiting step illustrated for the cyclodehydration of 2-phenyltriarylcarbinols [115].

Fig. 31. Brønsted relationship for attack of primary amines (○, aliphatic; ●, aromatic) on isocyanic acid [116].

The cause of the change in rate-limiting step is the variation of k_{-1} with amine structure.

A problem which occasionally arises in non-linear free-energy correlations is whether an observed rate constant *lower* than that extrapolated for a given series is *always* the result of a change in mechanism (Fig. 32). The low rate constant could result from a change in *mechanism* provided a change in rate-limiting step had occurred prior to the value of the parameter corresponding to the deviant point (Fig. 32).

9.4. Dependence on concentration as a free energy correlation

Solvent effects are essentially free-energy correlations [57] and we shall omit in this chapter those of strongly acidic media as these are clearly not pertinent to biochemistry. The effect of variation of solvent on a rate or equilibrium constant may be treated in the same way as a substituent effect. We may apply the same criteria to elucidate mechanistic or rate-determining step changes to free-energy relationships with solvent effects. Let us consider the hydrolysis of 5-phenyl-oxazoline-2-ones

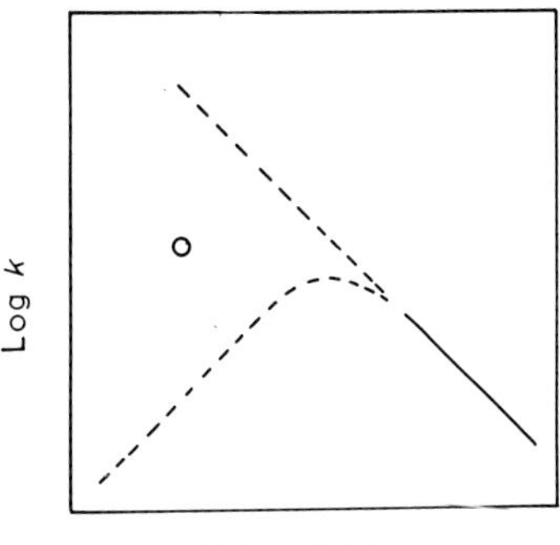

Fig. 32. Rate constant *lower* than predicted from an observed correlation does not *require* a change in rate-limiting step. The full line is the observed correlation.

(Fig. 33) where the reaction (Eqn. 131) has 3 zones [19] and is analyzed as follows:

$$\text{Ph-(oxazolin-5-one)} \longrightarrow \text{PhCONHCH}_2\text{CO}_2\text{H} \qquad (131)$$

the transitions A → B and B → C are "U-shaped" (concave upwards) and are thus effectively changes in mechanism. The "convex upwards" break in C is effectively a change in rate-limiting step and the mechanism involves an intermediate species; the break corresponds to either a change in rate-limiting step or the formation of a stoichiometric amount of conjugate base.

Changing buffer concentration can also change a rate-limiting step in consecutive

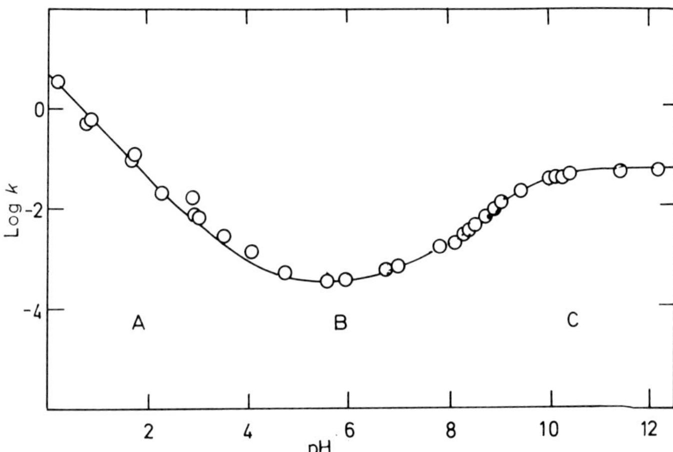

Fig. 33. Dependence on pH of the hydrolysis of 2-(4′-methylphenyl)oxazolin-5-one.

reactions. There are many examples where the change occurs at low concentration and Fig. 34 illustrates the results for the aminolysis of 4-nitrophenyl phenylmethanesulphonate by diethylamine (Eqn. 132) [117].

$$\mathrm{PhCH_2SO_2{-}O4NP} \underset{\mathrm{BH^+}}{\overset{\mathrm{B}}{\rightleftharpoons}} \mathrm{Ph\overline{C}HSO_2{-}O4NP} \xrightarrow{k_2} \text{sulphene} \xrightarrow{\mathrm{RNH_2}} \text{product} \qquad (132)$$

Many buffer curvature effects occur at relatively high molarities and in these cases extensive confirmatory tests must be carried out to make sure that a change in rate-limiting step is being observed and not a simple solvent effect [116,118,119]. Strictly the logarithm of the concentration must be employed in order to conform to a free energy. Mechanistic changes caused by concentration effects may also be detected in this way (Fig. 35); in the example the rate law is given by Eqn. 133 and

$$k = k_c[\mathrm{MeNH_2}]^2 + k_c'[\mathrm{MeNH_2}] + k_c'' \qquad (133)$$

the linear free-energy relationship possesses 3 sectors corresponding to different mechanisms. A good example of this phenomenon is the methylaminolysis of phenyl acetate [120a].

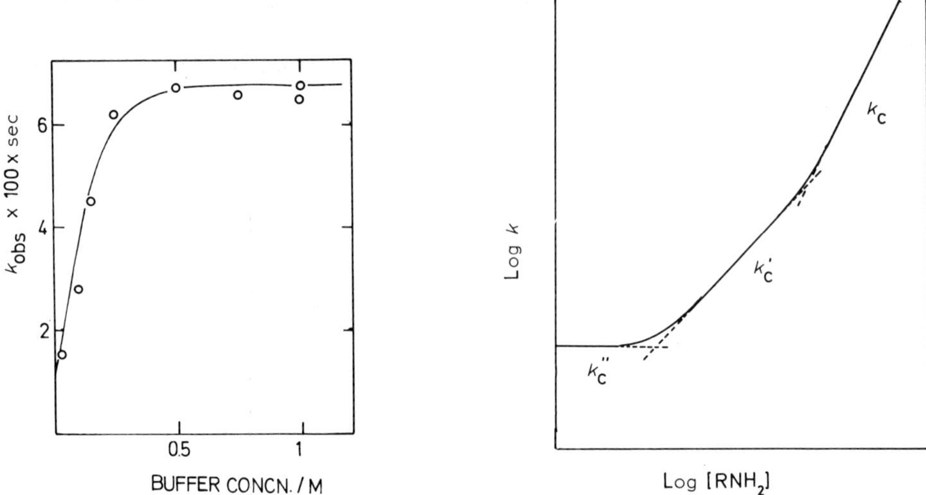

Fig. 34. Release of 4-nitrophenol from 4-nitrophenyl phenylmethanesulphonate in diethylamine buffer (fraction base 0.5) [117]. The intercept is calculated from data at zero buffer concentration.

Fig. 35. Change in mechanism caused by an increase in concentration (Eqn. 133) in the methylaminolysis of phenyl acetate [120a].

9.5. *Distinction between kinetic ambiguities*

A frequently occurring problem in determining mechanisms is kinetic ambiguity which arises because of the lability of the proton. For example specific acid-catalysed hydrolysis of amides has the rate law (Eqn. 134) which could arise from either oxygen (Eqn. 135) or nitrogen (Eqn. 136) protonation.

$$\text{Rate} = k_H[H^+][\text{amide}] \tag{134}$$

$$RCONHR'_2 \rightleftharpoons R\text{—}C(=\overset{+}{O}H)NR'_2 \longrightarrow \text{Products} \tag{135}$$

$$RCONR'_2 \rightleftharpoons R\text{—}C(=O)\overset{+}{N}HR'_2 \longrightarrow \text{Products} \tag{136}$$

A common method for differentiating between ambiguities due to labile protons is to substitute methyl for the proton. Assuming that no other factor is altered by the introduction of the methyl group then the comparison of rate constants with the models (methyl imidate in Eqn. 135 and acyl ammonium in Eqn. 136) should lead to the correct solution. This method has two major disadvantages [120b]: steric effect of the methyl as opposed to the hydrogen is different and the solvation of ground versus transition state will be different because the hydrogen will of necessity have a different "acidity" in each level while the methyl group will not be solvated. In order to obtain accurate models linear free-energy relationships must be involved in estimating these quantities. Owing to the errors involved in such estimates the lower level for a difference in rate constants consistent with a different mechanism is usually set at about 2 orders of magnitude.

9.6. *Proton transfer*

Free-energy correlations provide one of the most important approaches to the detailed study of proton transfer in transient intermediates. The formation of ureas from isocyanates and aromatic amines involves the transient formation of a zwitterionic intermediate (Eqn. 130) [121] whose decomposition is catalysed by both acids (HA) and bases (B). The third order rate constants k_{HA} and k_B fit non-linear Brønsted correlations (Figs. 36 and 37) for the reaction of aniline with isocyanic acid in the presence of acids and bases. The classical "Eigen type" pattern indicates a mechanism where proton transfer to or from a general base or acid becomes rate limiting at the pK of the donor or conjugate acid of acceptor ($Ph\overset{+}{N}H_2\text{—}CO\overset{-}{N}H$ or $Ph\overset{+}{N}H_2CONH_2$). For the mechanism to be consistent we need to calculate the pK of these acids using free-energy relationships in order to compare with the observed pK values. The acceptor has a calculated pK of 12 ± 2 [121] compared with the "weak"

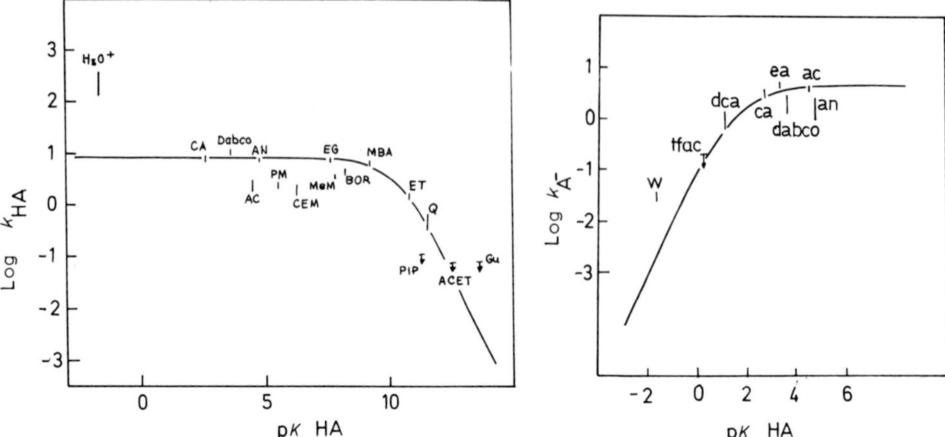

Fig. 36. General acid-catalysed aminolysis of isocyanic acid [121]; Eigen type curvature consistent with diffusion limiting proton transfer. CA, chloroacetic acid; Dabco, 1,4-diazabicyclo-(2,2,2)-octane; AC, acetic acid; AN, anilinium ion; PM, *N*-propargylmorpholinium ion; CEM, 2-chloroethylmorpholinium ion, MeM, *N*-methylmorpholinium ion; EG, ethyl glycinate; BOR, boric acid; MBA, methyl β-alaninate; ET, ethylammonium ion; Q, quinuclidinium ion; PIP, piperidinium ion; ACET, acetamidinium ion; Gu, guanidium ion.

Fig. 37. General base-catalysed aminolysis of isocyanic acid [121]. w, water; tfac, trifluoroacetic acid; dca, dichloroacetic acid; ea, ethoxyacetic acid; other identities are given in the legend to Fig. 36.

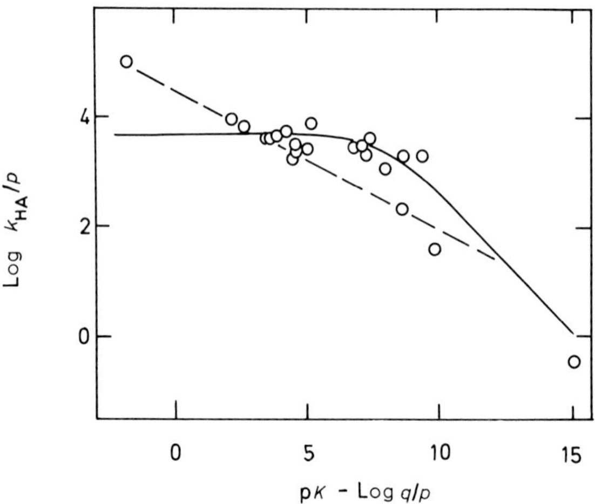

Fig. 38. Attack of semicarbazide on 4-chlorobenzaldehyde catalysed by acids (k_{HA}); the data can be interpreted as a poor linear Brønsted correlation or as a classical Eigen-type correlation with pK 9 [122].

value of 10.2 for acid catalysis; the donor has a calculated value 1 ± 2 comparing with the observed 1.8. A further piece of information consistent with the above conclusion is that proton catalysis is some 20–30-fold above the plateau line for general acid donors in accord with the greater mobility of this ion [122].

Barnett [122] has discussed Eigen type Brønsted plots as diagnostic tools for certain types of proton-transfer mechanisms. It is interesting that owing to current thought at the time 'sharp' Eigen curvature has been overlooked in a number of cases. An exceptionally good example of this phenomenon is acid-catalysed semicarbazone formation (Fig. 38) [93]; the third-order acid catalysis (k_{HA}) plot could fit either a poor linear correlation or a good "Eigen" curve with the proton above the plateau line. The criterion for choosing the Eigen mechanism is the existence of an intermediate with pK 9 [122]; estimation of pK values is clearly a prerequisite to the elucidation of such correlations.

References

1 Moore, W.J. (1962) Physical Organic Chemistry, 4th Edn., Prentice Hall, London, p. 296.
2 Fong, F.K. (1976) Accts. Chem. Res. 9, 433.
3a Bender, M.L. and Williams, A. (1966) J. Am. Chem. Soc. 88, 2502.
3b Bell, R.P. and Lidwell, O.M. (1940) Proc. Roy. Soc. (London) A176, 88.
3c Feather, J.A. and Gold, V. (1965) J. Chem. Soc. 1752.
4 Bell, R.P. (1973) The Proton in Chemistry, 2nd Edn., Chapman and Hall, London.
5 Moroney, M.J. (1956) Facts from Figures, Penguin.
6 Snedecor, G.W. (1946) Statistical Methods, 4th Edn., Iowa State College Press, Ames.
7a Brønsted, J.N. (1928) Chem. Rev. 5, 23.
7b Benson, S.W. (1958) J. Am. Chem. Soc. 80, 5151.
7c Schlag, E.W. (1963) J. Chem. Phys. 38, 2480.
7d Schlag, E.W. and Haller, G.L. (1965) ibid. 42, 584.
7e Bishop, D.M. and Laidler, K.J. (1965) ibid. 42, 1688.
7f Murrell, J.N. and Laidler, K.J. (1968) Trans. Far. soc. 64, 371.
7g Gold, V. (1964) ibid., 60, 739.
7h Pollak, E. and Pechukas, P. (1978) J. Am. Chem. Soc. 100, 2984.
8 Jencks, W.P. and Carriuolo, J. (1961) ibid. 83, 1743.
9 Knowles, J.R., Norman, R.O.C. and Prosser, J.H. (1961) Proc. Chem. Soc. 341.
10 Jencks, W.P. (1971) Cold Spring Harbor Symposia on Quantitative Biology, 26, 1.
11 Tommila, E. and Hinshelwood, C.N. (1938) J. Chem. Soc. 1801.
12 Pople, J.A. and Beveridge, D.L. (1970) Approximate Molecular Orbital Theory, McGraw-Hill, New York.
13 Alborz, M. and Douglas, K.T. (1980) J. Chem. Soc. Chem. Commun. 728.
14 Hupe, D.J. and Jencks, W.P. (1977) J. Am. Chem. Soc. 99, 451.
15 Deacon, T., Farrar, C.R., Sikkel, B.J. and Williams, A. (1978) ibid. 100, 2525.
16a Al-Rawi, H. and Williams, A. (1977) ibid. 99, 2671.
16b Leffler, J.E. (1953) Science 117, 340.
17 Sauers, C.K., Jencks, W.P. and Groh, S. (1975) J. Am. Chem. Soc. 97, 5546.
18 Hill, S.V., Thea, S. and Williams, A. (1981) unpublished work.
19 Curran, T.C., Farrar, C.R., Niazy, O. and Williams, A. (1980) J. Am. Chem. Soc. 102, 6829.
20 Brønsted, J.N. and Pedersen, K. (1924) Z. Phys. Chem. 108, 185.
21 Eigen, M. (1964) Angew. Chem. Int. Ed. 3, 1.

22a Kresge, A.J. (1975) Accts. Chem. Res. 8, 354.
22b Walters, E.A. and Long, F.A. (1969) J. Am. Chem. Soc. 91, 3733.
22c Hibbert, F., Long, F.A. and Walters, E.A. (1971) ibid. 93, 2829.
22d Hibbert, F. and Long, F.A. (1972) ibid. 94, 2647.
23 Marcus, R.A. (1956) J. Chem. Phys. 24, 966; (1960) Disc. Far. Soc. 29, 21; (1963) J. Phys. Chem. 67, 853, 2889; (1964) Annu. Rev. Phys. Chem. 15, 155; (1965) J. Chem. Phys. 43, 679.
24 Marcus, R.A. (1968) J. Phys. Chem. 72, 891 (1969) J. Am. Chem. Soc., 91, 7224; Cohen, A.O. and Marcus, R.A. (1968) J. Phys. Chem. 72, 4269.
25 Bell, R.P. (1976) J. Chem. Soc. (Far. 2) 72, 2088.
26 Kresge, A.J. (1973) Chem. Soc. Rev. 2, 475.
27a Kreevoy, M.M. and Konasiewich, D.E. (1971) Adv. Chem. Phys. 21, 241.
27b Kreevoy, M.M. and Oh, S.W. (1973) J. Am. Chem. Soc. 95, 4805.
28 Lewis, E.S., Shen, C.C. and More O'Ferrall, R.A. (1981) J. Chem. Soc. (Perkin 2), 10.
29a Bordwell, F.G., Boyle, W.J. and Hautala, J.A. (1969) J. Am. Chem. Soc. 91, 4002.
29b Bordwell, F.G. and Boyle, W.J. (1975) ibid. 97, 3447.
29c Bordwell, F.G. (1975) Far. Soc. Symp. 10, 100.
30 Jones, J.R. and Patel, S.P. (1974) J. Am. Chem. Soc. 96, 574.
31a Hammett, L.P. and Pfluger, H.L. (1933) ibid. 55, 571.
31b Hammett, L.P. (1935) Chem. Rev. 17, 125.
32 Burkhardt, G.N. (1935) Nature (London) 136, 684.
33 Williams, A. (1972) J. Chem. Soc. (Perkin 2), 808.
34a Williams, A. and Naylor, R.A. (1971) J. Chem. Soc. (B) 1967.
34b Ryan, J.J. and Humffray, A.A. (1966) ibid. 842.
35 Bruice, T.C. and Mayahi, M.F. (1960) J. Am. Chem. Soc. 82, 3067.
36 Williams, A. and Douglas, K.T. (1972) J. Chem. Soc. (Perkin 2) 1454.
37 Williams, A. and Douglas, K.T. (1974) ibid. 1727.
38 Humffray, A.A. and Ryan, J.J. (1969) J. Chem. Soc. (B) 1138.
39 Thea, S., Harun, M.G. and Williams, A. (1979) J. Chem. Soc. Chem. Commun. 717.
40 Douglas, K.T. and Williams, A. (1976) J. Chem. Soc. (Perkin 2), 515.
41 Stewart, R. and Yates, K. (1960) J. Am. Chem. Soc. 82, 4059.
42a Jaffe, H.H. (1953) Science 118, 246.
42b Jaffe, H.H. (1954) J. Am. Chem. Soc. 76, 4261.
43 Fersht, A.R. and Kirby, A.J. (1967) ibid. 89, 4853.
44 Wollenberg, R.H., Nimitz, J.S. and Gokcek, D.Y. (1980) Tetrahedron Lett. 2791.
45 Behinpour, K., Hopkins, A.R. and Williams, A. (1981) ibid. 279.
46 Bromilow, R.H. and Kirby, A.J. (1972) J. Chem. Soc. (Perkin 2), 149.
47 Graafland, T., Wagenaar, A., Kirby, A.J. and Engberts, J.B.F.N. (1979) J. Am. Chem. Soc. 101, 6981.
48a Yukawa, Y. and Tsuno, Y. (1959) Bull. Chem. Soc. Jpn. 32, 971.
48b Yukawa, Y., Tsuno, Y. and Sawada, M. (1966) ibid. 39, 2274.
49 Roberts, J.D. and Moreland, W.T. (1953) J. Am. Chem. Soc. 75, 2167.
50a Grob, C.A. and Schlageter, M.G. (1976) Helv. Chim. Acta 59, 264.
50b Grob, C.A. (1976) Angew. Chem. Int. Ed. 15, 569.
51 Taft, R.W. and Lewis, I.C. (1958) J. Am. Chem. Soc. 80, 2436.
52 Charton, M. (1964) J. Org. Chem. 29, 1222.
53a Taft, R.W. (1953) J. Am. Chem. Soc. 75, 4231.
53b Taft, R.W. (1956) in: M.S. Newman (Ed.), Steric Effects in Organic Chemistry, Wiley, New York, Ch. 13.
54 Ingold, C.K. (1930) J. Chem. Soc. 1032.
55 Pavelich, W.A. and Taft, R.W. (1957) J. Am. Chem. Soc. 79, 4935.
56 Wells, P.R. (1968) Linear Free Energy Relationships, Academic Press, New York, p. 38.
57 Grunwald, E. and Leffler, J.E. (1963) Rates and Equilibria of Organic Reactions, Wiley, New York.
58 McGowan, J.C. (1960) J. Appl. Chem. 10, 312.

59 Hancock, C.K., Meyers, E.A. and Yager, B.J. (1961) J. Am. Chem. Soc. 83, 4211.
60 Chapman, N.B. and Shorter, J. (1972) Advances in Free Energy Relationships, Plenum, New York, p. 95.
61a Charton, M. (1975) J. Am. Chem. Soc. 97, 1552.
61b Charton, M. (1973) Progr. Phys. Org. Chem. 10, 81.
62a Charton, M. (1975) J. Am. Chem. Soc. 97, 3691.
62b Charton, M. (1975) ibid. 97, 3694, 6159, 6472.
62c Charton, M. (1976) J. Org. Chem. 41, 2906, 2217.
62d Charton, M. and Charton, B. (1978) ibid. 43, 1161, 2383.
62e Charton, M. (1977) ibid. 42, 3531, 3535.
63 DeTar, D.F. (1980) J. Am. Chem. Soc. 102, 7988.
64 Charton, M. (1979) J. Org. Chem. 44, 903.
65 Fujita, T., Iwasa, J. and Hansch, C. (1964) J. Am. Chem. Soc. 86, 5175.
66 Hánsch, C. (1969) Accts. Chem. Res. 2, 232.
67a Miles, J.L., Robinson, D.A. and Canady, W.J. (1963) J. Biol. Chem. 238, 2932.
67b Hymes, A.J., Robinson, D.A. and Canady, W.J. (1965) ibid. 240, 134.
68 Hansch, C. and Coats, E. (1970) J. Pharm. Sci. 59, 731.
69 Blow, D.M. (1976) Accts. Chem. Res. 9, 145.
70 Hein, G.E. and Niemann, C. (1962) J. Am. Chem. Soc. 84, 4495.
71 Hansch, C., Deutsch, E.W. and Smith, R.N. (1965) ibid. 87, 2738.
72 Knowles, J.R. (1965) J. Theoret. Biol. 9, 213.
73 Pauling, L. and Pressman, D. (1945) J. Am. Chem. Soc. 67, 1003.
74 Williams, A., Lucas, E.C. and Rimmer, A.R. (1972) J. Chem. Soc. (Perkin 2), 621.
75 Hansch, C. and Calef, D.F. (1976) J. Org. Chem. 41, 1240.
76 Hansch, C., Leo, A. and Nikaitani, D. (1972) ibid. 37, 3090.
77 Leo, A., Hansch, C. and Elkins, D. (1971) Chem. Rev. 71, 525.
78a Hetnarski, B. and O'Brien, R.D. (1973) Biochemistry 12, 3883.
78b Charton, M. (1966) J. Org. Chem. 31, 2991.
79a Hansch, C. (1978) ibid. 43, 4889.
79b Murakami, Y., Aoyama, Y., Kida, M. and Nakano, A. (1977) Bull. Soc. Chem. Jpn. 50, 3365.
80 Grunwald, E. and Winstein, S. (1948) J. Am. Chem. Soc. 70, 846.
81 Scott, F.L., Flynn, E.J. and Fenton, D.F. (1971) J. Chem. Soc. (B) 277.
82 Kosower, E.M. (1958) J. Am. Chem. Soc. 80, 3253.
83 Reichert, C. (1965) Angew. Chem. Int. Ed. 4, 29.
84 Hille, M.B. and Koshland, D.E. (1967) J. Am. Chem. Soc. 89, 5945.
85 Sigman, D.S. and Blout, E.R. (1967) ibid. 89, 1747.
86 Kallos, J. and Avatis, K. (1966) Biochemistry 5, 1979.
87 Swain, C.G. and Scott, C.B. (1953) J. Am. Chem. Soc. 75, 141.
88 Edwards, J.O. and Pearson, R.G. (1962) ibid. 84, 16.
89 Edwards, J.O. (1954) ibid. 76, 1540; (1956) ibid. 78, 1819.
90a Hudson, R.F. (1962) Chimia 16, 173.
90b Hudson, R.F. (1965) Structure and Mechanism in Organophosphorus Chemistry, Academic Press, New York.
91 Johnson, C.D. (1973) The Hammett Equation, Cambridge.
92 Hammond, G.S. (1955) J. Am. Chem. Soc. 77, 334.
93 Cordes, E.H. and Jencks, W.P. (1962) ibid. 84, 4319.
94a Miller, S.I. (1959) ibid. 81, 101.
94b Grunwald, E. and Berkowitz, B.J. (1951) ibid. 73, 4939.
94c Gutbezahl, B. and Grunwald, E. (1953) ibid. 75, 559.
94d Taft, R.W. (1953) ibid. 75, 4231.
94e Hine, J. (1959) ibid. 81, 1126.
94f Ritchie, C.D., Saltiel, J.D. and Lewis, E.S. (1961) ibid. 83, 4601.

95 Thornton, E.R. (1967) ibid. 89, 2915.
96 Harris, J.C. and Kurz, J.L. (1970) ibid. 92, 349.
97 Jencks, D.A. and Jencks, W.P. (1977) J. Am. Chem. Soc. 99, 7948.
98 Critchlow, J.E. (1972) Trans. Far. Soc. 1774.
99 Dunn, B.M. (1974) Int. J. Chem. Kinetics 6, 143.
100 Jencks, W.P. (1972) Chem. Rev. 72, 705.
101 More O'Ferrall, R.A. (1970) J. Chem. Soc. (B) 274.
102 Arnett, E.M. and Reich, R. (1980) J. Am. Chem. Soc. 102, 5892.
103 Ritchie, C.D. (1972) Accts. Chem. Res. 5, 348.
104 Pross, A. (1975) Tetrahedron Lett. 1289; (1976) J. Am. Chem. Soc. 98, 776.
105a Pross, A. (1977) Adv. Phys. Org. Chem. 14, 69.
105b Buncel, E. and Chuaque, C. (1980) J. Org. Chem. 45, 2825.
105c Young, P.R. and Jencks, W.P. (1979) J. Am. Chem. Soc. 101, 3288.
105d Giesse, B. (1977) Angew. Chem. Int. Ed. 16, 125.
105e Farcasiu, D. (1975) J. Chem. Ed. 52, 76.
105f Agmon, N. (1981) Int. J. Chem. Kinetics 13, 333.
105g Poh, B.L. (1979) Aust. J. Chem. 32, 429.
105h McLennan, D.J. and Martin, P.L. (1979) ibid. 32, 2361.
106a Johnson, C.D. (1980) Tetrahedron 36, 3461.
106b Mclennan, D.J. (1978) ibid. 34, 2331.
107 Bell, R.P. and McTigue, P.T. (1960) J. Chem. Soc. 2983.
108a Fox, J.P. and Jencks, W.P. (1974) J. Am. Chem. Soc. 96, 1436.
108b Clark, J. and Perrin, D.D. (1964) Quart. Rev. Chem. Soc. 18, 295.
109 Barlin, G.B. and Perrin, D.D. (1966) ibid. 20, 75.
110 Holmquist, B. and Bruice, T.C. (1969) J. Am. Chem. Soc. 91, 2982, 2993, 3003.
111a Williams, A. and Salvadori, G. (1971) J. Chem. Soc. (B) 2401.
111b Fife, T.H. and Milstien, J.B. (1967) Biochemistry 6, 2901.
112a Milstien, J.B. and Fife, T.H. (1969) ibid. 8, 625.
112b Enriquez, P.M. and Gerig, J.T. (1969) ibid. 8, 3156.
112c Dupaix, A., Bechet, J.J. and Roucous, C. (1970) Bioch. Biophys. Res. Commun. 41.
113 Dupaix, A., Bechet, J.J. and Roucous, C. (1973) Biochemistry 12, 2559.
114 Williams, A. (1975) J. Chem. Soc. (Perkin 2) 947.
115 Hart, H. and Sedor, E.A. (1967) J. Am. Chem. Soc. 89, 2342.
116 Williams, A. and Jencks, W.P. (1974) J. Chem. Soc. (Perkin 2) 1753.
117 Davy, M.B., Douglas, K.T., Loran, J.S., Steltner, A. and Williams, A. (1977) J. Am. Chem. Soc. 99, 1196.
118 Thea, S., Kashefi-Naini, N. and Williams, A. (1981) J. Chem. Soc. (Perkin 2) 65.
119 Hand, E.S. and Jencks, W.P. (1975) J. Am. Chem. Soc. 97, 6221.
120a Jencks, W.P. and Gilchrist, M. (1966) ibid. 88, 104.
120b Williams, A. (1976) ibid. 98, 5645.
121 Williams, A. and Jencks, W.P. (1974) J. Chem. Soc. (Perkin 2) 1760.
122 Barnett, R.E. (1973) Accts. Chem. Res. 6, 41.

General references

Advances in Physical Organic Chemistry and Progress in Physical Organic Chemistry contain numerous articles related to the theory and practice of free energy correlations. The following references give general reading in the areas indicated.

General

1 Chapman, N.B. and Shorter, J. (1978) Correlation Analysis in Chemistry, Plenum, New York.
2 Leffler, J.E. and Grunwald, E. (1963) Rates and Equilibria of Organic Reactions, Wiley, New York.
3 Hine, J. (1975) Structural Effects on Equilibria in Organic Chemistry, Wiley, New York.
4 Chapman, N.B. and Shorter, J. (1972) Advances in Linear Free Energy Relationships, Plenum, New York.
5 Shorter, J. (1973) Correlation Analysis in Organic Chemistry, Oxford.

Brønsted relationship

6 Kresge, A.J. (1973) Chem. Soc. Rev. 2, 475.
7 Kresge, A.J. (1975) Accts. Chem. Res. 8, 354.
8 Caldin, E.F. and Gold, V. (1975) Proton Transfer Reactions, Chapman and Hall, London.
9 Jencks, W.P. (1969) Catalysis in Chemistry and Enzymology, McGraw-Hill, New York.
10 Bell, R.P. (1973) The Proton in Chemistry, 2nd Edn., Chapman and Hall, London.
11 Eigen, M. (1964) Angew. Chem. Int. Ed. 3, 1.

Hammett relationship

12 Johnson, C.D. (1973) The Hammett Equation, Cambridge.
13 Wells, P.R. (1968) Linear Free Energy Relationships, Academic Press, New York.
14 Jafffe, H.H. (1953) Chem. Rev. 53, 191.

Separation of inductive, resonance and steric effects

15 Taft, R.W. (1956) in: M.S. Newman (Ed.), Steric Effects in Organic Chemistry, Wiley, New York, ch. 13.

CHAPTER 6

Isotopes in the diagnosis of mechanism

ANDREW WILLIAMS

University Chemical Laboratories, Canterbury, Kent, Great Britain

1. Theoretical background

Although the detailed theoretical interpretation of primary isotope effects is a problem of great complexity and considerable uncertainty [1–3], many of the important aspects of the problem may be treated in a simplified and approximate manner with a precision which is adequate for most experimental work and which provides at least a qualitative insight into the mechanism of the observed effects.

In order to discuss the isotope effect we must make some fundamental assumptions regarding the passage from ground to product state. The potential energy surface is assumed to be independent of isotopic substitution in accord with the quantum-mechanical tenet that electronic and nuclear motions are not connected. We must further assume that the transition state theory is correct; in particular the assumption that ground and transition state are in equilibrium (the basic assumption of the transition-state theory) is very important. As we know, the equilibrium assumption is not strictly valid since the lifetime of the transition state may not be sufficient to allow all the vibrations to equilibrate. However, sophisticated theoretical treatments [4] have shown that the equations of the transition-state theory are substantially correct but the logic of their derivation according to the Eyring approach is wrong.

The transition-state theory relates the rate constant (k_1) for a reaction with the 'equilibrium constant' ($K_1^\ddagger$) for the equilibrium between reactants and the transition states (Eqn. 1) [5].

$$k_1 = K'(k \cdot T/h) K_1^\ddagger \tag{1}$$

$$\mathrm{B} + \mathrm{A} \quad \underset{}{\overset{K_1^\ddagger}{\rightleftharpoons}} \quad \mathrm{A}^\ddagger \quad \rightarrow \quad \text{products} \tag{2}$$

The K' is a transmission coefficient which expresses the fraction of transition-state 'species' going on to products relative to those returning to reactant and is usually considered, for no very good reason, to be insensitive to isotopic substitution. Boltzmann's constant, absolute temperature and Plancks constant are k, T and h respectively. The isotope effect is simply the ratio of the two rate constants (Eqn. 3).

$$k_1^{\mathrm{H}}/k_1^{\mathrm{D}} = K_1^{\ddagger\mathrm{H}}/K_1^{\ddagger\mathrm{D}} \tag{3}$$

Michael I. Page (Ed.), The Chemistry of Enzyme Action

The 'species' $A^{\ddagger}$ is not a molecule and $K_1^{\ddagger}$ is not a normal equilibrium constant. The transition state has one vibration less than a normal molecule ($3n - 7$ instead of $3n - 6$); the missing vibration mode is motion along the reaction coordinate.

The 'equilibrium constants' $K_1^{\ddagger}$ are not measurable and we must resort to statistical thermodynamics to estimate these values theoretically. The partition function (Q) is a quantity with no simple physical significance but it may be substituted for concentrations in the calculation of equilibrium constants (Eqns. 4 and 5) [5]. (It is assumed that there is no isotopic substitution in B.) Partition functions may be expressed as the product of contributions to the total energy from translational, rotational and vibrational motion (Eqn. 6).

$$K_1 = [A^{\ddagger}]/[A][B] = Q^{\ddagger}/Q_A \cdot Q_B \tag{4}$$

$$k_1^H/k_1^D = \left(Q_A^D/Q_A^H\right) \cdot \left({}^{H}Q^{\ddagger}/{}^{D}Q^{\ddagger}\right) \tag{5}$$

$$Q = Q_T \cdot Q_R \cdot Q_V \tag{6}$$

Rotational and translational contributions are essentially independent of the isotopic substitution as they refer to the molecule as a whole and the mass change on isotopic substitution is negligible. There will be one vibrational contribution for each mode and these will depend on isotopic substitution (Eqn. 7).

$$\left(Q^D/Q^H\right)_{vib} = \exp\left[\mathrm{h}(\nu_H - \nu_D)/2\mathrm{kT}\right] \frac{1 - \exp(-\mathrm{h}\nu_H/\mathrm{kT})}{1 - \exp(-\mathrm{h}\nu_D/\mathrm{kT})} \tag{7}$$

Since most of the vibrations are for non-substituted bonds considerable cancellation is possible; if the substitution is only in the vibrational mode which is destroyed in the transition state then Eqn. 7 reduces to Eqn. 8; the term $(1 - \exp(-h\nu_H/kT))/(1 - \exp(-h\nu_D/kT))$ is approximately unity.

$${}^{H}k_1/{}^{D}k_1 = \exp\left(h(\nu_H - \nu_D)/2kT\right) \tag{8}$$

In principle we may calculate the isotope effect provided we know the fundamental vibrational frequencies of reactant and transition state. Some of these frequencies are available from Raman or infrared spectra but with the advent of high-speed computers these may now be calculated.

We may divide isotope effects into the primary isotope effect where a bond to the isotopically substituted atom is made or broken in the course of the reaction and secondary isotope effects where the bonds to these atoms are not made or broken.

In order to obtain a pictorial idea of the origin of isotope effects we shall now consider potential-energy surfaces for reactions. The potential energy of the bond C–H varies with the length according to the Morse curve (Fig. 1). If the molecule

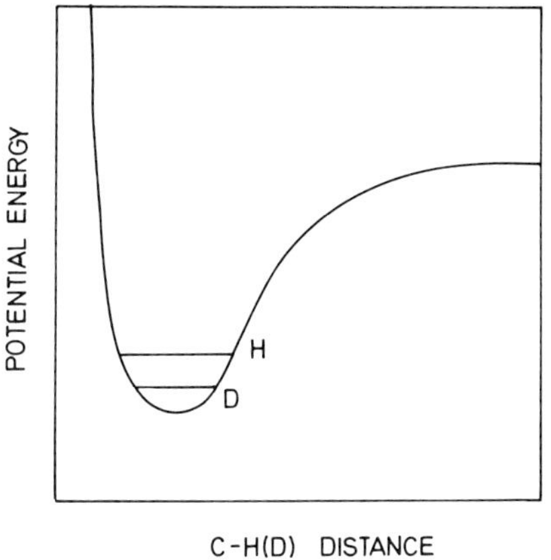

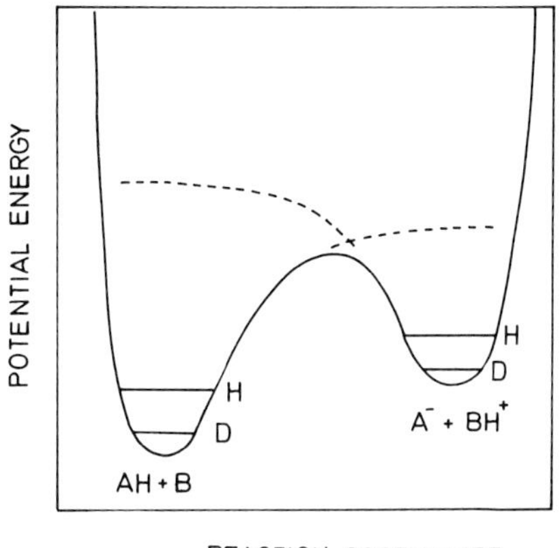

Fig. 1. Morse curve for the 'dissociation' of the bond C–H indicating zero-point energies for C–H and C–D.

Fig. 2. Combination of two Morse curves to produce a potential-energy curve for the transfer of a proton $AH + B \rightarrow A^- + HB^+$.

were isotopically labelled with D the potential-energy diagram would be identical but in the ground state the isotopically labelled species would have a different energy from the unlabelled owing to its zero-point energy. The species with the higher mass isotope will have the lower energy ($\frac{1}{2}h\nu$) because by Hooke's law $\nu \propto 1/\text{mass}^{1/2}$.

If the change of isotope has no effect on other vibrations the isotope effect is due to the difference in zero-point energies of the species in the ground state; the vibration in the transition state is of course lost. Equilibrium isotope effects result from the difference between the difference in zero-point energies in ground and product states. The energy diagram for the transfer of proton from AH to B (Fig. 2) can be considered to be composed of two Morse curves set end on end.

2. *Measurement of isotope effects*

The obvious way of measuring an isotope effect is to determine the rates of reaction of each pure isotopic species and is widely used to obtain primary and secondary deuterium effects. The use of the method depends on the availability of the pure isotopic species and on a relatively large difference in rate constants.

The first condition is usually met in deuterium compounds but the heavier isotopes such as ^{13}C, ^{15}N, and ^{18}O are more difficult to obtain pure and are consequently very expensive. It is also impractical to prepare isotopically pure compounds containing tritium or ^{14}C as apart from their expense and possible hazard these compounds would decompose from their own radiation.

The second condition limits us to deuterium (and tritium in principle) because a rate-determining proton transfer will have k_H/k_D at about 7 and a secondary effect of some 10–15%. Rate constants are seldom more accurate than 3–5% routinely and with good work to better than 1 or 2% so that isotope effects for the heavier atoms (^{13}C etc.) will be impossible to measure this way. There are reactions where the method for the rate constant allows accuracy of some 0.1% or better but these are not universal and in any case the effort involved in such work is extremely large.

Competition methods are employed for ratios of less than 5–10%; when two isotopic species react at different rate a mixture of the two will change composition (except at 100% reaction) and yield a product having a different isotopic composition from the reactant. This will give us a measure of the isotope effect. Radioactive isotopes give specific radiation (counts/sec/mole) proportional to the ratio of radioactive over stable isotope; radioactivity may be measured with a precision better than 1%.

Ratio of stable isotopes may be measured in a mass spectrometer; it is advisable to use relatively simple molecules (for example from degradation of one under study) so that the parent peaks are isolated from other possible ones. For example $^{12}CO_2$ and $^{13}CO_2$ have parent masses at 44 and 45 and the only nearby peaks are those from small amounts of ^{17}O and ^{18}O in the oxygen. A complex organic molecule will normally yield many smaller peaks immediately around the parent due to normal deuterium, ^{13}C and ^{18}O and determination of isotopic enrichments is only possible in uncomplicated mass spectra. Results good to 0.1–0.2% are easily obtained but simultaneous collection of ions from the two peaks of interest using a dual collector instrument gives precision to 0.01%.

The mathematics of the competition method is somewhat complex but part is easy to understand (Eqn. 9).

$$\mathrm{B + AH(D)} \xrightarrow{k_{\mathrm{H(D)}}} \mathrm{BH(D) + A} \tag{9}$$

The ratio [AH]/[AD] will vary as the reaction proceeds and [BH]/[BD] will then be a function of the isotope effect k_H/k_D and the changing [AH]/[AD] ratio. Let the concentration of AH and AD be respectively a_H and a_D and the concentration of B be b then Eqn. 10 holds.

$$\begin{aligned} -\mathrm{d}a_H/\mathrm{d}t &= k_H \cdot a_H \cdot b \\ -\mathrm{d}a_D/\mathrm{d}t &= k_D \cdot a_D \cdot b \end{aligned} \tag{10}$$

Rearranging the ratio of the rate constants and integrating from time zero to time t yields Eqns. 11–13.

$$\mathrm{d}a_H/\mathrm{d}a_D = (k_H \cdot a_H)/(k_D \cdot a_D) \tag{11}$$

$$k_D \cdot \mathrm{d}a_H/a_H = k_H \cdot \mathrm{d}a_D/a_D \tag{12}$$

$$k_H/k_D = \log(a_H/a_H^0)/\log(a_D/a_D^0) \tag{13}$$

Eqn. 13 may not be used directly because our normal methods of measurement give ratios of concentrations for the two isotopic species and not ratios of concentrations for species at time zero and t. Details on the recasting of Eqn. 13 into forms useful for specific situations are given in Melander's book [3]. A particularly simple example involves a very small extent of reaction where P_H and P_D are the concentrations of the two isotopic products at a given time. Eqn. 13 rearranges to Eqn. 14 which reduces to Eqn. 15 because P/a is small.

$$k_H/k_D = \log\left(1 - P_H/a_H^0\right)/\log\left(1 - P_D/a_D^0\right) \tag{14}$$

$$= \left(P_H/P_D\right)/\left(a_H^0/a_D^0\right) \tag{15}$$

3. Equilibrium isotope effects

It should in principle be possible to calculate the isotope effect for an equilibrium since the reactant and product are real compounds. The problem is rather more difficult for reactions in solution but has been solved successfully for a number of reactions in the gas phase. A useful example is the ionisation of acids in water (Eqn. 16).

$$XH(D) + H(D)_2O \rightleftharpoons X^- + H(D)_3O^+ \tag{16}$$

For all measurable dissociation constants XH will be a weaker acid than H_3O^+ and the proton will be bound more tightly; hence the dissociation will involve a decrease in zero point energy. This decrease will be smaller for the deuterated system hence deuteration should lower the tendency to dissociate ($K_H > K_D$). By the same argument K_H/K_D should increase with decreasing acid strength. These predictions are borne out by experiment (Fig. 3) and the slope of the line is of the order of magnitude to be expected in terms of zero-point energies.

If the vibration modes of reactant and product are known the equilibrium-isotope

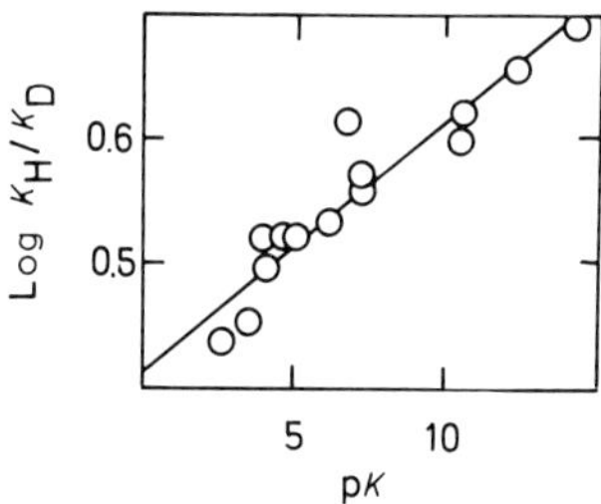

Fig. 3. Plot of log K_H/K_D versus pK. Data from R.P. Bell (1959) The Proton in Chemistry, 1st Edn., Methuen, London, p. 189.

effect should be calculable and Bunton and Shiner [6] considered the equilibrium (Eqn. 17); the frequencies are in general not known exactly but may be estimated from Badger's rule with sufficient accuracy to permit an estimate of their influence on the isotope effect (Table 1). The method predicts the solvent deuterium isotope effect tolerably well provided suitable estimates may be made about the hydrogen bonding in the transition state.

(17)

The lower acidity of deutero acids (in D_2O) compared with proto acids (in H_2O) provides us with a useful tool to diagnose the pre-equilibrium mechanism for acid-catalysed reactions (Eqn. 18).

$$S + H_3O^+ \underset{}{\overset{K_H}{\rightleftharpoons}} HS^+ \xrightarrow{k_2} \text{products} \qquad (18)$$

The isotope effect for this reaction is predicted by Eqn. 19 and since the k_2 step does not involve proton transfer the ratio k_2^H/k_2^D approximates to unity and the isotope effect is inverse because $K_H > k_D$.

$$k^H/k^D = \left(k_2^H/k_2^D\right) \cdot \left(K_D/K_H\right) \qquad (19)$$

Table 2 collects some data for acid-catalysed reactions involving pre-equilibrium protonations. The concerted mechanism for proton transfer ($S + H_3O^+ \rightarrow$ products) will of course lead to a normal isotope effect.

Base-catalysed reactions passing through an anionic species in a pre-equilibrium step (Eqn. 20) should also give inverse deuterium oxide solvent isotope effects.

$$SH + OH^- \overset{K_w/K_H}{\rightleftharpoons} S^- + H_2O \overset{k_2}{\rightarrow} \text{products} \qquad (20)$$

Table 1
Equilibrium isotope effects on the dissociation of acids in water and deuterium oxide [6]

Acid	pK_{obs}	K_{H_2O}/K_{D_2O}	K_{H_2O}/K_{D_2O} (calc.)
Water	15.74	6.5	4.6
β-Trifluoroethanol	12.4	4.4	4.5
Bicarbonate ion	10.25	4.4	3.8
4-Nitrophenol	7.21	3.67	3.3
Acetic acid	4.74	3.33	3.1
Chloroacetic acid	2.76	2.76	2.8

Table 2
Inverse isotope effects in specific acid-catalysed reactions

Reaction	k_H/k_D
Inversion of cane sugar	0.49
Bromination of acetone	0.48
Hydrolysis of acetal	0.38
Hydrolysis of acetamide	0.67

The equation governing the rate of reaction is Eqn. 21a and the isotope effect Eqn. 21b).

$$k = (K_H/K_w) \cdot k_2 \tag{21a}$$

$$k^H/k^D = (K_H/K_D)(k_2^H/k_2^D)(K_w^D/K_w^H) \tag{21b}$$

If we assume that k_2 is independent of the isotopic composition then k^H/k^D is less than unity because $K_H/K_D < K_w^H/K_w^D$ owing to water being the weaker acid. Of course this approach will not work for acids less acidic than water and since the deuteroxide ion in deuterium oxide is a stronger base than hydroxide we are unable to distinguish between the pre-equilibrium mechanism (Eqn. 20) and one involving direct hydroxide attack (Eqn. 22 – see Table 3).

$$SH + OH^- \quad \rightarrow \quad \text{products} \tag{22}$$

Neither of the above approaches should be used on their own to diagnose mechanism; of course an inverse isotope effect is clear evidence against rate-limiting proton transfer.

4. Primary isotope effects

Deuterium isotope effects have been widely used to demonstrate that bonds to hydrogen are broken in the rate-determining step of a reaction. The presence of a

Table 3
Inverse isotope effects in specific base-catalysed reactions

Reaction	k_{OH}/k_{OD}
Decomposition of diacetone alcohol	0.82
Alkaline hydrolysis of 4-nitrophenyl hippurate	0.7
Hydrolysis of 5-nitrocoumaranone	0.65

large effect is unequivocal evidence that this is the case. The maximum effect that can normally be expected for a single bond cleavage may be estimated from the stretching vibration for the X–H bond; since the H or D is attached to a much heavier atom $\nu^{H}/\nu^{D} = 2^{1/2}$ thus the difference in zero-point energies is given by $E_0 = \frac{1}{2}h\nu(1 - 2^{-1/2}) = 0.146h\nu$ (see Table 4) [7]. The calculated values are theoretical limits but recent work has shown that higher values might be obtained in the case of the hydrogen isotopes owing to the 'tunnelling' correction where the proton because of its small size is able to 'quantum-mechanically' tunnel through an energy barrier whereas deuterium with its larger size behaves classically [8]. We shall not deal with this subject further as it has little use at present in diagnosing mechanisms.

One of the oldest examples of the use of the deuterium isotope effect in diagnosis is the bromination of acetone; the large effect (see Table 5) confirms the hypothesis that enolisation is the slow step followed by a fast addition of bromine (Eqn. 23) [9].

$$CD_3COCD_3 \underset{\text{slow}}{\overset{\text{base}}{\rightarrow}} CD_2 = C(OD)CD_3 \underset{\text{fast}}{\overset{Br_2}{\rightarrow}} BrCD_2COCD_3 \qquad (23)$$

Decomposition of the chromate ester of isopropanol likewise involves rate-limiting proton transfer from carbon in the oxidation of isopropanol to acetone in the presence of chromic oxide [10].

Melander [11] found that tritiated toluene nitrated and brominated at the same rate as the unlabelled material arguing against decomposition of the Wheland intermediate as the rate-limiting step (Eqn. 24).

$$ArH(D) + NO_2^+ \rightleftharpoons \overset{+}{Ar}\begin{matrix} NO_2 \\ H(D) \end{matrix} \overset{\text{base}}{\rightarrow} ArNO_2 + BH(D) \qquad (24)$$

However, the intermediate is a high-energy species and its reaction with base might have a transition state where proton transfer has only just commenced (this is a use of the Hammond postulate) [12]; under these circumstances the zero-point energy difference for the two isotopic transition states might be close to that for the isotopic intermediate and a small isotope effect result. This is not strictly true as it assumes that the hydrogen 'vibrates' along the reaction coordinate in the transition state (see later).

A change in rate-limiting step is diagnosed by the primary deuterium isotope

Table 4
Zero-point energies, stretching vibrations and isotope effects for OH, CH and NH fission [7]

Bond	cm^{-1}	E^0 (cal/mole)	k_H/k_D
C–H	2800	1150	6.9
N–H	3100	1270	8.5
O–H	3300	1400	10.6

Table 5
Some primary isotope effects

Reaction	Catalyst	k_H/k_D
Bromination of acetone	Acetate ion	7.0
Bromination of methylacetylacetone	Acetate ion	5.8
Bromination of 2-carbethoxycyclopentanone	Chloroacetate ion	3.7
Bromination of nitromethane	Acetate ion	6.5
Bromination of acetone	Oxonium ion	7.7
Oxidation of isopropanol	Chromic oxide	6.7
Elimination of 2-phenylethyl-2,2-d_2-bromide	Ethoxide ion	7.1
Displacement at dimethylsulphonium bromide	Ethoxide ion	5.7
Nitration of tritiated toluene	Nitronium ion	1.0 [a]

[a] k_H/k_T measured.

effect on the diazo coupling (Eqn. 25) which falls from 6.5 to 3.6 as the pyridine concentration is raised from zero to 0.9 M. Proton transfer is rate-limiting in water but not quite in the high concentration of pyridine [13]. This effect can only be explained by a change in rate-controlling step as a change in reagent concentration will have no affect on zero-point energies.

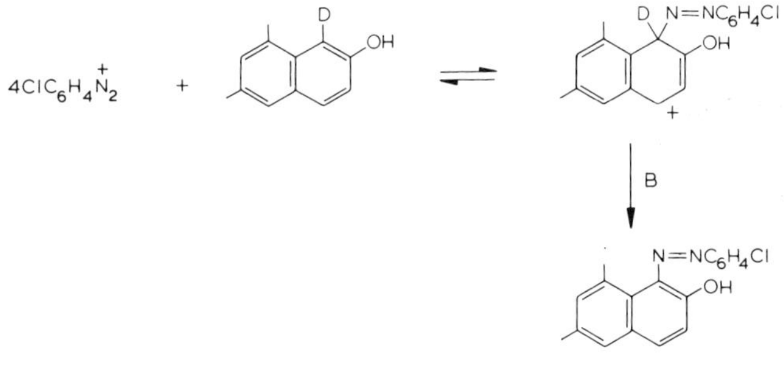

$4ClC_6H_4\overset{+}{N}_2$ + (25)

4.1. Variation in primary isotope effect

Perusal of Table 6 indicates that primary isotope effects vary from about 4 down to quite low values and this prompted considerable study concerning the origin of the variable effect. An early explanation was that the major part of the zero-point energy of the vibration along the reaction coordinate is retained in the transition state. For a linear 3-centre reaction in which hydrogen transfer occurs in the rate-limiting step this explanation is not compatible with transition-state theory because the basic tenet is that this stretching vibration no longer exists where the hydrogen is at the top of a potential barrier and is undergoing translation toward the acceptor (Eqn. 26).

$$\overleftarrow{A} - - - - - \overrightarrow{H} - - - - - \overleftarrow{B} \text{ (asymmetric)} \qquad (26)$$

Table 6
Primary isotope effects for ionisation catalysed by water

Substrate	k_H/k_D
Diethyl malonate	2.0
Nitromethane	3.8
Diacetyl	4.5
Malononitrile	1.5

This imaginary vibration has no potential minimum and therefore has no zero-point energy. Westheimer [2] showed how the hydrogen in the transition state may indeed possess vibrations leading to a diminution in isotope effect.

For a linear transition state we can consider the existence of both asymmetric and symmetric stretching vibrations. Although there is no zero-point energy from an asymmetric vibration (as above) the symmetric vibration (Eqn. 27) can still exist because it does not lead to either products or reactants.

$$\overleftarrow{A} - - - - - H - - - - - \overrightarrow{B} \text{ (symmetric)} \tag{27}$$

Fig. 4 illustrates the symmetric vibration in terms of the three-dimensional potential-energy diagram for the system. If the transition state is symmetrical (that is bond fission is half advanced) there will be no motion of the hydrogen atom in this mode which is perpendicular to the reaction coordinate in Fig. 4 and a normal isotope effect should be observed; there will be no difference in zero point energies in the transition state for hydrogen or deuterium. If the transition state is asymmetric (Eqn. 28) and the hydrogen lies close to either A or B the hydrogen atom may move in the transition state so that the zero-point energies change on substituting deuterium for hydrogen.

$$\overleftarrow{A} - - - \overleftarrow{H} - - - - - \overrightarrow{B} \tag{28}$$

To the extent that the difference in zero-point energy of the reagent is maintained in the transition state by this vibration the observed isotope effect will be reduced from the theoretical maximum [2].

The experimental verification of Westheimer's theory has been studied for the last decade or so and the explanation is now believed to be valid. More O'Ferrall [14] collected data for the value of k_H/k_D for the ionisation of ketones and nitroalkanes; he showed that at $\Delta pK = 0$ (where the pK of the attacking base equals that of the carbon acid) there is a maximum isotope effect. The transition state has a good chance of being symmetrical thus yielding a large isotope effect at $\Delta pK = 0$ where the energy of reactant is the same as that of product. Work from the laboratories of Kresge [15] and Jencks [16] confirms these notions for the case of heteroatoms (Fig. 5).

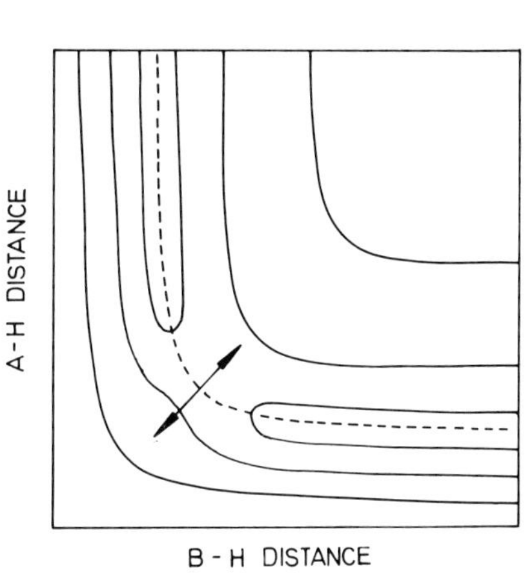

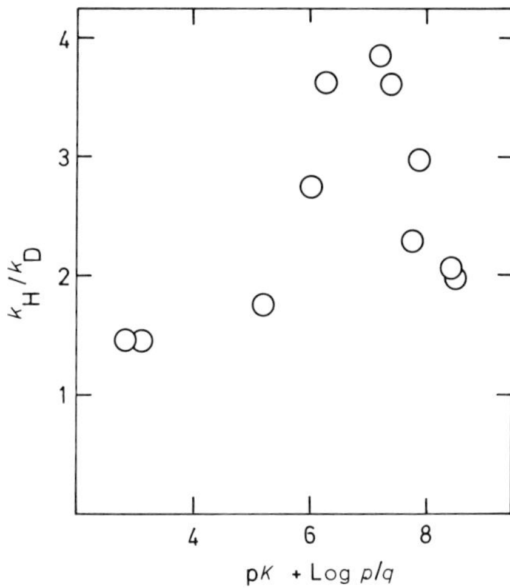

Fig. 4. Three-dimensional potential-energy diagram for the reaction $AH + B \rightarrow A^- + BH^+$. The dotted line represents the reaction coordinate and the double arrow the vibration mode of the hydrogen in the transition state (Eqns. 27 and 28).

Fig. 5. Effect of transition-state symmetry on the isotope effect. The value k_{H_2O}/k_{D_2O} as a function of pK for the reaction of methoxyamine with phenyl acetate catalysed by general acids; the rate constant essentially monitors the reaction $R\overset{+}{N}H_2$—C(OPh)—$O^-\cdot HA \rightarrow R\overset{+}{N}H_2$—C(OPh)—$OH\cdot A^-$ in the reaction complex [16].

4.2. *Solvent isotope effects*

(a) Nucleophilic versus general base catalysis

An often occurring mechanistic problem is the diagnosis of general base or nucleophilic catalysis which give identical kinetics. Imidazole is a well known catalyst for the hydrolysis of 4-nitrophenyl acetate in water and it is known to involve nucleophilic attack because *N*-acetylimidazole has been observed from ultraviolet spectral work [17]. The absence of a solvent deuterium isotope effect confirms the operation of the nucleophilic pathway (Table 7) because a primary isotope effect is expected for the general base mechanism.

general base HN⌒N ⌒ H—O ⌒ C(=O)—O4—NP (29)

nucleophilic HN⌒N ⌒ C(=O)—O 4NP (30)

This method is of use where it is not possible to observe an intermediate for

Table 7
Isotope effects for nucleophilic (n) and general base (g.b.) reactions

Reaction	Mechanism	k_{H_2O}/k_{D_2O}
Methoxyamine + *S*-ethyl-trifluorothioacetate	g.b.	4.4
Imidazole + 4-nitrophenyl acetate	n	1.0
Imidazole + ethyl dichloroacetate	g.b.	3.2
Acetic anhydride + acetate ion	g.b.	1.7
Acetylimidazole + imidazole	g.b.	3.6
Acetoxime acetate + imidazole	g.b.	2.0
Dichloroethyl acetate + imidazole	g.b.	1.9
4-Picoline + 4-nitrophenyl acetate	n	1.12
Methylamine + phenyl acetate	n	1.15
Acetate ion + 2,4-dinitrophenyl acetate	n	1.8
Hydroxylamine + phenyl acetate	n	1.5
Imidazole + phenyl acetate	n	1.8
Imidazole + β-propiolactone	n	1.2
Piperidine + phenyl acetate	n	1.19

This table is taken largely from a compilation in Johnson's review (Johnson, S.L. (1967) Adv. Phys. Org. Chem. 5, 237); persual of the data indicates that care must be taken in interpreting solvent isotope effects in the intermediate range 1.5–2.0.

example where its decomposition is so rapid as to leave only a minute steady-state concentration. Table 7 indicates that the full isotope effect is not reached for general base reactions at heteroatoms and the method is generally regarded as an empirical one where a value of about 2 divides rate-limiting proton transfer from non-proton transfer processes.

The cause of the low primary isotope effects compared with those predicted by Bell (Table 3) for heteroatoms has been a considerable puzzle. The work of Bergman et al. [15] and Cox and Jencks [16] indicates that the isotope effect depends sharply on ΔpK between attacking base and donor acid. Since all the ΔpK values for Table 7 differ markedly from zero it is not surprising that low isotope effects are observed. Of course the pK dependence of the isotope effect imparts considerable uncertainty and a low value (for example 1–1.5) may reflect simply a large pK difference: a large isotope effect (> 2.5) of course requires the proton transfer to be rate-limiting.

(b) Fractionation factors

Solvent isotope effects in mixtures of the isotopically enriched solvent can potentially lead to useful mechanistic data [18,19]. The isotope enrichment of the reactant species may not have the same value as that in the solvent despite completely free exchange and a fractionation factor must be defined (Eqn. 31)

$$\phi = ([SD]/[SH])/([ROD]/[ROH]) \tag{31}$$

where the solute SL is dissolved in a solvent ROL (where L is either D or H). The value ϕ quantifies the preference of the hydrogen site in SL for deuterium over protium relative that for the hydrogen site in ROL. Values of ϕ may be determined from NMR studies for various protonic species relative to water and typical values come between 0.4 and 1.3 [18,19].

For a simple proton transfer reaction (Eqn. 32) the equilibrium isotope effect is given by Eqn. 33.

$$\mathrm{RL} \rightleftharpoons \mathrm{PL} \tag{32}$$

$$K_H/K_D = ([\mathrm{PH}]/[\mathrm{RH}])/([\mathrm{PD}]/[\mathrm{RD}])$$
$$= \frac{[\mathrm{RD}]/[\mathrm{RH}]}{[\mathrm{ROD}]/[\mathrm{ROH}]} \Big/ \frac{[\mathrm{PD}]/[\mathrm{PH}]}{[\mathrm{ROD}]/[\mathrm{ROH}]} = \phi^R/\phi^P \tag{33}$$

In the case of the rate constant the isotope effect is given by Eqn. 34.

$$k_H/k_D = \phi^R/\phi^T \tag{34}$$

The above equations will hold for solvents and solutes with single proton exchangeable sites and for species with multiple sites it is assumed that the occupation of the other sites by D or H will not affect the deuterium preference for a particular site [18]. The equations governing the equilibria and rates are therefore the products of the simple equations (Eqns. 35 and 36).

$$K_H/K_D = \Pi_i^R \phi_i^R / \Pi_j^P \phi_j^P \tag{35}$$

$$k_H/k_D = \Pi_i^R \phi_i^R / \Pi_j^T \phi_j^T \tag{36}$$

The treatment of equilibria and reactions in mixtures of light and heavy solvents was investigated by Gross and Butler [21] who derived the following relationships (Eqns. 37 and 38) where there are n hydrogen sites.

$$K_x = K_0 \Pi_i^n (1 - x + x \cdot \phi_i^P) / \Pi_j^n (1 - x + x \cdot \phi_j^R) \tag{37}$$

$$k_x = k_0 \Pi_i^n (1 - x + x \cdot \phi_i^T) / \Pi_j^n (1 - x + x \cdot \phi_j^R) \tag{38}$$

For the dissociation of a monobasic acid SL in water (L_2O) the isotope effect is given by Eqn. 39.

$$K_H/K_x = (1 - x + x \cdot \phi_{SL}) / (1 - x + x \cdot \phi_{L_3O}) \tag{39}$$

The isotope effect at $x = 1$ $(K_H/K_D = \phi_{SL}/(\phi_{L_3O})^3)$ allows ϕ_{SL} to be computed and ϕ_{L_3O} may be taken from tables (0.69) [18] thus Eqn. 40 holds and the ratio K_H/K_x is not linear in x from $x = 0$ to 1.

The application of isotope effects in solvents with varying isotopic composition involves fitting the observed relationship between k_H/k_x and x to a theoretical equation. This procedure is known as the 'proton inventory' technique [18] as it effectively counts the protons which undergo a change between ground and transition state. The Gross–Butler equation for the reaction has a term $(1 - x + x \cdot \phi)$ for every exchangeable proton in the system and these cancel for those protons which do not change their state.

The simplest relationship between k_H/k_x and x is the linear equation (Eqn. 41) which arises from the involvement of a single proton site in the transition state and

$$k_x/k_H = 1 - x + x \cdot \phi \tag{40}$$

$$= 1 - x + x(k_D/k_H) \tag{41}$$

the fractionation factor ($\phi = k_D/k_H$) may be derived from the value of k_x/k_H at $x = 1$. Unfortunately it is possible to conceive of an apparently linear relationship between k_x/k_H and x arising from a more complicated scheme by fortuitous cancellation of factors [22]. An example of the linear plot is that for the decomposition of dichloroacetylsalicylate ion in water [23]. The Gross–Butler relationship for a possible transition state is Eqn. 42 from the mechanism (Eqn. 43).

$$k_x/k_H = \left(1 - x + x \cdot \phi_1^T\right)\left(1 - x + x \cdot \phi_2^T\right)/\left(1 - x + x \cdot \phi_{H_2O}^R\right)^2 \tag{42}$$

OCOCHCl$_2$; CO_2^- → [O$^{-\delta}$ CHCl$_2$; O ; O—H ; H ; O$^{-\delta}$; O]‡ (43)

The value of ϕ_{H_2O} is unity and ϕ_2 is taken as unity because ϕ undergoes a smooth transition from 1 (ϕ_{H_2O}) to 1(ϕ_{OH^-}). The equation reduces to a linear one where ϕ_1^T may be estimated from k_x/k_H at $x = 1$ ($\phi_1^T = 0.46$).

We should now concern ourselves with experimental errors because the curved plots from alternative formulations might be indistinguishable experimentally from the linear or vice versa. The maximum deviation between the different equations will occur mostly at $x = 0.5$ and we may calculate values of $(k_x/k_H)_{1/2}$ for the various formulations [18]. One proton yields a value 0.73 for the salicylate, two yields 0.704 and three 0.695. These figures are well outside the experimental error so that the observed linearity in the salicylate case is taken to be significant (Fig. 6). Other work indicates that the hydrolysis of salicylate ions involves a single proton transfer.

The approach employed by the Schowen laboratory [18,19] involves plotting k_x/k_H versus x but Albery and Davies [24] utilised the value of the ratio at $x = 0.5$

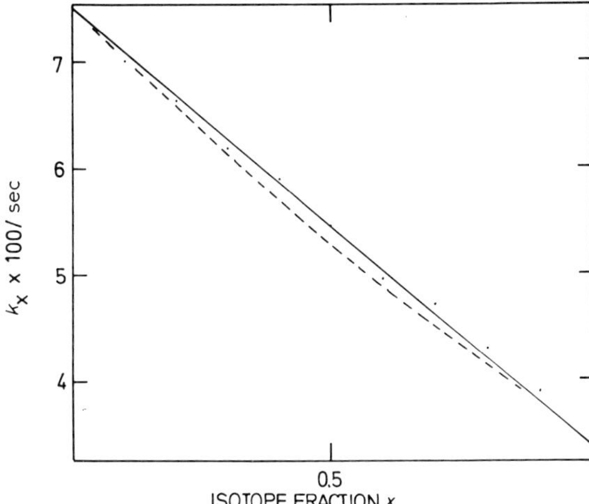

Fig. 6. The hydrolysis of dichloroacetylsalicylate anion in isotopically enriched water (x = fraction of deuterium oxide) [23]. The dashed line is the plot expected for a '2-proton' transition state.

where the maximal effect is expected and the value of the expression (Eqn. 44) is used to diagnose exchangeable protons. At $\gamma = 1$ the k_x/k_H plot is linear and at $\gamma = 0$ there are effectively infinite exchangeable protons in the transition state.

$$\gamma = 8 \ln(k_{1/2}/k_H)/(k_D/k_H)^{1/2}/\ln(k_D/k_H)^2 \tag{44}$$

The inventory technique is clearly a potentially useful tool for elucidating enzymatic transition states. Deacetylation of acetyl-α-chymotrypsin is consistent with a single proton transfer (Eqn. 45) [25] with a fractionation factor 0.42. It is thought that this result excludes the charge-relay mechanism (Eqn. 46) which would require a quadratic term for the dependence.

A similar dependence is observed for the acylation by *N*-acetyl-L-tryptophanamide [26]. Fortuitous cancellation as an explanation of the linearity [22] is effectively excluded by the result with the second substrate. Trypsin and elastase which are very similar enzymes compared with chymotrypsin also exhibit linear plots [26] consistent with the one proton transition state but asparaginase and glutaminase possess curved plots fitting square terms in x [26,27].

SER-195

HN N$^{+\delta}$---H---O(H)---C(O-SER)---O$^{-\delta}$ (45)

SER-195

-$CO_2^{-\delta}$---H---N N---H---O(H)---C(O-SER)---O$^{-\delta}$ (46)

A further interesting series of examples is the hydrolysis of acetyl imidazole which is known to involve a proton transfer from water (attacking) to a base. Hogg and co-workers [28,29] showed that the uncatalysed degradation involved 3 protons which changed their state whereas the imidazole-catalysed reaction had only 1 proton (Eqn. 47).

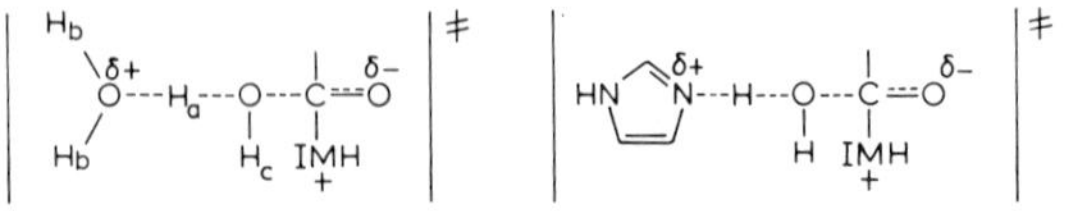

$$\left|\ \overset{H_b}{\underset{H_b}{}}\ \overset{\delta+}{O}{-}{-}{-}H_a{-}{-}{-}\underset{H_c}{O}{-}{-}{-}\underset{\overset{+}{IMH}}{C}{=}\overset{\delta-}{O}\ \right|^{\ddagger}\qquad \left|\ HN\langle\rangle\overset{\delta+}{N}{-}{-}H{-}{-}{-}\underset{H}{O}{-}{-}{-}\underset{\overset{+}{IMH}}{C}{=}\overset{\delta-}{O}\ \right|^{\ddagger} \tag{47}$$

Decomposition of the acetyl imidazole in urea solutions gave linear plots [29] versus x consistent with a single proton and a mechanism consistent with this is Eqn. 48 where the urea acts on the proton from the attacking water.

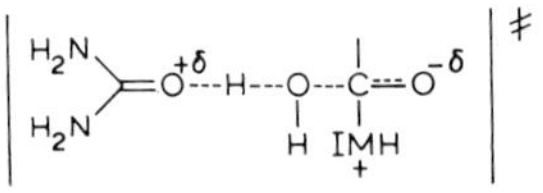

$$\left|\ (H_2N)_2C{=}\overset{+\delta}{O}{-}{-}{-}H{-}{-}{-}\underset{H}{O}{-}{-}{-}\underset{\overset{+}{IMH}}{C}{=}\overset{-\delta}{O}\ \right|^{\ddagger} \tag{48}$$

4.3. Heavy atom isotope effects

Next to the isotope effects of hydrogen the most studied element has been carbon where the effects range as high as 15% with ^{14}C; measurements of radioactivity are inherently of lower precision than mass spectrometric measurements making ^{13}C the preferable isotope. Results from isotope effects on carbon can be most informative as the majority of organic reactions involve carbon bond fission.

Carbon isotope effects can be used in a straightforward way to determine whether a bond cleaves in the rate-limiting step. For example the isotope effect on the decarboxylation of 2,4-dihydroxybenzoic acid [30a] varies with the nature and concentration of acid catalyst suggesting a change in rate-determining step (Eqn. 49).

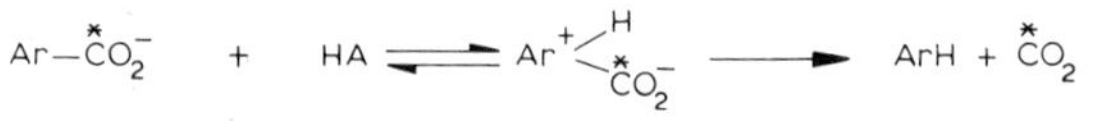

$$Ar{-}\overset{*}{C}O_2^- + HA \rightleftharpoons Ar^{+}\!\!<^{H}_{\overset{*}{C}O_2^-} \longrightarrow ArH + \overset{*}{C}O_2 \tag{49}$$

For a simple bond fission it is doubtful if comparison of the observed isotope effect with the theoretical maximum will give a trustworthy estimate of the position of the transition state; small effects are therefore ambiguous as diagnostic tools and must be coupled with other experiments. Another example involving decarboxylation is the bromine-catalysed decomposition of 3,5-dibromo-4-hydroxy-benzoic acid (Eqn. 50) [30b].

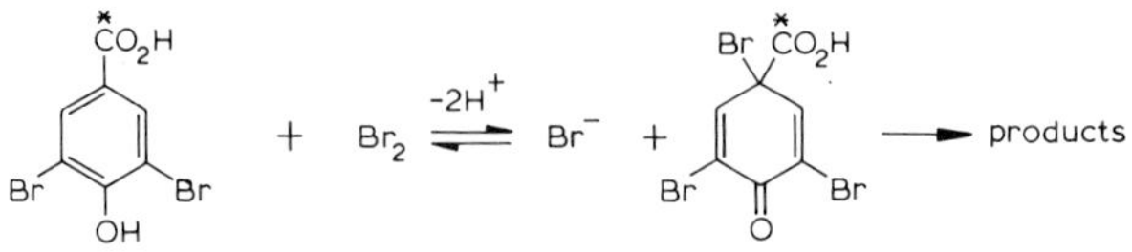

$$\text{3,5-}Br_2\text{-4-HO-}C_6H_2\overset{*}{C}O_2H + Br_2 \underset{}{\overset{-2H^+}{\rightleftharpoons}} Br^- + \text{4-Br-4-}\overset{*}{C}O_2H\text{-2,6-}Br_2\text{-cyclohexadienone} \longrightarrow \text{products} \tag{50}$$

In the absence of added bromide ion the ^{13}C effect is 1.00 ± 0.003 indicating no bond changes in the labelled carbon in the rate-limiting step; in the presence of 0.3 M Br^- the effect rises to 1.045 ± 0.001 as might be expected for full rupture consistent with rate-limiting decomposition of the intermediate (Eqn. 50).

5. *Secondary isotope effects*

As with most other isotope effects the literature on the subject is enormous; the reader is therefore referred to a comprehensive review by Halevi [30c] for further information. When an organic molecule ionises to yield a carbenium ion the adjacent C–H bond (the α-hydrogen) to the ionising atom weakens thus reducing the contribution of vibrational modes involving this bond to the zero-point energy of the transition state. An effect of some 10–15% is observed in an S_N1 reaction (Eqn. 51).

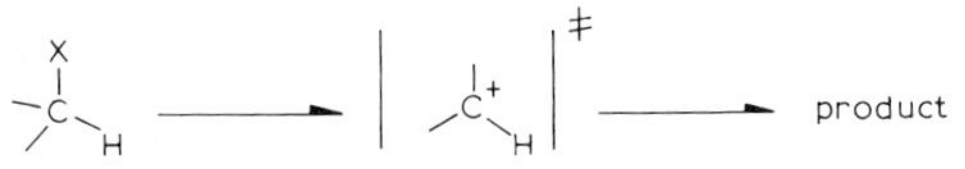

X–C(–)(–)–H ⟶ [>C⁺–H]‡ ⟶ product (51)

It may be argued from analogy with a stable molecule such as acetaldehyde that out of plane vibration of a bond from hydrogen to an sp^3 carbon is of higher frequency than that to an sp^2 carbon [31]. An interesting example of the use of the secondary isotope effect where hydrogen is attached to a reacting carbon is the hydrolysis of acetals and *ortho* acids [32]; the equilibrium isotope effect for the following reaction (Eqn. 52) has the value $K_H/K_D = 1.29$ so that this effect should be experienced in a carbenium-like transition state.

$$\text{Ph–C}(\text{H}^*)(\text{OH})(\text{Ph}) + \text{H}^+ \rightleftharpoons \text{Ph–C}^+(\text{H}^*)(\text{Ph}) + \text{H}_2\text{O} \quad (52)$$

General acid-catalysed hydrolysis of *ortho* esters (Eqn. 53) has the isotope effect 1.05 and acid-catalysed hydrolysis of acetals has $k_H/k_D = 1.19$ indicating that the former have less carbenium ion character in the transition state than the latter.

$$\left[\text{H}^*\text{–C(OEt)(OEt)–O(Et)----H–A}\right]^{\ddagger} \qquad \left[\text{H}^*\text{–C}^+\text{(R)(OEt)----O(Et)–H----A}\right]^{\ddagger} \quad (53)$$

These isotope effects are known for obvious reasons as α effects and β and γ effects are also known.

The α-seconday isotope effect has been applied to the hydrolysis of *O*-ethyl-*S*-phenyl benzaldehyde acetal hydrolysis by Faroz and Cordes (Eqn. 54) [33].

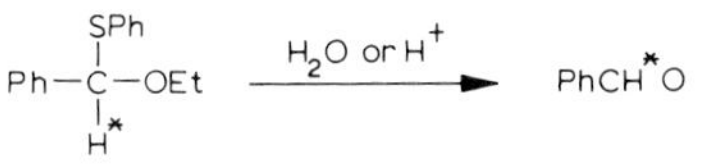

$$\text{Ph–C(SPh)(H}^*\text{)–OEt} \xrightarrow{\text{H}_2\text{O or H}^+} \text{PhCH}^*\text{O} \quad (54)$$

Acid-catalysed hydrolysis has an isotope effect of 1.038 ± 0.008 indicating little C–S cleavage in the transition state; water catalysis has an isotope effect of 1.13 ± 0.02 consistent with complete cleavage as suggested by Jensen and Jencks on other grounds [34]. doAmaral et al. have also shown that oxyanion attack on phenyl formates has an isotope effect of 1.12–1.24 as might be expected for full C–O bond formation [35].

$$\overset{*}{H}-C(=O)OAr \quad + \quad \bar{O}R \longrightarrow H^{*}-COOR + \bar{O}Ar \qquad (55)$$

The use of the α-deuterium isotope effects to diagnose changes in hybridisation and hence mechanism may be criticised on the grounds that changes in vibration of the α-hydrogen in ground and transition state may not arise from hybridisation changes. Westaway and Ali [36] indicate that crowding in the transition state will yield a 'large' α-deuterium isotope effect. Harris et al. [37] have also indicated that the effect is a sensitive probe of crowding at the transition state. Despite these criticisms the α-deuterium effect is very useful in determining the nature of the transition state relative to ground provided the model substrate is chosen carefully.

6. *Labelling techniques*

Labelling techniques have been used extensively to determine biogenetic paths and to the extent that determining stoichiometry resides in the province of mechanistic chemistry these techniques are covered. The philosophy of their use in this context is very simple and we shall discuss below some more sophisticated examples of labelling.

6.1. Position of bond cleavage

In a reaction where bond formation is to the same element as in bond cleaving simple product analysis does not usually yield the position of fission. The alkaline hydrolysis of 2′,4′-dinitrophenyl 3,5-dimethyl-4-hydroxybenzenesulphonate (Eqn. 56) can involve S–O or Ar–O fission and the problem is solved by labelling the solvent

$$ArSO_2OC_6H_3(NO_2)_2 \begin{cases} \xrightarrow[\text{acyl oxygen cleavage}]{H\overset{*}{O}^-} HOC_6H_3(NO_2)_2 \\ \xrightarrow[\text{aryl oxygen cleavage}]{H\overset{*}{O}^-} H\overset{*}{O}C_6H_3(NO_2)_2 \end{cases} \qquad (56)$$

water where the pattern of incorporation indicates the relative rates through each mechanism. The technique formerly applied to such a labelling problem involved decomposition of the products unambiguously to CO_2 which was measured using a dual collector mass spectrometer to compare the ratios of the various molecular ions. It is now possible to measure the products directly using the high-resolution mass

Table 8
Enrichment of 2,4-dinitrophenol from hydrolysis of 2′,4′-dinitrophenyl-3,5-dimethyl-4-hydroxybenzene-sulphonate [38]

Water enrichment	0% incorpn. [a]	Observed enrichment	100% incorpn. [a]
2.03	1.230	1.299	3.260
5.79	1.230	1.224	7.020

[a] Enrichment calculated from isotopic abundances.

spectrometer to select individual masses or to select all signals with a desired integral mass number. The latter method is usually employed as the former requires extremely high selectivity at the high mass numbers of interest. Table 8 illustrates the results obtained for the 2,4-dinitrophenol product (Eqn 56), where the calculated abundances are estimated using the known enrichments of the water oxygen and the published abundances of the other isotopes. Major contributors to the mass number $M + 2(C_6H_4N_2O_5 = 184)$ involve $^{14}C_4{}^{13}C_2H_4N_2O_5$, $C_6H_2D_2N_2O_5$, $C_6H_4{}^{15}N_2O_5$ and $C_6H_4N_2O_4{}^{18}O$.

Oxygen-18 incorporation may be used to eliminate more sophisticated problems than the bond cleavage one. The hydrolysis of *N*-sulphonylanthranilate (Eqn. 57)

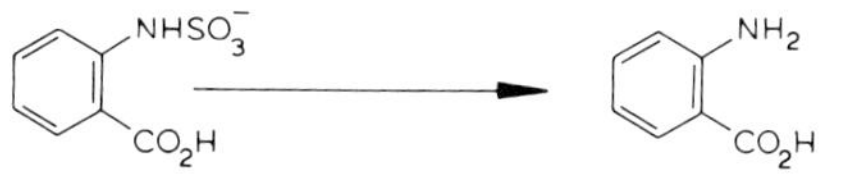

(57)

could involve the mixed sulphate–carboxylate anhydride (Eqn. 58) which will hydrolyse with C–O cleavage to give labelled anthranilic acid if the reaction is carried out in labelled water.

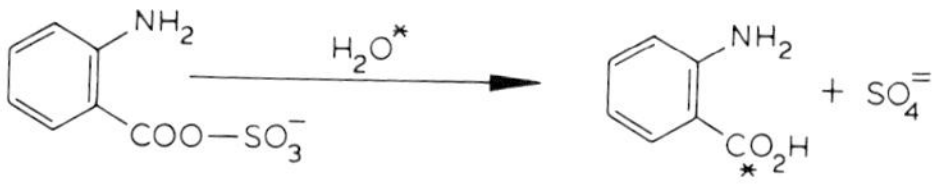

(58)

Table 9
Enrichment of anthranilic acid from the hydrolysis of *N*-sulphonylanthranilic acid [39]

Water enrichment	0% incorpn. [a]	Observed enrichment	100% incorpn. [a]
2.03	0.700	0.720	2.750
5.79	0.700	0.715	6.490

[a] Enrichment calculated from natural abundances.

Mass spectral analysis of the anthranilic acid will thus decide whether the anhydride path is involved and will strictly place an upper limit on the proportion of the reaction flux passing through this mechanism. Table 9 indicates the data and as before the contributions to the $M+2$ mass number from non-incorporation are calculated from tables of natural abundance.

From the above considerations it is clear that the larger the percentage enrichment of the labelling isotope the lower the level is at which we are able to define a mechanism. Clearly the use of 2% enriched water is adequate (and not expensive); critical experiments require larger enrichments which add considerably to the cost.

6.2. Proton transfer

Labelling by tritium or deuterium provides a tool for measuring whether a proton in an organic molecule is labile and also a method for measuring the speed of ionisation. With tritium this is a simple method for even the slowest process because of the accuracy of counting this particular isotope [40]. For example even the rate of ionisation of benzene may be followed in this way.

6.3. Detection of intermediates by isotope exchange

The discovery by Myron Bender [41] in 1951 that esters hydrolysing in water labelled with oxygen-18 undergo exchange at the carbonyl oxygen was a turning point in physical organic chemistry. The mere observation of exchange indicates the presence of a tetrahedral intermediate (Eqn. 59); it also tells us that the proton exchange reaction between the two forms of the intermediate must be faster than the decomposition of the intermediate to products. Moreover, the rate of exchange reaction coupled with the overall rate of the hydrolysis reaction allows us to measure the ratio of forward to reverse reactions.

$$\begin{array}{ccccccc}
\text{R—CO—OR}' & + & \text{OH}^- & \rightleftharpoons & \text{R—C(O}^-\text{)(OR}'\text{)—}\overset{*}{\text{O}}\text{H} & \longrightarrow & \text{product} \\
 & & & & \updownarrows \ \text{proton exchange} & & \\
\text{R—C}\overset{*}{\text{O}}\text{—OR}' & + & \text{OH}^- & \rightleftharpoons & \text{R—C(OH)(OR}'\text{)—}\overset{*}{\text{O}}{}^- & \longrightarrow & \text{product} \\
\text{exchanged ester} & & & & \text{intermediate} & &
\end{array} \tag{59}$$

An example from another area of chemistry shows the versatility of the exchange method. Alkanesulphonyl halides react with deuteroalcohols to produce a deuterated sulphonate where only one deuterium [42] is replaced (never more than one). This result shows that the mechanism is not a direct attack of alcohol on sulphonyl halide but involves an intermediate lacking one hydrogen (Eqn. 60).

$$RCH_2-SO_2Cl + DOR' \xrightarrow{Et_3N} RCHDSO_2-O-R'$$

Direct $$RCH_2-SO_2-Cl \;(\leftarrow {}^-OR') \longrightarrow RCH_2-SO_2-OR'$$ (60)

Sulphene $$RCH_2-SO_2-Cl \underset{}{\overset{Et_3N}{\rightleftharpoons}} RCH{=}SO_2 \xrightarrow{DOR'} RCHD-SO_2-OR'$$

6.4. Isoracemisation

The combination of hydrogen exchange and stereochemistry as tools for mechanism have elucidated a remarkable phenomenon. Let us consider the exchange of the α-proton in 1-phenyl-1-cyanopropane; if all the proton exchanges result in recemisation then the ratio of the rate of exchange to the rate of racemisation (k_e/k_α) = unity. Cram and Grosser [43] found this ratio to vary between 0.2 and 0.05 depending on the solvent and base present. The low exchange rate indicates that the planar carbanion necessary for racemisation must be retaining some of the original proton rather than allowing it to diffuse into the solvent. The explanation put forward by Cram [44] (Eqn. 61) is that the diffusion rate is much slower than the isomerisation or reprotonation rate from the loosely bound complex $(k_{-1}$ and $k_2 > k_3)$.

$$\text{Et(Ph)(CN)C–H} \underset{k_{-1}}{\overset{B\cdot k_1}{\rightleftharpoons}} \text{Et(Ph)(CN)C}^- \cdots \overset{+}{H}B \xrightarrow{D\overset{+}{B}\cdot k_3} \text{Et(Ph)(CN)C} \cdots D\overset{+}{B}$$

$$\text{Et(Ph)(CN)C}^- \cdots \overset{+}{H}B \;\overset{k_2}{\underset{k_2}{\rightleftharpoons}}\; \text{Et(NC)(Ph)C}^- \cdots \overset{+}{H}B \rightleftharpoons \text{Et(NC)(Ph)C–H}$$ (61)

In the above case the low k_e/k_α ratio indicates that the complex between acid and carbanion survives for a significant time prior to breakdown by diffusion. A further case involves tautomerism in the triphenylmethane system (Eqn. 62) in deuterium-containing solvents [45].

Table 10
Isoracemisation in triphenylmethane derivatives (Eqn. 62) [45]

Solvent	Base	H in product (%)
$(CD_3)_2SO$	NaOD	34
CH_3OD	NaOMe	47
Et_3COD	$N(isopr)_3$	98

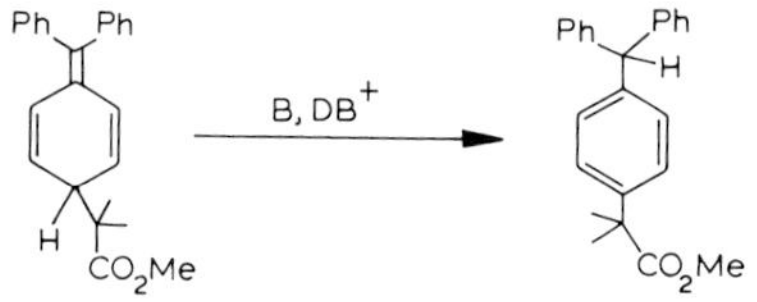

(62)

Table 10 indicates that under suitable conditions the proton is transmitted from one site to the other as the conjugate acid of the base without significant exchange with the solvent. The ion pair (Eqn. 63) has a significant lifetime compared with diffusion of BH^+ into the solvent. Both of these phenomena are termed isoracemisations – the base catalyst takes hydrogen or deuterium on a 'conducted tour' of the substrate from one site to another.

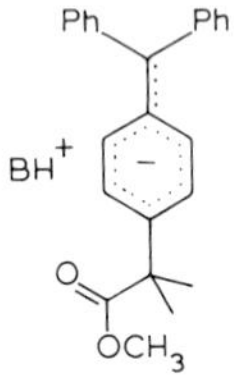

(63)

The factors leading to isoracemisation can lead to misinterpretation of isotope-exchange experiments. The lack of exchange of the proton with solvent protons is classical evidence for hydride transfer in the Cannizzaro reaction and Meerwein–Pondorff–Verley reduction and in the rearrangements involved in steroid biosynthesis. For some time the lack of deuterium exchange in the glyoxalase reaction was attributed to a hydride-transfer mechanism (Eqn. 64).

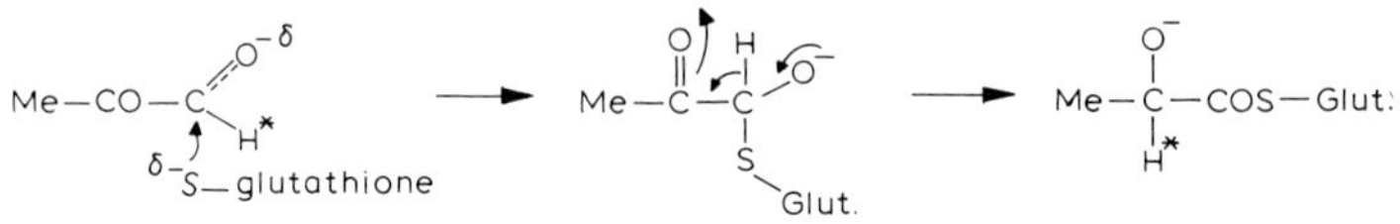

(64)

It was eventually shown by Hall et al. [46] that deuterium was exchanged to a small extent indicating that an isoracemisation mechanism was masking exchange in a *proton* – transfer process (Eqn. 65).

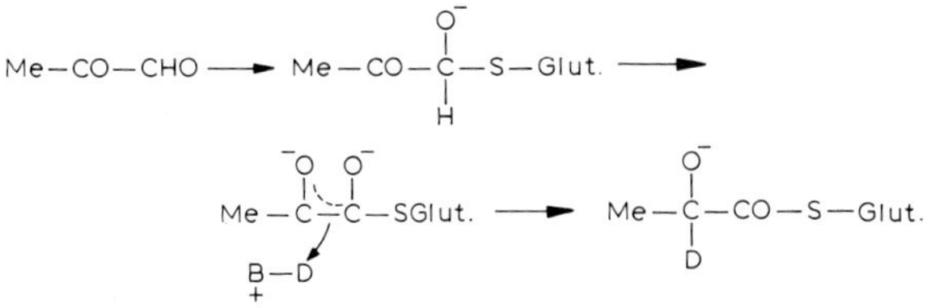

(65)

Another interesting example of the isoracemisation mechanism is in the triose phosphate isomerase pathway. Rieder and Rose [47] indicated that exchange of the proton in the triose phosphate occurred in the presence of the isomerase not consistent with the hydride shift. Later work by Herlihy et al. [48] indicated that some 3–6% of the hydrogen at C_1 of the product dihydroxyacetone phosphate was transferred to C_2 (Eqn. 66).

$$\begin{array}{c} CHO \\ | \\ CH^{*}OH \\ | \\ CH_2OPO_3H_2 \end{array} \underset{}{\overset{\text{isomerase}}{\rightleftharpoons}} \begin{array}{l} \quad H^{*} \\ \quad | \\ 1\ CHOH \\ \quad | \\ 2\ CO \\ \quad | \\ 3\ CH_2OPO_3H_2 \end{array} \tag{66}$$

The latter experiment indicates that the complex between carbanion and conjugate acid has a lifetime slightly greater than the diffusion or that the rearrangement step is commensurate with the diffusion rate (Eqn. 67).

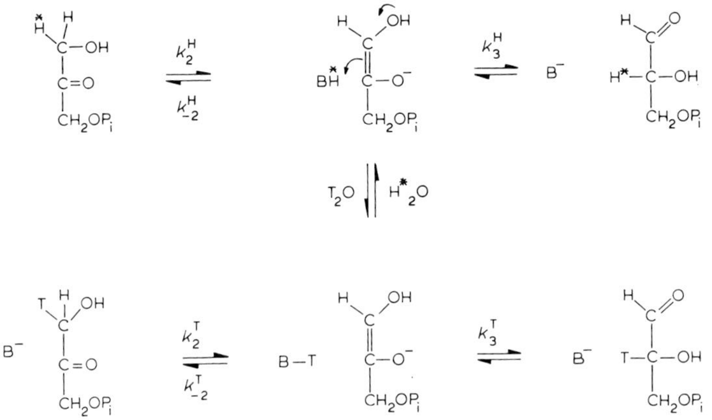

(67)

6.5. *Double-labelling experiments*

Detection of an intermediate by double labelling is well exemplified in the hydrolysis of phthalamic acid to phthalic acid [49]. One mechanism involves phthalic anhydride as an intermediate and this may be checked by labelling the amide carbon with ^{13}C and the water with ^{18}O (Eqn. 68).

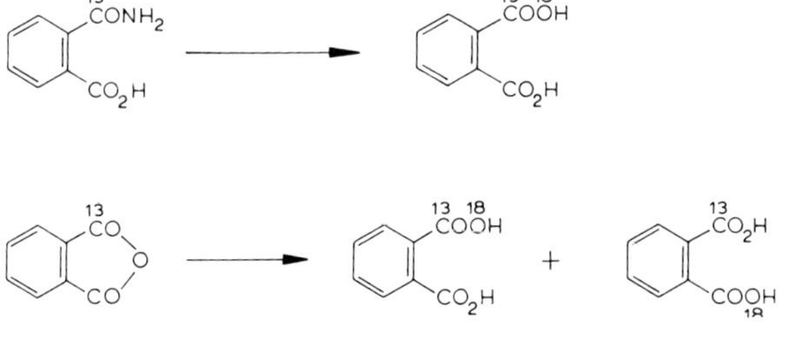

(68)

The product acid decarboxylates to CO^2 and the phthalic anhydride path should produce half the quantity of CO_2 of isotopic mass 47 ($^{13}C^{18}O^{16}O$) that the direct path yields. Knowing the isotopic composition of reactants the proportion of the reaction flux through each path is calculable from the ratio of 47 to 44 masses and the result is that only the anhydride path is taken.

Roberts et al.'s [50] classic demonstration of a symmetric intermediate in the reaction of chlorobenzene with potassium amide in liquid ammonia is an example of a double-labelling experiment where the second label is a heteroatom rather than an isotope. Direct attack will yield product labelled only on the carbon bearing the nitrogen whereas the route through the symmetrical intermediate (the benzyne) gives 50% of each of the possible products (Eqn. 69).

Cl NH2 NH2 + NH2 (69)

6.6. Isotopic enrichment

Enrichment of tritium is possible in reactions where the proton transfer is the rate-limiting step. This is caused by the primary isotope effect reducing the transfer rate for the tritium compared with that for hydrogen. Analysis of the specific activity of the reactant should in principle (as discussed earlier) give the isotope effect but the qualitative observation of an enrichment is sufficient to diagnose rate-limiting proton transfer. Tritiated ethanol (MeCHTOH) becomes enriched as it is oxidised by bromine because the proton is preferentially transferred [51] in the rate-limiting step.

References

1 Bigeliesen, J. and Wolfsberg, M. (1958) *Adv. Chem. Phys. 1*, 15.
2 Westheimer, F.H. (1961) *Chem. Rev. 61*, 265.
3 Melander, L. (1960) Isotope Effects on Reaction Rates, Ronald Press, New York,
4a Ross, J. and Mazur, P. (1961) *J. Chem. Phys. 35*, 19.
4b Eggers, D.F., Gregory, W.W., Halsey, G.D. and Rabinowitch, B.S. (1960) Physical Chemistry, John Wiley, New York, p. 439 ff.
5 Moore, W.J. (1962) Physical Chemistry, 4th Edn., Prentice Hall, London.
6 Bunton, C.A. and Shiner, V.J. (1961) *J. Am. Chem. Soc. 83*, 3207, 3214.
7 Bell, R.P. (1959) The Proton in Chemistry, Chapman and Hall, London.
8 Lewis, E.S. (1975) in: E.F. Caldin and V. Gold (Eds.), Proton Transfer Reactions, Chapman and Hall, London, p. 317.
9 Reitz, O. (1937) *Z. Phys. Chem. A179*, 119.
10 Westheimer, F.H. and Nicolaides, N. (1949) *J. Am. Chem. Soc. 71*, 211.
11 Melander, L. (1950) *Arkiv. Kemi 2*, 211.
12 Hammond, G.S. (1955) *J. Am. Chem. Soc. 77*, 334.
13 Zollinger, H. (1955) *Helv. Chim. Acta 38*, 1597, 1617.

14 More O'Ferrall, R.A. (1975) in: E.F. Caldin and V. Gold (Eds.), Proton Transfer Reactions, Chapman and Hall, London, p. 201.
15 Bergman, N.A., Chiang, Y. and Kresge, A.J. (1978) *J. Am. Chem. Soc. 100*, 5954.
16 Cox, M.M. and Jencks, W.P. (1981) *ibid. 103*, 572; (1978) *ibid. 100*, 5956.
17a Bruice, T.C. and Schmir, G.L. (1957) *ibid. 79*, 1663.
17b Bender, M.L. and Turnquest, B.W. (1957) *ibid. 79*, 1656.
18 Schowen, K.B.J. (1978) in: R.D. Gandour and R.L. Schowen (Eds.) Transition States of Biochemical Processes, Plenum, New York, p. 225.
19 Schowen, R.L. (1972) *Progr. Phys. Org. Chem. 9*, 275.
20 Albery, J. (1975) in: E.F. Caldin and V. Gold (Eds.) Proton Transfer Reactions, Chapman and Hall, London, p. 263.
21 For more detailed discussion of the Gross–Butler equation see refs. 7, 18 and 19.
22 Kresge, A.J. (1977) *J. Am. Chem. Soc. 95*, 2279.
23 Minor, S.S. and Schowen, R.L. (1973) *J. Am. Chem. Soc. 95*, 2279.
24 Albery, W.J. and Davies, M.H. (1972) *J. Chem. Soc. (Far. Trans. 1)* 167.
25 Pollock, E., Hogg, J.L. and Schowen, R.L. (1973) *J. Amer. Chem. Soc. 95*, 968.
26 Elrod, J.P., Gandour, R.D., Hogg, J.L., Kise, M., Maggiora, G.M., Schowen, R.L. and Venkatasubban, K.S. (1975) *Far. Soc. Symp. 10*, 145.
27 Quinn, D.M., Venkatasubban, K.S., Kise, M. and Schowen, R.L. (1980) *J. Am. Chem. Soc. 102*, 5365.
28a Patterson, J.F., Husky, W.P., Venkatasubban, K.S. and Hogg, J.L. (1978) *J. Org. Chem. 43*, 4935.
28b Hogg, J.L., Phillips, M.K. and Jergens, D.E. (1977) *J. Org. Chem. 42*, 2459.
28c Hogg, J.L. and Phillips, M.K. (1973) *Tetrahedron Lett.* 3011.
29 Gopalakrishnan, G., Hogg, J.L. and Venkatasubban, K.S. (1981) *ibid. 46*, 4356.
30a Lynn, K.R. and Bourne, A.N. (1963) *Chem. and Ind.* 782.
30b Grovenstein, E. and Ropp, C.A. (1956) *J. Am. Chem. Soc. 78*, 2560.
30c Halevi, E.A. (1963) *Progr. Phys. Org. Chem. 1*, 109.
31 Streitwieser, A., Jagow, R.H., Fahey, R.C. and Suzuki, S. (1958) *J. Am. Chem. Soc. 80*, 2326.
32 Cordes, E.H. (1970) *J. Chem. Soc. (Chem. Commun.)* 527.
33 Faroz, J.F. and Cordes, E.H. (1979) *J. Am. Chem. Soc. 101*, 1488.
34 Jensen, J.L. and Jencks, W.P. (1979) *ibid. 101*, 1476.
35 doAmaral, L., Bastos, M.P., Bull, H.G., Otiz, J.J. and Cordes, E.H. (1979) *ibid. 101*, 169.
36 Westaway, K.C. and Ali, S.F. (1979) *Can. J. Chem. 57*, 1089.
37 Harris, J.M., Paley, M.S. and Prasthope, T.W. (1981) *J. Am. Chem. Soc. 103*, 5915.
38 Thea, S., Guanti, G., Hopkins, A. and Williams, A. (1982) *ibid. 104*.
39 Hopkins. A. and Williams, A. (1982) *J. Org. Chem.*, 47, 1745.
40 Jones, J.R. (1973) The Ionisation of Carbon Acids, Academic Press, New York, p. 13.
41 Bender, M.L. (1951) *J. Am. Chem. Soc. 73*, 1626.
42 King, J.F. (1975) *Accts. Chem. Res. 8*, 10.
43 Cram. D.J. and Grosser, L. (1964) *J. Am. Chem. Soc. 86*, 5457.
44 Cram. D.J. (1965) Fundamentals of Carbanion Chemistry, Academic Press, New York.
45 Cram. D.J., Wiley, F., Fischer, H.P. and Scott, D.A. (1964) *J. Am. Chem. Soc. 86*, 5510.
46 Hall, S., Doweyko, A. and Jordan, F. (1976) *ibid. 98*, 7460.
47 Rieder, S.V. and Rose, I.A. (1959) *J. Biol. Chem. 234*, 1007.
48 Herlihy, J.M., Maister, S.G., Albery, W.J. and Knowles, J.R. (1976) *Biochemistry 15*, 5601.
49 Bender, M.L., Chow, Y.L. and Chloupek, F. (1958) *J. Am. Chem. Soc. 80*, 5380.
50 Roberts, J.D., Simmons, H.E., Carlsmith, L.A. and Vaughan, C.W. (1953) *ibid. 75*, 3290.
51 Kaplan, L. (1954) *ibid, 76*, 4645.

General references

1 Collins, C.J. and Bowman, N.S. (1971) Isotope Effects in Chemical Reactions A.C.S. Monograph 166, Van Nostrand Reinhold, New York.

2 Melander, L. (1960) Isotope Effects on Reaction Rates, Ronald Press, New York.
3 Cleland, W.W., O'Leary, M.H. and Northrop, D.B. (1977) University Park Press, Baltimore, MD.
4 Gandour, R.D. and Schowen, R.L. (1978) Transition States of Biochemical Processes, Plenum, New York.
5 Caldin, E. and Gold, V. (1975) Proton Transfer Reactions, Chapman and Hall, London.
6 Saunders, W.H. (1966) *Surv. Progr. Chem. 3*, 109.
7 Melander, L. and Saunders, W.H. (1980) Reaction Rates of Isotopic Molecules, Wiley-Interscience, New York.
8 Schowen, R.L. (1972) Progr. Phys. Org. Chem. 9, 275.
9 Bell, R.P. (1973) The Proton in Chemistry, Chapman and Hall, London.
10 Proton Transfer (1975) *Faraday Society Symposium,* Vol. *10*.

The mechanisms of chemical catalysis used by enzymes

MICHAEL I. PAGE

Department of Chemical Sciences, The Polytechnic, Huddersfield HD1 3DH, Great Britain

1. Introduction

Although enzymes may contain several hundred amino acid residues only two or three of them are usually chemically involved in the bond-making and -breaking steps in the transformation catalysed by the enzyme. The main purpose of the bulk of the amino acid residues is to provide the fairly rigid three-dimensional framework required to maximise the binding energy between substrate and enzyme (Ch. 1).

The majority of reactions occur by mechanisms involving a redistribution of electron density on going from the ground state to the transition state. The function of catalytic groups on the enzyme or coenzyme is to stabilise these electron-density changes or to provide an alternative pathway for the reaction. Charge complementarity is simply an extension of Pauling's electroneutrality principle. Electron-rich nucleophiles, bases and reducing agents will be paired with electron-deficient electrophiles, acids and oxidising agents, respectively. Charge may also be stabilised by delocalisation. Of course, this principle also extends to the resting enzyme. In the absence of substrate the electron distribution of groups within the enzyme will be "neutralised", as allowed by the three-dimensional framework, by complementary groups or solvent (Ch. 1).

The electron donor or acceptor properties of the side chains of amino acids are listed below:

Nucleophiles, electron donors, bases:

$-CH_2OH$, $-CH_2SH$, $-CH_2S^-$, $-CO_2^-$, CH_2NH_2, ArOH, imidazole (HN N ring)

Electrophiles, electron acceptors, acids:

$-CONH-$, $-CO_2H$, CH_2OH, CH_2SH, imidazole H^+, $CH_2NH_3^+$, M^{n+}

Substituents containing a hydrogen bonded to an electronegative atom can obviously act as a potential electron donor (hydrogen bond acceptor) via the heteroatom and as a potential electron acceptor (hydrogen bond donor) via the hydrogen.

Michael I. Page (Ed.), The Chemistry of Enzyme Action

Occasionally nature can teach us some chemistry (see Ch. 12), but generally the mechanisms used by enzymes are of the type that would be predicted by physical organic chemists. Sometimes the "chemical" mechanism itself makes a large contribution to the rate enhancement brought about by the enzyme compared with the rate of reaction in the absence of similar catalytic groups used by the enzyme. This is especially true in reactions formally involving carbanions where often the adduct-forming coenzymes provide routes to low-energy-stabilised carbanions derived from the substrate. However, even in such cases it is important to remember that the "non-reacting" parts of the protein and substrate contribute to the rate enhancement of the enzyme-catalysed reaction by utilising the binding energy to compensate for energetically unfavourable processes (Ch. 1).

In summary, chemical catalysis is efficient by avoiding the formation of unstable intermediates [1] which often simply means by-passing the development of charge on a particular atom. Of course, it is imperative that intermediates are not too stabilised by the enzyme to avoid the build-up of unwanted enzyme-bound derivatives.

2. *General acid base catalysis [2]*

It is useful to recall the definitions of pK_a and pK_b. The former is the negative logarithm of the dissociation constant, K_a, for an acid HA (Eqn. 1)

$$\mathrm{HA} \rightleftharpoons \mathrm{H^+ + A^-} \qquad K_a = (\mathrm{H^+})(\mathrm{A^-})/(\mathrm{HA}) \tag{1}$$

and pK_b is the negative logarithm of the equilibrium constant for protonation of a base, A^-, by water (Eqn. 2).

$$\mathrm{A^- + H_2O} \rightleftharpoons \mathrm{HA + OH^-} \qquad K_b = \frac{(\mathrm{HA})(\mathrm{OH^-})}{(\mathrm{A^-})(\mathrm{H_2O})} \tag{2}$$

The lower the pK_a the stronger is the acid HA and the weaker is its conjugate base, A^-. The lower the pK_b the stronger is the base A^- and the weaker is its conjugate acid, HA.

Many reactions involve the making and breaking of bonds to hydrogen attached to electronegative atoms. This invariably requires that the hydrogen is transferred as a proton or rather as a species in which the proton is coordinated to an electron-rich centre. Consequently many reactions are catalysed by acids and bases. Reactions in aqueous solution are commonly catalysed by the hydronium ion, H_3O^+, specific acid catalysis, or by hydroxide ion, OH^-, specific base catalysis. The function of the acid or the base is to facilitate the bond-making and -breaking processes. For example, substitution reactions at the acyl centre often involve expulsion of a leaving group X (I). This bond breaking may be very difficult if X is expelled as an unstable anion

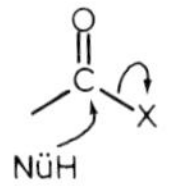

O C X NüH

I

X^-, but will be greatly facilitated if X is first protonated and then expelled as the more stable XH. Similarly the nucleophile often contains an ionisable hydrogen and proton removal will generate a more powerful nucleophile e.g. OH^- is a stronger nucleophile than H_2O.

In enzyme-catalysed reactions, of course, the powerful acid, H_3O^+, and base, OH^-, are not available and other acidic and basic groups must be used. The role of general acid and general base catalysts in increasing the rate of reactions may be exemplified by the hypothetical hydrolysis of peptides in the absence of H_3O^+ and OH^-. The reaction (Eqn. 3) must involve, at some stage during the reaction, the making of a bond between carbon and oxygen, the breaking of a bond between carbon and nitrogen and the transfer of several hydrogens as "masked" protons. If

$$RCONHR' + H_2O \rightleftharpoons RCO_2^- + R'NH_3^+ \qquad (3)$$

either of the heavy atom transfer steps occurred without proton transfer, high-energy-charged intermediates e.g. II would be formed with a large change in the acidity and basicity of the constituent atoms. For example the attacking water molecule is changed from a weak acid into a strong one, and both the amide nitrogen and the carbonyl oxygen are changed from very weak bases to strong ones.

$$\begin{array}{c} O^- \\ | \\ R-C-NHR' \\ | \\ {}^+OH_2 \end{array}$$

II

Formation of such unstable intermediates changes proton transfer between reactants and the solvent water ($pK_a \sim 14$, $pK_b \sim 0$) from being thermodynamically unfavourable to favourable. Catalysis can increase the rate of reaction by either (i) trapping such intermediates, (ii) stabilising the intermediate and the transition states leading to them, and (iii) providing an alternative mechanism not requiring the formation of such unstable intermediates. The mechanism of catalysis of many of these types of reactions is *enforced* by the lifetime and acid–base properties of the initially formed intermediate [3].

2.1. Catalysis by stepwise proton transfer (trapping)

Proton transfer between simple acids and bases occurs in a series of steps. Diffusion together is followed by proton transfer and then diffusion apart (Eqn. 4).

$$HA + B \underset{}{\overset{k_1}{\rightleftharpoons}} [AH \cdots B] \overset{k_H}{\rightleftharpoons} [A \cdots HB] \overset{k_2}{\rightleftharpoons} A^- + BH^+ \qquad (4)$$

If proton transfer is thermodynamically favourable i.e. the pK_a of the proton donor HA is less than that of the conjugate acid of the acceptor BH^+ ($pK_w - pK_b$ of the acceptor B) the rate-limiting step is diffusion together of the acid–base pair, k_1. If proton transfer is thermodynamically unfavourable i.e. the pK_a of the proton donor HA is greater than that of the conjugate acid of the acceptor BH^+ ($pK_w - pK_b$ of the acceptor B) the rate-limiting step is diffusion apart of the acid–base pair, k_2 (Eqn. 3).

Intermediate (II) is unstable with respect to proton transfer to or from the solvent water and with respect to breakdown back to reactants. Expulsion of water from II will occur much faster than expulsion of the unstable amine anion, $R'NH^-$. Removal of H^+ from the attacking water of II or proton addition to oxygen or nitrogen of II will generate relatively more stable intermediates. If these proton transfers occur to or from general acids or bases present rather than to solvent water a rate increase will be observed.

The extent of catalysis depends critically upon the stability of the intermediate II [3]. *If* the rate of expulsion of H_2O from II, k_{-1}, was slower than proton transfer to solvent water, the rate of formation of the intermediate, k_1, would be the rate-limiting step and no catalysis would be observed (Fig. 1). The rate of protonation of the amine nitrogen of II by solvent water, k_H (Eqn. 5, HA = H_2O) depends on the

$$\underset{\text{II}}{\mathrm{R{-}C(O^-)(^+OH_2){-}NHR'}} + \mathrm{HA} \underset{k_{-H}}{\overset{k_H}{\rightleftharpoons}} \underset{\text{III}}{\mathrm{R{-}C(O^-)(^+OH_2){-}\overset{+}{N}H_2R'}} + \mathrm{A^-} \qquad (5)$$

basicity of the nitrogen, K_b, and is given by $k_{-H}K_b = (k_{-H}K_w/K_a)$ where k_{-H} represents the diffusion-controlled abstraction of a proton by hydroxide ion with a value of approximately 10^{10}/M/sec. If the rate of expulsion of H_2O from II to regenerate reactants was faster or similar to the rate at which it is "trapped" [3] by proton abstraction from water the rate-limiting step occurs after the formation of

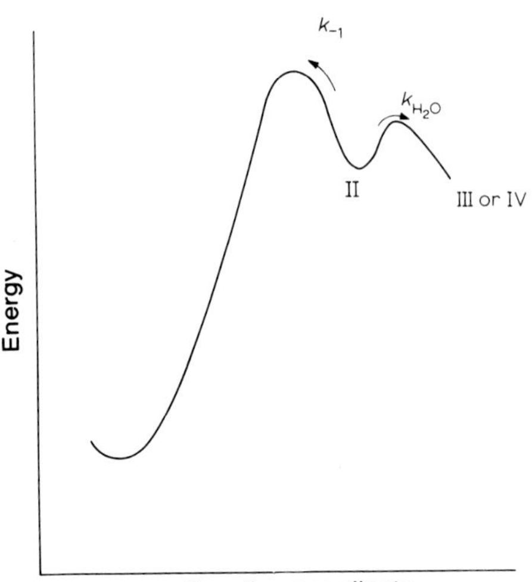

Fig. 1. Reaction coordinate diagram showing rate-limiting attack of water on an amide to give intermediate II which rapidly transfers a proton to solvent water to give III or IV.

intermediate II and the addition of general acids, i.e. other proton donors, may increase the observed rate.

If HA is a stronger acid than III (loss of proton from NH^+) the rate of proton transfer to II, k_H, would be diffusion controlled and the *observed rate would be independent of the acidity of HA,* i.e. making HA a strong acid would not increase the observed rate. If HA is a weaker acid than III (loss of proton from NH^+) the rate of proton transfer from general acids, HA, to the nitrogen of II (Eqn. 5) is given by $k_{-H}K_a^{HA}/K_a = k_{-H}K_a^{HA}K_b/K_w$, where K_a^{HA} and K_a are the acid dissociation constants of HA and III, respectively, and k_{-H} is the diffusion-controlled rate. *The observed rate would now be dependent upon the acidity of the catalyst HA and decrease with the pK_a of HA.*

In our hypothetical reaction the intermediate II could also be stabilised by proton transfer from oxygen to give IV (Eqn. 6).

$$\underset{\text{II}}{R\text{—}\overset{\overset{\displaystyle O^-}{|}}{\underset{\underset{\displaystyle {}^+OH_2}{|}}{C}}\text{—}NHR'} + B \underset{k_{-H}}{\overset{k_H}{\rightleftharpoons}} \underset{\text{IV}}{R\text{—}\overset{\overset{\displaystyle O^-}{|}}{\underset{\underset{\displaystyle OH}{|}}{C}}\text{—}NHR'} + BH^+ \tag{6}$$

The proton acceptor B could be solvent water or a general base catalyst. In this case also, the rate of reaction will only be catalysed if the rate of breakdown of II to regenerate reactants is faster than the rate of proton transfer. In this particular instance such catalysis would be independent of the basic strength of the catalyst B as proton transfer would invariably be thermodynamically favourable and hence occur at the maximum diffusion-controlled rate. If proton transfer to solvent is thermodynamically favourable such that proton donation to 55 M water is faster than, say 1 M added base any observed catalysis by base must represent transition-state stabilisation by a hydrogen-bonding (Ch. 7, Section 2.3) or a concerted (Ch. 7, Section 2.4) mechanism.

It is of interest to examine the general case when a nucleophile attacking a carbonyl group to eventually displace the group X contains a proton which becomes acidic when the intermediate is formed (Eqn. 7).

$$RC(=O)X + NuH \overset{K}{\rightleftharpoons} \underset{T^{\pm}}{R\text{—}\overset{\overset{\displaystyle O^-}{|}}{\underset{\underset{\displaystyle {}^+NuH}{|}}{C}}\text{—}X} \underset{k_{-H}}{\overset{k_H[B]}{\rightleftharpoons}} \underset{T^-}{R\text{—}\overset{\overset{\displaystyle O^-}{|}}{\underset{\underset{\displaystyle Nu}{|}}{C}}\text{—}X} + BH^+ \;\rightleftharpoons\; RC(=O)Nu + X^- \tag{7}$$

Again catalysis by the general base B will be observed when the intermediate $T^{\pm}$ breaks down to reactants faster than $T^{\pm}$ transfers a proton to water. The rate of formation of T^-, which may or may not represent the overall rate of reaction, is given by $k_H[B]K$, where K is the equilibrium constant for formation of the intermediate $T^{\pm}$ and k_H is the rate of proton transfer from $T^{\pm}$ to the catalyst B.

In such a mechanism, what is the effect of increasing the nucleophilicity of the nucleophile and the basicity of the catalyst? For a constant nucleophile changing the basicity of the catalyst would affect the rate as shown in the Eigen curve of Fig. 2a. When proton transfer from $T^{\pm}$ to B is thermodynamically favourable the rate of proton transfer is diffusion controlled and hence independent of the basicity of the catalyst B. When it is thermodynamically unfavourable the rate decreases proportionally to the decreased basicity of the catalyst, and is given by $Kk_{-H}K_a^{T^{\pm}}/K_a^{BH^+}$, where $K_a^{T^{\pm}}$ and $K_a^{BH^+}$ are the acid dissociation constants of $T^{\pm}$ and BH^+, respectively.

Increasing the nucleophilicity of the attacking nucleophile will increase the stability of $T^{\pm}$ and, in a constant series of nucleophiles generated, say, by changing the pK_b, the rate will be affected as shown in Fig. 2a, b, c. When proton transfer from $T^{\pm}$ to B is thermodynamically favourable the rate will increase with increasing nucleophilicity because the stability of $T^{\pm}$ is increased, i.e. K is increased. When proton transfer is thermodynamically unfavourable the overall rate of formation of T^- is independent of the basicity of the nucleophile. For a given general base catalyst increasing the basicity of the attacking nucleophile increases the stability of $T^{\pm}$ but this will be almost exactly compensated by a decrease in the rate of proton transfer, k_H, because $T^{\pm}$ will be a proportionally weaker acid.

An example will illustrate the typical rate enhancement brought about by general acid and base catalysts. The reaction of amines with penicillins requires the removal of a proton from the attacking amine after the initial formation of the tetrahedral

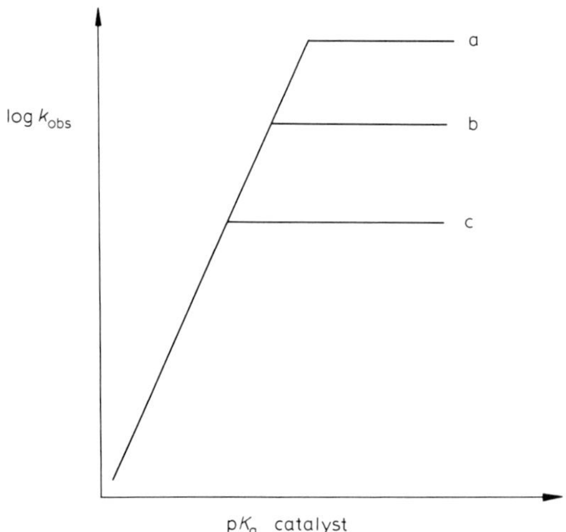

Fig. 2. Hypothetical Brønsted plot showing how the observed rate of the reaction outlined in Eqn. 7 of text varies with nucleophilicity of the attacking nucleophile and basicity of the catalyst: (a) constant nucleophile, increasing basicity of catalyst; when proton transfer from the intermediate is thermodynamically favourable increasing the nucleophilicity of the nucleophile increases the rate from (c) to (b).

intermediate (Eqn. 8) [4].

$$\mathrm{RNH_2} + \text{(β-lactam)} \underset{k_{-1}}{\rightleftharpoons} \mathrm{R\overset{+}{N}H_2}\text{–(tetrahedral intermediate, O}^-\text{)} \overset{B}{\rightleftharpoons} \mathrm{RNH}\text{–(tetrahedral intermediate, O}^-\text{)} \longrightarrow \mathrm{RNH{-}C(=O){-}CH_2CH_2{-}NH{-}} \qquad (8)$$

The rate of expulsion of RNH_2 to regenerate the reactants, k_{-1}, is 10^6–10^9/sec and much faster than proton transfer to water ($B = H_2O$), 10^{-1}–10^2/sec. When the general base catalyst is a second molecule of amine ($B = RNH_2$) proton transfer is much more efficient and a rate enhancement, based on the molar scale, of ca. 1000-fold is observed. Catalysis is due solely to the greater basicity of the amino group compared with water. Of course, compared with solvent water of 55.5 M the amine catalyst at 1 M is only about 20 times more efficient. In general, rate enhancements of 10–100 are brought about by general acids and bases. No greater further enhancement is obtained when the general acid or base catalyst of an enzyme acts within the enzyme–substrate complex [5].

2.2. Catalysis by preassociation [3]

When the rate-limiting step of a reaction such as that of Eqn. 7 is the diffusion-controlled encounter of two reagents an enzyme may increase the rate simply by having the nucleophile and catalyst preassociated i.e. they do not have to diffuse through solution before reaction can occur. This situation occurs [3] in simple intermolecular reactions when the rate of breakdown of the intermediate to regenerate reactants is faster than the rate of separation of the intermediate and catalyst, 10^{10}–10^{11}/sec, and when proton transfer between the intermediate and catalyst is thermodynamically favourable. The formation of the intermediate must take place by a preliminary association of the reactants and catalyst (Eqn. 9 and Fig. 3).

$$\mathrm{B + NuH + {>}C{=}O \rightleftharpoons B\cdot HNu\cdot {>}C{=}O \rightleftharpoons B\cdot H\overset{+}{N}u{-}\overset{|}{\underset{|}{C}}{-}O^-} \qquad (9)$$

The preassociation mechanism is preferred to the trapping mechanism because it generates an intermediate which is immediately trapped by an ultrafast proton transfer and thus avoids the diffusion-controlled step bringing catalyst and intermediate together. This mechanism is sometimes called a "spectator" mechanism because although the catalyst is present in the transition state it is not undergoing any transformation [7].

The rate enhancement obtainable by preassociation compared with "trapping" is given by the ratio of the rate of breakdown of the intermediate I to generate reactants to the rate of dissociation of the intermediate and catalyst I.C. (Fig. 3). The *maximum* lowering of the free energy of activation obtainable is the activation energy for diffusion of apart of the catalyst and intermediate i.e. ca. 15 kJ/mole, a rate enhancement of ca. 400. There can be no rate advantage from a preassociation mechanism when the proton transfer is thermodynamically unfavourable.

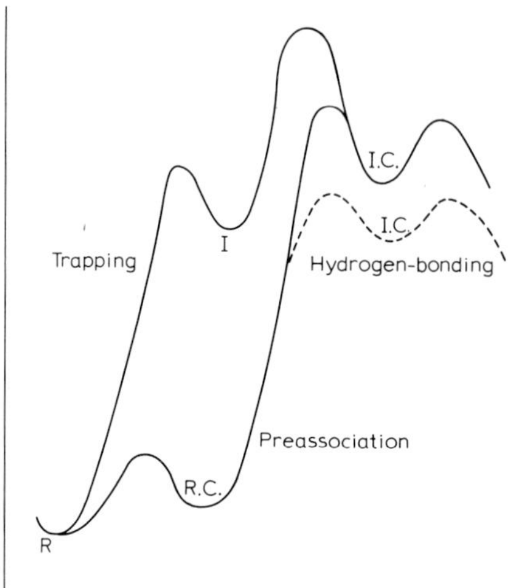

Fig. 3. Reaction coordinate diagram showing how a reaction will proceed through a preassociation complex (R.C.) when the associated intermediate and catalyst (I.C.) breaks down to reactants faster than it dissociates to separated I and C. Additional stabilisation of I.C. may occur through hydrogen-bonding.

2.3. Catalysis by hydrogen bonding

As an intermediate becomes progressively more *unstable* catalysis by hydrogen bonding becomes progressively more important [6]. In the preassociation mechanism for intermolecular reactions (Ch. 7, Section 2.2) the catalyst is in close proximity to the reactants because they are in the same solvent cage. If the catalyst is correctly located for subsequent proton transfer there is the possibility of stabilisation of the transition state by hydrogen bonding. This hydrogen bonding to the catalyst must be more favourable than that to water for there to be a rate enhancement. The dependence of the stability of hydrogen bonds upon acidity and basicity is small. But, for example, changing a proton donor hydrogen-bonding catalyst from water to one of pK_a 5 could increase the rate, compared with the preassociation mechanism, by ca. 150-fold if the dependence of hydrogen-bond strength upon pK_a had a typical Bronsted α value of 0.2 [6]. The importance of catalysis by hydrogen bonding increases sharply as the intermediate becomes less stable. The *maximum* rate advantage of a preassociation mechanism with hydrogen bonding compared with the trapping mechanism is proportional to the product of the free energy of dissociation of the catalyst–intermediate complex and the extra stabilisation of the intermediate brought about by hydrogen bonding to the catalyst *compared with that to water* (Fig. 3). This is sensibly a factor of ca. $400 \times 150 = 6 \times 10^4$.

2.4. Concerted catalysis

When the "intermediate" is so unstable that it cannot exist, i.e. it would have a lifetime of less than 10^{-14} sec, the reaction *must* proceed by a concerted mechanism.

The proton-transfer steps and other covalent bond-forming and -breaking processes occur simultaneously but with a varying degree of coupling between their motions. However, it is still not clear whether a concerted mechanism can occur when the intermediates, which would be formed in stepwise mechanism, have a significant lifetime. This is an important question for reactions catalysed by enzymes. If the sole criterion for a concerted mechanism is the stability of the intermediate then the nature of the intermediate itself will control whether the enzyme- and non-enzyme-catalysed mechanisms are forced to be similar.

If there is no, or very weak, coupling between the motions of bond making and breaking because of unfavourable geometry and orbital overlap a concerted mechanism can only occur when it is enforced i.e. when there is no activation barrier for one of the steps decomposing the hypothetical intermediate [1].

If there is an energetic advantage from coupling two steps of a reaction mechanism into one the mechanism may become concerted. This is unlikely to occur if the energy barriers to bond-making or -breaking steps are large. The barriers for proton transfer between electronegative atoms are usually smaller than those for carbon so the advantage for coupling is more likely to arise in the former type of reaction i.e. complex general acid–base catalysis [1].

2.5. Intramolecular general acid–base catalysis

It is often suggested that the transfer of a proton between electronegative atoms of substrates and those of acidic or basic groups in enzymes contributes to the large rate enhancement observed in enzyme-catalysed reactions [8]. Two aspects of the involvement of functional groups on the enzyme should be distinguished:

(1) the rate enhancement brought about by "chemical catalysis" relative to the uncatalysed or solvent-catalysed reaction i.e. the importance of catalysis (proton transfer) by acidic or basic species compared with water; and

(2) the rate enhancement brought about by chemical catalysis occurring within the enzyme–substrate complex and being of a lower kinetic order than an analogous intermolecular reaction [9].

Intramolecular reactions (Ch. 1) involving proton transfer generally show much smaller rate enhancements compared with intermolecular reactions proceeding by the same mechanism, than do those reactions involving the formation of a covalent bond between heavy atoms [5,10].

The rate enhancement that can be brought about by having a general acid or general base catalyst as part of the protein structure (Eqn. 10), as opposed to the chemically equivalent intermolecular mechanism (Eqn. 11), appears to be minimal.

$$\mathrm{S + H{-}A{-}Enzyme} \rightleftharpoons \underset{\mathrm{Enzyme}}{\underbrace{\mathrm{S \cdot HA}}} \longrightarrow \underset{\mathrm{Enzyme}}{\underbrace{\mathrm{S \cdots H \cdots A}^{\ddagger}}} \tag{10}$$

$$\mathrm{S + H{-}A} \longrightarrow \mathrm{S \cdots H \cdots A}^{\ddagger} \tag{11}$$

Of course, it may be necessary for the proton acceptor or proton donor to be at the active site as the equivalent intermolecular catalyst may be sterically prevented from reaching the enzyme-bound substrate and therefore in the absence of the general acid or general base on the enzyme a slower rate of reaction would result. However, the fact that the general acid or base is *part* of the enzyme apparently makes little contribution to the enormous rate enhancement brought about by the enzyme. The reason for this is that proton transfer reactions involve "loose" transition states (Ch. 1) so that in intermolecular reactions such as Eqn. 11 there is only a small loss of entropy in bringing the reactions together [5,9]. There is therefore little or no advantage in having the catalyst in close proximity to the substrate as in Eqn. 10.

Bimolecular reactions which proceed through "loose" transition states are entropically less unfavourable than those involving "tight" transition states (Ch. 1). This is because the entropy of low frequency motions in the transition state compensates for the large loss of translational and rotational entropy. Consequently, an analogous intramolecular reaction should show a smaller rate enhancement or effective concentration. Such a situation exists in proton-transfer reactions and all reactions involving intramolecular general acid or general base catalysis have low effective concentrations. This appears to be true for proton transfer to or from both electronegative atoms and carbon atoms [5]. Presumably this is because such reactions either have a transition state which is a very loose hydrogen-bonded complex or have a rate-limiting step which is diffusion controlled [2]. Thus there does not seem to be much of a rate advantage upon changing an intermolecular general acid- or general base-catalyzed reaction to an intramolecular or enzyme process. This appears to be true for mechanisms which occur by a stepwise or a coupled concerted process.

An example of a proton-transfer reaction occurring by the stepwise trapping mechanism (Ch. 7, Section 2.1) is the general base-catalysed aminolysis of benzylpenicillin. Amines react with penicillin to form an unstable tetrahedral intermediate which may be trapped by a diffusion-controlled encounter with a strong base as shown in Eqn. 8. Diamines also undergo this reaction but at a much faster rate than monoamines of the same basicity which is attributed to intramolecular general base catalysis i.e. the second amino group acts as a proton acceptor (V). However, the effect of intramolecularity itself is small and the effective concentration (Ch. 1)

H
H_2N ^+NH—C——N—
O^-

V

compared with the intermolecular general base-catalysed reaction is only about 1 M [4] i.e. the fact that the catalyst is held in close proximity to the reaction site makes it only as effective as a concentration of 1 M intermolecular catalyst, of the same basicity, freely diffusing through the solution.

An example of a general acid–base-catalysed reaction occurring by a concerted mechanism (Ch. 7, Section 2.4) is the hydrolysis of the hydroxy-amide (VI). The mechanism has been shown unequivocally to proceed by concerted general acid-catalysed breakdown of the tetrahedral intermediate (VII), in which there is coupling between proton transfer and carbon–nitrogen bond fission [11]. The presence of an intramolecular general acid catalyst such as the ammonium group of VIII causes a

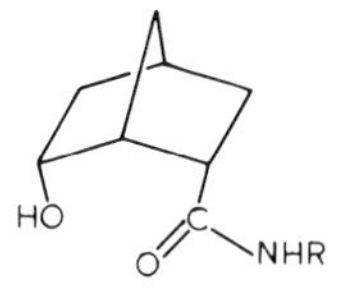

rate enhancement attributable to intramolecular catalysis. This rate enhancement is due almost entirely to the difference in acidities of the proton donors: water in the intermolecular reaction (VII, B = OH) and the ammonium ion in the intramolecular process (IX). However, the contribution of intramolecularity itself is small, the effective concentration (Ch. 1) is ca. 1 M i.e. the intramolecular general acid catalyst, although in close proximity to the reaction site, is only as efficient as an intermolecular catalyst of the same acid strength at a concentration of 1 mole/l.

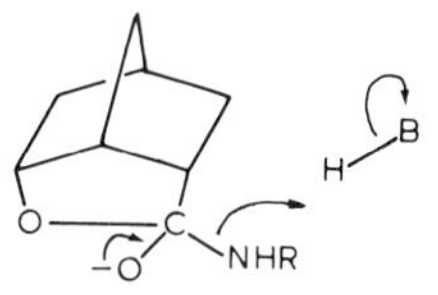

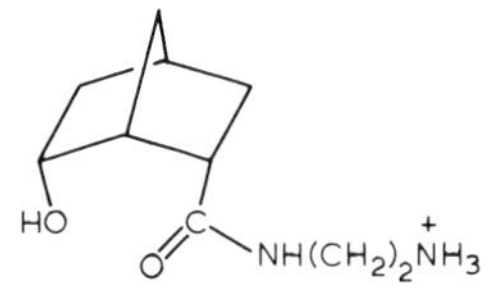

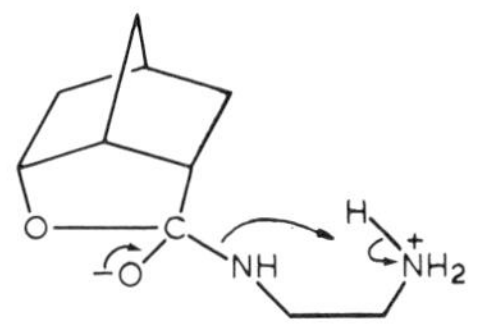

Most effective concentrations that have been reported for intramolecular general acid–base-catalysed reactions involve a stepwise mechanism in which a transport process is usually rate-limiting and the proton itself is in a potential-energy well in the transition state. The stepwise mechanism for acid–base catalysis represents the limit of a 'loose' transition state in which the reactant molecules are diffusing together and it is not surprising, therefore, that the entropy change associated with such bimolecular steps is small and that there is little rate advantage to be gained by covalently linking the reactant molecules together in an intramolecular reaction. It appears that the same is also true even if the proton is 'in flight' in the transition state of the intermolecular reaction, i.e. the reaction mechanism is one of concerted proton transfer. This is expected in view of the relatively small entropy changes associated with hydrogen-bonding equilibria [5].

The observations of low effective concentrations in intramolecular general acid–base-catalysed reactions may be extrapolated to determine the catalytic advantage of having a general acid or base catalyst as part of the protein structure in enzyme-catalysed reactions. The catalytic advantage appears to be minimal. Although it may be necessary for the proton acceptor or donor to be at the active site, as the equivalent intermolecular catalyst may be sterically prevented from reaching the enzyme-bound substrate, the fact that the general acid or base is *part* of the enzyme apparently makes little contribution to the enormous rate enhancement brought about by the enzyme [9,10].

There is an interesting contrast between the large contribution (a factor of ca. 10^8) to the rate enhancement of intramolecular and enzyme-catalysed reactions by nucleophilic catalysis and the much smaller contribution (ca. 1–10) of general acid–base catalysis.

3. Covalent catalysis

The advantage of covalent catalysis where an electrophilic or nucleophilic group on the peptide chain of the enzyme forms a covalent bond with the substrate, is immediately apparent by considering the difference in entropy changes between the equivalent intermolecular (Eqn. 12) and enzyme-catalysed mechanisms (Eqn. 13). In the latter, one of the reactants, B, is covalently bonded to the enzyme and a comparison of this reaction of the intermolecular reaction illustrates the advantage of binding the substrate to the enzyme *even if the chemical reactivity of B in the enzyme may be similar to that of B in intermolecular reaction.*

It is easily shown that the amount of catalysis by the enzyme is given directly by the binding energy between the enzyme and substrate in the transition state (Ch. 1).

$$A + B \longrightarrow A\text{–}B^{\ddagger} \qquad (12)$$

$$A + B\text{–Enzyme} \longrightarrow [A\text{–}B\text{–Enzyme}]^{\ddagger} \qquad (13)$$

Covalent catalysis is also effective because the transient chemical modification that occurs by forming a covalent bond to the substrate gives an intermediate which either facilitates the bond-making and -breaking processes or enables an active intermediate to be formed at a higher concentration than is possible in the absence of the catalyst. Most coenzymes function by one or both of these mechanisms.

An example of the pure entropic advantage of covalent catalysis is given by the hydrolysis of the hydroxy-amide (X), which is a good model for the serine proteases [12].

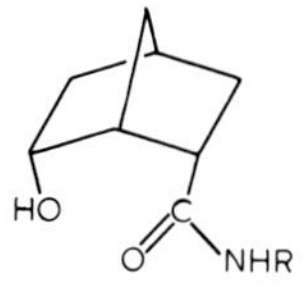

HO NHR

X

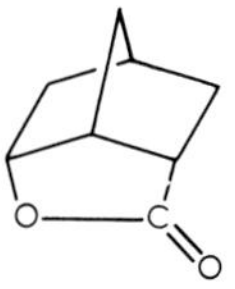

XI

The intramolecular nucleophilic hydroxy group causes a rate enhancement of ca. 10^8 mole/l compared with the equivalent intermolecular reaction. This is attributed to neighbouring group participation of the hydroxy group to form a lactone intermediate (XI) (equivalent to the acyl-enzyme in the serine proteases). The rate enhancement is due entirely to the entropy effect – the intramolecular reaction involves little change in strain energy on forming the lactone and solvation effects are minimal so that the nucleophilicity of the nucleophiles in the intra- and inter-molecular reactions are very similar [12].

The above example also illustrates another requirement for efficient nucleophilic catalysis, namely that the intermediate that is formed must react rapidly to prevent accumulation of the intermediate. The lactone intermediate is an ester which is much more reactive than the starting amide towards hydrolysis and is rapidly converted to the hydroxy-acid.

A simple illustration of the rate enhancement obtainable by chemical transformation is the conversion of carbonyl groups into imines (Schiff bases). This is dealt with at length in Ch. 8 and 9. Changing the oxygen of a carbonyl group for a nitrogen (Eqn. 14) has several profound effects upon the reactivity of the carbonyl carbon and adjacent groups.

$$>C{=}O + RNH_2 \rightleftharpoons >C{=}NR + H_2O \qquad (14)$$

Replacement of oxygen by nitrogen makes a molecule much more basic. For example the pK_b of acetone is ca. 21 compared with a value of ca. 7 for the corresponding imine. At pH 7 the fraction of the ketone protonated is ca. 10^{-14} (Eqn. 15) whereas half of the imine exists in the protonated form (Eqn. 16).

$$(H_3C)_2C{=}O + H^+ \rightleftharpoons (H_3C)_2C{=}\overset{+}{O}H \qquad (15)$$

$$(H_3C)_2C{=}NR + H^+ \rightleftharpoons (H_3C)_2C{=}\overset{+}{N}HR \qquad (16)$$

Unprotonated acetone is more susceptible to nucleophilic attack than the neutral imine group, because of the greater electronegativity of oxygen. However, the protonated imine is many millions times more reactive towards nucleophiles than the carbonyl group and the predominant reaction at neutral pH will occur by attack on the imine conjugate acid. Therefore, a significant rate enhancement of a reaction involving nucleophilic attack on a carbonyl carbon can result from simply converting the carbonyl to an iminium group.

Enamines are the nitrogen analogues of enols but their formation from imines is thermodynamically more favourable than enol formation from ketones (Table 1). The equilibrium constant for enol formation is ca. 10^{-8} compared with a value of 10^{-5} for enamine formation. However, at pH 7 half of the imine exists as the iminium ion and the proportion of enamine present is 10^8-fold greater than the proportion of enolate anion. In general, this implies that loss of an electrophile from

TABLE 1
Equilibrium constants for enol and enamine formation

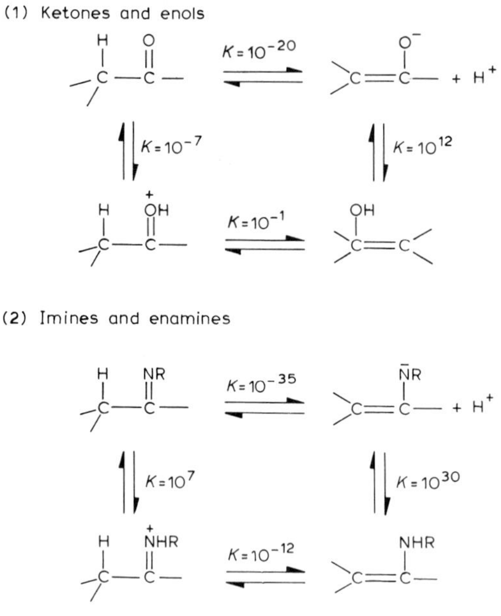

(1) Ketones and enols

H–C–C(=O)– $\underset{}{\overset{K=10^{-20}}{\rightleftharpoons}}$ >C=C(O$^-$)– + H$^+$

$K = 10^{-7}$ (ketone ⇌ protonated ketone); $K = 10^{12}$ (enolate ⇌ enol)

H–C–C(=$\overset{+}{O}$H)– $\overset{K=10^{-1}}{\rightleftharpoons}$ >C=C(OH)<

(2) Imines and enamines

H–C–C(=NR)– $\overset{K=10^{-35}}{\rightleftharpoons}$ >C=C($\bar{N}$R)– + H$^+$

$K = 10^{7}$ (imine ⇌ iminium ion); $K = 10^{30}$ (enamine anion ⇌ enamine)

H–C–C(=$\overset{+}{N}$HR)– $\overset{K=10^{-12}}{\rightleftharpoons}$ >C=C(NHR)–

the α-carbon of a charged iminium ion to form a neutral enamine (Eqn. 17) is much easier than that from a neutral ketone to form a charged enolate ion (Eqn. 18).

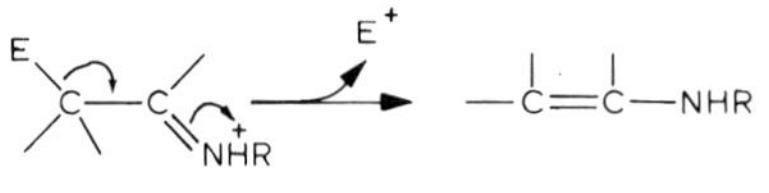

$$\text{E–C–C=}\overset{+}{\text{N}}\text{HR} \xrightarrow{\ \ E^+\ \ } \text{–C=C–NHR} \qquad (17)$$

$$\text{E–C–C=O} \xrightarrow{\ \ E^+\ \ } \text{–C=C–O}^- \qquad (18)$$

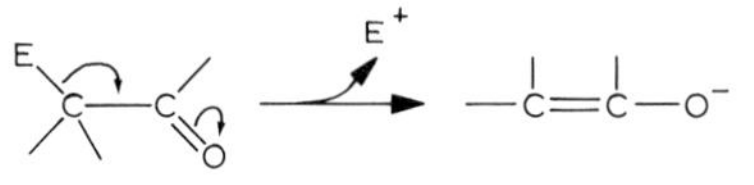

For example, the loss of a proton from the α-carbon (E = H, Eqns. 17 and 18) from the iminium ion is ca. 10^9 faster than that from the corresponding ketone [13].

The neutral enamine is, of course, a delocalised structure (XII) which readily explains the nucleophilic character of the enamine α-carbon.

$$>\text{C=C–NHR} \longleftrightarrow >\bar{\text{C}}\text{–C=}\overset{+}{\text{N}}\text{HR}$$

(XII)

4. Metal-ion catalysis

The coordination of electron donors to metal ions is an exergonic process. One role for metals in metalloenzymes is to act as electrophilic catalysts by stabilising the increased electron density or negative charge that is often developed during reactions [14]. Although this is undoubtedly true it has been frequently stated that because the metal ion may be multiply charged it is a better catalyst than a proton [15] and metal ions have been referred to as "superacids" [16]. There is little evidence to support this claim either for the binding or the activation of substrates. A proton binds more tightly to monodentate and even some bidentate ligands than do most mono-, di- or even tri-positively charged metal ions. The equilibrium of Eqn. 19 invariably lies more to the right than that of Eqn. 20 and this is often also true even if L is a bidentate ligand [17].

$$L + H_3O^+ \rightleftharpoons HL^+ + H_2O \tag{19}$$

$$L + M(H_2O)_x^{n+} \rightleftharpoons ML(H_2O)_{x-1}^{n+} + H_2O \tag{20}$$

A proton "binds" to an electron more tightly than any metal ion as indicated by its ionisation potential [18]. This is not surprising in view of the electron density surrounding the nucleus of a metal ion compared with that of the "bare" proton; the latter in fact could be called a "hyperacid".

A related consequence of the tight binding of a proton to basic sites is that such coordination increases the reactivity of adjacent bonds more so than does coordination of a metal ion. For example, coordination of a proton to a water molecule changes the acidity from a pK_a of 15.7 to -1.7 whereas coordination of a divalent metal ion changes it only to ca. 8 and even a tripositively charged ion changes it only to ca. 3 [19]. There is very little evidence that the binding of metal ions to substrates causes a larger rate enhancement than do protons. For example, although metal ions greatly increase the rate of enolisation of suitably structured ketones they are no more efficient than the hydronium ion. Zinc(II) ions greatly increase the rate of proton abstraction from 2-acetylpyridine (XIII) [20] but it is no more efficient than

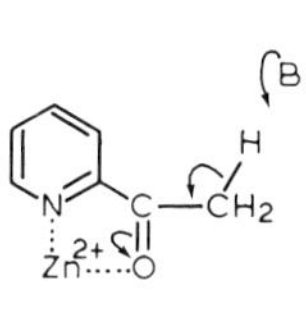

XIII 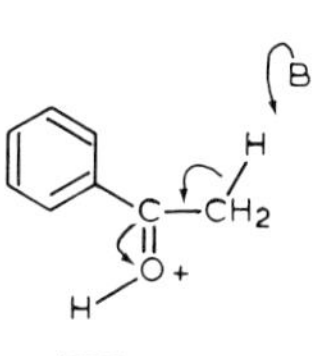 XIV

the coordination of a proton to the carbonyl group (XIV).

Metal ions are effective electrophilic catalysts for a wide variety of reactions [15] but they generally owe this efficiency to either:

(i) the model substrates invariably having a second coordination site available that is much more basic than the reactive site. If these two sites are suitably situated the substrate can act as a bidentate ligand. When reaction occurs this is usually accompanied by an increase in basicity of the reactive site coordinated to the metal ion which leads to more favourable binding and consequently a lowering of the activation energy. The metal ion binds more tightly to the transition state than it does to the ground-state structure of the substrate.

(ii) the limited concentration of hydronium ions in neutral aqueous solution. For example at pH 7 the hydronium-ion concentration is 10^{-7} M whereas the metal-ion concentration may be orders of magnitude higher than this.

Extrapolation of observations obtained from model systems to enzymes must be treated with caution. Metalloenzymes do not operate with high concentrations of metal ions and for those cases that have been studied the substrate usually acts as a monodentate ligand when it is directly coordinated to a metal ion e.g. the carbonyl oxygen of the amide link to be cleaved coordinates to the zinc atom of carboxypeptidase [21]. The protein itself is probably responsible for a large fraction of the binding energy resulting from the interaction of the substrate with the metalloenzyme (Ch. 1).

When the metal ion of the metalloenzyme acts as an electrophilic catalyst it serves an important function of stabilising the negative charges developed in the substrate. However, there is little evidence to suggest that it is markedly more efficient at this task than proton donors in the protein. For example ,an indication of the stabilisation of a negative charge on oxygen brought about by a metal ion can be estimated from the binding energies of this species resulting from coordination to the metal ion. The equilibrium constant for zinc(II) ion binding hydroxide ion (Eqn. 21) is ca. 10^5 estimated from K_a/K_w where K_a is the dissociation constant for the ionisation of zinc-bound water [19] and K_w is the dissociation constant of water.

$$Zn(H_2O)_n^{2+} + OH^- \rightleftharpoons Zn(H_2O)_{n-1}OH^+ + H_2O \quad (21)$$

$$H\text{–}A + OH^- \rightleftharpoons H_2O + A^- \quad (22)$$

The "stabilisation" of hydroxide ion by a proton donor HA may be estimated from Eqn. 22 where the equilibrium constant is given by K_a'/K_w where K_a' is the dissociation constant of the acid HA. For a pK_a' of 7 for HA the equilibrium constant for Eqn. 22 is ca. 10^7. Although the stabilisation of the negatively charged oxygen by the zinc(II) ion is considerable it is not better than a weak general acid catalyst.

In carboxypeptidase the carbonyl oxygen of the amide substrate is probably coordinated to the zinc(II) of the enzyme [21]. The metal ion will stabilise the tetrahedral intermediate by binding to the alkoxide anion (Eqn. 23).

$$R\text{—}\underset{\displaystyle NHR}{\underset{|}{C}}\text{=O}\cdots Zn^{2+} \xrightarrow{N\ddot{u}} \overset{+}{Nu}\text{—}\overset{\displaystyle R}{\overset{|}{\underset{\displaystyle NHR}{\underset{|}{C}}}}\text{—}O^-\cdots Zn^{2+} \quad (23)$$

Incidentally such stabilisation on oxygen would only definitely lead to a rate enhancement if formation of the intermediate were rate-limiting. If breakdown of the intermediate is the rate-limiting step the advantage of coordination is not so obvious. Although coordination would increase the concentration of the tetrahedral intermediate it would decrease its rate of breakdown (but not by as much as by coordination to a proton) because the electron density on oxygen is decreased by coordination which will reduce the ease of carbon–nitrogen bond fission (XV). The net effect would depend on the relative charge density on oxygen in the transition state [22].

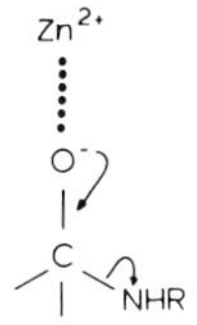

XV

The binding of the carbonyl oxygen of amides to zinc(II) of carboxypeptidase is based on X-ray studies of enzyme–inhibitor complexes. Of course, the amides studied may be inhibitors because they bind to the enzyme incorrectly. An alternative mechanism, which makes chemical sense, could involve metal-ion coordination to the amide nitrogen which would not only stabilise the tetrahedral intermediate (XVI) but also facilitate carbon–nitrogen bond cleavage. Such a mode of catalysis

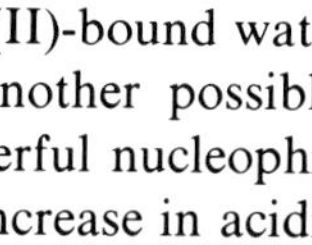

XVI 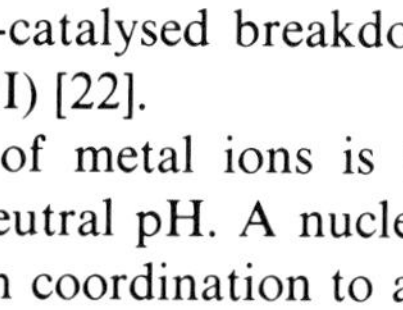 XVII

has been observed in the metal-ion-catalysed hydrolysis of penicillins [22]. Another possibility is general acid-catalysed breakdown of the tetrahedral intermediate by zinc(II)-bound water (XVII) [22].

Another possible role of metal ions is to provide a high concentration of a powerful nucleophile at neutral pH. A nucleophile with an ionisable proton shows an increase in acidity upon coordination to a metal ion (Eqn. 24).

$$M^{n+}(NuH) \rightleftharpoons M^{n+}(\bar{Nu}) + H^+ \tag{24}$$

The coordinated and deprotonated nucleophile will show a reactivity intermediate between that of the ionised and unionised nucleophile. At a given pH, if the increase in *concentration* of deprotonated nucleophile which results from coordination to a metal ion more than compensates for its decreased reactivity a rate enhancement will be obtained for a reaction of the coordinated nucleophile compared with the uncoordinated species. For example, at pH 7 there is 10^{-7} M hydroxide ion but as water bound to zinc(II) has a pK_a of ca. 9 [ref. 19] the amount of hydroxide ion bound to zinc is 10^{-2} M. As seen earlier coordination of hydroxide ion to a metal ion does not remove as much electron density from oxygen as does coordination to a proton. Generally the nucleophilicity of metal-coordinated hydroxide ion is only slightly less than that of the solvated species. For example, the relative rates of water (XVIII), zinc(II)-bound hydroxide ion (XIX) and hydroxide ion (XX) attack on acetaldehyde are 1, 5×10^5 and 3×10^8, respectively [24].

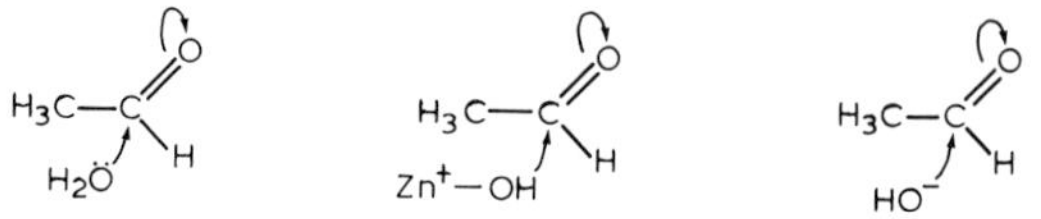

XVIII 1   XIX 5×10^5 XX 3×10^8

The role of metal ions in redox reactions is discussed in Ch. 7, Section 6.4.

5. *Catalysis by coenzymes*

5.1. *Pyridoxal phosphate coenzyme*

All living organisms use pyridoxal phosphate (PLP) a derivative of vitamin B_6 as a coenzyme to synthesise, degrade and interconvert amino acids. The chemical logic of this system is that pyridine ring can act as an electron sink and stabilise carbanions (see also Ch. 9).

Amino groups readily condense with the aldehyde of PLP to form a conjugated imine (XXI). The imines derived from amines may readily loose an electrophile, E,

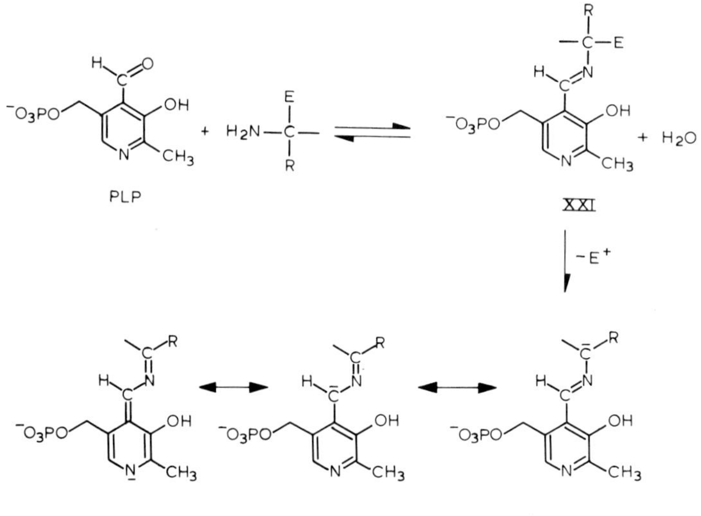

PLP   XXI $-E^+$ XXII

for example H^+, CO_2, to form a delocalised carbanion (XXII) which is especially stable if the pyridine nitrogen is protonated.

There has been much interest in the position of the hydroxyl proton in pyridoxal Schiff bases. Given that the imine nitrogen and phenolic hydroxyl are hydrogen-bonded it makes little difference to mechanistic conclusions whether the proton is nearer nitrogen or oxygen. With the proton nearer to oxygen the electronic structure is apparently simply that of a hydrogen-bonded imine (XXIII). The proton nearer to nitrogen gives a delocalised tautomer represented by the quinoid-enamine (a vinylogous amide) (XXIV) and the zwitterionic structure (XXV).

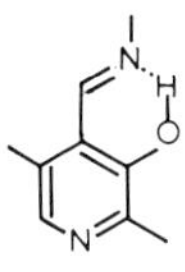

XXIII 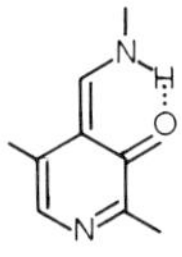 XXIV 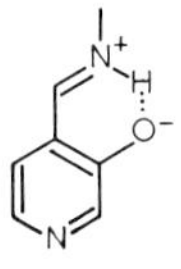 XXV

The 3-hydroxy group is very acidic for a phenol but this is not the result of orientation – all isomeric hydroxypyridines show pK_as of about 2 for ionisation of the hydroxy group. This effect is presumably due to formation of a zwitterionic structure rather than to any mesomerism. In pyridoxal the pK_a for OH is 4.2 and that for the pyridinium nitrogen 8.7. The pK_a of imines is normally 6–7 and although pyridoxal Schiff bases are tribasic the important species near neutrality are the protonated pyridinium zwitterion (XXVI) or its tautomer (XXVII) and these same structures with the pyridinium nitrogen unprotonated. The pK_a for loss of a proton from the intramolecularly hydrogen-bonded system is ca. 9 so the anion (XXVIII) must also exist to a significant extent.

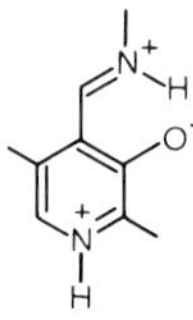

XXVI 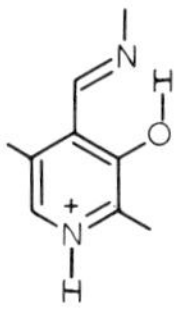 XXVII 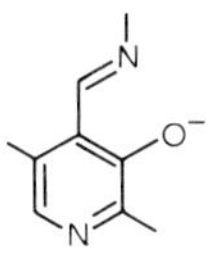 XXVIII

Although pyridoxal phosphate is involved in a wide variety of reactions the processes are all related mechanistically by the electron withdrawal toward the cationic or hydrogen-bonded imine nitrogen and into the electron sink of the pyridine/pyridinium ring.

Loss of the electrophile E from the imine (XXI) forms the carbanion (XXII) which may be attacked by another electrophile at two places. Addition to the α-carbon of the original amine generates a new imine which upon hydrolysis, gives a transformed amine derivative. Addition at the pyridoxal carbonyl carbon, usually by a proton, generates an isomeric imine (XXIX) which, upon hydrolysis, gives pyridoxamine phosphate (PMP) and the α-ketone derivative of the original α-amine (XXX). Because the steps are reversible this system allows a mechanism for nitrogen transfer (transamination) from one substrate to another, with PMP temporarily storing the nitrogen.

If the electrophile removed from the initially formed imine is a proton then deprotonation could give the racemic amino acid or that of inverted configuration.

If the electrophile removed from the initially formed imine is carbon dioxide then α-decarboxylation occurs. Protonation of the decarboxylated adduct at the α-carbon and subsequent hydrolysis would yield an amine and PLP. Protonation of the decarboxylated adduct at the pyridoxal carbonyl carbon and subsequent hydrolysis would yield an aldehyde and PMP.

If the electrophile removed from the initially formed imine is an aldehyde then a retroaldol condensation occurs. For example, serine may be converted to glycine by this mechanism.

Which electrophile is lost from the amino acid residue is, of course, controlled by the enzyme. One way this may occur is by the enzyme binding the PLP imine so that the electrophile is in close proximity to a suitable or base to aid abstraction and also so that the σ orbital of the bond to be broken is periplanar with the pπ acceptor system, i.e. orthogonal to the plane of the pyridine ring (XXXI). Maximal orbital overlap, stereoelectronic control, will lower the activation energy for the reaction. Aldol-type reactions can also occur with PLP; as in the laboratory the key to making carbon–carbon bonds is the formation of a stabilised carbanion. Proton abstraction from the initially formed imine gives a 'masked' carbanion which can nucleophilically attack electrophilic centres. For example, attack on coenzyme A is reminiscent of a Claisen ester condensation (XXXII).

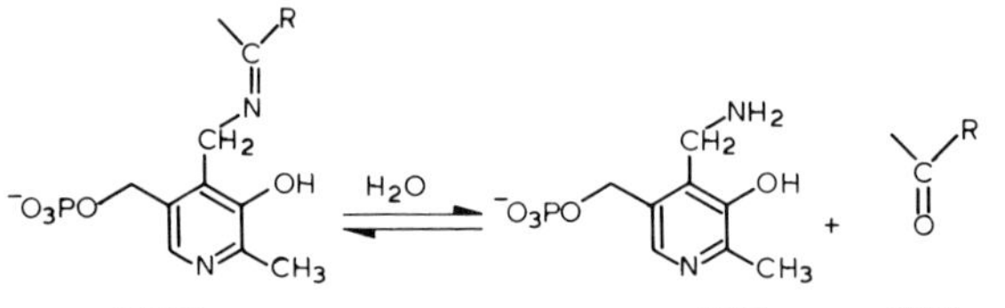

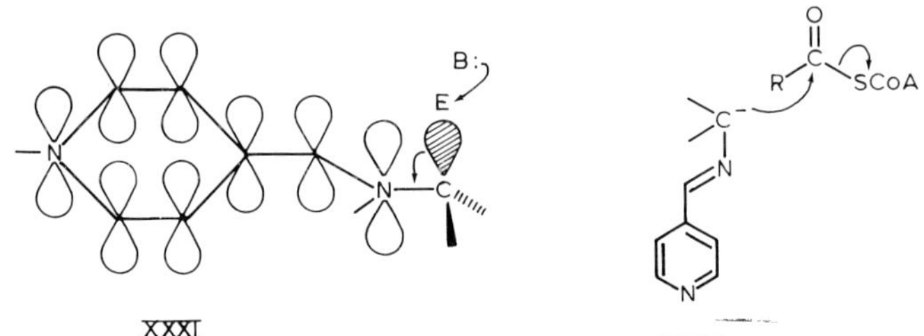

XXIX $\xrightarrow{H_2O}$ PMP + XXX

XXXI

XXXII

Another interesting feature of PLP catalysis is the stabilisation of both electron-poor and electron-rich centres generated at the β-carbon of the original amino acid. This is illustrated in Scheme 1. Pathway a involves α,β-elimination by proton

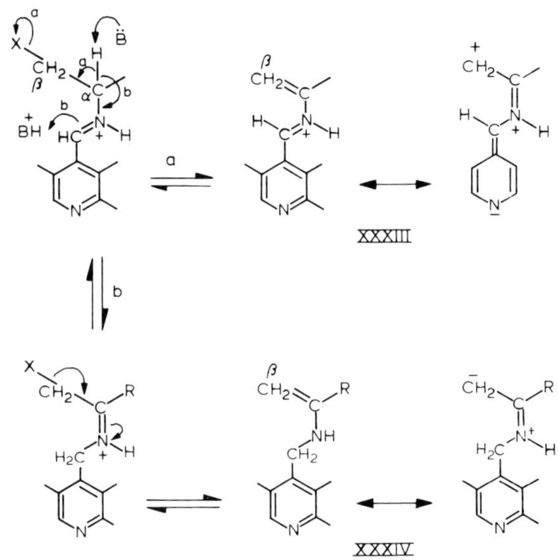

XXXIII

XXXIV

Scheme 1

abstraction at Cα and expulsion of the leaving group X at Cβ. This generates an electrophilic electron-deficient β-carbon which can be attacked by nucleophiles. Conversely, pathway b gives the isomeric imine seen earlier which can now loose an electrophilic X at Cβ to generate an enamine with a nucleophilic electron-rich β carbon (XXXIV), which can be attacked by electrophiles.

5.2. Thiamine pyrophosphate coenzyme

Thiamine pyrophosphate (XXXV) contains an *N*-alkylated thiazole and a pyrimidine. The function of this coenzyme is also to stabilise electron density and it is

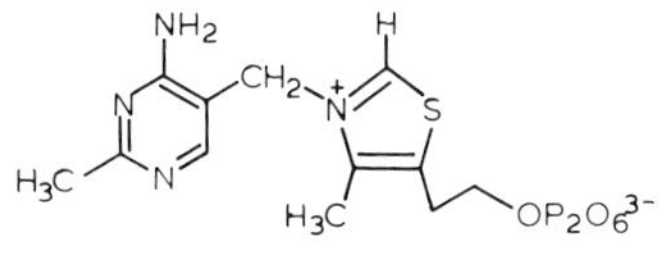

XXXV

active in the decarboxylation of α-keto acids, in the formation of α-ketols (acyloins) and in transketolase reactions.

Proton loss from C-2 of the *N*-alkyl thiazolium salt (XXXVI) gives an ylide-type carbanion (XXXVII) which is stabilised by the positive charge on nitrogen and by the electronegative sulphur – ionisation gives a non-bonded pair of electrons in an sp^2 orbital which cannot therefore be stabilised by pπ delocalisation. The acidity of C—H bonds is significantly enhanced by attached sulphur but, although this is usually explained by (d-p)π bonding between the carbanion lone pair and sulphur 3-d orbitals, theoretical calculations indicate that this is not significant and that sulphur carbanions owe their stability to the large polarisability of sulphur [25].

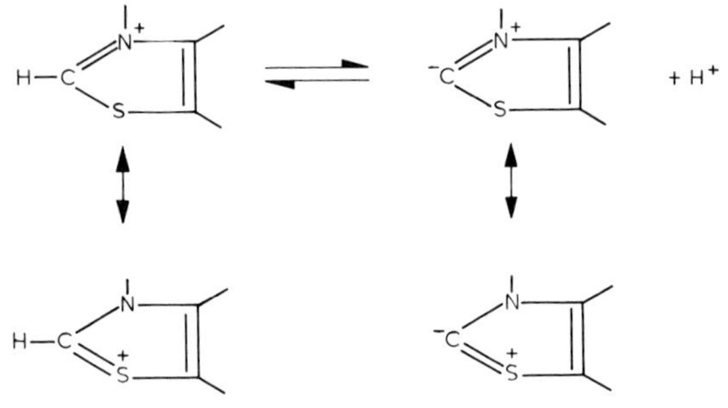

XXXVI XXXVII

As with all reactive intermediates it is important that they are not too stabilised to prevent facile further reaction. The thiazolium ylide is a potent carbon nucleophile but also a good leaving group. This is reminiscent of cyanide ion in the benzoin condensation and, in fact, the chemical logic of that reaction mechanism is similar to the thiamine-catalysed decarboxylations of α-keto acids (Scheme 2).

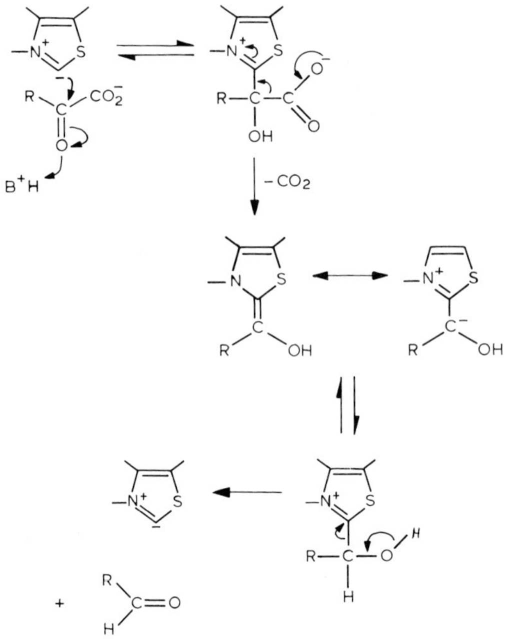

Scheme 2

Breaking the carbon–carbon bond of an α-keto acid to give carbon dioxide generates an unstable carbanion. However, nucleophilic attack of the thiamine carbanion on the α-carbonyl group gives an intermediate from which carbon dioxide is readily lost because the carbanion now generated is a delocalised system resembling an enamine (compare the amine-catalysed loss of α-electrophiles from carbonyl groups described earlier (Eqn. 17)). This intermediate, a hydroxy alkyl thiamine pyrophosphate, is in fact a form in which much of the coenzyme is formed in vivo. The process is completed by expulsion of the catalytic thiamine ylide.

The usefulness of thiamine pyrophosphate in carbon–carbon-bond-forming reactions is exemplified by the condensation of aldehydes and ketones to give α-hydroxy ketones (Scheme 3), and the mechanistically similar α-ketol transfers (Scheme 4).

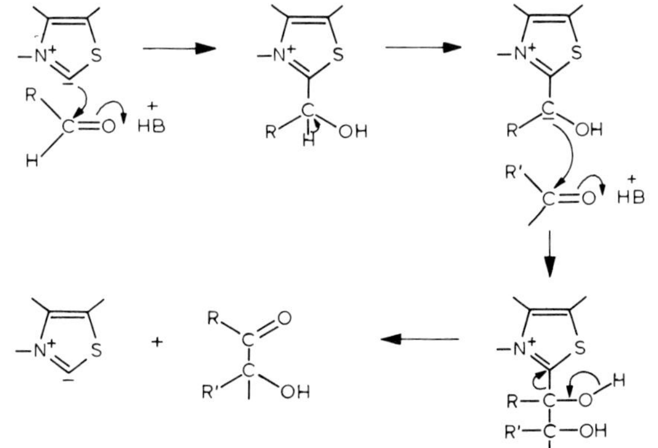

Scheme 3

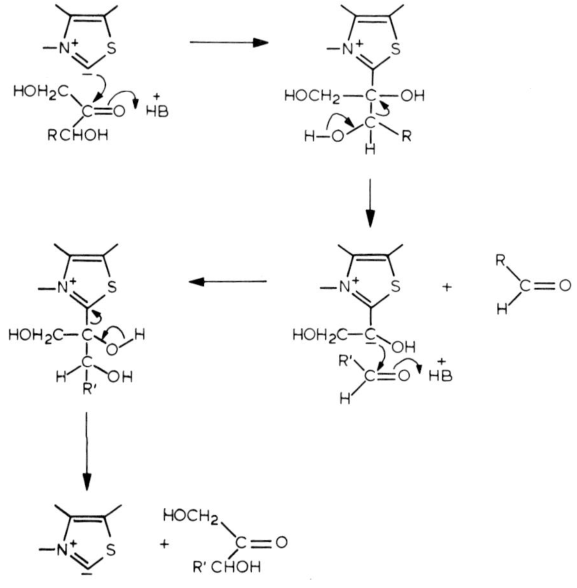

Scheme 4

In summary, the chemical usefulness of thiamine pyrophosphate depends on its ease of carbanion formation at C-2 which is not only a good nucleophile but also a reasonably stable leaving group. In addition the cationic imine stabilises the formation of a carbanion on the *adjacent* carbon bonded to C-2.

5.3. Adenosine triphosphate (ATP)

Adenosine triphosphate (ATP) is a purine ribonucleoside triphosphate which, at neutral pH, exists mainly as the tetra-anion:

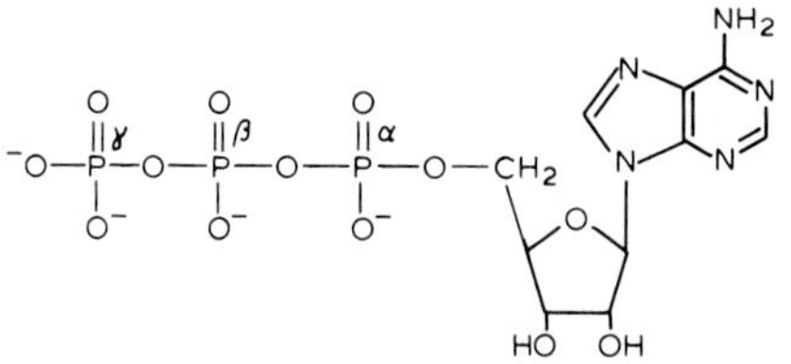

ATP

All reactions involving ATP usually require Mg^{2+} or other divalent cation to reduce the negative charge density on the molecule and to control the site of nucleophilic attack. ATP allows many reactions to take place by phosphorylation of the substrate which provides a good leaving group for carbon–carbon bond synthesis and elimination reactions. Although the role of ATP is often discussed in terms of its very exergonic hydrolysis to adenosine diphosphate and monophosphate these reactions rarely occur directly and it is more useful to consider their mechanistic role.

The two terminal linkages of the triphosphate are phosphoric acid anhydrides i.e. they are activated phosphoric acid derivatives and nucleophilic attack at the electrophilic γ or β phosphorus is thermodynamically very favourable. The advantage of phosphoric anhydride derivatives over acyl derivatives is that the former are kinetically stable in neutral aqueous solution unless a suitable enzyme is present to catalyse its reactions.

An example will readily illustrate the chemical logic of using ATP as a phosphorylating agent to generate substrates with a good leaving group. The direct expulsion of OH groups from saturated or unsaturated carbon is a difficult process because hydroxide ion is a strong base and a poor leaving group:

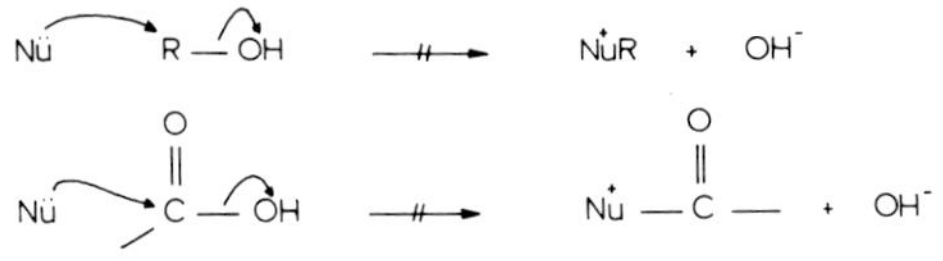

$$\mathrm{Nu}^{-} \quad \mathrm{R{-}OH} \not\longrightarrow \overset{+}{\mathrm{Nu}}\mathrm{R} + \mathrm{OH}^{-}$$

$$\mathrm{Nu}^{-} \quad \mathrm{C(=O){-}OH} \not\longrightarrow \overset{+}{\mathrm{Nu}}{-}\mathrm{C(=O)}{-} + \mathrm{OH}^{-}$$

Of course, the carboxylate anion is an unactivated, low-energy resonance-stabilised system that is even less susceptible to nucleophilic attack. However, if the hydroxyl group is first converted to a phosphate ester expulsion will occur readily because the leaving group is now a resonance-stabilised phosphate anion which is weakly basic:

$$ATP + ROH \xrightarrow{\quad} R{-}OPO_3^{2-} \xrightarrow{Nu} \overset{+}{Nu}R + HPO_4^{2-}$$
ADP

The formation of the phosphate ester is a phosphoryl transfer in which the real or hypothetical *meta*phosphate, PO_3^-, is transferred to the hydroxyl group.

Expulsion of oxygen from carboxylic acids can occur if ATP is used to convert the non-activated substrate into an activated one by phosphoryl transfer:

$$R{-}C(=O){-}O^- + P(=O)(O^-)_2{-}O{-}ADP \longrightarrow R{-}C(=O){-}O{-}P(=O)(O){-}O^-$$

The product is an acyl phosphate which is itself a mixed acid anhydride and readily attacked by nucleophiles:

$$R{-}C(=O){-}O{-}P(=O)(O^-){-}O^- + \ddot{N}u \longrightarrow R{-}C(=O){-}\overset{+}{Nu} + HPO_4^{2-}$$

Acyl phosphates are too reactive to be acyl-transfer agents themselves, but are used for example to make acetyl coenzyme A, the intermediate used for acetyl transfer.

$$H_3C{-}CO_2^- + ATP \xrightarrow[\text{kinase}]{\text{acetate}} H_3C{-}C(=O){-}OPO_3^{2-} \xrightarrow[\text{phospho-transacetylase}]{\text{CoASH}} H_3CCOSCoA + HPO_4^{2-}$$

Here ATP is effectively acting as a dehydrating agent by activating the carboxylate oxygen anion for elimination by intermediate conversion to a phosphate ester from which phosphate can be displaced by the thiolate anion nucleophile of coenzyme A.

5.4. Coenzyme A

Coenzyme A, CoASH, is ubiquitously found in living organisms and consists of a phosphodiester linking adenosine and pantetheine residue with a reactive terminal thiol group:

$$\text{Adenosine(3'-}OPO_3^{2-}\text{)-5'-}CH_2O{-}P(=O)(O^-){-}O{-}P(=O)(O^-){-}OCH_2CMe_2CH(OH)CONH(CH_2)_2CONHCH_2CH_2SH$$

CoASH

Thiols are much more acidic than alcohols and the pK_a of coenzyme A is 8 so there is a reasonable concentration of the powerfully nucleophilic thiolate anion present at neutral pH.

$$\text{CoASH} \rightleftharpoons \text{CoAS}^- + \text{H}^+$$

Acylated coenzyme A are thioesters, RCOSCoA, which are common intermediates used to transfer acyl groups and as a source of carbanions.

Non-enzyme-catalysed bond-forming and -breaking reactions, such as the aldol and Claisen condensations, require strongly basic conditions to form the carbanion intermediates. In living systems, the carbanion formed in Claisen-type condensation is the α-anion of an acylthioester, usually acyl coenzyme A, rather than the usual acyl oxygen ester.

$$\text{RCH}_2\text{COSR}' \rightleftharpoons \text{R}\bar{\text{C}}\text{H}{-}\text{C}(=\text{O}){-}\text{SR}' \longleftrightarrow \text{RCH}{=}\text{C}(\text{O}^-){-}\text{SR}$$

$$\xrightarrow{\text{R}-\text{C}(=\text{O})-\text{R}} \text{R}{-}\text{CH}(\text{COSR}'){-}\text{C}(\text{OH})(\text{R}){-}\text{R}$$

$$\xrightarrow{\text{R}-\text{C}(=\text{O})-\text{SR}'} \text{RCH}(\text{COSR}'){-}\text{C}(=\text{O}){-}\text{R}$$

α-Hydrogens of acylthioesters are more acidic than those of oxygen esters which makes the corresponding carbanion more easily formed and less reactive but more selective. The pK_as of acylthioesters are 2–4 units lower than the corresponding oxygen derivative. Replacement of the alcoholic oxygen of a carboxylic ester by sulphur reduces the carbonyl stretching frequency by ca. 40/cm. The latter does not imply a great contribution from the delocalisation of the sulphur lone pair as occurs with amides compared with esters. The effect of changes in mass and bond lengths

$$\text{R}{-}\text{C}(=\text{O}){-}\text{SR} \qquad \text{R}{-}\text{C}(\text{O}^-){=}\overset{+}{\text{S}}\text{R}$$

causes the reduced carbonyl stretching frequency of acylthioesters and the force constant for the carbonyl group is probably the same in O- and S- esters. This is substantiated by the basicities of O- and S- esters being similar. On the other hand, there is little evidence to support the common suggestion that resonance in thioesters is reduced compared with an oxygen analogue. This would be a nice explanation of the increased acidity of α-hydrogens of acylthioesters because it would indicate that the carbonyl group is more ketone-like than ester-like which would not only help stabilise the carbanion but also make the carbonyl group more susceptible to nucleophilic attack. It is not obvious why sulphur is good at stabilising carbanions. An analogous situation is seen with thiamine pyrophosphate (Section 5.2):

thiol ester carbanions

$$\rangle\bar{C}-C(=O)SR \longleftrightarrow \rangle C=C(O^-)SR$$

α-carbanion derivatives of thiamine pyrophosphate

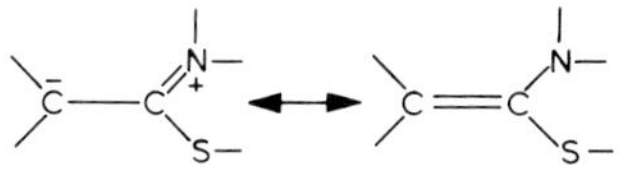

$$\rangle\bar{C}-\overset{+}{C}(=N-)S- \longleftrightarrow \rangle C=C(N-)S-$$

In all the acetyl coenzyme-A-utilising enzymes which catalyse Claisen-type condensations the reaction involves the conversion of the acetyl methyl into a methylene group. A simple example illustrates the use of thiol esters both as carbanion-stabilising systems and as readily hydrolysable esters. The conversion of glyoxalate to malate uses acetyl coenzyme. A probable mechanism is outlined below:

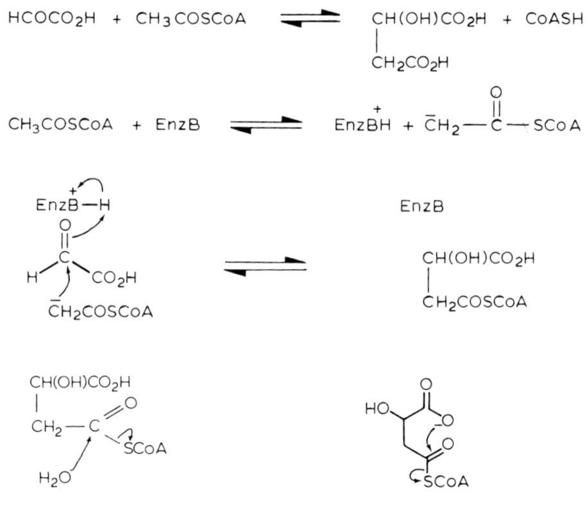

$$HCOCO_2H + CH_3COSCoA \rightleftharpoons CH(OH)CO_2H\text{–}CH_2CO_2H + CoASH$$

$$CH_3COSCoA + EnzB \rightleftharpoons Enz\overset{+}{B}H + \bar{C}H_2-\overset{O}{\overset{\|}{C}}-SCoA$$

$$Enz\overset{+}{B}-H,\ O=C(H)(CO_2H) + \bar{C}H_2COSCoA \rightleftharpoons EnzB + CH(OH)CO_2H\text{–}CH_2COSCoA$$

XXXVIII

XXXIX

It is not clear at present whether the malyl-CoA intermediate is hydrolysed directly (XXXVIII) or by intramolecular catalysis, by the neighbouring carboxylate group to give an intermediate anhydride (XXXIX). In either case the important point is that the malyl-CoA intermediate does not accumulate because the thiol anion of CoA is readily expelled by nucleophilic attack. The thiol anion is a good leaving group because it is the conjugate base of relatively strong acid, i.e. compare the ease of expulsion of thiol anion RS^- (pK_a of RSH = 8) with alkoxide ion RO^- (pK_a of ROH = 15).

6. *Oxidation and reduction*

Oxidation is the removal of electrons and oxidising agents are electrophiles, conversely reduction is the addition of electrons and reducing agents are nucleophiles. There are no obvious oxidising or reducing groups in the side chains of amino acids found in proteins. Enzymes carrying out redox transformations use either organic coenzymes or transition metals as the electron-transfer agents.

The standard reduction potentials of some systems used by enzymes in redox reactions are as follows:

$NAD^+ \rightleftharpoons NADH - 320$ mV

Riboflavin $\rightleftharpoons$ Dihydroflavin $-$ 200 mV

Ubiquinone $\rightleftharpoons$ Dihydroubiquinone + 100 mV

Cytochrome c—$Fe^{3+} \rightleftharpoons$ Cytochrome c—Fe^{2+} + 300 mV

$O_2 \rightleftharpoons H_2O + 810$ mV

As the free-energy change accompanying a redox reaction is given by Eqn. 25, where E_0 is the standard reduction potential, n is the number of electrons transferred per mole and F is Faraday's constant, a reaction is thermodynamically feasible if the difference in reduction potentials is positive. Thus in the preceding list NADH will reduce flavins but cytochrome c cannot.

$$\Delta G = -nF\Delta E_0 \qquad (25)$$

Single electron transfer generates radicals and although this mechanism is now more common than once thought in non-biological redox reactions, its prevalence in enzyme-catalysed reactions is limited to coenzymes with quinoid-type structures e.g. flavins, coenzyme Q, vitamins C, E and K and to enzymes containing transition metals. Of course, there is a growing interest in metabolic disorders initiated by radical reactions. Reduction by 2-electron transfer can take place by either (a) hydride, H^-, transfer or (b) discrete electron, e^-, and proton, H^+, addition.

6.1. Hydride transfer

The hydride ion is a very strong base (pK_a of hydrogen > 40) and hydride transfer to electrophilic centres in non-enzyme-catalysed reactions usually requires H^- either to be "pushed off", as in metal hydride reductions or the Cannizzaro reaction, or to be "pulled off" by strong electrophiles, such as carbonium ions or bromine. The Cannizzaro reaction is another good example of how reaction at one centre in a molecule may completely modify the properties of an adjacent potential reaction site. Hydride transfer from an aldehyde group is thermodynamically unfavourable but addition of hydroxide ion to the carbonyl group, as in the Cannizzaro reaction, makes the tetrahedral adduct a powerful hydride donor to another aldehyde molecule.

$$\mathrm{RCHO} + OH^- \rightleftharpoons \mathrm{R{-}CH(OH)O^-}$$

$$\mathrm{R{-}CH(OH)O^-} + \mathrm{RCHO} \longrightarrow \mathrm{RCOOH} + \mathrm{RCH_2{-}O^-}$$

$$\mathrm{RCOOH} + \mathrm{RCH_2{-}O^-} \rightleftharpoons \mathrm{RCOO^-} + \mathrm{RCH_2OH}$$

Possibly, glyoxylase, which catalyses the conversion of β-keto aldehydes to β-hydroxy thiol esters, acts by an analogous sort of intramolecular hydride transfer.

$$\mathrm{R{-}C({=}O){-}CHO + GSH \longrightarrow RCH(OH){-}COSG}$$

However, the mechanism is not unambiguous and an alternative pathway involves proton abstraction to give an enediolate which can then be reprotonated at the adjacent carbon.

6.2. Nicotinamide coenzymes

A major coenzyme of redox enzymes is nicotinamide adenine dinucleotide (NAD) (Table 2). The cofactor is composed of an adenosine monophosphate residue linked by a phosphodiester to a 5-phosphoribosyl-1-nicotinamide.

TABLE 2
Examples of enzymes dependent upon nicotinamide coenzymes

Reaction	Enzyme
(1) $>C{=}O + H_2 \rightleftharpoons >CHOH$	Alcohol dehydrogenase Lactate dehydrogenase Malate dehydrogenase
(2) $>C{=}C< + H_2 \rightleftharpoons >CH{-}CH<$	Steroid reductase
(3) $>C{=}N{-} + H_2 \rightleftharpoons >CH{-}NH{-}$	Dihydrofolate reductase
(4) $-C({=}O)OH + H_2 \rightleftharpoons -CH{=}O$	Aldehyde dehydrogenase
(5) $>C{=}O + H_2 + NH_3 \rightleftharpoons >CH{-}NH_2$	Glutamate dehydrogenase

The chemically redox reactive part of the coenzyme is the nicotinamide residue. The formal addition of hydride to the pyridinium ring gives 1,4-dihydronicotinamide, NADH, selectively in biological systems although the 1,2- and 1,6-dihydro isomers are also formed by simple chemical reduction. Hydride-ion transfer takes place directly between the nicotinamide system and the substrate, as shown by

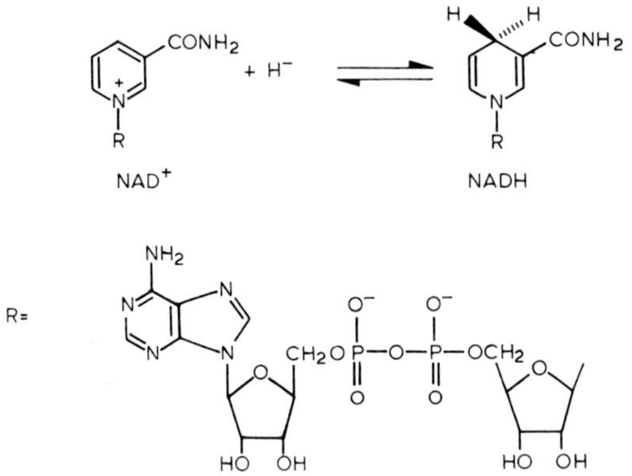

NAD$^+$ + H^- ⇌ NADH

R=

isotopic labelling experiments. C-4 of NADH is prochiral and, because of the asymmetric centres in the molecule, the methylene hydrogens are diastereotopic and described as proR (H_R) and proS (H_S) (XXXX). Class A enzymes, e.g. alcohol dehydrogenase, use the proR hydrogen of NADH and the class B enzymes, e.g. glutamate dehydrogenase use the proS hydrogen. Although NAD is an oxidising

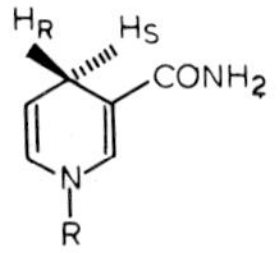

XXXX

agent and NADH a reducing agent the reduction potential of most redox substrates is more positive than that of NAD and, metabolically, most enzymes using nicotinamide coenzymes function as reductases. The dihydropyridine ring is stable in air and is not oxidised by molecular oxygen.

Dihydropyridine is a powerful source of hydride ion because conversion of NADH to NAD is accompanied by a large change in the resonance energy (~ 62 kJ/mole) as the system becomes aromatic. Although there is a possibility that hydrogen transfer takes place by a radical pathway, the mechanism of reduction by NADH is most simply understood in terms of hydride transfer. The receptor site for the hydride ion must be electrophilic. Many reactions involve hydride transfer to the carbonyl group and the negative charge developed on oxygen in the transition state is stabilised/neutralised by an electrophile (XXXXI). This may be either a metal ion, such as Zn(II) in alcohol dehydrogenase or a general acid, such as the ion in lactate dehydrogenase.

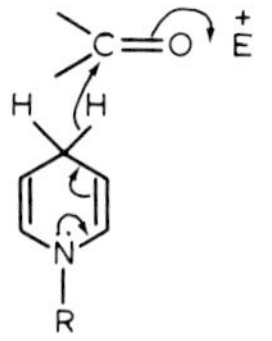

XXXXI

6.3. Flavin coenzymes

These yellow redox systems all contain the conjugated tricyclic isoalloxazine ring system. Riboflavin, vitamin B_2, is an isoalloxazine which is phosphorylated and then adenylated to two active redox coenzyme forms – FMN (flavin adenine mononucleotide, R = phosphorylated ribose) and FAD (flavin adenine dinucleotide, adenylated FMN).

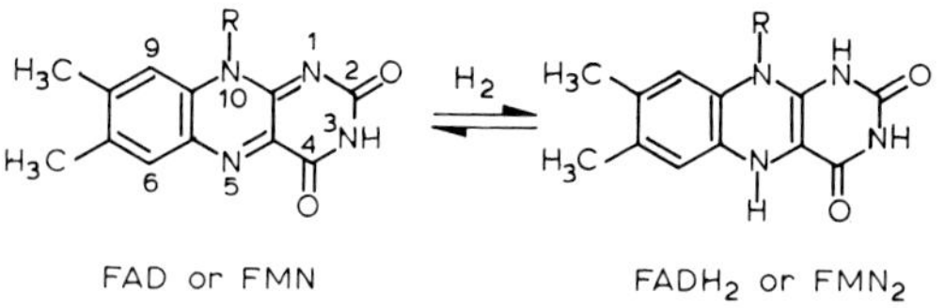

FAD or FMN $FADH_2$ or FMN_2

Reduction of the two formal imine residues gives 1,5-dihydroflavin ($FADH_2$ or $FMNH_2$) with a free-energy change of +38.6 kJ/mole reflecting the decrease in resonance energy. FAD is a 14π electron system whereas the reduced form contains 16π electrons and thus would be antiaromatic if the molecule was planar, but, in fact, the reduced flavin is butterfly shaped. Dihydroflavins are weaker reducing agents than dihydronicotinamide and NADH is used to reduce enzyme-bound flavins. Together with the nicotinamide coenzymes the flavins are the primary acceptors of electron pairs from molecules undergoing oxidation such as aldehydes, amines, amino acids, alcohols, carboxylic acids, dithiols, hydroxy acids and ketones. However, the nicotinamides apparently undergo only 2-electron transfer reactions but the flavins are involved in both 2-electron and 1-electron pathways. Flavins form relatively stable radicals (XXXXII) and are used to couple the 2-electron redox reactions of dihydropyridines with 1-electron redox reaction.

The pK_a of dihydroflavin is 6.7 and therefore at neutral pH nearly half of the molecules will be ionised (XXXXIII).

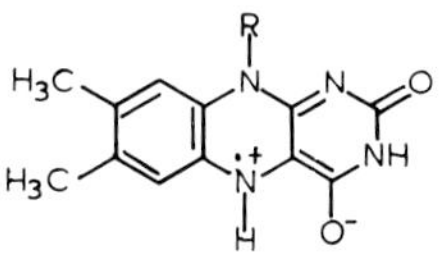

XXXXII 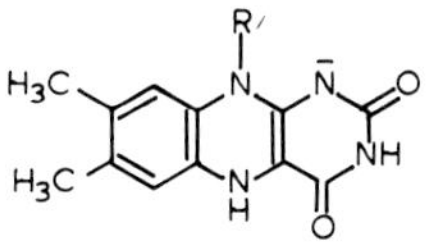   XXXXIII

Another difference between the redox systems is that the monocyclic dihydronicotinamides are inert to oxygen but the tricyclic dihydroflavins readily reduce oxygen to hydrogen peroxide.

A final distinction from nicotinamides is that the flavin coenzymes generally form tight non-dissociable non-covalent complexes with the apoenzyme. Nicotinamides are released at the end of each catalytic cycle and so are consumed as substrate as part of the redox stoichiometry. Because flavins are tightly bound to the apoprotein ($K_D = 10^{-7}$–10^{-11} M) the coenzyme must be oxidised/reduced at the end each turnover before the enzyme complex again becomes catalytically active. Differential binding of flavin and dihydroflavin is responsible for the wide range of redox potentials for flavoproteins so that oxidation or reduction can be thermodynamically favourable. For example, D-amino acid oxidase binds FAD with a dissociation constant of 10^{-7} M but $FADH_2$ with one of 10^{-14} M which changes the reduction potential from −200 for the FAD/$FADH_2$ couple free in solution to 0 mV when bound to the enzyme.

(a) Oxidation of amino and hydroxy acids and carboxylic acids

Unsaturation α, β to a carboxyl group is introduced by flavoenzymes (Eqn. 26, X=NH, O or CR_2).

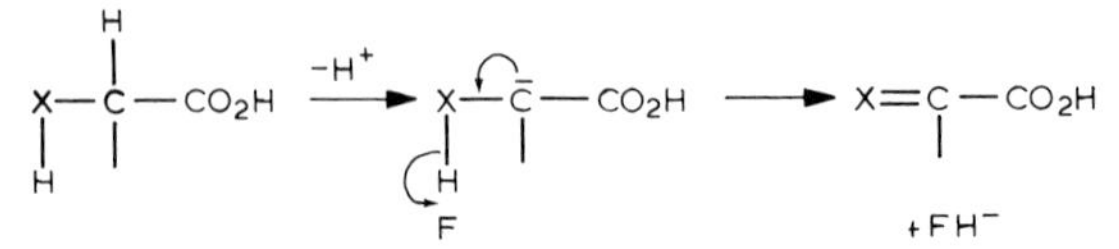

$$\mathrm{H{-}X{-}\overset{H}{\underset{|}{C}}{-}CO_2H} \xrightarrow{-H^+} \mathrm{H{-}X{-}\bar{C}{-}CO_2H} \;(\mathrm{F}) \longrightarrow \mathrm{X{=}C{-}CO_2H} + \mathrm{FH^-} \tag{26}$$

It is thought that the mechanism involves abstraction of the proton α to the carboxyl group followed by oxidation of the resulting carbanion. The exact nature of the oxidation step – 1-electron vs. 2-electron transfer and the importance of flavin N_5 or C_{4a} adducts – is presently unknown. Nucleophilic addition can take place by hydride or carbanion attack at N_5 or C_{4a} of the isoalloxazine (Scheme 5). However, there is radical trapping and CIDNP evidence that carbanion oxidation can take place by 1-electron transfer. 1-Electron transfer from a carbanion to the electron-deficient, but aromatic, oxidised flavin has been observed in model systems (Scheme 5).

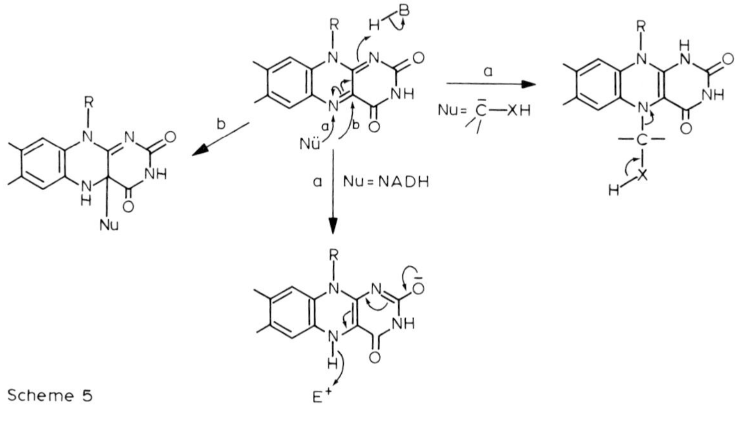

Scheme 5

(b) Oxidation of thiols

Thiols may be oxidised to disulphides by the intermediate formation of covalent 4a flavin–sulphur adducts; subsequent nucleophilic attack by another thiol anion displaces dihydroflavin as a good leaving group (XXXXIV).

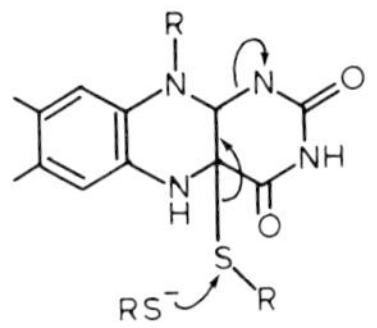

XXXXIV

The above reactions have been described in terms of substrate oxidation which is the predominant metabolic pathway. However, as enzymes are catalysts the reaction in the reverse direction – reduction of substrate by dihydroflavin – will occur by the microscopically reversed mechanisms.

(c) Reductive activation of oxygen by dihydroflavins

Reduced flavins may be reoxidised by molecular oxygen with concomitant production of hydrogen peroxide and transfer of one or two oxygen atoms to a substrate (Eqn. 27).

$$\text{Enz—FH}_2 + \text{O}_2 + \text{S} \longrightarrow \text{Enz—F} + \text{H}_2\text{O}_2 + \text{SO} \qquad (27)$$

The two classes of flavoenzymes which reduce oxygen are the oxidases and the monooxygenases. Because of the deleterious nature of hydrogen peroxide these enzymes are often in proximity to catalase which converts peroxide into oxygen and water.

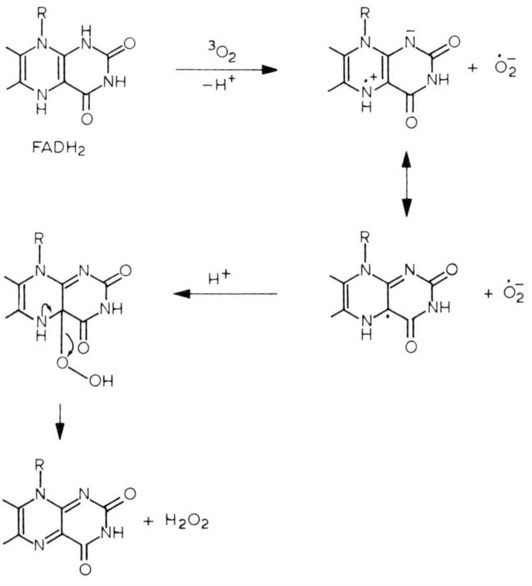

FADH2 $\xrightarrow[-H^+]{^3O_2}$ + $\dot{O}_2^-$ ⟷ + $\dot{O}_2^-$ $\xrightarrow{H^+}$ → FAD + H_2O_2

Because the ground-state electronic configuration of oxygen is the triplet diradical it seems likely that 1-electron transfer takes place to yield the 1-electron oxidised dihydroflavin, a semiquinone-type radical, and the peroxide anion radical. Combination of these two radicals would give a covalent flavin 4a-hydroperoxide, which is thought to be a common intermediate for both flavomonooxygenases and flavooxidases. Hydrogen peroxide is rapidly released from this intermediate, although it is not certain that this is the route used in vivo.

(d) Flavomonooxygenases

Hydroxylation of electron-rich aromatic rings, *N*-oxidation of amines, *S*-oxidation of alkyl sulphides and the conversion of aldehyde and ketones to acids and esters (apparent Baeyer–Villiger reactions) are amongst reactions catalysed by this class of flavoenzymes.

The electron deficiency of the flavin 4a-hydroperoxide is enhanced by the electron-withdrawing isoalloxazine ring which increases its susceptibility to nucleophilic attack. *N*- and *S*-oxidation of amine and sulphides, respectively, occurs efficiently and 10^4–10^5 times more effectively than using alkyl peroxides or hydrogen peroxide as the oxidant (Eqn. 28).

$$\text{FH—O—O—H} + \ddot{\text{N}}\text{u} \longrightarrow \text{FH—OH} + \text{NuO} \qquad \text{Nu} = \text{N}\langle \text{ or } \text{S}\langle \tag{28}$$

Hydroxylation of aromatic rings probably occurs by a similar mechanism i.e. the electron-rich aromatic ring nucleophilically attacks the flavin peroxide (electrophilic attack by the peroxide on the aromatic ring at activated positions) (XXXXV).

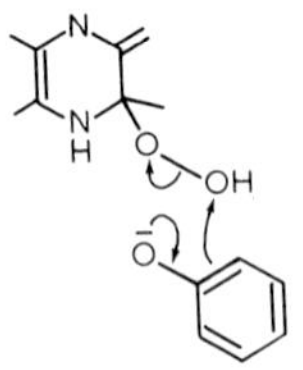

XXXXV

6.4. Electron transfer with metals

The ability of a metal to act as an electron donor or acceptor is very dependent upon the type and arrangement of the ligands surrounding the metal. This is true thermodynamically and kinetically. The electron density at the metal is ligand-dependent as is the overall stability of the complex. The ease of electron-transfer process itself is dependent not only upon the reduction potential but also the geometry of the complex because of Franck–Condon restrictions.

(a) Thermodynamic stability of metal complexes

By convention metals are assigned an oxidation number or state, but, of course, the charge density of the metal itself in a complex is dependent upon the nature of the surrounding ligands and upon the solvent. A consequence of this convention is that it is common to discuss the change in redox potential of a given oxidation state of a metal *brought about by the ligands.* It is important to realise that this, too, then becomes a convention and in reality one should discuss the thermodynamic stability of the whole complex. One could equally well bias a discussion of changes in redox potentials towards the stabilising influence of the metal *on* the ligands.

There is no single generalisation relating the values of formation constants of complexes but a number of useful correlations exist. For a given ionic size increase in charge almost invariably results in a substantial increase in complex stability. If a metal exists in two different oxidation states the more highly charged ion is the smaller and complexes with the metal in the higher oxidation states are normally more stable. There are exceptions to this rule, for example complexes of 1,10-phenanthroline, 2,2-bipyridyl, carbon monoxide and cyanide are more stable with the metal in the lower oxidation state. These ligands have relatively low-energy antibonding orbitals available to accept electrons "back" from the metal.

'Hard' acids or class a cations bind more strongly to 'hard' bases and 'soft' acids or class b cations more strongly to 'soft' bases – the principle of hard and soft acids and bases. This principle, however, lacks a satisfactory quantitative basis.

If, for the half-reaction:

$$M^{n+} + e^- \rightleftharpoons M^{(n-1)+}$$

the concentration of M^{n+} is reduced, say by formation of a stable complex or by precipitation as an insoluble salt, the reduction potential will become less positive and the oxidation state of M^{n+} has been apparently stabilised. Complexation tends to stabilise higher oxidation states. For example, the reduction potentials of cobalt(III) shows that the overall formation constant for $[Co(NH_3)_6]^{3+}$ is ca. 10^{31}-fold greater than that for $[Co(NH_3)_6]^{2+}$.

$$[Co(H_2O)_6]^{3+} + e^- \rightleftharpoons [Co(H_2O)_6]^{2+} \qquad E^0 = +195 \text{ mV}$$

$$[Co(NH_3)_6]^{3+} + e^- \rightleftharpoons [Co(NH_3)_6]^{2+} \qquad E^0 = +10 \text{ mV}$$

(b) Kinetic effects

If the metal complex is inert with respect to ligand-exchange, electron transfer usually takes place by a tunnelling or an *outer sphere* mechanism. The act of electron transfer takes place much faster ($\sim 10^{-15}$ sec) than the rate of change of nuclear configuration ($> 10^{-13}$ sec). The nuclei effectively remain frozen in position during electron transfer so that there are restrictions on changes in spin angular momentum and in geometry (Franck–Condon principle).

The rates of outer-sphere reactions between metal complexes vary widely. For example the rate of electron transfer between two octahedrally coordinated low-spin complexes which differ from one another only by the presence of an extra electron in the t_{2g} orbitals of the complex with the metal in the lower oxidation state tend to be very fast ($\sim 10^3$/M/sec). Transfer of an electron from low-spin Fe(II) t_{2g}^6 to low-spin Fe(III) t_{2g}^5 involves the transfer of an electron from one t_{2g} orbital to another, although the electron lost is not necessarily the same as the one gained. Low-spin metal complexes of this kind have very similar metal–ligand bond lengths in their two oxidation states. For example, Fe—C in $[Fe(CN)_6]^{4-}$ is 1.92 Å and in $[Fe(CN)_6]^{3-}$ is 1.95 Å. If one complex is high spin and the other low spin a slower rate of electron transfer results. Cobalt(III) complexes are usually low spin and those of cobalt(II) are high spin so the electron configuration of the atoms will be: Co(III) t_{2g}^6, Co(II) $t_{2g}^5\,e_g^2$. Transfer of an electron from the Co(II) e_g orbital to the empty e_g orbital of Co(II) leaves both complexes electronically excited and these higher-energy states will contribute to the activation energy.

If the metal–ligand bond distances change considerably with oxidation state this can also retard the rate of electron transfer. For example the Fe—O distance in $[Fe(H_2O)_6]^{2+}$ is 2.21 Å but that in $[Fe(H_2O)_6]^{3+}$ is 2.05 Å so if electron transfer took place from $[Fe(H_2O)_6]^{2+}$ to $Fe(H_2O)_6^{3+}$, with both complexes in their ground states, the product would be a compressed Fe(II) ion and a stretched Fe(III) ion. It is therefore necessary that the molecules become vibrationally excited *before* electron transfer takes place i.e. the Fe(III) and Fe(II) geometries must be expanded and compressed, respectively.

Outer-sphere reactions between complexes of different metals are often very fast, and are, of course, assisted by the decrease in free energy accompanying chemical reaction.

(c) Redox metalloproteins and oxygen

We are concerned in this section with redox metalloproteins and it is thought that many of these systems involve oxygen bound to a metal ion. Oxygen is an oxidising agent and as such is an electron acceptor. However, ground-state oxygen is triplet and contains two electrons in antibonding π orbitals; electron donation into one of these orbitals is thus unfavourable. Superoxide radical anion, O_2^-, is a weak base

$$O_2 \xrightarrow{e^-} O_2^- \xrightarrow{e^-} O_2^{2-} \text{ peroxide ion}$$

$$O_2^- \underset{}{\overset{H^+}{\rightleftharpoons}} HO_2^\bullet \quad pK_a\ 4.7$$

$$O_2^{2-} \overset{H^+}{\rightleftharpoons} HO_2^- \overset{H^+}{\rightleftharpoons} H_2O_2 \quad pK_a\ 11.6$$

$$H_2O_2 \xrightarrow{2e^-\ 2H^+} 2\,H_2O$$

and nucleophile in aqueous solution but is a powerful nucleophile in aprotic conditions. Superoxide itself is not a powerful oxidant but it rapidly abstracts protons from weakly acidic substrates and then disproportionates to peroxide and oxygen, both of which are strong oxidants.

Oxygen can be bound to metal ions with several geometries: side-on (to form a triangle), end-on linear and end-on bent. The electronic possibilities include a simple electron-pair donation from oxygen, 1-electron acceptance by oxygen so that there is effectively a coordinated superoxide ion and the charge on the metal is increased one unit ($M \cdot O_2 \leftrightarrow M^+O_2^-$) and 2-electron acceptance to give effectively coordinated peroxide with the oxidation state of the metal increased by two ($M \cdot O_2 \leftrightarrow M^{2+} \cdot O_2^{2-}$) [26].

One of the effects of ligation to metal ions on the properties of superoxide ion is to stabilise it with respect to reduction to oxygen (the reduction potential of O_2 is made more positive).

(d) Iron-containing proteins and enzymes

Both Fe(II), d^6, and Fe(III), d^5, usually form octahedral complexes which are high spin unless there is a strong ligand field such as imine nitrogen donors. The affinity of Fe(III) for saturated nitrogen ligands is very low but that for ligands that coordinate by oxygen is high. As usual, the nature of the ligands and solvent can greatly affect the reduction potential of Fe^{3+} to Fe^{2+}. For example, the lower oxidation state is usually favoured by a polar solvent and by ligands which are electron acceptors through inductive effects, π orbitals e.g. pyridine-type or through orbitals e.g. sulphur. Good σ electron donor ligands e.g. imidazole favour the higher oxidation state.

Low-spin Fe(II) or Fe(III) is more compact than high-spin iron. Consequently if the geometry of the ligands is 'fixed' e.g. as in porphyrin system or by a rigid protein structure the binding energy and coordination geometry of the metal may be different in its two spin states.

The quite remarkable chemistry of iron proteins is reflected by the fact that the square planar tetradentate porphyrin (XXXXVI) is coordinated to Fe in many different proteins but these molecules show such diverse behaviour as oxygen transport, electron transport and catalytic reactivity in metalloenzymes. The protein must be responsible, by providing the other ligand(s) and by general environmental (cf. solvent) effects, for the differences in reduction potential and chemical reactivity of complexes of the various iron porphyrin systems.

The macrocyclic tetrapyrrole ring provides 4 nitrogens to ligate to iron in equatorial positions. To complete the octahedral complex one of the axial positions is filled by a ligand from the apoprotein – a histidine imidazole in the case of haemoglobin and a cysteine thiolate anion in cytochrome P450. The remaining axial position may be occupied by a protein ligand, water or substrate or left vacant.

Low-spin Fe can fit into the cavity of the tetrapyrrole macrocycle (XXXXVII) but high-spin Fe takes up the unusual square pyramidal stereochemistry such that Fe(III) is ~ 0.3 Å and Fe(II) is ~ 0.7 Å out of the plane defined by the 4 nitrogens

(XXXXVIII). When haemoglobin binds oxygen there is a shift from low-spin Fe(II)

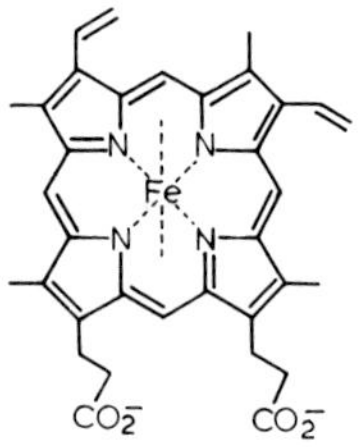

XXXXVI

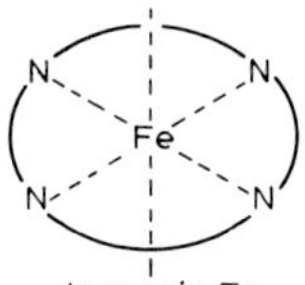

low spin Fe

XXXXVII

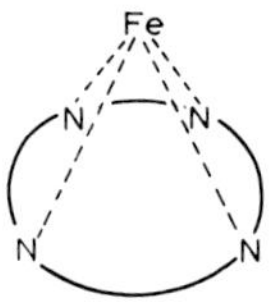

high spin Fe

XXXXVIII

to high-spin Fe(II) causing a movement of Fe towards the equatorial plane of the macrocycle. However, this movement is not as large as has been assumed in the past [25].

Microsomal cytochrome P450 is an iron-dependent oxygenase which reduces one atom of oxygen to water whilst the other is transferred to the substrate. The general outcome of this reaction is hydroxylation of the substrate which may be important for its consequent breakdown or solubilisation if it is toxic.

Substrate binding to low-spin Fe(III) P450 is followed by electron transfer from an iron–sulphur cluster protein to generate high-spin Fe(II). Oxygen can bind reversibly as an axial ligand to give a diamagnetic complex. This is followed by rate-limiting addition of a second electron to give, formally, superoxide ion bound to Fe(II). It is not known if this is the active hydroxylating agent or if a protonated derivative is involved. The function of the metal ion is presumably to generate an electrophilic oxygen.

Catalase, which catalyses the conversion of hydrogen peroxide to oxygen and water (Eqn. 29), and peroxidases, which catalyse the conversion of alkyl peroxides to alcohol and water with concomitant dehydrogenation of a substrate (Eqn. 30), are Fe(III) haemoproteins.

$$2\,H_2O_2 \longrightarrow O_2 + 2\,H_2O \qquad (29)$$

$$AH_2 + ROOH \longrightarrow ROH + H_2O + A \qquad (30)$$

Singlet molecular oxygen is not produced in the decomposition of hydrogen peroxide by catalase or by horseradish peroxidase although it is formed in the base-catalysed reaction [27]. The free enzyme contains high-spin out-of-plane Fe(III). Coordination of the peroxide to Fe(III) again generates an electrophilic oxygen which is more susceptible to nucleophilic attack by hydride transfer (IL). This mechanism could also explain the oxidation of ethanol to acetaldehyde by the enzyme bound peroxide (L).

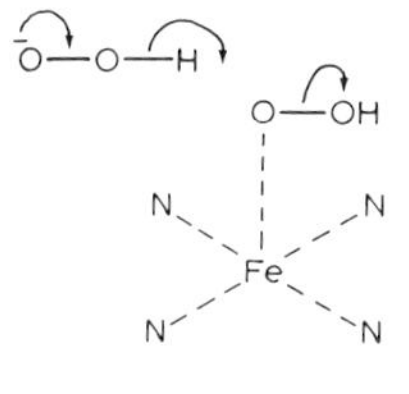

IL

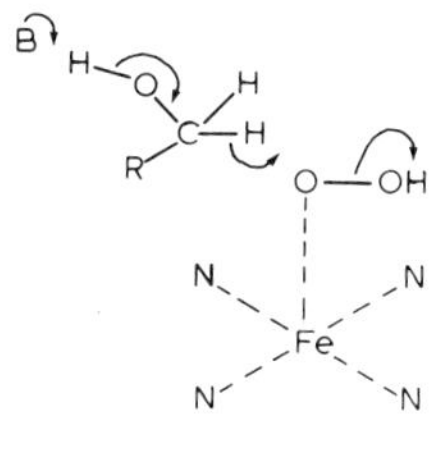

L

(e) Copper-containing oxidases and monooxygenases

The reduction potential of Cu^{2+} to Cu^{+} is 153 mV, but the relative stabilities of Cu(I) and Cu(II) depend very strongly on the nature of the ligands, the solvent or the nature of neighbouring atoms in a crystal. In aqueous solution only low-equilibrium concentrations ($< 10^{-2}$ M) of Cu^{+} can exist and the only stable simple derivatives are insoluble. This is because of the higher solvation and lattice energies and greater complex formation constants for complexes of Cu(II). There are various Cu aggregates in which there is a short Cu · · · · Cu distance of 2.4–2.8 Å. Although the Cu · · · · Cu distance in metallic Cu is ~ 2.6 Å, Cu—Cu bonding is thought not to be significant in the aggregates and the short distance is probably a result of ligand stereochemistry.

The d^9 configuration of Cu(II) makes it subject to Jahn–Teller distortion in a regular octahedral or tetrahedral environment, and its stereochemistry, except in very rare situations, is therefore never regular. The typical distortion of "octahedral" complexes gives four short Cu—L bonds and two long axial ones, which, in the limit, is indistinguishable from square coordination. In general, Cu(II) does not bind the fifth and sixth ligands strongly.

In redox Cu proteins the nature of the ligands is not generally known. Some are intensely blue due to charge transfer arising in a Cu—S σ bond. Oxidation of Cu(I) to Cu(II) is usually accompanied by electron transfer to oxygen, which can either be reduced to hydrogen peroxide or water with or without oxygenation of a substrate.

Two-electron oxidations of substrates by copper enzymes reduce O_2 to H_2O_2:

$$SH_2 + O_2 \longrightarrow S + H_2O_2$$

An example is galactose oxidase which catalyses the oxidation of a primary alcohol to an aldehyde. However, the mechanism of this reaction is not clear although it has been suggested that the active enzyme contains Cu(II) and hydride transfer from the alcohol to the metal (LI) is followed by two 1-electron transfers

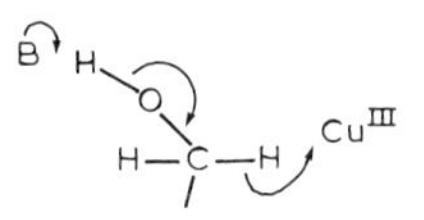

LI

from Cu(I) to O_2.

Some copper enzymes, containing 4 or more copper atoms, carry out a 4-electron reduction of both oxygen atoms of O_2 to H_2O and do not oxygenate the substrate. For example, ascorbate oxidase converts the enediol of vitamin C to an α,β-diketone in dehydroascorbate with oxygen acting as an external electron acceptor. Ascorbate (LII) is a good bidentate ligand and coordination to Cu(II) could facilitate electron transfer. It has been suggested that two Cu(II) are coupled thus allowing the electron transfer from the substrate to the metal. Four Cu^+ then undergo rapid 4-electron transfers to oxygen in a poorly understood process. Two electrons are removed from ascorbate to give dehydroascorbate (LIII).

LII $\underset{}{\overset{-2H^+}{\rightleftharpoons}}$ [ascorbate dianion] $Cu^{2+}\cdots Cu^{2+}$ $\longrightarrow$ LIII $+ Cu^+\cdots Cu^+$

A third category of copper enzymes are monooxygenases which, in addition to reducing O_2 to H_2O, add oxygen to the substrate. For example dopamine-β-hydroxylase hydroxylates the benzylic position of dopamine (LIV) to give noradrenalin, one

LIV $+ {}^{18}O_2 +$ ascorbate $\xrightarrow{Enz-Cu^{2+}}$ [noradrenalin] $+ H_2{}^{18}O$ + dehydroascorbate

of the neurotransmitters. As we have seen previously, there is often a need for an external agent to produce an active coenzyme from an inactive precursor. In this case, ascorbate is required to reduce Cu(II) to Cu(I) which can then, presumably, in turn activate oxygen by acting as an electron acceptor generating an electrophilic oxygenating agent:

$$Enz \cdot Cu^+ + O_2 \longrightarrow \left[Enz \cdot Cu^+ \cdot O_2 \leftrightarrow Enz \cdot Cu^{2+} \cdot O_2^-\right]$$

Copper needs to be electron-rich (in a low oxidation state) to bind oxygen but if binding does occur directly the degree of electron transfer from Cu(I) to oxygen is not known. Formally transfer of one electron would generate a bound superoxide anion and metal ion complexes of this species are known. Although the superoxide radical anion contains an odd number of electrons it is not a reactive electron-transfer oxidant unless the resulting peroxide anion is stabilised by protonation or coordination to a metal ion. Copper-bound oxygen may thus act as an electrophile but at present the exact mechanism remains speculative.

References

1 Jencks, W.P. (1981) Chem. Soc. Rev. 10, 345.
2 Jencks, W.P. (1972) Chem. Rev. 72, 705.
3 Jencks, W.P. (1976) Acc. Chem. Res. 9, 425.
4 Morris, J.J. and Page, M.I. (1980) J. Chem. Soc. Perkin II 212.
5 Page, M.I. (1973) Chem. Soc. Rev. 2, 295.
6 Jencks, W.P. and Gilbert, H.F. (1977) Pure and Appl. Chem. 49, 1021; Young, P.R. and Jencks, W.P. (1977) J. Am. Chem. Soc. 99, 1026.
7 Kershner, L.D. and Schowen, R.L. (1971) J. Am. Chem. Soc. 93, 2014.
8 Wang, J.H. and Parker, L. (1967) Proc. Natl. Acad. Sci. (U.S.A.) 58, 2451.
9 Page, M.I. (1979) Int. J. Biochem. 10, 471.
10 Page, M.I. (1977) Angew. Chem. Int. Edn. 16, 449.
11 Morris, J.J. and Page, M.I. (1980) J. Chem. Soc. Perkin II 685.
12 Morris, J.J. and Page, M.I. (1980) J. Chem. Soc. Perkin II 679.
13 Bender, M.L. and Williams, A. (1966) J. Am. Chem. Soc. 88, 2502.
14 Mildvan, A.S. (1970) Enzymes 2, 246.
15 Bender, M.L. (1971) Mechanisms of Homogeneous Catalysis from Proteins to Proteins, Wiley, New York, p. 212.
16 Westheimer, F.H. (1955) Trans. N.Y. Acad. Sci. 18, 15.
17 Stability Constants of Metal-Ion Complexes, Chem. Soc. Special Publ., No. 17, 1964; No. 25, 1971.
18 Cotton, F.A. and Wilkinson, G. (1966) Advanced Inorganic Chemistry, Interscience, 2nd Edn.
19 Jencks, W.P. and Regenstein, J. (1970) in: H.A. Sober (Ed.), Handbook of Biochemistry, CRC, Cleveland, OH, 2nd Edn. p. J-150.
20 Cox, B.G. (1974) J. Am. Chem. Soc. 96, 6823.
21 Lipscomb, W.N. (1974) Tetrahedron 30, 1725.
22 Gensmantel, N.P., Proctor, P. and Page, M.I. (1980) J. Chem. Soc. Perkin II 1725.
23 Lipscomb, W.N. (1970) Accts. Chem. Res. 31, 81; Kaiser, E.T. and Kaiser, B.L. (1972) Accts. Chem. Res. 5, 219.
24 Prince, R.H. and Wooley, P.R. (1972) J. Chem. Soc. Dalton 1548.
25 White, D.K., Cannon, J.B. and Traylor, T.G. (1979) J. Am. Chem. Soc. 101, 2443.
26 Sakaki, S., Hori, K. and Okyoshi, A. (1978) Inorg. Chem. 17, 3183.
27 Smith, L.C. and Kulig, M.L. (1976) J. Am. Chem. Soc. 98, 1027.
28 Hamilton, G.A. (1978) J. Am. Chem. Soc. 100, 1899.

Enzyme reactions involving imine formation

DONALD J. HUPE

Merck Sharp and Dohme Research Laboratories, Box 2000, Rahway, NJ 07065, U.S.A.

1. Introduction

There are a substantial number of enzymatic reactions which involve the formation of a carbon nitrogen double bond between substrate and enzyme which do not involve pyridoxal, pyridoxamine or other cofactors. The most common type of enzyme found is that in which a condensation occurs between a lysine amino group from the protein and a carbonyl group from the substrate [1,2]. These enzymes, which include aldolases, decarboxylases, dehydratases, and others, have many mechanistic similarities. A common feature in these enzymes is that they allow the activation of a carbonyl group in a manner that stabilizes the formation of an α-carbanion. Despite the fact that the formation of an imine bond is mechanistically complex, requiring a number of bond-forming and -breaking reactions including protonations and deprotonations, evolutionary pressure has dictated the generation of a variety of enzymes utilizing this pathway. In this review we will attempt to show the value of formation of this adduct in terms of physical organic studies that demonstrate the amount of enhancement of reactivity obtained by conversion of a carbonyl group to an iminium ion.

Another important class of proteins involving imine formation between lysine amino groups and substrate carbonyl groups are the pigment proteins binding retinal [3]. These include opsin which is the pigment involved in the vision, and bacterhodopsin, the pigment found in purple membrane of halobacteria. The imine linkage in these proteins is responsible not only for binding the pigment, but also contributes substantially to the chromophore that allows light absorption. These will not be considered in detail in this review.

A completely different class of imine-forming enzyme than those described above is typified by histidine decarboxylase, which operates via the formation of an imine bond between substrate and enzyme but does so by using the amino group of the substrate and a carbonyl group attached to the protein [4]. The carbonyl group in this case is that associated with a pyruvamide moiety covalently attached to the protein. In this instance the pyruvamide acts mechanistically in a manner similar to pyridoxal.

The review presented here will focus initially on the physical organic chemistry of imine formation, and the chemical reactivity of the imine adduct thus formed. The

Michael I. Page (Ed.), The Chemistry of Enzyme Action

most well characterized examples of each class of enzyme involving imine formation will then be outlined.

2. *Iminium ion formation*

The formation of iminium ions from basic amines and carbonyl compounds is a 2-step process involving addition of the amine to the carbonyl group to form a tetrahedral intermediate, followed by dehydration of this carbinolamine to form the new C—N double bond. At higher pH values the dehydration of the carbinolamine is the rate-determining step [5]. The rate expression, for example, of the reaction of methyl amine with acetone is dominated above pH 4 by a term proportional to methyl amine which reflects rate-determining dehydration of the carbinolamine formed in a rapid pre-equilibrium [6] ($k_{imine} = 11.4 \times 10^{-2}$ [CH_3NH_2]/M-sec). The direct observation of this reaction is difficult because of the small equilibrium constant for imine formation. This type of reaction is therefore commonly studied by trapping the imine as it is formed with hydroxylamine, which reacts rapidly to form an oxime. Because the equilibrium constant for formation of the imine between methyl amine and acetone is so small, the equilibrium is established very rapidly. (The observed rate constant for a reaction proceeding to an equilibrium position is larger than the first-order rate constant for the forward reaction [7].) Thus the addition of methylamine and acetone to an aqueous solution results in the establishment of an equilibrium concentration of imine (and iminium ion) in several seconds. In several studies described below wherein reactions subsequent to imine formation occur, it is common to find a presumption of rapid imine equilibria prior to the slower α-proton abstraction or decarboxylation events that occur subsequently.

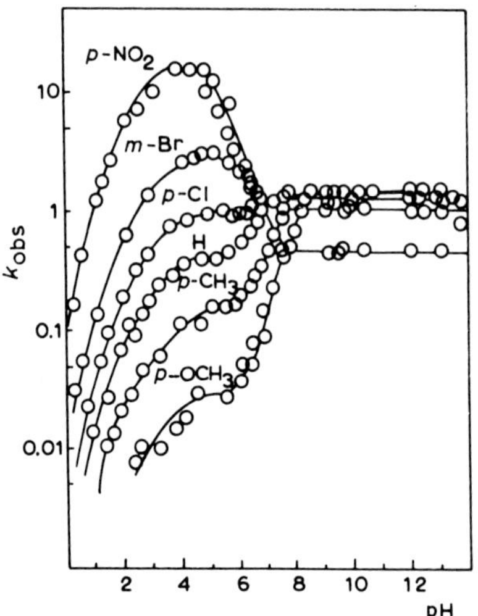

Fig. 1. pH rate profiles for the hydrolysis of substituted benzylidene-1,1-dimethylethylamines [8].

The change in rate-determining step from carbinolamine dehydration to carbinol amine formation occurs as the pH decreases [5,8,9]. As shown in Fig. 1 for the hydrolysis of substituted benzylidene-1,1-dimethylethyl amines, the pH rate profiles vary dramatically at lower pH values. The high pH rates are due to attack of hydroxide ion on protonated imine, which is the reverse of the dehydration of carbinolamine. At lower pH values the term due to water addition to protonated imine becomes substantial for those imines with electron withdrawing substituents. At even lower pH values the rate drops because of the change in rate determining step to the decomposition of the carbinolamine intermediate, which is the reverse of carbinolamine formation. In those cases where the zwitterionic intermediate formed by amine addition to a carbonyl compound (k_1 in Scheme 1) is very unstable, the required proton-transfer steps (k_h in Scheme 1) may become rate-determining. This is particularly true in those cases where the attacking amine is very weakly basic [10].

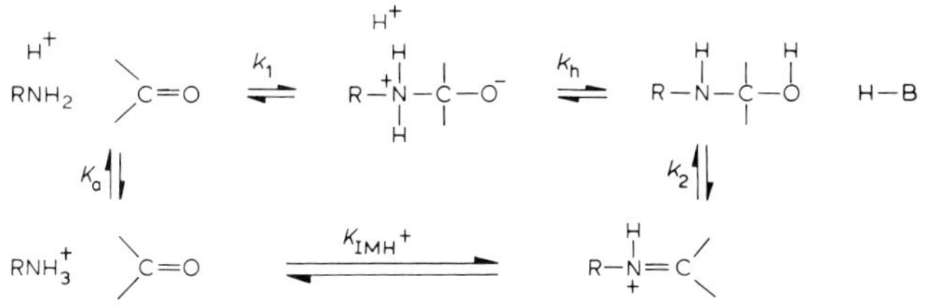

Scheme 1

A generalized pathway for imine-forming reactions is shown Scheme 1. The protonated amine in equilibrium with the protonated imine (K_{IMH^+}) must first ionize to form unprotonated amine (K_a) in order for the reaction to occur. The unprotonated amine adds (reversibly at high pH) to form the zwitterionic intermediate (k_1) which undergoes proton transfer (k_h) to become the neutral carbinolamine.

The subsequent dehydration reaction is general acid-catalyzed ($\alpha = 0.25$) [5] and this results in the formation of the iminium ion product. In enzymatic reactions that require imine formation, most mechanistic attention has been paid to those steps subsequent to the imine formation. The data presented above however suggest some basic requirements for a catalyzed process within an enzyme-active site that results in the formation of a substrate–enzyme imine complex.

It is clear from Scheme 1 that aside from the formation of a new carbon nitrogen bond and the breaking of an old carbon oxygen bond a number of proton transfers must be accomplished in order to form an iminium ion. Aside from the essential lysine group observed at the active site of imine-forming enzymes, it is common to find other functional groups that can play the role of proton donor or acceptor in such processes. For example, a second lysine group has been found at the active site of both muscle aldolase and acetoacetate decarboxylase. In the case of the muscle aldolase Hartman and Brown [11] have found that an essential lysine (146) is modified by the active-site-directed reagent BrAcNHEtOP. The modification of this second active-site lysine prevents the binding of the natural substrate. In the case of acetoacetate decarboxylase a second lysine is adjacent to the imine-forming lysine in the primary sequence and may be the positively charged group responsible for lowering the pK_a of the imine-forming lysine [12,13]. Aside from the ability of these secondary lysines to lower the pK_a of the amino groups in a manner that enhances the reactivity of the iminium ions, these groups could also conceivably play a role in catalyzing the formation of the iminium ions. As shown in Scheme 2 a reasonable mechanism can be drawn for the condensation reaction of a substrate carbonyl with

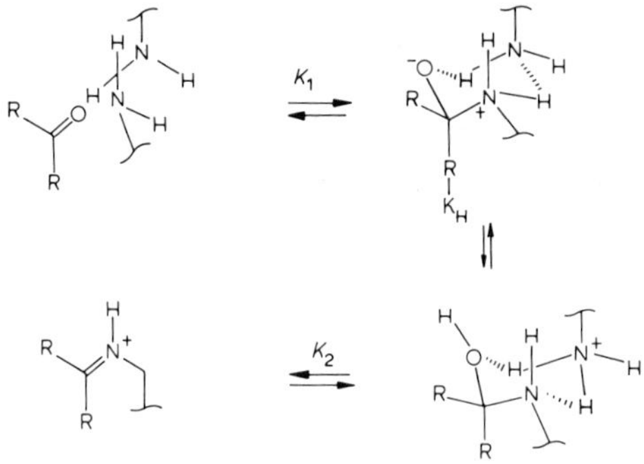

Scheme 2

the essential lysine at the active site of an enzyme in a manner in which the proton-transfer steps (k_h) are accomplished as well as the general acid catalysts of the loss of OH^- (k_2). Hine and Chou [14] have demonstrated that diamines form imines more rapidly than do monoamines as a result of just such intramolecular catalysis.

The equilibrium constants for formation of imine from carbonyl and amine (K_{IM}) are in general larger than the equilibrium constants for formation of iminium ion from carbonyl and ammonium ion (K_{IMH^+}). As shown in Scheme 3 the equilibria for these processes are related and the ratio of K_{imh^+}/K_{im} is equal to the ratio $K_a^{RNH_3^+}/K_a^{IMH^+}$. Since the sp^2 iminium ions are more acidic than the sp^3

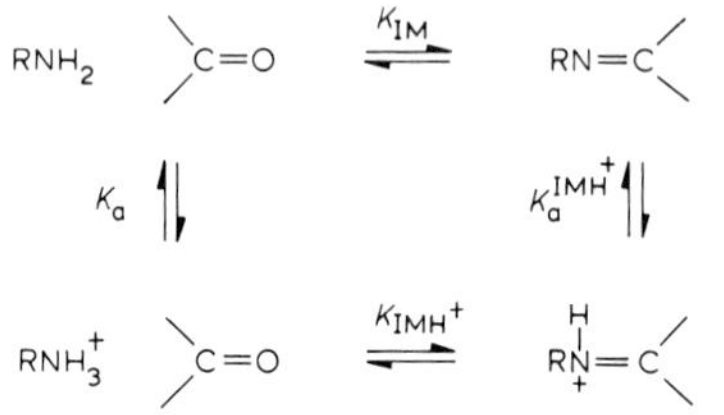

Scheme 3

hybridized ions by a factor of approximately 10^3 [9,15–17] the equilibrium concentration of iminium ions is considerably smaller than the equilibrium concentration of imine adduct at high pH. Only a few values of equilibrium constants for

TABLE 1
Ionization constants for iminium ions, $K_a^{IMH^+}$

Iminium ion	pK	Ref.
$C_6H_5CH{=}\overset{+}{N}HC(CH_3)_3$	6.7	9
$(C_6H_5)_2C{=}\overset{+}{N}HCH_3$	7.2	9
p-$CH_3OC_6H_4CH{=}\overset{+}{N}HCH(CH_3)_2$	7.1	9
$(CH_3)_2CH{=}\overset{+}{N}HCH_3$	7.0	17

imine formation have been measured directly for carbonyl compounds with aliphatic amines, and some are listed in Table 1. A typical situation, therefore, for a ketone and primary aliphatic amine might be one in which the $pK_a^{RNH_3^+} = 10$, $pK_a^{IMH^+} = 7$, $K_{IM} = 10^{-1}$, and $K_{IMH} = 10^{-4}$.

Shown in Fig. 2 is a plot of the concentrations of the various entities in equilibrium calculated using these values. At arbitrarily chosen concentrations of 0.1 M ketone and 0.01 M amine, the concentration vs. pH profiles are as shown. Only a very tiny fraction of either the amine or ketone is tied up a imine or iminium ion at any pH value. The concentration of imine is greatest at high pH values and decreases with decreasing amine concentration. The concentration of iminium ion is linked to the concentration of ammonium ion. Thus the concentration of iminium ion does not decrease above the pK_a for the iminium ion, since it is dependent upon the concentration of protonated amine. This is a critical point for understanding the kinetics of reactions involving the iminium ion since the reasonable initial expectation is that the concentration of this species should decrease above the pK_a for the

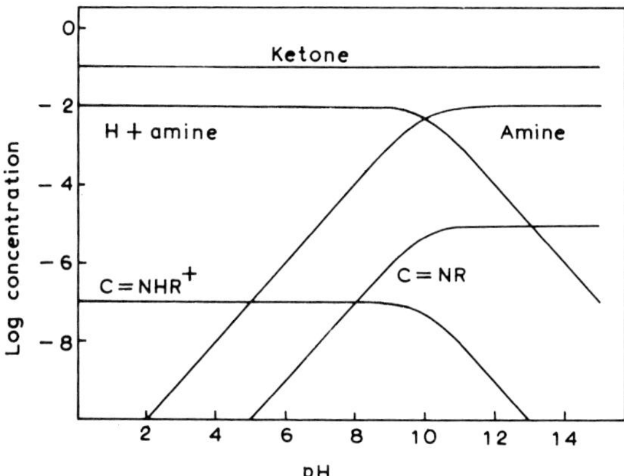

Fig. 2. Typical equilibrium concentrations of ketone, amine, imine and iminium ion (see text).

ionization. It is found, therefore, that reactions involving iminium ion formation in which that formation is rapid and reversible, demonstrate kinetic dependence upon the concentration of protonated amine.

The ability to form imine or iminium ion adducts differs for primary secondary and tertiary amines. Ammonia and primary amines may form any of the sp^2 or sp^3 adducts, whereas secondary amines may form iminium ions but not imines. The fact that secondary amines are as effective as primary amines in catalyzing the α-depro-

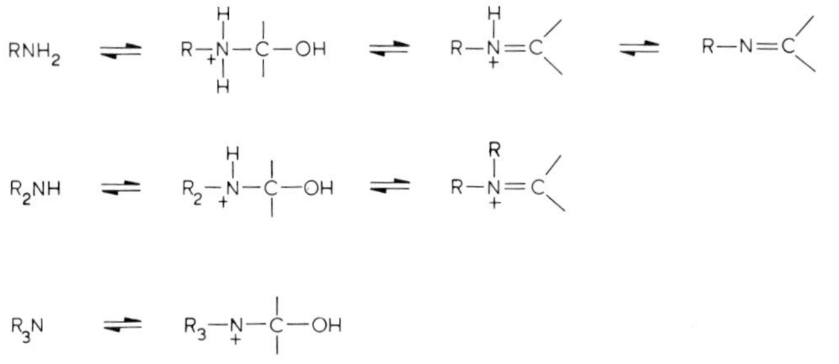

$$RNH_2 \rightleftharpoons R{-}\overset{+}{N}H_2{-}C{-}OH \rightleftharpoons R{-}\overset{+}{N}H{=}C\!< \rightleftharpoons R{-}N{=}C\!<$$

$$R_2NH \rightleftharpoons R_2{-}\overset{+}{N}H{-}C{-}OH \rightleftharpoons R{-}\overset{+}{N}R{=}C\!<$$

$$R_3N \rightleftharpoons R_3{-}\overset{+}{N}{-}C{-}OH$$

Scheme 4

tonation reactions described below [18], indicates the importance of the iminium ion in this process, rather than the imine present in higher concentration. Tertiary amines are incapable of forming either iminium ions or imines and the lack of catalytic activity of tertiary amines is a typical pattern used as evidence for the presence of an imine-forming pathway in catalysis. Westheimer and Cohen [19] found in an early study of a retroaldol reaction, for example, that it was catalyzed by primary and secondary amines but not by tertiary amines.

It is interesting to compare the equilibrium constants for the formation of the iminium ion from simple aliphatic carbonyl compounds and amines with the same reaction catalyzed by an enzyme. Values of K_m for substrate DHAP with aldolase has been measured [20] in the micromolar range so that for the enzyme substrate iminium ion $K_{IMH^+} = 10^6$. This contrasts with K_{IMH^+} values of 10^{-4} described above. It is apparent therefore that a considerable fraction of the binding energy comes from the usual non-covalent associations between substrate and enzyme site [21] rather than because of the stability of the covalent imine bond being formed. It is clear also that this binding may be disrupted, since, for example, a change in substrate from DHAP to dihydroxyacetone sulfate causes a 10^2 decrease in binding, even though the same imine bond is formed with aldolase [22].

3. *Activation of carbonyl groups by iminium ion formation*

The most important classes of carbonyl group reactions are those which generate an α-carbanion and include the retroaldol reaction, enolization reactions, and α-decarboxylation reactions of β-keto acids. Each of these reactions is catalyzed by primary amines in model systems by virtue of iminium ion formation. Studies of these processes have contributed to the understanding of the enhancement of the carbonyl group activity and the factors that make it useful for enzyme systems to operate via this pathway. The fundamental question answered by these studies is how much can the reaction be enhanced by conversion of the carbonyl group to an iminium ion, and upon what factors does this enhancement depend. The methods used include measuring the rates of exchange of α-protons for labeled protons [24], rates of iodination of intermediate enolate anion or enamine [16], or rates of elimination of β-leaving groups [18]. In each of these cases the proton abstraction is the slow step and the general notions concerning iminium ion reactivity derived from each of these methods is comparable.

In model system studies on the loss of acetic acid from I to form the chromophoric product II (Scheme 5), it was found that primary amines were much more effective as catalysts than were comparably basic tertiary amines [18]. In addition to

OAc

I II

Scheme 5

the simple proton abstraction mechanism available to tertiary amines, primary amines could also condense with the substrate carbonyl rapidly to form a small equilibrium concentration of iminium ion, and that iminium ion could be deprotonated in a rate-determining step by general base catalysts present in solution. The assignment of rate-determining step was assisted by isotope affect studies. As shown in Fig. 3, for a typical primary amine (cyanomethylamine, pK_a 5.34) a bell shaped pH rate profile is obtained. No such pH profile is found for tertiary amines which are incapable of forming iminium ions with the substrate. The rate expression for the overall reaction due to amine was: $k_{obs} = k_a\,[RNH_3^+] + k_{ab}\,[RNH_3^+]\,[RNH_2] + k_b\,[RNH_2]$. The individual terms in the kinetic rate expression were ascribed to the three different types of transition states drawn below (Scheme 6), in which the iminium ion is deprotonated by water, another mole of free amine, or hydroxide ion, respectively.

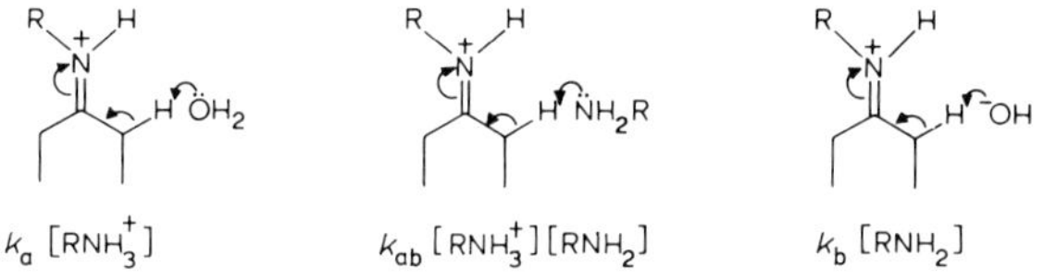

R H N H ÖH$_2$ R H N H ṄH$_2$R R H N H $^-$OH

$k_a\,[RNH_3^+]$ $k_{ab}\,[RNH_3^+][RNH_2]$ $k_b\,[RNH_2]$

Scheme 6

As noted above, the dependence of reaction rates on the concentration of

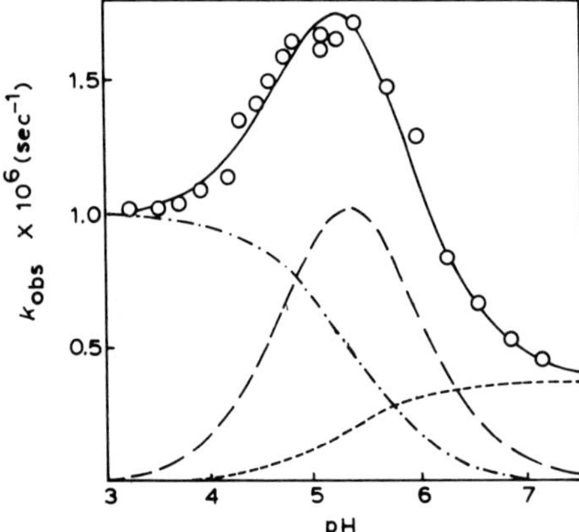

Fig. 3. pH rate profile for the conversion of I to II catalyzed by cyanomethylamine. The individual contributing terms are also shown [18].

iminium ion gives an apparent dependence upon the protonated amine concentration as long as the formation of the iminium ion is rapid (so that the iminium ion is in equilibrium with ammonium ion). There is no observable change in the pH rate profile associated with the pK_a of the iminium ion.

Since this system allowed the measurement of individual terms in the rate expression for iminium ion catalysis it was possible to vary the proton-abstraction base and the imine-forming amine in order to determine the effects of these changes on the overall rate [15]. When the iminium ion was held constant and the base doing the proton abstraction was varied, a Brønsted plot with a slope, $\beta = 0.5$ was found. When the proton-abstracting base was held constant and the imine-forming amine was varied, the rate increased with a decrease in the pK_a of the amine. The slope of a plot of the rate constant versus pK_a of the imine-forming amine was -0.5. Thus increasing the pK_a of the proton-abstracting base by 2 units increased the rate of the reaction by 1 order of magnitude. Decreasing the pK_a of the imine-forming amine by 2 units increased the rate by 1 order of magnitude. These two effects are perfectly reasonable since β values for proton abstraction from ketones are of this order of magnitude, and increasing the electrophilicity of the iminium ion by decreasing the pK_a of the parent amine should increase the rate proton abstraction.

The balance between these two changes is demonstrated in Fig. 4 in which the calculated pH rate profiles for a variety of amines are displayed. The bell shaped term for the catalysis by a series of primary amines varying 10000-fold in basicity are approximately the same size. Since these bell shaped terms are due to the primary amine acting as both the imine-forming amine and as the proton abstractor, the magnitude of these terms reflects a change in both the pK_a of the proton-abstracting base and the pK_a of the imine-forming amine. The fact that the terms remain essentially constant over a wide range of pK_a reflects the balance of these two independent electronic effects.

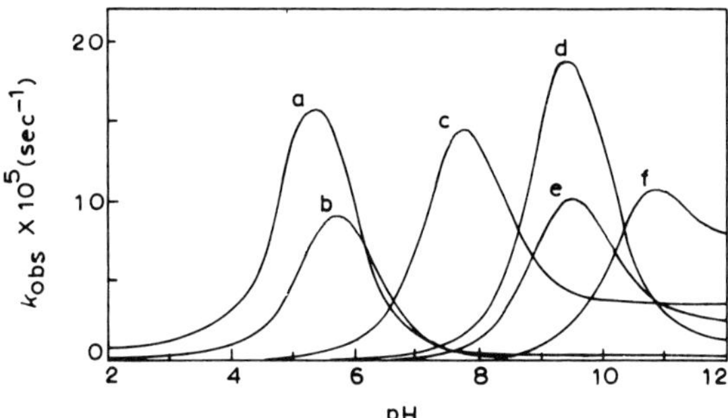

Fig. 4. Calculated pH rate profiles for amines of widely varying pK_a, demonstrating bell shaped curves of similar magnitude [15].

It might initially seem that the best catalyst for doing a proton abstraction via the iminium ion pathway might be one which had a very low pK_a imine-forming amine, and a high pK_a proton-abstracting base. Although this combination would produce a very high *rate constant* for proton abstraction, the observed *rate* at any pH would be small. The low pK_a amine would be mostly unprotonated, and the high pK_a base would be mostly protonated, and therefore be unable to catalyze the reaction. An analysis of these factors leads to the conclusion that the highest turnover of substrate would be obtained from an enzyme which has an imine-forming amine with a pK_a equivalent to the pK_a of a proton-abstracting base [24]. Another conclusion is that the optimal rate is obtained when the pH of the solution is equivalent to the pK_a of the amine and the pK_a of the base. The absolute value of the maximal rate is independent of the pH chosen. One might expect, therefore, that evolutionary pressure would select enzymes with lysine amino groups responsible for iminium ion formation which have pK_a values around neutrality, and that the proton-abstracting base would also have a pK_a around neutrality. Frey et al. [12] and Schmidt and Westheimer [13] have shown that the imine-forming lysine amino group in acetoacetate decarboxylase has a pK_a of 6.0, consistent with the notion that a lower pK_a amino group would create a more reactive iminium ion. It has been proposed that the proton-abstracting base in rabbit muscle aldolase is the anion of the substrate phosphate which would have a pK_a of approximately 7 [25]. Thus there is at least some experimental confirmation of the notions described above.

4. *Aldolases*

Rabbit muscle aldolase is the archetypical type 1 (imine-forming) aldolase and constitutes 3% of the soluble protein obtained from this source [26]. It is responsible for catalyzing the cleavage or synthesis of the carbon–carbon bond which joins carbon 3 and 4 in fructose 1,6-diphosphate (F-1,6P_2) to form dihydroxyacetone phosphate (DHAP) and glyceraldehyde-3-P (G-3P). The mechanism involves (in the synthesis direction) the binding of DHAP as an iminium ion first (1) (Scheme 7). An

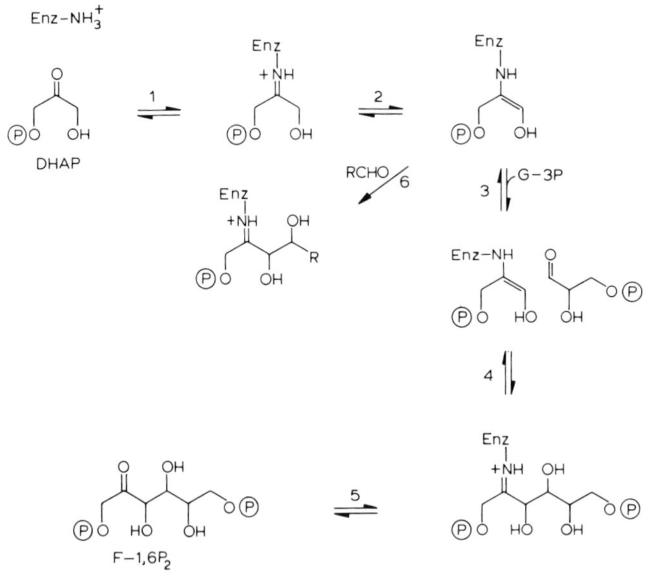

Scheme 7

α-proton abstraction then occurs stereospecifically (2), and after this G-3P binds (3). The condensation reaction then occurs to form the new C—C bond (4), resulting in the iminium ion of F-1,6P_2 which can then hydrolyze (5). The evidence for this mechanism is presented below.

The primary amino acid sequence of rabbit muscle aldolase has been determined [27]. The molecular weight of the protein is 160000, and it has 4 equivalent subunits each bearing an active site. Some aldolases are found as tetramers of α and β subunits which differ only by the conversion of Asn 358 to an aspartate residue. Because this enzyme is essential to glycolysis, because it is readily available, and because it is mechanistically interesting it has been the subject of a large number of studies. These include determinations of active-site residues, order of substrate addition, stereochemistry, and chemical mechanism which will be integrated below in an attempt to generate a picture of the active-site structure.

Following the study by Westheimer and Cohen [19] of the primary and secondary amine catalyzed dealdolization of diacetone alcohol which was proposed to occur through an essential imine intermediate, Speck and Forist [28] suggested that the same class of reaction may occur in aldolase. The first characterization of a stable covalent entity formed between substrate and enzyme was done by Horecker et al. [29] on transaldolase. This enzyme, which forms a stable complex with dihydroxyacetone formed from the cleavage of fructose-6-P, was found to be susceptible to reduction by sodium borohydride. It was demonstrated that this adduct was formed irreversibly in the presence of substrate and borohydride, and that irreversible enzymatic inactivation occurred at the same rate as incorporation of [^{32}P]DHAP. Grazi et al. [30] were responsible for demonstrating that a similar adduct was formed between DHAP and rabbit muscle aldolase in the presence of borohydride, and by using [^{14}C]DHAP isolated the radioactive derivative by hydrolysis followed by chromatography [31]. The adduct was identified as N^6-β-glyceryllysine, and its

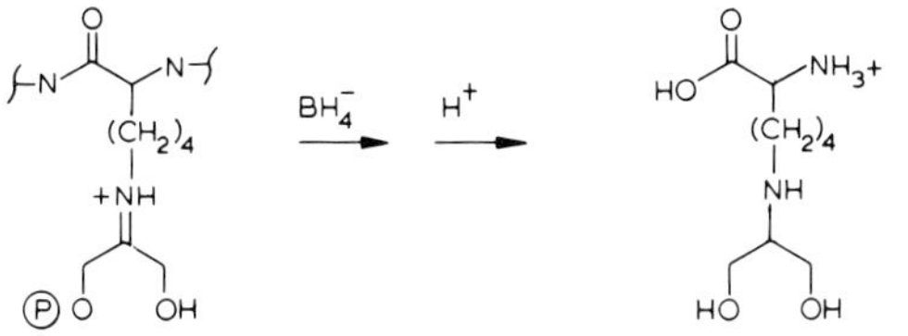

Scheme 8

structure was confirmed by synthesis [32]. The conclusion was that both transaldolase and muscle aldolase formed an imine linkage between the ϵ-amino group of an active site lysine and the carbonyl group of substrate. Further evidence for the formation of this linkage, and its presence in the mechanistic pathway arose from the inhibition of enzyme in the presence of substrate with cyanide, because of an adduct formed by addition of cyanide to the imine [33]. Also experiments demonstrating ^{18}O exchange during the course of substrate turnover provided further supporting evidence [34]. These experiments described above are the archetypical experimental means of demonstrating an imine intermediate in an enzymatic reaction. Dilasio et al. [35] have also shown that the reduction reaction with sodium borohydride is, in the case of liver aldolase, a stereospecific process resulting in the addition of hydride to to the si face of the iminium ion.

The thiol groups present in aldolase have been studied extensively. Four of the total of 8 cysteine thiols are particularly reactive toward thiol reagents. These are CYS-72, 237, 287, and 336 whereas the 4 other cysteines, CYS-134, 149, 177, and 199, are buried in the native protein and are modified only after denaturation. This behavior is exemplified by the modification of aldolase with substituted sulfenyl sulfoxides [36]. In this case, the rapid modification occurred to CYS-72 and 336 in the absence of substrate but did not in the presence of substrate. CYS-237 and 287 were modified in the presence or absence of substrate. In a similar manner, phenanthroline and copper oxidized CYS-72 and 336 to form an inactive disulfide [37], and this modification was prevented by the presence of substrate. Also, bromoacetate attacks an active-site thiol with substrate protection, whereas other thiol groups are not protected by substrate. Other reagents that have been used to quantify, and separate thiol groups into active-site or non-active-site categories are chlorodinitrobenzene, *p*-mercuri-benzoate, *N*-ethyl-maleimide, and carboxyethyldisulfide [38]. The last of these is interesting because it is reversed by mercaptoethanol through a thiol disulfide interchange reaction. Thus one can make an inactive derivative of aldolase which is a mixed disulfide of the active-site thiol and

β-mercaptopropionic acid and reactivate this protein by treatment with a low molecular weight thiol.

The α-haloacetol phosphates were created as potential active-site-directed inhibitors of enzymes using DHAP since alkylative trapping of the base required for proton abstraction is likely [39]. Whereas this did occur with triosephosphate isomerase [40], the corresponding trapping of the proton-abstracting base in aldolase did not occur. α-Iodoacetol phosphate was responsible for inhibition of the enzyme by oxidizing thiol groups [41]. α-Chloroacetol phosphate was found to react with an active-site thiol group at high pH values [42]. Interestingly, in this study as in others, it has not been possible to prove that modification of this thiol group alone results in complete abolition of activity. Thus the thiol group may have an auxiliary function in catalysis, rather than some fundamental catalytic goal. The inactivation of rabbit muscle aldolase by 2-keto-3-butenylphosphate has been proposed to occur by reaction with both the essential lysine amino group and the active site thiol group as shown below in Scheme 9 [20,25,43].

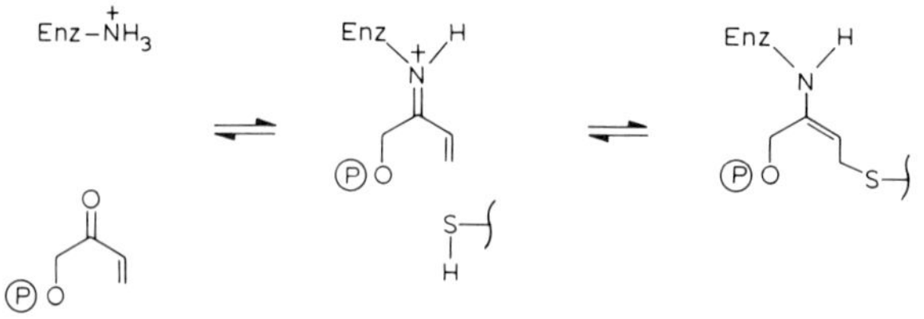

Enz$-\overset{+}{N}H_3$ ⇌ Enz, H, N+ ... ⇌ Enz, H, N ... S

Scheme 9

This inhibition is prevented by converting the enzyme to the mixed disulfide described above. The summary of the data on thiol groups suggests that there are therefore 8 per subunit, but 4 of them are buried within the central portion of the protein and not exposed. Two thiols are on the surface of the protein and 2 are in a position to be protected by the substrate. There is no solid evidence that the protectable thiols are involved in the mechanism of aldolase, but it is likely that they are close to the active-site lysine.

A variety of basic amino acid side chains are involved in the active-site structure of aldolase. At least 3 lysine groups are important at the active site, including the critical Lys-227 which is involved in iminium ion formation [27]. In addition to this one, however, there is apparently another lysine important for binding of the 6-phosphate group on the substrate, since protection against inactivation by pyridoxal phosphate is afforded by fructose 1,6-bisphosphate to a greater extent than either DHAP or fructose 1-phosphate [44]. Another lysine, identified as Lys-146, was trapped by *N*-bromoacetylethanolamine phosphate, as mentioned above [11]. Alkylation of this active-site lysine prevents the binding of substrate. The pK_a value of this amino group is probably lower than typical lysine values of 10.2, since alkylation of this amino group occurs at pH 8.5, where it is presumably unprotonated, but not at pH 6.5 where it is presumably protonated. At the lower pH value the reagent alkylates another basic amino acid side chain at the active site, histidine-359 [45]. It had been proposed that the lowering of the pK_a of Lys-146 might be accomplished by the juxtaposition of an arginine residue which is known to be at the active site, and which would be positively charged throughout the pH range [44]. The presence of this side chain was demonstrated by inhibition of aldolase with phenylglyoxal in the presence of borate, with protection by substrate or phosphate [46]. It is an interesting fact that despite the presence of these numerous basic catalysts which would be capable of carrying out the proton abstraction from substrate, none of these is trapped by reagents such as the α-haloacetol phosphates specifically designed to alkylate the base.

The modification of histidine residues and tyrosine residues produces a similar, and interesting, change in the enzyme activity. Photo-oxygenation of the 40 histidine residues in aldolase, sensitized by Rose Bengal, results in the modification of approximately one-half of the imidazole rings [47]. The activity is reduced to a small fraction of its initial value by this operation. The modified enzyme can still form the imine with substrate, and in fact has little change in the rate of cleavage of fructose 1,6-bisphosphate. The most dramatic change in the enzyme is in the rate of proton transfer to the eneamine. As shown above in Scheme 7, the change occurs in step 2 in which proton transfer occurs, rather than in the cleavage step. Interestingly, the rate of cleavage of F-1,6-P is stimulated by the addition of other acceptor aldehydes (step 6), because the eneamine is trapped by the aldehyde even though the protonation reaction is prevented by the modification. Removal of the C-terminal tyrosine, Tyr-361, results in very similar behavior [48]. Aldolase that has been treated with carboxypeptidase is modified in a manner that results in much lowered F-1,6-P cleavage, rate-determining proton transfer, and stimulation by aldehydes [49,50]. The enzyme is still capable of forming an iminium ion, as evidenced by borohydride trapping experiments [51]. Acetylation of tyrosine residues by acetylimidazole results in very similar modification of enzymatic activity [52]. The net effect, therefore, of carboxypeptidase treatment, acetylation, or photochemical oxygenation is to convert the aldolase into an enzyme having transaldolase-like activity.

Hanson and Rose [53] have pointed out that all aldolases, regardless of mechanism, operate with a retention mechanism. As shown in Scheme 10 a proton is removed from the face of an incipient enolate anion or eneamine and then the base doing the proton abstraction must swing out of the way, an aldehyde is bound to that same site, and C—C formation occurs. In muscle aldolase, for example, it is

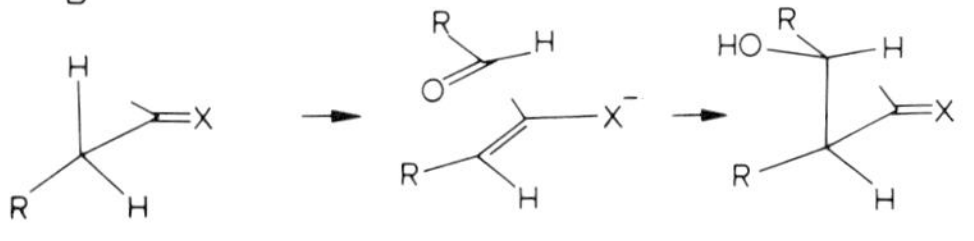

Scheme 10

known that the pro-S proton is removed and replaced with a carbon on the same face of the eneamine in order to create the appropriate stereochemistry for fructose.

The fact that all aldolases operate via this same mechanism suggests that this stereochemically cryptic pathway is an advantageous one that has been selected through evolutionary pressure because it serves a useful purpose. Clearly an inversion mechanism can be envisioned which would lead to the same product from the same starting materials and yet natural selection has avoided that pathway in all cases. This contrasts dramatically with Claissen-type condensations, which invariably operate via an inversion mechanism. One rationale proposed for this behavior is that this might allow proton recycling, so that the same proton abstracted from C-3 could be used to protonate the oxygen on C-4 during the condensation reaction. In enzymes with rapid turnover the loss or acquisition of a proton from solvent may be slow. In this case, however, the relatively slow turnover of aldolase would seem to preclude this explanation. Also, since modified aldolase can carry out the cleavage reaction, including the C-4 OH proton transfer, but not carry out the C-3 proton transfer, this suggests that the same base is not involved in both processes. We have proposed a rationale for this generally observed behavior of aldolases and decarboxylases [25]. If an enzyme is required to cleave a carbon–carbon bond in the same position in space where the subsequent protonation reaction occurs, then this automatically affords protection against the enolization of ketones other than the intended substrate. Thus the more complex retention mechanism, requiring the movement of the proton-abstracting base, provides insurance against the indiscriminant enolization, dehydration or tautomerization of ketonic metabolites other than the substrate. The evolutionary pressure, therefore, selecting this pathway arises from a need to avoid unwanted reactions, rather than a drive to produce a more rapid rate of the desired reaction.

Rose et al. [54] have demonstrated by isotope exchange experiments that aldolase operates with an ordered mechanism, as outlined in Scheme 7, in which the formation of an iminium ion with DHAP is followed by the abstraction at C-3 of the pro-S proton. Only after this event may G-3-P bind in a productive manner. A condensation step then can occur to produce the iminium ion of the fructose. In the reverse reaction the cleavage step of the fructose creates an eneamine, but that eneamine may not be protonated before the G-3-P has left the enzyme surface. Any acceptable proposed mechanism for aldolase must include this order of addition of substrates.

Grazi et al. [22] have found that dihydroxyacetone sulfate (DHAS) is capable of forming an iminium ion at the active site which reacts with borohydride or with cyanide, but which is incapable of undergoing proton abstraction. We have shown that the rate of irreversible inhibition of aldolase by an enone-type inhibitor bound as the iminium ion (as shown in Scheme 9) is independent of whether the ionized group is phosphate or sulfate. Thus it appears that the change from phosphate to sulfate does not disrupt the electron-sink activity of the iminium ion. It has been demonstrated in a model system that an intramolecular proton abstraction by α-keto-phosphates through a 7-member transition state occurs readily. Because of this data, and because of the fact that active-site-directed alkylating agents are incapable of trapping the proton-abstracting base as they do in other enzymes, we

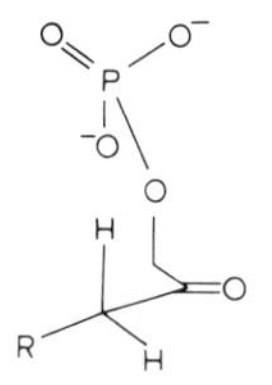

Scheme 11

have proposed that the substrate phosphate is responsible for the proton abstraction rather than a base on the enzyme [25]. This would account for the fact that DHAS does not undergo proton exchange since the lower pK_a oxygens on sulfate would not catalyze proton abstraction even after formation of the iminium ion.

Presented in Fig. 5 is a proposed mechanism for the action of aldolase which incorporates all of the data described above. Although some features of this proposed mechanism are speculative it does provide a working model which incorporates the stereochemical, chemical, kinetic, and other features elucidated by the many experiments described above. In this figure the protein cleft is approximated by the box drawn over the atoms which is open to solvent water on the top and far sides. In step A the conformation drawn is consistent with the fact that only the pro-S proton on C-3 is exchanged with solvent [55]. The pro-S C—H has been drawn so that it may overlap with the π orbital of the iminium ion during proton

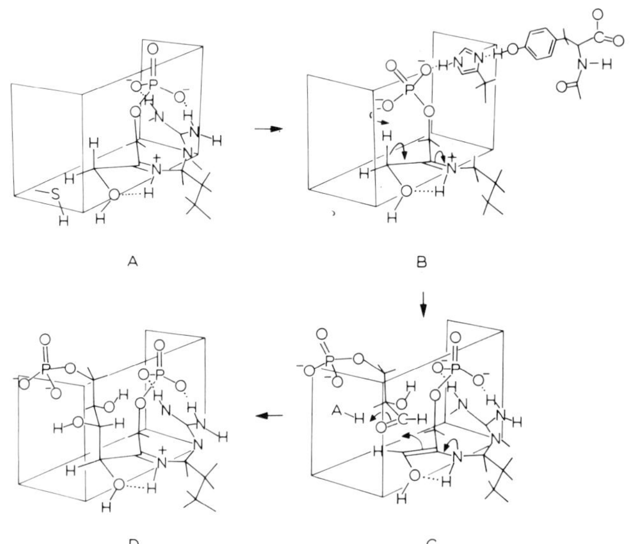

Fig. 5. A proposed active site structure for muscle aldolase [25].

transfer. The face of the iminium ion exposed to solvent has been shown by Dilasio et al. [35] to be the side attacked by borohydride, and therefore also by water. The E isomer of the iminium ion has been drawn because this allows hydrogen bonding between the C-3 hydroxyl and the iminium ion. This would explain the fact that the iminium ion generated by the condensation of hydroxyacetone phosphate (with no OH on C-3) with the lysine amino group is less stable than that formed with DHAP [56,57]. The C-1 phosphate has been drawn in a position hydrogen-bonded with an arginine residue. The iminium ion is sufficiently exposed to solvent so that borohydride, cyanide or tetranitromethane [58] reactions with the bound substrate are possible. A thiol group has been drawn in a position where the enone phosphate inhibitor could react with it.

In step B the phosphate group has been rotated to a position where proton abstraction may occur in an intramolecular molecular fashion. The pK_a of the phosphate (approx. 7) is ideal for proton abstraction as discussed earlier, and the rotation of the phosphate around the C-1 C—O bond provides the two positions absolutely required for the base doing the proton abstraction if a retention mechanism is occurring. The positioning of the phosphate is maintained by His-359 and Tyr-361, in such a manner that disruption of either of these residues would produce a lesion in the proton-transfer rate. In the cleavage direction, with modified enzyme, the eneamine would be protonated very slowly and reversal of the cleavage reaction could occur in the presence of other aldehydes, giving the transaldolase behavior noted above. In step C, the acceptor aldehyde binds after the phosphate has rotated back to its original position, thus insuring the retention mechanism. The aldehyde carbonyl must be protonated by a stereospecific catalyst rather than by water since tagatose 1,6-bisphosphate (which is identical to fructose 1,6-bisphosphate except for an inverted configuration at C-4) is not cleaved [59]. Following step D after formation of the iminium ion of fructose 1,6-bisphosphate, water would attack the re face to form the carbinolamine which can then eliminate the lysine amine to generate the keto form of F-1,6-P_2. It is clear from the geometrical arrangement described that this catalyst would be extremely efficient for enolizing DHAP, but would be completely ineffective at enolizing any other substrate, since C-4 would take the place required for the base during the proton abstraction.

Although this review has focused on muscle aldolase intensively because it is the most well studied enzyme of this kind, it is only one of a variety of aldolases catalyzing similar reactions using an iminium ion pathway. Many of the mechanistic features of these enzymes, including retention mechanism, borohydride reaction etc., are similar [60–63].

5. *Transaldolase*

Transaldolase is a critical enzyme in the metabolism of carbohydrates directed through the pentose phosphate pathway [63]. It is responsible for transferring a 3-carbon unit between various acceptor aldehydes so that an equilibrium of various chain length sugar phosphates may be obtained. As demonstrated in Scheme 12, transaldolase removes the first 3 carbons from sedoheptulose-7-P and transfers them to G-3-P [64,65]. This results in the formation of F-6-P and erythrose-4-P. The

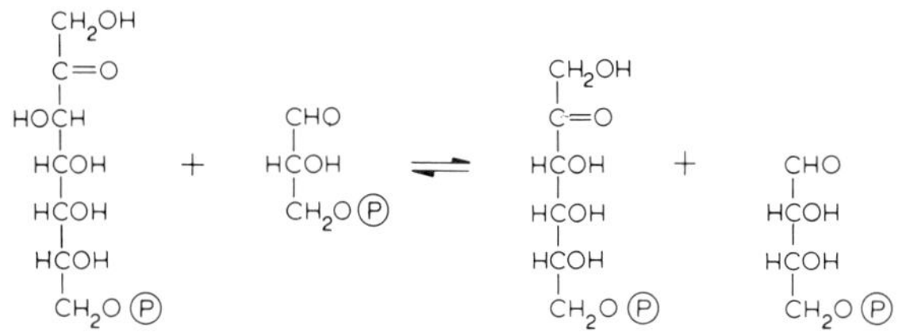

$$\begin{array}{l}CH_2OH\\ C{=}O\\ HOCH\\ HCOH\\ HCOH\\ HCOH\\ CH_2O\text{Ⓟ}\end{array} + \begin{array}{l}CHO\\ HCOH\\ CH_2O\text{Ⓟ}\end{array} \rightleftharpoons \begin{array}{l}CH_2OH\\ C{=}O\\ HCOH\\ HCOH\\ HCOH\\ CH_2O\text{Ⓟ}\end{array} + \begin{array}{l}CHO\\ HCOH\\ HCOH\\ CH_2O\text{Ⓟ}\end{array}$$

Scheme 12

mechanism of the enzyme is based on iminium ion chemistry and is analogous to muscle aldolase, but with substantial and interesting differences. Most of the mechanistic work on transaldolase has been carried out on the enzyme from the yeast *Candida utilis*, because of the large amount of activity found in this source. The activity derived from this source can be separated into 3 isoenzymes, I, II, III, so-labeled because of their order of elution from DEAE Sephadex [66]. Isoenzyme III has been crystallized and is the subject of most of the careful mechanistic studies. It is a dimer of two identical subunits each with a molecular weight of 37 000 [67]. Isoenzyme I is a dimer of smaller molecular weight subunits and isoenzyme II is a mixture of the monomers forming isoenzymes I and III. A variety of donor substrates, both phosphorylated and not, and having a variety of chain lengths had been found. An array of acceptor substrates both phosphorylated and not have also been elucidated [63].

No other cofactors are required for enzymatic catalysts of this reaction. Amino acid analysis on isolated isoenzymes I and III indicate that the lower molecular weight monomers from I are not proteolytic degradation products of the monomers from III [68]. When dimers of I ($\alpha\alpha$) are mixed with dimers of III ($\beta\beta$) an equilibrium concentration of II ($\alpha\beta$) is formed.

There is substantial evidence indicating the formation of an iminium ion with substrate. The adduct formed between lysine amino group of the enzyme and the substrate is much more stable than the corresponding adduct of aldolase, since the enamine that is formed does not protonate, as it does in aldolase. Horecker and Smyrniotis [69] demonstrated, for example, that sedoheptulose does not incorporate labeled dihydroxyacetone when incubated in the presence of transaldolase. The implication is that the dihydroxyacetone adduct formed on the enzyme does not

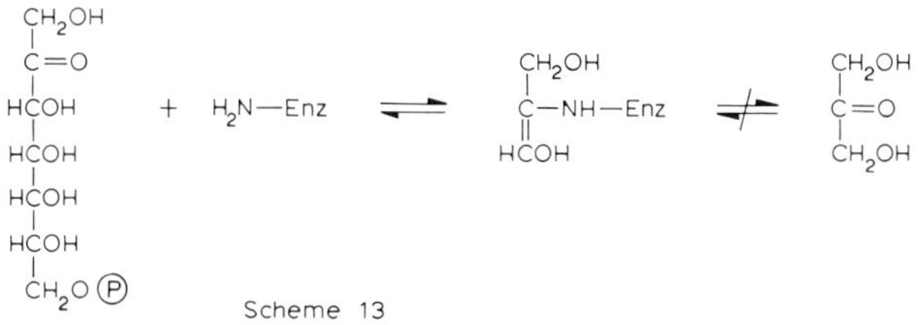

$$\begin{array}{l}CH_2OH\\ C{=}O\\ HCOH\\ HCOH\\ HCOH\\ HCOH\\ CH_2O\text{Ⓟ}\end{array} + H_2N{-}Enz \rightleftharpoons \begin{array}{l}CH_2OH\\ C{-}NH{-}Enz\\ \| \\ HCOH\end{array} \not\rightleftharpoons \begin{array}{l}CH_2OH\\ C{=}O\\ CH_2OH\end{array}$$

Scheme 13

exchange with free dihydroxyacetone in solution. The same adduct may be formed from labeled fructose-6-P and isolated as a stable entity, bearing one dihydroxyacetone per dimeric enzyme [70,71]. The chromatographically isolated adduct can then be allowed to react with acceptor aldehydes such as G-3-P to form product. Reduction of the adduct with borohydride causes the expected loss of activity associated with irreversible modification of the essential lysine amino group. Hydrolysis of this reduced adduct results in the production of the expected *N*-*β*-glyceryl-lysine [72]. The presence of an iminium ion intermediate was also demonstrated by inhibition with cyanide [73]. The enamine adduct bound to the active site is reactive toward electrophiles and oxidants as well as the acceptor aldehydes. For example, tetranitromethane reacts readily with inhibition of the enzyme [74]. Hexocyanoferrate III is also a specific irreversible inhibitor in the presence of substrates [75].

Evidence has been accumulated that shows that transaldolase is a "half the sites" enzyme, in which binding of substrate to one site almost completely inactivates the other active site. As mentioned above, only 1 mole of substrate is bound to the active site, even though the enzyme is a dimer. In the presence of excess F-6-P a burst of one equivalent of G-3-P is produced followed by a slower conversion of F-6-P to G-3-P plus DHA. It appears therefore, that the blocking of one active site by adduct formation with dihydroxyacetone results in a modified second active site which is able to slowly cleave F-6-P [76]. This second site behavior is identical to that found in transaldolase in which one active site has been modified by reducing the dihydroxyacetone adduct with borohydride. Photo-oxidation of one of the two histidine residues in isoenzyme III results in the loss of activity of the enzyme, as is consistent with half-sites reaction [77]. The dye 1-anilino-8-naphthalene sulfonate on the other hand is bound with 2 moles of dye per mole of enzyme, indicating two binding sites for F-6-P [78].

The accumulated evidence, therefore, suggests that transaldolase is a dimer which is capable of binding 1 mole of dihydroxyacetone obtained from a variety of donors at one of its two active sites. This binding event converts the second active site into one which is incapable of transaldolase activity but does have an aldolase-like activity with a K_m for F-6-P which is essentially not altered, but which has a substantially reduced rate of cleavage. Unlike the first active site this second active site allows protonation of the enamine to complete the aldol-like reaction and does not exhibit reduction with borohydride. The major reaction at the first active site is completed by binding of any of a variety of acceptor aldehydes to create a new C-3–C-4 bond.

6. *Acetoacetate decarboxylase*

By far the most well studied enzyme catalyzing a decarboxylation reaction in which iminium ion formation is involved, is acetoacetate decarboxylase [79]. This enzyme obtained from *Clostridium acetobutylicium* [80] has an apparent molecular weight of approximately 340 000, and is composed of subunits with a molecular weight of 29 000 [81]. It is apparently composed of 12 subunits, and these may be dissociated into dimers of approximately 60 000 molecular weight. These dimers are capable of reforming the native structure after removal of denaturant, whereas the 29 000 MW monomers are not. Modification and inactivation of the enzyme is caused by acetylation of the active lysine with acetic anhydride or by reaction with acetopyruvate to form eneamine [83]. The stoichiometry of these reactions requires 1 mole of reagent per 60 000 MW dimer, even though the monomers are apparently identical. In this sense acetoacetate decarboxylase is reminiscent of transaldolase.

Many carefully performed studies on this enzyme have generated the proposed mechanism shown in Scheme 14, involving the imine-forming condensation of the β-carbonyl group of the substrate. The generation of the more active electron sink

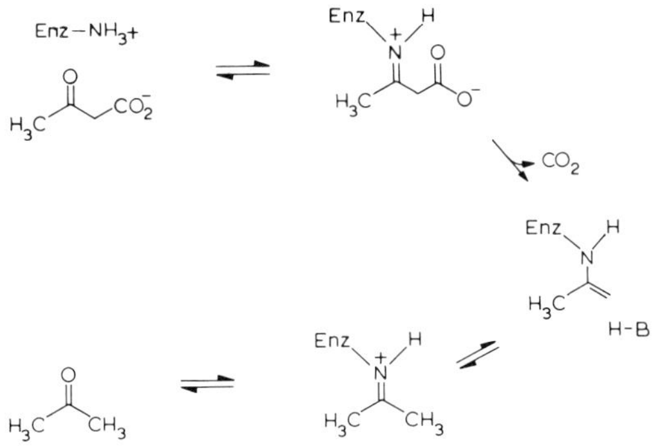

Scheme 14

then assists in the decarboxylation event, leaving the eneamine of acetone. Protonation of this intermediate followed by hydrolysis results in the formation of acetone. The evidence for this mechanism includes many of the methods described above for the aldolases. These include the observation of ^{18}O exchange with the oxygen of water during the catalytic event, as required by the mechanism [84], and inhibition by cyanide [85]. Irreversible inactivation of the enzyme in the presence of both substrate and borohydride was found, whereas borohydride had no effect on the enzyme alone [85]. When this operation was carried out on [3-^{14}C]acetoacetate, the label was incorporated into the protein at a rate of one label per approximately 50 000 MW. Acid hydrolysis led to the identification of the modified lysine derivative ϵ-N-isopropyl lysine. This is consistent with reduction of the intermediate formed after decarboxylation and protonation of the substrate. A single lysine amino group was labeled during these operations out of the total of 20/subunit, suggesting a unique lysine ϵ-amino group involved in the catalytic event. The sequence of peptide in the region containing the active site lysine is [96: —Glu—Leu—Ser—Ala—Tyr—Pro—Lys*—Lys—Leu—. The catalysis of the loss of carbon dioxide from substrate was demonstrated in model system reactions to involve essentially the same mechanism, but the process was catalyzed at a rate approximately 3 orders of magnitude less than the enzymatic reaction [87].

The cleaver introduction of a reporter group was used to determine the pK_a of the essential lysine amino group at the active site [88,89]. 5-Nitrosalicylaldehyde condensed with the active-site lysine group to form an iminium ion which was then reduced with borohydride. This covalent adduct was shown to have a phenolic pK_a

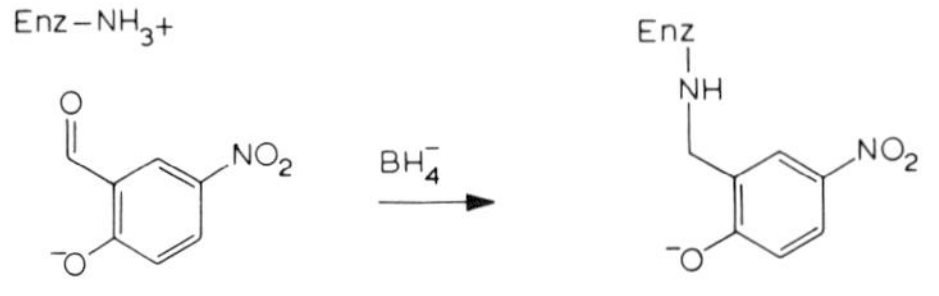

Scheme 15

of 2.4, which is 3.5 pK_a units below the expected value for this phenol. The spectrophotometric properties of this reporter group also allowed the measurement of the pK_a of the lysine amino group which was found to be approximately 6.0, 4.7 units below the expected value. The pK_a value predicted for the active-site lysine on this basis was 6.0, and compared favorably with the value of 5.9 determined by kinetic methods involving the pH dependence of acetylation of the lysine. We have discussed above the merits of lowering the pK_a of an imine-forming amine in terms of the greater ability of the iminium ion formed to act as an electron sink. The lowering of pK_a of the lysine group could be due to either a more hydrophobic environment or to the juxtaposition of a positively charged group. The fact that the phenol pK_a is also lowered suggests that the latter is a more likely explanation.

The decarboxylation of β-keto acids by both primary amines and by acetoacetate decarboxylase has been studied by the method of carbon isotope effects [90,91]. A typical isotope effect for the amine-catalyzed reaction is 1.03 for the cyanomethylamine-catalyzed decarboxylation of acetoacetate at pH 5.0. This demonstrates that the carbon–carbon bond is being cleaved in the rate-determining decarboxylation step. The comparable carbon isotope effect for the enzymatic decarboxylation is $k^{12}/k^{13} = 1.018$ and is pH-independent over the range pH 5.3–7.2. This data suggests that in the enzymatic reaction the decarboxylation event is at least partially rate-determining.

A variety of β-diketones have been demonstrated to be very effective inhibitors of acetoacetate decarboxylase [83]. Although they bind tightly to the active site they are not reduced by borohydride. Corresponding α- or γ-diketones are not inhibitory [84]. Acetopyruvate, for example, inhibits competitively with respect to acetoacetate with $K_i = 10^{-7}$ M. This compound has been shown to react with model amino compounds to yield enamines which are not reducible by borohydride. The steric requirements for this class of inhibitors has been studied in some detail, demonstrating the considerable latitude of structure available for binding at the active site.

Other interesting inhibitors developed for the enzyme include 2-oxopropane sulfonate [93], and 2-oxopropane phosphonate [94]. The sulfonate binds with K_i approximately equal to K_m for acetoacetate, and is reduced by borohydride. The phosphonate is a competitive inhibitor also, and a more effective inhibitor of the enzyme than the sulfonate when in the monoanionic form, but a much worse inhibitor in the dianionic form. The phosphonate is not reducible with borohydride, which is explained on the basis of charge interaction.

It had been recognized that the catalyst providing the proton to the α-carbon in the last step of the decarboxylase reaction might also be capable of enolizing ketones. It had been shown that acetone is enolized by acetoacetate decarboxylase [95] in a sequential manner such that the imine-forming reaction must be rapid [96]. It was also shown that 2-butanone is a substrate, and that only one of the methylene α-protons was exchanged, indicating a stereospecific reaction. An interesting recent study has demonstrated the stereochemistry of the reaction at the active site of this enzyme [97,98]. Analogs of acetoacetate which were conformationally restricted and stereochemically defined were used to determine the absolute stereochemistry of the event occurring on the natural substrate. A preference for *R*-2-carboxycyclohexanone over its enantiomer was established. The *R*-2-carboxycyclohexanone had values of $K_m = 7.5$ mM and $V_{max} = 60$ μM/min/mg, whereas the values for acetoacetate were 10 mM, and 740 μM/min/mg. Thus the conformationally restricted substrate was bound as well as normal substrate, but had a slower turnover. The V_{max} value for the *S*-2-carboxycyclohexanone was at least 12 times smaller than the V_{max} for its enantiomer.

Acetoacetate decarboxylase also catalyzes the proton exchange of a variety of ketonic substrates, with a high degree of stereospecificity. For example, the 2R protons on cyclohexanone are exchanged by the enzyme, whereas the 2S protons

$^{-}O_2C$, D — AAD → H, D

Scheme 16

exchange only at much slower rate. The combination of these two stereospecific processes, the decarboxylation and the proton transfer, results in a net retention of stereochemistry during turnover of acetoacetate. Thus as shown in Scheme 16, *R*-2-deutero-carboxycyclohexanone is stereospecifically converted to *S*-2-deutero-cyclohexanone. This requires a net retention stereochemistry and is therefore analogous to the behavior of aldolases, in that the loss of the group involved in the carbon–carbon bond cleavage is followed by replacement on that site with the proton. As with aldolase this requires a movement of the base doing the proton abstraction.

7. *Pyruvate-containing enzymes*

Decarboxylations of α-amino acids are some of the most widely studied enzymatic reactions, and had been, at one time, presumed to be exclusively associated with pyridoxal-dependent catalysis. The reactivity of an enzyme of this type with carbonyl group reagents such as hydrazines, cyanide or hydroxylamine, was therefore considered acceptable evidence for the presence of a pyridoxal coenzyme. Thus it was a striking discovery when several enzymes were found which exhibited these properties, but did not contain pyridoxal. A covalently bound pyruvamide moiety was discovered to be essential for catalysis in proline reductase [99] and in a bacterial histidine decarboxylase [100]. Since these initial important discoveries a growing list of pyruvate-containing enzymes has been found [4]. These include *S*-adenosylmethionine decarboxylase from *E. coli* [101], yeast [102] and rat liver [103], phosphatidyl serine decarboxylase from *E. coli* [104] and aspartate decarboxylase from *E. coli* [105]. Because the histidine decarboxylase studied by Riley and Snell [100] has been the most intensively examined, its properties will be reviewed here.

It is possible to write a simple mechanism for pyruvamide containing enzymes catalyzing decarboxylation reactions of α-amino acids, and it was therefore suspected that the pyruvamide was behaving in a manner analogous to that shown in Scheme 17. This reaction scheme bears many similarities to pyridoxal containing

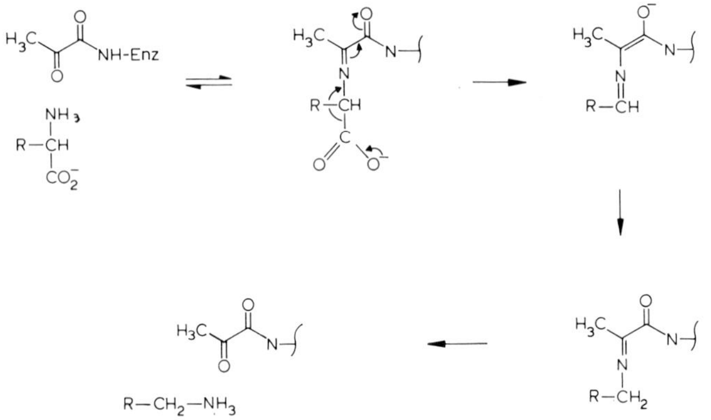

Scheme 17

enzymatic decarboxylation, since the condensation of the α-amino group with a carbonyl function results in the creation of an electron sink such that the carbon dioxide may be lost. Subsequent protonation of the same carbon that had borne the carbon dioxide results in the completion of the exchange followed by hydrolysis to yield product.

In an interesting model system study Owen and Young [105] have shown that pyruvamide itself is capable of catalyzing analogous reactions under surprisingly mild conditions. For example, treatment of phenylglycine in tetrahydrofuran solution with pyruvamide results in the production of both benzylamine, and the subsequent deamination product, benzaldehyde. The pyruvamide, therefore, appears to be able to catalyze both the decarboxylation, and the deamination reactions normally associated with pyridoxal catalysis. The corresponding deamination of aliphatic amines under comparable conditions did not occur. The imine did form under these conditions, however.

Rosenthaler et al. [106] purified histidine decarboxylase from Lactobacillus 30A and demonstrated that there was no pyridoxal phosphate, as had been suggested by Rodwell [107]. Treatment with [^{14}C]phenylhydrazine labeled the protein, but did not if the protein was first reduced with borohydride. Chymotrypsin digestion of the [^{14}C]phenylhydrazone treated enzyme resulted in a labeled fragment identified as *N*-pyruvoylphenylalanine [100]. Borohydride reduction of the native enzyme resulted in lactate production after hydrolysis. Thus it was established that a pyruvoyl group is covalently bound as an amide to the NH_2-terminal phenylalanine. As is consistent with this proposed mechanism the enzyme is also inhibited by cyanide and by hydroxylamine. The iminium ion predicted by the mechanism above was trapped with borohydride in the presence of substrate and identified [108].

The question of the source of the pyruvamide group had been answered beautifully by Snell and coworkers. A mutant of Lactobacillus 30A was obtained which was deficient in histidine decarboxylase [109], and which contained an enzymatically inactive protein which was immunologically cross-reactive with active histidine decarboxylase [110]. Preparation of this protein showed only 3–5% of the wild-type activity. The activity of this protein, however, increased 20-fold at pH 7.5 during final purification. The inactive protein did not react with phenylhydrazine, and was a single polypeptide of approximately 37 000 MW. The active protein was composed of two peptide chains of approximately 28 000 and 9000 MW with the larger subunit containing the pyruvamide moiety. During the activation of the inactive protein the cleavage occurred to form the two smaller subunits found in the wild-type enzyme. During the activation process, 1 mole of ammonia was generated at the same time that one pyruvamide group was produced. Sequencing data from both inactive pro-enzyme and from the α and β subunits of the active enzyme demonstrated that the protein is produced as a single peptide and then cleavage occurs between two serine residues. This cleavage is coupled with a conversion of one serine residue to pyruvamide. Although little information is available concerning the details of this conversion, it is clear that it is a non-enzymatic process, in which the pro-enzyme conversion occurs spontaneously at 37°C and pH 7.6 with a half-time of 3 h.

The structure of the enzyme appears to be a hexamer, with a dumbell shape 60 Å in width and 120 Å in length. The native enzyme has a molecular weight of approximately 280 000, with a subunit composition $(\alpha\text{-}\beta)_6$, and the pro-enzyme has a subunit composition $(\pi)_6$ [111].

Aside from the reduction experiments with or without substrate, and other carbonyl group reactions, evidence has accrued also for the presence of a thiol group at the active site of histidine decarboxylase [12]. It has been demonstrated, for example, that approximately one thiol group per α–β subunit is titrated with iodoacetamide or *p*-chloromercuribenzoate, with concomitant inactivation of the enzyme [113]. This same active-site thiol group is titrated with DTNB with loss of activity. Interestingly, the competitive inhibitors histamine and imidazole enhance the reactivity of these thiol residues toward DTNB. Upon denaturation, the enzyme demonstrates the presence of one additional thiol group per α–β subunit. The non-competitive inhibitor histidine methyl ester protects against the titration of this group. It is possible that the histamine and imidazole are enhancing the thiol activity by providing a counter ion, since this type of behavior has been observed in other cases [114]. It is possible that the thiol group, which has an apparent pK_a of 9.2, may be participating in the enzymatic process as the proton-transfer catalyst required in the mechanism.

The stereochemistry of the overall process has recently been studied. When [S-α-^{3}H]histidine is used as substrate, the product is [S-α-^{3}H]histamine. This observation requires retention of configuration, and is therefore analogous to the stereochemistry of decarboxylations catalyzed by pyridoxal, or by imine formation.

8. Dehydratases

A number of enzymes use iminium ion formation not only to activate the substrate carbonyl group for α-proton abstraction, but also to enhance the rate of elimination of a group from the iminium thus formed. In Section 2 above, an elimination reaction similar to that shown in Scheme 18 was used as an example of primary amine catalyzed α-proton abstraction, and the data was also unambiguous since the

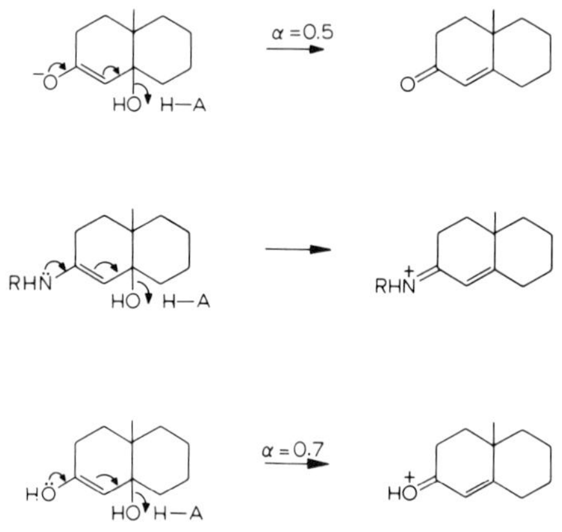

$\alpha = 0.5$

HO H—A

RHN

HO H—A

RHN

$\alpha = 0.7$

HO

HO H—A

HO

Scheme 18

subsequent elimination step was rapid. Thus, no experimental evidence on catalysis of the loss of OH^- from the eneamine was available because this step was faster than the rate-determining proton abstraction that preceded it. It has been shown, however, that in the simple elimination reaction catalyzed by acids and bases in aqueous solution, with no imine formation, the loss of the hydroxyl group is catalyzed by general acids for both the enolate anion ($\alpha = 0.5$), and the enol ($\alpha = 0.7$) [115]. The clear implication is that the eneamine, which is intermediate in reactivity between these two, would undergo general acid catalyzed loss of OH^-, with $\alpha = 0.6$. As described above, the utility of iminium ion formation lies in the fact that it activates the carbonyl group to the maximum extent with entities that are present in high concentration at neutral pH values. Thus even though the iminium ion does not undergo proton abstraction as rapidly as the protonated ketonc, for example, it is present in higher concentration, and therefore the overall reaction is much faster. Much the same is true for the eneamine formation in elimination reactions, since the eneamine is very reactive toward elimination of the β leaving group, but is less so than the enolate anion. However, the eneamine concentration can be much greater at neutral pH than the enolate anion and therefore this pathway is a worthwhile one for enzymes to use.

Dehydroquinase is an enzyme in the pathway responsible for forming aromatic amino acids from acyclic precursors [116]. It is part of a 5-enzyme complex in *Neurospora crassa* which exhibits the property of co-ordinate activation by the first substrate [117]. The enzyme *Aerobacter aerogenes* was used by Hanson and Rose [118] to determine the absolute stereochemical course of citric acid biosynthesis. These workers also demonstrated that the elimination of water proceeds in a SYN manner with the prochiral R-proton removed followed by elimination of hydroxide. This was in contrast to the commonly observed anti-elimination of water from most

Scheme 19

substrates by enzymatic dehydratases. Since anti-elimination is commonly observed in non-enzymatic concerted elimination reactions, Hanson and Rose proposed that a stable carbanionic intermediate was involved in this enzymatic process.

Working with dehydroquinase isolated from *E. coli* 83-2, Butler et al. [119] demonstrated that the reaction was catalyzed via iminium ion formation, in a manner analogous with the model system shown in Scheme 19. They demonstrated that the enzyme was inhibited in the presence of both substrate and sodium borohydride. The inhibited enzyme did not regain activity upon dialysis. The enzymatic SYN elimination is contrasted with the non-enzymatic, base-catalyzed, trans-elimination found in aqueous solution, in which the pro-S proton is stereo-selectively removed (Scheme 20). The stereochemistry of proton abstraction in the non-enzymatic case arises from the pro-S proton being in an axial position such that

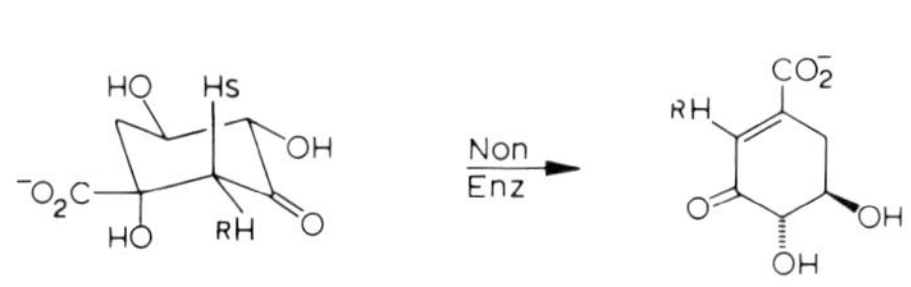

Scheme 20

overlap with the carbonyl group will occur when the molecule is in its most stable conformation. The enzymatic SYN elimination implies a substrate conformational change during the enzyme-catalyzed process. It has been demonstrated that the carboxylate anion is not responsible for non-enzymatic intramolecular proton abstraction, since the methyl ester gives the preferred pro-S proton abstraction also [121,122]. In the enzymatic reaction the methyl ester is neither a substrate nor an inhibitor. This suggests that the carboxylate anion is involved in the binding process. An isopropylidene derivative has been prepared which has the stereochemistry shown in Scheme 21. This ketal derivative is a substrate for dehydroquinase and reacts at about one-half the rate of natural substrate. This demonstrates that the hydroxyl groups are not involved in binding and also restricts the conformations that are possible for the substrate when bound to the active site. Those conformations in which the hydroxyl groups are in diaxial positions could not be achieved by the derivatized substrate, and therefore cannot be involved in catalysis. On the basis of this information it has been proposed that the skew-boat conformation drawn below is the form present at the enzyme-active site. Since, as described above, the loss of hydroxide from the eneamine is general acid catalyzed, it is possible that the

Scheme 21

SYN elimination involves proton abstraction by a base which is then in the correct position to also act as a general acid catalyst.

There are several enzymes which embody not only the elements of the proton abstractions, and dehydrations described above but which also carry out reactions making further use of eneamine and iminium ion chemistry. For example, the complex reaction catalyzed by the enzyme converting 2-keto-3-deoxy-L-arabonate to α-ketoglutarate semialdehyde involves a condensation step with a lysine amino group, loss of α-proton, and dehydration in a manner analogous to the reactions seen above. This enzyme, however, uses the vinylogous iminium ion created by this process to catalyze a proton abstraction at the γ-position as shown (Scheme 22), creating an enol. This enol can tautomerize to give the aldehyde product, resulting from hydrolysis of the iminium ion. Portsmouth et al. [122] demonstrated the incorporation of two labeled protons from water, as expected from this mechanism. They also showed that borohydride inactivated the enzyme in the presence of substrate.

Borohydride trapping was also used to demonstrate the presence of an iminium ion in the pathway involved in pyrrole biosynthesis from two molecules of Δ-aminolevulinic acid [123]. This reaction, shown in Scheme 23, involves iminium ion formation with 1 mole of substrate, eneamine formation by deprotonation, and then

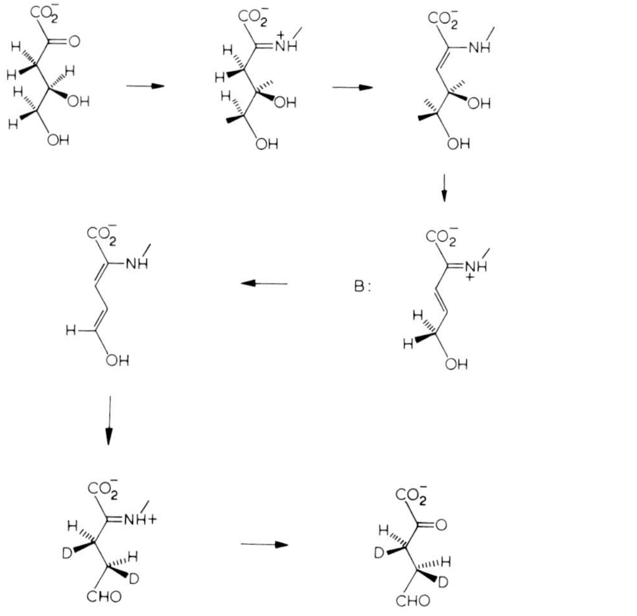

Scheme 22

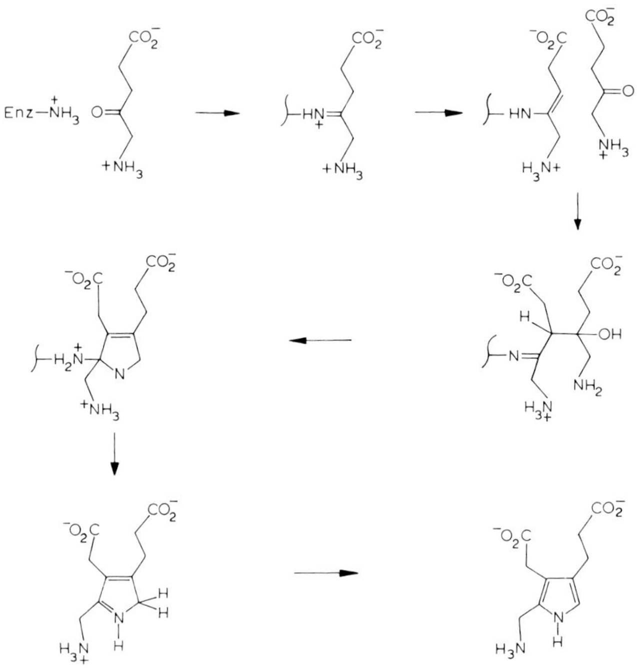

Scheme 23

an aldol condensation. This adduct is then dehydrated using another α-deprotonation followed by eneamine expulsion of hydroxide ion. The iminium ion generated by these events is attacked by an intramolecular nucleophilic amine, followed by expulsion of the lysine amino group. The overall driving force for the reaction is provided, in part, by the aromatic stability of the pyrrole ring produced. This mechanism is particularly interesting because it embodies most of the known reactions that iminium ion formation enhances. These include: α-proton abstraction, α–β-dehydration, aldol condensation, and activation of carbonyl groups towards nucleophilic attack.

9. Conclusions

Although this review concerns those reactions catalyzed by iminium ion formation, it is important to note that there are enzymatic reactions that could logically be catalyzed by iminium ion formation but which are not. Yeast aldolase, for example, is the best known case [26] (see Ch. 6). This enzyme is metal ion dependent, does not demonstrate the loss activity in the presence of both substrate and borohydride, and is sensitive to inhibition by EDTA. The reaction catalyzed by this enzyme is identical to that catalyzed by the imine-forming enzyme, and even has evolved to exhibit the same retention stereochemistry. Another example is Δ^4-3-oxosteroid reductase which is responsible for the NADPH-dependent reduction of the enone double bond to the corresponding dihydrosteroid [124]. Even though iminium ion formation would increase the reactivity of this substrate toward the β-hydride addition, a demonstrated lack of the required oxygen exchange proves that this does not occur.

Iminium ion formation and conversion to eneamine are useful, therefore, in the activation of carbonyl groups for α-carbanion formation. This pathway has often been selected by evolution even though other practical methods of accomplishing the same enzyme chemistry at neutral pH exist, and even though the pathway involves a complex succession of events. The primary value of this adduct formation appears to be that it allows the formation of the highest concentration of the most reactive intermediate under neutral pH conditions. The contribution of imine formation to binding energy does not seem to be a substantial reason for the selection of this pathway. The consistency of the retention mode of stereochemistry in those reactions involving carbon–carbon bond cleavage is striking, and suggests that a fundamental reason for this behavior exists.

Acknowledgement

Ruth Hagerman provided invaluable assistance in the preparation of this review.

References

1 Boyer, P. (Ed.) (1972) The Enzymes, 3rd edn., Vol. III, Academic Press, New York.
2 Boyer, P. (Ed.) (1972) The Enzymes, 3rd edn., Vol. VI, Academic Press, New York.
3 Stoeckenius, W. (1980) Acts. Chem. Res. 13, 337; Florkin, M. and Stotz, E. (Eds.) Comprehensive Biochemistry, Vol. 27, Ch. II.
4 Snell, E. (1977) Trends Biochem. Sci. 2, 131.
5 Jencks, W.P. (1969) Catalysis in Chemistry and Enzymology, McGraw-Hill, New York, pp. 490–496.
6 Williams, A. and Bender, M. (1966) J. Am. Chem. Soc. 88, 2508.
7 Jencks, W.P. (1969) Catalysis in Chemistry and Enzymology, McGraw-Hill, New York, p. 587.
8 Cordes, E.H. and Jencks, W.P. (1963) J. Am. Chem. Soc. 85, 2843.
9 Patai, S. (Ed.) (1970) The Chemistry of the Carbon–Nitrogen Double Bond, Interscience, New York, Ch. 10.
10 Rosenberg, S., Silver, S., Sayer, J. and Jencks, W.P. (1974) J. Am. Chem. Soc. 96, 7986.
11 Hartman, F. and Brown, J. (1970) J. Biol. Chem. 251, 3057.
12 Frey, P., Kokesh, F. and Westheimer, F. (1971) J. Am. Chem. Soc. 93, 7266.
13 Schmidt, D. and Westheimer, F. (1970) Biochemistry 10, 1249.
14 Hine, J. and Chou, Y. (1981) J. Org. Chem. 46, 649.
15 Hupe, D., Kendall, M. and Spencer, T. (1972) J. Am. Chem. Soc. 95, 2271.
16 Bender, M. and Williams, A. (1966) J. Am. Chem. Soc. 88, 2502.
17 Hine, J., Menon, B., Jensen, J. and Mulders, J. (1966) J. Am. Chem. Soc. 88, 3367.
18 Hupe, D., Kendall, M. and Spencer, T. (1972) J. Am. Chem. Soc. 94, 1254.
19 Westheimer, F. and Cohen, H. (1938) J. Am. Chem. Soc. 60, 90.
20 Motiu-DeGrood, R., Hunt, W., Wilde, J. and Hupe, D. (1979) J. Am. Chem. Soc. 101, 2182.
21 Jencks, W.P. (1975) Adv. Enzymol. 43, 219.
22 Grazi, E., Sivieri-Pecorari, C., Gagliano, R. and Trombetta, G. Biochemistry 12, 2583.
23 Ref. 14, and many earlier papers in this series by Hine.
24 Van Tamelen, E. (Ed.) (1977) Bio-Organic Chemistry, Academic Press, New York, Ch. 13.
25 Periana, R.A., Motiu-DeGrood, R., Chiang, Y. and Hupe, D. (1980) J. Am. Chem. Soc. 102, 3923.
26 Boyer, P. (Ed.) (1972) The Enzymes, 3rd edn., Vol. VII, Academic Press, New York, Ch. 6.
27 Lai, C., Nakai, N. and Chang, D. (1974) Science 137, 1204.
28 Speck, J. and Forist, A. (1957) J. Am. Chem. Soc. 79, 4659.
29 Horecker, B., Pontremoli, S., Ricci, C. and Cheng, T. (1961) Proc. Natl. Acad. Sci. (U.S.A.) 47, 1949.
30 Grazi, E., Cheng, T. and Horecker, B. (1962) Biochem. Biophys. Res. Commun. 7, 250.
31 Grazi, E., Rowley, P., Cheng, T., Tchola, O. and Horecker, B. (1962) Biochem. Biophys. Res. Commun. 9, 38.
32 Speck, J., Rowley, P. and Horecker, B. (1963) J. Am. Chem. Soc. 85, 1012.
33 Cash, D. and Wilson, I. (1966) J. Biol. Chem. 241, 4290.
34 Model, P., Ponticarvo, L. and Rittenberg, D. (1978) Biochemistry 7, 1339.
35 Dilasio, A., Trombetta, G. and Grazi, E. (1977) FEBS Lett. 73, 244.
36 Steinman, H. and Richards, F. (1970) Biochemistry 9, 4360.
37 Kobashi, K. and Horecker, B. (1967) Arch. Biochem. Biophys. 121, 178.
38 Kowal, J., Cremona, T. and Horecker, B. (1965) J. Biol. Chem. 240, 2485.
39 Hartman, F. (1970) Biochemistry 9, 1776.
40 Hartman, F. (1971) Biochemistry 10, 146.
41 Hartman, F. (1970) Biochemistry 9, 1783.
42 Paterson, M., Norton, I. and Hartman, F. (1972) Biochemistry 11, 2070.
43 Wilde, J., Hunt, W. and Hupe, D. (1979) J. Am. Chem. Soc. 101, 2182.
44 Shapiro, S., Esner, M., Pugh, E. and Horecker, B. (1968) Arch. Biochem. Biophys. 128, 554.
45 Hartman, F., Suh, B., Welch, M. and Barker, R. (1973) J. Biol. Chem. 248, 8233.
46 Lobb, R., Stokes, A., Hill, H. and Riordan, J. (1976) Eur. J. Biochem. 70, 517.
47 Hoffee, P., Lai, C., Pugh, E. and Horecker, B. (1967) Proc. Natl. Acad. Sci. (U.S.A.) 57, 107.

48 Drescher, E., Boyer, P. and Kowalsky, A. (1960) J. Biol. Chem. 235, 604.
49 Rose, I., O'Connell, E. and Mehler, A. (1965) J. Biol. Chem. 240, 1758.
50 Spolter, P., Adelman, R. and Weinhouse, S. (1965) J. Biol. Chem. 240, 1327.
51 Kobashi, K., Lai, C. and Horecker, B. (1966) Arch. Biochem. Biophys. 117, 437.
52 Pugh, E. and Horecker, B. (1967) Arch. Biochem. Biophys. 122, 196.
53 Hanson, K. and Rose, I. (1975) Accts. Chem. Res. 8, 1.
54 Rose, I., O'Connell, E. and Mehler, A. (1965) J. Biol. Chem. 240, 1758.
55 Rieder, S. and Rose, I. (1959) J. Biol. Chem. 234, 1007.
56 Rose, I. and O'Connell, E. (1969) J. Biol. Chem. 244, 126.
57 Pratt, R. (1977) Biochemistry 16, 3988.
58 Healy, M. and Christen, P. (1972) J. Am. Chem. Soc. 94, 7911.
59 Tung, T., Ling, K., Byrne, W. and Lardy, H. (1954) Biochim. Biophys. Acta, 14, 488.
60 Boyer, P. (Ed.) (1970) The Enzymes, 3rd edn. Vol, VII, Academic Press, New York, Ch. 8.
61 Mavridis, I. and Tulinsky, A. (1976) Biochemistry 15, 4410.
62 Meloche, H.P. (1981) Trends Biochem. Sci. 6, 38.
63 Van Tamelen, E. (Ed.) (1973) Bio-Organic Chemistry, Academic Press, New York, Ch. 3.
64 Horecker, B. and Smyrniotis, P. (1953) J. Am. Chem. Soc. 75, 2021.
65 Horecker, B., Smyrniotis, P., Hiatt, H. and Marks, P. (1955) J. Biol. Chem. 212, 827.
66 Tchola, O. and Horecker, B. (1966) Methods Enzymol. 9, 499.
67 Schutt, H. and Brand, K. (1975) Arch. Biochem. Biophys. 169, 287.
68 Tsolas, O. and Horecker, B. (1970) Arch. Biochem. Biophys. 136, 303.
69 Horecker, B. and Smyrniotis, P. (1955) J. Biol. Chem. 212, 811.
70 Venkataraman, R. and Racker, E. (1961) J. Biol. Chem. 236, 1883.
71 Horecker, B., Pontrenoli, S., Ricci, C. and Cheng, T. (1961) Proc. Natl. Acad. Sci. (U.S.A.) 47, 1949.
72 Speck, J., Rowley, P. and Horecker, B. (1963) J. Am. Chem. Soc. 85, 1012.
73 Brand, K. and Horecker, B. (1968) Arch. Biochem. Biophys. 123, 312.
74 Kobashi, K. and Brand, K. (1972) Arch. Biochem. Biophys. 148, 169.
75 Christen, P., Cogoli-Grentner, M., Healy, M. and Lubini, D. (1976) Eur. J. Biochem. 63, 223.
76 Tsolas, O. and Horecker, B. (1976) Arch. Biochem. Biohys. 173, 577.
77 Brand, K., Tsolas, O. and Horecker, B. (1969) Arch. Biochem. Biophys. 130, 521.
78 Brand, K. (1970) FEBS Lett. 7, 235.
79 Boyer, P. (Ed.) (1972) The Enzymes, 3rd Edn. Vol. VI, Academic Press, New York, Ch. 8.
80 Zerner, B., Coutts, S., Lederer, F., Waters, H. and Westheimer, F. (1966) Biochemistry 5, 813.
81 Tagaki, W. and Westheimer, F. (1968) Biochemistry 7, 895.
82 O'Leary, M. and Westheimer, F. (1968) Biochemistry 7, 913.
83 Tagaki, W., Guthrie, J.P. and Westheimer, F. (1968) Biochemistry 7, 905.
84 Hamilton, G. and Westheimer, F. (1959) J. Am. Chem. Soc. 81, 6332.
85 Fridovich, I. and Westheimer, F. (1962) J. Am. Chem. Soc. 84, 3208.
86 Laursen, R. and Westheimer, F. (1966) J. Am. Chem. Soc. 88, 3426.
87 Guthrie, J. and Westheimer, F. (1967) Fed. Proc. 26, 562.
88 Frey, P., Kokesh, F. and Westheimer, F. (1971) J. Am. Chem. Soc. 93, 7266.
89 Kokesh, F. and Westheimer, F. (1971) J. Am. Chem. Soc. 93, 7270.
90 O'Leary, M. and Baughn, R. (1972) J. Am. Chem. Soc. 94, 626.
91 Guthrie, J.P. and Jordan, F. (1972) J. Am. Chem. Soc. 94, 9136.
92 Autor, A. and Fridovich, I. (1970) J. Biol. Chem. 245, 5214.
93 Fridovich, I. (1968) J. Biol. Chem. 243, 1043.
94 Kluger, R. and Nakaoka, K. (1974) Biochemistry 13, 910.
95 Tagaki, W. and Westheimer, F. (1968) Biochemistry 7, 901.
96 Hammons, G., Westheimer, F., Nakaoka, K. and Kluger, R. (1975) J. Am. Chem. Soc. 97, 1568.
97 Benner, S. and Morton, T. (1981) J. Am. Chem. Soc. 103, 991.
98 Benner, S., Rozzell, J. and Morton, T. (1981) J. Am. Chem. Soc. 103, 993.
99 Hodgins, D. and Abeles, R. (1967) J. Biol. Chem. 242, 5158.

100 Riley, W. and Snell, E. (1968) Biochemistry 7, 3520.
101 Wickner, R., Tabor, C. and Tabor, H. (1970) J. Biol. Chem. 245, 2132.
102 Cohn, M., Tabor, C. and Tabor, H. (1977) J. Biol. Chem. 252, 8212.
103 Pegg, A. (1977) FEBS Lett. 84, 33.
104 Satre, M. and Kennedy, E. (1978) J. Biol. Chem. 253, 479.
105 Owen, T. and Young, P. (1974) FEBS Lett. 43, 308.
106 Rosenthaler, J., Guirard, B., Chang, G. and Snell, E. (1965) Proc. Natl. Acad. Sci. (U.S.A.) 54, 152.
107 Rodwell, A. (1953) J. Gen. Microbiol. 8, 233.
108 Recsai, P. and Snell, E. (1970) Biochemistry 9, 1492.
109 Recsai, P. and Snell, E. (1972) J. Bacteriol. 112, 624.
110 Recsai, P. and Snell, E. (1973) Biochemistry 12, 365.
111 Hackert, M., Meador, W., Oliver, R., Salmon, I., Recsai, P. and Snell, E. (1981) J. Biol. Chem. 256, 687.
112 Lane, R. and Snell, E. (1976) Biochemistry 15, 4175.
113 Chang, G. and Snell, E. (1968) Biochemistry 7, 2012.
114 Wilson, J., Wu, D., Motiu-DeGrood, R. and Hupe, D. (1980) J. Am. Chem. Soc. 102, 359.
115 Hupe, D.J., Kendall, M., Sinner, G. and Spencer, T. (1973) J. Am. Chem. Soc. 95, 2260.
116 Mitsuhashi, S. and Davis, B. (1954) Biochim. Biophys. Acta 15, 54.
117 Welch, G. and Gaertner, F. (1975) Proc. Natl. Acad. Sci. (U.S.A.) 72, 4218.
118 Hanson, K. and Rose, I. (1963) Proc. Natl. Acad. Sci. (U.S.A.) 50, 981.
119 Butler, J., Alworth, W. and Nugent, M. (1974) J. Am. Chem. Soc. 96, 1617.
120 Haslam, E., Turner, M., Sargent, D. and Thompson, R. (1971) J. Chem. Soc. C, 1489.
121 Vaz, A., Butler, J. and Nugent, M. (1975) J. Am. Chem. Soc. 97, 5914.
122 Portsmouth, D., Stoolmiller, A. and Abeles, R. (1967) J. Biol. Chem. 242, 2751.
123 Shemin, D. and Nandi, D. (1967) Fed. Proc. 26, 390.
124 Wilton, D. (1976) Biochem. J. 155, 487.

Pyridoxal phosphate-dependent enzymic reactions: mechanism and stereochemistry *

MUHAMMAD AKHTAR [1], VINCENT C. EMERY [1] and JOHN A. ROBINSON [2]

[1] *Department of Biochemistry and* [2] *Department of Chemistry, University of Southampton, Bassett Crescent East, Southampton, S09 3TU, Great Britain*

1. Historic background: Braunstein–Snell hypothesis

The original discovery leading to our present interest in pyridoxal-P-dependent biological reactions was made in 1934 by Paul György [1] in the United States (Western Reserve University, Cleveland). He identified a nutritional factor that could cure a specific dermatitis produced in young rats when they were reared on a deficient diet. György called this factor vitamin B_6 which was subsequently shown to be pyridoxine (Fig. 1, 1). The term vitamin B_6 is currently used to include a number of closely related substances of dietary origin (Fig. 1, 1, 2 and 3), which, in animals, are metabolised to produce the enzymically active form, pyridoxal phosphate (Fig. 1, 5). The pathways for these transformations [2] are shown in Fig. 1.

Pyridoxal phosphate and its close relative, pyridoxamine phosphate (Fig. 1, 6) participate in a multiplicity of different biological processes, most noteworthy amongst these being the reaction involved in the metabolism of amino acids [3–6]. Several dozen enzymes are known whose activities are dependent on the presence of pyridoxal phosphate in their active sites. Our current view regarding the mechanisms of pyridoxal-P-dependent transformations is based on the inspired suggestion originally made in 1953, independently by Braunstein and Shemyakin [7] in the Soviet Union, and Snell and his coworkers [8] in the United States. Before considering the basic tenets of the Braunstein–Snell hypothesis we examine the nature of the chemical problems involved in the metabolism of amino acids. In the later sections we shall note that the key event in the metabolism of amino acids is either the

* In recognition of his numerous contributions to bio-organic chemistry, this paper is dedicated to Professor E. Lederer on the occasion of his 75th birthday, which was on 5th June, 1983.

Michael I. Page (Ed.), The Chemistry of Enzyme Action

cleavage of the C_α–COOH or C_α–H bond. With the unmodified amino acid skeleton such cleavage reactions will generate intermediary species (Fig. 2, 1 and 2), whose conjugate carbon acids may be estimated to have pK_a values in excess of 30, rendering these types of cleavages thermodynamically most unfavourable. The Braunstein–Snell hypothesis provided the solution to this mechanistic problem by suggesting that a prerequisite for such a cleavage reaction is the formation of a Schiff

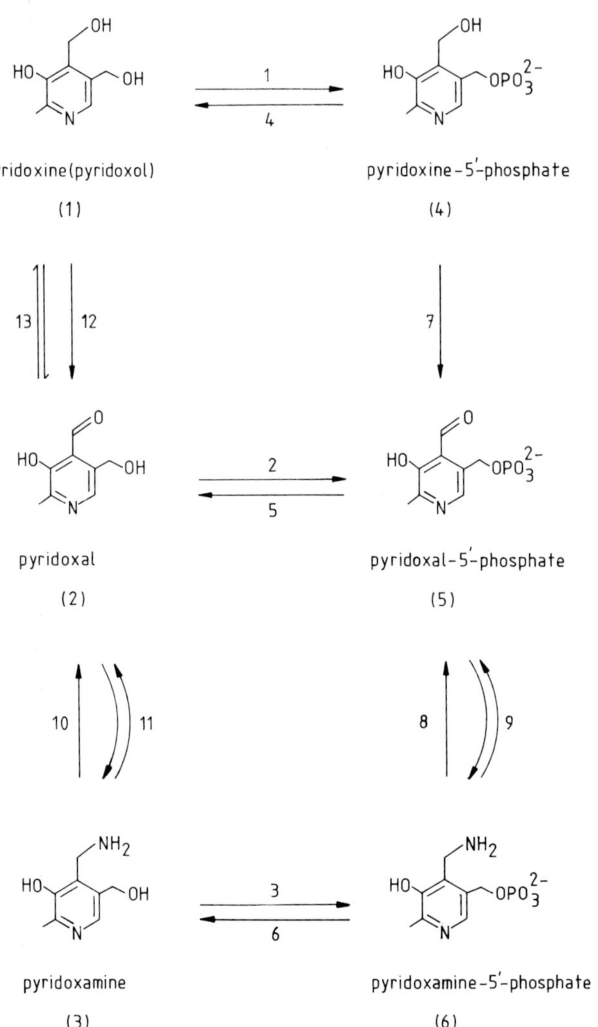

Fig. 1. The metabolic interconversion of various forms of vitamin B_6. Reactions 1, 2 and 3 are catalysed by pyridoxal kinase, reactions 4, 5 and 6 by various phosphatases, reactions 7, 8 and 10 by pyridoxal-P oxidases, reactions 9 and 11 by certain aminotransferases, and reactions 12 and 13 by various pyridoxal dehydrogenases (taken from ref. 2).

base between the amino acid and the enzyme-bound pyridoxal-P̄ moiety, presumably via a transimination process (Fig. 2, 3 → 4) *. The cleavage of the bond to C_α is now stereoelectronically assisted and leads to the formation of the putative carbanionic species which is considerably resonance stabilised by delocalisation of the electron pair on the positively charged pyridinium nitrogen atom (Fig. 2). In other words the pK_a of the relevant carbon acid is thus greatly reduced – by how much we could not say.

In the light of this background we consider an enormous body of experimental work on pyridoxal-P-dependent enzymic reactions which has accumulated since the enunciation of the Braunstein–Snell hypothesis. Broadly speaking, at a crucial stage in these reactions, the cleavage of either the C_α–COOH or C_α–H bond is involved. In Sections 2–5 transformations belonging to these two classes are discussed sequentially with major emphasis on the description of events which occur on the

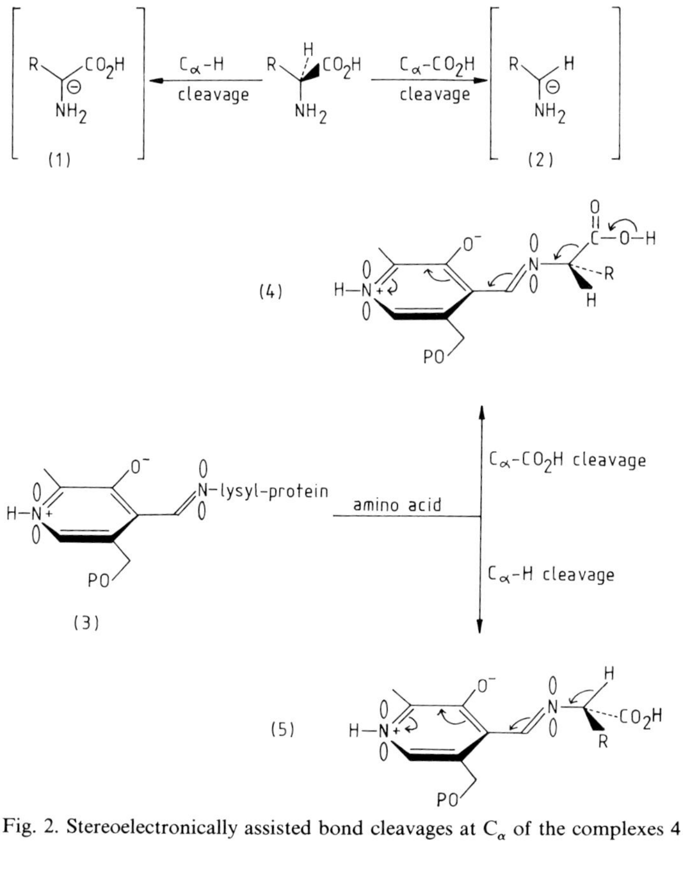

Fig. 2. Stereoelectronically assisted bond cleavages at C_α of the complexes 4 and 5.

* For the arrangement of substituents around the pyridinium ring, see comments on p. 367.

substrate. In the following sections we examine the structural and stereochemical details of coenzyme–enzyme (binary) and substrate–coenzyme–enzyme (ternary) complexes and also evaluate the nature of chemical events which may occur within these complexes.

2. *Pyridoxal phosphate-dependent reactions involving C_α–CO_2H bond cleavage*

Amino acid decarboxylases catalyse the generic reaction, 1 → 5, shown in Fig. 3. They are widely distributed in nature and are found in most living organisms [9]. In animals the enzymes of this class participate in the biosynthesis of a wide variety of pharmacologically active amines some of which exert a profound influence on the central nervous system. For example, a crucial reaction in the biosynthesis of both adrenalin (epinephrine) and γ-aminobutyric acid (GABA) (Fig. 4) is the involvement of a specific amino acid decarboxylase. Despite their importance to clinical medicine, decarboxylases from mammalian sources have not been purified in sufficient amounts for detailed kinetic and mechanistic studies. Consequently, in the pioneering studies by Hanke and coworkers [10a] enzymes from bacterial sources [11] were used to establish the basic mechanistic framework for the reaction [10a]. The general mechanism to emerge * from these studies is shown in Fig. 3.

The importance of stereoelectronic assistance in the cleavage of the C_α–CO_2H bond within the coenzyme–substrate Schiff base complex has already been emphasised. Protonation of the transient intermediate (Fig. 3, 3) at C_α can then proceed with retention, inversion or racemisation. It is worth bearing in mind that during this conversion a chiral centre of an α-amino acid ** is transformed into a prochiral

* The mechanism for the decarboxylation process in Fig. 3 is in fact due to F.H. Westheimer (Harvard) [10b].

** According to the Cahn–Ingold–Prelog nomenclature system [12a], L-amino acids have (*S*) absolute configuration, and D-amino acids have (*R*) configuration. The system may be exemplified by reference to an amino acid. The 4 substituents around C_α are assigned the following decreasing order of priority $-NH_2 > -COOH > -alkyl > -H$. If the molecule is now viewed with the group of lowest priority, –H, away from the observer then the remaining three groups in the order $-NH_2$, –COOH, –alkyl make an anti-clockwise arrangement for an L-amino acid as in structure 1 (Fig. 3) and clockwise for a D-amino acid. These centres thus have (*S*) and (*R*) absolute configuration respectively (these descriptors are derived from the Latin words, *rectus* = right and *sinister* = left).

centre * of an amine (Fig. 3, 5). The stereochemical outcome of the reaction therefore has no physiological implication since the same chemical entity (5) is produced irrespective of which steric course is followed. A knowledge of the stereochemical course, however, has a profound mechanistic relevance and may be determined only by judicious labelling with isotopes of hydrogen, which due to their presence alone generate a centre of chirality at C_α of the amine.

The high stereospecificity of decarboxylations catalysed by a series of bacterial decarboxylases was first demonstrated by Hanke [10a], who carried out the reactions in ^2H_2O, isolated the corresponding chiral [1-$^2H_1$]amines and showed that they were enantiomerically pure. Since then the steric courses of a number of decarboxylases have been more fully defined. Most of these act on L-amino acids and by appropriate isotopic labelling these have all been shown to proceed with overall retention at C_α ** (see Fig. 5).

A good illustration of the methods involved in the study of the stereochemistry of the decarboxylation process is seen in the elucidation of the steric course of tyrosine decarboxylase [13]. In a recent investigation two enantiomeric [1-^{3}H]tyramines were produced, one by decarboxylation of (L-[2-^{3}H]tyrosine in H_2O and the other by decarboxylation of unlabelled tyrosine in [^{3}H]H_2O (Fig. 6). The absolute configura-

* In this article the *pro*chiral hydrogen atoms are designated as H_{Re} and H_{Si} according to the system suggested by Prelog and Helmchen [12b]. The structure (a) is viewed with *one* of the *pro*chiral hydrogen

(a) alkyl—C—NH2 with *H and ■H

(b) ■H(Re), *H(Si), alkyl, NH2

(c) *H(Si), ■H(Re), H2N, alkyl

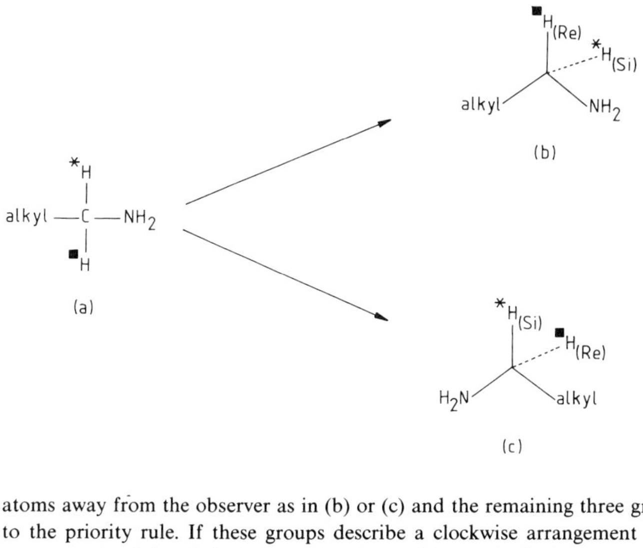

atoms away from the observer as in (b) or (c) and the remaining three groups are then ordered according to the priority rule. If these groups describe a clockwise arrangement (b) then the prochiral hydrogen atom contained therein is designated as *Re*, and hence the hydrogen atom placed away from the observer becomes H_{Si}. The converse case yielding the same answer is shown in (c).

** In accordance with the requirements of the mechanism outlined in Fig. 3, the incubation [10a] of amine 5 with the corresponding decarboxylase led to the stereospecific exchange of one of the hydrogen atoms at C_α of 5 with the protons of the medium, due to a partial reaction.

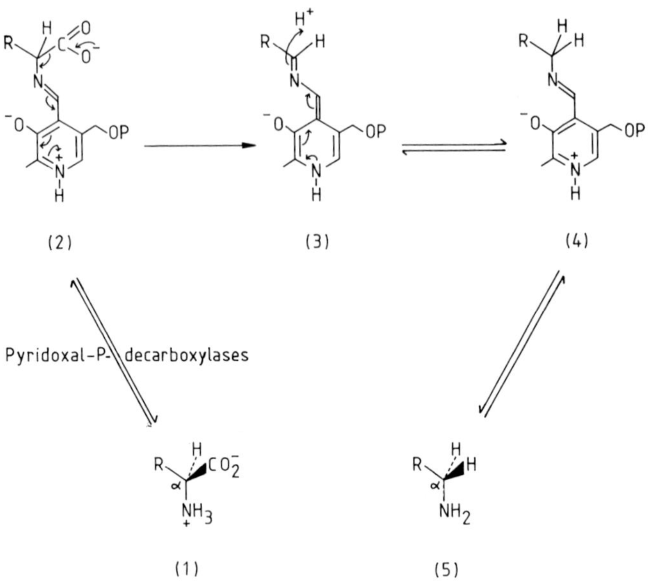

Fig. 3. A general mechanism for an amino acid α-decarboxylase.

tions of these two samples were established in the following way: enantiomerically pure (R)- and (S)-[1-^{3}H]tyramine were prepared by stereospecific synthesis and each was separately oxidised by either pea-seedling diamine oxidase, or by rat-liver monoamine oxidase. During oxidation with the former enzyme, tritium was lost to the medium from the (S) enantiomer and retained in the product during oxidation of the (R) enantiomer, whereas the latter enzyme exhibited the opposite stereo-

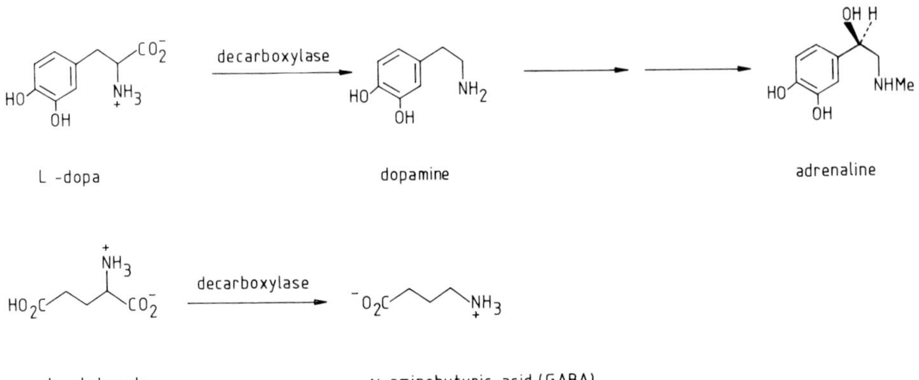

decarboxylase

L -dopa

dopamine

adrenaline

decarboxylase

L -glutamate

γ-aminobutyric acid (GABA)

Fig. 4. The involvement of amino acid decarboxylases in the biosynthesis of pharmacologically active compounds.

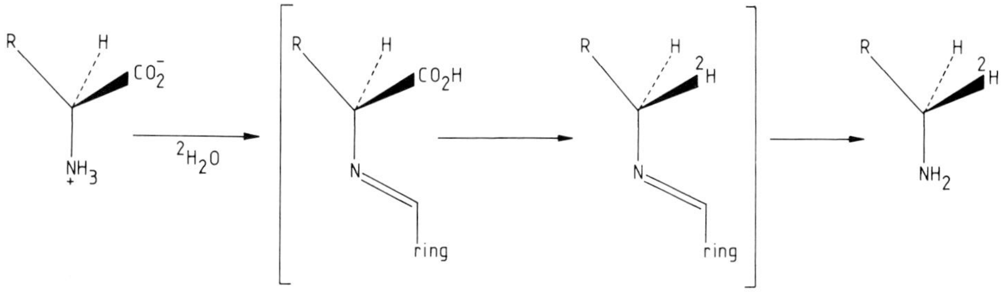

Fig. 5. Scheme showing the retention of stereochemistry in the reaction catalysed by decarboxylases.

specificity (Fig. 7). These two amine oxidases could then be used to establish the absolute configuration and enantiomeric purity of the [1-^{3}H]tyramines produced enzymically by decarboxylation of tyrosine. Tyrosine decarboxylases from plant, microbial and mammalian origin were tested in this way and in every case the reaction was shown to proceed with complete retention (Fig. 5). The same conclusion had also been reached earlier by a different group of workers [14,15] using deuterium labelling, and by related methods, the decarboxylations of L-lysine [16–18], L-glutamate [19–21] and L-histidine [22], catalysed by their respective decarboxylases, have also been shown to proceed with retention. This stereochemical consistency, manifested over a range of different enzymes, may reflect a hidden mechanistic or catalytic advantage associated with the retention mode, but this is made less likely by the recent observation [23] that the bacterial enzyme, α,ω-*meso*-diaminopimelate decarboxylase, which converts α,ω-*meso*-diaminopimelate into L-lysine, proceeds with inversion at C_ω (Fig. 8). This enzyme, therefore, is not only the sole pyridoxal-P-dependent decarboxylase acting at a chiral centre of (*R*) absolute

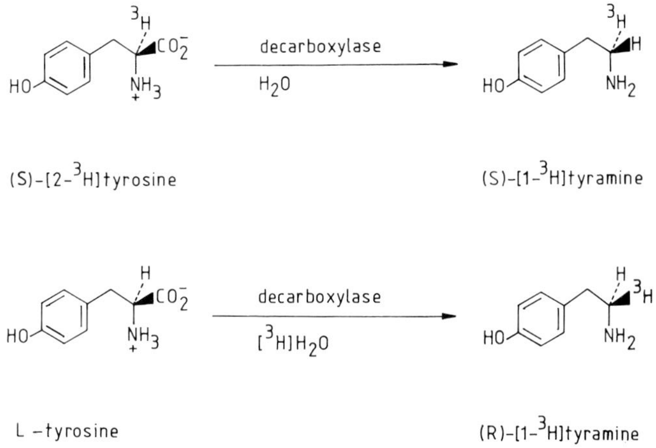

(S)-[2-^{3}H]tyrosine → (S)-[1-^{3}H]tyramine (decarboxylase, H_2O)

L-tyrosine → (R)-[1-^{3}H]tyramine (decarboxylase, [^{3}H]H_2O)

Fig. 6. The stereospecific preparation of enantiomeric [1-^{3}H]tyramines.

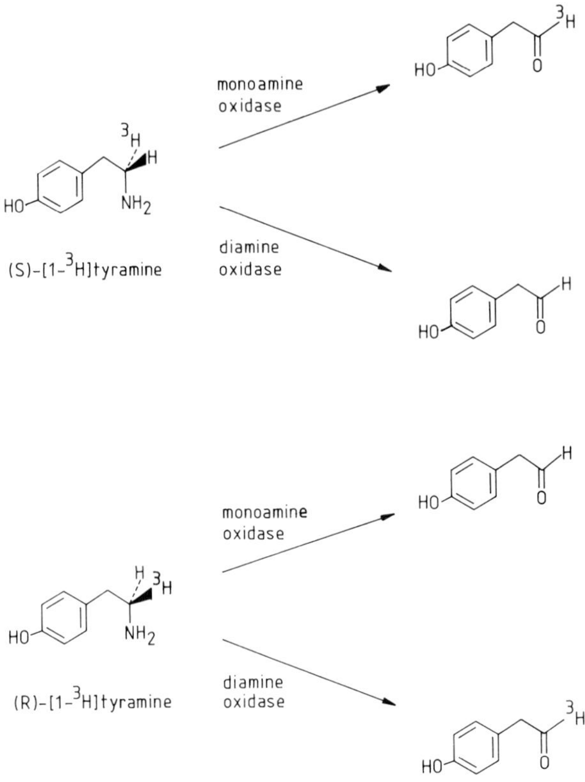

Fig. 7. Use of amine oxidases to establish the absolute configuration of enantiomeric [1-^{3}H]tyramines.

configuration to have been studied to date, it is the only one known to proceed with inversion * at a chiral centre.

The decarboxylation of L-aspartic acid to L-alanine is catalysed by a pyridoxal-P-dependent β-decarboxylase whose reaction mechanism is clearly different from that of the α-decarboxylases since the initial step probably involves C_α–H bond cleavage. The steric course at C_β during the normal decarboxylation reaction has recently been shown [23b] to be *inversion.* In addition to this, however, the enzyme will also catalyse the decarboxylation of amino-malonic acid to glycine and Meister and coworkers [24,25] have shown that this process involves loss of the *Si* carboxyl group with overall retention at C_α.

A list of some of the decarboxylases is included in Table 1.

Several amino acid α-decarboxylases of bacterial origin are known that do not

* Note, there are other pyridoxal-P-dependent enzymes that have an inversion step (ALA-synthetase, racemases).

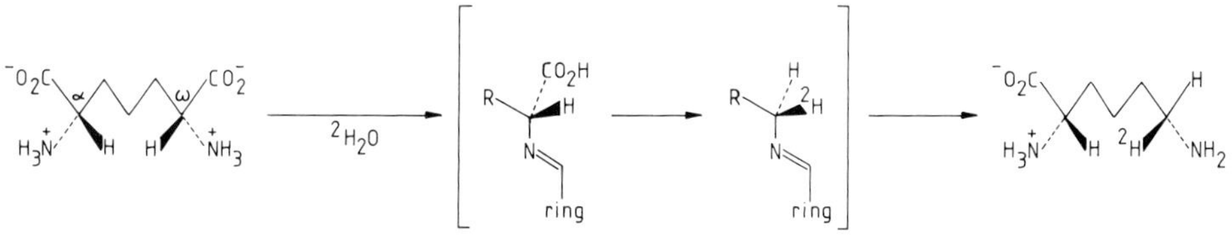

Fig. 8. An example of a decarboxylation reaction occurring via an inversion process.

require pyridoxal phosphate as a coenzyme, but instead use a pyruvyl residue of the enzyme [26]. The mechanism of these reactions again involves formation of a substrate–pyruvyl Schiff base from which CO_2 can be lost by decarboxylation (Fig. 9), the electron sink this time being the carbonyl group, which is activated presumably by the presence of a Schiff base at the α-position. The steric courses of two such pyruvyl-decarboxylases, L-histidine decarboxylase [27] and *S*-adenosyl-L-methionine decarboxylase [28], have been investigated and shown to proceed with overall retention at C_α of the amino acid.

Decarboxylation of glycine. A multienzyme complex consisting of a pyridoxal-P protein, a flavin-linked dehydrogenase and requiring tetrahydrofolate has been

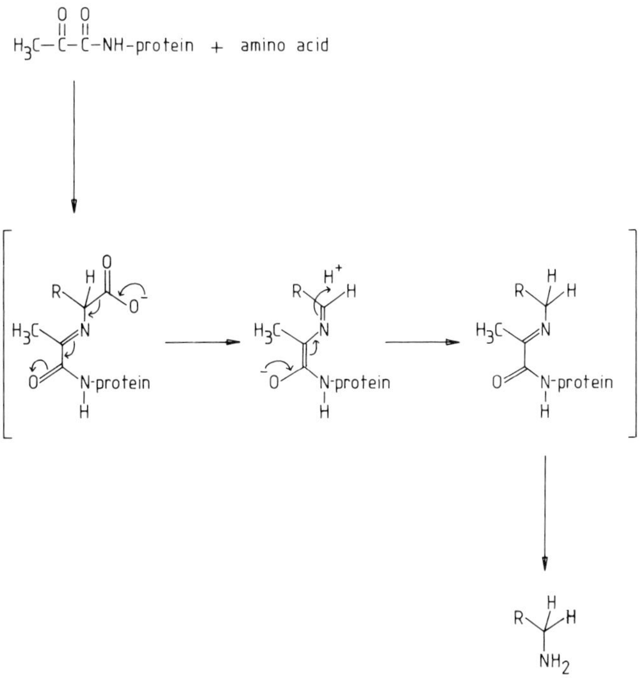

Fig. 9. Mechanistic pathway for the decarboxylation of amino acids by enzymes containing a pyruvyl residue.

TABLE 1

Decarboxylases

This list is primarily based on the previously published tables [30,31]. It is to be noted that the previous compilations [30,31] also give a brief account of the physical properties of the enzymes.

Decarboxylase	Source	Substrate utilised	Amine produced
Bacterial			
1. Aspartate α [112]	*Rhizobium trifoli*	Aspartate	β-Alanine
	Rhizobium leguminosarum		
	E. coli		
	Azobacter vinelandii		
2. Aspartate β [113–115]	*Clostridium perfringens*	Aspartate	α-Alanine
	Clostridium welchii		
	Archromobacter sp.		
3. Arginine [116–118]	*E. coli*	Arginine	Agmatine
4. Aminobenzoate [119]	*E. coli*	*p*-Aminobenzoate	Aniline
5. α-Aminoisobutyrate [120]	Pseudomonas sp.	α-Aminoisobutyrate	Isopropylamine
6. αα′-Diaminopimelate [121,122]	*E. coli*	αα′-Diamino-pimelate	Lysine
	Bacillus sphaericus asporogenous		
7. Glutamate [123–125]	*E. coli*	Glutamate	Glutamine
		γ-Hydroxyglutamate	γ-Hydroxyglutamine
8. Histidine [116,126,127]	*E. coli*	Histidine	Histamine
	Clostridium welchii		
	Lactobacillus 30a		
9. Leucine [128,129]	*Proteus vulgaris*	Leucine	Isoamylamine
		Valine	Isobutylamine
		Norvaline	2-Methylbutylamine
10. Lysine [116,130,131]	*E. coli*	Lysine	Cadaverine
11. Ornithine [116,132]	*E. coli*	Ornithine	Putrescine
	Lactobacillus 30a		
12. Phenylalanine [133]	*Streptococcus faecalis*	Phenylalanine	Phenylethylamine
		[illegible]	Tyramine

Enzyme	Source	Substrate	Product
13. Tryptophan [134]	Streptococci	Tryptophan	Tryptamine
14. Tyrosine [135]	*Streptococcus faecalis*	Tyrosine	Tyramine
		3,4-Dihydroxy-phenylalanine	3,4-Dihydroxy-phenylethylamine
Mammalian and plant			
15. Aminomalonate [136,137]	Liver	Aminomalonate	Glycine
16. Aromatic L-amino acid [138–141]	Kidney, brain, liver	3,4-Dihydroxy-phenylalanine	3,4-Dihydroxy-phenylethylamine
		Tyrosine	Tyramine
		5-Hydroxytryptophan	Serotonin
		Tryptophan	Tryptamine
		Phenylalanine	Phenylethylamine
		Histidine	Histamine
17. Cysteine sulphinate [142–144]	Liver	Cysteine sulphinate	Hypotaurine
		Cysteate	Taurine
18. 3,4-Dihydroxyphenylalanine [145]	*Sparticum scoparium*	3,4-Dihydroxy-phenylalanine	3,4-Dihydroxy-phenylethylamine
19. Glutamate [142,146,147]	Brain	Glutamate	γ-Aminobutyrate
		Cysteine sulphinate	Hypotaurine
		Cysteate	Taurine
20. Glutamate [148–150]	Squash, avocado, green peppers etc.	Glutamate	γ-Aminobutyrate
	Barley	γ-Methylene-glutamate	γ-Amino-α-methylene-glutamate
21. Histidine [151,152]	Human mast cells	Histidine	Histamine
	Rat hepatoma		
	Normal mast cells		
	Rat foetal tissues		
22. Histidine [145]	Spinach and other higher plants	Histidine	Histamine
23. Ornithine [153]	Calf liver	Ornithine	Putrescine
24. *S*-Adenosylmethionine [154,155]	Rat liver	*S*-Adenosyl-methionine	*S*-Methyladenosyl-L-homocysteamine
	Mouse mammary		

isolated from *Peptococcus glycinophilus* and catalyses the decarboxylation of glycine [9,29]. The mechanism of the process is, however, not yet fully understood.

3. *Pyridoxal phosphate-dependent enzymic reactions involving C_α–H bond cleavage*

3.1. Aminotransferases

(a) Metabolic background

Evolutionary biology must hold a special place for L-glutamate, since this is the only amino acid whose carbon–amine bond is produced from 'first principles' using ammonia. The enzyme responsible for this process in most organisms is glutamate dehydrogenase that catalyses the reductive amination of α-oxoglutarate using NADH or NADPH as the hydride donor:

$$\begin{aligned} &{}^{-}O_2C\text{–}CH_2\text{–}CH_2\text{–}CO\text{–}CO_2^- + NH_3 + NAD(P)H + H^+ \\ &\rightleftharpoons {}^{-}O_2C\text{–}CH_2\text{—}CH_2 - CH(NH_2)\text{–}CO_2^- + NAD(P)^+ \end{aligned} \qquad (1)$$

In some bacteria and plants, however, the formation of L-glutamate is catalysed by L-glutamate synthetase, in a somewhat wasteful * process in which ammonia is produced at the active site of the enzyme from L-glutamine:

$$\begin{aligned} &H_2N\text{–}OC\text{–}CH_2\text{–}CH_2\text{–}CH(NH_2)\text{–}CO_2^- \rightarrow {}^{-}O_2C\text{–}CH_2\text{–}CH_2\text{–}CH(NH_2)CO_2^- \\ &\qquad\qquad \alpha\text{-oxoglutarate} + NADPH \xrightarrow{\text{“}NH_3\text{”}} \text{L-Glu} + NADP^+ \end{aligned} \qquad (2)$$

The α-amino function of other L-amino acids is then produced from the corresponding α-oxo-acids through a transamination process [30,31] in which L-glutamate is almost without exception the amino donor **:

$${}^{-}O_2C\text{–}CO\text{–}R + \text{L-Glu} \rightleftharpoons \text{2-oxoglutarate} + R\text{–}CH(\overset{+}{N}H_3)\text{–}CO_2^- \qquad (3)$$

When the reaction occurs as shown in Eqn. 3, the enzyme nomenclature system currently in vogue uses only the name of the unique amino acid to specify an aminotransferase, e.g. aspartate aminotransferase, alanine aminotransferase, etc. etc.

---

* Wasteful, because the formation of L-glutamine from L-glutamate and ammonia requires ATP. This investment of ATP is reflected in the fact that the glutamate synthetase reaction is irreversible, whereas the glutamate dehydrogenase reaction is reversible.

** Readers should note that the coupling of aminotransferases and glutamate dehydrogenase in the reverse direction is involved in the catabolism of amino acids.

(b) Transaminations at the C_α of amino acid

Aspartate aminotransferase has been available in a highly purified state, and in abundance from several sources [31] for more than 2 decades. The primary amino acid sequence [32] of the enzyme as well as its tertiary structure at 2.8 Å resolution [33] is now known. It is therefore not surprising that our current view on the mechanism of action of pyridoxal-P-dependent reactions in general and aminotransferases in particular is *primarily* based on the extensive physicochemical studies which have been performed on aspartate aminotransferase [34].

The studies culminating in the proposal of the mechanism of action of aminotransferases as outlined in Fig. 10, include the early observation that the transformation occurs without intermediate formation of ammonia [35] and with the acquisition of one proton from the solvent which is located at C_α in the new amino acid [36]. It was also shown that the β-hydrogen of the amino acid is not a participant * [37–40].

The general mechanism shown in Fig. 10 is supported by a considerable body of evidence from both enzymic and model studies [30,31,34] and the key step after the formation of the L-amino acid-pyridoxal-P Schiff base (Fig. 10, 1) is a reversible tautomerisation reaction involving C_α of the substrate and C-4′ of the coenzyme (Fig. 10, 1 → 3). Hydrolysis of the intermediate (Fig. 10, 3) produces the pyridoxamine-P–enzyme complex (Fig. 10, 4) releasing the α-oxo-acid. The reversal of the same sequence but using a new α-oxo-acid produces the corresponding L-amino acids (Fig. 10). There are some aminotransferases particularly those involved in biosynthesis of bacterial cell walls, which are specific for D-amino acids. A list of more important aminotransferases is included in Table 2. Full details of the methodology used in the elucidation of the structure of intermediates involved in reactions catalysed by aminotransferases and experimental approaches used to shed light on the steric course of these reactions will be discussed in a later section.

(c) Mechanistic studies on miscellaneous transaminations

A closely related group of pyridoxal-P-dependent aminotransferases act not at the chiral α-carbon atom of an amino acid, but instead at the prochiral ω-aminomethylene carbon of amino acids such as lysine [42], ornithine [43] and γ-aminobutyrate [44] (GABA). The mechanism of these processes is essentially analogous to that shown in Fig. 10 for transamination at C_α. The steric course with respect to the substrate, however, is only discernible if the ω-aminomethylene group is first rendered chiral by isotopic substitution. This has been achieved [21a,b,45a] in the case of GABA by starting from glutamate and making use of the known steric course of the glutamate decarboxylase reaction (vide supra) (Fig. 11). Thus (4*R*)-[4-^{3}H]GABA was prepared [45a] from unlabelled glutamate using glutamate decarboxylase in [^{3}H]H_2O, whereas (4*S*)-[4-^{3}H]GABA was generated in normal water

* It is though interesting to draw attention to the fact that incubation of L-alanine with alanine aminotransferase results not only in the expected exchange of its α-hydrogen but also of the β-hydrogen [41] atoms with the protons of the medium.

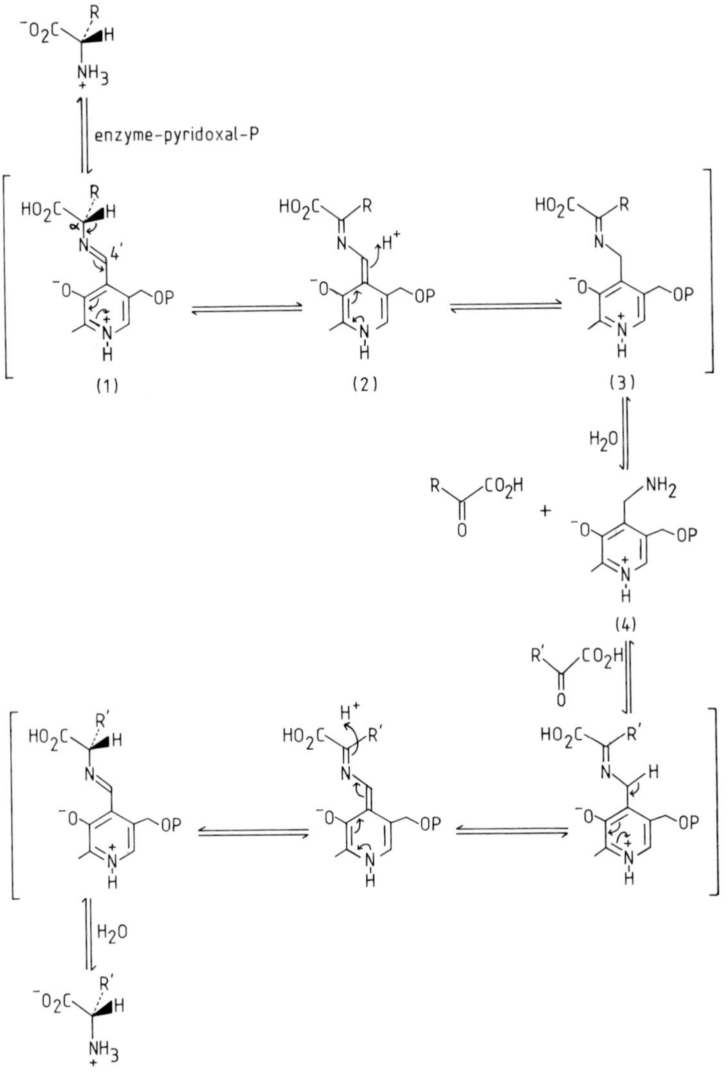

Fig. 10. A general mechanistic scheme for aminotransferases.

from L-[2-^{3}H]glutamate. When each labelled GABA was then transformed by a bacterial GABA aminotransferase into succinate semialdehyde, tritium was lost preferentially from the (4*S*) isomer to the medium. A similar result has also been obtained [21b] using a mammalian brain GABA-ω-aminotransferase *. In contrast, another aminotransferase [45a], ω-amino acid: pyruvate aminotransferase from Pseudomonas sp F-126, acting at the γ-carbon of GABA, stereospecifically removed tritium from the (4*R*)-[4-^{3}H]GABA (see Fig. 11). These two different bacterial

TABLE 2

Aminotransferases

This list is primarily based on the previously published tables [30,31]. It is to be noted that the previous compilations [30,31] also give a brief account of the physical properties of the enzymes.

Aminotransferase	Source	Utilisation of amino acids other than that of column 1	Oxoacid utilised
1. N^2-Acetylornithine [156]	*E. coli*	–	α-Oxoglutarate
2. L-Alanine [157]	Pig heart (cyt)	α-Aminobutyrate α-Aminoadipate	α-Oxoglutarate
3. L-Alanine [158,159]	Rat liver	–	α-Oxoglutarate
4. L-Alanine [160]	Human liver	–	Glyoxylate
5. D-Alanine [161]	*Bacillus subtilis*	D-Glutamate D-Aminobutyrate D-Aspartate D-Asparagine D-Norvaline	α-Oxoglutarate
6. β-Alanine [162]	*Pseudomonas fluorescens*	γ-Aminobutyrate δ-Aminovaline β-Aminoisobutyrate ε-Aminocaproate	Pyruvate α-Oxobutyrate Glyoxylate
7. α-Aminoadipate [163]	Rat kidney	L-Kyneurenine	α-Oxoglutarate
8. γ-Aminobutyrate [164]	Rat brain	β-Alanine D,L-β-Aminoisobutyrate	α-Oxoglutarate
9. γ-Aminobutyrate [165]	*Pseudomonas fluorescens*	–	α-Oxoglutarate
10 5-Aminolevulinic acid [166]	*Rhodopseudomonas spheroides*	γ-Aminobutyrate δ-Aminovaline ε-Aminocaproate	Pyruvate
11. Aspartate [167–171]	Pig heart (cyt)	L-Alanine	α-Oxoglutarate
12. 7,8-Diaminopelargonic acid [172]	*E. coli*	–	*S*-Adenosyl-2-oxomethyl-thiobutyrate

TABLE 2 (continued)

Aminotransferase	Source	Utilisation of amino acids other than that of column 1	Oxoacid utilised
13. Glycine [173–175]	*Micrococcus denitrificans*	Serine Asparagine	Oxaloacetate
14. Glycine [176]	Human liver	Alanine Glutamate Methionine Arginine Glutamine	α-Oxoglutarate
15. Glutamine [177–180]	Rat liver (cyt) Rat kidney	Most monocarboxylic α-amino acids	Any monocarboxylic α-oxoacid
16. L-Histidinol-1-phosphate [181–183]	*Salmonella typhimurium*	Histidine	α-Oxoglutarate
17. Leucine [184]	Pig heart (cyt)	Isoleucine Valine Aminobutyrate Methionine	α-Oxoglutarate
18. Lysine [185]	*Achromobacter liquidum*	Ornithine	α-Oxoglutarate
19. Ornithine [186]	Pig kidney (mit)	–	Any α-Oxoacid
20. Phenylalanine [187]	*Achromobacter euridycea*	Broad specificity: tyrosine tryptophan aspartate	α-Oxoglutarate Oxaloacetate Phenylpyruvate
21. Pyridoxamine [188–190]	Pseudomonas sp.	Analogues of pyridoxamine	Pyruvate
22. Putrescine [191–193]	*E. coli* B	Other diamines (C_4–C_7)	α-Oxoglutarate Pyruvate
23. Tyrosine [194,195]	Rat liver (cyt)	Phenylalanine Tryptophan	α-Oxoglutarate

aminotransferases therefore exhibit different steric courses with respect to C_γ of GABA.

More recently, lysine and ornithine ω-aminotransferase have also been investigated [45b] and each was shown to remove stereospecifically the H_{Si} atom from their respective substrates at the ω centre.

A bacterial aminotransferase [46] promotes a decarboxylative transamination reaction with α-aminoisobutyrate in the presence of pyruvate. The reaction occurs via the sequence of Fig. 12 involving an initial cleavage of the C_α–CO_2H bond in the substrate pyridoxal-P Schiff base complex (Fig. 12, 1) followed by reprotonation at C-4′ of the coenzyme to give the pyridoxamine-P–enzyme complex (Fig. 12, 4) that participates in the transamination of pyruvate. However, the enzyme will also transform L-alanine at a significant rate by a half-transamination reaction into pyruvate, thereby implying that it is now the C_α–H bond of the amino acid that is first broken.

3.2. Racemases

Amino acid racemases catalyse the formation of a racemic mixture from either L- or D-amino acids. The enzymes of this class have only been detected in bacteria where they are involved in the formation of D-amino acids required for cell-wall biosynthesis. A crucial step in the racemase catalysed reactions must involve the cleavage of C_α–H bond to give a quinonoid intermediate of type 2 (Fig. 10) which is then protonated in a non-stereospecific fashion. Detailed mechanistic studies on these

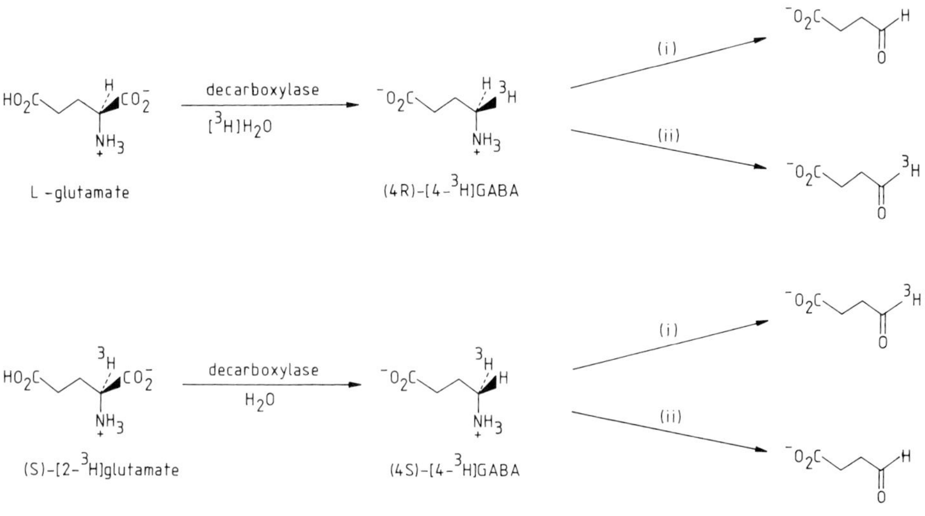

Fig. 11. Steric course of aminotransferases acting on an ω-aminomethylene group of enantiomeric [4-^{3}H]GABA. (i) = ω-amino acid pyruvate aminotransferase; (ii) = GABA aminotransferase.

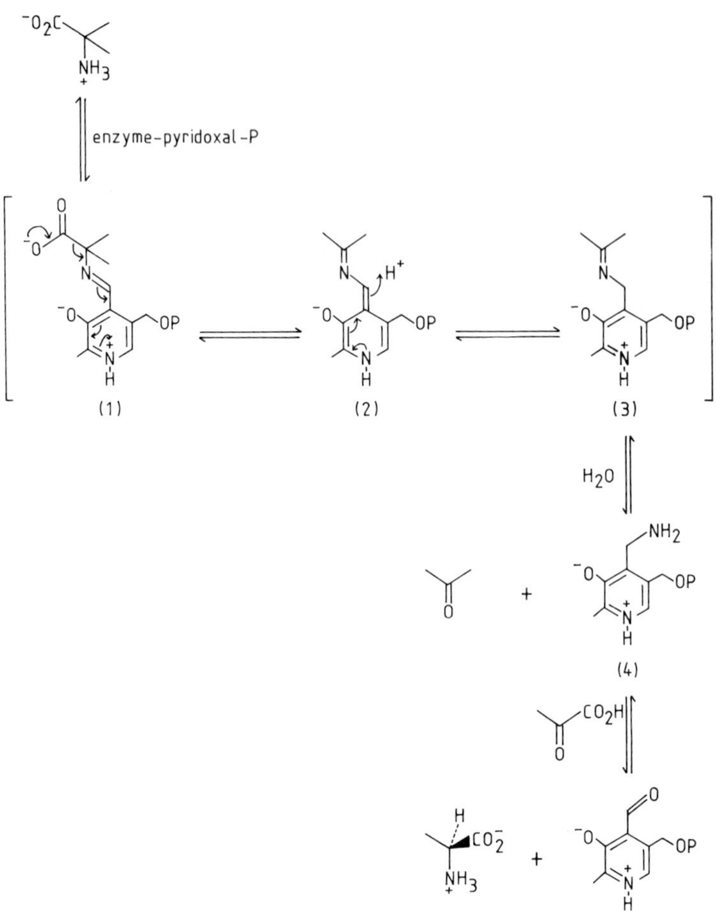

Fig. 12. Mechanism of a decarboxylative transamination reaction involving α-aminoisobutyrate and pyruvate as substrates.

reactions have however not been performed. A comprehensive review on racemases and other related enzymes is available [4]. Suffice to mention here that two racemases, alanine racemase and arginine racemase have been purified to homogeneity [4].

3.3. Serine hydroxymethyltransferase (SHMT) *

Pyridoxal-P-dependent serine hydroxymethyltransferase (SHMT; 5,10-methylene-

* We have included the SHMT reaction in the C_α–H bond cleavage category; although this is true only for the conversion of glycine into serine. In the reverse direction the reaction falls in a third category, not used in the present article, that is, cleavage of the C_α–R bond.

tetrahydrofolate-glycine-hydroxymethyltransferase) catalyses the interconversion * of glycine and serine according to the reaction of Eqn. 1 in Fig. 13 in which the C_1 unit is shuttled via tetrahydrofolate [47,48]. The enzyme also possesses activity for the cleavage of allothreonine and threonine which is independent of tetrahydrofolate [49,50] (Fig. 13, Eqn. 2) (threonine aldolase activity). In addition, SHMT participates in the exchange of one of the α-hydrogen atoms of glycine with the protons of the medium [51,52]. The exchange process is stimulated several-fold (20–50) in the presence of tetrahydrofolate [50,51]. On the basis of these considerations and other studies performed with cytoplasmic as well as mitochondrial SHMT from rabbit liver **, the catalytic mechanism for the enzyme as shown in Fig. 14 has been proposed [50,52]. This involves, in the initial step, the formation of a glycine–pyridoxal-P–enzyme complex that undergoes C_α-deprotonation giving rise to the glycyl anion equivalent (Fig. 14, 2) which then reacts with formaldehyde transiently generated at the active site from N^5,N^{10}-methylenetetrahydrofolate, to produce the Schiff base of serine–pyridoxal-P–enzyme (Fig. 14, 3) from which the hydroxy amino acid is released by the usual hydrolytic process. In the threonine aldolase reaction the glycyl anion equivalent (Fig. 14, 2) reacts with acetaldehyde in an analogous fashion ***. The support for the general mechanism of Fig. 14 comes from several quarters. For example, that SHMT catalyses the formation of serine from glycine and formaldehyde in the *absence* [50,53] of tetrahydrofolate, even though the rate of this reaction is only 2% of that in the presence of tetrahydrofolate, is in accordance with the basic tenet of the mechanism. The possibility that during catalysis the aldehydes may bind to the active site by Schiff base linkage has been considered [48], but is eliminated by the demonstration that in the cleavage of threonine, containing ^{18}O at the hydroxyl group, the acetaldehyde produced quantitatively retained the label [54].

The steric course of this reaction with respect to C_α of glycine has been deduced by two groups and crucial to each was the availability of a method for the configuration analysis of stereospecifically tritiated glycine. For this purpose two enzymic reactions each of known steric course were utilised. One was glycolic acid oxidase which was shown [55,108] to remove specifically the H_{Re} atom from glycolic acid. Thus, when L-serine was cleaved in $[^3H]H_2O$ with SHMT, $[2\text{-}^3H]$glycine was formed and this upon conversion into glycolic acid and then oxidation using the

* Although SHMT catalyses a freely reversible reaction, in vivo the enzyme is predominantly involved in the cleavage of serine. The C_1 unit transferred to tetrahydrofolate is used in a number of biosynthetic processes.

** SHMT has also been purified from rat [56a] and sheep [56b] liver. All these preparations show a number of activities, particularly for the cleavage of allothreonine and β-phenylserine but the activity for the cleavage of threonine and aminomalonate is dependent on the source [56a]. It should, though, be noted that SHMT catalysed reactions with amino acids other than L-serine, are unlikely to have any physiological significance.

*** The reaction produces allothreonine and threonine in the ratio [48,53,56a] of 98:2, while in the reverse direction at saturating concentration of the relevant amino acid, allothreonine is cleaved about 3 times faster than threonine with the rabbit liver cytoplasmic enzyme [49].

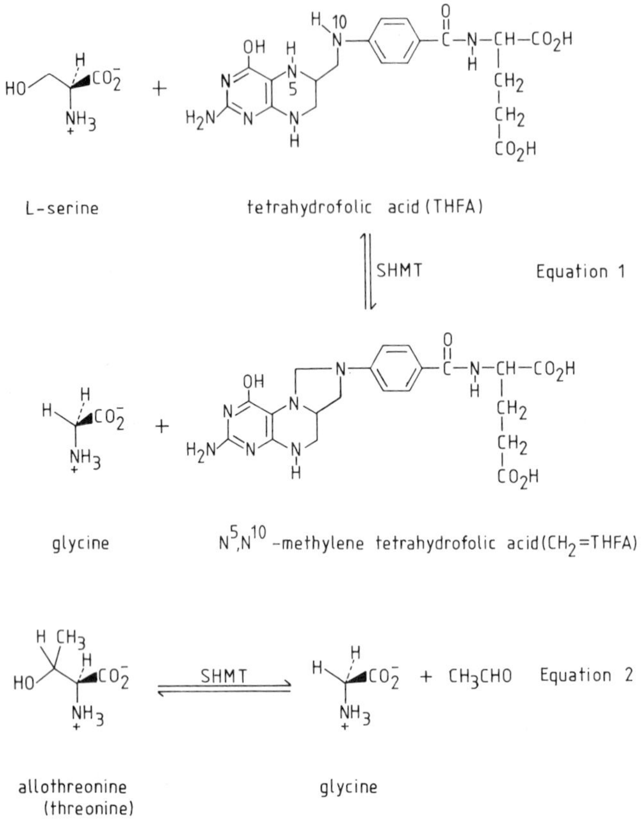

Fig. 13. Reactions catalysed by serine hydroxymethyltransferase (SHMT). It should be noted that the reaction of Eqn. 1 is the only one with an established physiological role.

oxidase retained most of the tritium label. Its absolute configuration was therefore (*S*) (see Fig. 15).

In a different approach [52,57,58] (Fig. 16) the known property of SHMT to catalyse the exchange of one of the hydrogen atoms at C_α of glycine with protons from the medium was used to prepare both enantiomers of [2-^{3}H]glycine, the sample shown to be (*R*)-[2-$^3H_2$]glycine was obtained by equilibrating (*RS*)-[2-$^3H_2$]glycine with SHMT in H_2O, whereas (*S*)-[2-^{3}H]glycine was obtained correspondingly from SHMT in [^{3}H]H_2O. The configurational analysis this time involved the oxidation of each tritiated glycine to glyoxylic acid using D-amino acid oxidase. Since this enzyme is specific for D-amino acids it was assumed that the 2-H_{Si} atom of glycine would be removed stereospecifically. The tritiated glycine prepared from (*RS*)-[2-^{3}H]glycine in H_2O retained most of the tritium label upon conversion into glyoxylic acid, whereas its counterpart lost most of the label. Both samples were therefore of high enantiomeric purity and possessed the (*R*) and (*S*) absolute configurations, respec-

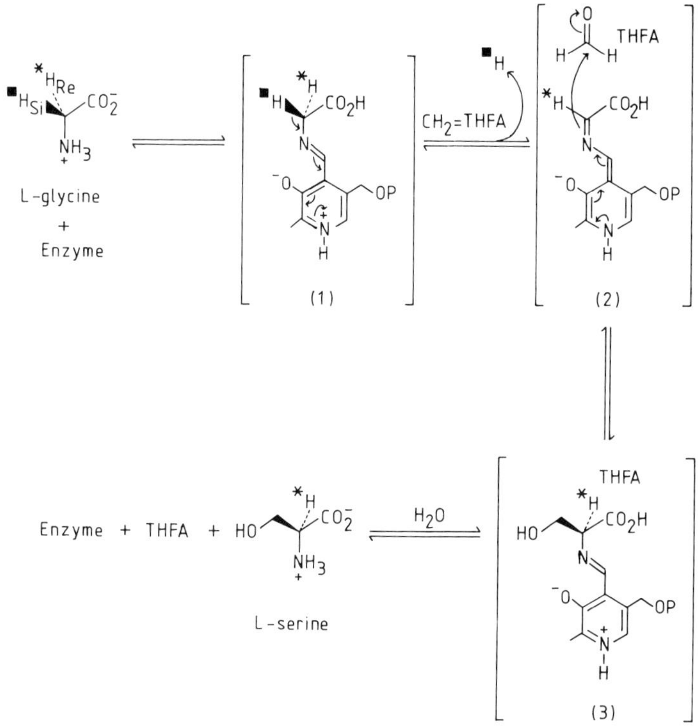

Fig. 14. Mechanism for the action of SHMT. An important feature of the mechanism is the transient generation of formaldehyde at the active site.

tively (see Fig. 16). Finally, these (R)- and (S)-[2-^{3}H]glycines were transformed using rabbit liver SHMT into L-serine and en route (R)-[2-^{3}H]glycine retained most of the tritium label while the (S) isomer lost most of the radioactivity.

From these results we see that the 2-H_{Si} atom of glycine is replaced during the SHMT reaction and since L-serine has the (S) absolute configuration, the overall steric course at C_α is retention (see Fig. 14). A similar retention at C_α was shown in the interconversion of glycine into allothreonine (threonine).

A second stereochemical feature of the SHMT reaction to have been studied concerns the stereochemistry of the events which occur at C-3 (C_β) of serine or the

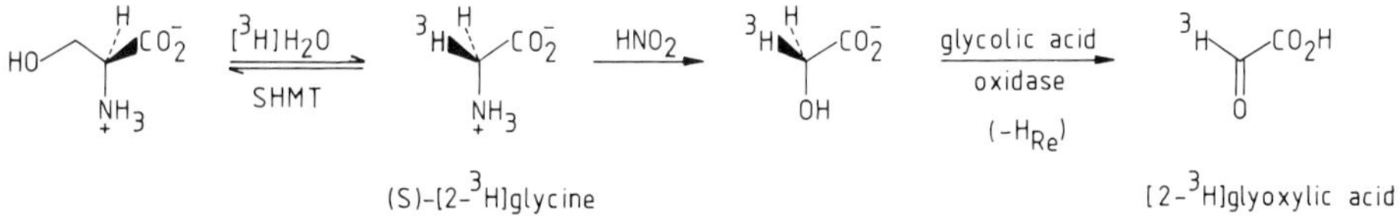

Fig. 15. The configurational assignment of (S)-[2-^{3}H]glycine prepared from serine using SHMT.

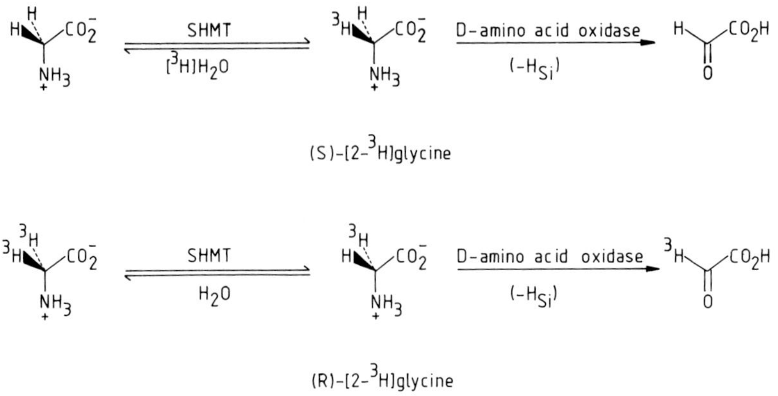

Fig. 16. Configurational analysis of enantiomeric glycines prepared via an exchange reaction catalysed by SHMT.

methylene carbon of N^5,N^{10}-methylenetetrahydrofolate.

Biellmann and Schuber [59] used [^{3}H]HCO_2H as a source of the C_1 unit for the conversion of glycine into serine in rat liver slices. The labelled formate is presumably converted into N^5,N^{10}-[^{3}H]methylenetetrahydrofolate by the sequence shown in Fig. 17 and its labelled methylene group then used to generate (*S*)-[3-^{3}H]serine. The

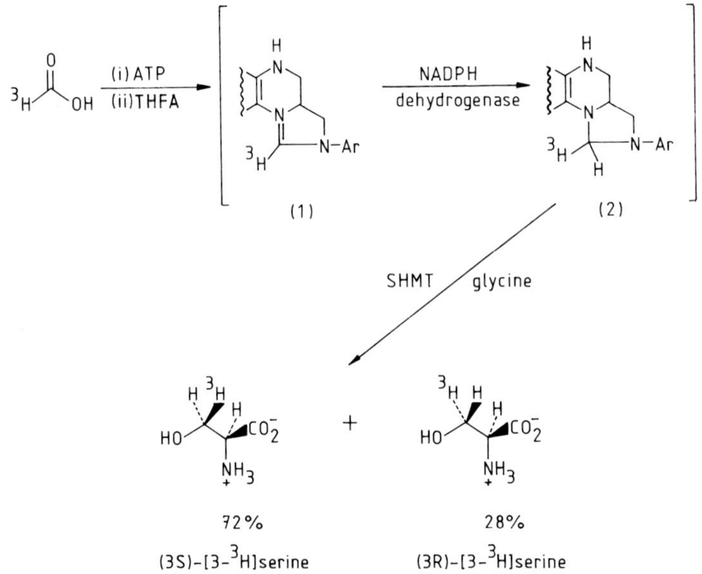

Fig. 17. The incorporation of [1-^{3}H]HCOOH into C-3 of serine using rat liver slices.

(S)-[3-^{3}H]serine formed was degraded into ethanol in such a way that C-3 of serine becomes C-1 of ethanol, and the alcohol dehydrogenase reaction was then used to establish the enantiomeric purity and the absolute configuration of the labelled centre. The sample was in fact 72% (S)-[1-^{3}H]ethanol and therefore the tritiated serine contained 72% of the label in the ($3S$) and the remainder in the ($3R$) isomer (see Fig. 17).

A similar result was obtained by Tatum et al. [60] using a purified SHMT from rabbit liver. They converted stereospecifically labelled [3-^{3}H]serines of known absolute configuration with the enzyme and determined the distribution of tritium label between the two diastereotopic positions at the N^5,N^{10}-CH_2 group in the N^5,N^{10}-methylenetetrahydrofolate produced during the reaction. This was achieved by stereospecific oxidation * using N^5,N^{10}-methylenetetrahydrofolate dehydrogenase which transfers one of the methylene bridge hydrogen atoms to $NADP^+$ affording NADPH. This latter was then converted using NADase into nicotinamide as shown in Fig. 18. The relative tritium contents of the N^5,N^{10}-methenyltetrahydrofolate and the nicotinamide will then be a direct reflection of the stereospecificity of the hydroxymethyl group transfer. Fig. 18 illustrates this approach using ($3R$)-[3-^{3}H]serine, of 98% configurational purity, and shows that the N^5,N^{10}-methylenetetrahydrofolate had tritium distributed between the two diastereotopic methylene bridge positions in the ratio of 68 to 32.

The results from two groups of investigators are therefore mutually reinforcing and indicate a partial loss of stereospecificity during the process involved in the transfer of the C-1 unit. One way in which this might arise is by the transient formation of formaldehyde at the active site (as is implied in the mechanism [50,52] of Fig. 14), which in at least 24 turnovers out of 100 is able to rotate before combination with tetrahydrofolate. An alternative explanation [60] is that when bound at the active site the hydroxymethyl group of serine is able to react in two different conformations (Fig. 19, 1 and 2) the proportions of which reflect the final tritium distribution in the N^5,N^{10}-methylenetetrahydrofolate. We regard the latter explanation less likely since this requires a crucial binding group, –OH, of the substrate, serine, to occupy two entirely different sites on the enzyme. The two main arguments [60] presented in support of the suggestion may be scrutinised as follows. First, that SHMT catalyses the cleavage of allothreonine as well as threonine was suggested to indicate some flexibility in the steric requirement of the binding sites in the enzyme. If flexibility is the key to the argument, it would be preferable to assume that the binding sites for the two α-hydrogen atoms of serine are capable of accommodating other alkyl substituents with varying degrees of efficiency. Secondly, attention was drawn to a non-stereospecific decarboxylation of aminomalonate to glycine catalysed by SHMT from rat liver mitochondria. It is our view that when an artificial substrate for an enzyme displays strict stereochemical discipline, it is

---

* Since a method is not yet available for the identification of the two diastereotopic N^5,N^{10}-methylene hydrogen atoms, neither the orientation of ^{3}H in intermediate 1 (Fig. 18), nor the steric course of the dehydrogenase reaction is known.

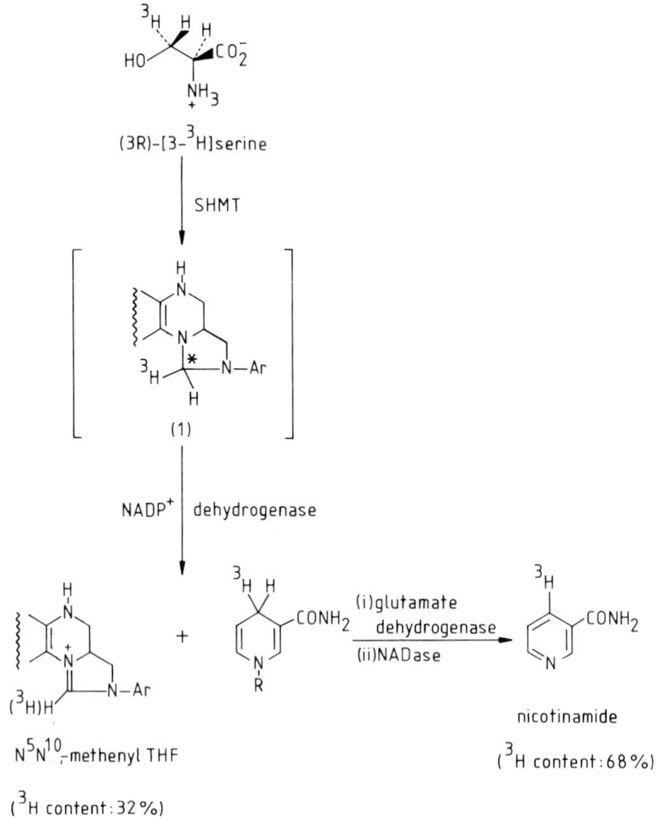

Fig. 18. The steric course of the C_1 unit transfer process as studied using serine stereospecifically labelled at C-3. For convenience the pathway is shown using the (3*R*) enantiomer only. Attention is drawn to the fact that a method is not at present available for the assignment of absolute stereochemistry to ^3H at the C* of structure 1.

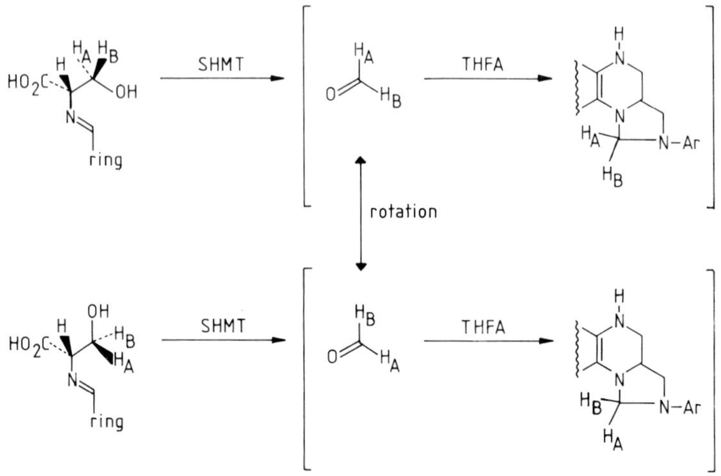

Fig. 19. The partial loss of stereospecificity in the transfer of the C_1 unit. The proposal makes the unlikely assumption that the hydroxyl group of serine occupies two different binding domains at the active site.

justified to use the results for mechanistic interpretation; when it does not then the result is merely an observation.

*3.4. 5-Aminolevulinate synthetase * (ALA synthetase)*

Pyridoxal-P-dependent 5-aminolevulinate synthetase catalyses the formation of 5-aminolevulinic acid from glycine and succinyl-CoA, and this constitutes the first irreversible step on the main pathway leading to the biosynthesis of tetrapyrroles. ALA synthetase [61] has been purified from several bacterial, avian and mammalian sources, but the amount of enzyme obtained in these preparations is relatively small. Light-grown *Rhodopseudomonas spheroides* is at present the best source of the

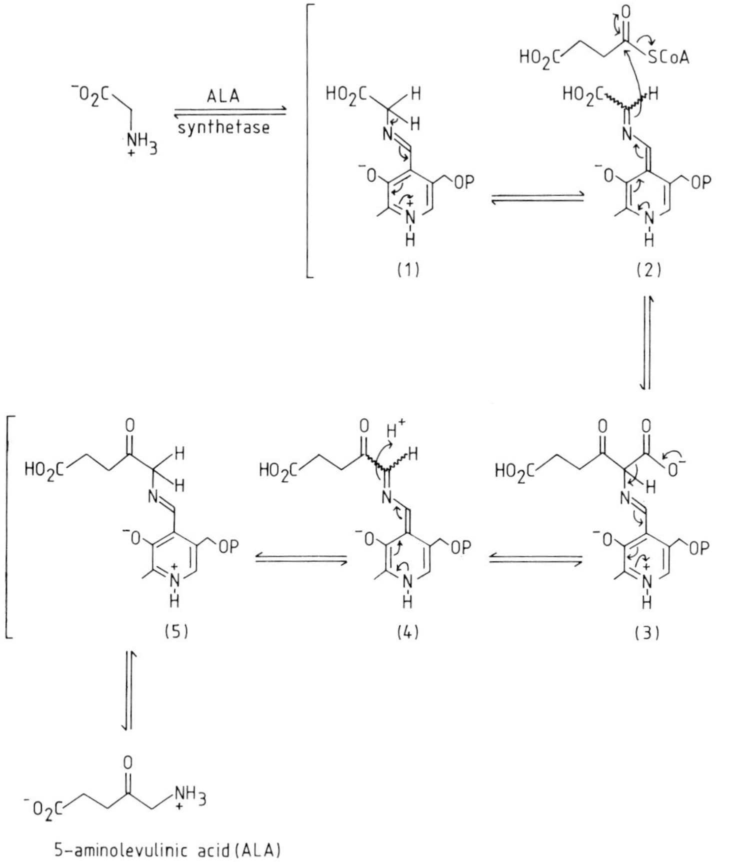

ALA synthetase

(1) (2) (3) (4) (5)

5-aminolevulinic acid (ALA)

Fig. 20. A mechanism for the 5-aminolevulinate synthetase catalysed reaction.

---

* Throughout this chapter the more conventional "synthetase" is used for enzymes which according to recent nomenclature will be regarded as "synthases".

enzyme, and therefore most of the mechanistic studies have been performed using purified ALA synthetase from this organism. A mechanism for the reaction has been proposed [62,63] (Fig. 20) that involves, first the formation of the glycyl-anion equivalent (Fig. 20, 2) and then its acylation by succinyl-CoA to afford transiently the 1-amino-2-oxoadipate–pyridoxal-P–enzyme complex (Fig. 20, 3). Two alternatives are now available for the manipulation of the complex (Fig. 20, 3). First a decarboxylation process (Fig. 20, 3 → 4) promoted by the dual electron withdrawing effects of the carbonyl group and the pyridinium ring in structure 3 can give the complex (Fig. 20, 4) from which ALA is formed by protonation and Schiff base hydrolysis. Secondly, hydrolysis of the intermediate (Fig. 20, 3) releases 1-amino-2-oxoadipate that then decarboxylates non-enzymically to give ALA. The latter possibility has, however, been eliminated [64].

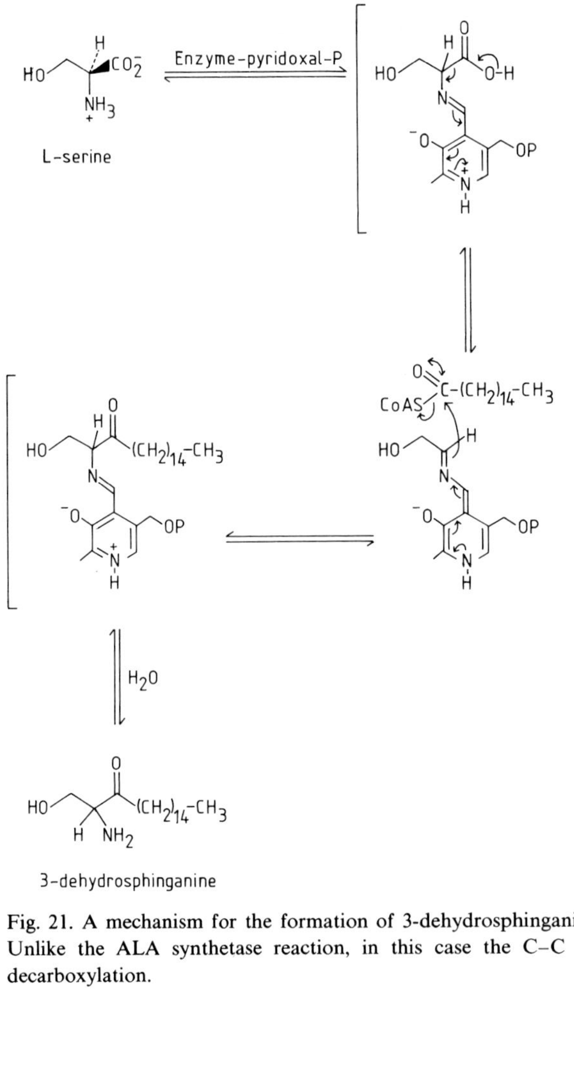

Fig. 21. A mechanism for the formation of 3-dehydrosphinganine by a pyridoxal-P-dependent enzyme. Unlike the ALA synthetase reaction, in this case the C–C bond formation only occurs following decarboxylation.

The stereochemistry of the hydrogen lost from C_α of glycine during the ALA synthetase catalysed reaction was identified [62,64] by using (R)- and (S)-[2-^{3}H]glycine. The ALA produced from these species was reduced immediately with sodium borohydride and then cleaved with periodate to release the important C-5 methylene group of ALA as formaldehyde, which could be trapped for radiochemical analysis. Since the formaldehyde produced in this way from (R)-[2-^{3}H]glycine had lost most of the tritium label, whereas that from (S)-[2-^{3}H]glycine had retained most of the label, it follows that the enzyme stereospecifically removes the 2-H_{Re} atom.

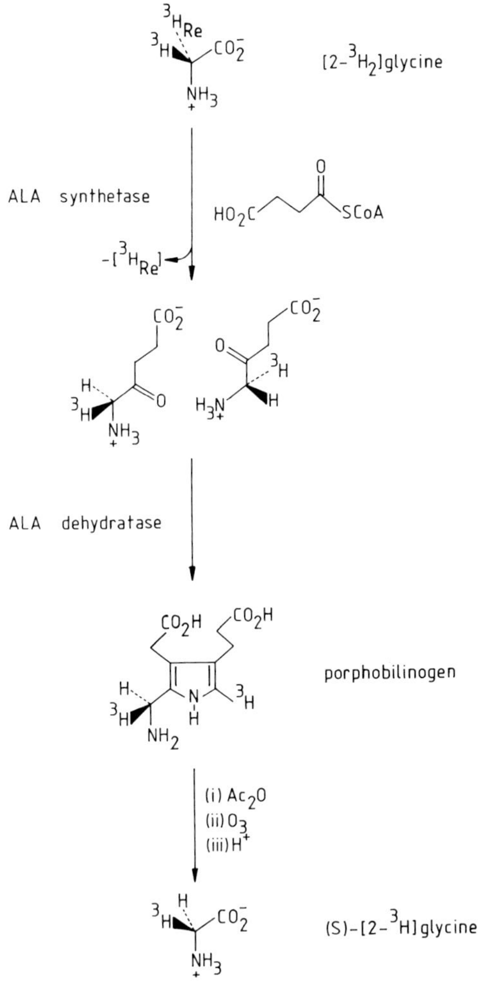

[2-$^3H_2$]glycine

ALA synthetase

HO_2C–CH_2CH_2–CO–SCoA

–[$^3H_{Re}$]

ALA dehydratase

porphobilinogen

(i) Ac_2O
(ii) O_3
(iii) H^+

(S)-[2-^{3}H]glycine

Fig. 22. The elucidation of the stereochemistry at C-5 of ALA. The analysis involves the isolation of this position as glycine via the sequence shown. A point to note is that C-5 of ALA cannot be directly analysed due to a rapid loss of stereochemistry through a facile enolisation process.

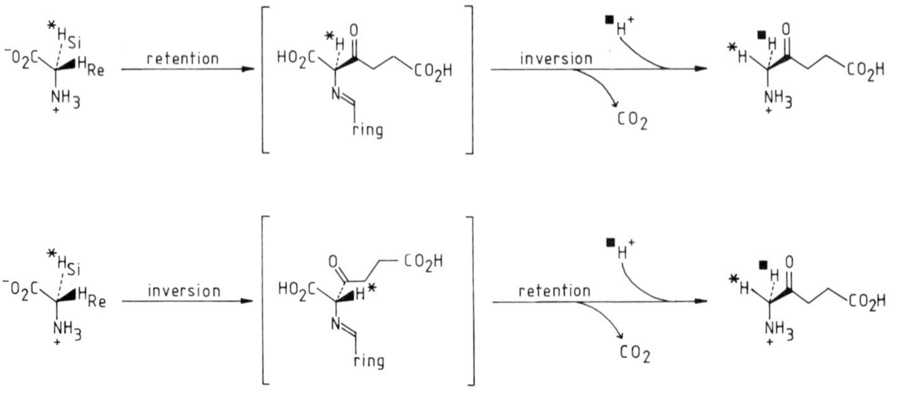

Fig. 23. Two possible steric courses for the ALA synthetase reaction. In the upper sequence the replacement of a C–H bond by a C–C bond occurs with retention and the decarboxylation process with inversion. The converse is the case in the lower sequence.

This is an important result because the conceptually simpler reaction pathway involving an initial C_α–CO_2H bond cleavage followed by condensation with succinyl-CoA to afford enzyme bound ALA in a single step, would require the intact transfer of both enantiotopic protons at C-2 of glycine into C-5 of ALA. Indeed in an analogous pyridoxal-P-dependent reaction in which 3-dehydro-sphinganine is produced from palmitoyl-CoA and serine, the reaction occurs without labilisation of the C-2 hydrogen of serine, thus implicating the participation of a mechanism involving the cleavage of the C_α–CO_2H bond [65] (Fig. 21).

The absolute configuration of the [5-^{3}H]ALA produced from (*S*)-[2-^{3}H]glycine was deduced [63] by converting the ALA directly into porphobilinogen (PBG; Fig. 22) via ALA-dehydratase. This enzymic reaction is in fact the second biosynthetic step in porphyrin, chlorin and corrin biosynthesis. The key aminomethylene group of PBG, which is derived unchanged from ALA, could then be extruded by an ozonolytic pathway to afford [2-^{3}H]glycine which after incubation with SHMT lost most of the radioactivity to the medium *. Its absolute configuration is therefore predominantly (*S*).

With these key pieces of information in hand the overall steric course of the ALA synthetase reaction can be deduced. Since the transformation affects two different bonds at C_α it is necessary that the intermediate (Fig. 20, 3) should undergo a substantial conformational change at the active site if the cleavage of both are to be stereoelectronically assisted by the π-system of the Schiff base. Of course it is not known whether the cleavages of the C_α–CO_2H bond and the C_α–H bond occur on the same face or different faces of the π system. Nevertheless the overall steric course

* The 5-H_{Re} atom is lost stereospecifically from that ALA unit that affords C-2 in PBG [66] (see Fig. 22).

must be inversion because (S)-[2-^{3}H]glycine is converted into (S)-[5-^{3}H]ALA and this means that either the replacement of the 2-H_{Re} atom by the succinyl moiety occurs with retention and decarboxylation with inversion or vice versa (Fig. 23).

4. *Pyridoxal phosphate-dependent reactions involving modifications at C_β*

4.1. *General introduction*

A large group of pyridoxal-P-dependent enzymes catalyse reactions that affect C_β as well as C_α of the amino acid and these can be divided into two main groups (Fig. 24): (a) those that involve the substitution of a group at C_β (β replacement) and (b) those that involve the overall elimination of the α-hydrogen atom and the C_β substituent with the formation of α-oxoacid and ammonia (β elimination–deamination).

The mechanisms of these processes are closely related and appear to have in common the formation on the enzyme of an aminoacrylate–pyridoxal-P complex (Fig. 25, 1). Addition of a new nucleophile in a Michael fashion followed by hydrolysis of the Schiff base would represent an overall β replacement reaction, whereas hydrolysis of the aminoacrylate intermediate affords α-oxoacid and ammonia in a β elimination reaction (see Fig. 25). In these reactions the crucial elimination process is straightforward when X is a good leaving group, but when X is an aromatic system (for example indolyl) a prior protonation of the aromatic nucleus is a prerequisite for elimination.

4.2. *β Replacement reactions*

(a) Tryptophan synthetase

Tryptophan synthetase is the best known and most extensively studied example of the enzymes catalysing β replacement reactions. Our current understanding of the genetics and biochemistry of tryptophan synthetase is due to the painstaking and elegant studies of Yanofsky and coworkers [67] carried out during the last 2 decades at the University of Stanford. Tryptophan synthetase catalyses the last reaction in the biosynthesis of tryptophan:

Indole-3-glycerolphosphate + L-serine → L-tryptophan

+ D-glyceraldehyde-3-phosphate (4)

The enzyme is also capable of catalysing the following two half-reactions:

Indole-3-glycerolphosphate → indole + 3-glyceraldehyde-3-phosphate (5)

Indole + L-serine → L-tryptophan (6)

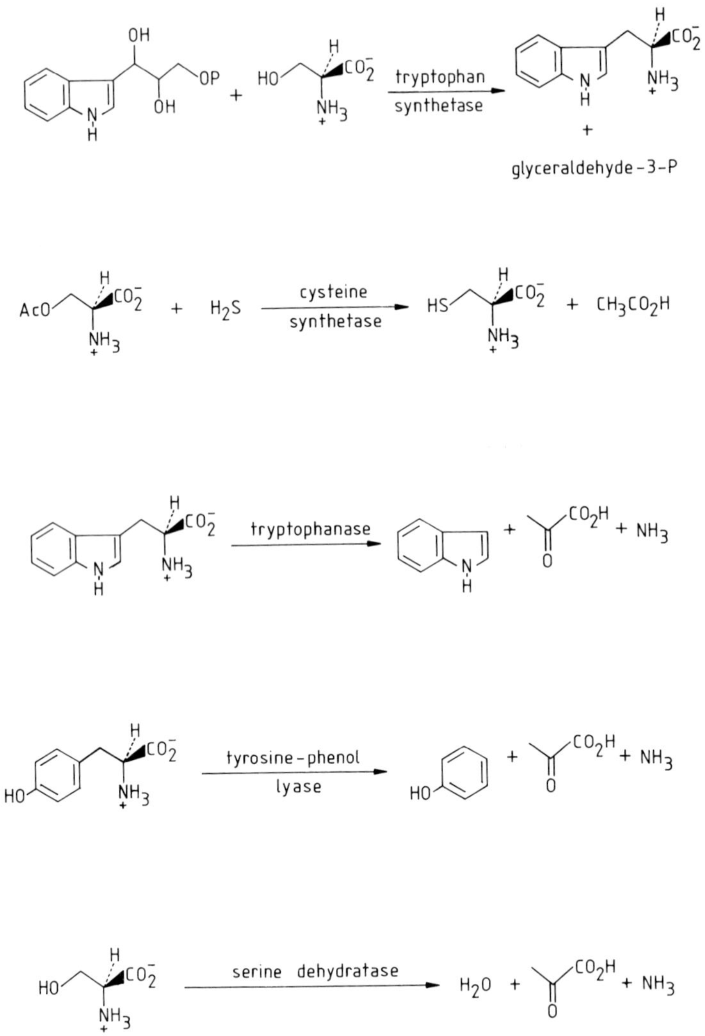

tryptophan synthetase

glyceraldehyde-3-P

cysteine synthetase

AcO + H_2S → HS + CH_3CO_2H

tryptophanase

tyrosine-phenol lyase

serine dehydratase

H_2O + NH_3

Fig. 24. Examples of β replacement and β elimination–deamination reactions.

Tryptophan synthetase from *E. coli* contains two different types of polypeptides which are referred to as α and β subunits. The physiologically functional tryptophan synthetase complex consists of two α and two β subunits and may be represented by the stylised illustration shown in Fig. 26. Upon dilution the complex dissociates to furnish the α subunits as monomers of mol. wt. 29 000 and a dimer consisting of two β subunits, mol. wt. 100 000. The β dimer contains one pyridoxal phosphate per polypeptide chain. These two components have been separated and used in the study of partial reactions.

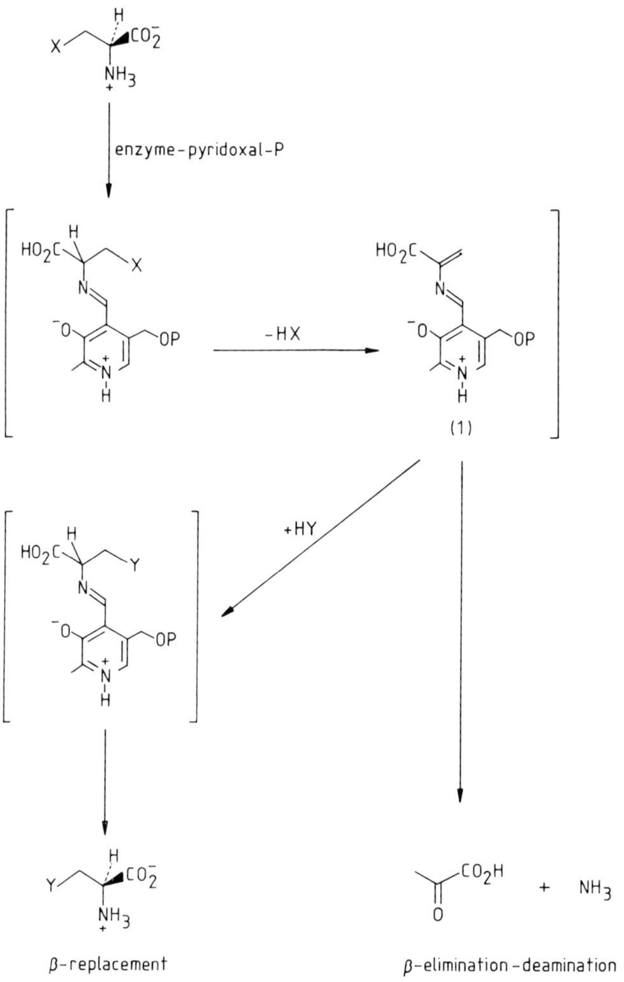

enzyme-pyridoxal-P

−HX

(1)

+HY

β-replacement

β-elimination-deamination

Fig. 25. A general mechanism of β replacement and β elimination–deamination reactions involving a common aminoacrylate intermediate (1).

In the presence of both the substrates, indole-3-glycerolphosphate and L-serine, tryptophan synthetase, $\alpha_2\beta_2$, catalyses two parallel reactions. The α subunits in the complex carry out the cleavage of indole-3-glycerolphosphate releasing glyceraldehyde-3-phosphate in the medium but the indole remains bound to the complex. Concomitantly, the β_2 subunits promote the conversion of serine into the aminoacrylate-pyridoxal-P species (Fig. 27, 1) which acts as a Michael acceptor for the indole producing the L-tryptophan–pyridoxal-P complex from which the amino acid is released by hydrolysis (Fig. 27). That the aforementioned molecular events can be described in such vivid detail is due to an ingenious series of partial reactions

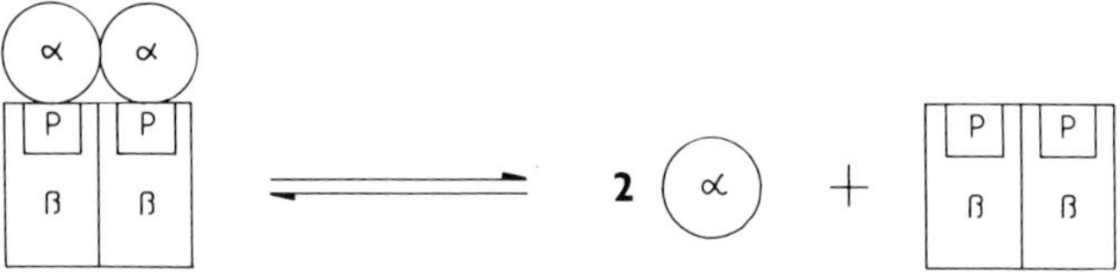

tryptophan synthetase
$\alpha_2\beta_2$ protein

P = pyridoxal phosphate

Fig. 26. A stylised model of tryptophan synthetase, $\alpha_2\beta_2$, and the two components, α protein and β_2 protein, derived therefrom. The model shows an intimate interaction between the α and β subunits at the pyridoxal-P binding sites.

discovered by Yanofsky and Crawford [67] using either the intact complex, $\alpha_2\beta_2$, or the components isolated therefrom. Thus when the intact complex, $\alpha_2\beta_2$, is incubated with indole-3-glycerolphosphate, but in the *absence* of L-serine, the cleavage reaction of Eqn. 5 occurs. The release of indole in this reaction provides evidence for the existence of this species in an enzyme bound form during the operation of the physiological reaction. This conclusion is further supported by the demonstration that the complex, $\alpha_2\beta_2$, does in fact produce L-tryptophan from exogenously added indole in the presence of L-serine (Eqn. 6).

We now turn to studies with the isolated α protein and β_2 protein. The α protein promotes the cleavage reaction of Eqn. 5 at about 1% of the rate exhibited by the complex, $\alpha_2\beta_2$, for this conversion. On the other hand the pyridoxal-P containing β_2 protein catalyses the condensation reaction between indole and L-serine (Eqn. 6). The efficiency of this process is about 5% of that observed with the intact complex. The β_2 protein in the *absence* of indole also catalyses the deamination reaction:

$$\text{L-serine} \rightarrow \text{pyruvate} + NH_3 \qquad (7)$$

but this activity is completely abolished by the addition of the α protein. In our view this observation reveals that the aminoacrylate-pyridoxal-P moiety (Fig. 27, 1) produced at the active site of the β_2 protein is exposed to the medium for decomposition via the reaction of Eqn. 7, and that this region of the protein is shielded by the α subunit in the intact complex, $\alpha_2\beta_2$ (Fig. 26). In the physiological reaction the delivery as well as the utilisation of indole must occur at α subunit–β subunit interface.

Tryptophan synthetase has also been purified from several other sources, and *Neurospora crassa* appears to be a particularly convenient source for the large scale preparation of the enzyme [67]. The enzyme from the latter source has been used for detailed stereochemical studies particularly with respect to the events which occur at

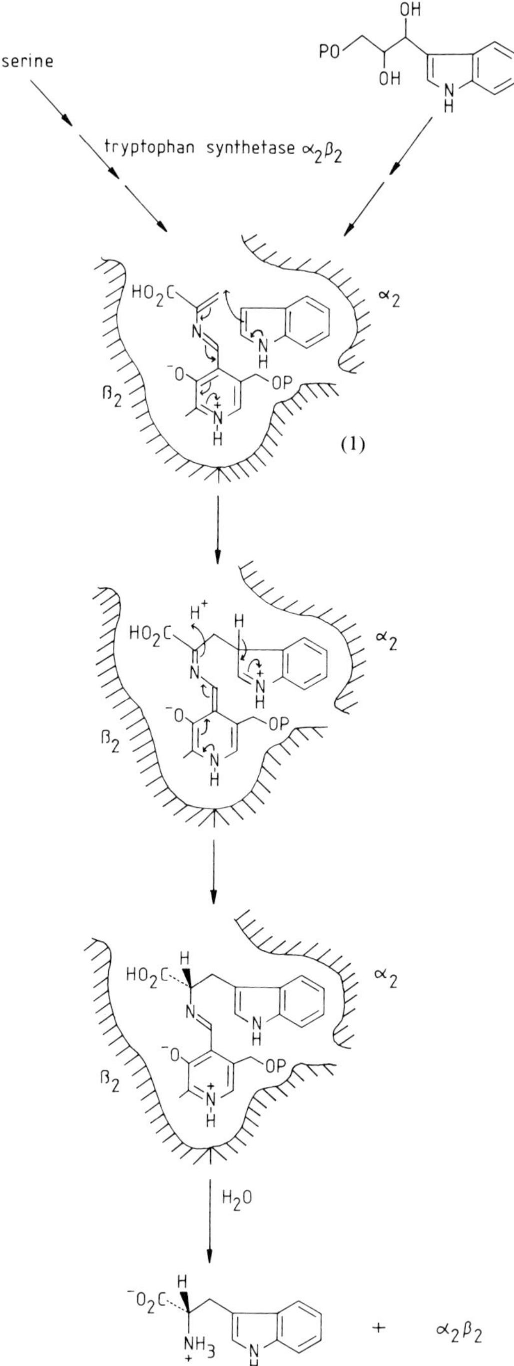

Fig. 27. A mechanism for the tryptophan synthetase, $\alpha_2\beta_2$, catalysed reaction. The scheme shows that the indole formed in the reaction remains bound to the enzyme for condensation with the aminoacrylate intermediate.

C_β. In order to study this aspect stereospecific isotopic labelling of L-serine at C_β together with an appropriate stereochemical analysis of the reaction product is required. Serine stereospecifically tritiated at C_β has been prepared [68] from either [1-^{3}H]glucose or [1-^{3}H]mannose using a sequence of enzymic reactions of known steric courses (see Fig. 28). These samples of labelled serine were converted with tryptophan synthetase from *Neurospora crassa* to give two species of tritiated tryptophan [69]. The configuration of each tritiated tryptophan was determined by oxidative degradation to furnish first aspartic acid and then malic acid, followed by equilibration of the latter with fumarase. The fumarase reaction proceeds with the loss of the 3-H_{Re} atom from malate. It was found that during the equilibration the [3-^{3}H]malate obtained from the degradation of tryptophan that had been biosynthesised from L-(3*R*)-[3-^{3}H]serine, lost most of the tritium, whereas the sample originating from L-(3*S*)-[3-^{3}H]serine retained most of the tritium. The replacement of the β-OH by β-indole must therefore proceed with retention (Fig. 29). In other words, the indole unit adds to the same stereoheterotopic face of the π system as that from which the –OH group departed. Certainly this result is inconsistent with direct S_N2 displacement at the β-carbon atom.

The same result has also been obtained by Fuganti and coworkers [70] using stereospecifically labelled serines of synthetic origin. In fact, by related methods all of the β replacement reactions studied to date have been shown to proceed with retention at C_β (Table 3).

4.3. β Elimination–deamination reactions

In common with β replacement reactions, β elimination–deaminations also proceed via the formation of an aminoacrylate species (Fig. 25, 1) protonation of which at C_β followed by hydrolysis leads ultimately to the release of pyruvate and ammonia. However, the exact timing and mechanism of these final steps are at present unclear.

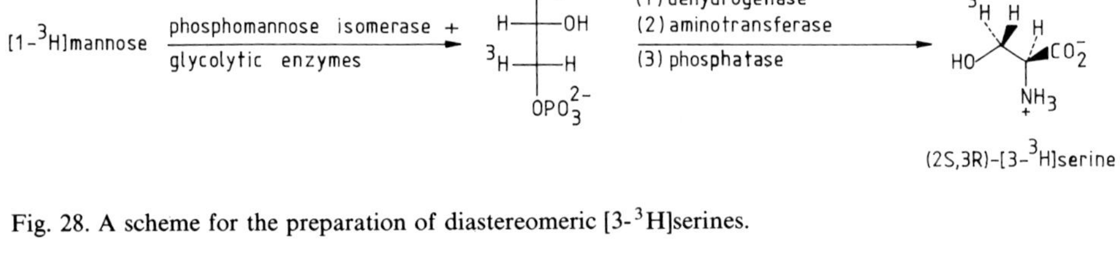

[1-^{3}H]glucose
phosphoglucose isomerase + glycolytic enzymes
(1) dehydrogenase
(2) aminotransferase
(3) phosphatase
(2S,3S)-[3-^{3}H]serine

[1-^{3}H]mannose
phosphomannose isomerase + glycolytic enzymes
(1) dehydrogenase
(2) aminotransferase
(3) phosphatase
(2S,3R)-[3-^{3}H]serine

Fig. 28. A scheme for the preparation of diastereomeric [3-^{3}H]serines.

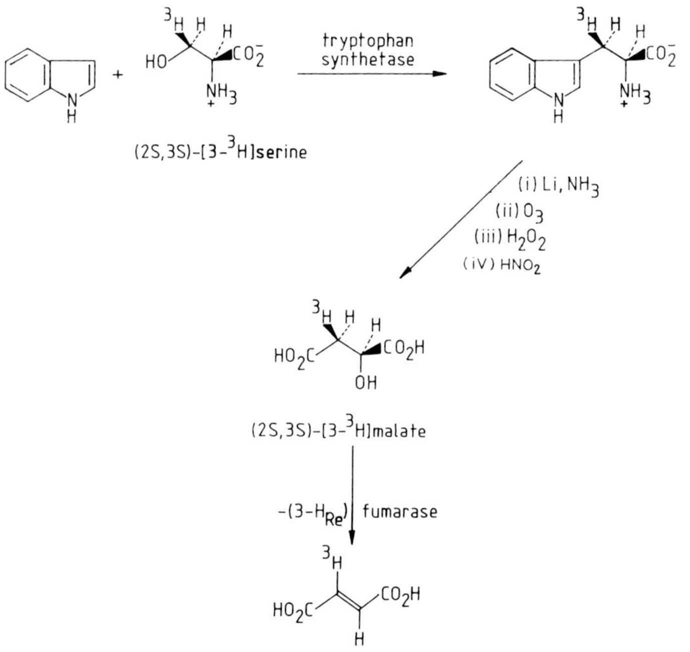

Fig. 29. The retention of stereochemistry in the tryptophan synthetase catalysed reaction. The stereochemical assignment was made at the level of malate.

(a) Tryptophan synthetase–β_2 protein

Attention has already been drawn to the fact that the β_2 protein catalyses the deamination of L-serine. Although this reaction has no physiological significance it has been utilised for mechanistic studies pertinent to β elimination–deamination reactions. The conversion of L serine into pyruvate in ^2H_2O has been investigated by Floss using an NMR method [69]. It was found that during the conversion there was no incorporation of deuterium at C-β thus implying an intramolecular transfer of hydrogen from C_α to C_β during the formation of pyruvate. The stereochemistry of this process was elucidated [69] using L-serine stereospecifically deuterated and tritiated at C-3. The elegant method developed by the schools of Cornforth [71] and Arigoni [72] was used for the chirality analysis of the methyl group and it was shown that the conversion 1 → 5 (Fig. 30) had occurred with retention of configuration at C_β. This feature, taken together with the demonstration of an intramolecular transfer of hydrogen from C_α to C_β, suggests that a single catalytic group is involved in both the removal of C_α–H in the conversion 1 → 2 and reprotonation of the aminoacrylate species 3 (Fig. 30). Furthermore, all three processes, (i) cleavage of C_α–H, (ii) elimination of –OH and (iii) reprotonation at C_β must be mediated by catalytic group(s) located on the same side of the active site.

In fact, all the β elimination–deamination reactions studied to date have been shown to occur with retention of configuration at C_β (Table 4).

TABLE 3

Stereochemical course of β replacement reactions:

X–CH$_2$–C(H)(CO$_2^-$)(N$^+$H$_3$) ⟶ Y–CH$_2$–C(H)(CO$_2^-$)(N$^+$H$_3$)

X	Y	Enzyme	Steric course at C_β
CH_3–C(=O)–O–	HS–	*O*-Acetylserine sulphydrase (cystine synthetase)	Retention [68]
HS–	NC–	β-Cyanoalanine synthetase	Retention [196]
HO–	indol-3-yl	Tryptophan synthetase Tryptophan synthetase β_2 protein Tryptophanase	Retention [69] Retention [69] Retention [104]
HO–	4-hydroxyphenyl	Tyrosine phenol lyase	Retention [197]
4-hydroxyphenyl	3,4-dihydroxyphenyl	Tyrosine phenol lyase	Retention [198]
HO–	^-O_2C–CH(N$^+$H$_3$)–CH$_2$–CH$_2$–S–	Cystathionine-β-synthetase	Retention [199]

(b) Tryptophanase

The bacterial enzyme tryptophanase consists of 4 identical subunits and its major physiological role is the catabolism of tryptophan to indole, pyruvate and ammonia (Fig. 24). In vitro, the enzyme is also capable of catalysing a variety of β replacement reactions as well as the deamination of serine. Unlike the 1,2-intramolecular hydrogen transfer described above for the β_2 protein catalysed reaction, Floss and coworkers [74,104] have shown that during the cleavage of appropriately labelled tryptophan a 1,3-intramolecular hydrogen transfer from C_α to C-3 of the indolyl moiety occurs. Since protonation of the indolyl moiety is a prerequisite for facile β elimination, the intramolecular transfer must occur prior to β elimination, and this accounts for the fact that no intramolecular transfer from C_α to C_β is noted during the overall transformation. These observations are rationalised in the mechanism shown in Fig. 31, in which the C_α–H removed in the conversion 1 to 2 is the one used to protonate C-3 of the indolyl moiety to yield structure 3. If the same mechanistic pattern is followed for the non-physiological cleavage of serine by tryptophanase then the proton transferred from the C_α of the amino acid will be used to protonate the C-3 hydroxyl group prior to elimination (Fig. 32).

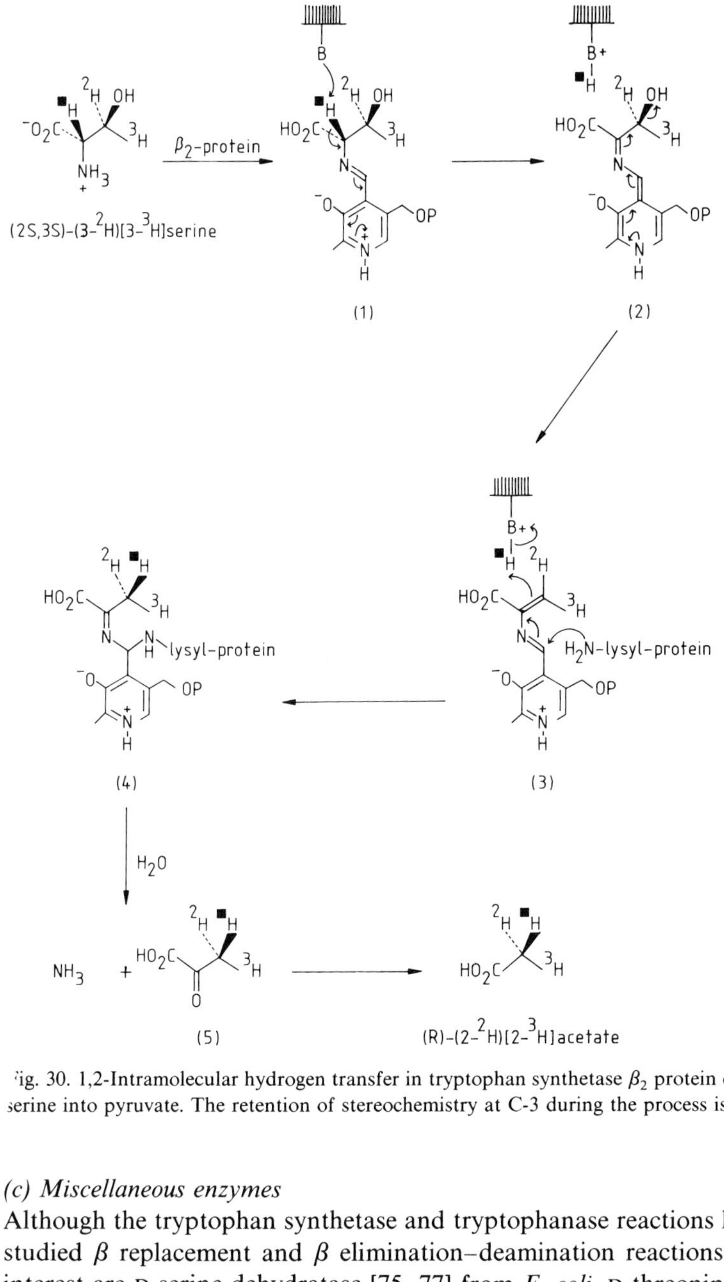

Fig. 30. 1,2-Intramolecular hydrogen transfer in tryptophan synthetase β_2 protein catalysed conversion of serine into pyruvate. The retention of stereochemistry at C-3 during the process is also shown.

(c) Miscellaneous enzymes

Although the tryptophan synthetase and tryptophanase reactions have been the best studied β replacement and β elimination–deamination reactions, others of special interest are D-serine dehydratase [75–77] from *E. coli*, D-threonine dehydratase and L-threonine dehydratase from *Serratia marcescens* [78]. The only information available on the above enzymes is that in these cases also, the events at C_β occur with retention of configuration.

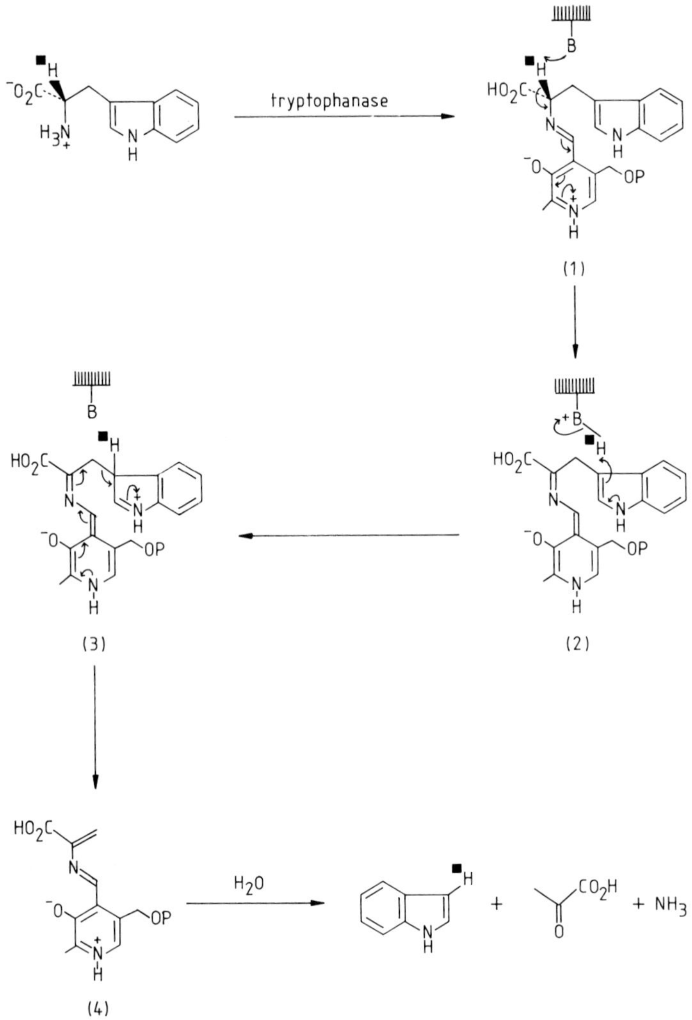

Fig. 31. 1,3-Intramolecular hydrogen transfer in the tryptophanase catalysed reaction.

An unusual variant of a β elimination reaction is shown by *kyneurinase* which catalyses the production of L-alanine and 3-hydroxyanthranilate from 3-hydroxy kyneurine (Fig. 33), an intermediate in tryptophan biosynthesis. Although the reaction has not been subjected to mechanistic investigation a pathway consistent with the previously demonstrated properties of pyridoxal-P to stabilise a β-carbanion is illustrated in Fig. 34.

TABLE 4

Stereochemical course of β-elimination–deamination reactions:

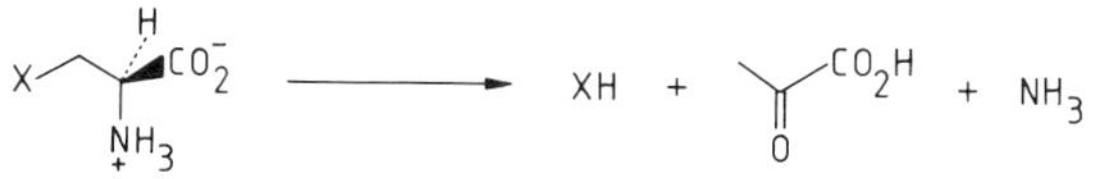

$X-CH_2-CH(\overset{+}{N}H_3)-CO_2^- \longrightarrow XH + CH_3COCO_2H + NH_3$

X	Medium	Substrate configuration	Enzyme	Steric course at C_β
HO–	^2H_2O	(2*S*,3*S*)-[3-^{3}H]Serine (2*S*,3*R*)-[3-^{3}H]Serine	Tryptophan synthetase β_2 protein	ND [73]
HO–	H_2O	(2*S*,3*S*)-[3-$^2H_1$]-[3-^{3}H]Serine (2*S*,3*R*)-[3-$^2H_1$]-[3-^{3}H]Serine (2*S*,3*S*)-[2-$^2H_1$]-[3-^{3}H]Serine	Tryptophan synthetase β_2 protein	Retention [69]
HO–	^2H_2O	(2*S*,3*S*)-[3-^{3}H]Serine (2*S*,3*R*)-[3-^{3}H]Serine	Tryptophanase	Retention [74,104]
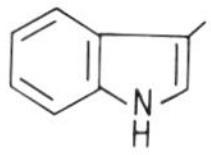 indol-3-yl (N–H)	^2H_2O	(2*S*,3*S*)-[3-^{3}H]Tryptophan (2*S*,3*R*)-[3-^{3}H]Tryptophan	Tryptophanase	Retention [104]
HO–	^2H_2O	(2*R*,3*R*)-[3-^{3}H]Serine (2*R*,3*S*)-[3-^{3}H]Serine	D-Serine dehydratase	Retention [75]
HO–	^2H_2O	(2*S*,3*R*)-[3-^{3}H]Serine (2*S*,3*S*)-[3-^{3}H]Serine	Tyrosine phenol lyase	Retention [200]
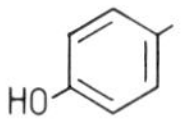 HO–C_6H_4–	[^{3}H]H_2O	(2*S*,3*R*)-[3-$^2H_1$]Tyrosine (2*S*,3*S*)-[3-$^2H_1$]Tyrosine	Tyrosine phenol lyase	Retention [200]
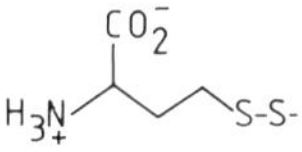 $H_3\overset{+}{N}CH(CO_2^-)CH_2$–S–S–	^2H_2O	(2*S*,3*S*)-[3-^{3}H]Cystine (2*S*,3*R*)-[3-^{3}H]Cystine	*S*-Alkylcysteine lyase	Retention [196,201]

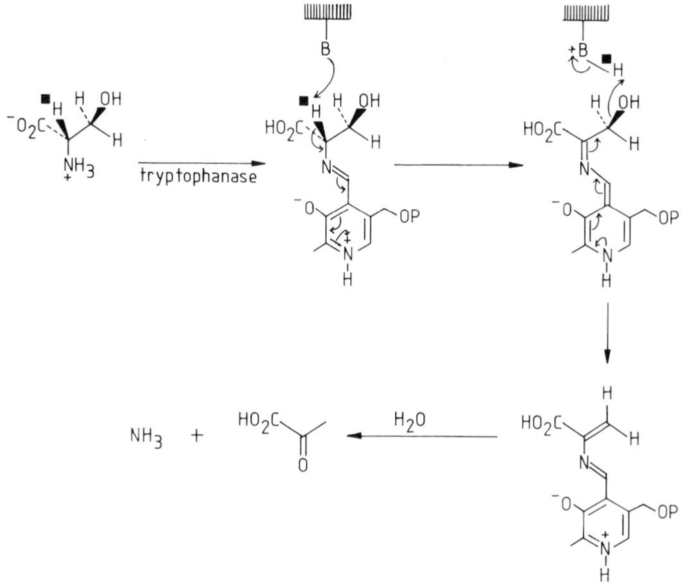

Fig. 32. Scheme rationalising the lack of intramolecular transfer in the non-physiological β-elimination–deamination of serine catalysed by tryptophanase.

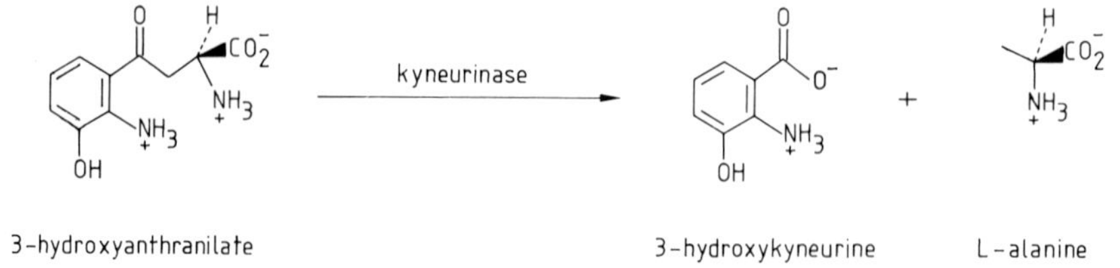

Fig. 33. Reaction catalysed by kyneurinase.

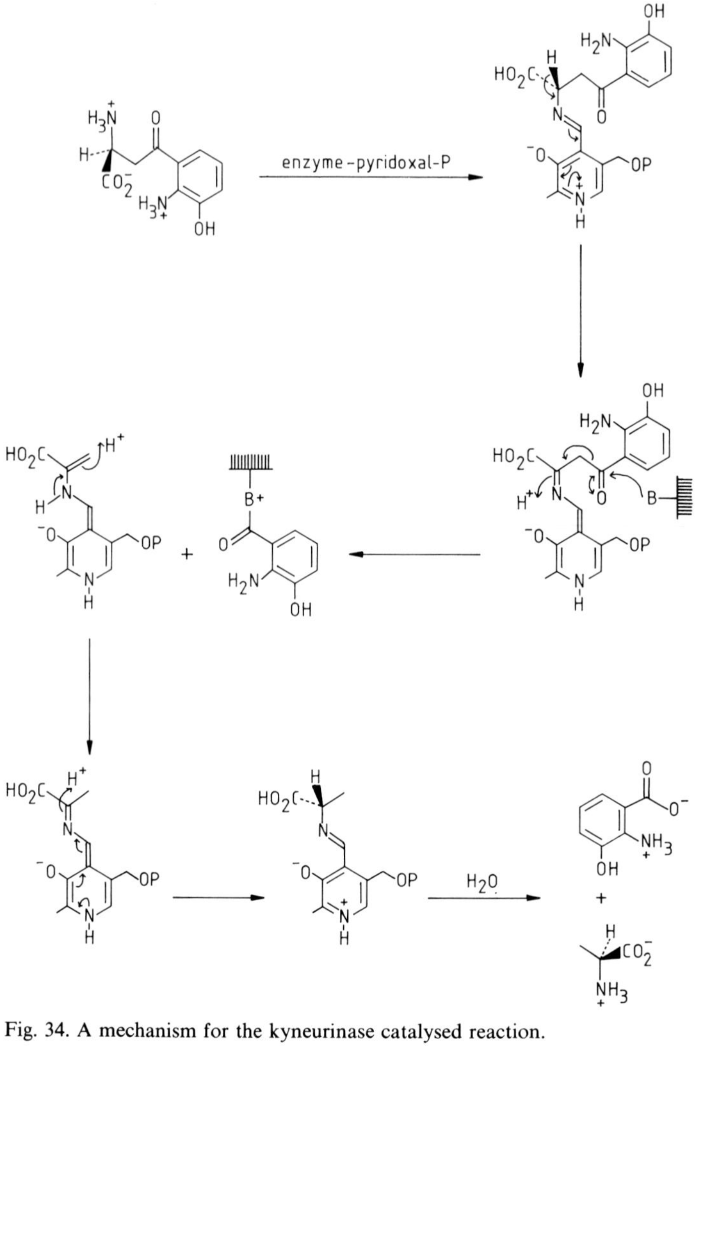

Fig. 34. A mechanism for the kyneurinase catalysed reaction.

5. Pyridoxal phosphate-dependent reactions occurring at C_γ

5.1. Enzymic aspects

The versatility of pyridoxal-P for providing stereoelectronically assisted bond cleavages is further illustrated by γ replacement and γ elimination–deamination reactions where mechanistic and stereochemical interest now focusses on the changes which occur at the three centres (C_α, C_β and C_γ) in the α-amino acid substrate. Representative examples of this class of enzyme catalysed reactions are shown in Fig. 35.

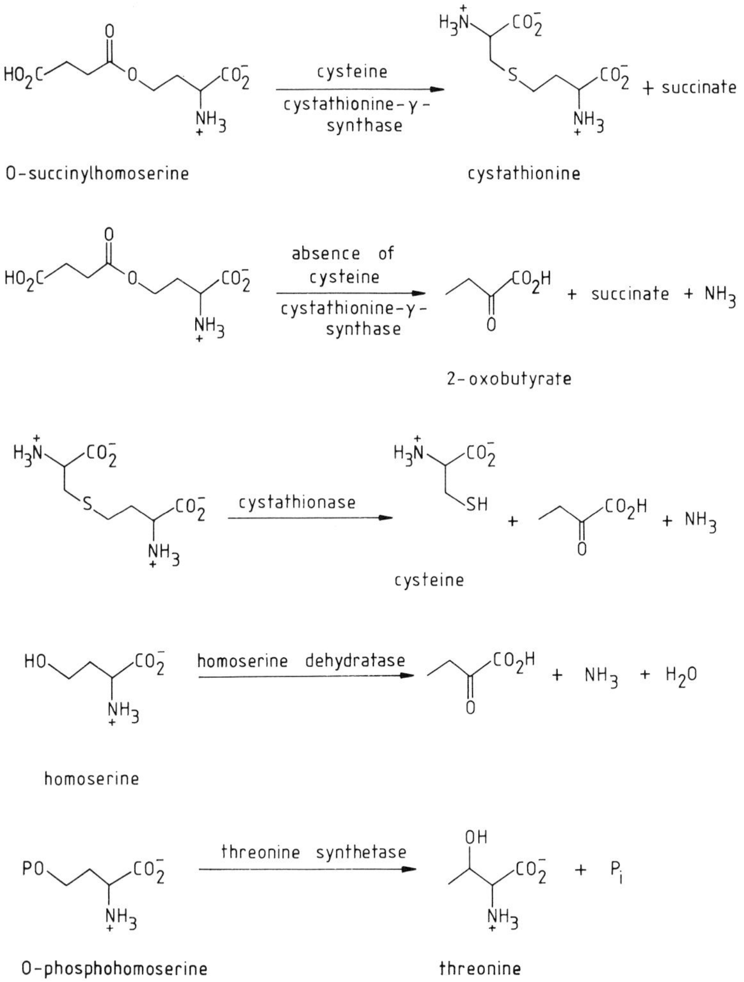

O-succinylhomoserine —cysteine / cystathionine-γ-synthase→ cystathionine + succinate

O-succinylhomoserine —absence of cysteine / cystathionine-γ-synthase→ 2-oxobutyrate + succinate + NH_3

cystathionine —cystathionase→ cysteine + 2-oxobutyrate + NH_3

homoserine —homoserine dehydratase→ 2-oxobutyrate + NH_3 + H_2O

O-phosphohomoserine —threonine synthetase→ threonine + P_i

Fig. 35. Examples of reactions occurring at C_γ.

Cystathionine-γ-synthase isolated from *Salmonella typhimurium* is a tetramer (molecular weight 160 000) and catalyses, in vivo, the γ-replacement of *O*-succinylhomoserine with cysteine [79] to yield cystathionine. The latter, by way of homocysteine, is involved in the biosynthesis of methionine. In other species of bacteria and plants the succinyl moiety may be replaced by acetyl, phosphoryl, or malonyl moieties [80]. In the absence of cysteine the enzyme catalyses an abnormal reaction resulting in the formation of α-oxobutyrate. The latter reaction has been utilised for mechanistic investigations pertinent to the γ-elimination–deamination process (vide infra).

γ-Cystathionase has a molecular weight of 190 000 and has been found in mammalian liver and the mold *Neurospora crassa* [81]. Its physiological role in *N. crassa* is the catabolism of cystathionine to cysteine, α-oxobutyrate and ammonia; however, in vitro, it will also effect the γ elimination of homoserine to yield α-oxobutyrate [82].

Threonine synthetase has been purified from *N. crassa* by Flavin [83] and catalyses the synthesis of L-threonine from *O*-phosphohomoserine and H_2O via a γ–β replacement process.

A general mechanism encompassing each of these processes can be formulated which involves the formation at the active site of the common intermediate (Fig. 36, 1) which arises via a multistep process discussed later (cf. Fig. 39).

5.2. Stereochemical aspects

A stereochemical investigation of these reactions is complicated, not only because changes occur at three centres and for each two stereochemical outcomes are possible * – retention or inversion – but also because of the possibility of proton shifts from one centre to another via a basic group in the enzyme active site. Indeed, it has been observed that in the abnormal γ-elimination–deamination reaction catalysed by cystathionine-γ-synthase, the conversion of *O*-succinylcystathionine into α-oxobutyrate, is attended by an intramolecular hydrogen transfer from C_α, as well as C_β, to C_γ [85]. Information on the stereochemistry of events occurring at C_β was provided by studies on homoserine dehydratase [84] and cystathionine-γ-synthase [85] catalysed reactions performed in ^2H_2O (Fig. 37, 1 → 2 and 3–4). In each case α-oxobutyrate was formed containing deuterium at C_β. Oxidative degradation of these β-deuterated α-oxobutyrates to the corresponding deuterated propionates followed by the examination of their ORD spectra showed that in each case they possessed the (*S*) absolute configuration. These results were confirmed and extended by Fuganti and Coggiola [86]. The salient features of the conclusions drawn from these experiments are illustrated in Fig. 37 and highlight the fact that in the overall reaction the removal of the C_β-H_{Re} atom and its subsequent replacement occurs with overall retention. The stereochemical status of C_β in the threonine synthase reaction

* With the exception of C_α.

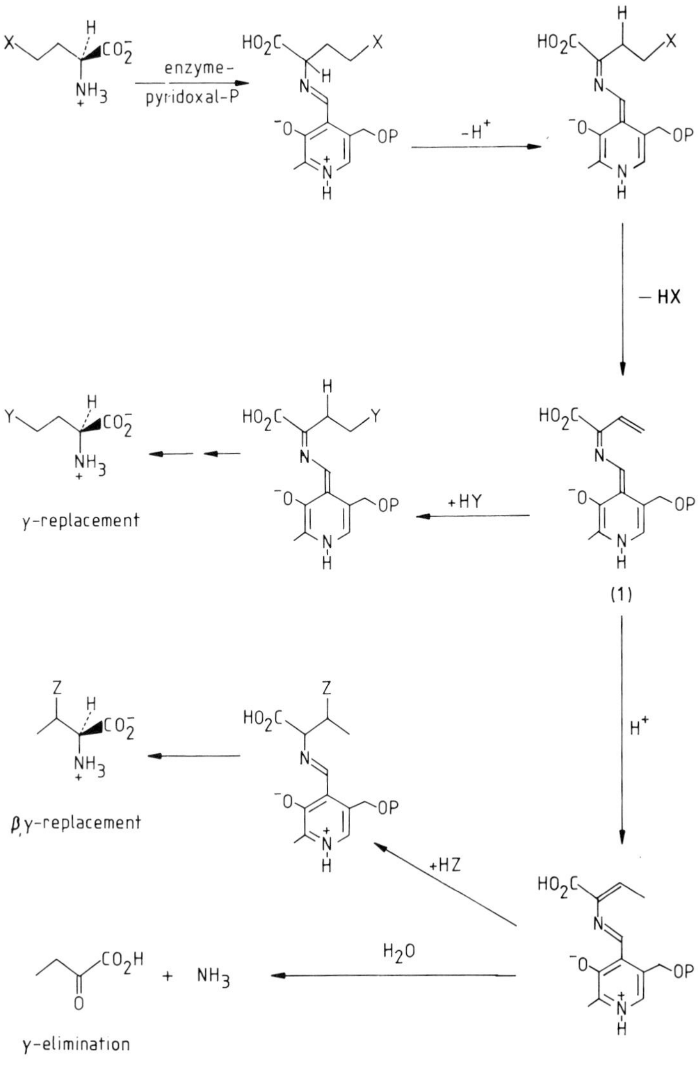

Fig. 36. A general mechanism for reactions occurring at C_γ via the intermediacy of a vinylglycine species 1.

has also been investigated [87]. This reaction differs from the examples mentioned above since it is the C_β-H_{Si} atom which is labilised in the overall reaction.

The steric course at C_γ during the normal γ-replacement and abnormal γ-elimination–deamination reactions catalysed by cystathionine-γ-synthase has been deduced in an elegant series of experiments described by Walsh and coworkers. An important aid in this stereochemical investigation was the observation that L-vinylglycine [88]

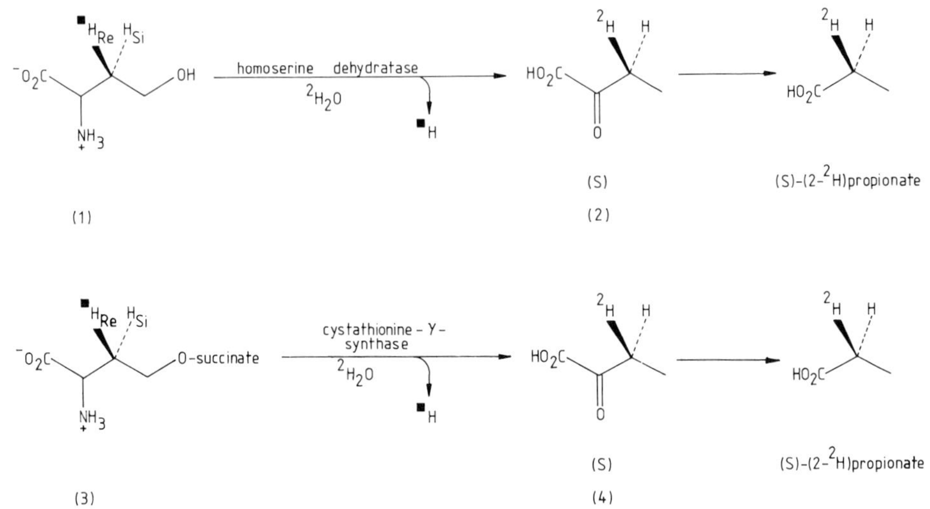

Fig. 37. The loss of a hydrogen at C_β and its replacement by a hydrogen from the medium with retention of configuration during the γ-elimination–deamination process.

(Fig. 38, 2) can also serve as a substrate in either the γ-replacement (in the presence of L-cysteine) or γ-elimination–deamination (absence of L-cysteine) modes. The vinylglycine enters into the normal catalytic cycle by Schiff base formation at the active site followed by loss of the C_α hydrogen to afford the same intermediate generated from the natural substrate. With this background, information on the stereochemistry of the events occurring at C_γ in the biosynthesis of cystathionine may be obtained by comparing the stereochemistry of the products obtained from the conversion of (i) *O*-succinyl homoserine labelled stereospecifically at the strategic C_γ with deuterium and (ii) vinylglycine containing deuterium at only one of the C_γ vinylic positions. Experiments based on this approach were performed [89] and it was shown that in the reaction catalysed by cystathionine-γ-synthase, both (4*S*)-[4-^{2}H]-*O*-succinyl homoserine and (*Z*)-[4-^{2}H]vinylglycine * were converted into (4*S*)-[4-^{2}H]cystathionine (Fig. 38). This result is only possible if both transformations proceed through the same intermediate (Fig. 38, 3). These experiments thus show that in the enzymic conversion of (4*S*)-[4-^{2}H]-*O*-succinyl homoserine (Fig. 38, 1) to products, the terminal double bond of the complex (Fig. 38, 3) has the (*Z*) configuration. This observation when considered in conjunction with the earlier demonstration that in the conversion the C_β-H_{Re} atom is removed [84–86], allows

* (*Z*), from the German word, *zusammen* = together. In the present context this refers to the two groups of highest priority on the same side of the double bond (^{2}H and C chain on the same side of 2: Fig. 38). (*E*), from entgegen = opposite, referring to the highest priority groups on the opposite side.

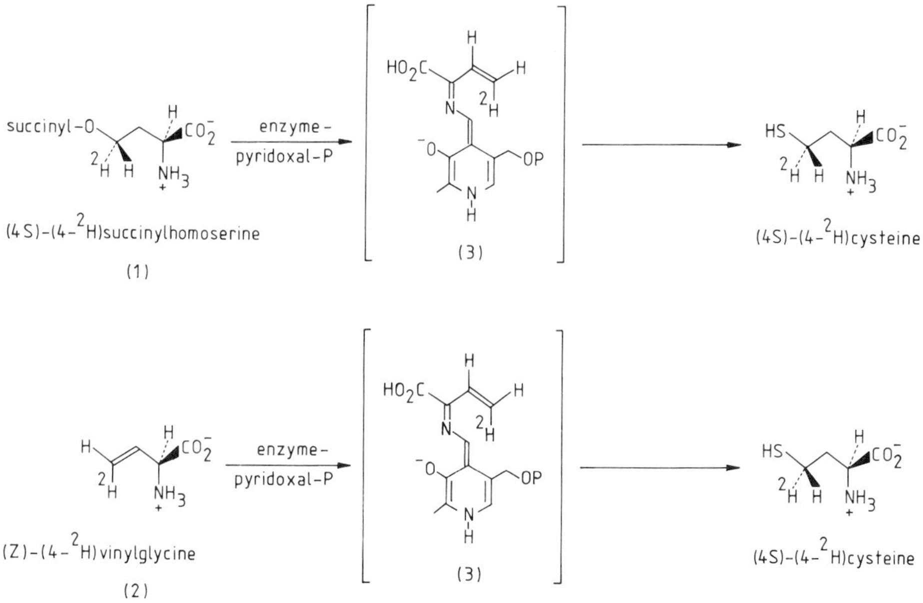

(4S)-(4-^{2}H)succinylhomoserine (1) → enzyme-pyridoxal-P → [(3)] → (4S)-(4-^{2}H)cysteine

(Z)-(4-^{2}H)vinylglycine (2) → enzyme-pyridoxal-P → [(3)] → (4S)-(4-^{2}H)cysteine

Fig. 38. The demonstration that in the reaction catalysed by cystathionine-γ-synthetase with (4*S*)-(4-^{2}H)succinylhomoserine, the (*Z*)-vinylglycyl species (3) is involved, and that the overall transformation occurs with retention of stereochemistry at C_{γ}.

the conclusion that the formation of the vinyl double bond of 3 as well as its subsequent saturation must occur via a net *syn* process. Cumulatively, these experiments suggest that during both the normal γ replacement and the 'artificial' γ-elimination–deamination reactions catalysed by cystathionine-γ-synthase all the stereoelectronically assisted bond breaking and forming events at C_{α}, C_{β} and C_{γ} occur in a *syn* process. It is well known in solution chemistry that concerted 1,2 eliminations proceed by an *anti* pathway because the corresponding *syn* eliminations are energetically unfavourable. If the same mechanistic constraint is applied to the enzyme catalysed reaction discussed above, then the argument may be developed that in these cases the bond cleavage and formation events are *syn* only in a superficial sense; in fact, the processes occur in a step-wise fashion using a single catalytic group or multiple groups organised on the same side of the protein. A schematic representation of this process is outlined in Fig. 39, in which first the C_{α}, then the C_{β} and finally the C_{γ} bonds are aligned perpendicularly to the electron withdrawing centre(s) of the coenzyme for stereoelectronically assisted bond manipulations eventually producing the intermediate 5 by an apparent *syn* pathway. The formation of the product from 5 then involves the reversal of the same reactions permitting the retention of stereochemistry at all the centres.

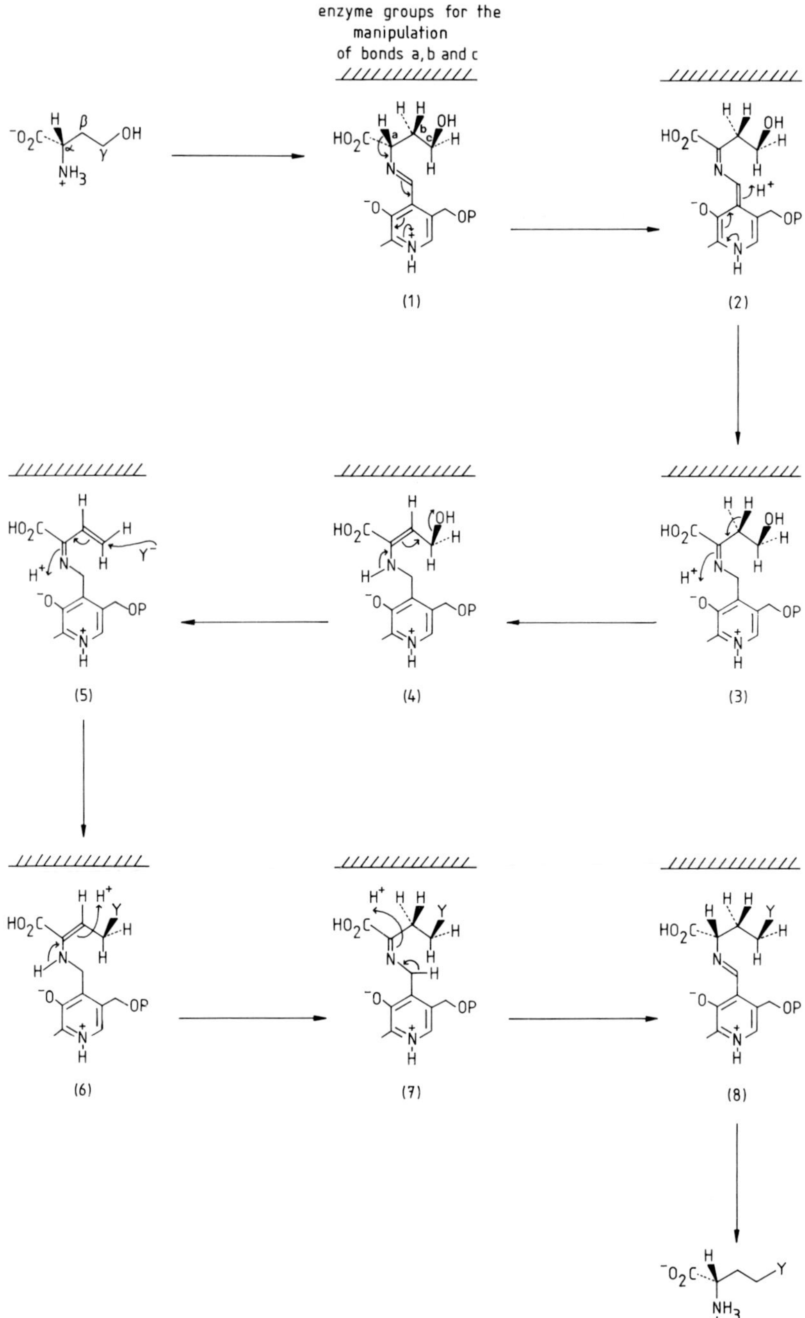

Fig. 39. Illustration showing the stepwise cleavage and formation of bonds at C_α, C_β and C_γ via an overall *syn* arrangement.

6. *Structure and molecular dynamics within the binary and ternary complexes*

Above we have used concepts and drawn detailed structures for various species involved in pyridoxal-P-dependent enzymes without presenting experimental evidence. This section gives an account of spectroscopic and chemical studies which have formulated our views on the structure of the binary (coenzyme–enzyme) and ternary (substrate–coenzyme–enzyme) complexes and the nature of chemical events which occur within them.

6.1. Electronic spectrum of the coenzyme chromophore

Spectroscopic studies have shown that the chromophoric system of the coenzyme absorbs at 390 nm, in a region clearly separated from the 280 nm peak of the protein moiety. This feature has been exploited for a variety of structural and kinetic studies [31,34,90]. Pyridoxal phosphate, whether in its free form or when bound in the binary or ternary complexes, can exist, depending upon the pH, in a number of ionised states dictated by the pK_a values for the various ionisable groups within the molecule. Metzler and coworkers [91] have ascertained spectrophotometrically the pK_as for 3-hydroxypyridine (pK_a: 5.1, 8.6) and *N*-methyl-3-hydroxypyridine (pK_a: 4.96) and concluded the pK_a of the 3-hydroxy function to be 5.1 and the pyridinium nitrogen to be 8.6 (Fig. 40). This information allows one to interpret the pK_a values of 3.1 and 8.3 obtained by Morozov et al. [92] for pyridoxal, as those corresponding to the 3-hydroxy and N-1 groups respectively (Fig. 41). It was noted [92] that the presence of a 5′-phosphate group on pyridoxal had practically no effect on the shape, half-width, or position of the absorption bands and hence pK_as as noted above.

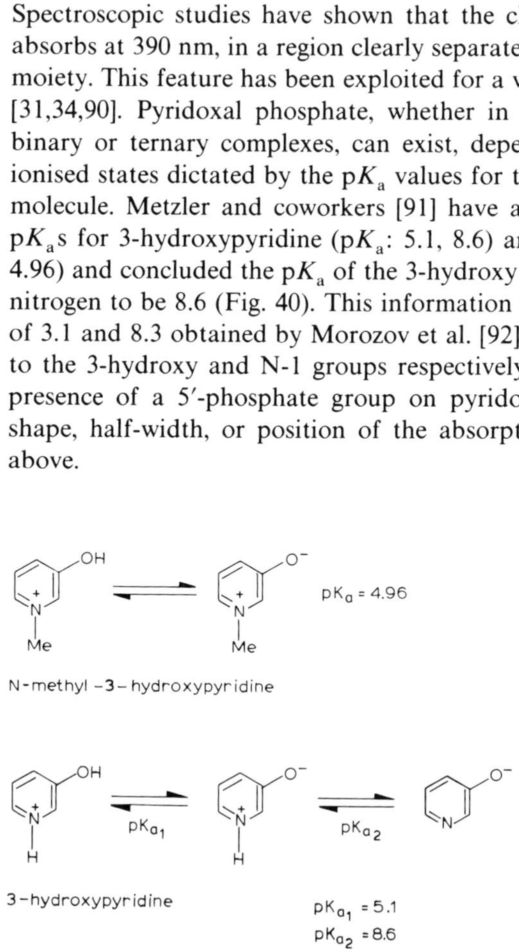

OH O⁻ pKa = 4.96
Me Me
N-methyl-3-hydroxypyridine

OH O⁻ O⁻
pKa1 pKa2
H H
3-hydroxypyridine
pKa1 = 5.1
pKa2 = 8.6

Fig. 40. The pK_a values for the 3-hydroxy and N-1 groups of 3-hydroxypyridine as inferred by reference to the pK_a of *N*-methyl-3-hydroxypyridine.

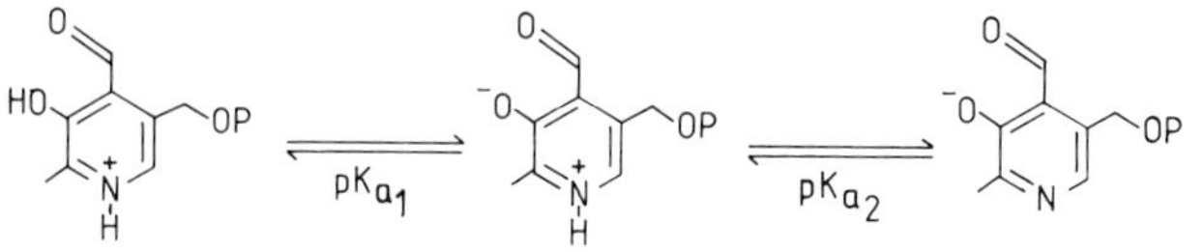

(1) $\underset{pK_{a_1}}{\rightleftharpoons}$ (2) $\underset{pK_{a_2}}{\rightleftharpoons}$ (3)

$pK_{a_1} = 3{\cdot}1$

$pK_{a_2} = 8{\cdot}3$

Fig. 41. The pK_a values for 3-hydroxy and N-1 groups of pyridoxal-P. The state of ionisation of the 5′-phosphate moiety is not shown.

From this information it may be inferred that in the physiological pH range 5.0–7.5 pyridoxal phosphate exists as a Zwitterion with the 3-hydroxy group in its anionic form and the pyridinium nitrogen in its cationic form (Fig. 41, 2). The recognition of this feature has two implications: first, that a Zwitterion is a more appropriate structure for the enzyme bound pyridoxal phosphate moiety, and second, that the spectral changes observed during catalysis may, with some confidence, be attributed to the covalency changes occurring at the strategic C-4′ position, rather than being due to alterations in the state of ionisation of the groups in the pyridinium ring.

In close analogy to 3-hydroxypyridine, the chromophores in pyridoxamine phosphate and other related species of types 2 and 3 (Fig. 42) possessing a tetrahedral centre at C-4′, absorb at about 330 nm. The presence of an aldehyde function at this position, as in pyridoxal phosphate (Fig. 42, 1) shifts the absorption spectrum to 390 nm. The conversion of the aldehyde into a Schiff base is attended by a slight hypsochromic shift, but the protonation of the Schiff base produces a dramatic bathochromic shift and species of type 5 (Fig. 42) absorb at approximately 420–440 nm. Guided by the data from chemical model systems we examine the spectroscopic information obtained in the formation of binary and ternary complexes with several B_6 enzymes.

The binary complex can exist as an equilibrium mixture of a number of species as shown in Fig. 43. The major species apparent in a number of binary complexes, e.g. glutamate decarboxylase and aspartate aminotransferase [31,90] appear to be those absorbing at 420 and 333 nm which are attributed to structures 2 and 3 (Fig. 43) respectively; as expected the ratio of these species is pH dependent. On the other hand, the ternary complex may exist as an equilibrium mixture composed of species from both binary and ternary complexes, thus producing a composite electronic absorption profile; however, the ternary complex of aspartate aminotransferase exhibits only two major absorption maxima at 430 and 340 nm due to substrate–coenzyme Schiff base and enzyme bound pyridoxamine-P respectively [90]. It is interesting to note the spectra observed for aspartate aminotransferase

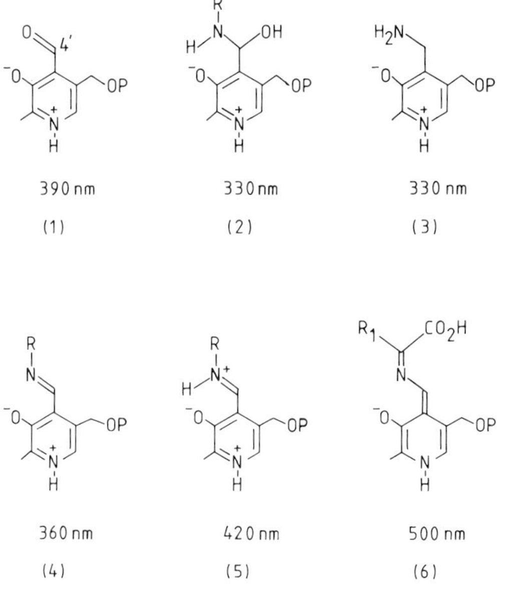

Fig. 42. Long wavelength electronic absorption maxima of pyridoxal-P and related species (R = lysyl residue in binary complex or amino acid in ternary complex).

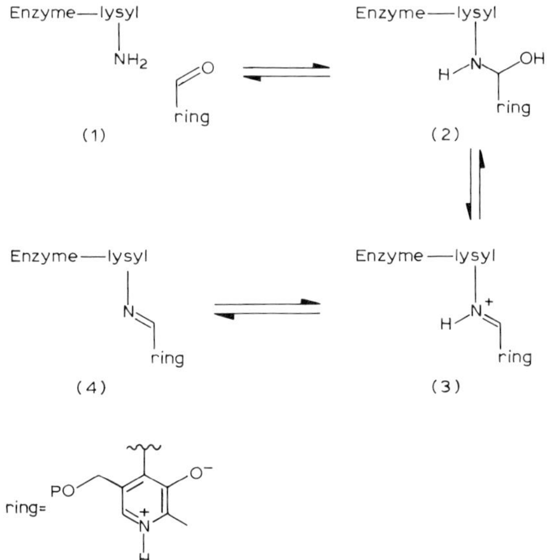

Fig. 43. Possible constituent species of the binary (coenzyme–enzyme) complex.

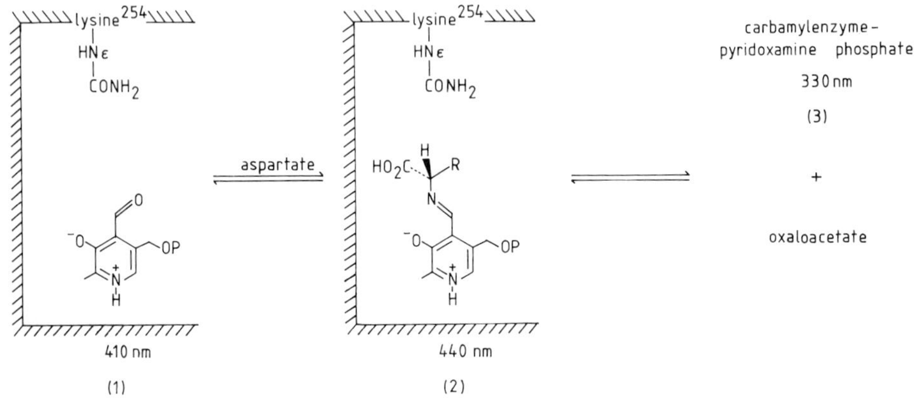

Fig. 44. Half-transamination reaction catalysed by carbamylated aspartate aminotransferase.

carbamylated at the active site lysine 254; the binary complex now binds pyridoxal phosphate non-covalently (Fig. 44) and shows an absorption maximum centred at 410 nm * (Fig. 45) [93]. Addition of aspartate to yield a ternary complex is attended by the characteristic maxima at 330 and 430 nm (Fig. 45).

Depending upon the relative rates of the reactions participating in the interconversion of various species in the ternary complex, one may expect to see spectrophotometrically the presence of the quinonoid intermediate of type 6 (Fig. 42) in a number of pyridoxal-P-dependent enzymic reactions; such an intermediate would be expected to possess a bathochromically shifted long wavelength absorption maximum. This has indeed been clearly viewed for the ternary complex of serine hydroxymethyltransferase with glycine as shown in Fig. 46, the absorption at 495 nm being attributed to the quinonoid intermediate [94].

Studies with tryptophan synthetase [95] and tryptophanase [96] highlight the strength as well as the weakness of the spectroscopic approach as a tool for the identification and kinetic analysis of intermediary species produced in pyridoxal-P-dependent reactions. The involvement of the species of types 1, 2 and 4 (Fig. 31) in tryptophanase catalysed reactions and the kinetics of their formation have been extensively studied spectroscopically [96], but the quinonoid intermediate (2) and the acrylyl pyridinium species (4) could not be discerned from each other because the two chromophores absorbed in the same region at about 500 nm.

* The absorption maximum of pyridoxal phosphate is red-shifted by about 20 nm and this may be due to the effect of the environment on the chromophore. This feature points to the need for caution in the application of spectroscopic techniques to the identification of enzyme bound intermediates since a change in the position of a signal could be due either to covalancy changes in the probe or its presence in an environment different from the one used for obtaining the reference data.

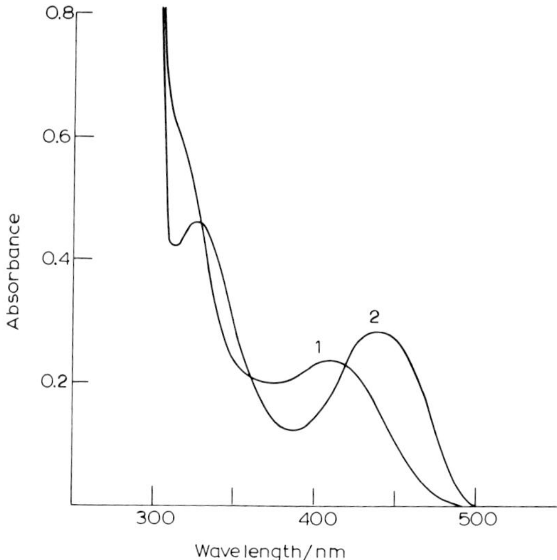

Fig. 45. Electronic absorption spectrum for carbamylated aspartate aminotransferase in the presence of pyridoxal-P (curve 1) and following the addition of aspartate to the latter (curve 2).

6.2. Chemical studies on binary (coenzyme–enzyme) complexes

The spectroscopic evidence for the existence of a Schiff base linkage between the coenzyme and the protein in the binary complex and between the substrate and the

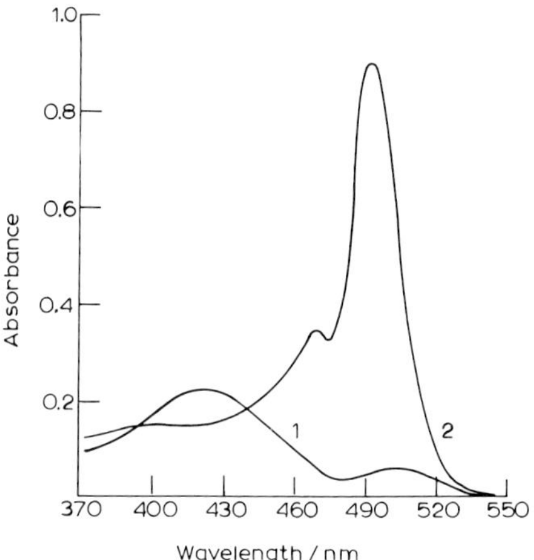

Fig. 46. Electronic absorption spectrum for SHMT in the presence of glycine (curve 1). Note the formation of the quinonoid intermediate absorbing at 495 nm, the extent of which is greatly increased in the presence of THFA (curve 2) (see section on SHMT).

coenzyme in the ternary complex has received independent support from chemical modification studies. Following the use of $NaBH_4$ by Fischer and coworkers [97] in 1958 to stabilise the Schiff base linkage in phosphorylase *b*, an enzyme in which pyridoxal phosphate does not play a direct catalytic role, the technique was extended to aspartate aminotransferase [98,99] and cystathionase [100]. It was shown that the treatment of these two enzymes with $NaBH_4$ resulted in the loss of their catalytic activities and acid hydrolysis of the modified enzymes gave ϵ-pyridoxyl-lysine [98]. This suggested that in the native enzyme the ϵ-amino group of a lysyl residue must be involved in the binding of the coenzyme via a Schiff base. In principle, any of the four rapidly interconverting chemical entities may constitute a binary complex. A successful trapping of the Schiff base intermediate with $NaBH_4$ thus depends on the presence of 3 and 4 (Fig. 43) as predominant species under steady-state conditions, and also on whether the tertiary structure of the enzyme allows the entry of the reagent into the active site. It is therefore not surprising that these favourable combinations of circumstances have been found in only a few cases.

6.3. Stereochemical aspects of the reduction of Schiff base at C-4′ with $NaBH_4$

In 1973 work in the laboratories of Duilio Arigoni at Eidgenössischen Technischen Hochschule (ETH) Zurich, culminated in the development of an elegant approach [101] for the delineation of the stereochemistry of $NaBH_4$ reduction of the Schiff base linkage, thus adding a new interest to the study of this group of enzymes. As a prelude to this discussion, let us consider some of the stereochemical conventions whose comprehension is necessary for the full appreciation of the story. In the Schiff base complex of type 1 (Fig. 47) the stereochemical outcome of the reduction reaction would depend on the face of the trigonal C atom at C-4′ which is accessible to the incoming reagent. According to the convention of Hanson [102] the two faces are defined as *Re* or *Si* * depending on whether the groups, =N–, –Ring, –H, at C-4′ make a clockwise or an anticlockwise arrangement. Thus the face viewed by the observer from the top in 1 (Fig. 47) is *Re* while that from the bottom is *Si*. For the readers who do not have the agility of a yogi to view standing on their heads, the illustration in structure 2 (Fig. 47) shows the arrangement as it would appear from the bottom.

Stereospecific reduction at the *Re* and *Si* faces using the isotopically labelled reagent, NaB^2H_4 or NaB^3H_4, will produce the (*R*)- and (*S*)-enantiomers respectively (Fig. 47). The contribution of the ETH group was not merely the recognition of this principle, but more importantly the development of a protocol [101] for the configurational analysis of the resulting species of types (*R*)-3 and (*S*)-3 (Fig. 47).

The coenzyme–enzyme Schiff base in aspartate aminotransferase was reduced with NaB^3H_4, and the resulting derivative hydrolysed to give [4′-^{3}H]pyridoxyl-lysine. The latter on *N*-chlorination, followed by base catalysed dehydrohalogenation gave

* *Re* and *Si* are derived from the Latin words: *rectus* = right, and *sinister* = left and pronounced as 'ray' and 'sigh' respectively.

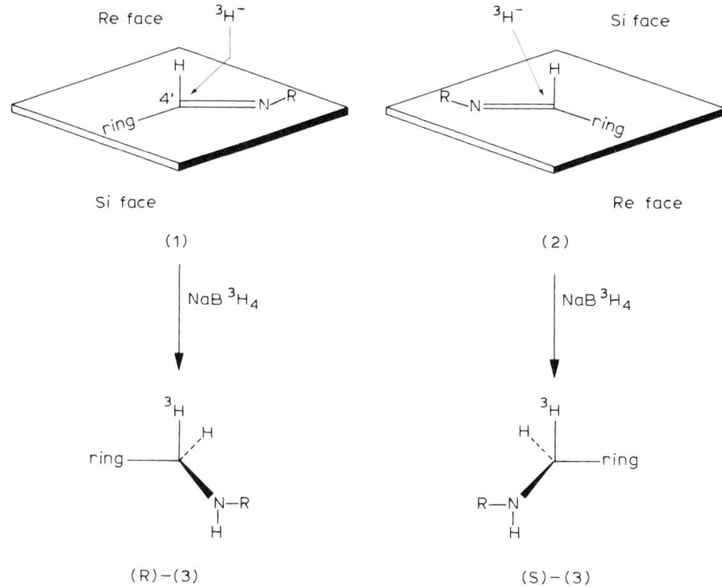

Fig. 47. Reduction of the C-4′ Schiff base in the binary complex with NaB^3H_4 showing that *Re* and *Si* face attack yields the (*R*)-[4′-^{3}H] and (*S*)-[4′-^{3}H] enantiomers respectively.

[4′-^{3}H]pyridoxamine which was converted by aspartate aminotransferase *apo*enzyme * in the presence of α-oxoglutarate into pyridoxal with retention of ^{3}H. A knowledge of the stereochemical course of the latter reaction which removes the C-4′ H_{Si} atom allowed the conclusion that the [4′-^{3}H]pyridoxamine had the (*R*) configuration (Fig. 48). The reduction by NaB^3H_4 must have involved [101] attack on the *Re* face of the π-electron system at C-4′ of the species (Fig. 48, 1). The conclusion for aspartate aminotransferase has recently been confirmed by Zito and Martinez-Carrion [93] and the same approach when applied to tyrosine decarboxylase [103], also showed the hydride attack to occur on the *Re* face at C-4′ of the coenzyme–enzyme imine bond. In both cases, therefore, the *Re* face at C-4′ must be exposed to the 'solvent side' in the binary complex.

6.4. Structure and stereochemistry of the substrate–coenzyme bond in ternary complexes

The above approach has also been extended to the study of the 'solvent sidedness' of the face of the substrate–coenzyme Schiff base, in a ternary complex. During the normal functioning of a pyridoxal-P-dependent enzyme several types of ternary complex will be present, although only two species represented by structures 1 and 2

* The apoenzyme obtained by the removal of pyridoxal-P from aspartate aminotransferase catalyses the half-transamination: pyridoxal + amino acid $\rightleftharpoons$ pyridoxamine + α-oxo-acid.

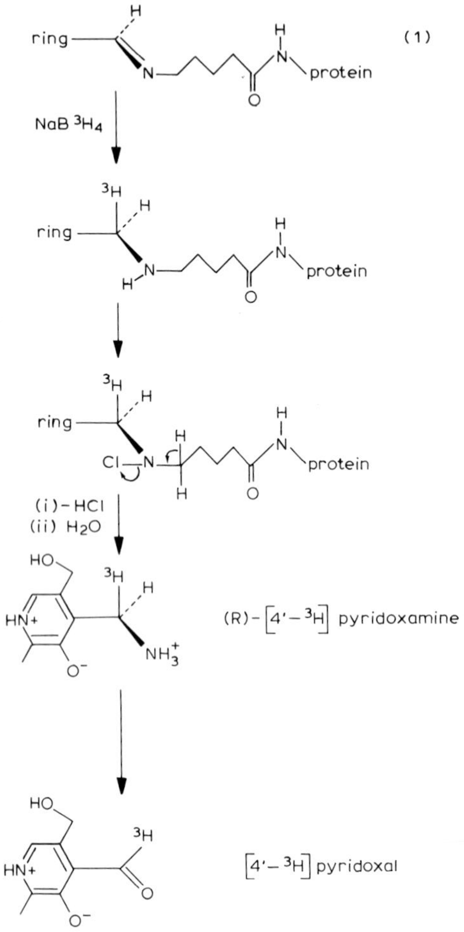

Fig. 48. Chirality analysis at C-4′ of pyridoxamine obtained following reduction of the binary complex of aspartate aminotransferase with NaB^3H_4. It should be noted that at the dehydrohalogenation step only half the reaction will occur via the pathway shown (cf. reaction of 4: Fig. 49).

(Fig. 49) are relevant to the present discussion. These, on treatment with NaB^3H_4 will produce the respective dihydro derivatives arising from either the trapping of the Schiff base at C-4′ or C_α respectively. It is worth remembering that under steady-state conditions, in the presence of its physiological substrate, an enzyme will partition itself between at least a dozen binary and ternary complexes. The concentration of any one of these intermediates, therefore, is likely to be extremely small, hence posing a major experimental challenge to the trapping of an intermediate of types 1 and 2 (Fig. 49) in sufficient quantities for detailed stereochemical analysis. Notwithstanding this difficulty the ETH group quenched an incubation of aspartate

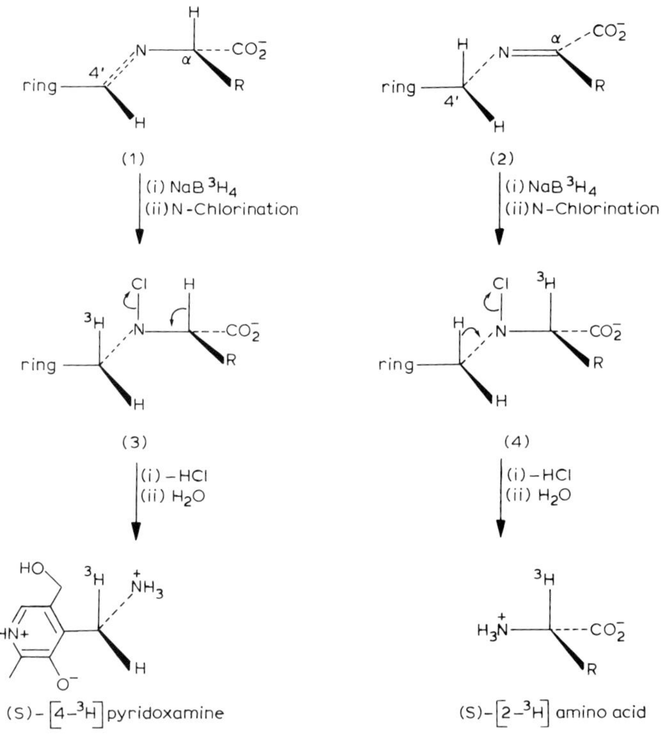

Fig. 49. Structures present in the ternary complex which will be amenable to NaB^3H_4 reduction and subsequent configurational analysis of C-4′ and C_α.

aminotransferase and L-aspartate with NaB^3H_4 and using the degradative sequence of Fig. 49 established that the [4′-^{3}H]pyridoxamine, as well as the amino acid, had the (*S*) absolute configuration [101]. The experiments show that hydride attack at C-4′ of 1 and C_α of 2 occurred from the direction *corresponding* to the *Si* face at C-4′ of the ternary complexes * (Fig. 49) [101].

Similar experimental approaches have been extended to the reduction of substrate–coenzyme Schiff bases in two other ternary complexes. In the case of tyrosine decarboxylase [103], the ternary complex produced from the physiological substrate, L-tyrosine, was used in the trapping, while with tryptophanase [104], a reducable ternary complex was generated using L-alanine, a competitive inhibitor for the enzyme. In these two cases, once again it was the *Si* face at C-4′ of the substrate–coenzyme Schiff base in the ternary complex that was available for attack by $NaBH_4$. These results provide an interesting contrast with the *Re* face attack

* Readers should note that according to the convention the face at C-4′ of 1 (Fig. 49) is *Si* whereas that at C_α of 2 is *Re*.

established in the preceding section for the reduction of the Schiff bases in the corresponding binary complexes.

The pattern recorded above raises the question whether the change of face at C-4′ to the 'solvent side' is a mandatory requirement in the transformation of binary into ternary complexes in pyridoxal-P-dependent reactions. That this may be so was the view beginning to prevail until a timely reminder, or perhaps an undue caution, came from a more recent report by Zito and Martinez-Carrion [93]. As has already been cited, these workers repeated the earlier experiments of the Zurich School on aspartate aminotransferase confirming the *Re* face hydride attack at C-4′ in the binary complex. However, aspartate aminotransferase carbamylated at the active site Lys-258 was used to produce the substrate–coenzyme Schiff base linkage in the ternary complex. Since the modified enzyme catalysed the half-transamination reaction:

$$\text{Aspartate} + \text{carbamylenzyme–pyridoxal phosphate} \rightleftharpoons \text{oxaloacetate}$$
$$+ \text{carbamylenzyme–pyridoxamine phosphate}$$

relatively slowly, a sufficiently high concentration of the ternary complex of type 2 (Fig. 44) was available for trapping. Reduction of the complex followed by the usual work-up showed that in this case the substrate–coenzyme Schiff base in the ternary complex was attacked by hydride at the *Re* face. This result for the ternary complex is thus at variance with those from the other two examples cited earlier where the *Si* face was attacked. We must, though, bear in mind two features; first, that the carbamylenzyme has less than 1% of the activity of the native enzyme and, secondly, that in the carbamylenzyme the substrate–coenzyme Schiff base is formed by a direct condensation reaction (see Fig. 44) rather than by transamination which is the case with the native enzyme.

Leaving the last anomaly aside, the other two examples considered above clearly show that as far as the accessibility to external reagents is concerned, two different faces at C-4′ of the coenzyme are available for reaction; *Re* in the binary and *Si* in the ternary complex. At first sight this change of steric course appears dramatic pointing to a fundamental structural reorganisation upon binding of the substrate to the binary complex to produce the ternary complex. This need not be so, experience from the field of natural product chemistry tells us that a high degree of preference for a given steric course is often the result of small but subtle structural differences; so subtle that organic chemistry had to reach its zenith before such subtleties were clearly comprehended by Barton [105] (London; now Gif-Sur-Yvette).

Notwithstanding this, several explanations have been, or can be, offered to rationalise the opposite stereochemical modes of the reduction of the Schiff bases in the binary and ternary complexes and these are considered below.

(a) A rotation of the C-4–C-4′ bond of the coenzyme placing the substrate–nitrogen atom in the ternary complex in an orientation that is opposite to that occupied by the ε-amino nitrogen of the lysyl residue in the binary complex (Fig. 50). Such an outcome may be the result of quite diverse courses including, for example, the

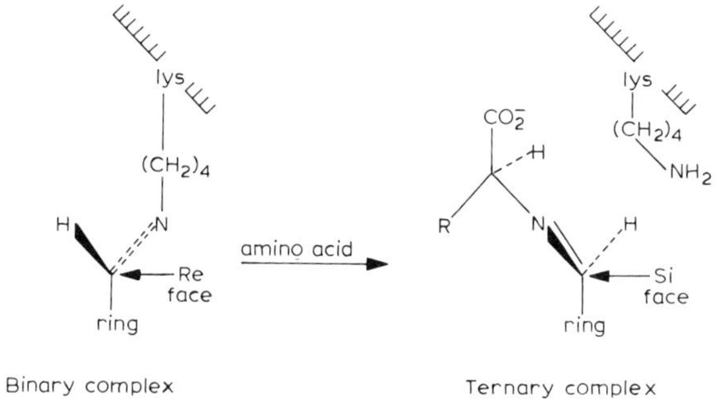

Fig. 50. A stylised diagram showing the change of face in the rearrangement of a binary into a ternary complex. The incoming reagent in both cases enters from the same direction.

rotation of the entire skeleton of the coenzyme [103,106].

(b) A conformational change in the protein moiety shielding one of the faces of the coenzyme and exposing the other.

(c) Finally, it is likely that the reaction of $NaBH_4$ with the Schiff base linkage of the binary complex is analogous to the physiological process involving the addition of an amino acid to the same linkage during the formation of the ternary complex. If this were the case, it follows that the presence of the amino acid moiety in the ternary complex would obstruct the entry of $NaBH_4$ from the direction originally available in the binary complex. $NaBH_4$ is now forced to attack the Schiff base from the direction previously less favourable, thus at a new face (Fig. 51).

6.5. Stereochemical and mechanistic events at C_α of the substrates and at C-4′ of the coenzyme during catalysis

The structural details of the binding interactions around C-4′ of the coenzyme in various complexes have been presented in the preceding section. We now consider

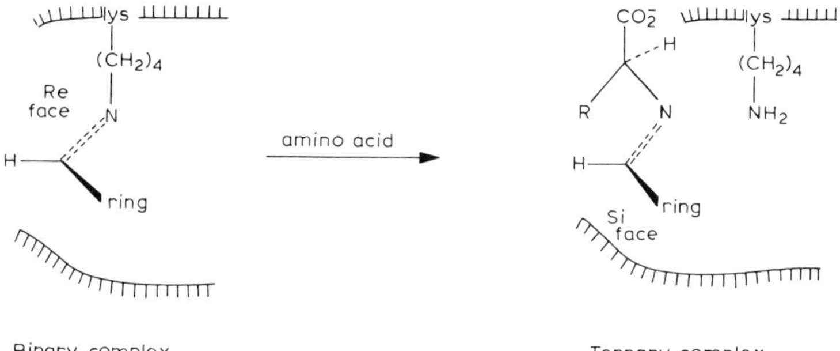

Fig. 51. An illustration showing that the *Re* face attack in the ternary complex is blocked by the presence of the amino acid moiety.

the nature of the stereoelectronic factors which operate during bond forming and breaking processes in the ternary complexes. Our current knowledge in this area is predominantly based on the elegant studies of Dunathan [107] performed on the half-reaction catalysed by transaminases:

$$\text{Enzyme–pyridoxal phosphate} + \text{amino acid}$$
$$\rightleftharpoons \text{enzyme–pyridoxamine phosphate} + \alpha\text{-oxoacid}$$

The main chemical feature of the transformation is the allylic rearrangement in which the C_α–H bond in the amino acid is broken and a new C–H bond at C-4′ of the coenzyme is constructed (Fig. 52). Questions pertinent to this type of conversion are: (i) Is the formation and cleavage of bonds in the process *syn* (as in 1 → 3) or *anti* (as in 2 → 4) (Fig. 52) (ii) What is the geometry of the C_α–H bond with respect to the two C-4′ faces of the coenzyme? In other words, is the C_α–H bond oriented on the side of the *Re* or *Si* face at C-4′ (Fig. 53)?

The answer to the first question was originally provided [107] using pyridoxamine-pyruvate aminotransferase, an enzyme involved in the degradation of pyridoxamine by a soil bacterium. The enzyme catalyses the reaction:

$$\text{Pyridoxal} + \text{L-alanine} \xrightleftharpoons[]{\text{pyridoxamine-pyruvate aminotransferase}} \text{pyridoxamine} + \text{pyruvate}$$

in which the formation of pyridoxamine occurs in a turnover process, thus enabling

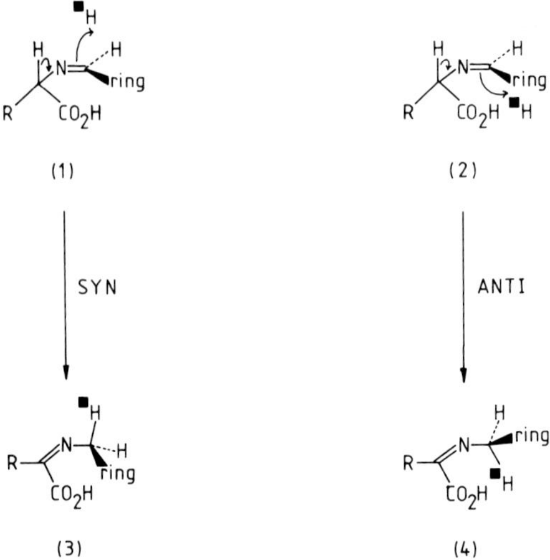

Fig. 52. The two possible stereochemical courses for the C_α–C-4′ allylic rearrangement observed in transamination reactions.

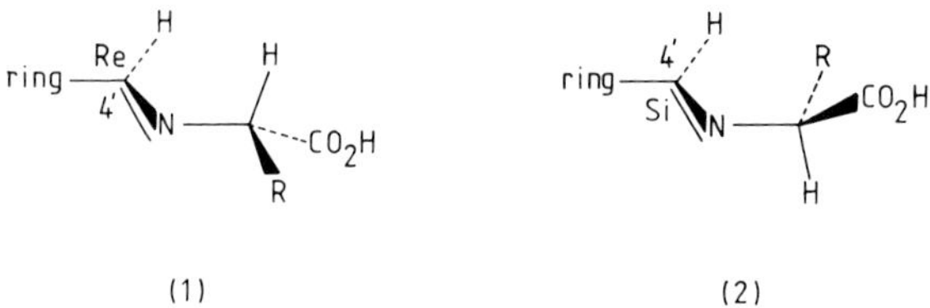

Fig. 53. Geometry of the C_α–H bond with respect to the faces of the Schiff base at C-4′.

the availability of this compound in amounts required for isotopic analysis. Incubations of the aminotransferase in ordinary H_2O with [2-^{2}H]alanine and pyridoxal gave pyridoxamine containing a small but significant amount of deuterium at C-4′, thus indicating that there was a component of intramolecular hydrogen transfer from C_α to C-4′ during the tautomerisation process. The view was strengthened by performing the above experiment with unlabelled alanine in D_2O when the analysis of the pyridoxamine, thus produced, revealed that the intramolecular hydrogen transfer had now occurred to the extent of 50%. The conclusion was confirmed by studying the reaction in the reverse direction when the transfer of hydrogen from C-4′ of pyridoxamine into the biosynthetic alanine could also be demonstrated.

Although these experiments have been performed using an untypical enzyme, they provided a deeper insight into the chemistry of the prototropic shift than was required by the question above. The demonstration of an intramolecular hydrogen transfer proves beyond doubt that the tautomerisation reaction occurs via a *syn* elimination–addition process in which the bond to C_α that is broken lies orthogonally * to the plane of the Schiff base linkage, as indeed had been expected on theoretical considerations [107]. The results also indicate (but do not prove) that a single catalytic group is involved, both in the cleavage of the C_α–H bond of the amino acid moiety and the formation of the new bond at C-4′ of the coenzyme (Fig. 54). Since the extent of intramolecular transfer under the normal conditions of enzymic reaction is low, this indicates that the rate at which the overall tautomerisation reaction occurs is considerably slower than the rate of exchange of hydrogen between the catalytic group, –X–H, of the enzyme and the protons of the medium (Fig. 54, reactions a and b). The last point further suggests that in the case of pyridoxamine-pyruvate aminotransferase the catalytic group, –X–H, in the ternary complex faces the ‘solvent side’ to allow the exchange.

We now consider the second question; the orientation with respect to the Schiff base face at C-4′, of the bond to C_α cleaved in the tautomerisation reaction. In 1971 Dunathan [107] described the first example of a study shedding light on this aspect by showing that in the reaction catalysed by the *apoenzyme* of *aspartate* aminotransferase in ^2H_2O in the forward direction, one deuterium was incorporated

* i.e. perpendicular. This is equivalent to the statement that the bond to C_α cleaved is aligned parallel to the *p*-orbitals of the conjugated system.

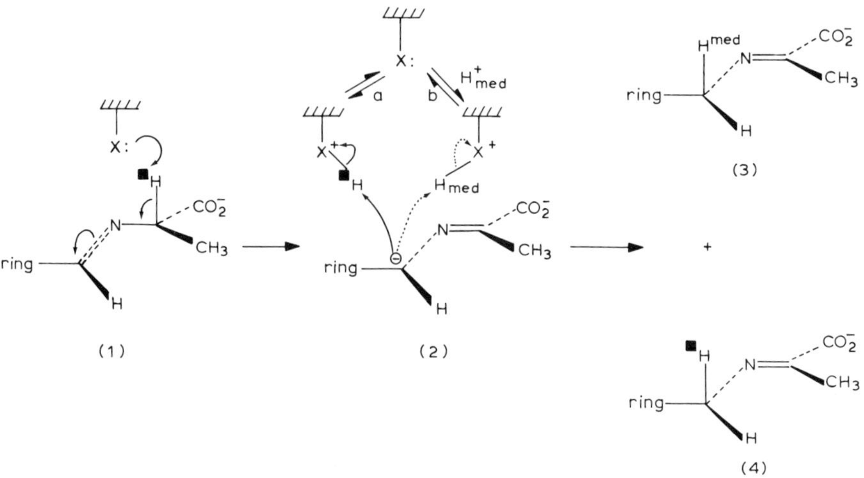

Fig. 54. 1,3-Intramolecular hydrogen transfer catalysed by a single base. According to this proposal the extent of intramolecular transfer depends on the rate of the tautomerisation process versus the exchange process (reaction a plus b).

at C-4′ of the pyridoxamine, while in the reverse direction the same deuterium was removed. This demonstrated that shuttling of hydrogen to and from C-4′ of the coenzyme occurs in a stereospecific fashion. Dunanthan's experiment also indicated that the [4′-^{2}H]pyridoxamine produced above had the (*S*) chirality (hydrolysis product of 3, $H^{med} = {}^2H$, Fig. 54). "The stereochemical deduction was", writes Dunathan, "confirmed in an elegant way by Besmer and Arigoni [108]. [4′-^{3}H]Pyridoxamine was degraded to glycine and thence to glycolate. Enzymatic oxidation of the glycolate led to the assignment of (*S*) symmetry to glycolate and thus *S* symmetry to the original [4′-^{3}H]pyridoxamine (Fig. 55)".

The experiments conclusively prove that the addition of hydrogen to C-4′ of the coenzyme occurs at the *Si* face of the Schiff base. Evidence has already been provided for the *syn* nature of the tautomeric process in the reaction catalysed by pyridoxamine-pyruvate aminotransferase [107]. If the same precedent is extended * to aspartate aminotransferase it then follows that the bond to C_α that is formed and broken in this case must also be located on the *Si* face at C-4′, in the catalytic complex, as shown in structure 2 (Fig. 53). In other words, the alternative arrangement for *syn* proton transfer shown in 1 (Fig. 53) is precluded by these experiments. The direction of hydrogen addition to C-4′ of the coenzyme in the half-reaction has also been studied using several other L-amino acid requiring aminotransferases and in every case the medium hydrogen was shown to add to the *Si* face at C-4′ (Table 5). These experiments have led to the generalised view that in B_6-dependent reactions

* It is regrettable that such crucial experiments, with far reaching implications, have not been performed on a single enzyme.

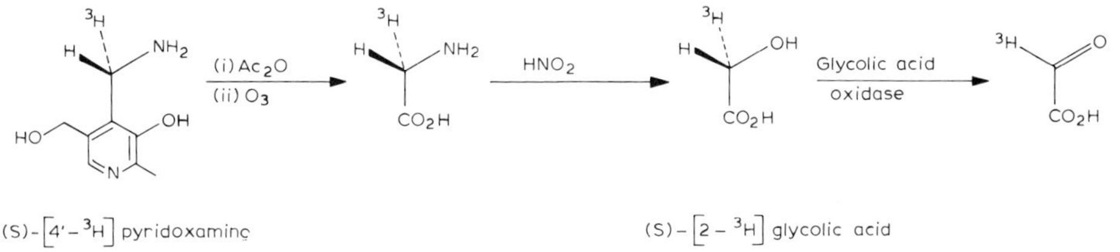

(i) Ac_2O (ii) O_3 → HNO_2 → Glycolic acid oxidase →

(S)-[4'-^{3}H] pyridoxamine (S)-[2-^{3}H] glycolic acid

Fig. 55. Configurational analysis of [4'-^{3}H]pyridoxamine obtained by the half-transamination reaction catalysed by the *apo* protein of aspartate aminotransferase in [^{3}H]H_2O.

bond breaking and forming events occur in a *syn* fashion on the *Si* face at C-4′, giving rise to the disposition of various groups in the ternary complex as shown in structure 1 (Fig. 56). What is behind this conformity? Can the specific arrangement of substituents at C_α of amino acids of L configuration, with the carboxylate in a position to hydrogen bond with the protonated Schiff base Fig. 56, 1, be responsible for steering the cleavable C_α–H bond to project on the *Si* face at C-4′? We cannot answer this question with absolute certainty at this stage since parallel studies with aminotransferases which utilise D-amino acids as their physiological substrates have not yet been reported. Notwithstanding this, that the key to *Si* face preference may

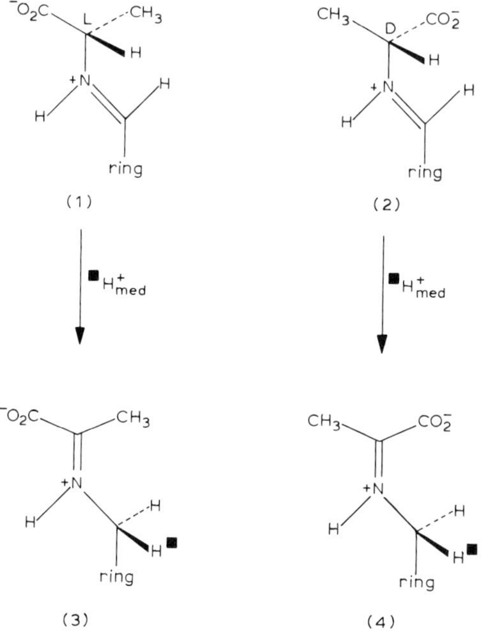

Fig. 56. *Syn* tautomerisation process using the *Si* face at C-4′. Note that the arrangement of groups around C_α is different for amino acids of L and D configuration.

TABLE 5

The stereochemistry of reactions at the C-4′ Schiff base faces in the binary (coenzyme–enzyme) and ternary (substrate–coenzyme) complexes

Enzyme	Face of coenzyme at C-4′ attacked by $NaBH_4$ in the binary complex	Face of coenzyme at C-4′ attacked by $NaBH_4$ in the ternary complex		Face of coenzyme at C-4′ protonated in the half-transamination reaction		Intramolecular transfer of hydrogen from C_α of substrate to C-4′ of the coenzyme
		Face	Substrate	Face	Substrate	
1. Alanine aminotransferase [107]	–	–	–	*Si*	Alanine	–
2. Aspartate aminotransferase [101,108,202]	*Re*	*Si*	L-Aspartate	*Si*	Aspartate	–
3. Carbamylated aspartate aminotransferase [93]	*Re*	*Re*	L-Aspartate	*Si*	Aspartate	–
4. Dialkyl aminotransferase [203]	–	–	–	*Si*	α-Aminoisobutyrate	–
5. Glutamate decarboxylase [110]	–	*Si*	α-Methylglutamate	*Si*	α-Methylglutamate	–
6. Pyridoxamine-pyruvate aminotransferase [107,204]	–	–	–	*Si*	Pyruvate L-Alanine	Yes
7. Serine hydroxymethyl transferase [109]	–	–	–	*Si*	D-Alanine	–
8. Tryptophanase [104]	–	*Si*	L-Alanine	–	–	–
9. Tryptophan synthetase β_2 protein [69,111]	–	*Si*	L-Serine + 2-mercaptoethanol	*Si*	Serine + 2-mercaptoethanol	–
10. Tryptophan synthetase $\alpha_2\beta_2$ [205]	*Re*	*Si*	L-Serine			
11. Tyrosine decarboxylase [103]	*Re*	*Si* *Si*	L-Tyrosine L-Tyramine	– –	– –	– –

lie elsewhere is suggested by three indirect examples of "artificial" transamination reactions. Serine hydroxymethyltransferase in the presence of its competitive inhibitor, D-alanine, promotes a transamination to afford pyridoxamine phosphate [48]. The steric course of the transamination process has once again been shown to involve the addition of a medium proton to the *Si* face at C-4′ [109]. Assuming that the group on the enzyme that is involved in the removal of H_{Si} from glycine in the physiological reaction, is the one which catalyses in a *syn* fashion the 'artificial transamination' with D-alanine, then the geometric arrangement of the carboxylate group at C_α will have to be opposite to that attained with L-amino acids (compare 1 with 2: Fig. 56). Another example of *Si* face preference is provided by L-glutamate decarboxylase which in the presence of the substrate analogue, α-methyl-D,L-glutamate, catalyses the conversion of pyridoxal phosphate to pyridoxamine phosphate once again in an 'artificial transamination' process. When the reaction was performed in $[^3H]H_2O$, the $[4'\text{-}^3H]$pyridoxamine phosphate produced was shown to have the (*S*) absolute configuration [110] thereby indicating that during the reaction tritium is added to the *Si* face of the Schiff base at C-4′. Assuming that the only significant effect of the methyl group at C_α in the ternary complex is to cause the protonation at C-4′ instead of C_α, during what is otherwise the normal catalytic cycle of the enzyme, leads to the conclusion that the group involved in the C_α protonation in the physiological substrate must also be located towards the *Si* face. This feature, when taken in conjunction with the knowledge that amino acid decarboxylases catalyse the physiological reactions with the retention of stereochemistry suggests that in the transition state for the decarboxylation process, C_α–COOH bond must also be oriented toward the *Si* face at C-4′ in the ternary complex (Fig. 57).

In all the examples studied to date, during half-transaminations, physiological or 'artificial', the medium derived proton has been shown experimentally to enter the *Si* face at C-4′ of the coenzyme. This surprising stereochemical consistency in diverse types of pyridoxal-P-dependent enzymes has been interpreted by Dunathan and Voet [111] as evidence for the evolution of this family of enzymes from a common ancestor whose distribution of binding groups at the active site has remained essentially unaltered throughout the course of evolution.

The argument that the bond to C_α that is formed or cleaved is also located orthogonally on the *Si* face at C-4′ is however based on extrapolation. In those cases where the substrate–coenzyme Schiff base linkage was reduced with $NaBH_4$, the hydride attack was also found to occur on the *Si* face at C-4′. These facts have been reconciled by assuming that the catalytic group(s) participating in reactions at C_α are located on the solvent side of the ternary complexes from which $NaBH_4$ also attacks. However, it should be borne in mind that all the 3 parameters, (i) stereochemistry of $NaBH_4$ attack, (ii) stereochemistry of protonation at C-4′ and (iii) use of the *Si* face in the manipulation of the C_α bond of the substrate, have not yet been determined for a single enzyme (for example, see Table 5).

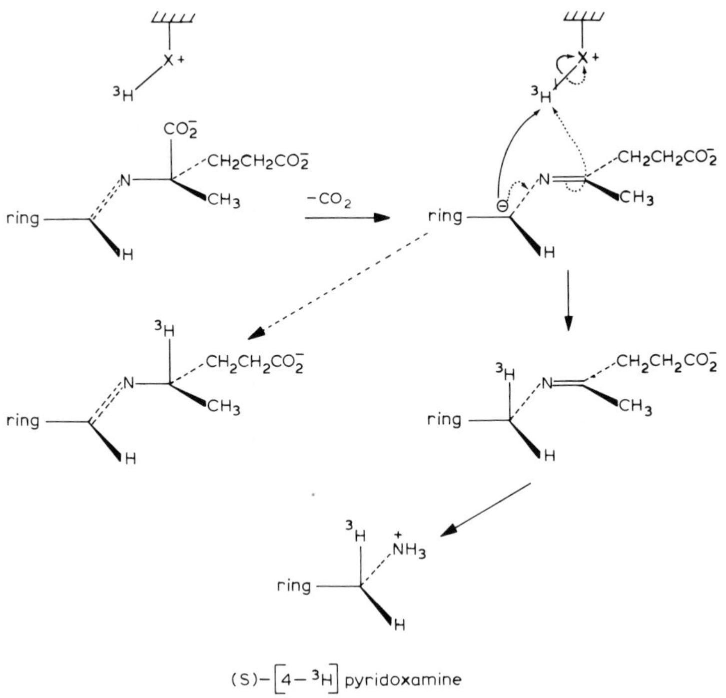

(S)-[4-^{3}H] pyridoxamine

Fig. 57. The formation of (S)-[4′-^{3}H]pyridoxamine in the reaction catalysed by glutamate decarboxylase with α-methyl-D,L-glutamate in [^{3}H]H_2O. Dotted arrows show the course, had the reaction with the analogue followed the pathway expected with L-glutamate. The heavy arrows show the observed transamination reaction.

6.6. Orientation of the pyridinium ring of the coenzyme in the binary and ternary complexes

Throughout the discussion on the stereochemistry of the coenzyme face at C-4′ we have omitted the full structure of the pyridinium ring. This is because it is not known for certain which of the two arrangements of substituents as shown in structures 1 and 2 (Fig. 58) are involved in the organisation of the binary and ternary complexes. Ivanov and Karpeisky [106a] had previously hypothesised the orientation of groups in a binary (coenzyme–enzyme) complex similar to that shown in structure 1 because this permits the interaction of the phenolic oxygen with the protonated Schiff base. Indeed the X-ray data on the coenzyme–enzyme complex of aspartate aminotransferase [33] have been interpreted to support this view. Whether this structural feature represents a general phenomenon must await the availability of additional examples. The X-ray data on aspartate aminotransferase [33] have also supported the view originally [101] highlighted by the $NaBH_4$ reduction experiments that the *Re* face at C-4′ of the Schiff base is exposed to the solvent side in the

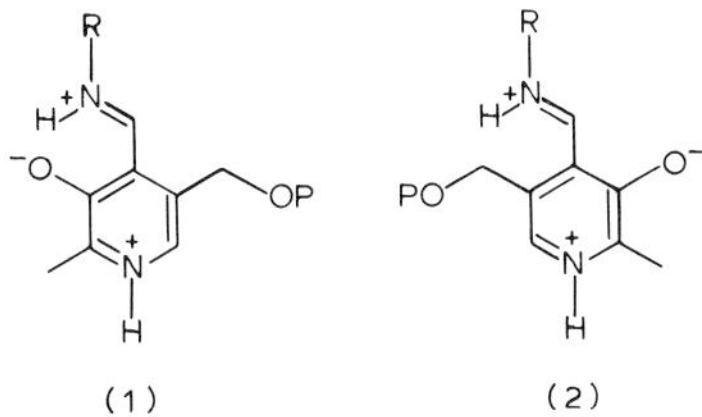

Fig. 58. Two possible orientations of the C-4′ Schiff base with respect to various substituents in the pyridine ring. Note that only in strcture 1 is the protonated Schiff base able to hydrogen bond with the 3-hydroxy anion.

binary complex (the face obtained by rotating structure 1 (Fig. 58) along N-1–C-4′ axis by 180°) Corresponding X-ray diffraction studies on a ternary complex have not yet been reported, though a hypothetical model for the interaction of the substrate with the coenzyme–enzyme complex based on difference electron density maps of various enzyme derivatives has been proposed. Influenced by the tacit assumptions made in the model and the involvement of the *Si* face at C-4′ in bonding processes revealed by the chemical studies (see above), we have conformed and used, as far as possible, the arrangement of structure 1 for ternary complexes in mechanistic illustrations in Sections 1–5.

References

1 György, P. (1934) Nature (London) 133, 498.
2 Snell, E.E. and Haskell, B.E. (1970) in: M. Florkin and E.H. Stotz (Eds.), Comprehensive Biochemistry, Vol. 21, Elsevier, Amsterdam, pp. 47–71.
3 Meister, A. (1965) Biochemistry of the Amino Acids, 2nd Edn., Vol. 1, Academic Press, New York.
4a Adams, E. (1972) in: P.D. Boyer (Ed.), The Enzymes, Vol. VI, Academic Press, New York, p. 479.
b Adams, E. (1976) Adv. Enzymol. 44, 69.
5 Vederas, J.C. and Floss, H.G. (1980) Accts. Chem. Res. 13, 455.
6 Retey, J. and Robinson, J.A. (1982) Stereospecificity in Organic Chemistry and Enzymology, Verlag Chemie, Weinheim.
7 Braunstein, A.E. and Shemyakin, M.M. (1953) Biokhimiya 18, 393.
8 Metzler, D.E., Ikawa, M. and Snell, E.E. (1954) J. Am. Chem. Soc. 76, 648.
9 Boeker, E.A. and Snell, E.E. (1972) in: P.D. Boyer (Ed.), The Enzymes, Vol. VI, Academic Press, New York, p. 217.
10a Mandeles, S., Koppelman, R. and Hanke, M.E. (1954) J. Biol. Chem. 209, 327.
b Westheimer, F.H. (1959) in: P.D. Boyer, H. Lardy and K. Myrback (Eds.), The Enzymes, Vol. I, Academic Press, New York, p. 259 (whole review).
11 Gale, E.F. (1946) Adv. Enzymol. 6, 1.
12a Cahn, R.S., Ingold, C. and Prelog, V. (1966) Angew. Chem. Int. Ed. 5, pp. 385.
b Prelog, V. and Helmchen, G. (1972) Helv. Chim. Acta 55, 2581.
13 Battersby, A.R., Chrystal, E.J.T. and Staunton, J. (1980) J. Chem. Soc. Perkin I, 31.
14 Belleau, B. and Burba, J. (1960) J. Am. Chem. Soc. 82, 5751.
15 Belleau, B., Fang, M., Burba, J. and Moran, J. (1960) J. Am. Chem. Soc. 82, 5752.
16 Spenser, I.D. and Leistner, E. (1975) J. Chem. Soc. Chem. Commun. 378.

17 Battersby, A.R., Murphy, R. and Staunton, J. (1982) J. Chem. Soc. Perkin I, 449.
18 Gerdes, H.J. and Leistner, E. (1979) Phytochemistry 18, 771.
19 Yamada, H. and O'Leary, M.H. (1978) Biochemistry 17, 669.
20 Santaniello, E., Kienle, M.G., Manzocchi, A. and Bosisio, E. (1979) J. Chem. Soc. Perkin I, 1677.
21a Battersby, A.R., Staunton, J. and Tippett, J. (1982) J. Chem. Soc. Perkin I, 455.
b Bouclier, M., Jung, M.J. and Lippert, B. (1979) Eur. J. Biochem. 98, 363.
22 Battersby, A.R., Joyeau, R. and Staunton, J. (1979) FEBS Lett. 107, 231.
23a Asada, Y., Tanizawa, K., Sawada, S., Suzuki, T., Misoni, H. and Soda, K. (1981) Biochemistry 20, 6881.
b Chang, C.C., Laghai, A., O'Leary, M.H. and Floss, H.G. (1982) J. Biol. Chem. 257, 3564.
24 Palekar, A.G., Tate, S.S. and Meister, A. (1970) Biochemistry 9, 2310.
25 Palekar, A.G., Tate, S.S. and Meister, A. (1971) Biochemistry 10, 2180.
26 Recsei, P.A. and Snell, E.E. (1970) Biochemistry 9, 1492.
27a Battersby, A.R., Nicoletti, M., Staunton, J. and Vleggar, R. (1980) J. Chem. Soc. Perkin I, 43.
b Santaniello, E., Manzocchi, A. and Biondi, P.A. (1981) J. Chem. Soc. Perkin I, 307.
28 Allen, R.R. and Klinman, J.P. (1981) J. Biol. Chem. 256, 3233.
29 Klein, S.M. and Sagers, R.D. (1967) J. Biol. Chem. 242, 301, and references cited therein.
30 Guirard, B.M. and Snell, E.E. (1964) in: M. Florkin and E.H. Stotz (Eds.), Comprehensive Biochemistry, Vol. 15, Elsevier, Amsterdam, p. 138.
31 Braunstein, A.E. (1973) in: P.D. Boyer (Ed.), The Enzymes, Vol. IXB, Academic Press, New York, p. 379.
32a Ovchinnikov, Yu.A., Egorov, C.A., Aldanova, N.A., Feigina, M.Yu., Lipkin, V.M., Abdulaev, N.G., Grishin, E.V., Kiselev, A.P., Modyanov, N.N., Braunstein, A.E., Polyanovsky, O.L. and Nosikov, V.V. (1973) FEBS Lett. 29, 31.
b Kagamiyama, H., Sakakibara, R., Wada, H., Tanase, S. and Morino, Y. (1977) J. Biochem. (Tokyo) 82, 291.
c Barra, D., Bossa, F., Doonan, S., Fahmy, H.M.A., Hughes, G.J., Kakoz, K.Y., Martini, F. and Petruzzelli, R. (1977) FEBS Lett. 83, 241.
d Shlyapnikov, S.V., Myasnikov, A.N., Severin, E.S., Myagkova, M.A., Torchinsky, Yu.M. and Braunstein, A.E. (1979) FEBS Lett. 106, 385.
e Hausner, U., Christen, P., Wilson, K.J., Eichele, G., Ford, G.C. and Jansonius, J.N. (1979) Experientia 35, 935.
33 Ford, G.C., Eichele, G. and Jansonius, J.N. (1980) Proc. Natl. Acad. Sci. (U.S.A.) 77, 2559.
34 Fasella, P. (1967) Annu. Rev. Biochem. 36, 185–210.
35 Tanenbaum, S.W. (1956) J. Biol. Chem. 218, 733.
36 Hilton, M.A., Barnes Jr., F.W. and Enns, T. (1956) J. Biol. Chem. 219, 833.
37 Meister, A. (1951) Nature (London) 168, 1119.
38 Sprinson, D.B. and Rittenberg, D. (1950) J. Biol. Chem. 184, 405.
39 Grisolia, S. and Burris, R.H. (1954) J. Biol. Chem. 210, 109.
40 Tamiya, N. and Oshima, T. (1962) J. Biochem. (Jpn.) 51, 78.
41 Babu, U.M. and Johnston, R.B. (1974) Biochem. Biophys. Res. Commun. 58, 460.
42 Soda, K., Misono, H. and Yamamoto, T. (1968) Biochemistry 7, 4102.
43 Strecker, H.J. (1965) J. Biol. Chem. 240, 1225.
44 Scott, E.M. and Jakoby, W.B. (1959) J. Biol. Chem. 234, 932.
45a Burnett, G., Walsh, C., Yonaha, K., Toyama, S. and Soda, K. (1979) J. Chem. Soc. Chem. Commun. 826.
b Tanizawa, K., Yoshimura, T., Asada, Y., Sawada, S., Misono, H. and Soda, K. (1982) Biochemistry 21, 1104.
46 Bailey, G.B., Chottamangsa, O. and Vutivej, K. (1970) Biochemistry 9, 3243.
47 Blakley, R.L. (1969) in: A. Neuberger and E.D. Tatum (Eds.), Frontiers in Biology, Vol. 13, North-Holland, Amsterdam, p. 190.
48 Schirch, L. (1982) Adv. Enzymol. 53, 83.

49 Schirch, L. and Gross, T. (1968) J. Biol. Chem. 243, 5651.
50 Akhtar, M., El-Obeid, H.A. and Jordan, P.M. (1975) Biochem. J. 145, 159.
51 Schirch, L. and Jenkins, W.T. (1964) J. Biol. Chem. 239, 3801.
52 Jordan, P.M. and Akhtar, M. (1970) Biochem. J. 116, 277.
53 Chen, M.S. and Schirch, L. (1973) J. Biol. Chem. 248, 7979.
54 Jordan, P.M., El-Obeid, H.A., Corina, D.L. and Akhtar, M. (1976) J. Chem. Soc. Chem. Commun. 73.
55 Weber, H. (1965) Ph.D. Dissertation No. 3591, Eidgenössische Technische Hochschule, Zürich.
56a Palekar, A.G., Tate, S.S. and Meister, A. (1973) J. Biol. Chem. 248, 1158.
b Ching, W.M. and Kallen, R.G. (1979) Biochemistry 18, 821.
57 Akhtar, M. and Jordan, P.M. (1968) J. Chem. Soc. Chem. Commun. 1691.
58 Akhtar, M. and Jordan, P.M. (1969) Tetrahedron Lett. 875.
59 Biellmann, J.F. and Schuber, F. (1967) Biochem. Biophys. Res. Commun. 27, 517.
60 Tatum, C.M., Benkovic, P.A., Benkovic, S.J., Potts, R., Schleicher, E. and Floss, H.G. (1977) Biochemistry 16, 1093.
61 Akhtar, M. and Jordan, P.M. (1979) in: D.H.R. Barton and W.D. Ollis (Eds.), Comprehensive Organic Chemistry, Vol. 5, Pergamon, Oxford, p. 1121.
62 Zaman, Z., Jordan, P.M. and Akhtar, M. (1973) Biochem. J. 135, 257.
63 Abboud, M.M., Jordan, P.M. and Akhtar, M. (1974) J. Chem. Soc. Chem. Commun. 643.
64 Akhtar, M., Abboud, M.M., Barnard, G., Jordan, P.M. and Zaman, Z. (1976) Phil. Trans. R. Soc. Lond. B 273, 117.
65a Weiss, B. (1963) J. Biol. Chem. 238, 1953.
b Stoffel, W. (1971) Annu. Rev. Biochem. 40, 57–82.
66 Abboud, M.M. and Akhtar, M. (1976) J. Chem. Soc. Chem. Commun. 1007.
67 Yanofsky, C.D. and Crawford, I.P. (1972) in: P.D. Boyer (Ed.), The Enzymes, Vol. 7, Academic Press, New York, pp. 1–31.
68 Floss, H.G., Schleicher, E. and Potts, R. (1976) J. Biol. Chem. 251, 5478.
69 Tsai, M.D., Schleicher, E., Potts, R., Skye, G.E. and Floss, H.G. (1978) J. Biol. Chem. 253, 5344.
70 Fuganti, C., Ghiringhelli, D., Giangrasso, D., Graselli, P. and Amisano, A.S. (1974) Chim. Industr. 56, 424.
71 Cornforth, J.W., Redmond, J.W., Eggerer, H., Buckel, W. and Gutschow, C. (1969) Nature (London) 221, 1212.
72 Lüthy, J., Retey, J. and Arigoni, D. (1969) Nature (London) 221, 1213.
73 Skye, G.E., Potts, R. and Floss, H.G. (1974) J. Am. Chem. Soc. 96, 1593.
74 Schleicher, E., Mascaro, K., Potts, R., Mann, D.R. and Floss, H.G. (1976) J. Am. Chem. Soc. 98, 1043.
75 Cheung, Y. and Walsh, C. (1976) J. Am. Chem. Soc. 98, 3397.
76 Yang, I.Y., Huang, Y.Z. and Snell, E.E. (1975) Fed. Proc. 34, 496.
77 Also see: Walsh, C., Pascal, R.A., Johnston, M., Raines, R., Dikshit, D., Krantz, A. and Honma, M. (1981) Biochemistry 20, 7509.
78 Crout, D.H.G., Gregorio, M.V.M., Müller, U.S., Komatsubara, S., Kisumi, M. and Chibata, I. (1980) Eur. J. Biochem. 106, 97.
79 Kaplan, M. and Flavin, M. (1966) J. Biol. Chem. 241, 4463.
80 Datko, A., Giovanelli, J. and Mudd, S.H. (1974) J. Biol. Chem. 249, 1139.
81 Flavin, M. and Slaughter, C. (1970) Methods Enzymol. 17B, 433–439.
82 Davis, L. and Metzler, D. (1972) in: P. Boyer (Ed.), The Enzymes, 3rd Edn., Vol. 7, Academic Press, New York, p. 33.
83 Flavin, M. (1968) Methods Enzymol. 5, 951–959.
84 Krongelb, M., Smith, T.A. and Abeles, R.H. (1968) Biochim. Biophys. Acta 167, 473.
85 Posner, B.I. and Flavin, M. (1972) J. Biol. Chem. 247, 6412.
86 Fuganti, C. and Coggiola, D. (1977) Experientia 33, 847.
87 Fuganti, C. (1979) J. Chem. Soc. Chem. Commun. 337.

88 Johnston, M., Marcotte, P., Donovan, J. and Walsh, C. (1979) Biochemistry 18, 1729.
89 Chang, M.N.T. and Walsh, C. (1981) J. Am. Chem. Soc. 103, 4921.
90 Fasella, P., Carotti, D., Giartosio, A., Riva, F. and Turano, C. (1978) in: N. Seiler, M.J. Jung and J. Koch-Weser (Eds.), Enzyme-Activated Irreversible Inhibitors, Elsevier, Amsterdam, pp. 87–108.
91 Metzler, D.E., Harris, C.M., Johnson, R.J., Siano, D.B. and Thompson, J.A. (1973) Biochemistry 12, 5377.
92 Morozov, Yu.V., Bazhulina, N.P. and Karpeisky, M.Ya. (1968) in: E.E. Snell, A.E. Braunstein, E.S. Severin and Yu.M. Torchinsky (Eds.), Pyridoxal Catalysis: Enzymes and Model Systems, John Wiley, New York, pp. 53–66.
93 Zito, S.W. and Martinez-Carrion, M. (1980) J. Biol. Chem. 255, 8645.
94 Schirch, L. and Jenkins, W.T. (1964) J. Biol. Chem. 239, 3801.
95 Martell, A.E. (1982) Adv. Enzymol. 53, 163.
96 Snell, E.E. (1975) Adv. Enzymol. 42, 287.
97 Fischer, E.H., Kent, A.B., Snyder, E.R. and Krebs, E.G. (1958) J. Am. Chem. Soc. 80, 2906.
98 Hughes, R.C., Jenkins, W.T. and Fischer, E.H. (1962) Proc. Natl. Acad. Sci. (U.S.A.) 48, 1615.
99 Turano, C.P., Fasella, P., Vecchini, P. and Giastosio, A. (1961) Atti. Acad. Nazl. Lincei. Sci. Fis. Mat. Nat. 30, 532.
100 Matsuo, Y. and Greenberg, D.M. (1959) J. Biol. Chem. 234, 507.
101 Austermühle-Bertola, A. (1973) Ph.D. Dissertation No. 5009, Eidgenössische Technische Hochschule, Zürich.
102 Hanson, K.R. (1966) J. Am. Chem. Soc. 88, 2731.
103 Vederas, J.C., Reingold, I.D. and Sellers, H.W. (1979) J. Biol. Chem. 254, 5053.
104 Vederas, J.C., Schleicher, E., Tsai, M.D. and Floss, H.G. (1978) J. Biol. Chem. 253, 5350.
105 Barton, D.H.R. (1950) Experientia 6, 316.
106a Ivanov, V.I. and Karpeisky, M.Ya (1969) Adv. Enzymol. 32, 21.
b Tumanyan, V.G., Mamaeva, O.K., Bocharov, A.L., Ivanov, V.I., Karpeisky, M.Ya. and Yakovlev, G.I. (1974) Eur. J. Biochem. 50, 119.
107 Dunathan, H.C. (1971) Adv. Enzymol. 35, 79–134.
108 Besmer, P. and Arigoni, D. (1968) Chimia (Switz.) 22, 494.
109 Voet, J.G., Hindelang, D.M., Blanck, T.J.J., Ulevitch, R.J., Kallen, R.G. and Dunathan, H.C. (1973) J. Biol. Chem. 248, 841.
110 Sukharera, B.S., Dunathan, H.C. and Braunstein, A.E. (1971) FEBS Lett. 15, 241.
111 Dunathan, H.C. and Voet, J.G. (1974) Proc. Natl. Acad. Sci. (U.S.A.) 71, 3888.
112 Billen, D. and Lichstein, H.C. (1949) J. Bacteriol. 58, 215.
113 Meister, A., Sober, H.A. and Tice, S.V. (1951) J. Biol. Chem. 189, 577–591.
114 Novogrodsky, A., Nishimura, J.S. and Meister, A. (1963) J. Biol. Chem. 238, 1903.
115 Wilson, E.M. and Kornberg, H.L. (1963) Biochem. J. 88, 578.
116 Gale, E.F. (1940) Biochem. J. 34, 392.
117 Melnykovych, G. and Snell, E.E. (1958) J. Bacteriol. 76, 518.
118 Boeker, E.A., Fischer, E.H. and Snell, E.E. (1969) J. Biol. Chem. 244, 5239.
119 McGullough, W.G., Piligian, J.G. and Daniel, I.J. (1957) J. Am. Chem. Soc. 79, 628.
120 Bailey, G.B. and Dempsey, W.B. (1967) Biochemistry 6, 1526.
121 Dewey, D.L., Hoare, D.S. and Work, E. (1954) Biochem. J. 58, 523.
122 Meadow, P. and Work, E. (1958) Biochim. Biophys. Acta 29, 180.
123 Homola, H.D. and Dekker, E.E. (1967) Biochemistry 6, 2626.
124 Shukuya, R. and Schwert, G.W. (1959) J. Biol. Chem. 235, 1649.
125 Umbreit, W.W. and Gunsalus, I.C. (1945) J. Biol. Chem. 159, 333.
126 Rodwell, A.W. (1953) J. Gen. Microbiol. 8, 224, 233.
127 Epps, H.M.R. (1945) Biochem. J. 39, 42.
128 Ekladius, L., King, H.K. and Sutton, C.R. (1957) J. Gen. Microbiol. 17, 602.
129 Sutton, C.R. and King, H.K. (1962) Arch. Biochem. Biophys. 96, 360.
130 Gale, E.F. and Epps, H.M.R. (1944) Biochem. J. 38, 232.

131 Maretzki, A. and Mallette, M.F. (1962) J. Bacteriol. 83, 720.
132 Taylor, E.S. and Gale, E.F. (1945) Biochem. J. 39, 52.
133 McGilvery, R.W. and Cohen, P.P. (1948) J. Biol. Chem. 174, 813.
134 Mitoma, C. and Udenfriend, S. (1960) Biochim. Biophys. Acta 37, 356.
135a Gale, E.F. (1940) Biochem. J. 34, 846.
b Epps, H.M.R. (1944) Biochem. J. 38, 242–249.
136 Matthew, M. and Neuberger, A. (1963) in: E.E. Snell, P.M. Fasella, A.E. Braunstein and A. Rossi-Fanelli (Eds.), Chemical and Biological Aspects of Pyridoxal Catalysis, Pergamon, Oxford, p. 243.
137 Shimura, K., Nagayama, H. and Kikuchi, A. (1956) Nature (London) 177, 935.
138 Lovenberg, W., Weissbach, H. and Udenfriend, S. (1962) J. Biol. Chem. 237, 89.
139 Buxton, J. and Sinclair, H.M. (1956) Biochem. J. 62, 27P.
140 Werles, E. and Aures, D. (1960) Z. Physiol. Chem. 316, 124.
141 Smith, S.E. (1960) Br. J. Pharmacol. 15, 319.
142 Davison,-A.N. (1956) Biochim. Biophys. Acta 19, 66.
143 Hope, D.B. (1955) Biochem. J. 59, 497.
144 Awapara, J. and Wingo, W.J. (1953) J. Biol. Chem. 203, 189.
145 Werle, E. and Raub, A. (1948) Biochem. Z. 318, 538.
146 Roberts, E. and Frankel, S. (1950) J. Biol. Chem. 187, 55.
147 Roberts, E. and Simonsen, D.G. (1963) Biochem. Pharmacol. 12, 113.
148 Beevers, H. (1951) Biochem. J. 48, 132.
149 Anderson, J.A., Cheldelin, V.H. and King, T.E. (1961) J. Bacteriol. 82, 354.
150 Fowden, L. and Done, J. (1953) Biochem. J. 55, 548.
151 Werle, E. and Koch, W. (1949) Biochem. Z. 319, 305.
152 Weissbach, H., Lovenberg, W. and Udenfriend, S. (1961) Biochim. Biophys. Acta 50, 177.
153 Haddox, M.K. and Russell, D.H. (1981) Biochemistry 20, 6721.
154 Feldman, M.J., Levy, C.C. and Russell, D.H. (1972) Biochemistry 11, 671.
155 Sakai, T., Hori, C., Kano, K. and Oka, T. (1979) Biochemistry 18, 5541.
156 Vogel, H.J. and Jones, E.E. (1971) Methods Enzymol. 17A, 260.
157 Jenkins, W.T. and Saier, M.H. (1970) Methods Enzymol. 17A, 159.
158 Segal, H.L. and Matsuzawa, T. (1970) Methods Enzymol. 17A, 153.
159 Gatehouse, P.W., Hopper, S., Schatz, L. and Segal, H.L. (1967) J. Biol. Chem. 242, 2319.
160 Richardson, K.E. and Thompson, J.S. (1970) Methods Enzymol. 17A, 163.
161 Martinez-Carrion, M. and Jenkins, W.T. (1970) Methods Enzymol. 17A, 167.
162 Hayaishi, O., Nishizuka, Y., Tatibana, M., Takeshita, M. and Kuno, S. (1961) J. Biol. Chem. 236, 781.
163 Tobes, M.C. and Mason, M. (1977) J. Biol. Chem. 252, 4591.
164 Sytinsky, I.A. and Vasilijev, V.Y. (1970) Enzymologia 39, 1.
165 Scott, E.M. and Jakoby, W.B. (1959) J. Biol. Chem. 234, 932.
166 Turner, J.M. and Neuberger, A. (1970) Methods Enzymol. 17A, 188.
167 Jenkins, W.T. (1961) J. Biol. Chem. 236, 478, 1121.
168 Green, D.E., Leloir, L.F. and Nocito, V. (1945) J. Biol. Chem. 161, 559.
169 Jenkins, W.T. and Sizer, I.W. (1957) J. Am. Chem. Soc. 79, 2655.
170 Fasella, P., Lis, H., Turano, C. and Vecchini, P. (1960) Biochim. Biophys. Acta 45, 529.
171 Meister, A., Sober, H.A. and Peterson, E.A. (1954) J. Biol. Chem. 206, 89.
172 Stoner, G.L. and Eisenberg, M.A. (1975) J. Biol. Chem. 250, 4029.
173 Kornberg, H.L. and Morris, J.G. (1965) Biochem. J. 95, 577.
174 Gibbs, R.G. and Morris, J.G. (1966) Biochem. J. 99, 27.
175 Gibbs, R.G. and Morris, J.G. (1970) Methods Enzymol. 17A, 981.
176 Thompson, J.S. and Richardson, K.E. (1966) Arch. Biochem. Biophys. 117, 599.
177 Braunstein, A.E. and Ting-Seng, H. (1960) Biochim. Biophys. Acta 44, 187.
178 Cooper, A.J.L. and Meister, A. (1972) Biochemistry 11, 661.

179 Braunstein, A.E. and Hsü, T.S. (1960) Biokhimiya 25, 758.
180 Kupchik, H.Z. and Knox, W.E. (1970) Methods Enzymol. 17A, 951.
181 Martin, R.G. (1970) Arch. Biochem. Biophys. 138, 239.
182 Albritton, W.L. and Levin, A.P. (1969) Biochem. J. 114, 662.
183 Martin, R.G. and Goldberger, R.F. (1967) J. Biol. Chem. 242, 1168.
184 Taylor, R.T. and Jenkins, W.T. (1966) J. Biol. Chem. 241, 4396.
185 Soda, K. and Misono, H. (1971) Methods Enzymol. 17B, 222.
186 Jenkins, W.T. and Tsai, H. (1970) Methods Enzymol. 17A, 281.
187 Fujioka, M., Morino, Y. and Wada, H. (1970) Methods Enzymol. 17A, 585.
188 Ayling, J.E. and Snell, E.E. (1968) Biochemistry 7, 1616 and 1626.
189 Kolb, H., Cole, R.D. and Snell, E.E. (1968) Biochemistry 7, 2946.
190 Snell, E.E. (1970) Vitam. Horm. 28, 265.
191 Kim, K.H. (1964) J. Biol. Chem. 239, 783.
192 Kim, K.H. and Tchen, T.T. (1971) Methods Enzymol. 17B, 812.
193 Michaels, R. and Kim, K.H. (1966) Biochim. Biophys. Acta 115, 59.
194 Valeriote, F.A., Auricchio, F., Riley, D. and Tomkins, G.M. (1969) J. Biol. Chem. 244, 3618.
195 Granner, D.K. and Tomkins, G.M. (1970) Methods Enzymol. 17A, 633.
196 Tsai, M.D., Weaver, J., Floss, H.G., Conn, E.E., Creveling, R.K. and Mazelis, M. (1978) Arch. Biochem. Biophys. 190, 553.
197 Fuganti, C., Ghiringhelli, D., Giangrassio, D. and Grasselli, P. (1974) J. Chem. Soc. Chem. Commun. 726.
198 Sawada, S., Kumagai, H., Yamada, H. and Hill, R.K. (1975) J. Am. Chem. Soc. 97, 4334.
199 Borcsok, E. and Abeles, R.H. (1982) Arch. Biochem. Biophys. 213, 695.
200 Kumagai, H., Yamada, H., Sawada, S., Schleicher, E., Mascaro, K. and Floss, H.G. (1977) J. Chem. Soc. Chem. Commun. 85.
201 Mazelis, M. and Creveling, R.K. (1975) Biochem. J. 147, 485–491.
202 Dunathan, H.C., Davis, L., Kury, P.G. and Kaplan, M. (1968) Biochemistry 7, 4532.
203 Bailey, G.B., Kusamrarn, T. and Vultivej, K. (1970) Fed. Proc. 29, 857.
204 Ayling, J.E., Dunathan, H.C. and Snell, E.E. (1968) Biochemistry 7, 4537.
205 Miles, E.W., Houck, D.R. and Floss, H.G. (1982) J. Biol. Chem. 257, 14203.

CHAPTER 10

Transformations involving folate and biopterin cofactors

S.J. BENKOVIC and R.A. LAZARUS

Department of Chemistry, Pennsylvania State University, University Park, PA 16802, U.S.A.

1. Introduction

This short review examines the reactions of enzymes that employ tetrahydrobiopterin or derivatives of tetrahydropteroyl poly-γ-glutamates for promoting transformations involving hydroxylation and 1-carbon unit transfer. This paper is not meant to be comprehensive but to outline the main types of reactions in which these cofactors participate against a background of their biological importance. A major objective is to foster an appreciation of their chemical diversity and its relation to structural elements within the cofactors. The mechanism of action of 3 enzymes: serine hydroxymethylase, thymidylate synthetase, and phenylalanine hydroxylase is discussed in greater depth reflecting in part the degree of attention they have received. For other more comprehensive accounts the reader is referred to the referenced articles [1–4].

2. Structure

Folic acid cofactors (1) are derived from tetrahydropteroic acid linked to the α-amino function of poly-γ-glutamyl peptides of varying chain length ($n = 1–8$). The absolute configuration at C-6 for tetrahydrofolate (H_4-folate) which should also include the pteroyl polyglutamates, has been defined as (S) by X-ray studies on 5,10-methenyltetrahydrofolate [5].

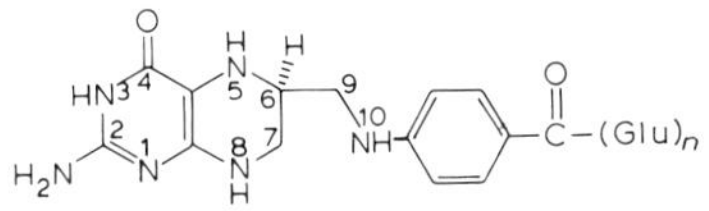

(1)

Stereospecific replacement of the p-aminobenzoyl$(Glu)_n$ by the L-erythro-1,2-dihydroxypropyl side chain furnishes tetrahydrobiopterin (BPH_4, 2) [6].

Michael I. Page (Ed.), The Chemistry of Enzyme Action

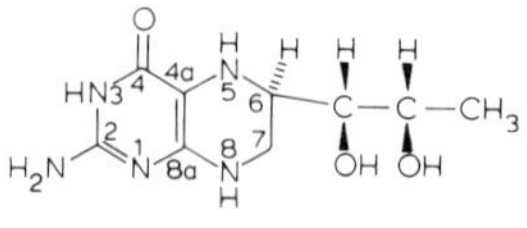

(2)

It has now been established that pteroylpoly(γ-glutamates) are the major, and sometimes only, intracellular forms of folate [7]. Within a particular source, a single length normally predominates but a distribution of lengths usually is found. In rat liver the major and minor conjugates are penta- and tetra-glutamates respectively, whereas in a Corynebacterium species they are tetra- and tri-glutamates [8,9]. Many in vitro studies have shown that the pteroylpolyglutamates often are more efficient substrates than the corresponding monoglutamates for folate-requiring enzymes including methionine synthetase, thymidylate synthetase, 5′-phosphoribosyl-5-amino-4-imidazole-carboxamide (AICAR) transformylase, formyl tetrahydrofolate synthetase, serine hydroxymethylase and methylenetetrahydrofolate reductase [10–16]. Their increased efficiency relative to the monoglutamate derivative stems mainly from a decrease in their dissociation from the cofactor enzyme binary complex or a decrease in the dissociation of the other substrate from the ternary complex owing to synergistic binding [16]. An additional advantage of these conjugates is that they do not readily cross cell membranes and are retained preferentially, whereas the monoglutamate derivatives appear to be the transport forms [17]. The synthesis of the pteroyl polyglutamate is carried out by ATP synthetases, that after isolation and partial purification can faithfully reproduce synthesis as it occurs in vivo [18].

The conformations in solution of H_4-folate, 5,10-methylene-H_4-folate, and the BPH_4 analogues, 6-methyl- and 6,7-dimethyl-5,6,7,8-tetrahydropterin ($6MPH_4$ and $DMPH_4$) have been examined by proton NMR. The H_4-folate exists as a roughly equal mixture of two half-chair conformations, with the C-6 proton axial and the other with the same proton equatorial [19]. The tetrahydropyrazine ring for both $6MPH_4$ and $DMPH_4$, likewise, is in a half-chair conformation with the 6-methyl being in an equatorial position for $6MPH_4$ but in rapid exchange between the axial and equatorial positions for $DMPH_4$ [20,21]. 5,10-Methylene-H_4-folate, which mediates 1-carbon unit transfer at the oxidation level of formaldehyde, exhibits a solution conformation with the tetrahydropyrazine ring also adopting a half-chair conformation and with the aromatic side chain most likely extended away from pteridine moiety [22].

Crystallographic structures derived from X-ray analysis of various pterins generally have confirmed these assignments of conformation and the expected planarity of the pyrimidine ring, and have verified the double bond character of the bond between N-8 and C-8a owing to its participation in a vinylogous amide grouping [23]. This resonance is responsible for the negligible basicity of N-8 in an aqueous medium relative to N-5. The dissociation constants for the 2-amino-4-hydroxypterin are only slightly altered by the substitution at N-10 or C-6 required for the cofactor structures so that their pK_a values are approximately 1.0–1.4 (N-1), 4.8–5.5 (N-5) and 9.5–10.5 (N-3) [24]. The pK_a of N-10 in H_4-folate is ca. −1.3 owing to electrostatic effects arising from prior protonation at N-5.

3. Reduction

Dihydrofolate reductase catalyzes the NADPH-dependent reduction of dihydrofolate to H_4-folate. The reductase appears to be the major intracellular receptor for the action of 4-amino analogues of folic acid including methotrexate and trimethoprim. Inhibition of the enzyme by methotrexate depletes the H_4-folate pool resulting in a decreased synthesis of purines, pyrimidines and amino acids that are dependent on H_4-folate prompting the wide and effective use of methotrexate in chemotherapy [25]. In addition a differential sensitivity to certain drugs, such as trimethoprim, by mammalian and bacterial reductases has led to the development of a class of compounds with potent antibacterial activity [26]. Since differential inhibitory effects must reside in subtle changes in the active site topography, the elucidation of the primary and tertiary structure of the reductase from both bacterial and animal sources in the presence and absence of substrates, cofactors and inhibitors has been a major focus of research activity. Amino acid sequences of the enzymes from *L. casei, Streptococcus faecium, E. coli* (RT500), *E. coli* (MB1428), mouse L1210 and chicken liver have been reported [27–32]. Particularly significant is the resolution of the X-ray structure of the *L. casei* reductase–NADPH–methotrexate ternary complex at 2.5 Å [33] and the recent discovery that the reductase and thymidylate synthetase activities exist on a single polypeptide in the protozoa, *Crithidia fasciculata* [34].

The stereochemistry of the reduction of dihydrofolate to H_4-folate by the enzyme requires transfer of hydrogen at the re face of dihydrofolate, 3, to generate the correct stereochemistry at C-6. This transfer involves the 4-pro-R hydrogen of NADPH for the reductases from chicken liver, L1210 cells, *S. faecalis* and *E. coli* [35–38]. The reductase from *L. casei* will also reduce folic acid, albeit at a slower

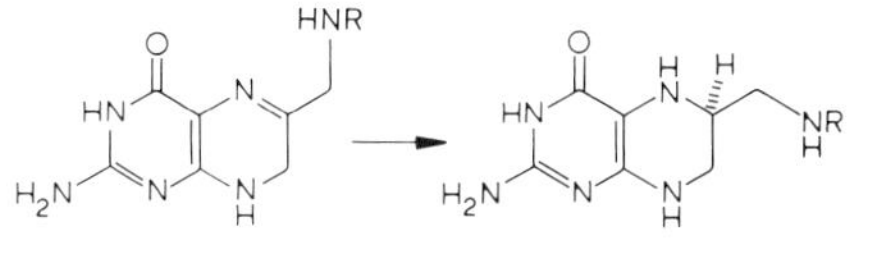

(3) (4)

rate. This reduction also proceeds through the transfer of hydrogen from the pro-R position of NADPH to yield the 7-pro-S hydrogen so that the net reduction at both C-6 and C-7 involves the same side of NADPH and the same face of folic acid [39]. This demonstration coupled to inferences stemming from the X-ray analysis of the ternary methotrexate–NADPH complex suggests that the antifolate methotrexate is bound with its pteridine ring rotated 180° from the position occupied by folate and dihydrofolate.

Oxidation of the tetrahydropterin cofactor during the hydroxylase-catalyzed reactions produces a dihydropterin quinonoid intermediate that may exist as two distinct (*para* and *ortho*) neutral, tautomeric forms (5, 6). The same intermediate can be generated non-enzymatically by various oxidizing agents including in-

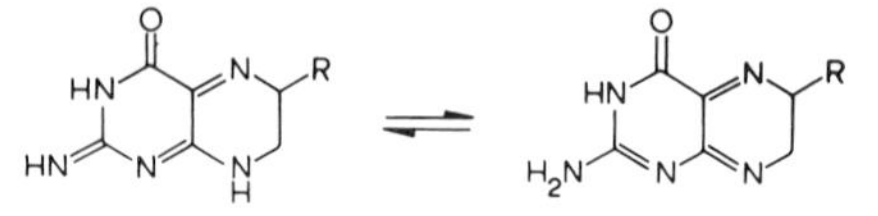

(5) (6)

dophenol, and ferricyanide, as well as electrochemically [40–42]. This 2-electron oxidized material is unstable and rearranges to the 7,8-dihydropterin in a buffer-catalyzed process. Reductases have been isolated from various mammalian sources, which in general demonstrate a marked preference for NADH as a cofactor, that catalyze the conversion of "quinonoid" dihydropteridines to tetrahydropteridines. The dihydrofolate reductase is capable of reducing 7,8-dihydrobiopterin [43] although the isomerization of quinonoid dihydrobiopterin to 7,8-BH_2 may not be significant in vivo [44].

4. *Methylene transfer*

The formation of 5,10-methylene tetrahydrofolate (5,10-CH_2-CH_4-folate) is mediated by serine hydroxymethylase, a pyridoxal-dependent enzyme, and constitutes a principal route into the 1-carbon unit pool at the oxidation level of formaldehyde. The enzyme participates in a surprising diversity of reactions: the α, β cleavage of β-hydroxy-L-amino acids in the absence of tetrahydrofolate; the transamination of D-alanine; the exchange of the pro-2S hydrogen of glycine in the presence of tetrahydrofolate; and the decarboxylation of aminomalonate [45–51]. In the first three reactions the stereochemistry of the C-α bond cleavage is conserved; in contrast the enzyme shows no specificity for the heterotopic carboxyl groups in the decarboxylation of aminomalonate [51]. Unexpectedly, the enzyme is able to catalyze the removal of the α proton from several L-amino acids, e.g. phenylalanine and alanine [46]. If stereoelectronic control is followed in forming the carbanionic intermediates stabilized by pyridoxal phosphate, such control requires the perpendicular orientation of the bond to be broken to the plane of the extended conjugated system [52]. Thus, for serine hydroxymethylase these perplexing results demand the presence of two conformations for the substrate–pyridoxal phosphate enzyme complex (7, 8) that can actively partition to product through their respective carbanionic forms.

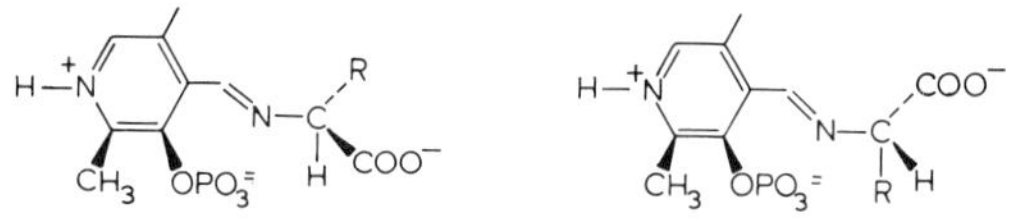

(7) (8)

With the above evidence in mind, a similar rationale may be applied to the results of stereochemical studies on the transfer of the 1-carbon, formaldehyde equivalent, between serine and H_4-folate. Utilizing 3*R*- and 3*S*-[3-^{3}H[serines, the 5,10-[^{3}H]CH_2-H_4-folate formed under one turnover conditions shows a 25% crossover of label from each serine stereoisomer into the 5,10-methylene-H_4-folate containing a methylene unit of opposite configuration [53]. This result is in excellent agreement with an earlier one performed with intact rat-liver slices where [^{3}H]formate was converted to [3-^{3}H]serine, presumably via 5,10-methenyl- and 5,10-methylene-H_4-folate [54]. The observed partial stereospecificity for the transfer process, thus, may arise from two different conformations of the initial serine at the β-carbon and is consistent with the observed cleavage of L-allothrenonine/L-threonine and *erythro-/threo-*β-phenylserine in a 3–7 ratio by the same enzyme in the absence of H_4-folate [46]. Alternatively, crossover may occur through a formaldehyde intermediate generated at the active site by addition of water to a N-5 iminium H_4-folate that is trapped with only partial stereospecificity by the glycyl-pyridoxal phosphate anion.

The evidence for an enzyme-bound formaldehyde equivalent is based mainly on the observations of: (1) the utilization of α-methylserine as a substrate (V is about 10% that for serine) and (2) the cleavage of β-hydroxy-L-amino acids including serine in the absence of H_4-folate. Hence a covalent adduct involving H_4-folate in order to break the C_β–C_α bond [55] is not required. Statement 1 opposes the anticipated addition–elimination of H_2O from the seryl–pyridoxal phosphate complex prior to its trapping by H_4 folate (9) so as to remove after tautomerization the C-3 unit of serine and form 5,10-CH_2-H_4-folate. Statement 2 suggests that the role

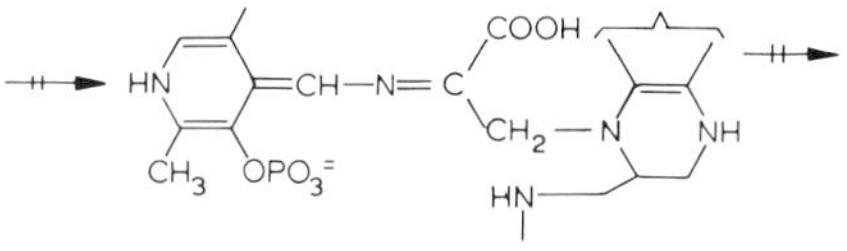

(9)

of H_4-folate in accelerating by ca. 10^3 serine cleavage may be in part synergistic [56], indeed the folate analog 2-amino-6-(4-carbethoxybenzylamino)-4-hydroxyquinazoline stimulates the cleavage of β-phenylserine yet lacks both the N-5 and N-10 nitrogens [57]. The trapping of formaldehyde by tetrahydrofolate likewise occurs with the folate-binding protein of mitochondria which also displays dimethylglycine dehydrogenase activity producing glycine and 5,10-methylene H_4-folate [58].

A key enzyme in pyrimidine biosynthesis, thymidylate synthetase, catalyzes the reductive methylation of 2′-deoxyuridylate (dUMP) to thymidylate (dTMP) with the concomitant conversion of 5,10-methylene-H_4-folate to 7,8-dihydrofolate. The cofactor serves both as a 1-carbon carrier and a reductant. There is substantive evidence based on direct and indirect studies with the inhibitor 5′-fluoro-2′-deoxyuridylate, which will not be reviewed here, that the dUMP is covalently bound to the enzyme, possibly through a thiol group [59]. The intermediate ternary complex is hypothesized to have the following structure (10), where the CH_2 unit is attached through the N-5 of H_4-folate. The ring opening of the 5,10-methylene-H_4-folate is thought to proceed through the electrophilic iminium cation at N-5 by analogy to the products trapped during the decomposition of acyclic and cyclic geminal

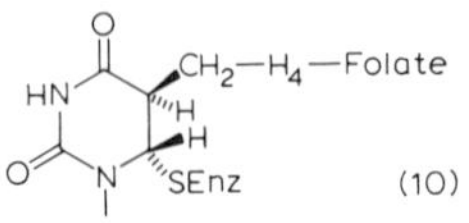

(10)

diamines derived from formaldehyde, and kinetic studies on the general acid-catalyzed ring opening of symmetrical and unsymmetrical imidazolidines [60,61]. Collectively the data were in accord with the preferential opening of the ring under kinetic control to give the most stable carbocation. Thus, the difference between the N-5 and N-10 basicity directly influences the direction of ring opening.

It is known that the hydrogen at C-6 of H_4-folate is transferred to the methyl group of thymine in the enzyme-catalyzed process [62]. Studies of the steric course of this process employing a chiral methylene unit in the cofactor, i.e., 5,10-^{3}H,^{2}HC-H_4-folate generated from (3*R*)-[3-^{3}H,3-^{2}H]- or (3*S*)-[3-^{3}H,3-^{2}H]serine and H_4-folate with serine hydroxymethylase, yield dTMP with a chiral methyl group [63]. Analysis of the methyl group configurations through conversion to their respective acetates indicated that the dTMP derived from (3*S*)-serine yielded a chiral acetate of *S* configuration, whereas dTMP formed from (3*R*)-serine gave acetate of *R* configuration. Thus the reaction catalyzed is highly stereospecific and consistent with the absence of a freely rotating methylene group. However, in order to unequivocally establish the course of the hydrogen transfer, the absolute stereochemistry of 5,10-^{3}H,^{2}HC-H_4 folate must still be established.

Unlike the steps to form the ternary complex (10), there have been no convincing chemical models for this stereospecific hydrogen-transfer step. An attempt to generate thymine through the retro-ene reaction of 11 to give 13 that might then undergo a 1,3 proton shift to complete the process yielded thymine (14) in ca. 47% yield, but the hydrogen transferred (traced by using ^{2}H at C-6) was exclusively in the methyl group of thymine [64]. Obviously the C-6 of 14 should have been labeled if a

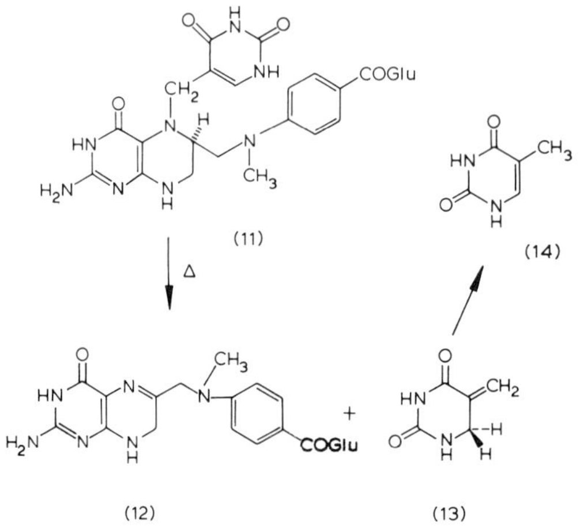

(11)

(14)

(12)

(13)

retro-ene mechanism were operative. A radical cage mechanism involving a 1,3 hydride shift (a concerted 1,3 shift is not allowed) or an ion pair mechanism involving a cationic folate (at the quinonoid level of oxidation) and thymidylate anion would seem to best account for this finding as well as that observed earlier in the pyrolysis of dihydroquinoline adducts [65].

The quinonoid form of the pteridine might also participate in the flavin requiring methylation of ribothymidine by 5,10-methylene-H_4-folate in the tRNA of *Streptococcus faecalis* [66]. In this case the methyl group derives its third hydrogen from solvent and not the C-6 position of the folate, so that the reduced flavin may trap the quinonoid cofactor of the ion pair before internal hydrogen transfer from C-6.

5. *Formyl transfer*

The transfer of 1-carbon units at this oxidation level originally was thought to involve two derivatives of H_4-folate, 10-formyl-H_4-folate and 5,10-methenyl-H_4-folate which acted as cofactors for the two transformylases in de novo purine biosynthesis [67–69]. However, recent work has shown that the glycinamide ribonucleotide transformylase (GAR TFase) from *E. coli* as well as from avian liver utilize 10-formyl-H_4-folate as the actual cofactor [70,71]. The preference for 10-formyl-H_4-folate was masked by the presence of the opposite, unreactive diastereomer (R at C-6 in H_4-folate) which is an excellent competitive inhibitor of the enzyme. The apparent reactivity of the 5,10-methenyl-H_4-folate in the same assay arose because of a contaminating cyclohydrolase activity capable of selectively hydrolyzing it to the correct diastereomer of 10-formyl-H_4-folate.

The further examination of both GAR TFase and AICAR TFase with 5- and/or 8-deazafolate analogs formylated at N-10 has eliminated mechanisms requiring cyclization to the 5,10-methenyl species prior to 1-carbon unit transfer [72]. For example with GAR TFase, analog 15 manifests 77% *V* and with AICAR TFase compound 16 exhibits 47% *V* relative to 10-formyl-H_4-folate. Since neither *N*-formyl compound can cyclize to a stable methenyl species either off or on the enzymes, participation by the latter is eliminated. It is surprising that GAR TFase and AICAR TFase "read" the pterin residue, since 15 is an effective inhibitor of AICAR

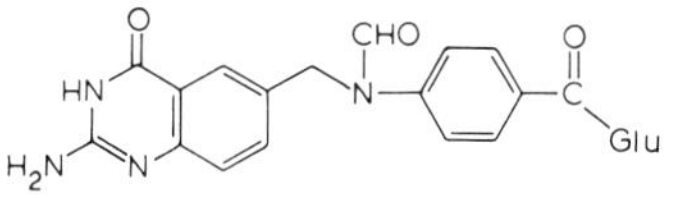

(15) 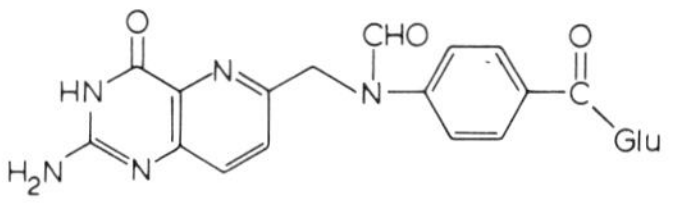 (16)

TFase and vice versa. Furthermore, the transfer of the 10-formyl group does not appear to proceed through a dehydration–hydration sequence involving an amine function on either enzyme since ^{18}O is not exchanged into the 1-carbon unit during transfer [73].

The formation of 10-formyl tetrahydrofolate is catalyzed by multifunctional

enzymes. A trifunctional, single protein has been isolated from ovine, rabbit and porcine liver [74–76] and from yeast [77] that catalyzes the following 3 reactions:

$$5,10\text{-}CH_2\text{—}H_4\text{—folate} + NADP^+ \underset{\text{(dehydrogenase)}}{\rightleftharpoons} 5,10\text{-}\overset{+}{C}H\text{—}H_4\text{—folate} + NADPH$$

$$5,10\text{-}\overset{+}{C}H\text{—}H_4\text{—folate} + H_2O \underset{\text{(cyclohydrolase)}}{\rightleftharpoons} 10\text{-}CHO\text{—}H_4\text{—folate} + H^+$$

$$10\text{-}CHO\text{—}H_4\text{—folate} + ADP + P_i \underset{\text{(synthetase)}}{\rightleftharpoons} H_4\text{—folate} + ATP + HCOO^-$$

A bifunctional protein containing the dehydrogenase and cyclohydrolase activities has been purified from *E. coli.* Whether these activities reside at a single or multiple site is not yet resolved unequivocally. In the case of the trifunctional protein from porcine liver, probes employing tryptic digests and assays for intermediate species, e.g. the level of the intermediate 5,10-methenyl-H_4-folate, implicate a proximal or single dehydrogenase–cyclohydrolase site [78–80].

Note that the hydrolytic ring opening of 5,10-methenyl-H_4-folate proceeds to the N-10 formyl isomer which is the product of kinetic control. In this case (recall 5,10-methylene) the higher basicity of N-5 relative to N-10 insures an increased degree of protonation, so that it is the preferred leaving group in the decomposition of any tetrahedral intermediate arising from nucleophilic addition to C-11 [81].

Recently the trifunctional protein, GAR TFase and AICAR TFase from chicken liver have been found to copurify from AICAR and GAR ligand affinity columns and the GAR TFase and trifunctional protein have been cross-linked [82]. Based on the activities residing on the trifunctional enzyme, a reason for their association is readily imagined. There is a hint that these enzymes in turn may be part of a larger macromolecular complex containing all the enzymes necessary for purine biosynthesis [83]. Whether this hypothesized interaction is important might be demonstrated by experiments designed to demonstrate more efficient product formation arising from "channeling" or in the coordinate regulation of the actual synthesis of the enzymes themselves.

With respect to mechanism of action, the most extensive kinetic and equilibrium exchange studies have been carried out on monofunctional 10-formyl-H_4-folate synthetase from *Cl. cylindrosporum* [84]. The data support a random sequential mechanism that does not involve the formation of freely dissociable intermediates. The most likely mechanism, however, is not concerted but probably involves the formation of a formyl phosphate intermediate, since the synthetase catalyzes phosphate transfer from carbamyl phosphate but not acetyl phosphate to ADP with H_4-folate serving as an activator. Carbamyl phosphate is an inhibitor of 10-formyl-H_4-folate synthesis – an inhibition that can be eliminated only when both ATP and formate are present in accord with the concept that it spans both sites [85]. It would be of considerable interest to attempt to demonstrate positional isotope exchange employing [β,γ-^{18}O]ATP for this enzyme in order to further implicate an enzyme-bound formyl phosphate species [86].

6. *Methyl transfer*

Insight into the mechanism of methyl group transfer may be emerging from recent studies on methylenetetrahydrofolate reductase which catalyzes the reaction:

$$\mathrm{NADPH + H^{+} + 5,10\text{-}CH_2{—}H_4{—}folate \rightarrow NADP^{+} + 5\text{-}CH_3{—}H_4{—}folate}$$

The enzyme contains FAD as a prosthetic group and the steady-state kinetics implicate a ping-pong mechanism. Since dihydrofolate is a potent competitive inhibitor of NADPH–methylenetetrahydrofolate oxidoreduction, it was thought that 5-*N*-methyl-7,8-dihydrofolate might serve as an enzyme-bound intermediate. However, its formation is precluded by the failure to observe ^3H loss from C-6 of the pteridine ring during reduction [87]. Significantly the enzyme catalyzes the reduction of quinonoid dihydropterins (BH_2 and $DMPH_2$) with V values comparable to that for CH_2-H_4-folate [88]. This result lends support to the proposal that the reduction of methylenetetrahydrofolate proceeds by tautomerization of the 5-iminium cation 17 to form quinonoid 5-methyldihydrofolate 18 which is then reduced. Species 18 is an attractive but speculative candidate as the methyl donor in reactions involving methyl transfer from 5-CH_3-H_4-folate to acceptors such as homocysteine. Its formation would require an internal oxidation by the transferase, since these enzymes do not appear to require an external redox agent [2]. Methyl group formation or transfer catalyzed by H_4-folate thus may represent the case where both the transfer and redox properties of the cofactor are exploited.

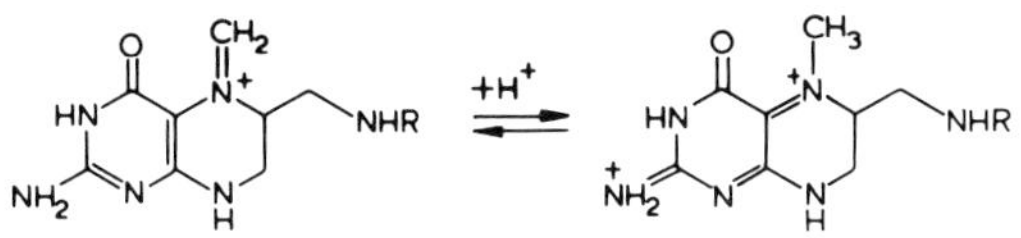

(17) (18)

7. *Hydroxylation*

Tetrahydrobiopterin (BPH_4) is the natural cofactor required for the mammalian aromatic amino acid monooxygenases: phenylalanine, tyrosine and tryptophan hydroxylase [4,89]. During the course of the reaction catalyzed by these enzymes, a molecule of oxygen is cleaved in order to hydroxylate the respective amino acid substrate. The remaining atom of oxygen is reduced to water at the expense of the cofactor, which is oxidized to the quinonoid form. Despite the many studies on the pterin-dependent hydroxylases, their precise mechanism of action is not well understood. This discussion will focus on mammalian phenylalanine hydroxylase (PAH), which has been favored for investigation due to its relative stability and ease of purification [90,91]. Since in addition all 3 enzymes either require or are stimulated by iron [92–95] PAH probably serves as a prototype for the others. Two fundamental questions may be posed with regard to the mechanism of action: what is the site of oxygen activation – i.e. the tetrahydropterin and/or another locus such as the metal ion on the enzyme – and what is the structure of the hydroxylating species.

A structural requirement of the reduced pterin (other reducing agents are inactive) necessary for activity is the 2-amino-4-hydroxy substitution pattern at the pyrimidine ring [96]. Either diastereoisomer of BPH_4 is active with identical K_m values (23 μM), although the natural isomer has a 4-fold greater V and induces substrate inhibition by phenylalanine [97]. Similar results are observed for both tyrosine hydroxylase and tryptophan hydroxylase, however, the K_m values for the natural isomer (6L)-BPH_4 are lower than those for the unnatural one [6]. The use of other tetrahydropterin analogs such as 6-methyltetrahydropterin ($6MPH_4$), 6,7-dimethyltetrahydropterin ($DMPH_4$), or 6-phenyltetrahydropterin ($6PPH_4$), which serve as reasonably good cofactors [98,99], demonstrates the enzyme's insensitivity to the nature of the side chain on the pyrazine ring although some of the regulatory responses are different [100].

This tolerance to structural changes has culminated in the demonstration that various amino-substituted pyrimidines can function as cofactors [98,101]. They exhibit K_m values similar to those found with the corresponding tetrahydropterin analogs, although the relative V_{max} values are only 1–5%. Thus the minimal structural requirements for cofactor activity are found in the 2,4,5-triamino-6-hydroxypyrimidines. Groups at the 5-position other than amino yield inactive compounds [102]. The requirement for amino substitution is repeated in the tetrahydropterins, since neither the 5- nor the 8-deaza-6-methyltetrahydropterin support hydroxylation [102,103].

The apparent requirement for amino substitution is manifest in the limited electrochemical reduction potential range for active compounds. Their reduction potential versus the normal hydrogen electrode at pH 7 is +0.15 V for $DMPH_4$ and $6MPH_4$ and +0.18 V [104] for 2,4,5-triamino-6-hydroxypyrimidine (TAP) respectively (C.L. Luthy, personal communication) whereas 5-deaza-6-methyltetrahydropterin, a competitive inhibitor of $DMPH_4$ and $6MPH_4$ ($K_I = 50$ μM), has a much higher reduction potential (+0.97 V) [103]. The interpretation of this activity–redox potential correlation, however, is ticklish. It does not necessarily follow that the cofactor's mode of action is to reduce some residue, perhaps iron, on the enzyme, but may be an index of the redox potential difference necessary for efficient combination of the cofactor with molecular oxygen.

Support for the argument that the cofactor is the site for oxygen activation stems from two sets of experiments. Kaufman [105] first observed the accumulation of a transient species during PAH turnover that decays non-enzymatically or, in the presence of a "stimulator" protein, enzymatically to quinonoid-BPH_2. Since the rate of tyrosine formation was more rapid than decay of this species, the transfer of oxygen must already be complete. This deduction led to the postulate that the intermediate is comprised of elements from both BPH_4 and the remaining oxygen atom, possibly a 4*a*-hydroxyl adduct. Secondly the products derived from PAH turnover of TAP and 5-benzylamino-2,4-diamino-6-hydroxypyrimidine (BTAP) result from cleavage at the 5-amino substituent yielding 2,4-diamino-5,6-dihydroxypyrimidine (divicine) – after reduction with mercaptoethanol – and in the case of BTAP, benzylamine [106].

The extension to $6MPH_4$ (19) of the observation of a PAH-catalyzed transient tetrahydropterin-derived intermediate that is a precursor of the quinonoid dihydropterin (22) has firmed the argument for the existence of a specific 4*a*-hydroxy adduct (21). The proof of structure rests on a striking similarity between its UV spectrum and that of the chemically synthesized 4*a*-hydroxy adduct derived from 5-deaza-6-methyltetrahydropterin. As for BPH_4, oxygen transfer has already occurred since the rate of tyrosine formation is identical to the rate of disappearance of the tetrahydropterin [102]. In addition the pH dependence of the rate of decay of this intermediate ($21 \rightarrow 22$) follows the expected rate-determining hydronium ion-catalyzed dehydration of a carbinolamine species [107]. Electrophilic oxygen attack at the 4*a* position is also consistent with observed structural rearrangements arising as a consequence of 4*a*-peroxide intermediates generated in non-enzymatic oxidations of tetrahydropterins [108,109]. The observation of the adduct from both $6MPH_4$ and BPH_4 eliminates any possible artifactual products arising from cyclization on the quinonoid of the dihydroxypropyl side chain in BPH_4. Thus the accumulated evidence infers but does demand a precursor 4*a*-peroxy intermediate (20) as being directly involved in oxygen atom transfer. The possible transformations of the pterin cofactor are summarized in the following scheme.

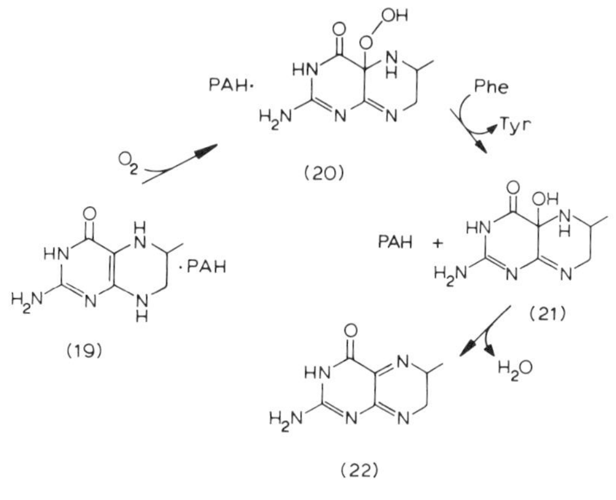

Although the above data suggest that the pterin is the initial site of oxygen attachment, the requirement of Fe for activity increases the mechanistic possibilities. It has recently been demonstrated that 1.0 mole of iron per 50 000 subunit molecular weight is required for maximal activity with a direct correlation between iron content and specific activity [92]. EPR data of native resting PAH reveals a signal with a *g* value at 4.3, which is characteristic of high spin (5/2) Fe^{3+} [92,93]. Therefore a possible role of the cofactor could be to prereduce the enzyme to the

Fe^{2+} state which could then react with molecular oxygen. However in the absence of O_2 the tetrahydropterin is not oxidized by a stoichiometric amount of PAH, thus reducing the likelihood of this mechanism [4,102].

The question as to the precise structure of the hydroxylating agent remains unresolved. The 3 most likely possibilities presuming an initially formed 4*a*-peroxy adduct are: (1) the 4*a*-peroxy adduct itself; (2) an activated form of the adduct such as a carbonyl oxide type, oxenoid-type species; and (3) an iron–oxygen adduct. In the first two cases the role of the iron may reside in catalyzing both the formation of the 4*a*-peroxy adduct and the resultant oxygen transfer; in the third case the high reactivity of the iron–oxygen adduct is required for the aromatic hydroxylation. In all these cases the oxidizing agent must be capable of generating the intermediate arene oxide required for the NIH shift [110].

A 4*a*-hydroperoxy adduct (23) has been observed directly with flavoprotein monooxygenases [111,112]. These enzymes catalyze aromatic hydroxylation; however their substrates are phenols rather than an unactivated phenyl ring. They do not

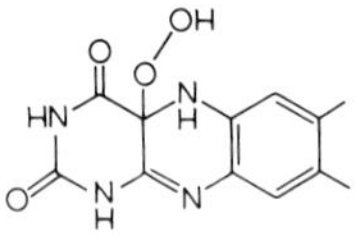

(23)

contain metal ions so that transfer of oxygen occurs directly or from an activated form of the peroxide, e.g. a carbonyl oxide [113] or an *N*-oxide [114]. Moreover, the reactivity of dihydroflavins with molecular oxygen is nearly instantaneous ($t_{1/2} < 1$ sec) compared to the sluggish reactivity of the tetrahydropterins ($t_{1/2} \sim 5$ min). Thus one may deduce that iron may act in the dual role noted earlier, possibly through a structure such as 24.

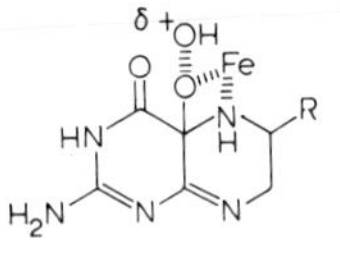

(24)

Support for the intermediacy of the carbonyl oxide mechanism stems mainly from the observation of stoichiometric 5-amino group expulsion from the pyrimidine cofactors during PAH turnover regardless of the extent of coupling [106]. However, an attempt to demonstrate PAH-catalyzed cyclization of 25 to 26 proved unsuccessful, despite the requirements for such a process if the tetrahydropterin follows a similar reaction course [102]. Thus the case for their intermediacy is flawed. Carbonyl oxides are rather poor electrophilic reagents so that the hydroxylation of an aromatic ring probably proceeds via a radical species [115].

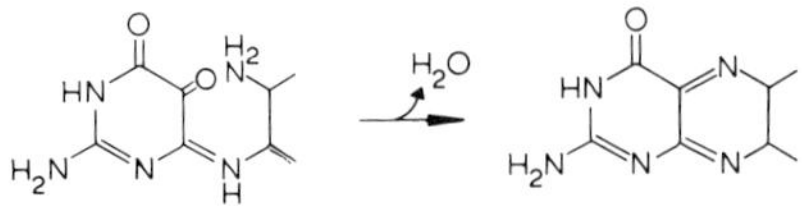

(25) (26)

On chemical grounds, the most active hydroxylating agent would be an oxoiron species. Model studies on non-enzymic oxygen transfer reactions have suggested that a type of oxoiron species may be the active hydroxylating species in monooxygenases [116,117]. Recent evidence supports the concept that heme–Fe^{3+} complexes of cytochrome P-450 are readily oxidized to higher oxidation states such as the ferryl ion (Fe^{4+}=O)–heme radical cation [118,119]. Model reactions of non-heme iron with peroxides can result in aromatic hydroxylation, furthermore via intermediates that undergo the NIH shift [120]. However, it remains to be conclusively demonstrated that the iron atom and pterin are in close enough proximity so that any of these combined mechanisms may have validity.

References

1 Benkovic, S.J. (1980) Annu. Rev. Biochem. 49, 227–251.
2 Blakley, R.L. (1969) The Biochemistry of Folic Acid and Related Pteridines, Elsevier, New York, p. 569.
3 Kisliuk, R.L. and Brown, G.M. (Eds.) (1979) Chemistry and Biology of Pteridines, Elsevier, Amsterdam, p. 713.
4 Kaufman, S. and Fisher, D.B. (1974) in: O. Hayaishi (Ed.), Molecular Mechanisms of Oxygen Activation, Academic Press, New York, pp. 285–369.
5 Fontecilla-Camps, J., Bugg, C.E., Temple, C., Rose, J.D., Mongomery, J.A. and Kisliuk, R.L. (1979) in: R.L. Kisliuk and G.M. Brown (Eds.), Chemistry and Biology of Pteridines, Elsevier, Amsterdam, pp. 235–240.
6 Hasegawa, H., Shunsuke, I., Ichiyama, A., Sugimoto, T., Matsumura, S., Oka, K., Kato, T., Nagatsu, T. and Akino, M. (1979) in: R.L. Kisliuk and G.M. Brown (Eds.), Chemistry and Biology of Pteridines, Elsevier, Amsterdam, pp. 183–188.
7 Baugh, C.M. and Krumdieck, C.L. (1971) Ann. N.Y. Acad. Sci. 186, 7–28.
8 Connor, M.J. and Blair, J.A. (1980) Biochem. J. 186, 235–242.
9 Shane, B. (1980) J. Biol. Chem. 255, 5649–5654.
10 Coward, J.K., Parameswaran, K.N., Cashmore, A.R. and Bertino, J.R. (1974) Biochemistry 13, 3899–3903.
11 Coward, J.K., Chello, P.L., Cashmore, A.R., Parameswaran, K.N., DeAngelis, L.M. and Bertino, J.R. (1975) Biochemistry 14, 1548–1552.
12 Kisliuk, R.L., Gaumont, Y., Lafer, E., Baugh, C.M. and Montgomery, J.A. (1981) Biochemistry 20, 929–934.
13 Baggott, J.E. and Krumdieck, C.L. (1979) Biochemistry 18, 1036–1041.
14 Curthoys, N.P. and Rabinowitz, J.C. (1972) J. Biol. Chem. 247, 1965–1971.
15 Matthews, R.G., Ross, J., Baugh, C.M., Cook, J.D. and Davis, L. (1982) Biochemistry 21, 1230–1234.
16 Matthews, R.G. and Baugh, C.M. (1980) Biochemistry 19, 2040–2045.
17 Shane, B. and Stokstad, E.L.R. (1976) J. Biol. Chem. 251, 3405–3410.
18 McGuire, J.J., Hoieh, P., Coward, J.K. and Bertino, J.R. (1980) J. Biol. Chem. 255, 5776–5788.
19 Poe, M. and Hoogsteen, K. (1978) J. Biol. Chem. 253, 543–546.
20 Bieri, J.H. and Viscontini, M. (1977) Helv. Chim. Acta 60, 447–453.
21 Furrer, H.J., Bieri, J.M. and Viscontini, M. (1978) Helv. Chim. Acta 61, 2744–2751.
22 Poe, M., Jackman, L.M. and Benkovic, S.J. (1980) Biochemistry 19, 4576–4582.
23 Bieri, J. (1979) in: R.L. Kisliuk and G.M. Brown (Eds.), Chemistry and Biology of Pteridines, Elsevier, Amsterdam, pp. 19–24.
24 Kallen, R.G. and Jencks, W.P. (1966) J. Biol. Chem. 241, 5845–5850.
25 Bertino, J.R. and Johns, D. (1972) in: I. Brodsky (Ed.), Cancer Chemotherapy, Grune and Stratton, New York, pp. 9–22.

26 Burchall, J.J. and Hitchings, G.H. (1965) Mol. Pharmacol. 1, 126–136.
27 Freisheim, J.H., Bitar, K.G. Reddy, A.V. and Blankenship, D.T. (1978) J. Biol. Chem. 253, 6437–6444.
28 Gleisner, J.M., Peterson, D.L. and Blakley, R.L. (1974) Proc. Natl. Acad. Sci. (U.S.A.) 71, 3001–3005.
29 Stone, D., Phillips, A.W. and Burchall, J.J. (1977) Eur. J. Biochem. 72, 613–624.
30 Bennett, C.D., Rodkey, J.A., Sondey, J.M. and Hirschmann, R. (1978) Biochemistry 17, 1328–1337.
31 Stone, D., Paterson, S.J., Raper, J.H. and Phillips, A.W. (1979) J. Biol. Chem. 254, 480–488.
32 Kumar, A.A., Blankenship, D.T., Kaufman, B.T. and Freisheim, J.H. (1980) Biochemistry 19, 667–678.
33 Matthews, D.A., Alden, R.A., Freer, S.T., Xuong, N. and Kraut, J. (1979) J. Biol. Chem. 254, 4144–4151.
34 Ferone, R. and Roland, S. (1980) Proc. Natl. Acad. Sci. (U.S.A.) 77, 5802–5806.
35 Pastore, E.J. and Friedkin, M.J. (1962) J. Biol. Chem. 237, 3802–3810.
36 Blakley, R.L., Ramasastri, B.V. and McDougall, B.M. (1963) J. Biol. Chem. 238, 3075–3079.
37 Pastore, E.J. and Williamson, K.L. (1968) Fed. Proc. 27, 764.
38 Pastore, E.J., Friedkin, M. and Jardetsky, O. (1963) J. Am. Chem. Soc. 85, 3058–3059.
39 Charlton, P.A., Young, D.W., Birdsall, B., Feeney, J. and Roberts, G.C.K. (1979) J. Chem. Soc. Chem. Commun., 922–924.
40 Craine, J.E., Hall, E.S. and Kaufman, S. (1972) J. Biol. Chem. 247, 6082–6091.
41 Archer, M.C. and Scrimgeour, K.G. (1970) Can. J. Biochem. 48, 278–287.
42 Lund, H. (1975) in: W. Pfleiderer (Ed.), Chemistry and Biology of Pteridines, Gruyter, Berlin, pp. 645–670.
43 Webber, S., Deits, T.L., Snyder, W.R. and Whitely, J.M. (1978) Anal. Biochem. 84, 491–503.
44 Stone, K.J. (1976) Biochem. J. 157, 105–109.
45 Schirch, L. and Chen, M. (1973) J. Biol. Chem. 248, 7979–7984.
46 Ulevitch, R.J. and Kallen, R.G. (1977) Biochemistry 16, 5342–5360.
47 Hansen, J. and Davis, L. (1979) Biochim. Biophys. Acta 568, 321–330.
48 Schirch, L. and Jenkins, W.T. (1964) J. Biol. Chem. 239, 3797–3800.
49 Besmer, P. and Arigoni, D. (1968) Chimia 23, 190.
50 Akhtar, M., El-Obeid, H.A. and Jordan, P.M. (1975) Biochem. J. 116, 159–168.
51 Palekar, A.G., Tate, S.S. and Meister, A. (1973) J. Biol. Chem. 248, 1158–1167.
52 Dunathan, H.C. (1966) Proc. Natl. Acad. Sci. (U.S.A.) 55, 713–716.
53 Tatum, C.M., Benkovic, P.A., Benkovic, S.J., Potts, R., Schleicher, E. and Floss, H.G. (1977) Biochemistry 16, 1093–1102.
54 Biellmann, J.F. and Schumber, J.F. (1970) Bell. Soc. Chim. Biol. 52, 211–227.
55 Schirch, L. and Diller, A. (1971) J. Biol. Chem. 246, 3961–3966.
56 Chen, M.S. and Schirch, L. (1973) J. Biol. Chem. 246, 7979–7984.
57 Jones, C.W. and Priest, D.G. (1976) Arch. Biochem. Biophys. 174, 305–311.
58 Wiltwer, A.J. and Wagner, C. (1980) Proc. Natl. Acad. Sci. (U.S.A.) 77, 4484–4488.
59 Pogolotti, A.L. and Santi, D.V. (1977) in: E.E. vanTamelen (Ed.), Bioorganic Chemistry, Academic Press, New York, pp. 277–311.
60 Moad, G. and Benkovic, S.J. (1978) J. Am. Chem. Soc. 100, 5495–5499.
61 Fife, T.H. and Pellino, A.M. (1980) J. Am. Chem. Soc. 102, 3062–3071.
62 Lorenson, M.Y., Maley, G.F. and Maley, F. (1967) J. Biol. Chem. 242, 3332–3344.
63 Tatum, C., Vederas, J., Schleicher, E., Benkovic, S.J. and Floss, H. (1977) J. Chem. Soc. Chem. Commun. 7, 218–220.
64 Charlton, P.A. and Young, D.W. (1980) J. Chem. Soc. Chem. Commun. 614–616.
65 Wilson, R.S. and Mertes, M.P. (1973) Biochemistry 12, 2879–2886.
66 Delk, A.S., Nagle Jr., D.P. and Rabinowitz, J.C. (1979) J. Biol. Chem. 255, 4387–4390.
67 Warren, L. and Buchanan, J.M. (1957) J. Biol. Chem. 229, 613–626.
68 Warren, L., Flaks, J.G. and Buchanan, J.M. (1957) J. Biol. Chem. 229, 627–640.
69 Hartman, S.C. and Buchanan, J.M. (1959) J. Biol. Chem. 234, 1812–1816.

70 Dev, I.K. and Harvey, R.J. (1978) J. Biol. Chem. 253, 4242–4244.
71 Smith, G.K., Benkovic, P.A. and Benkovic, S.J. (1981) Biochemistry 20, 4034–4036.
72 Smith, G.K., Mueller, W.T., Benkovic, P.A. and Benkovic, S.J. (1981) Biochemistry 20, 1241–1245.
73 Smith, G.K., DeBrosse, C., Mueller, T.W., Sleicker, L. and Benkovic, S.J. (1982) Biochemistry 2870–2874.
74 Schirch, L. (1978) Arch. Biochem. Biophys. 189, 283–290.
75 Tan, L.U.L., Drury, E.J. and Mackenzie, R.E. (1977) J. Biol. Chem. 252, 1117–1122.
76 Paukert, J.L., Straus, L.D. and Rabinowitz, J.C. (1976) J. Biol. Chem. 251, 5104–5111.
77 Paukert, J.L., Williams, G.R. and Rabinowitz, J.C. (1977) Biochem. Biophys. Res. Commun. 77, 147–154.
78 Dev, I.K. and Harvey, R.J. (1978) J. Biol. Chem. 253, 4245–4253.
79 Tan, L.U.L. and Mackenzie, R.E. (1977) Biochim. Biophys. Acta 485, 52–59.
80 Cohen, L. and Mackenzie, R.E. (1978) Biochim. Biophys. Acta 522, 311–317.
81 Benkovic, S.J. (1978) Accts. Chem. Res. 11, 314–320.
82 Smith, G.K., Mueller, W.T., Wasserman, G.F., Taylor, W.T. and Benkovic, S.J. (1980) Biochemistry 19, 4313–4321.
83 Rowe, P.B., McCairns, C., Madsen, B., Sauer, D. and Elliott, H. (1978) J. Biol. Chem. 253, 7711–7721.
84 Hines, R.H. and Hamony, J.A.K. (1973) CRC Rev. Biochem. 1, 501–535.
85 Buttlaire, D.H., Balfe, C.A., Wendland, M.F. and Himes, R.H. (1979) Biochim. Biophys. Acta 567, 453–463.
86 Midelfort, C.F. and Rose, I.A. (1976) J. Biol. Chem. 251, 5881–5887.
87 Matthews, R.G. and Haywood, B.J. (1974) Biochemistry 18, 4845–4851.
88 Matthews, R.G. and Kaufman, S. (1980) J. Biol. Chem. 255, 6014–6017.
89 Youdim, M.B.H. ed. (1979) Aromatic Amino Acid Hydroxylase and Mental Disease, John Wiley, New York, p. 340.
90 Shiman, R., Gray, D.W. and Pater, A. (1979) J. Biol. Chem. 254, 11300–11306.
91 Kaufman, S. and Fisher, D.B. (1970) J. Biol. Chem. 245, 4745–4750.
92 Gottschall, D.W., Dietrich, R.F., Benkovic, S.J. and Shiman, R. (1982) J. Biol. Chem. 845–849.
93 Fisher, D.B., Kirkwood, R. and Kaufman, S. (1972) J. Biol. Chem. 247, 5161–5167.
94 Kuhn, D.M., Ruskin, B. and Lovenberg, W. (1980) J. Biol. Chem. 255, 4137–4143.
95 Hoeldtke, R. and Kaufman, S. (1977) J. Biol. Chem. 252, 3160–3169.
96 Ayling, J.E., Boehm, G.R., Textor, S.C. and Pirson, R.A. (1973) Biochemistry 12, 2045–2051.
97 Bailey, S.W. and Ayling, J.E. (1978) J. Biol. Chem. 253, 1598–1605.
98 Bailey, S.W. and Ayling, J.E. (1978) Biochem. Biophys. Res. Commun. 85, 1614–1621.
99 Storm, C.B. and Kaufman, S. (1968) Biochem. Biophys. Res. Commun. 32, 788–793.
100 Ayling, J.E. and Helfand, G.D. (1975) in: W. Pfleiderer (Ed.), Chemistry and Biology of Pteridines, de Gruyter, Berlin, pp. 305–319.
101 Kaufman, S. (1979) J. Biol. Chem. 254, 5150–5154.
102 Lazarus, R.A., Dietrich, R.F., Wallick, D.E. and Benkovic, S.J. (1981) Biochemistry 20, 6834–6841.
103 Moad, G., Luthy, C.L., Benkovic, P.A. and Benkovic, S.J. (1979) J. Am. Chem. Soc. 101, 6068–6076.
104 Archer, M.C. and Scrimgeour, K.G. (1970) Can. J. Biochem. 48, 526–527.
105 Kaufman, S. (1975) in: W. Pfleiderer (Ed.), Chemistry and Biology of Pteridines, de Gruyter, Berlin, pp. 291–304.
106 Bailey, S.W. and Ayling, J.E. (1980) J. Biol. Chem. 255, 7774–7781.
107 Jencks, W.P. (1964) Prog. Phys. Org. Chem. 2, 63–128.
108 Mager, H.I.X. (1975) in: W. Pfleiderer (Ed.), Chemistry and Biology of Pteridines, de Gruyter, Berlin, pp. 753–773.
109 Yamamoto, H. and Pfleiderer, W. (1973) Chem. Ber. 106, 3194–3202.
110 Guroff, G., Daly, J.W., Jerina, D.M., Rensen, J., Witkop, B. and Udenfriend, S. (1967) Science 157, 1524–1530.
111 Ballou, D.P. (1982) in: V. Massey and C.H. Williams (Eds.), Symposium on Flavins and Flavoproteins, Elsevier, Amsterdam, pp. 301–310.

112 Massey, V. and Hemmerich, P. (1975) in: P.D. Boyer (Ed.), The Enzymes, Vol. 12, Academic Press, New York, pp. 191–252.
113 Hamilton, G.A. (1974) in: O. Hayaishi (Ed.), Molecular Mechanisms of Oxygen Activation, Academic Press, New York, pp. 405–451.
114 Frost, J.W. and Rastetter, W.H. (1981) J. Am. Chem. Soc. 103, 5242–5245.
115 Sawaki, Y., Kato, H. and Ogata, Y. (1981) J. Am. Chem. Soc. 103, 3832–3837.
116 Sharpless, K.B. and Flood, T.C. (1971) J. Am. Chem. Soc. 93, 2316–2318.
117 Hamilton, G.A. (1964) J. Am. Chem. Soc. 86, 3391–3392.
118 Chin, D.H., LaMar, G.N. and Balch, A.L. (1980) J. Am. Chem. Soc. 102, 5945–5947.
119 Groves, J.T. and Van Der Puy, M. (1976) J. Am. Chem. Soc. 98, 5290–5297.
120 Castle, L. and Lindsay Smith, J.R. (1978) J. Chem. Soc. Chem. Commun., 704–705.

CHAPTER 11

Glycosyl transfer

MICHAEL L. SINNOTT

University of Bristol, School of Chemistry, Cantock's Close, Bristol BS8 1TS, Great Britain

1. Introduction

Perusal of any textbook of general biochemistry will reveal how widespread and important are glycosyl-transfer reactions in living systems. The simple hydrolysis of oligo- and poly-saccharides is a common first step in the processing of nutrients, and is complemented by the phosphorolysis of the cell's own storage polysaccharides and oligosaccharides in response to the energy demands made upon it. These storage carbohydrates are themselves built up by other glycosyl transfers from nucleotide diphosphoryl sugars; and the same donors (with CMP neuraminic acid) also serve to glycosylate proteins for recognition and translocation. Glycosyl transfers are also critically involved in the conservation of genetic information; nucleotides are biosynthesised via the reaction of phosphoribosyl pyrophosphate with nitrogen nucleophiles and DNA glycosylates excise altered bases. Even the redox coenzyme NAD can also serve as a glycosylating agent for proteins.

All these reactions are nucleophilic substitutions at a carbon adjacent to oxygen as shown, the nature of X and Y, their degree of protonation, deprotonation or metal coordination, and the stereochemistry of the reaction (retention or inversion), varying with the individual process.

The main effect of the oxygen atom next to a reaction centre in chemical terms is to stabilise the buildup of positive charge on the central atoms: the lone pairs of electrons on the oxygen atom release electrons into the vacant p orbital of an adjacent carbonium ion centre. This means that, compared with nucleophilic substitutions at purely carbon centres, the C—X bond can break to a much greater extent before there is much C—Y bond formation. Indeed, in the limit, for most processes in water the C—X bond completely breaks first and the oxocarbonium ion is a discrete intermediate.

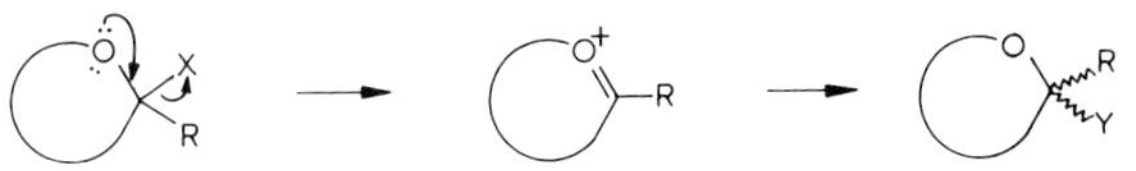

Michael I. Page (Ed.), The Chemistry of Enzyme Action

Few mechanistic studies have however been made of biological glycosyl derivatives themselves. Quantitative structure–reactivity correlations are one of the main tools of the physical organic chemist, and most of the biologically important glycosyl-transfer reactions involve substrates which are dauntingly difficult to synthesise. Furthermore, kinetic data on substrates with many ionisation states – and many tautomers of each individual ionisation state – are notoriously difficult to interpret. It should therefore be no surprise that most of our knowledge of the chemistry of these processes comes from studies of the hydrolysis of simple acetals, ketals and glycosides. These are treated in a critical and constructive review, from a mechanistic standpoint, of the whole of carbohydrate chemistry before 1969 [1]. Therefore the bases on which mechanistic points were established before that date will be described as briefly as is compatible with coherence, and more space will be given to developments since then. There are additional, later reviews on various aspects of acetal and ketal hydrolysis [2–6]. Glycosyl transfer is also treated in the chapters by Capon in an annual review of organic reaction mechanisms [7].

2. *Effects of the direction of the lone pairs of oxygen in space*

It is now well established that hydrogen-like atomic orbitals can be used in molecular orbital or valence bond calculations to give an accurate picture of the energies, geometries and charge distributions of isolated small molecules containing only first-row elements. Such calculations produce pictures like Fig. 1a of the wave functions (4) describing the orbitals containing the oxygen lone pairs in water or dimethyl ether: these stick out like rabbits' ears on either side of the molecular plane. If the R—ÔR angle has the tetrahedral value of 109°, then the two orbitals are sp^3

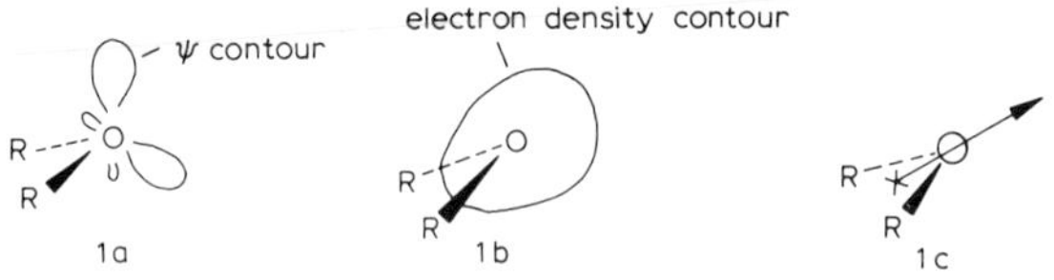

Fig. 1

hybrids, i.e. composed of 3 parts hydrogen-like p orbital and one part hydrogen-like s orbital. The two orbitals give rise to a smooth distribution of electronic charge in which individual rabbits' ears are not discernible (Fig. 1b) [8]. The net effect, if nuclear charge is taken into account, is a powerful dipole in the plane of the molecule, bisecting the ROR angle. So long as all four lone-pair electrons are present and the symmetry of the system is preserved, however, it is entirely arbitrary whether these two lone pairs are represented as being in two equivalent sp^3 hybrid orbitals, or in a p and an sp orbital, or any intermediate arrangement: the same charge distribution and the same energy result [9,10].

However, photoelectron spectroscopy of dimethyl ether [11] and tetrahydropyran

[12] indicated *two* lone-pair ionisation energies. This difference has its origin in the differing electronic states of the cation radical left behind after ejection of the photoelectron. If the p-type orbital is only singly occupied, the positive charge avoids the nucleus more effectively than if it inhabits the (orthogonal) n-type (sp) orbital. The two peaks with tetrahydropyran are emphatically not due to axial and equatorial lone pairs.

If the symmetry of the system is broken, as in glycosyl derivatives, it is best to adopt the nonequivalent (p-type and n-type) representation of the lone pairs on oxygen. This is illustrated with α-D-xylopyranosyl fluoride in Fig. 2. With this representation it is important to note that the p-type lone pair is "tilted" at a dihedral angle of ca. 30% to the axial C—F bond.

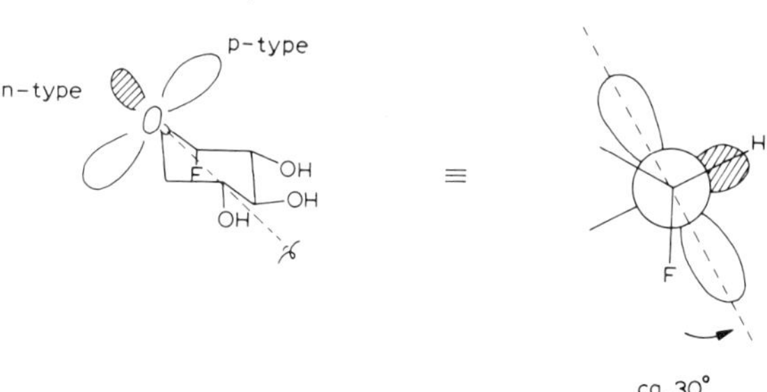

Fig. 2

2.1. *The anomeric effect*

The preference of electronegative substituents at C1 of pyranose rings for the axial orientation has been known for 30 years [13]. This preference is the reverse of what would be expected in cyclohexane chemistry, and is termed the anomeric effect. It is responsible for the exclusive production of α-glycopyranosyl halides when these are equilibrated under strongly acidic conditions (Fig. 3, I).

HBr in HOAc

I

Fig. 3, I

If β-glycopyranosyl halides are made by other routes they can be shown in a number of cases to adopt normally disfavoured conformations (Fig. 3, II).

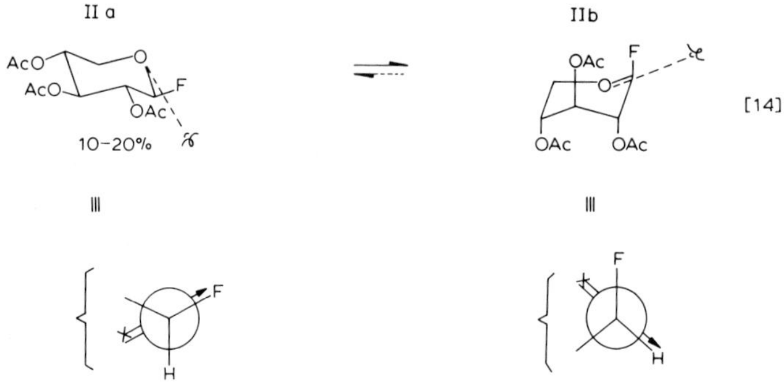

II a

IIb

AcO

AcO

OAc

F

10–20%

OAc

F

OAc

OAc

[14]

F

H

F

H

Fig. 3, II

The effect can be simply rationalised by considering the electrostatic interactions between the C1—F dipole and the concentration of negative charge associated with the lone pairs of electrons (Fig. 3). In the case of the ${}^4C_1$ conformer, a, the negative end of the C1—F dipole exactly eclipses the concentration of highest negative charge density, but is removed from it when the normally disfavored ${}^1C_4$ conformation, in which all four substituents are axial, is adopted.

In the case where the electronegative substituent at C1 is not torsionally symmetrical – as with pyranoses and pyranosides – an exactly similar effect is observed with respect to rotation about the O1—C1 bond. This is termed the exo-anomeric effect [15], and ensures that the preferred conformation of an α-glycoside is as shown: the other rotamer not disfavoured by the exo-anomeric effect has R exactly under the pyranose ring, where there are severe non-bonded interactions (Fig. 4).

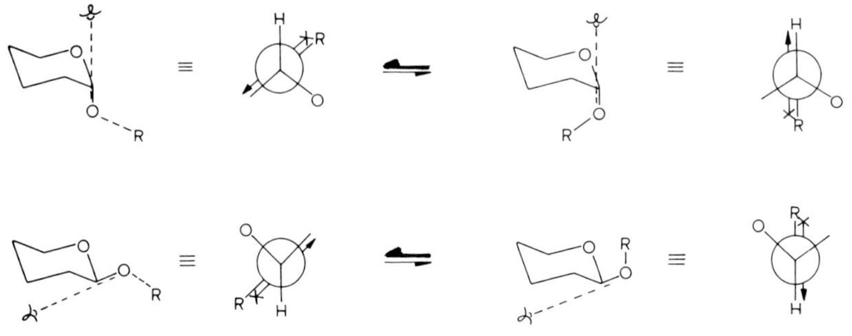

Fig. 4

In the case of β-glycosides there are again two rotamers which are not disfavoured by the anomeric effect (Fig. 4), the rotamer involving fewer ordinary steric interactions being the one that is exclusively observed.

As might be anticipated, positively charged substituents at C1 of a pyranose ring have a more than ordinary preference for the equatorial orientation, or other orientation in which the positive charge can eclipse the concentration of negative charge on oxygen. This "reverse anomeric effect" is so powerful that 2,3,4,6-tetra-*O*-acetyl-α-D-glucopyranosyl-4-methyl pyridinium ion adopts the $B_{2,5}$ conformation (III).

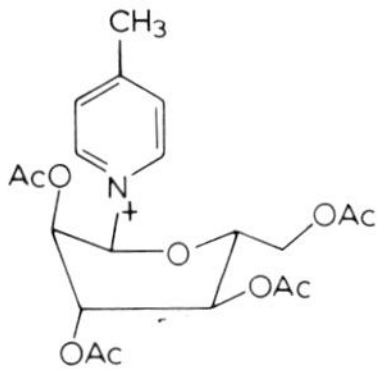

III

That this normally strongly disfavoured conformation is not a consequence of the simple bulk of the pyridine was elegantly demonstrated using the imidazole derivative IV, which adopts the ^4C chair conformation but flips to the boat on protonation (V) [16].

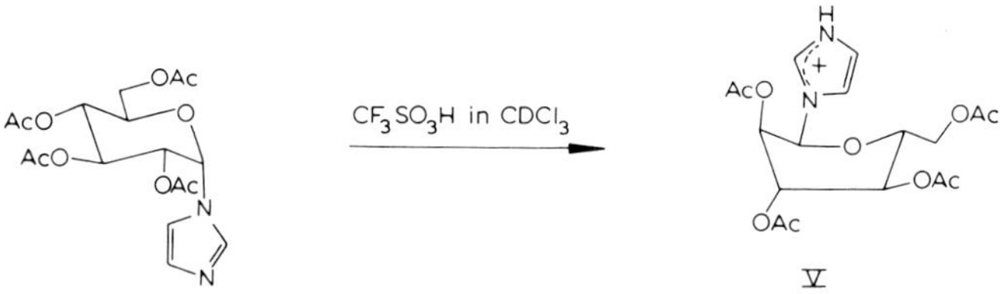

IV

V

As would be anticipated for a phenomenon which is fundamentally electrostatic, this effect is attenuated in polar solvents: α-glucosylimidazole in water does not change conformation on protonation [17]. There is evidence that the anomeric effect itself is likewise solvent sensitive. Table 1 gives representative data on the strength of the anomeric effect (defined as the additional stabilisation of the axial substituent in the glucose series, compared with cyclohexane).

Electrostatic interactions with the oxygen lone-pair electrons are not, exclusively, the sole cause of the anomeric effect. Increasing the positive charge on C5 by substitution with electronegative groups increases the strength of the dipole and hence the anomeric effect (Fig. 3) [18].

The operation of the anomeric effect in furanosyl derivatives is less clear. A vast amount of data has been accumulated on nucleotides and deoxy nucleotides, but the bulky substituents at C4 and C1 of the ribofuranose ring ensure that the conformational choice for these compounds is between N (VI) and S (VII), in both of which the aglycone is eclipsing one lone pair on oxygen (regarding them as equivalent) [24].

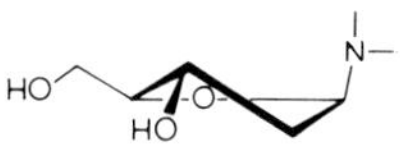

VI

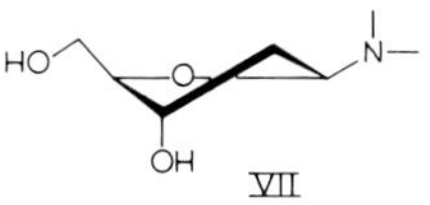

VII

TABLE 1
Strength of the anomeric effect (kcal/mole)

Substituent	Solvent	Effect (kcal/mole)	Reference
	CCl_4	1.5	
	C_6D_6	1.5	
	CS_2	1.4	
	$CDCl_3$	1.1	
OMe	$(CD_3)_2C{=}O$	1.2	[19] [a]
	CH_3OD	1.1	
	$(CD_3)_2SO$	1.2	
	CD_3CN	1.0	
	D_2O	0.6	
OH	H_2O	0.55	[20]
Cl	2-Chlorotetrahydropyran	2.7	
Br	2-Bromotetrahydropyran	> 3.2	[21]
I	2-Iodotetrahydropyran	> 3.1	
—OAc	Acetic anhydride	1.3	[18]
$—N_3$	(Same as OAc in $CDCl_3$)		
$—NH_2$	$CDCl_3$	0	[22]

[a] Calculated using an A value of 0.6 for OCH_3 in cyclohexane [23].

Another explanation of the anomeric effect has come to the fore in recent years, although the electrostatic explanation still receives theoretical support [25]. This is based on ideas of frontier molecular orbital theory and envisages overlap of the p-type lone pair on the ring oxygen atom overlapping with the σ* orbital of the bond between C1 and an electronegative aglycone. This is feasible only in the axial case (Fig. 5), the σ* orbital of the equatorial C—X bond being orthogonal to the p-type

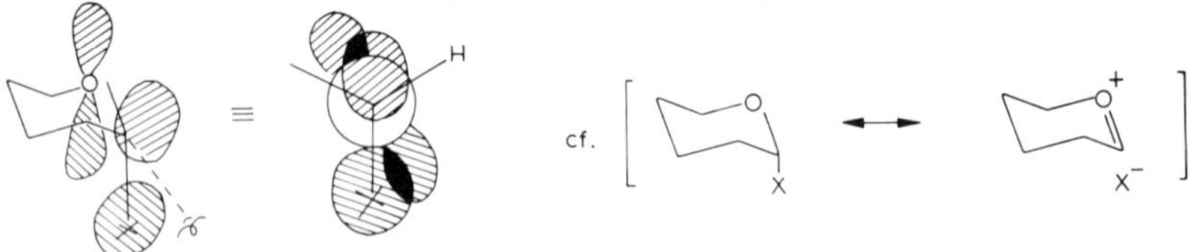

Fig. 5

lone pair. This lone pair σ* overlap preferentially stabilises the axial orientation. A valence bond representation of this explanation involves "no-bond resonance", and in this representation it is clear why axial C—X bonds are longer than equatorial C—X bonds [26], and also why, if the C—X bond is lengthened, the O5—C1 bond is shortened [27]. Variations in ^{35}Cl and ^{79}Br nuclear quadrupole coupling constants are also accounted for on this picture [28].

The main flaw in the frontier orbital explanation is that it fails to predict the existence of the reverse anomeric effect: on the "no-bond resonance" picture there is no reason why a C—$\overset{+}{N}$ bond, polarised in exactly the same sense as a C—Hal bond, should not likewise prefer the axial orientation. A couple of theoretical studies have taken into account both electrostatic and no-bond resonance interactions [29,30] but no satisfactory reconciliation of the reverse anomeric effect with the no-bond resonance picture has emerged. Implicit redefinition of the reverse anomeric effect as the tendency of electron donating substituents – in a π [30] or a σ [31] sense – to be equatorial is unhelpful, not only because of the confusion it sows, but also because the effect as so redefined has no experimental basis.

A possible reconciliation of the two opposing theories might be that the "no-bond resonance" model adequately rationalises changes of a few hundredths of an Ångström in bond lengths, but that these small changes are not reflected in the energetics of the system. Thus, although bond a shortens and bond b lengthens in compounds VIII and IX as the pK_a of the leaving group decreases [32], in accord with the "no-bond resonance picture" the same effect is not seen with carbohydrates [33], in which (presumably) crystal forces are stronger.

VIII

IX

2.2. Geometrical changes on oxocarbonium ion formation

In a 6-membered, pyranose ring the p-type lone pair is tilted at ca. 30° to the axial substituent at C1. If there are only three substituents at C1 – i.e. if a carbonium ion is formed – then the empty p orbital at C1 in a perfect chair is at a dihedral angle of 60° to the p-type oxygen lone pair. For this ion to be maximally stabilised by overlap between the empty p orbital on C1 and the p-type lone pair the system must distort so that the two p orbitals are exactly eclipsed (Fig. 6). One way of doing this is for the cation to adopt a half-chair conformation in which C5, O5, C1 and C2 are coplanar.

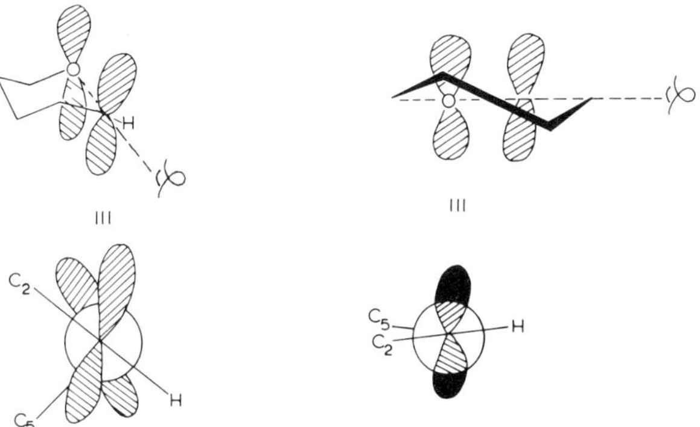

Fig. 6

Exactly similar considerations apply to δ lactones: in order for the p-type lone pair on oxygen to conjugate with the carbonyl group, the lactone function must be planar. Indeed, aldonolactones are bound very tightly to a number of glycosidases and this is considered to be by virtue of their resemblance to transition states leading to oxocarbonium ion intermediates [34].

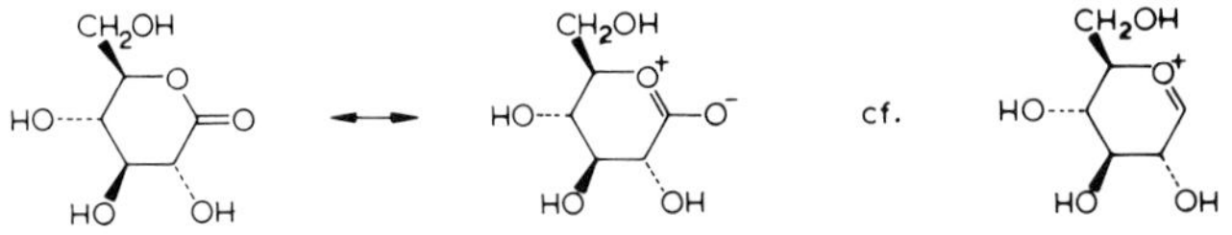

X

XI

However, the half-chair conformation is not the only one in which planarity can be achieved. The classical boat conformation also fulfils the stereoelectronic requirements for maximum lone pair–carbonyl overlap, and is quite close to the half-chair in energy; microwave investigations, and concordant force-field calculations indicate that in the gas phase the classical boat conformer of δ valerolactone (XII) lies only ca. 0.6 kcal above the half-chair [35]. In $CDCl_3$ and C_6H_6 solution, 2,3,4-tri-*O*-acetyl-D-xylonolactone adopts predominantly a boat conformation [36]. The possibility of a glycopyranosyl cation at an enzyme active site being in a boat conformation could therefore profitably be borne in mind.

XII

Since ordinary envelope conformations of saturated 5-membered rings have 4 of the 5 atoms coplanar anyway, there is little manifestation of these effects in the reactions of furanosides.

2.3. Stereoelectronic control of reactions of acetals?

Deslongchamps has proposed that in the reactions of tetrahedral intermediates in the hydrolysis of esters and amides the group which leaves is that which has two antiperiplanar lone pairs of electrons; in the case of oxygen the two lone pairs are considered as identical sp^3 hybrids, the antiperiplanar one making a dihedral angle of 180° with the bond to be cleaved [37]. The theory as originally formulated considered the lifetime of these intermediates to be short compared to the time of rotation about a C—O or C—N single bond, which has since been shown to be demonstrably incorrect [38,39].

Despite the untenability of the theory in its original form in the area for which it was first proposed, the concordance of the idea with simple, pictorial frontier molecular orbital theory has led to its extension to acetals. Indeed, it is possible to find statements in this vein about the catalytic power of glycosyl-transferring enzymes [40]. It must be emphasised that the limited data on the hydrolysis of simple acetals directly refute the idea of an antiperiplanar lone pair of electrons assisting departure of the leaving group. Departure of $H_2PO_4^-$ from the equatorial position of an aldopyranosyl phosphate is faster than from the axial position [41]. This is precisely what would be expected on the basis of the destabilisation of the equatorial substituent by the anomeric effect. The anomeric effect is augmented in mannose derivatives by the "$\Delta 2$ effect" (additional electrostatic repulsion between the electronegative substituent at C1 and the C2 OH [42]): accordingly, the equatorial/axial ratio is highest for mannose (Fig. 7).

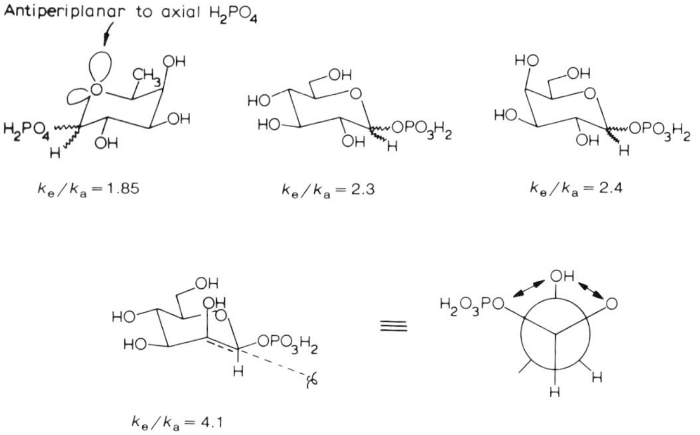

Fig. 7

Likewise the equatorial epimer of compound VIII loses *p*-nitrophenolate 3.3 times faster than VIII itself [43].

The only convincing demonstration of the reality of stereoelectronic control in the acetal function pertains to a fragmentation, rather than a simple leaving-group departure [44].

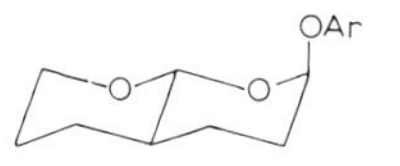

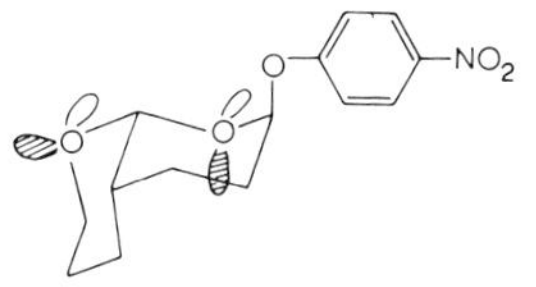

Finally, it should perhaps be noted that the catalytic efficiency of lysozyme can be accounted for quantitatively by valence-bond methods which envisage the pyranose ring as merely changing from $^4C_1$ chair to half-chair [45], and that recent thorough theoretical studies of nucleophilic additions to carbonyl compounds (reactions closely resembling the reverse of departure from an acetal centre) indicate that the frontier orbital description is "inadequate in general" [46].

3. *The chemistry of processes occurring with electrophiles or acids*

3.1. Lifetimes of oxocarbonium ions

Many glycosyl-transfer reactions, and similar reactions involving acetals and ketals, are written as involving a discrete oxocarbonium ion. Proof positive of the reality of such an intermediate is its direct observation, and the demonstration that it has a sensible lifetime under the conditions of the reaction. One can apply two criteria for the "sensibleness" of the lifetime of reactive intermediate. An absolute, conceptual one is that the lifetime must be longer than the period of a bond vibration, $\sim 10^{-13}$ sec. A less exacting one is that the intermediate lives long enough for any other part of the precursor molecule to diffuse away: in water at 25°C this gives a lower limit of $\sim 10^{-10}$ sec [47,48].

Some exceptionally stable oxocarbonium ions can be observed in neutral aqueous solutions: others can be observed in aqueous sulphuric acid and their lifetimes extrapolated to pure water. Some results are given in Table 2. By the normal standards of organic chemists' intuition, *O*-methylated acetophenone is a fairly stable oxocarbonium ion, yet its lifetime is approaching the region in which its existence could be called into question. A linear free-energy relationship exists between the lifetimes of oxocarbonium ions derived (formally) by *O*-methylation of ketones, and rates of attack of SO_3^{2-} on the parent aldehyde or ketone [50]. This relationship was extrapolated to formaldehyde, and gave an estimated lifetime of $CH_3O\overset{+}{C}H_2$ of 10^{-15} sec. The very long extrapolation could be questioned in detail, but there was no avoiding the firm prediction that the methoxymethyl cation was too unstable to exist, and that all reactions putatively involving it are in truth bimolecular. Note that it is the lifetime in the solvent concerned which is the crucial

TABLE 2
Lifetimes of oxocarbonium ions in water (sec) at 25° [49]

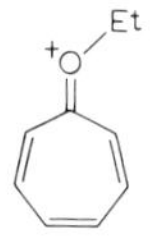

Ion	Lifetime (sec)
Et, +O (cycloheptatrienyl)	59
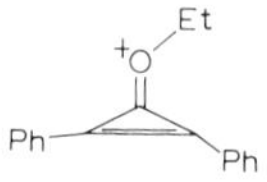 Et, +O, Ph, Ph (cyclopropenyl)	3 X 10^{-4}
+O–CH_3, OCH_3 (phenyl)	7 X 10^{-7}
+O–CH_3 (phenyl)	2 X 10^{-8}
+O–CH_3, NO_2 (phenyl)	8 X 10^{-9}

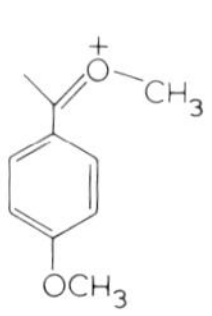

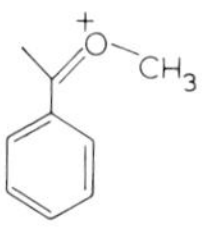

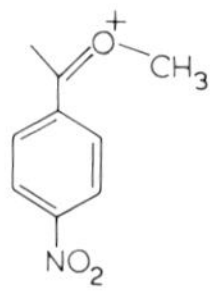

parameter – $CH_3OCH_2^+$ has been observed as a stable species in highly acid media [51].

3.2. Preassociation mechanisms

The prediction that reactions of methoxymethyl derivatives in water are all, necessarily, bimolecular is verified by studies of the reaction of the 2,4-dinitrophenolate [52] and methoxymethyl *N,N*-dimethylanilinium ions in water [53], the reactions of which are unequivocally bimolecular. However, the sensitivity to the nucleophilicity of the attacking group is low: if logarithms of the second-order rate constants for attack of nucleophiles on the methoxymethyl system are plotted against logarithms of second-order rate constants for attack of the same nucleophiles on methyl bromide on methanol, the points are badly scattered about a trend line of slope 0.2–0.3. The rate of reaction of amines with methoxymethyl *N,N*-dimethylanilinium ions is likewise very insensitive to their basicity [53]. α-Deuterium kinetic isotope effects – which measure the extent to which the central atom has approached sp^2 hybridisation at the transition state – are generally in the region normally associated with S_N1 reactions (k_H/k_D values of 1.1–1.2 per deuteron) although small, non-polarisable nucleophiles attacking the anilinium ions give rise to effects below this (in the case of fluoride, even inverse) [53]. β_{lg} values are around -1 [53,54]. All these data indicate a transition state in which both entering nucleophile and leaving group are still present, but the charge on the central carbon atom is nearly as fully developed as in a true S_N1 reaction.

Attack on the methylene group of methoxymethyl derivatives is sterically unrestricted. The position with regard to the more hindered sugar derivatives in water is not very clear cut. The pH-independent hydrolysis of aryloxytetrahydropyrans show higher $\Delta S^\dagger$ values, a more negative β_{lg} value, and different sensitivities to added salts and organic solvents, than neutral methoxymethyl derivatives [54]; they are unambiguously unimolecular. As electron-withdrawing hydroxyl substituents are added to the tetrahydropyranyl ring the intermediate oxocarbonium ion will become less stable – at what point it will become too unstable to exist is not clear. pH-independent hydrolyses of D-glucosyl and D-xylosyl derivatives are subject to solute effects bigger than those experienced with aryloxytetrahydropyrans, but much less than those in the methoxymethyl system, and with the effects parallelling basicity rather than nucleophilicity [55]. It seems as if in water reactions of 2-deoxyaldosides, and of ketosides – where the oxocarbonium ion is tertiary, not secondary – are unambiguously unimolecular, but that reactions of aldosides might just – but only just – go through intermediates so unstable that the presence of any nucleophile is enforced in some circumstances.

In solvents of nucleophilicity similar to water, but of lower polarity, quite unambiguously the transition states for the formation of the products of nucleophilic displacement reactions also contain the leaving group. Reactions of 2,3,4,6-tetra-*O*-methyl α-D-glucopyranosyl and α-D-mannopyranosyl halides with various nucleophiles in methanol [56] gave different products, the glucosyl derivatives yielding largely the β compound, the mannosyl a 50 : 50 mixture. The results were rationalised by a version of the Winstein solvolysis scheme [57] since modified [58] to include a solvent-separated ion pair.

$$
\begin{array}{ll}
\alpha\text{-GlyCl} \rightleftharpoons \alpha\text{-Gly}^{+}\text{Cl}^{-} \xrightarrow{\text{ROH}} \beta\text{-GlyOR} & \\
\quad \alpha\text{-Gly}^{+}\text{Cl}^{-} \rightleftharpoons \alpha,\beta\text{-Gly}//\text{OR} \rightarrow \alpha,\beta\text{-GlyOR} & \\
\beta\text{-GlyCl} \rightleftharpoons \beta\text{-Gly}^{+}\text{Cl}^{-} \xrightarrow{\text{ROH}} \alpha\text{-GlyOR} & \\
\quad \beta\text{-Gly}^{+}\text{Cl}^{-} \rightleftharpoons \alpha,\beta\text{-Gly}//\text{OR} &
\end{array}
$$

The main difficulty with this scheme lies in the concept of an intimate ion pair as a discrete species. Solvent-separated ion pairs can be detected by several physical techniques, as can intimate ion pairs where there is an obvious barrier to recombination [59]. In the absence of a barrier to recombination, the idea of an "ion pair" as an intermediate with a real lifetime (rather than as the extremum of a highly excited bond vibration), seems paradoxical – what is keeping the two halves of the pair apart?

Certainly, the idea of an ion pair (or an ion–dipole complex) cannot accommodate the results of solvolysis of a series of glucosyl derivatives in mixtures of ethanol and trifluoroethanol [60]. There was discrimination amongst solvent components no matter whether the configuration of the product was retained or invented. Acid-catalysed solvolysis of phenyl glucosides showed much the same pattern as solvolysis of compounds with more basic leaving groups. Since phenol is less nucleophilic than the solvent components, ion–dipole complexes cannot be invoked to explain observed specificities. Finally, although the leaving groups used did not lend themselves to direct observation of internal return (the production of products of retained configuration by apparent recombination of an ambident carbonium ion or ambident leaving group with its partner at a different site), the very high proportions of retained product from acidic solvent components when fluoride was the leaving group could be regarded as "internal return" of a leaving group whose nucleophilic atoms are connected by hydrogen bonds (XV).

XV: glucosyl fluoride (HO, HO, OH, CH₂OH) with F···H–O–CH_2CF_3 —heat CF_3CH_2OH→ XVI: glycoside OCH_2CF_3, 88.5% (+ 11.5% β)

It is apparent that nucleophilic displacements at glycosyl centres (as at other centres where formation of a solvent-equilibrated carbonium ion is just not favourable enough) has the following characteristics. (i) Low sensitivity to the nucleophilicity of the attacking nucleophile (low β_{nuc} value if a series of nucleophiles with the same nucleophilic atom is considered). (ii) High sensitivity to leaving group ability (strongly negative β_{lg}). (iii) High charge density at the reaction centre, resulting in rates varying much as one would expect for formation of a free carbonium ion. Common observation of high α-deuterium kinetic isotope effects. (iv) Internal return under appropriate conditions. (v) Stereochemical lack of discrimination (simple electrostatic effects will commonly lead to an excess of inversion). The idea of very weak nucleophilic assistance, even of an electrostatic nature, from the same side as the leaving group, may seem to go against a fundamental principle of physical organic chemistry, but internal return would not happen without it. Further, the occurrence of nucleophilic substitution reactions at bridgeheads (e.g. the solvolysis of 1-adamantyl derivatives [61]) requires the postulation of frontside nucleophilic assistance, if the view is taken that all reactions going through carbonium ions too unstable to exist are necessarily bimolecular.

It seems to the author that characteristics (iv) and (v) sharply distinguish this type of preassociation reaction from a classical Ingold S_N2 reaction. One can provide the following physical picture. The reaction is, energetically, overwhelmingly unimolecular: substrate molecules get within a couple of kcal/mole of the transition state, without a solvent molecule or molecule of nucleophile becoming immobilised. What then happens to these molecules nearly at the transition state – to molecules with the fissile bond containing nearly enough energy to break – is that any species in the immediate solvent shell (including different atoms of the same leaving group) that can provide electrons to neutralise the positive charge, does so wherever it is with respect to the leaving group. This picture is similar to that advanced by Gregoriou for organic reactions in general [62], but has applicability only to reactions which can be shown to proceed via preassociation mechanisms.

There are strong analogies between "metaphosphate" as an intermediate in phosphoryl transfer reactions and glycosyl cations as intermediates in glycosyl-transfer reactions [47]. Both types of reaction proceed by preassociation reactions which are, energetically, overwhelmingly unimolecular. Sound evolutionary reasons can be perceived, why Nature chooses as working materials structural moieties which ordinarily are transferred by preassociation mechanisms. In a classical, Ingold S_N2 reaction the structure of the transition state is determined as much by the nucleophile as by the substrate. If it is accepted that enzymes work by stabilising the transition state of the reaction, then an enzyme that has evolved to catalyse an Ingold S_N2 reaction with one nucleophile will have to be drastically altered to catalyse an S_N2 reaction with another nucleophile. However, if the reaction catalysed is a preassociation reaction – i.e. is energetically overwhelmingly an S_N1 reaction, whose outcome is determined by the nucleophiles which happen to be in the vicinity of electrophile – then minor changes in the nucleophile binding site will suffice to alter the course of the reaction, and evolution of diverse specificities will be rapid.

There is experimental support for this speculation. The natural inducer of the enzymes of the *lac* operon of *E. coli* is allolactose, formed from lactose by *lacZ* β-galactosidase without the glucose becoming free of the enzyme [63].

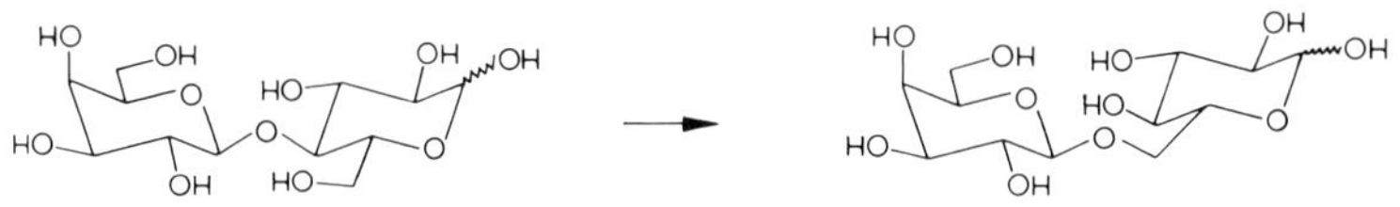

XVII → XVIII

Arguments have been advanced that this production of allolactose occurs via an enzyme-catalysed internal return mechanism [64]. The wild-type form of the second β-galactosidase of *E. coli* (*ebg*^{0}) cannot synthesise allolactose from lactose. However, if a *lacZ*$^-$ strain of the bacterium is put under intense selection pressure such that it has to (a) find a way of hydrolysing lactose fast enough to be able to grow on it, and (b) produce enough allolactose to induce the *lac* operon, and produce enough *lac* permease to allow lactose to enter the cell, then two spontaneous point mutations are sufficient to produce an *ebg* enzyme which performs both tasks. As with the *lacZ* enzyme, the glucose moiety does not become free of the enzyme in the lactose ⇌ allolactose interconversion [65].

3.3. Chemical synthesis of glycosides

With rare exceptions the reactions of synthetic organic chemistry are not, unlike enzymic reactions, performed with the substrates non-covalently bond to a template in the correct orientation for reaction. Therefore the reactions which are widely used in synthetic chemistry tend to be those in which the reaction coordinate lies at the bottom of a deep valley in the free-energy surface: small changes in substrate structure or conditions then have little effect. Preassociation reactions, by contrast, are represented by plateaux on free-energy surfaces. Most glycoside synthesis however involve nucleophilic displacements of suitable nucleophiles at C1 of sugars. It should therefore be no surprise that the literature on the preparation of glycosides and nucleosides provides recipes ever more desperate and baroque [65–68], as attempts are made to direct these preassociation reactions along predictable lines. Equatorial aryl glycopyranosides with *trans* substituents on C2 can be made in partially aqueous solution by reaction of the phenolate with the axial protected glycosyl halide (itself accessible in a thermodynamically controlled reaction – Section 2.1). High stereospecificity is observed possibly by virtue of the ability of the *trans*-2-acyloxy group to act as an intramolecular *syn* nucleophile. Certainly, when it is desired to make equatorial alkyl glycosides, and less polar solvents and anhydrous conditions have to be resorted to, orthoesters can be formed from *cis* acyloxy halides [69–71].

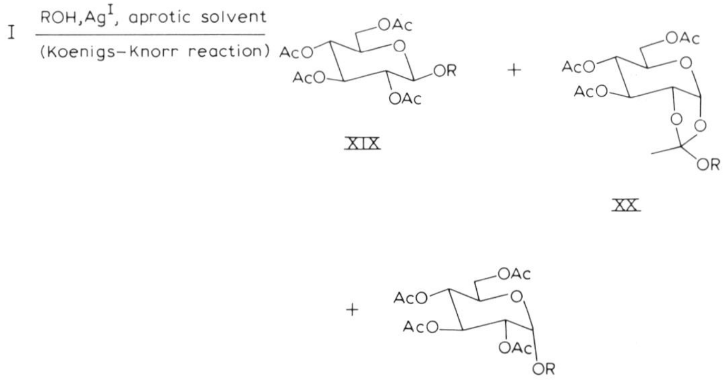

I $\xrightarrow[\text{(Koenigs–Knorr reaction)}]{\text{ROH, Ag}^{\text{I}}\text{, aprotic solvent}}$ XIX + XX + XXI

XIX

XX

XXI

It is usual to add a heavy metal as an electrophilic catalyst and an acid acceptor (e.g. Ag_2CO_3, $Hg(CN)_2$), as well as a drying agent such as calcium sulphate.

The extreme lability of acetobromoglucosamine (XXII) [72] – it readily converts to the amine hydrobromide (XXIII) in the presence of traces of moisture – can be attributed to the ease with which the acetamido group acts as an intramolecular *syn* preassociation nucleophile.

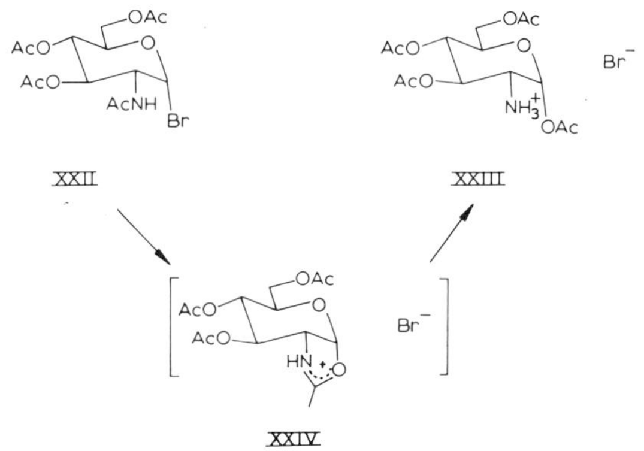

Notwithstanding the ability of *cis* acyloxy and acylamido groups to act as *syn* preassociation nucleophiles, a *trans* disposition is far more effective. Thus acetobromomannose (the C2 epimer of I) gives largely a mixture of α-glycoside and orthoester on attempted glycosylation.

The one rational synthesis of axial glycopyranosides (with a *cis*-1,2-configuration) is due to Lemieux and involves reaction of an axial glycopyranosyl halide protected with non-participating groups (e.g. benzyl) with a tetra-alkyl ammonium halide and the appropriate alcohol in a dipolar aprotic solvent. Lemieux' rationalisation of the success of his procedure is based upon ion pairs, ion triplets and stereoelectronic control of the decomposition thereof [73]. However, the tetra-alkylammonium halide epimerises the glycosyl halide at a rate comparable with the time of reaction, and it would appear that a simpler rationalisation of the success of the method is based upon the faster rate of departure of the equatorial halide and the greater capacity of a neutral, rather than an anionic, nucleophile to act as a preassociation nucleophile in the axial position (cf. [66]): halide ion, which would reconvert the equatorial halide to the axial one, is repelled by the negative end of the dipole of the substituent on C2.

Syntheses of glycofuranosyl derivatives – including nucleosides – via glycofuranosyl halides are as yet entirely empirical.

3.4. Effect of oxocarbonium ion structure

Table 3 gives relative rates of pH-independent departure of alkoxides or phenoxides from oxocarbonium ion centres in aqueous solution. It is seen that these features which are anticipated to stabilise the ion (e.g. aromatic character in the case of the ethoxy tropylium ion) do in fact do so. The effect of the electron-withdrawing

TABLE 3

Effect of structure on pH-independent hydrolyses of acetals: relative rates at 39°C, at constant leaving group acidity

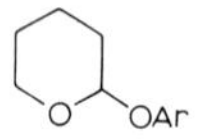

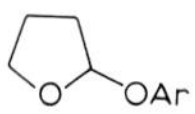

Compound	Relative rate
2-aryloxytetrahydropyran (OAr)	1 [54]
2-aryloxytetrahydrofuran (OAr)	2 [74]
CH_3OCH_2OAr	2×10^{-4} [52,54] [a]
aryl glucopyranoside (OAr)	3×10^{-6} [75]
7-OEt, 7-OR cycloheptatriene	9×10^{12} [b]

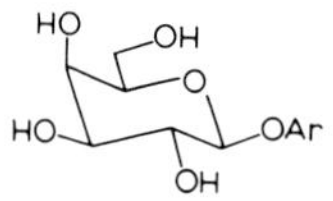

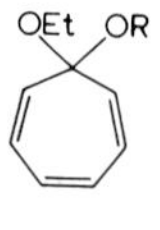

[a] Bimolecular preassociation mechanism with water as a nucleophile.

[b] Comparison made at 15°C with R = Et [76], data for 2-(*p*-nitrophenoxy) tetrahydropyran [77] extrapolated to this temperature, and then extended to a leaving group pK_a of 16 using a β_{lg} value of -1.18 for aryloxytetrahydropyrans obtained at 39°C [54].

hydroxyl groups in the pyranosyl series is very marked; even substituents at C5 have a pronounced effect, the pH-independent hydrolyses of the 2,4-dinitrophenolates (XXV) giving a ρ_I of -6.3 ± 0.8 in the *gluco* and -5.4 ± 0.7 in the *galacto* series.

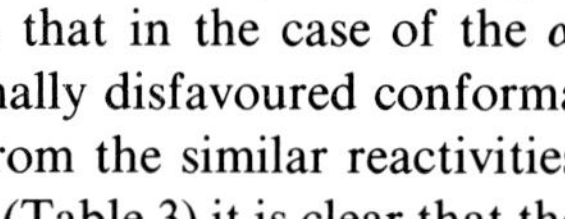

XXV

$Y = H, CH_2OH, CH_2Cl$ or CH_3

$R_1 = H, R_2 = OH$, *gluco* series

$R_2 = OH, R_1 = H$, *galacto* series

Data on nitrogen leaving groups are less extensive and are summarised in Table 4. Note that in the case of the *α-gluco* and *α-xylo* compounds the sugar rings adopt normally disfavoured conformations.

From the similar reactivities of tetrahydropyranyl and tetrahydrofuranyl derivatives (Table 3) it is clear that the difference between a 5-membered and a 6-membered ring per se cannot account for the increased reactivity of furanosyl derivatives, as exhibited for example in the departure of nicotinamide from NAD^+ (Table 4), and the 200–400-fold faster departure of $H_2PO_4^-$ from α-D-ribofuranosyl phosphate than from α-D-glucopyranosyl phosphate [183]. It appears rather that relief of non-bonded interactions on forming an oxocarbonium ion is the main cause of the difference in the reactivity of the various glycosyl derivatives here considered. Conversion of an equatorial hydroxyl of a pyranoside to an axial hydroxyl group –

TABLE 4

Effect of structure on pH-independent hydrolyses of *N*-(alkoxyalkyl) quaternary ammonium ions at 25°C (corrected to constant leaving-group basicity)

$CH_3OCH_2\overset{+}{N}(CH_3)_2Ar$ [a]

4.9×10^4 [53]

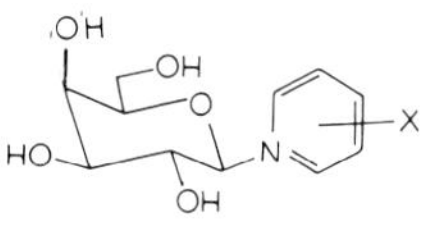

1 [79]

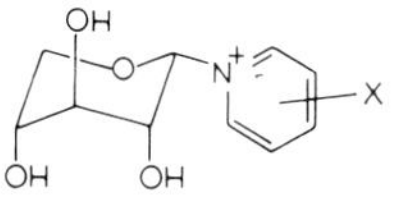

4.7 [80]

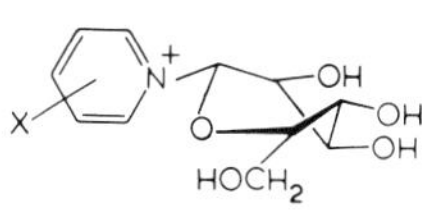

20 [80]

(adenosine diphosphate)

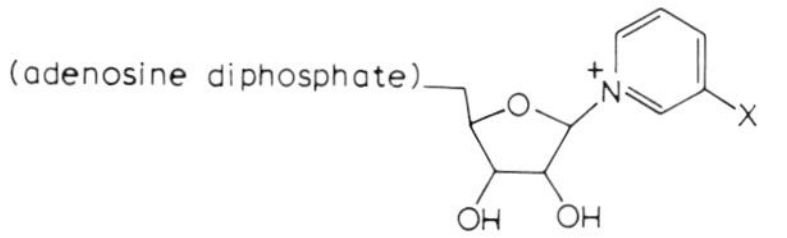

24 [81]

[a] Bimolecular preassociation mechanism.

as with the change from glucose to galactose – increases the rate by a factor of about 3 [75]. Axial substituents in a chair conformation favour the formation of the half-chair conformation in which 1,3-diaxial interactions are relieved [1]. Similar relief of strain may well account for the enhanced rates of furanosyl derivatives.

It is noteworthy that the α-glucopyranosyl and α-xylopyranosyl pyridinium salts are not hydrolysed very much faster than their β-anomers. If an antiperiplanar lone pair of electrons were a requirement for pyridine departure, one would expect the two salts – which could put all substituents except the leaving group equatorial in the $^4C_1$ conformation – to be hydrolysed much faster than their anomers [80].

If rates for the water reaction of $CH_3OCH_2\overset{+}{N}Me_2Ar$ [53] and CH_3OCH_2OAr [52] are extrapolated to the pK_a of hydrogen fluoride, and compared with the rate of hydrolysis of CH_3OCH_2F [82], the rates of departure of nitrogen, oxygen and fluorine leaving groups, at constant pK_a, are seen to be roughly constant. A similar procedure with β-D-galactopyranosyl nitrophenolates, pyridinium salts and fluoride yields the order $N \ll O < F$ – i.e. the order of ground-state stabilisation by the anomeric effect (although the stability of the azide is a puzzle) [83].

The leaving ability of leaving groups attached by sulphur is much lower than that of oxygen-leaving groups of the same pK_a – it has been estimated that the departure of arenethiolate from PhCH(OMe) SAr is some 10^5 times slower than the departure of phenolate from PhCH(OMe)OAr [48,84]. Likewise, it can be estimated that

β-D-galactopyranosyl picrate would have a lifetime in water of about a minute [75,85], but aqueous solutions of the thiopicrate can be handled quite readily [86].

3.5. Intramolecular nucleophilic assistance

Much data of a qualitative, preparative nature exist on the way neighbouring acyloxy groups bedevil the synthesis of glycosides by participating in nucleophilic substitution reactions at C1 of glycosyl halides (see Section 3.3). The acquisition of quantitative data had for the most part to wait upon the determination of the tertiary structure of the glycosidase lysozyme, and the construction of supposed model compounds for the action of this enzyme. The pH-independent hydrolysis of *p*-nitrophenyl 2-acetamido-2-deoxy-β-D-glucopyranoside was reported in 1967 [87], but convincing demonstration that the enhanced rate of aglycone liberation was due to neighbouring group participation by the acetamido group, leading to oxazoline formation had to wait a decade [88], the oxazoline was detected with X = F, and methanolysis was demonstrated to give the methyl β-glycoside quantitatively.

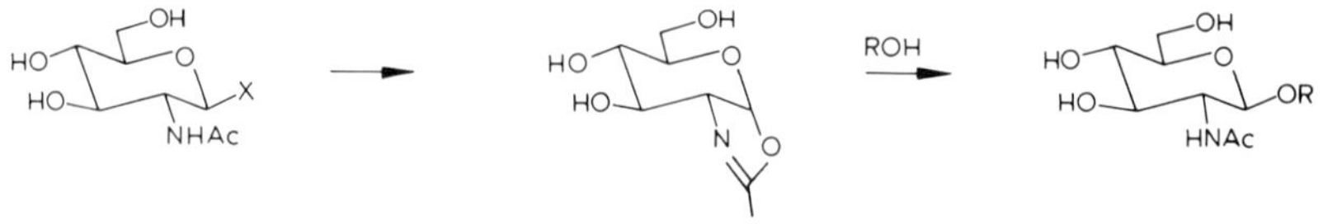

XXVI → XXVII —ROH→ XXVIII

With X = 2,4-dinitrophenolate, this participation gave a rate enhancement over the glucoside of 40 in hydrolysis, and about 500 in acetolysis [89]. In the transition state for amide group participation, the charge is more dispersed than in the transition state for reaction without participation, and accordingly participation is more important in less polar solvents. The more effective participation in the hydrolysis of the fluoride (a rate enhancement of around 400 being estimated) may be attributable to the anomeric effect of fluoride rendering conformations in which acetamido group is better placed to participate more accessible.

Whilst there is no evidence for kinetically significant participation by unionised hydroxyl groups in reactions of glycosyl derivatives, participation by ionised hydroxyl groups is responsible for the well known base lability of aryl glycopyranosides. In the case of *trans*-1,2-glycosides an epoxide is transiently formed, which with hexoses usually gives 1,6-anhydrosugars [1].

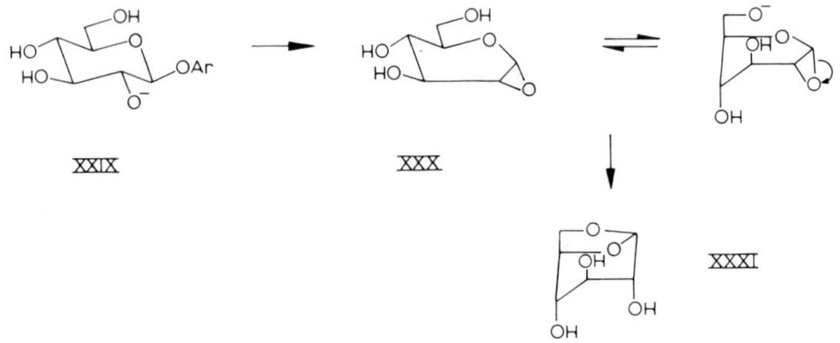

XXIX → XXX ⇌ → XXXI

α-Glucopyranosides are much less labile – 70° higher temperatures are required for a comparable rate to the β compound – but the reaction probably involves direct attack by the ionised 6-hydroxyl although some bimolecular attack is claimed [90]. Nitrophenyl glycosides can react as well by intramolecular and intermolecular S_N2_{Ar} reactions on the benzene ring [91,92].

Aryl furanosides with *trans* 2-OH groups are also base labile, probably because of an analogous formation of a 1,2-epoxide [93].

3.6. Electrostatic stabilisation?

The determination of the tertiary structure of the hens' egg white lysozyme, and the subsequent proposal that the side-chain carboxylic acid group of Glu 35 act as a general acid, and the ionised side chain of Asp 52 acted to electrostatically stabilise an oxocarbonium ion intermediate, led to a search for a manifestation of these phenomena in model systems [6]. In the scramble electrostatic stabilisation was invoked where there was no compelling reason to distinguish it from ordinary nucleophilic participation. Thus, 3,5-dichlorophenolate departs from compound XXXII 100 times faster than from its *p*-isomer [94]: a similar 22-fold acceleration of the *ortho* compared to the *para* isomer was observed with compound XXXIII. The rate acceleration in both cases was attributed to electrostatic stabilisation, but since the 3-methoxyphthalide was in both cases the product, it is not clear why humdrum neighbouring group participation was not invoked.

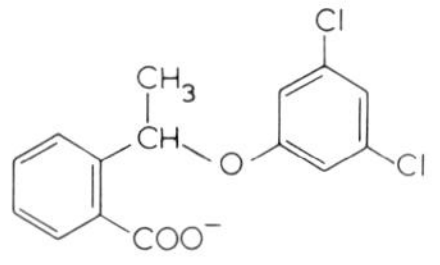

XXXII 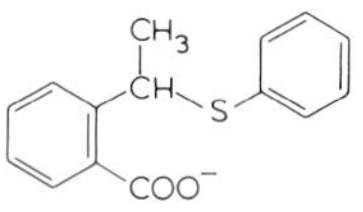 XXXIII

A very carefully thought out test of the reality of electrostatic stabilisation of a developing oxocarbonium ion was applied by Loudon and Ryono [95] using compounds XXXIV and XXXV. (The oxocarbonium ion is generated by rate-determining proton transfer to the vinyl ether function, rather than by departure of a leaving group.) Neither carboxylate can participate nucleophilically; any lactone from XXXIV would be unconscionably strained. Furthermore, water molecules in XXXIV cannot be accommodated between the carboxylate and the oxocarbonium ion centre: the *endo* face of the system is hydrophobic.

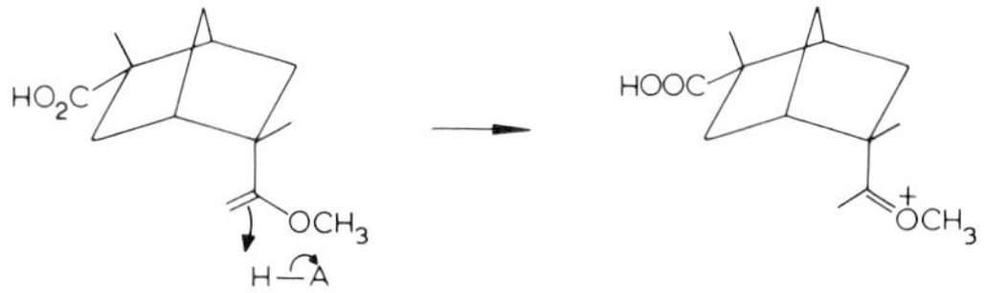

XXXIV 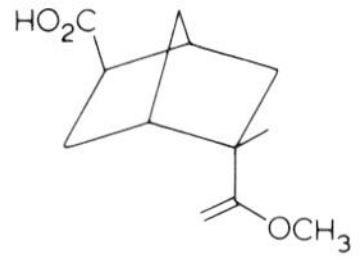 XXXV

Only insignificant differences on the rate of hydrolysis were observed on ionisation of the carboxylic acid group of XXXIV – about the same as those observed with XXXV.

The idea of intramolecular electrostatic, as distinct from covalent, stabilisation of an oxocarbonium ion must therefore be considered as yet unproven.

4. Processes occurring via the application of acidic or electrophilic assistance to the departure of oxygen-leaving groups

When the leaving group is of low acidity – as it is in common acetals, ketals, and glycosides – then protonation (or coordination with a metal) can increase its leaving-group ability. In the case of protonation, two cases can be discussed. In the first, the conjugate acid is a species with a real lifetime, so the proton can be transferred to and from the oxygen of the leaving group many times before the C—O bond is broken. In such circumstances the rate depends only on the ability of the medium to protonate a neutral molecule, rather than on the nature of the individual acids present, and specific acid catalysis is observed. This is the situation with run of the mill acetals and ketals, and with all glycosides. The second case describes situations where the protonated substrate is too unstable to exist. In this case the protonation reactions are necessarily bimolecular: the rate depends on the identity of the individual acids present and general acid catalysis is observed. General acid catalysis (but not specific acid catalysis) can be intramolecular as well as intermolecular, and both can be made synchronous with nucleophilic participation.

4.1. Specific acid catalysis of acetals, ketals and glycosides

In the case where both alkoxyl groups are identical, only one mechanism can be written *:

$$R_1R_2C(OR_3)_2 + H_3O^+ \longrightarrow R_1R_2C(OR_3)(\overset{+}{O}HR_3) \xrightarrow{\text{slow}} R_1R_2C{=}\overset{+}{O}R_3 + R_3OH$$

$$\longrightarrow R_1R_2C(OH)(OR_3) \xrightarrow{\text{fast}} R_1R_2C{=}O + R_3OH$$

Evidence for this mechanism includes [4]:

(i) Solvent deuterium kinetic isotope effects, k_{D_2O}/k_{H_2O}, of around 2. D_2O is less basic than H_2O, and therefore the same stoichiometric concentration of strong acid

* In an achiral environment.

will protonate a higher fraction of the substrate in D_2O than in H_2O. (The rate-determining C—O fission is not subject to a solvent isotope effect.)

(ii) Positive entropies of activation. Bimolecular reactions involving the solvent, which necessarily require the partial immobilisation of a solvent molecule, tend to have lower entropies of activation than immolecular reactions, although the position is clouded by other phenomena which may increase order.

(iii) Inverse dependence of the rate in moderately concentrated acids on the activity of water. We can write

$$\log k_{ob} + H_O = \omega \log a_{H_2O} + c$$

where H_O is the Hammett acidity function, a measure of the ability of the solution to protonate a neutral base. The H_O scale and the pH scale become identical in dilute aqueous acid. ω values for unimolecular reactions are usually negative, as are ω values for the hydrolysis of acetals and ketals (including acetals of formaldehyde, which rather casts doubt on the criterion).

(iv) The reaction is strongly accelerated by electron-donating substituents in Ar or R_1, Hammett ρ or Taft ρ^* * values being

$ArCH(OR)_2$	−3.3 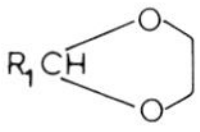
$R_1CMe(OR)_2$	−3.5
$R_1CH(OR)_2$	−3.4
R_1CH (cyclic acetal, 1,3-dioxane)	−4.0

(v) Substituents in *both* potential leaving groups of symmetrical formals ($CH_2(OR)_2$) give a ρ^* value of −1.66 [96]. This is a combination of three separate effects: electron-donating substituents in the two potential leaving groups will increase the concentration of the conjugate acid, decrease the leaving-group ability of the protonated group, but increase the stability of the oxocarbonium ion by virtue of the group remaining.

(vi) Volumes of activation for acetal and ketal hydrolysis are near zero, in accord with what is commonly observed for A1 reactions [97].

The buildup of the intermediate simple hemiacetal is known in a couple of cases [98,99], but not where the only pathway for hydrolysis is specific acid catalysis. In

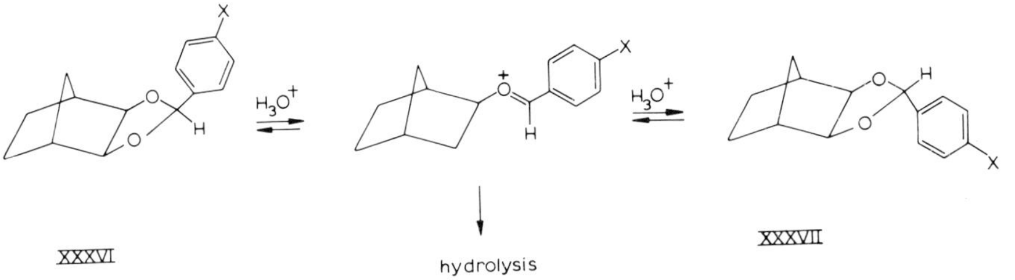

XXXVI $\underset{}{\overset{H_3O^+}{\rightleftharpoons}}$ (oxocarbonium ion) $\overset{H_3O^+}{\rightleftharpoons}$ XXXVII

↓ hydrolysis

X = NO_2	hydrolysis ≫ epimerisation	hydrolysis > epimerisation
X = H	hydrolysis > epimerisation	hydrolysis ~ epimerisation
X = MeO	hydrolysis ~ epimerisation	hydrolysis ≪ epimerisation

* $\sigma^* = 0.45\,\sigma^I$. ρ^* values are on the same scale as Hammett ρ values for aromatic compounds.

the case of cyclic acetals, the leaving group is physically attached to the oxocarbonium ion, and, as the stability, and hence the lifetime of the oxocarbonium ion increases, so does the probability of recombination. This was demonstrated elegantly using compounds XXXVI and XXXVII [100].

When the two groups attached to the *pro* acyl carbon atoms are not identical, then a mechanistic ambiguity arises since productive protonation can take place, in principle, on either oxygen. In the case of glycosides this protonation could give rise to either a cyclic or an acyclic oxocarbonium ion (with such nucleophilic assistance as is enforced). The cyclic oxocarbonium ion reacts with water to give aldose directly, but the acyclic ion gives a hemiacetal or hemiketal which then reacts further. In principle, too, an acyclic ion could react with a different hydroxyl group of the sugar, or with the same hydroxyl group in a different sense, so anomerisation, and change of ring size concomitant with hydrolysis is firm evidence for an acyclic pathway.

There is no doubt that simple aldopyranosides react via exocyclic protonation (top pathway). The evidence for this includes:

(i) Strongly positive entropies of activation, indicating a dissociative reaction [1].

(ii) Aglycone ^{18}O kinetic isotope effects (k_{16}/k_{18}) of 1.03 for methyl α-D-glucopyranoside [101] and 1.035 for *p*-nitrophenyl β-D-galactopyranoside [102].

(iii) A linear relationship (slope 1.0) between logarithms of rates of acid hydrolysis of methyl β-D-glycopyranosides and logarithms of rates of pH-independent hydrolyses of the corresponding 2,4-dinitrophenyl glycosides (XXV) [75].

(iv) Pronounced retarding effect of electron-withdrawing substituents at C5, in accord with the substituents being in the oxocarbonium ion, rather than the leaving group part of the molecule [75,103]. The ρ_I value calculated by Capon and Ghosh

[104] for the hydrolysis of 2-naphthyl β-glucoside, glucuronide, and glucuronide anion is -6.5, in close agreement with that obtained with compounds XXV and the corresponding methyl glycosides. The slope of 1.0 obtained in the linear free-energy relationship described in iii above means that the effect of substituents on C5 on the reversible protonation step is negligible.

(v) Small effect of substituents in aglycone. This is in accord with the aglycone being the leaving group in the rate-determining transition state, since the effect of substituents on the concentration of the conjugate acid, and on the leaving ability of the protonated aglycone, are in opposite senses. From the literature the following β_{lg} values for the acid-catalysed hydrolyses of aryl glycopyranosides can be calculated: $+0.27 \pm 0.04$ for β-glucosides [105], $+0.01 \pm 0.02$ for α-glucosides [106], $+0.09 \pm 0.02$ for β-xylosides [107], and $+0.026 \pm 0.012$ for α-mannosides [108]. The slightly bigger β_{lg} values for the β-glycosides could well be real: if so they imply more bond breaking in the transition state for α-glycoside hydrolysis.

Acid-catalysed hydrolysis of glycopyranosides is very sensitive to the substituent at C2 – a ρ_I value of -8 to -9 can be calculated for the methyl β-glucosides when the substituent is —OH, —NH_3^+, and H [1]. Bulky aglycones result in significant steric acceleration: the faster rates of hydrolysis of tertiary alkyl glycosides are due as much to steric strain as to aglycone–oxygen cleavage [109], although alkyl–oxygen fission does take place, as it does with ferrocenylcarbinyl β-glucoside (XXXVIII), which gives the very stable ferrocenylcarbinyl cation [110].

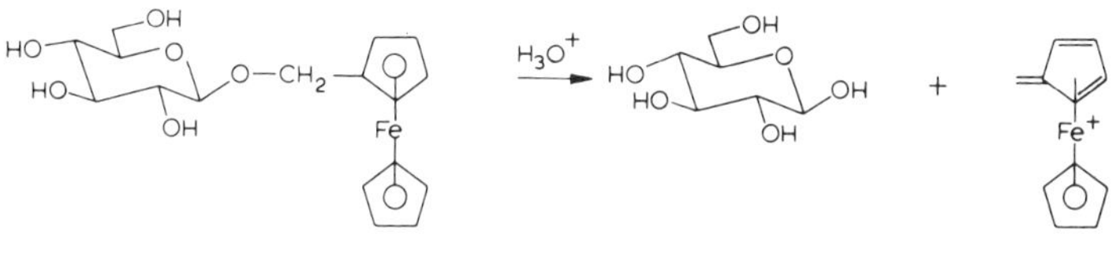

XXXVIII

The reverse anomeric effect will act in a complex way upon α- and β-glycosides: it will disfavour protonation of the α compound but increase the leaving ability once the aglycone is protonated. In the event methyl β-glycosides are hydrolysed twice as fast as methyl α-glycosides, but because of the differing β_{lg} values, this order is reversed with aryl glycosides, so that phenyl α-glucoside is hydrolysed about 4 times, and the p-nitrophenyl glucoside about 10 times, faster than their β anomers [1].

Alkyl aldofuranosides normally react faster than their pyranoside isomers [111]. The reactions are normally associated with negative entropies of activation, and this combined with ω values more indicative of an A2 reaction, has led to the suggestion that the reaction of furanosides is A2 with ring opening [1]. It was also suggested that negative entropies of activation could be accounted for by reversible ring opening and closing [111], but one would have thought such a process would have led to the accumulation of the more stable pyranosides. Given the dominance of the translational component of the partition function, one might expect that reactions giving two fragments (aglycone and glycosyl cation) would have more positive entropies of activation than reactions in which the two fragments remain attached.

Acid-catalysed hydrolyses of glycosides proceeding through aldofuranosyl cations, moreover, can be observed [112]. Electron-withdrawing substituents in the aglycone, as is clear from the data on pyranosides, do not appreciably affect reaction through the cyclic glycosyl cation; but they strongly decelerate reaction through the acyclic cation, in which the aglycone is part of the oxocarbonium ion. Therefore electron-withdrawing substituents favour the cyclic mechanism: in the case of β-D-xylofuranosides, the 2-methoxy ethyl derivative reacts through an acyclic cation ($\Delta S^{\dagger} = -2.6$ eu) whereas the 2-chloroethyl compound reacts through a cyclic cation ($\Delta S^{\dagger} = +2.6$ eu).

The aglycone electronegativity at which the change from acyclic to cyclic mechanism occurs has been examined for various aldofuranosides: the data are given in

TABLE 5

Aldofuranosides with the most electron-withdrawing aglycone compatible with a ring-opening mechanism for acid-catalysed hydrolysis (parentheses: most electron-donating aglycone compatible with reaction via an aldofuranosyl cation) [113,114]

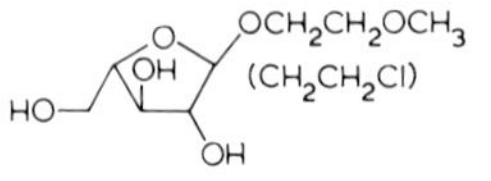

HO, O, OH, OH; $OCH_2CH_2OCH_3$ (CH_2CH_2Cl) β-D-Xylo	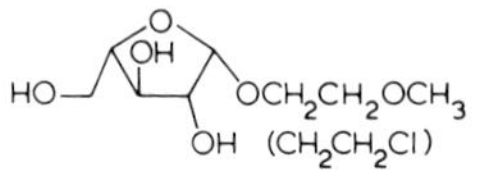 HO, O, OH, OH; $OCH_2CH_2OCH_3$ (CH_2CH_2Cl) α-D-Xylo
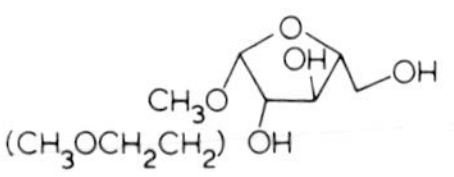 O, OH, OH, OH; CH_3O ($CH_3OCH_2CH_2$) β-D-ara	$(CH_3)_2CHO$ (CH_3CH_2); O, OH, OH, OH α-D-ara
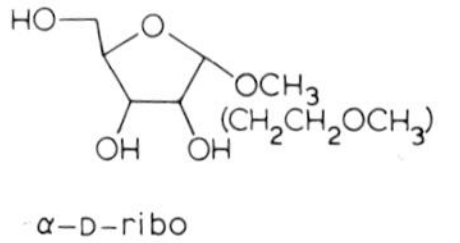 HO, O, OH, OH; OCH_3 ($CH_2CH_2OCH_3$) α-D-ribo	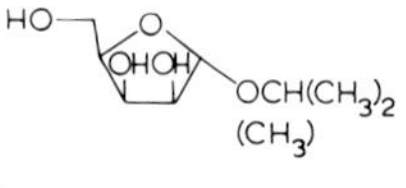 HO, O, OH, OH; $OCH(CH_3)_2$ (CH_3) α-D-lyxo

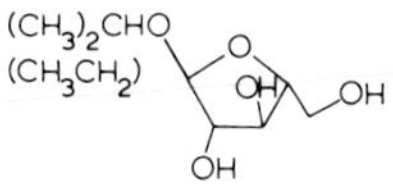

Ring-opening is observed in the β-D-ribo series even with

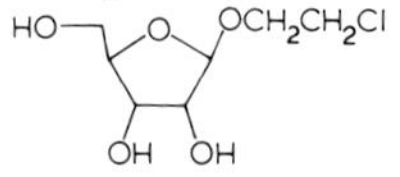

HO, O, OH, OH; OCH_2CH_2Cl

Table 5. The precise factors which determine the relative energies of the ring-opening and cyclic pathways in each individual furanoside are not readily understood, but the intuitive expectation that lengthening the C1—O4 bond would further increase the crowding of *cis* substituents on C3 and C4, which would therefore favour a cyclic mechanism, does appear to be fulfilled. Replacement of a 2-hydroxyl group by an NH_3^+ group, as in the pyranoside series results in a sharp deceleration in the rate of acid-catalysed hydrolysis [115].

Considering that the acid-catalysed hydrolysis ("inversion") of sucrose has been studied perfunctorily for the last century and a half, surprisingly little is known about the hydrolysis of ketosides. The positive entropies of activation of methyl β-D-fructofuranoside (+13 eu) and sucrose (+8 eu) argue in favour of an A-1 mechanism [1] involving a fructofuranosyl cation.

4.2. Intramolecular nucleophilic assistance in specific acid-catalysed processes

There is surprisingly little data on this. The acetamido group participates in the acid-catalysed hydrolysis of methyl 2-acetamido-2-deoxy-β-D-glucopyranoside [1] to the same extent as in the spontaneous reaction of the dinitrophenolate [75], but apart from that examples are scarce. Capon and Thacker [116] estimated that the ring-closure reactions of the straight-chain dimethyl acetals of sugars were anchimerically assisted, but a careful search failed to turn up other examples [117]. Interestingly, in the compounds XXXIX and XL, neighbouring group participation by the double bond or the cyclopropane ring contributes rate accelerations of 2.5 and 121, respectively [118], compared to factors of 10^{11} and 10^{14} for the solvolysis of the corresponding 7-norbornyl *p*-toluenesulphonates.

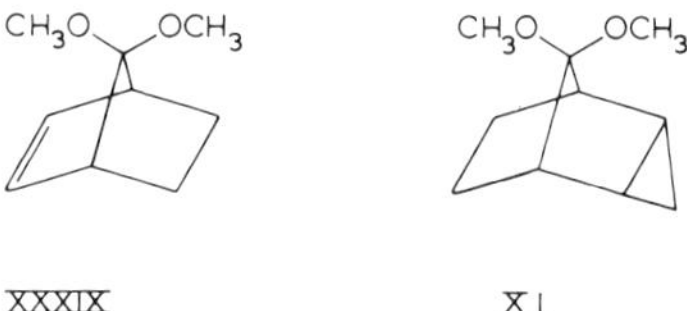

4.3. Intermolecular general acid catalysis of the hydrolysis of acetals and ketals

Fig. 8 shows, in general terms, the reaction profile for the specific acid-catalysed hydrolysis of an acetal or a ketal.

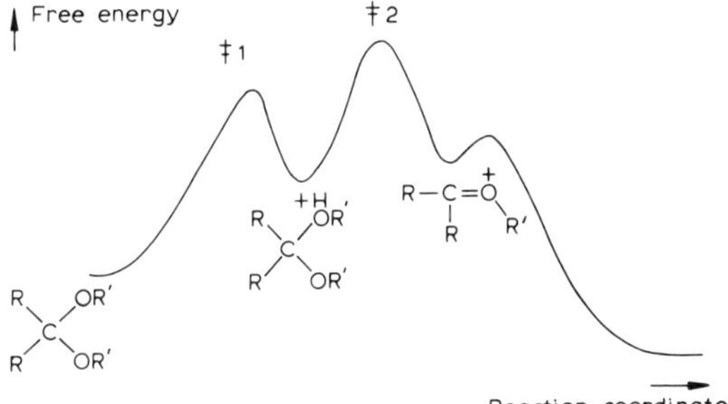

Fig. 8

If transition state 1 – because of the low basicity of the acetal * – becomes of higher energy than transition state 2, then rate-determining proton transfer, and general acid catalysis, will be observed. This was the reasoning adopted by Kankaanperä and Lahti [120,121] to observe general acid catalysis in the hydrolysis of $CH_3CH(OEt)OCH_2Cl_3$. Anderson and Capon [122] considered the requirements for the merging of transition states 1 and 2. Applying the Hammond postulate, they predicted that this was most likely to happen when the leaving group was of low basicity – when both transition state 1 and transition state 2 would move towards the protonated acetal – or when the oxocarbonium ion was very stable, in which case transition state 2 would move towards the protonated acetal. Accordingly, intramolecular general acid catalysis was observed in the hydrolysis of compounds XLI–XLVI, in which either the leaving group is of low basicity, or the oxocarbonium ion is particularly stable: in the case of XLVI, it is not an ion at all.

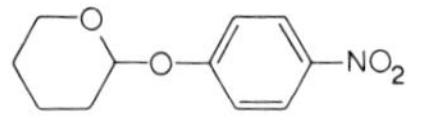

XLI [77] 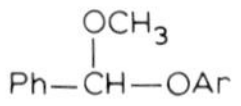 XLII [123] 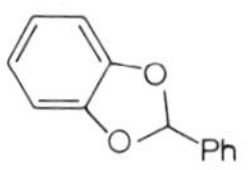 XLIII [126]

XLIV [76] 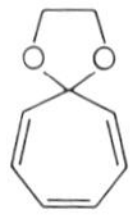 XLV [125] 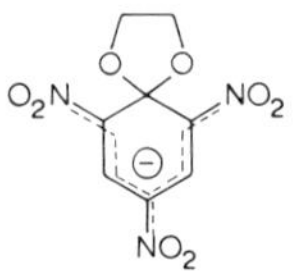 XLVI [124]

Steric strain – whether due to a small ring, which is opening at the transition state, or to crowding, which is relieved at the transition state – also facilitates the observation of general acid catalysis in the hydrolysis of compounds XLVII–L.

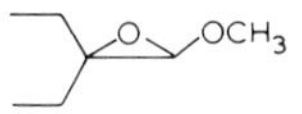

XLVII [127] 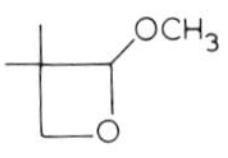 XLIX [128] 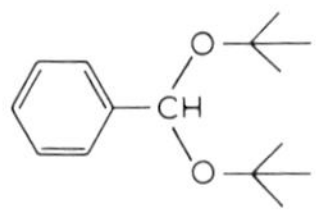 L [129]

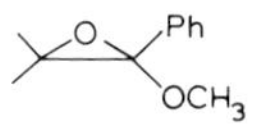

XLVIII [127]

On the Anderson–Capon picture this strain could be regarded as the equivalent of stabilising the oxocarbonium ion: it is the free-energy difference between the ion

* The barrier to proton transfer between oxygen atoms in the thermodynamically favourable direction is almost invariably diffusional [119].

and the neutral substrate which is important, rather than the absolute position of either on the energy scale. However, a conceptually simpler way of rationalising those structures which lend themselves to the observation of general acid catalysis is that this is observed when the protonated acetal or ketal is too unstable to exist. At 30°C compound XLI decomposes in 80% aqueous dioxan at a rate of 9×10^{-5}/sec. If we assume that the pK_a difference of $Ar\overset{+}{O}H_2$ and ArOH is the same as that between $ArNH_3^+$ and $ArNH_2$, then data on *p*-nitroaniline indicate that this is around 17.3 pK units [130]. Using a β_{lg} value of -1.18 for aryloxytetrahydropyrans [54], it is possible to estimate the rate of decomposition of exocyclically protonated acetal (XLI) of around 2×10^{16}/sec, i.e. the protonated compound is too unstable by far to exist.

In general acid catalysis of acetals two processes – proton transfer and C—O cleavage – are occurring simultaneously, and therefore the simple 2-dimensional free-energy diagram of Fig. 8 is an oversimplification. In a More O'Ferrall–Jencks diagram the free-energy axis is perpendicular to the plane of the paper (free energy commonly being represented by contours), and the degrees of the two concurrent processes are represented by displacements in orthogonal directions. Thus, Fig. 9a represents specific acid catalysis, and Fig. 9b spontaneous departure of the leaving group: in both cases the reaction coordinate lies round the edges of the diagram. Fig. 9c represents a hypothetical general acid-catalysed reaction in which C—O breakage

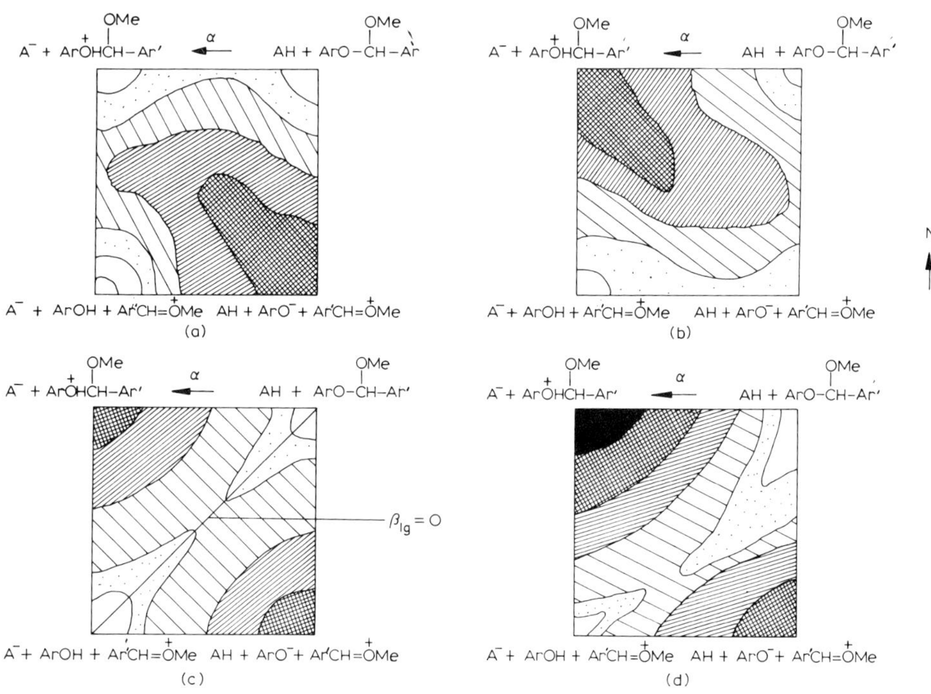

Fig. 9

and proton transfer are synchronised, so that $\beta_{lg} = 0$. This type of diagram can be used to interpret the results of Capon and Nimmo [123] on substituent effects on the general acid catalysis of ArCH(OMe)OAr and of Jensen et al. [182] on similar reactions of the alkyl derivatives pXC$_6$H$_5$—CH(OR)$_2$ and PhCH(OEt)OR. (The aryloxy group leaves in the first step.) In the case of substituted benzaldehyde methyl phenyl acetals, as the substituents become more electron donating, i.e. as the intermediate cation XC$_6$H$_4$CH=$\overset{+}{\text{O}}$Me becomes more stable the Brønsted α for catalysis by a series of carboxylic acid decreases (from 1.0 for X = mNO$_2$ to 0.68 for X = pMeO). Making the cation more stable corresponds to lowering the south edge of the free-energy diagram: this has the effect of moving the transition state east. In the same way decreasing the strength of the acid raises the whole western edge: this has the effect of making ρ more negative – i.e. increasing the extent of C—O cleavage and moving the transition state south. In the case of benzaldehyde methyl-substituted phenyl acetals, PhCH(OMe)OAr, as the leaving group becomes more acidic rates of general acid-catalysed hydrolysis pass through a minimum and Brønsted α values decrease. Making the leaving group more acidic corresponds to tilting the diagram about the line for $\beta_{lg} = 0$, depressing the south-east and raising the north-west corner. This has the effect of moving the transition state east, crossing the $\beta_{lg} = 0$ line in the process.

The characterisation of transition states by 2-dimensional structure–reactivity correlations has been dealt with rigorously [131]. The type of behaviour described above characterises a transition state in which the reaction coordinate is in an ENE—SSW direction. The observation of Brønsted α values of > 0.5, and non-linear variation of rate with aglycone acidity, with most substituents giving a negative β_{lg} value, suggests that qualitatively the free-energy profile looks like Fig. 9d.

It is clear from this work that acidic assistance to aglycone departure in the active sites of enzymes can result in principle in a large range of β_{lg} values, or even non-linear plots, in a way which depends on the aglycones used, the acidic group concerned, and the extent of nucleophilic assistance to aglycone departure (corresponding, formally, to stabilisation of the oxocarbonium ion). This is in addition to the normal problems of obtaining structure–reactivity parameters for enzymic reactions – "chemical noise" caused by the bad fit of unnatural substrates, and possible change of rate-limiting step. Structure–reactivity correlations should therefore be used only with the utmost caution in studies of glycosidase mechanism.

It is noteworthy that there is a pH-independent "water" reaction in the hydrolysis of compound XLIII [126]. That this involves general acid catalysis by a water molecule is shown by the strongly negative entropy of activation (−21 eu) and a solvent deuterium kinetic isotope effect ($k_{H_2O}/k_{D_2O} = 1.61$). Effective general acid catalysis can therefore be applied by an acid of higher pK_a (15.7) than the leaving group (~ 10): it quite clearly however must involve only hydrogen bonding rather than complete proton transfer.

4.4. Intramolecular general acid catalysis of the hydrolysis of acetals, ketals and glycosides

The lability of salicyl β-D-glucopyranoside (LI) compared with its *para* isomer was tentatively attributed to intramolecular general acid catalysis in 1963 [132], in

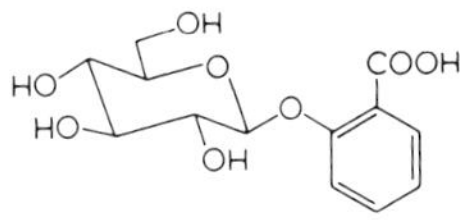

LI

addition to the acid-catalysed hydrolysis, a pH-independent reaction was observed apparently governed by the ionisation of the aromatic carboxylic acid group. Following upon the proposal that the unionised carboxylic acid group of Glu 35 in hens' egg white lysozyme acted as a general acid catalyst, the kinetics of the hydrolyses of a whole series of salicyl derivatives (LII–LV) were investigated. Participation of the *ortho* carboxylic acid group was demonstrated by comparison either with *para* isomer or the methyl ester.

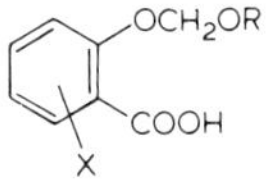

LII [133–137] 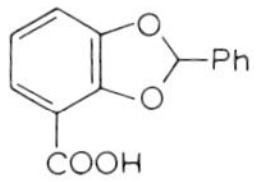   LIII [138] 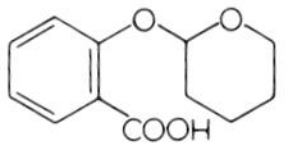   LIV [139]

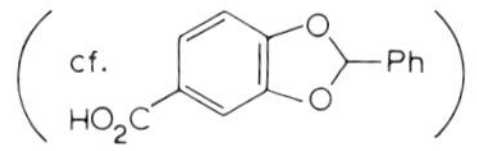

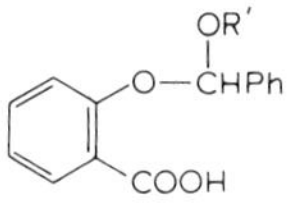

(R′ = Et [140]; R′ = Me [139])

LV

That the rate acceleration due to the carboxylic acid was intramolecular general acid catalysis, and was not the kinetically equivalent A_1 reaction of the anion, in which the transition state was stabilised by electrostatic catalysis (or *syn* nucleophilic attack) was established by the following pieces of evidence.

(i) Compound LIII, in which the ionised carboxylate group would be remote from the oxocarbonium ion centre behaves similarly to LII and LV.

(ii) Compounds LVI and LVII were prepared and shown to be stable under the conditions of, but absent during, the hydrolysis of LII (X = H, R = Me) [135].

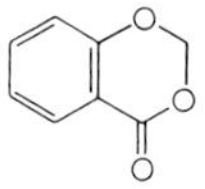

LVI 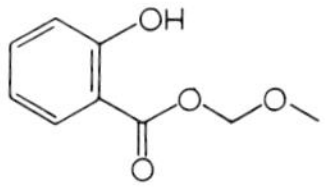   LVII

(iii) Solvent isotope effects on the intramolecular reaction ($k_{H_2O}/k_{D_2O} = 1.43$ for LII and 1.28 for LIII, were similar to those observed in the intermolecular general acid catalysis of benzylidene catechol (XLIII) (1.4–1.5) [138].

(iv) The effects of substituents in the salicyl moiety, when correctly analysed, indicate a low sensitivity to acid strength and a high sensitivity to leaving-group ability. Derivatives of salicylic acid, in which this moiety behaves as a phenol, are an unfortunate choice for the investigation of neighbouring group reactions since the phenolic group and the carboxylic acid groups are conjugated and substituents affect both functions. If data for LII (R = Me) [141] and (LV) R = Et) [140] are analysed according to

$$\log(k_1/k_0) = \rho_1\sigma_1 + \rho_2\sigma_2$$

ρ_1 (governing the phenol) was 0.89 for the methoxymethyl system and 0.93 for substituted analogues of compound LV (R = Et); ρ_2 (governing the carboxylic acid) was 0.02 for the methoxymethyl system and ~ 0.2 for the benzaldehyde derivatives (LV). The similarity of these effects is some reassurance that the requirement for a nucleophile to participate in the reactions of methoxymethyl derivatives does not have very significant effects on the consequences of the application of acidic catalysis.

Since compounds LIII, LIV, and LV are the salicyl analogues of compounds XLIII, XLI and XLII, in which intermolecular general acid catalysis is observed, it is possible to estimate the effective molarity of the carboxylic acid group acting as a general acid in these salicyl acetals. Figures of around 10^2 M for LIII [138], and $10^{3.7}$, $10^{4.5}$ and 10^4 M are estimated for LIV, LV (R = Et) and LV (R = Me), respectively [142]. The effective molarity is not sensibly altered by substituents in the salicyl moiety of LV.

The stereochemical requirements for observation of intramolecular general acid catalysis seem quite severe – less effective participation is observed in the hydrolysis of 8-methoxymethoxy 1-naphthoic acid and none at all in the hydrolysis of (2-methoxymethoxyphenyl) acetic acid. Hydrolysis of 2-methoxymethoxy-3-methyl benzoic acid is quite strongly sterically accelerated [136] but this is also true of its specific acid-catalysed reactions.

Putting an acid catalytic group in the oxocarbonium ion portion of a hydrolysing acetal or ketal has resulted in the observation of intramolecular general acid catalysis only in the aromatic systems LVIII and LIX, but not in the aliphatic system (LX). Even so, the effective molarity of the carboxylic acid group in the hydrolysis of

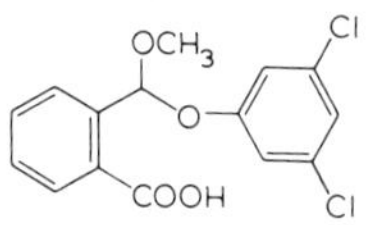

OCH3 Cl O COOH Cl

L VIII [94]

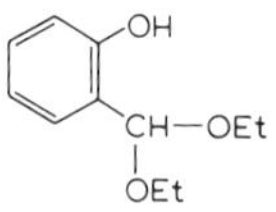

OH CH—OEt OEt

L IX [143]

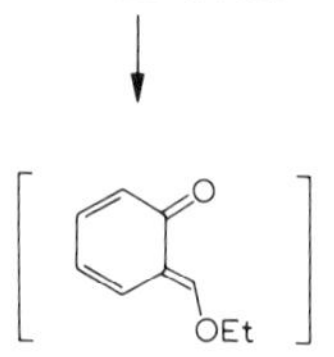

O OEt

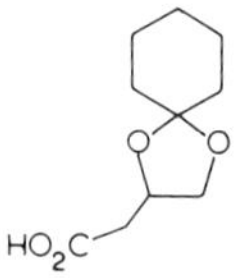

HO2C O O

L X [144]

LVIII is only 38 M [142] and the reaction of LIX may well be brought about by the non-ionic nature of the product.

4.5. Intramolecular general acid catalysis concerted with intramolecular nucleophilic (or electrostatic) assistance

Most of the quantitative work in this area has been done with putative models for lysozyme action. Thus, synchronous acetamido group participation and intramolecular general acid catalysis was held to be responsible for the modest rate enhancement of compound LXI compared with the *p*-nitrophenyl compound or salicyl glucoside [145].

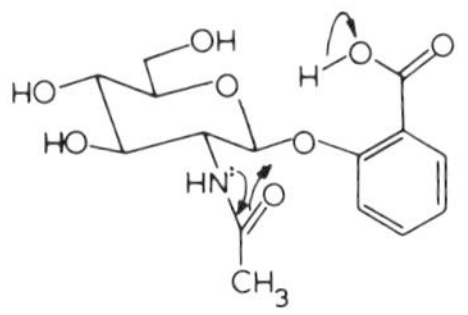

LXI

The construction of molecules designed to show the synchronous operation of intramolecular general acid catalysis and electrostatic (or nucleophilic) catalysis has met with mixed success. Thus, although compounds LXII and LXIII exhibit bell-shaped pH profiles – indeed at the maximum compound LXIII hydrolyses 3×10^9 faster than its dimethyl ester – it is difficult to disentangle any possible "electrostatic" or nucleophilic effect brought into play by the ionisation of the carboxyl from the ordinary inductive and mesomeric substituent effects on oxocarbonium ion stability caused by changing $^{-}$COOH for COO^{-}. Interestingly, participation by the ionised carboxylate of LXIV in the intermolecular general acid-catalysed reaction with water (but not with carboxylic acids) results in a rate-enhancement of about 30 [146].

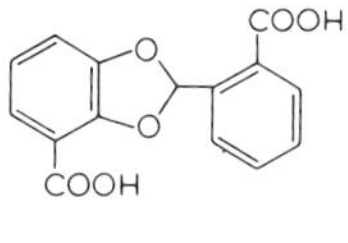

LXII [146] 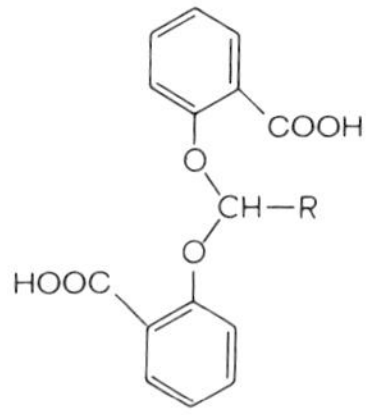 LXIII [147] 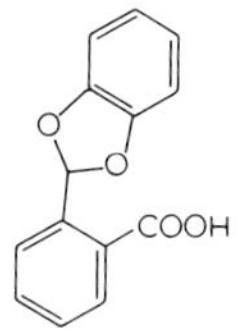 LXIV [146]

There is some evidence that in the hydrolysis of compound LXV the two carboxylate groups act in a synchronous fashion. A bell-shaped pH-rate profile is observed, and even though at the maximum the rate is only 4×10^4 times faster than the diethyl ester, the compound with only one carboxyl group (LXVI) shows no evidence of intramolecular carboxyl participation: intramolecular general acid catalysis only appears when the cationic centre has extra stabilisation [148].

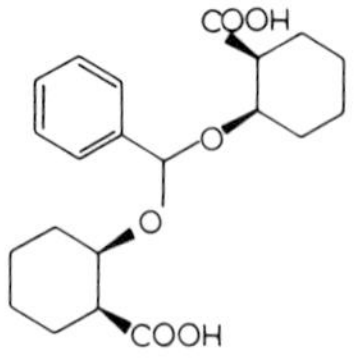

LXV LXVI

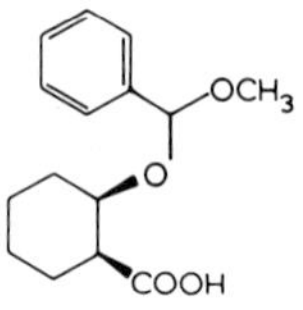

To the author, the extreme difficulty with which nucleophilic (or electrostatic) participation by carboxylate is made manifest in model systems makes it likely that the carboxylate group identified in the active site, not only of lysozyme but also a number of other glycosyl-transferring enzymes, [149] contributes little or nothing to the energetics of the bond-breaking process, and serves merely as a preassociation nucleophile to catch the glycosyl cation and allow the aglycone to diffuse away.

4.6. Electrophilic catalysis

Electrophilic catalysis of leaving group departure is most effective when both the leaving group and the metal are "soft" – thus Koenigs–Knorr glycosylations using glycosyl chlorides and bromides use Ag^{I} or Hg^{II} (Section 3.3). However, the association of many glycosyl-transferring enzymes with metal ions encourages speculation that coordination to a metal ion could increase the leaving-group ability and constitute a possible source of enzymic rate-enhancement. Support for this idea was first obtained with the glycoside (LXVII), whose hydrolysis is markedly accelerated by divalent ions of the first transition series, in the order $Cu^{2+} \gg Ni^{2+} > Co^{2+}$ [150]. With acetal (LXVIII) a similar order of metal ion efficacies was obtained ($Cu^{2+} >$

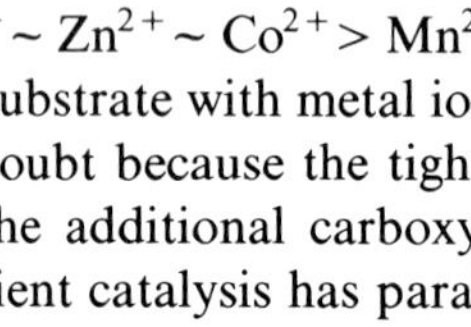

LXVII

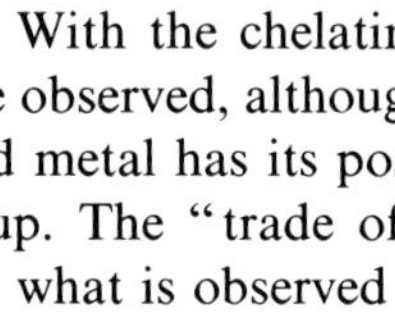

LXVIII X=R=H

LXIX R=COO^-, X= 3,4 Cl_2

$Ni^{2+} \sim Zn^{2+} \sim Co^{2+} > Mn^{2+}$) [151]. With the chelating agent (LXIX) saturation of the substrate with metal ions can be observed, although the rate enhancement is less, no doubt because the tightly bound metal has its positive charge partly neutralised by the additional carboxylate group. The "trade off" between tight binding and efficient catalysis has parallels with what is observed with enzymic catalysis [152].

5. Acid- and electrophile-assisted departure of nitrogen, sulphur and fluorine from oxocarbonium ion centres

The biological significance of glycosyl transfer to and from nitrogen is enormous. Even a much applied technique depends on it – S_N1 and A_1 departure of methylated guanine and adenine, respectively, from a 2-deoxy aldofuranosyl centre plays a key role in the Maxam–Gilbert DNA-sequencing procedure [153]. Despite this biological importance, surprisingly little work of mechanistic significance exists. A contrasting situation obtains in respect of glycosyl transfer to and from sulphur: a fair amount is known about the chemistry of the process, even though the biological significance of the reactions is not great. (Thiols can be detoxified in mammalian liver by conjugation with glucuronic acid [154] and unnatural thioglycosides can be hydrolysed by various glycosidases [86,154–156].) The only natural thioglycosides of any importance are the glucosinolates (LXX) ("mustard-oil glucosides") which are hydrolysed by "myrosinase" when tissues of the plant family Cruciferae are crushed [157].

LXX

"myrosinase" → Glucose + [R–C=N–O–SO_3^-, S⁻] → R—N=C=S + SO_4^{2-}

Glycosyl fluorides are apparently not found in nature, but their hydrolysis by glycosidases is rapid, and they are useful tools for the investigation of glycosidase mechanisms. Hydrolysis of glycosyl fluorides of the same anomeric configuration as the substrate exhibits Michaelis–Menten kinetics and, because F is isoelectronic with —OH, optical rotation changes on hydrolysis are dominated by the configuration of the anomeric centre of the product. They are thus the substrates of choice for determining the initial products of glycosidase action [158]. The hydrolysis of glycosyl fluorides of the *opposite* anomeric configuration to the substrate by "inverting" glycosidases has neatly established a number of mechanistic points about these enzymes [159]. The hydrolysis of β-D-glucopyranosyl fluoride by say, glucoamylase [160] requires two molecules of the fluoride to be bound at the active site, although the second molecule can be replaced by other glucosyl derivatives. β-D-Glucopyranose is the product. The only reasonable mechanism for this transformation is that the β-fluoro group mimics the attacking water molecule in the normal transition state for action of the enzyme on an α-glucoside. However, with the fluoride the tautomeric state of the acidic and basic groups is reversed, and the enzyme synthesises a glycosidic link which is then hydrolysed in the usual manner (Fig. 10).

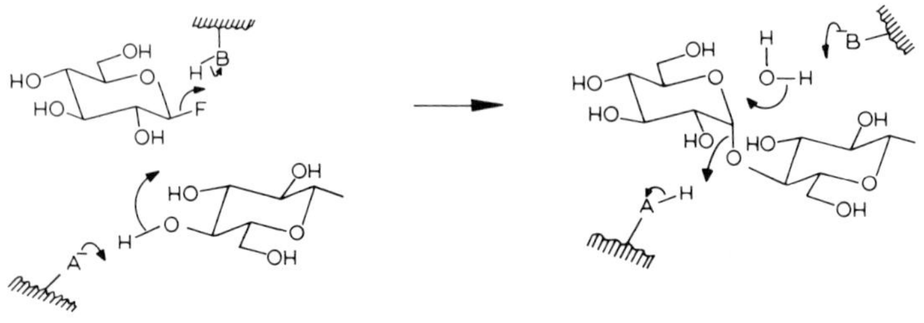

Fig. 10

5.1. Hydrolysis of glycosylamines

The hydrolysis of glycosylamines derived from primary amines and ammonia proceeds via a ring-opening mechanism [1]. A detailed polarimetric and photometric investigation of the hydrolysis of anomerically pure *N*-aryl glucosylamines [161,162] indicated that rapid anomerisation preceded hydrolysis. This hydrolysis was subject to general acid catalysis. The rates, extrapolated to zero buffer concentration went through maxima as the pH was lowered, the position of the maximum shifting to lower pH (or H_0) as the amine became less basic. This behaviour was attributed to a change in rate-determining step from attack of water on the Schiff base in neutral solution to decomposition of the carbinolamine intermediate in acid (Fig. 11). Similar pH dependence is observed in the hydrolysis of *N*-glucosyl piperidine [163].

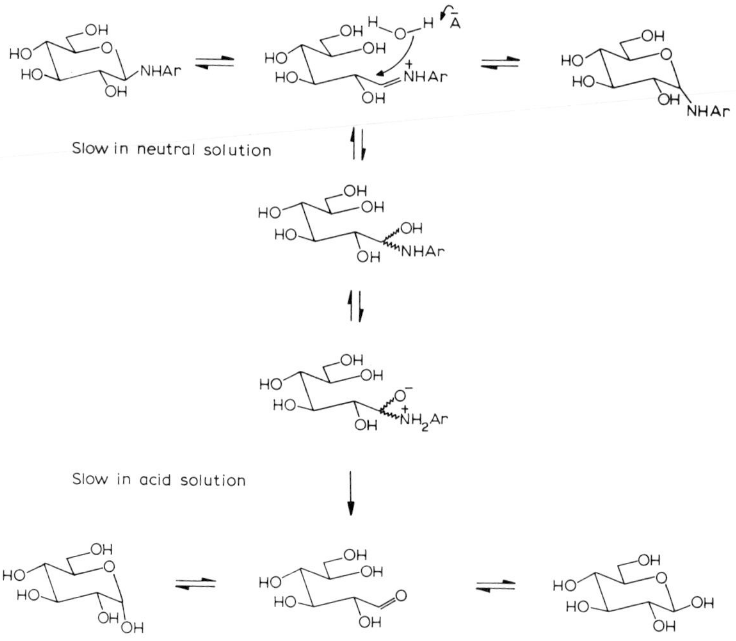

Slow in neutral solution

Slow in acid solution

Fig. 11

5.2. *Hydrolysis of nucleosides and deoxy nucleosides*

The structures of the 5 common nucleosides and deoxy nucleosides are given below, together with the numbering system of the heterocyclic base.

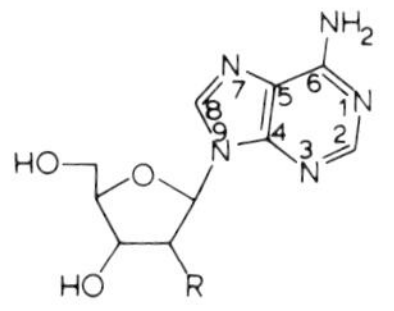

Adenosine 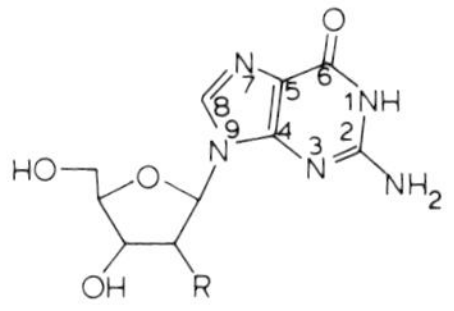 Guanosine

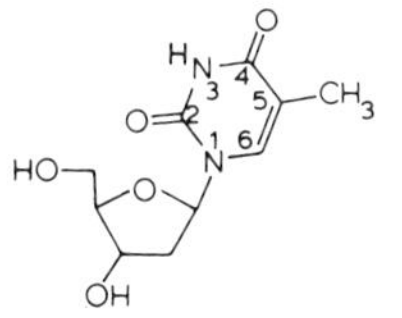

Thymidine 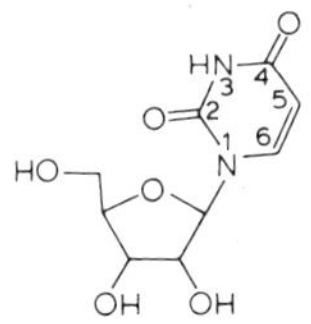 Uridine 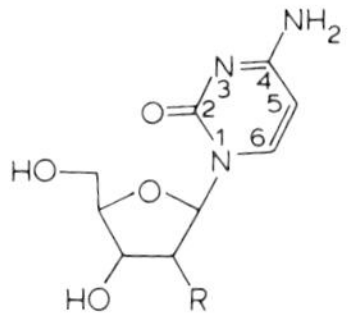 Cytidine

R = H or OH

Contrary to statements in the early literature, the weight of evidence is that in hydrolysis of the above compounds, the C—N bond breaks to give an aldofuranosyl cation and a purine or pyrimidine base as an anion, neutral molecule or cation. Analogies to the mechanism of hydrolysis of glycosylamines are misconceived since the lone pair on N1 of a pyrimidine, or N9 of a purine participating in a ring system with some aromatic character, is not as available as the lone pair on a glycosylamine. The only evidence favouring a glycosylamine-like hydrolysis mechanism is the observation of ring expansion and anomerisation reactions during the acid-catalysed hydrolysis of deoxythymidine and deoxyuridine (but not 5-bromodeoxyuridine and deoxycytidine) [164], and even this occurred to a comparatively minor extent.

Regarded as glycosyl-transfer reactions, various features of these hydrolyses are readily understood.

(i) C—N cleavage is much readier in deoxynucleosides than nucleosides: this is to be expected by analogy with reactions of aldopyranosides, and is accounted for by the destabilising effect of an electron-withdrawing hydroxyl group on the oxocarbonium ion centre. This electron-withdrawing effect slows down the acid-catalysed departure of adenine by a factor of 1000 [165] and of guanine by a factor of 500 [166]. The effect of removing the sugar 3-OH is much less [165], as expected for a more remote substituent.

(ii) The purines are fairly readily converted to good leaving groups by protonation on N7 and thus purine nucleosides are more readily cleaved in acid than pyrimidine nucleosides. This results in the well known depurination of DNA in acid solution.

(iii) The combination of the low basicity of thymidine, deoxyuridine and 5-bromodeoxyuridine, and the ease of generation of 2-deoxy aldosyl cations, has enabled a pH-independent S_N1 departure of pyrimidine anions to be observed [167]. If literature pK_a values are corrected for the fraction of the pyrimidine anion present as the N3-protonated tautomer, a β_{lg} value of -1.0 for this reaction can be calculated.

(iv) In the case of deoxycytidine and its derivatives, the replacement of the electron-withdrawing oxygen at C4 of deoxyuridine by an electron-donating NH_2 suppresses departure of the pyrimidine anion. The reaction is now (specific) acid-catalysed, reaction through the monoprotonated and diprotonated species being observed. Log K_{obs} versus pH (or H_O) for deoxycytidine and its 5-methyl and 5-bromo derivatives plots have two sections of gradient ~ -1, joined by a pH-independent region around pH 2, corresponding to reaction of the completely monoprotonated substrate [168].

Plots of log k_{obs} against pH (or H_O) for acid-catalysed hydrolysis of guanosine, deoxyguanosine, and deoxyadenine are linear (gradient 1) over the entire observable region, which encompasses a pK value of the substrate. This is to be expected if

$$\begin{array}{ccccc} S + H^+ & \overset{K_{a1}}{\rightleftharpoons} & SH^+ & \overset{K_{a2}}{\rightleftharpoons} & SH_2^{2+} \\ & & \downarrow & & \downarrow \\ & & k_1 & & k_2 \end{array}$$

the compounds are reacting via both monoprotonated, and diprotonated forms at rates such that $k_1/K_{a1} = k_2/K_{a2}$ [166]. The hypothesis was confirmed by a study of 7-methyl guanosine, 1,7-dimethyl guanosine, and 7-methyl deoxyguanosine. All three quaternary compounds hydrolysed in a pH-independent fashion until high acidity. At these high acidities the rates were comparable with those of the unmethylated compounds. In very concentrated perchloric acid a levelling off of the rate, corresponding to pK_a of ~ -2.5 and probable complete diprotonation of the substrate could be observed for both guanosine and its 7-methyl derivative [169].

In accord with the aldosyl cation mechanism for nucleoside hydrolysis, positive entropies of activation are observed for the hydrolysis of deoxyadenosine and deoxyguanosine: the oxygenated purine nucleosides have slightly negative $\Delta S^{\neq}$ values [170]. Also in accord with the steric acceleration anticipated on this mechanism, 7-ribofuranosyl purines, except 7-ribofuranosyl guanine, are hydrolysed faster than their naturally occurring 9-isomers. In the hydrolysis of adenosine and its analogue inosine (in which the 6-NH_2 is replaced by a keto group) α-deuterium kinetic isotope effects of around 1.2 are found [171].

The antibiotic psicofuranine, a ketoside analogue of adenosine in which the anomeric hydrogen is replaced by CH_2OH, also reacts via both monocationic and dicationic forms, although in this case the log k_{obs} pH plots are not strictly linear [172].

Pyrimidine nucleosides decompose in acid in other ways before the C—N bond is cleaved.

5.3. Hydrolysis of hemithioacetals, hemithioketals, and thio- and thia-glycosides

Two factors dominate the title process – the low basicity of dicovalent sulphur, which makes protonation and hydrogen bonding more difficult, and the lower ability of sulphur to stabilise an adjacent carbonium ion centre. This is conventionally attributed to poor overlap between the 3p orbital containing the sulphur lone pair and the vacant 2p orbital on carbon (Fig. 12), and is manifested in carbohydrate

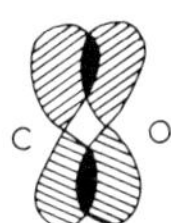

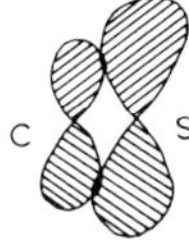

Fig. 12.

chemistry by the 40-fold slower rate of methanolysis of compound LXXI compared to its analogue with oxygen in the ring [173].

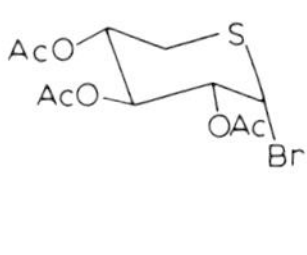

In the case of acid-catalysed hydrolysis of hemithioacetals and hemithioketals these two factors are in opposition, the first favouring initial C—O cleavage and the second initial C—S cleavage. The factors influencing the balance between C—O and C—S cleavage in the case of the specific acid catalysis of the cleavage of *O,S*-acetals of benzaldehyde have been analysed by Jensen and Jencks [48]. In accord with the lesser ability of sulphur to stabilise an adjacent positive charge, electron-donating substituents in the benzene ring, which make such stabilisation less necessary, favour C—O cleavage. Approximate structure–reactivity parameters are, for C—O cleavage of ArCH(OEt)SEt, $\rho = -2$; of PhCH(SEt)OR, $\beta_{lg} = 0.3$; and of ArCH(SR)OEt $0.5 < \beta_{rg*} < 1.0$; for C—S cleavage of ArCH(OEt)SPh, $\rho = -1.4$; of PhCH(OMe)SAr $\beta_{lg} = 0.57$; and of PhCH(OR)SPh, $\beta_{rg} = 0.62$. Apart from the differing sensitivity to carbonium ion stability the two processes are fairly similar. Other things being equal, as with PhCH(OEt)SEt, or oxathiolanes (LXXII), C—O cleavage is slightly favoured over C—S cleavage [174].

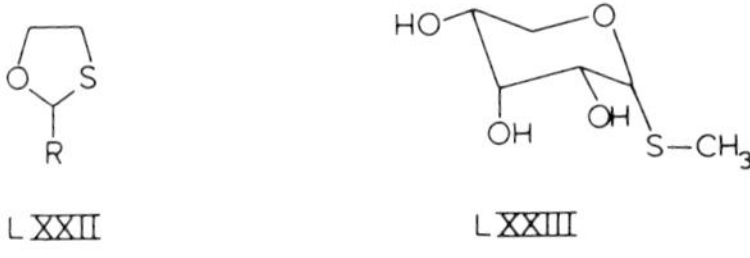

* rg = remaining group.

Therefore, the ring-opening mechanism should be more readily observable in the hydrolysis of 1-thioglycopyranosides than in the hydrolysis of the corresponding fully oxygenated compounds. Accordingly, ring contraction and anomerisation accompany the acid-catalysed hydrolysis of compound LXXIII [175]. Where the sugar is glucose, and ring opening, in the absence of axial substituents, is presumably less likely on steric grounds, the cyclic mechanism is still favoured [176]: phenyl β-D-glucopyranoside is hydrolysed 25 times slower than its oxygen analogue, but the ethyl 1-thio-β-D-glucopyranoside is hydrolysed 4 times faster than its oxygen analogue.

Interestingly, the replacement of the ring oxygen of methyl α- and β-D-xylopyranosides by sulphur increases the rate of acid hydrolysis by an order of magnitude [173]: apparently the lower electronegativity of sulphur, resulting in an increased concentration of the conjugate acid, is not totally offset by the lower ability of sulphur to stabilise an adjacent cationic centre.

As expected from the low hydrogen-bonding ability of sulphur, attempts to observe either intra- or inter-molecular general acid catalysis in the hydrolysis of hemithioacetals have met with little success. The only convincing example is compound LXXIV, in which acid catalysis is concerted with nucleophilic attack by the second carboxylate and contributes a rate enhancement of about 10^2 [177].

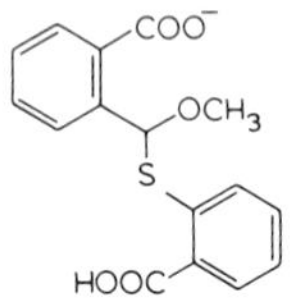

LXXIV

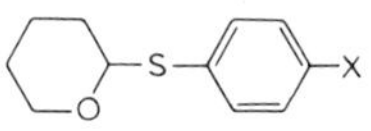

LXXV

The softer nature of sulphur, compared to oxygen, suggests that hemithioacetals would be more susceptible to electrophilic catalysis. This expectation is fulfilled in the case of compounds LXXV. Bimolecular rate constants for catalysis by Hg^{2+} are around 10^3 bigger than those for catalysis by protons, but otherwise the two reactions are fairly similar ($\rho = 0.88$ and 0.96 for Hg^{2+} and H_3O^+, respectively [178]).

5.4. *Hydrolysis of glycosyl fluorides*

In addition to their spontaneous hydrolyses, glycosyl fluorides [83] undergo acid- and base-catalysed reactions. The pattern of reactivity shown by various glycosyl fluorides in the base-catalysed process resembles that shown in the base-catalysed hydrolysis of aryl glycosides, including the much lower reactivity of *cis*-α-fluorides [114]. Both reactions proceed by the same mechanism (Section 2.3). The acid-catalysed reactions, although written as examples of specific acid catalysis, may well show general acid catalysis [114]: species involving fully protonated covalent fluorine may well be too unstable to exist in water.

6. *Envoi*

From the foregoing description it should be apparent that a reasonable understanding of the reactions of *O*-glycosides is available by diligent and sceptical reading of the literature of physical organic chemistry. Considerable caution must be exercised, however, since in this area it is not the things we don't know which cause difficulty, it is the things we know which ain't so. Particular scepticism is called for in two areas: the postulation of kinetic effects on the basis of static protein X-ray crystallographic pictures, and the interpretation of the output of molecular orbital programs with drastic, unstated approximations – and, worse, the simple pictorial representation of the supposed effects.

Experimentally, the major gap in our knowledge concerns attack on, and departure from, glycosyl centres by nitrogen. There is additionally always the possibility that more careful analysis of the products of enzyme reactions may turn up unusual reactions – such as the production of 1,6-anhydropyranose units by a glucosyl transferase [180] or phage λ lysozyme [181].

References

1 Capon, B. (1969) Chem. Rev. 69, 407.
2 Cordes, E.H. and Bull, H.G. (1978) in: R.D. Gandour and R.L. Schowen (Eds.), Transition States in Biochemical Processes, Plenum, New York, p. 429.
3 Cordes, E.H. (1967) Prog. Phys. Org. Chem. 4, 1.
4 Cordes, E.H. and Bull, H.G. (1974) Chem. Rev. 74, 581.
5 Fife, T.H. (1972) Accts. Chem. Res. 5, 264.
6 Dunn, B.M. and Bruice, T.C. (1973) Adv. Enzymol. 37, 1.
7 Capon, B. (1965–1980) Chapters on Reactions of Aldehydes and Ketones and their Derivatives, in annual volumes of Organic Reaction Mechanisms (various editors), John Wiley/Interscience, New York.
8 Streitweiser, A. and Owens, P.H. (1973) Orbital and Electron Density Diagrams, Macmillan, New York, p. 120.
9 McWeeny, R. (1979) Coulson's Valence, 3rd Edn., Oxford University Press, London, p. 211.
10 Murrell, J.N., Kettle, S.F.A. and Tedder, J.M. (1965) Valence Theory, John Wiley, New York, p. 184.
11 Cradock, S. and Whiteford, R.A. (1972) J. Chem. Soc. Faraday II 68, 281.
12 Kobayashi, T. and Nagakura, S. (1973) Bull. Chem. Soc. Jpn. 46, 1558.
13 Edward, J.T. (1955) Chem. Ind., 1102.
14 Paulsen, H., Luger, P. and Heiker, F.R. (1979) in: W.A. Szarek and D. Horton (Eds.), Anomeric Effect: Origin and Consequences, American Chemical Society, Washington, DC, p. 63.
15 Lemieux, R.U., Koto, S. and Voisin, D., ibid, p. 17.
16 Lemieux, R.U. (1971) Pure Appl. Chem. 27, 527.
17 Finch, P. and Nagpurkar, A.G. (1976) Carbohydrate Res. 49, 275.
18 Lemieux, R.U. (1964) in: P. de Mayo (Ed.), Molecular Rearrangements, Interscience, New York, p. 740.
19 Lemieux, R.U., Pavia, A.A., Martin, J.C. and Watanabe, K.A. (1969) Can. J. Chem. 47, 4427.
20 Angyal, S.J. (1969) Angew. Chem. (Int. Edn.) 8, 157.
21 Anderson, C.B. and Sepp, D.T. (1967) J. Org. Chem. 32, 607.
22 Paulsen, H., Györgydeák, Z. and Friedmann, M. (1974) Chem. Ber. 107, 1590.
23 Eliel, E.L. (1962) Stereochemistry of Carbon Compounds, McGraw-Hill, New York, p. 236.

24 Altona, C. and Sundaralingam, M. (1972) J. Am. Chem. Soc. 94, 8205.
25 Tvaroška, I. and Kožár, T. (1980) J. Am. Chem. Soc. 102, 6929.
26 Romers, C., Altona, C., Buys, H.R. and Havinga, E. (1969) Top. Stereochem. 4, 73.
27 Jeffrey, G.A. (1979) in: W.A. Szarek and D. Horton (Eds.), Anomeric Effect: Origin and Consequences, American Chemical Society, Washington, DC, p. 50.
28 David, S., ibid, p. 1.
29 Jeffrey, G.A., Pople, J.A. and Radom, L. (1972) Carbohydrate Res. 25, 117.
30 Wolfe, S., Whangbo, M.H. and Mitchell, D.J. (1979) Carbohydrate Res. 69, 1.
31 David, S., Eisenstein, O., Hehre, W.J., Salem, L. and Hoffmann, R. (1973) J. Am. Chem. Soc. 95, 3806.
32 Jones, P.G. and Kirby, A.J. (1979) J. Chem. Soc. Chem. Commun., 288.
33 Kirby, A.J. (1982) The Anomeric Effect and Related Stereoelectronic Effects at Oxygen, Springer Verlag, Berlin/Heidelberg, p. 60.
34 Leaback, D.H. (1968) Biochem. Biophys. Res. Commun. 32, 1025.
35 Philip, T., Cook, R.L., Malloy, T.B., Allinger, N.L., Chang, S. and Yuh, Y. (1981) J. Am. Chem. Soc., 103, 2151.
36 Nelson, C.R. (1979) Carbohydrate Res. 68, 55.
37 Deslongchamps, P. (1975) Tetrahedron 31, 2463.
38 Caswell, M. and Schmir, G.L. (1979) J. Am. Chem. Soc. 101, 7323.
39 Capon, B. and Ghosh, A.K. (1981) J. Am. Chem. Soc. 103, 1765.
40 Gorenstein, D.G., Findlay, J.B., Luxon, B.A. and Kar, D. (1977) J. Am. Chem. Soc. 99, 3473.
41 O'Connor, J.V. and Barker, R. (1979) Carbohydrate Res. 73, 227.
42 Stoddard, J.F. (1971) Stereochemistry of Carbohydrates, Wiley-Interscience, New York, p. 70.
43 Chandrasekar, S. and Kirby, A.J. (1978) J. Chem. Soc. Chem. Commun., 171.
44 Kirby, A.J. and Martin, R.J. (1979) J. Chem. Soc. Chem. Commun., 1079.
45 Warshel, A. and Weiss, R.M. (1980) J. Am. Chem. Soc. 102, 6218.
46 Stone, A.J. and Erskine, R.W. (1980) J. Am. Chem. Soc. 102, 7185; cf. Stone, A.J. (1978) Chem. Soc. Specialist Rep. Theoret. Chem. 3, 39.
47 Jencks, W.P. (1981) Chem. Soc. Rev. 10, 345.
48 Jensen, J.L. and Jencks, W.P. (1979) J. Am. Chem. Soc. 101, 1476.
49 McClelland, R.A. and Ahmad, M. (1978) J. Am. Chem. Soc. 100, 7027.
50 Young, P.R. and Jencks, W.P. (1977) J. Am. Chem. Soc. 99, 8238.
51 Fărcaşiu, D., O'Donnell, J.J., Wiberg, K.B. and Matturro, M. (1979) J. Chem. Soc. Chem. Commun., 1124.
52 Craze, G.-A., Kirby, A.J. and Osborne, R. (1978) J. Chem. Soc. Perkin II, 357.
53 Knier, B.L. and Jencks, W.P. (1980) J. Am. Chem. Soc. 102, 6789.
54 Craze, G.-A. and Kirby, A.J. (1978) J. Chem. Soc. Perkin II, 354.
55 Bennet, A.J. and Sinnott, M.L. (1983) unpublished.
56 Rhind-Tutt, A.J. and Vernon, C.A. (1960) J. Chem. Soc., 4637.
57 Harris, J.M. (1974) Progr. Phys. Org. Chem. 11, 89.
58 Schuerch, C. (1979) in: W.A. Szarek and D. Horton (Eds.), Anomeric Effect: Origin and Consequences, American Chemical Society, Washington, DC, p. 80.
59 Grunwald, E., Highsmith, S. and I, T.-P. (1974) in: M. Szwarc (Ed.), Ions and Ion-Pairs in Organic Reactions, John Wiley, p. 447.
60 Sinnott, M.L. and Jencks, W.P. (1980) J. Am. Chem. Soc. 102, 2026.
61 Ando, T. and Tsukamoto, S. (1977) Tetrahedron Lett. 2775.
62 Gregoriou, G. (1979) Chim. Chron. 8, 219.
63 Jobe, A. and Bourgeois, S. (1972) J. Mol. Biol. 69, 397.
64 Sinnott, M.L. (1978) FEBS Lett. 94, 1.
65 Hall, B.G. (1982) J. Bacteriol. 150, 132.
66 Hough, L. and Richardson, A.C. (1979) in: D.H.R. Barton and W.D. Ollis (Eds.), Comprehensive Organic Chemistry, Vol. 5 (E. Haslam (Vol. Ed.)), Pergamon, Oxford, p. 714.

67 Walker, R.T., ibid, p. 59.
68 Paulsen, H. (1982) Angew. Chem. (Int. Edn.) 21, 155.
69 Wulff, G., Röhle, G. and Krüger, R. (1972) Chem. Ber. 105, 1097.
70 Wulff, G., Röhle, G. and Schmidt, U. (1972) Chem. Ber. 105, 1111.
71 Wulff, G. and Röhle, G. (1972) Chem. Ber. 105, 1122.
72 Micheel, F. and Petersen, H. (1959) Chem. Ber. 92, 298.
73 Lemieux, R.U., Hendricks, K.B., Stick, R.V. and James, K. (1975) J. Am. Chem. Soc. 97, 4056.
74 Lönnberg, H. and Pohjola, V. (1976) Acta Chem. Scand. 30A, 669.
75 Cocker, D. and Sinnott, M.L. (1975) J. Chem. Soc. Perkin II, 1391.
76 Anderson, E. and Fife, T.H. (1969) J. Am. Chem. Soc. 91, 7163.
77 Fife, T.H. and Brod, L.H. (1970) J. Am. Chem. Soc. 92, 1681.
78 Lowry, T.H. and Richardson, K.S. (1981) Mechanism and Theory in Organic Chemistry, 2nd Edn., Harper and Row, New York, p. 139.
79 Jones, C.C., Sinnott, M.L. and Souchard, I.J.L. (1977) J. Chem. Soc. Perkin II, 1191.
80 Hosie, L., Marshall, P.J. and Sinnott, M.L. (1983) J. Chem. Soc. Perkin II, submitted.
81 Bull, H.G., Ferraz, J.P., Cordes, E.H., Ribbi, A. and Apitz-Castro, R. (1978) J. Biol. Chem. 253, 5186.
82 Kokesh, F.C. and Hine, J. (1976) J. Org. Chem. 41, 1976.
83 Jones, C.C. and Sinnott, M.L. (1977) J. Chem. Soc. Chem. Commun., 767.
84 Fife, T.H. and Anderson, E. (1970) J. Am. Chem. Soc. 92, 5464.
85 Page, I.D., Pritt, J.R. and Whiting, M.C. (1972) J. Chem. Soc. Perkin II, 906.
86 Sinnott, M.L., Withers, S.G. and Viratelle, O.M. (1978) Biochem. J. 175, 539.
87 Piszkiewicz, D. and Bruice, T.C. (1967) J. Am. Chem. Soc. 89, 6237.
88 Ballardie, F.W., Capon, B., Dearie, W.M. and Foster, R.L. (1976) Carbohydrate Res. 49, 79.
89 Cocker, D. and Sinnott, M.L. (1976) J. Chem. Soc. Perkin II, 618.
90 Lai, Y.-Z. and Ontto, D.E. (1979) Carbohydrate Res. 75, 51.
91 Horton, D. and Luetzow, A.E. (1971) J. Chem. Soc. Chem. Commun., 79.
92 Tsai, C.S. and Reyes-Zamora, C. (1972) J. Org. Chem. 37, 2725.
93 Lönnberg, H. (1978) Finn. Chem. Lett. 213.
94 Fife, T.H. and Przystas, T.J. (1977) J. Am. Chem. Soc. 99, 6693.
95 Loudon, G.M. and Ryono, D.E. (1976) J. Am. Chem. Soc. 98, 1900.
96 Aftalion, F., Hellin, M. and Cousseman, F. (1965) Bull. S.C. France, 1497.
97 Kohnstam, G. (1970) Progr. Reaction Kinetics 5, 335.
98 Mori, A.L., Porzio, M.A. and Schalager, L.L. (1972) J. Am. Chem. Soc. 94, 5034.
99 Jensen, J.A. and Lenz, P.A. (1978) J. Am. Chem. Soc. 100, 1291.
100 Capon, B. and Page, M.I. (1970) J. Chem. Soc. Chem. Commun., 1443.
101 Banks, B.E.C., Meinwald, Y., Rhind-Tutt, A.J., Sheft, I. and Vernon, C.A. (1961) J. Chem. Soc., 3240.
102 Rosenberg, S. and Kirsch, J.F. (1981) Biochemistry 20, 3196.
103 Timmell, T.E., Enterman, W., Spencer, F. and Soltes, E.J. (1965) Can. J. Chem. 43, 2296.
104 Capon, B. and Ghosh, B.C. (1971) J. Chem. Soc. (B), 739.
105 Nath, R.L. and Rydon, H.N. (1954) Biochem. J. 57, 1.
106 Hall, A.N., Hollingshead, S. and Rydon, H.N. (1961) J. Chem. Soc., 4290.
107 Van Wijnendaele, F. and De Bruyne, C.K. (1969) Carbohydrate Res. 9, 277.
108 De Bruyne, C.K. and De Bock, A. (1979) Carbohydrate Res. 68, 71.
109 Cocker, D., Jukes, L.E. and Sinnott, M.L. (1973) J. Chem. Soc. Perkin II, 190.
110 Collins, P.M., Overend, W.G. and Rayner, B.A. (1973) J. Chem. Soc. Perkin II, 310.
111 Capon, B. and Thacker, D. (1967) J. Chem. Soc. (B), 185.
112 Lönnberg, H., Kankaanperä, A. and Haapakka, K. (1977) Carbohydrate Res. 56, 277.
113 Lönnberg, H. and Valtonen, L. (1978) Finn. Chem. Lett., 209.
114 Lönnberg, H. and Kulonpää, A. (1977) Acta Chem. Scand. 31A, 306.
115 Morgan, D.M.L. and Neuberger, A. (1977) Carbohydrate Res. 53, 167.

116 Capon, B. and Thacker, D. (1967) J. Chem. Soc. (B), 1010.
117 Anderson, E. and Capon, B. (1972) J. Chem. Soc. Perkin II, 515.
118 Lamaty, G., Malaval, A., Roque, J.-P. and Geneste, P. (1972) Bull. S.C. France, 4563.
119 Bell, R.P. (1973) The Proton in Chemistry, 2nd Edn., Chapman and Hall, London, p. 111.
120 Kankaanperä, A. and Lahti, M. (1969) Acta Chem. Scand. 23, 2465.
121 Kankaanperä, A. and Lahti, M. (1969) Acta Chem. Scand. 23, 3266.
122 Anderson, E. and Capon, B. (1969) J. Chem. Soc. (B), 1033.
123 Capon, B. and Nimmo, K. (1975) J. Chem. Soc. Perkin II, 1113.
124 Crampton, M.R. and Willison, M.J. (1974) J. Chem. Soc. Perkin II, 1686.
125 Fife, T.H. and Anderson, E. (1971) J. Org. Chem. 36, 2357.
126 Capon, B. and Page, M.I. (1972) J. Chem. Soc. Perkin II, 522.
127 Mori, A.L. and Schalager, L.L. (1972) J. Am. Chem. Soc. 94, 5039.
128 Atkinson, R.F. and Bruice, T.C. (1974) J. Am. Chem. Soc. 96, 819.
129 Anderson, E. and Fife, T.H. (1971) J. Am. Chem. Soc. 93, 1701.
130 Cox, R.A. and Stewart, R. (1976) J. Am. Chem. Soc. 98, 488.
131 Jencks, D.A. and Jencks, W.P. (1977) J. Am. Chem. Soc. 99, 7948.
132 Capon, B. (1963) Tetrahedron Lett., 911.
133 Dunn, B.M. and Bruice, T.C. (1970) J. Am. Chem. Soc. 92, 2410.
134 Dunn, B.M. and Bruice, T.C. (1970) J. Am. Chem. Soc. 92, 6589.
135 Capon, B., Smith, M.C., Anderson, E., Dahm, R.H. and Sankey, G.H. (1969) J. Chem. Soc. (B), 1038.
136 Capon, B., Anderson, E., Anderson, N.S., Dahm, R.H. and Smith, M.C. (1971) J. Chem. Soc. (B), 1963.
137 Dunn, B.M. and Bruice, T.C. (1971) J. Am. Chem. Soc. 93, 5725.
138 Capon, B., Page, M.I. and Sankey, G.H. (1972) J. Chem. Soc. Perkin II, 529.
139 Fife, T.H. and Anderson, E. (1971) J. Am. Chem. Soc. 93, 6610.
140 Buffet, C. and Lamaty, G. (1976) Rec. Trav. Chim. 95, 1.
141 Craze, G.-A. and Kirby, A.J. (1974) J. Chem. Soc. Perkin II, 61.
142 Kirby, A.J. (1980) Adv. Phys. Org. Chem. 17, 183.
143 Kankaanperä, A., Oinonen, L. and Lahti, M. (1974) Acta Chem. Scand. 28A, 440.
144 Buffet, C. and Lamaty, G. (1976) Bull. S.C. France, 1887.
145 Piszkiewicz, D. and Bruice, T.C. (1968) J. Am. Chem. Soc. 90, 2156.
146 Capon, B. and Page, M.I. (1972) J. Chem. Soc. Perkin II, 2057.
147 Anderson, E. and Fife, T.H. (1973) J. Am. Chem. Soc. 95, 6437.
148 Fife, T.H. and Przystas, T.J. (1979) J. Am. Chem. Soc. 101, 1202.
149 Legler, G. (1973) Mol. Cell. Biochem. 2, 31.
150 Clark, C.R. and Hay, R.W. (1973) J. Chem. Soc. Perkin II, 1943.
151 Przystas, T.J. and Fife, T.H. (1980) J. Am. Chem. Soc. 102, 4391.
152 Jencks, W.P. (1975) Adv. Enzymol. 43, 219.
153 The chemistry is described by G.M. Blackburn (1979) in: D.H.R. Barton and W.D. Ollis (Eds.), Comprehensive Organic Chemistry, Vol. 5 (E. Haslam, Vol. Ed.) Pergamon, Oxford, p. 156.
154 Dutton, G.J. and Illing, H.P.A. (1972) Biochem. J. 129, 539.
155 Fedor, L.R. and Murty, B.S.R. (1973) J. Am. Chem. Soc. 95, 8410.
156 Lowe, G., Sheppard, G., Sinnott, M.L. and Williams, A. (1967) Biochem. J. 104, 893.
157 Benn, M. (1977) Pure Appl. Chem. 49, 197.
158 Barnett, J.E.G. (1971) Biochem. J. 123, 607.
159 Hehre, E.J., Brewer, C.F. and Genghof, D.S. (1979) J. Biol. Chem. 254, 5942.
160 Kitahata, S., Brewer, C.F., Genghof, D.S., Sawai, T. and Hehre, E.J. (1981) J. Biol. Chem. 256, 6017.
161 Capon, B. and Connett, B.E. (1965) J. Chem. Soc. 4492.
162 Capon, B. and Connett, B.E. (1965) J. Chem. Soc. 4497.
163 Simon, H. and Palm, D. (1965) Chem. Ber. 98, 433.
164 Cadet, J. and Teoule, R. (1974) J. Am. Chem. Soc. 96, 6517.
165 Garrett, E.R. and Mehta, P.J. (1972) J. Am. Chem. Soc. 94, 8532.

166 Zoltewicz, J.A., Clark, D.F., Sharpless, T.W. and Grahe, G. (1970) J. Am. Chem. Soc. 92, 1741.
167 Shapiro, R. and Kang, S. (1969) Biochemistry 8, 1806.
168 Shapiro, R. and Danzig, M. (1972) Biochemistry 11, 23.
169 Zoltewicz, J.A. and Clark, D.F. (1972) J. Org. Chem. 37, 1193.
170 Hevesi, L., Wolfson-Davidson, E., Nagy, J.B., Nagy, O.B. and Bruylants, A. (1972) J. Am. Chem. Soc. 94, 4715.
171 Romero, R., Stein, R., Bull, H.G. and Cordes, E.H. (1978) J. Am. Chem. Soc. 100, 7620.
172 Garrett, E.R. (1960) J. Am. Chem. Soc. 82, 827.
173 Whistler, R.L. and van Es, T. (1963) J. Org. Chem. 28, 2303.
174 Guinot, F. and Lamaty, G. (1972) Tetrahedron Lett. 2569.
175 Clayton, C.J., Hughes, N.A. and Saeed, S.A. (1967) J. Chem. Soc. (C), 644.
176 Bamford, C., Capon, B. and Overend, W.G. (1962) J. Chem. Soc., 5138.
177 Fife, T.H. and Przystas, T.J. (1980) J. Am. Chem. Soc. 102, 292.
178 Fedor, L.R. and Murty, B.S.R. (1973) J. Am. Chem. Soc. 95, 8407.
179 Barnett, J.E.G. (1969) Carbohydrate Res. 9, 21.
180 Pazur, J.H., Tominaga, Y., DeBrosse, C.W. and Jackman, L.M. (1978) Carbohydrate Res. 61, 279.
181 Bienkowska-Szewczyk, K. and Taylor, A. (1980) Biochim. Biophys. Acta 615, 489.
182 Jensen, J.L., Herold, L.R., Lenz, P.A., Trusty, S., Sergi, V., Bell, K. and Rogers, P. (1979) J. Am. Chem. Soc. 101, 4672.
183 Bunton, C.A. and Humeres, E. (1969) J. Org. Chem. 34, 572.

CHAPTER 12

Vitamin B_{12}

KENNETH L. BROWN

University of Texas at Arlington, Department of Chemistry, P.O. Box 19065, Arlington, TX 76019 (U.S.A.)

1. Introduction and scope of this chapter

Vitamin B_{12} (cyanocobalamin, R = —CN in Fig. 1) was first isolated from liver as a red crystalline substance in 1948 [1,2], the result of a vigorous search for the so-called anti-pernicious anemia factor of raw liver. Within the next 15 years from a combination of biochemical studies, natural product isolations and X-ray crystallography, it became apparent that both of the coenzymatically active species (methylcobalamin, R = CH_3 and 5′-deoxyadenosylcobalamin, R = 5′-deoxyadenosine in Fig. 1) of B_{12} contained surprisingly stable carbon–cobalt bonds. This organometallic nature of the B_{12} coenzymes makes them unique among the coenzymes (and very possibly among natural products in general) and determines the general perspective of this chapter. Thus, although numerous B_{12} derivatives with alterations in every conceivable part of the molecule (including the metal atom) can either be prepared chemically or isolated from natural sources (see, for instance, ref. 3, pp. 24–32 and ref. 4, pp. 280–295), this chemistry will not be significantly reviewed in this work. Instead we will take a broad perspective of organocobalt chemistry with a view toward providing a chemical background from which the mechanism of action of the B_{12} coenzymes can, hopefully, eventually be understood. In this regard it is important to point out that much of this chemistry has been worked out in the last 25 years using organocobalt model chelates which, for the most part, are vastly simpler in structure than the B_{12} coenzymes. Examples of some of these chelates are shown in Fig. 2 along with the chelate abbreviations which will be used in this chapter [5]. Many of the examples of reactions of organocobalt reagents referred to will be taken from the vast body of literature on such organocobalt chelates. In very many cases, similar, and often strictly analogous chemistry can be demonstrated for organocobalt derivatives of B_{12}.

2. Structure

The structures of the relevant B_{12} species are shown in Fig. 1 (R = —CN, cyanocobalamin or vitamin B_{12}, R = —CH_3, methylcobalamin, R = 5′-deoxyadenosyl, 5′-deoxyadenosylcobalamin or coenzyme B_{12}). Owing to the large number of known

Michael I. Page (Ed.), The Chemistry of Enzyme Action

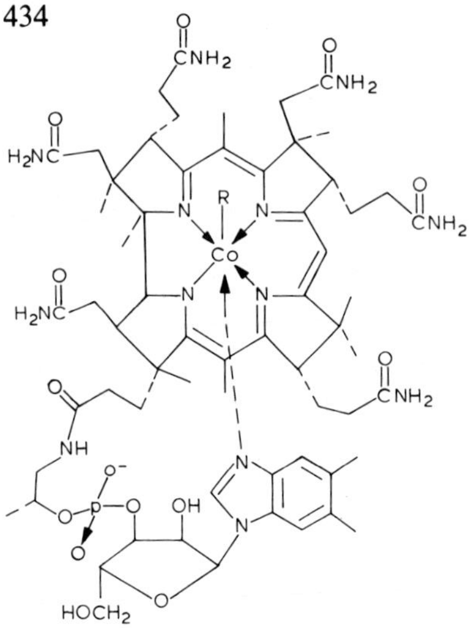

Fig. 1. Structure of the coenzymic derivatives of vitamin B_{12}. R = —CN, cyanocobalamin or vitamin B_{12}; R = $—CH_3$, methylcobalamin; R = 5′-deoxyadenosyl, 5′-deoxyadenosylcobalamin or coenzyme B_{12}.

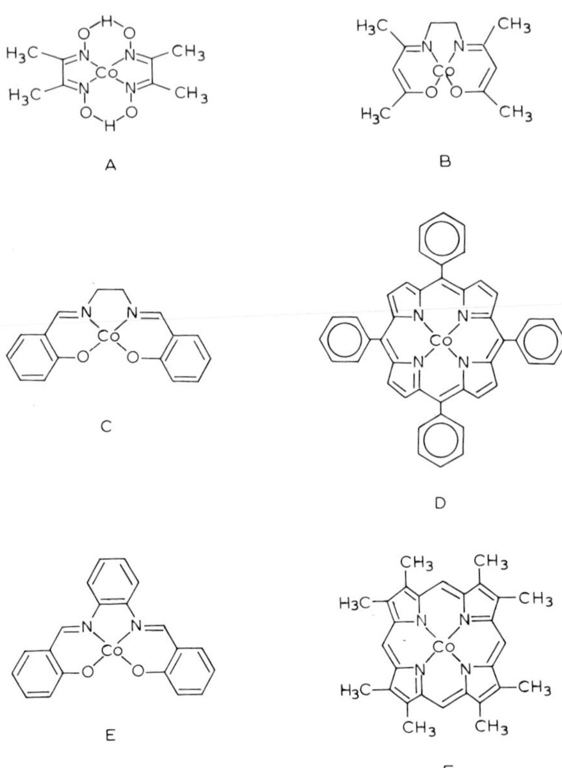

Fig. 2. Examples of cobalt chelates capable of stabilizing carbon–cobalt bonds. For additional examples see ref. 40. A, bis(dimethylglyoximato)cobalt = cobaloxime, $Co(D_2H_2)$; B, bis(acetylacetone)ethylenediiminecobalt, Co(BAE); C, bis(salicylaldehyde)ethylenediiminecobalt, Co(SALEN); D, tetraphenylporphyrincobalt, Co(TPP); E, bis(salicylaldehyde)-*o*-phenylenediiminecobalt, Co(SALOPH); F, octamethylporphyrincobalt, Co(OMP).

structural derivatives, nomenclature in this area is horridly complex. The interested reader is directed to Chapter 2 of ref. 4 and ref. 6. These structures are well established from X-ray crystallographic work, mostly by Prof. Hodgkin and co-workers [7,8]. Such structures are basically composed of 4 elements: (i) the central cobalt atom, (ii) the corrin macrocycle (and attached peripheral side chains and methyl groups) providing 4 nearly planar equatorial nitrogen ligands, (iii) the pendant 5,6-dimethylbenzimidazole nucleotide covalently attached to the corrin macrocycle and providing the lower (or α) axial nitrogen ligand, and (iv) the upper (or β) ligand, an organic group in the coenzymic B_{12} derivatives. Such structures are conveniently abbreviated as shown in 1.

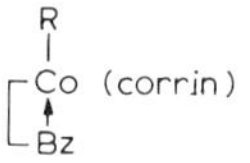

```
  R
  |
┌Co (corrin)
│ ↑
└Bz              (1)
```

The corrin macrocycle is derived biochemically from the porphyrin precursor uroporphyrinogen III [9]. However, it differs in several important respects from the porphyrin macrocycle. Perhaps the most important difference is the lack of the methene bridge between the A and D rings (carbon 20 in the porphyrin numbering system) which leads to a break in the conjugation, a loss of aromaticity, and a consequent buckling or non-planarity of the corrin macrocycle. Such buckling is readily apparent in the X-ray structure of 5′-deoxyadenosylcobalamin [10].

Other differences between the corrin macrocycle and its porphyrinogen precursors include decarboxylation of the acetate side chain at C-12, amidation of the other carboxylate side chains and addition of 7 "extra" methyl groups, all of which are now known to be derived from *S*-adenosylmethionine [11–13]. In addition, the C-17 propionate side chain is attached as an amide to isopropanolamine which in turn is esterified to the phosphate of a 3′-α-ribonucleotide of 5,6-dimethylbenzimidazole. The benzimidazole nitrogen not involved in the *N*-glycosidic bond is available as an axial ligand for the central cobalt atom (the so-called pendant axial ligand), making the cobalamins a pentadentate ligand system.

As has been pointed out (ref. 4, Chapter 6), the cobalamins are characterized by extreme steric crowding due to the extensive peripheral substitution, coordination of the bulky pendant ligand, and strain caused by the missing methene bridge (i.e. direct attachment of C-1 to C-19). Hence, while the Co—N bond lengths are similar to those of cobalt(III)-ammine [10], two of the N—Co—N angles in the equatorial plane are substantially different from 90° and the 4 equatorial nitrogens are slightly tetrahedrally displaced above and below their least squares plane. Although such structural features have long been thought to contribute in some crucial way to the action of the B_{12} coenzymes, the key to this particular puzzle has not yet been found.

3. *Oxidation states*

The cobalt atom in the cobalamin pentadentate ligand system, as well as in all the other simpler model systems capable of stabilizing carbon–cobalt bonds, may exist in the usual +3, +2, and +1 oxidation states. All organocobalt complexes for which measurements have been made are diamagnetic, regardless of coordination state, and formally contain cobalt in the +3 oxidation state and a carbanion ligand. However, it is equally reasonable to consider these species to be complexes of cobalt(II) with an organic radical (with antiferromagnetic spin coupling) or of cobalt(I) with a carbonium ion. As we shall see, the cobalt atom in organocobalt complexes frequently does not behave chemically like cobalt(III). Nonetheless, a number of cobalamins with inorganic β ligands are well known (e.g. aquo- and hydroxy-cobalamin) and are clearly d^6, low spin cobalt(III) complexes with octahedral (or nearly octahedral) geometries.

Cobalt(II) complexes (such as cob(II)alamin, formerly called B_{12r}) can be obtained by several methods including controlled potential reduction [14,15], anaerobic photolysis of some organocobalt species [16] (see Section 5.2(b)), partial oxidation of cobalt(I) complexes, in some cases acidification of solutions of cobalt(I) complexes (see below) and chemical reduction. Chemical reduction of cobalt(III) cobalamins to the +2 oxidation state has been achieved with hydrogen over platinum oxide [16], with neutral and acidic solutions of vanadium(III) [17], with chromous acetate at pH 3 [18] and with amalgamated zinc in 0.1 M aqueous perchloric acid [19]. All such cobalt(II) species are probably 5-coordinate, and are low spin d^7 systems containing an unpaired electron and thus displaying an ESR spectrum [20].

Many such cobalt(II) complexes reversibly bind diatomic oxygen in solution and have been extensively studied in this regard. However, all such cobalt(II) complexes are more or less labile toward air oxidation and in general cannot be handled in air without suffering some oxidation. Evidently the lability of cobalt(II) complexes toward air oxidation is quite sensitive to the nature of both the axial and equatorial ligands.

Cobalt(I) complexes (such as cob(I)alamin formerly called B_{12s}) can be obtained both by electrolytic reduction [14,21] and chemical reduction of both cobalt(II) and cobalt(III) complexes. Although borohydride ion is by far the most commonly used reducing agent for most cobalt chelates, sodium, sodium amalgam, and potassium have been used [22] as has zinc dust in acetic acid [23] and in aqueous ammonium chloride [24] as well as chromous ion [25].

Such cobalt(I) complexes are extremely air-sensitive, low spin d^8 systems, some of which are believed to be 4-coordinate square-planar and all of which are highly nucleophilic reagents due to the lone pair behavior of the d_z^2 electrons. These nucleophilic cobalt(I) chelates are the conjugate bases of hydrido–cobalt complexes (Eqn. 1) [23,26] which in turn are unstable, decomposing to hydrogen and cobalt(II) (Eqn. 2) [23,27,28], a reaction which is known to be reversible, at least in some cases.

$$\text{HCo(Chelate)L} \rightleftharpoons \text{H}^+ + \text{Co(Chelate)L}^- \tag{1}$$

$$\text{HCo(Chelate)L} \rightleftharpoons 1/2\text{H}_2 + \text{Co}^{(\text{II})}\text{(Chelate)L} \tag{2}$$

The latter reaction provides another route to cobalt(I) complexes via reduction of cobalt(II) species with hydrogen, a spontaneous reaction at least for some complexes. Although it has also been reported that cobalt(II) complexes, including cob(II)alamin, undergo disproportionation in alkaline media [29,30], these reports have recently been refuted [31,32].

One of the principal differences among the cobalt chelates which can stabilize carbon–cobalt bonds appears to be the acidity of the hydridocobalt species (Eqn. 1). pK_a values as disparate as 1.0 for hydridocobalamin [28], 10.5 for hydrido(tri-*n*-butylphosphine)cobaloxime [33] and ca. 20 for hydridopentacyanocobaltate [34] have been reported.

4. *Enzymology of the B_{12} coenzymes*

Although the enzymology of the B_{12} coenzymes is not the subject of this chapter, it is instructive to briefly consider what is known about the mechanism of action of these coenzymes in order to guide our subsequent discussion of organocobalt chemistry. Adenosylcobalamin is known to be a cofactor for about a dozen enzymatic intramolecular 1,2-rearrangements [35] in which an electronegative group and a hydrogen on adjacent carbons exchange places (Eqn. 3).

$$-\underset{X}{\overset{|}{\underset{|}{C_1}}}-\underset{H}{\overset{|}{\underset{|}{C_2}}}- \longrightarrow -\underset{H}{\overset{|}{\underset{|}{C_1}}}-\underset{X}{\overset{|}{\underset{|}{C_2}}}- \qquad (3)$$

The current state of knowledge regarding the mechanisms of these reactions [36,37] is summarized diagrammatically in Fig. 3.

It is now well established for a number of these enzymes that upon binding substrate to the holoenzyme the coenzyme is "activated", leading to formation of a 5′-deoxyadenosyl radical and cob(II)alamin at the active site (step 1, Fig. 3). Although it is generally assumed that a homolytic carbon–cobalt bond cleavage

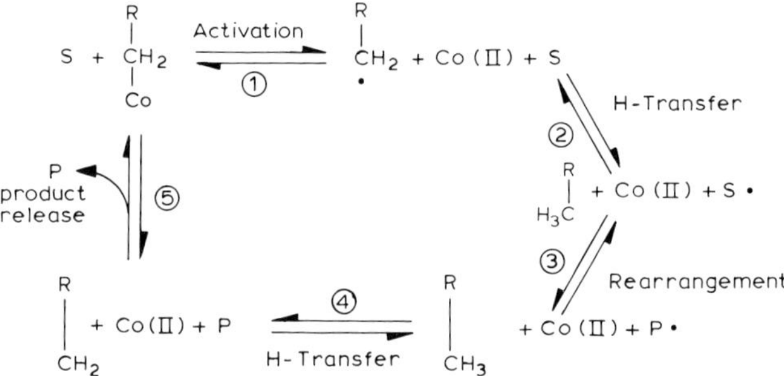

Fig. 3. Schematic representation of the catalytic cycle of a 5′-deoxyadenosylcobalamin-requiring enzyme. See text for descriptions of the individual steps.

occurs, there is no evidence against a heterolytic cleavage mechanism followed by a 1-electron transfer. In the second step, the migrating hydrogen atom of the substrate is abstracted by the 5′-deoxyadenosyl radical and hence becomes equivalent with the two 5′ hydrogens of the coenzyme in the 5′-deoxyadenosine moiety thus formed. In step 3, the substrate radical produced in the hydrogen-transfer step somehow undergoes rearrangement to become a product radical. Virtually nothing is known about the mechanism of this step although there has been considerable speculation. The substrate radical itself could presumably rearrange, it could undergo a 1-electron transfer with cob(II)alamin and the resulting carbonium ion or carbanion could rearrange, or it could couple with cob(II)alamin to produce an organocobalamin intermediate which rearranges and then homolytically dissociates to regenerate cob(II)alamine and produce the product radical. At any rate, following formation of the product radical, it abstracts one of three equivalent hydrogens from the 5′-methyl group of 5′-deoxyadenosine to form the product and regenerate the 5′-deoxyadenosyl radical. Release of the product (step 5) and recombination of the 5′-deoxyadenosyl radical with cob(II)alamin regenerates the coenzyme and completes the catalytic cycle.

Methylcobalamin is known to participate in several enzymatic methyl group transfer reactions [38], including the synthesis of methionine via N^5-methyltetrahydrofolate-homocysteine transmethylase (methionine synthetase), an enzyme found widely distributed in the biosphere. It is now well established that this reaction proceeds via transfer of the N^5-methyl group of N^5-methyltetrahydrofolate to enzyme-bound B_{12} (probably cob(I)alamin) to produce a methylcobalamin–enzyme intermediate. In a second step the cobalt-bound methyl group is transferred to homocysteine to generate methionine. Although this step is often postulated to occur via nucleophilic displacement of cob(I)alamin by the thiolate anion of homocysteine, the purported model system evidence for such nucleophilic displacements of cobalt(I) is weak and other mechanisms are consistent with the known enzymology.

Methylcobalamin also participates as a substrate for an enzymatic reaction in the formation of methane in certain methanogenic bacteria. In this reaction, which can be considered as a model for the second half-reaction of methionine synthetase, the cobalt-bound methyl group is transferred to 2-mercaptoethanesulfonic acid (coenzyme M) [39], the *S*-methyl sulfide of which is subsequently reduced to yield methane. Additionally, methylcobalamin participates in the rather obscure process of acetate synthesis in certain bacteria in which it can be shown that the cobalt-bound methyl group becomes the acetate methyl carbon. The details of the involvement of methylcobalamin in this process are, however, unknown.

5. Chemical reactivity of organocobalt complexes

Organocobalt chemistry is a broad, diverse area which has been approached both from the traditional organometallic viewpoint and in terms of modeling the action of the B_{12} coenzymes. For the purpose of discussing those aspects of organocobalt chemistry most likely to be relevant to the biochemistry of vitamin B_{12} we can conveniently divide the subject into 4 areas: reactions in which carbon–cobalt bonds are formed, reactions in which carbon–cobalt bonds are cleaved, axial ligand substitution reactions, and reactions of the cobalt-bound organic group.

5.1. Reactions in which carbon–cobalt bonds are formed

Organocobalt synthetic routes have recently been reviewed extensively [40,41]. Routes are available starting from cobalt(I), cobalt(II) and cobalt(III). These will be briefly described in order to give the reader an idea of the scope of this synthetic area.

(a) Synthesis of organocobalt complexes via cobalt(I) reagents

Despite the sensitivity of cobalt(I) reagents to air oxidation, the extremely high nucleophilicity of these reagents makes them attractive substrates for alkylation reactions. Since cobalt(I) reagents must always be prepared in situ from complexes in a higher oxidation state, this route is commonly referred to as reductive alkylation.

All cobalt(I) nucleophiles (Eqn. 4) as well as their conjugate acids (Eqn. 5) react rapidly with alkyl halides to produce the appropriate alkyl cobalt complexes.

$$\mathrm{Co(Chelate)L^- + RX \rightarrow RCo(Chelate)L + X^-} \tag{4}$$

$$\mathrm{HCo(Chelate)L + RX \rightarrow RCo(Chelate)L + HX} \tag{5}$$

Alkyl tosylates, alkyl phosphates and even alkenyl halides are effective alkylating agents. Both primary and secondary alkyl halides are effective alkylating agents in all chelate systems, although secondary alkylcobalamins are notably unstable [42]. Unstrained tertiary alkyl halides fail to produce tertiary alkyl–cobalt complexes in all ligand systems, presumably due to the inherent instability of such species owing to steric hindrance. However, several strained tertiary alkyl–cobalt complexes have been synthesized including 1-methyl-2,2-diphenylcyclopropyl(pyridine)cobaloxime [43] and 1-adamantyl- and 1-norbornyl(pyridine)cobaloximes [44].

Kinetic studies of the reactions of cobalt(I) nucleophiles with alkyl halides [45] are consistent with an S_N2 mechanism for these reactions. Indeed, inversion of configuration at the displacement center has been observed both for secondary [46] and primary [47,48] alkylating agents. However, for some alkylating agents for which inversion of configuration is not possible [44,49] as well as in certain other instances [50], the electron-transfer mechanism of Eqns. 6 and 7 is probably operative.

$$\mathrm{Co^{(I)}(Chelate)L^- + RX \rightarrow Co^{(II)}(Chelate)L + R\cdot + X^-} \tag{6}$$

$$\mathrm{Co^{(II)}(Chelate)L + R\cdot \xrightarrow{\text{fast}} RCo(Chelate)L} \tag{7}$$

Both cobalt(I) supernucleophiles (Eqn. 8) and hydridocobalt species (Eqn. 9) will add to unsaturated electrophiles although the mode of addition is different for the two species as indicated.

$$Co^{(I)}(Chelate)L^- + XCH{=}CH_2 \rightarrow XCH_2CH_2Co(Chelate)L \quad (8)$$

$$HCo(Chelate)L + XCH{=}CH_2 \rightarrow CH_3CHXCo(Chelate)L \quad (9)$$

With the possible exception of hydridocobalamin [42], olefins must be suitably activated with electron-withdrawing substituents to undergo these reactions. Hence, acrylonitrile ($X = CN^-$) and methylacrylate ($X = COOCH_3$) are active alkylating agents [51] but allyl alcohol ($X = CH_2OH$) is not [52]. The mechanism of these reactions is not as well studied, but Schrauzer et al. [53] have presented spectral evidence that the alkylation of cobalt(I) nucleophiles with activated olefins proceeds via a π complex intermediate (Eqns. 10 and 11).

$$Co^{(I)}(D_2H_2)L^- + XCH{=}CH_2 \rightleftharpoons \underset{Co^{(I)}(D_2H_2)L^-}{\overset{XCH{=}CH_2}{\uparrow}} \quad (10)$$

$$\underset{Co^{(I)}(D_2H_2)L^-}{\overset{XCH{=}CH_2}{\uparrow}} \xrightleftharpoons{\pm H^+} XCH_2CH_2Co(D_2H_2)L \quad (11)$$

Gaudemer and coworkers [54] have studied the reaction of deuteridocobaloxime with *trans*-phenylpropene (Eqn. 12) and presented ^{1}H-nmr evidence as well as

$$DCo(D_2H_2)py + (\Phi)(H)C{=}C(H)(CH_3) \longrightarrow \Phi{-}\underset{Co(D_2H_2)py}{\underset{|}{CH}}{-}CHD{-}CH_3 \quad (12)$$

conformational arguments that a single deuterium atom is incorporated stereospecifically into the product via a *cis* addition of the hydridocobalt species to the double bond. This result suggests that the reaction proceeds via a 4-center transition state which is consistent with the kinetic studies of Halpern and Wong [55] on addition of hydridopentacyanocobaltate to olefins.

Cobalt(I) nucleophiles and hydridocobalt complexes also add to acetylene and substituted alkynes to yield alkenyl cobalt complexes (Eqns. 13 and 14), again with a change in addition mode, although mixtures of products are obtained for some X [56].

$$Co^{(I)}(Chelate)L^- + XC{\equiv}CH \longrightarrow (H)(X)C{=}C(H)(Co(Chelate)L) \quad (13)$$

$$HCo(Chelate)L + XC{\equiv}CH \longrightarrow (H)(H)C{=}C(X)(Co(Chelate)L) \quad (14)$$

The reaction of pyridine cobaloxime(I) with phenylacetylene has been shown to occur via direct attack of the cobalt(I) nucleophile on the unsubstituted sp carbon with concerted or stepwise addition of a proton from solvent [56,57] while the analogous reaction of hydrido(pyridine)cobaloxime occurs with *cis* addition [57].

Cobalt(I) nucleophiles and hydridocobalt reagents also add to epoxides to form 2-hydroxyalkyl cobalt species (Eqn. 15)

$$\left.\begin{matrix} Co^{(I)}(D_2H_2)py^- \\ \text{or} \\ HCo(D_2H_2)py \end{matrix}\right\} + H_3C{-}\overset{O}{\widehat{CH{-}CH_2}} \longrightarrow H_3CCHOHCH_2Co(D_2H_2)py \quad (15)$$

although without a change in the mode of addition [52]. Inexplicably, addition of both types of cobalt reagent to styrene oxide also forms the same product (Eqn. 16) but in this case addition occurs in the opposite sense to that of propylene oxide [57].

$$\left.\begin{matrix} Co^{(I)}(D_2H_2)py^- \\ \text{or} \\ HCo(D_2H_2)py \end{matrix}\right\} + \Phi{-}\overset{O}{\widehat{CH{-}CH_2}} \longrightarrow HOCH_2(\Phi)CHCo(D_2H_2)py \quad (16)$$

The mechanisms of these reactions are not known.

Finally, some successful reductive arylations have been reported via reaction of aryl halides with cobalt(I) nucleophiles. Generally speaking these reactions are only successful when the aryl halide has an electron-withdrawing substituent and yields are generally less than 15% [58,59]. Thus, pyridinecobaloxime(I) reacts with *p*-bromoacetophenone, methyl-*m*- and methyl-*p*-bromobenzoate, and *p*-bromo-α,α,α-trifluorotoluene, to produce the expected substituted phenyl(pyridine)cobaloximes in yields of 11.5, 10, 7, and 5% respectively.

(b) Synthesis of organocobalt complexes via cobalt(II) reagents

Cobalt(II) reagents in several equatorial ligand systems are also known to react with alkyl halides according to the overall stoichiometry of Eqn. 17 [16,60,61].

$$2Co^{(II)}(\text{Chelate})L + RX \rightarrow RCo(\text{Chelate})L + XCo(\text{Chelate})L \quad (17)$$

The kinetics of this reaction have been shown to be first order in cobalt(II) chelate and first order in alkyl halide and the dependence of the rate constants on leaving group and alkyl halide structure suggests the atom-transfer mechanism of Eqns. 18 and 19, in which the rate-determining step is abstraction of a halogen atom from the alkyl halide by cobalt(II). Although the synthetic utility of this reaction is limited by the stoichiometry to a theoretical yield of 50% (based on cobalt), Roussi and Widdowson [62] recently showed that yields of alkylcobaloximes as high as 95% can be obtained when the reaction is run in the presence of excess zinc. Presumably zinc reduces the halocobalt(III) complex formed in the atom-transfer step (Eqn. 18) to cobalt(II) permitting further reaction.

$$Co^{(II)}(\text{Chelate})L + RX \rightarrow XCo(\text{Chelate})L + R\cdot \quad (18)$$

$$Co^{(II)}(Chelate)L + R\cdot \rightarrow RCo(D_2H_2)L \quad (19)$$

As exemplified by Eqn. 19, cobalt(II) complexes can efficiently trap organic radicals to form organocobalt complexes. Hence, organic radicals generated by means other than the atom-transfer step of Eqn. 18 may also be used to alkylate cobalt(II) complexes. Thus, Roche and Endicott [63] generated organic radicals by photolysis of carboxylatopentaammine–cobalt(III) complexes in acidic media (Eqn. 20)

$$Co(NH_3)_5OOCR^{2+} \underset{H^+}{\overset{h\nu}{\rightarrow}} Co^{2+} + 5NH_4^+ + CO_2 + R\cdot \quad (20)$$

in the presence of cobalt(II) complexes of several macrocyclic equatorial ligands and obtained the appropriate organocobalt products in yields as high as 40%. This work included the first successful preparation of organocobalt complexes with completely saturated equatorial ligand systems. Similarly, Espenson and Martin [19] decomposed organic hydroperoxides in the presence of a 2-fold excess of cobalt(II) reagent to produce alkoxy radicals (Eqn. 21)

$$Co^{(II)}(Chelate)L + RC(CH_3)_2COOH \rightarrow RC(CH_3)_2O\cdot + HOCo(Chelate)L \quad (21)$$

which lose acetone to form alkyl radicals (Eqn. 22) which are then trapped by excess cobalt(II) complex (Eqn. 23).

$$RC(CH_3)_2O\cdot \rightarrow CH_3\overset{\overset{\displaystyle O}{\|}}{C}CH_3 + R\cdot \quad (22)$$

$$R\cdot + Co^{(II)}(Chelate)L \rightarrow RCo(Chelate)L \quad (23)$$

Reaction kinetics were first order in cobalt(II) chelate and first order in hydroperoxide and the second-order rate constants were surprisingly insensitive to the nature of the cobalt chelate equatorial ligand system over a wide variety of cobalt chelates.

Finally, Goedken and coworkers [64] have shown that cobalt(II) complexes react with organohydrazines in the presence of molecular oxygen to produce organocobalt products, in some cases in very high yield (Eqn. 24).

$$Co^{(II)}(Chelate)L + RNHNH_2 \overset{O_2}{\rightarrow} RCo(Chelate) + N_2 \quad (24)$$

Although it has not been very thoroughly studied, this exothermic reaction occurs for cobalt(II) complexes in a number of equatorial ligand systems. Brown and coworkers [59] have attempted to use this procedure for the preparation of substituted arylcobaloximes with limited success. Thus, while *p*-carbomethoxyphenylhydrazine does react with pyridinecobaloxime(II) to produce the appropriate aryl(pyridine)cobaloxime, the yield was not significantly greater than that obtained via reductive arylation with *p*-carbomethoxyphenylbromide [58]. Although the mechanism of this interesting reaction (Eqn. 24) is not known, Goedken et al. [64] have speculated that oxidation of the organohydrazine to a diazene by oxygen is catalyzed by the cobalt(II) complex. Subsequent decomposition of the diazene (Eqn. 25) produces nitrogen and organic radicals which are trapped by the cobalt(II) reagent. Observation of the formation of nitrogen during this reaction (Eqn. 24) supports this view.

$$R{-}N{=}N{-}R \rightarrow 2R\cdot + N_2 \tag{25}$$

(c) Synthesis of organocobalt complexes via cobalt(III) reagents

Reactions of cobalt(III) complexes with organomagnesium halides (Eqn. 26)

$$RMgX + XCo^{(III)}(\text{Chelate})L \rightarrow RCo^{(III)}(\text{Chelate}) + MgX_2 \tag{26}$$

or organolithium reagents were among the earliest of methods used to produce organocobalt complexes [65,66]. With the possible exception of equatorial ligand systems whose cobalt complexes are difficult to reduce, this route is rarely used for the synthesis of alkylcobalt complexes, the simpler reductive alkylation route being preferred. However, it is ideal for preparation of Grignard and organolithium compatible aryl cobalt complexes. Unfortunately, because of the extremely low solubility of most cobalt(III) chelates in ethereal solvents, an inconvenient reverse addition of organometallic reagent to cobalt complex is required and the organometallic reagent must often be employed in substantial excess in order to allow for its reaction with acidic sites on the cobalt equatorial and/or axial ligands.

Hydroxy- and alkoxy-cobalt(III) complexes are also known to react with reagents which can form stabilized carbanions on loss of a proton [67], such as malononitrile, to form organocobalt complexes (Eqn. 27, $R = H$ or CH_3, $R' = CN$, $R'' =$ CN, $CONH_2$, etc.).

$$ROCo(\text{Chelate})L + R'CH_2R'' \rightarrow R'CH(R'')Co(\text{Chelate})L + ROH \tag{27}$$

Both malononitrile and phenylacetonitrile were shown to react with aquocobalamin in this manner [68]. Similarly, cobalt(III) complexes react with monosubstituted alkynes to produce alkynylcobalt products [69].

Recently, Callot and Schaeffer [70] found that cobalt(III) porphyrins react with diazoalkanes to produce substituted vinyl cobalt products in good yield (Eqn. 28).

$$R(R')CH{-}C({=}N_2){-}R'' + XCo^{(III)}(\text{Porphyrin}) \longrightarrow (R)(R')C{=}C(R'')Co(\text{Porphyrin}) \tag{28}$$

This reaction is believed to proceed via insertion of a carbene formed from the diazoalkane into one of the Co—N bonds of the cobalt(III) complex followed by elimination of HX to form the organocobalt complex.

Finally, cobalt(III) cobaloximes and cobalamins have been found to react with vinyl ethers in nucleophilic solvents in the presence of a weak base to form β-substituted ethylcobalt complexes [71] (e.g. Eqn. 29).

$$XCo(D_2H_2)py + CH_2{=}CHOCH_2CH_3 \xrightarrow[N(CH_2CH_3)_3]{CH_3CH_2OH} (CH_3CH_2O)_2CHCH_2Co(D_2H_2)py \tag{29}$$

Although there is still some debate about the mechanism of these reactions, they are thought to proceed via intermediate formation of a cationic cobalt(III)–olefin π complex (Eqn. 30)

$$XCo(D_2H_2)py + CH_2{=}CHOCH_2CH_3 \rightleftharpoons H_2C \overset{\overset{+}{Co}(D_2H_2)py}{\cdots\cdots} C(OCH_2CH_3)(H) \tag{30}$$

which is subsequently trapped by attack of the nucleophilic solvent (ethanol) on the substituted carbon of the olefinic moiety to form the organocobalt product.

5.2. Reactions in which carbon–cobalt bonds are cleaved

The area of carbon–cobalt bond cleavage reactions is of perhaps the greatest direct relevance to the mechanism of action of the B_{12} coenzymes. Thus, while it is not yet clear if new organocobalt intermediates are formed (and decomposed) from substrates and the cobalt-containing moiety of adenosylcobalamin during the enzyme-catalyzed intramolecular rearrangements (see Section 4), it is clear that the activation of adenosylcobalamin, as well as the coenzymic action of methylcobalamin require that the carbon–cobalt bond of the coenzymes be cleaved.

In a general sense, carbon–cobalt bond cleavage can occur by any one of three mechanisms: mode I cleavage (Eqn. 31)

$$R{-}Co(Chelate) \rightarrow [R^+] + Co^{(I)}(Chelate) \tag{31}$$

in which both electrons of the organometallic bond remain on cobalt to generate cobalt(I) product and a carbonium ion (or incipient carbonium ion), mode II, or homolytic cleavage to produce a cobalt(II) complex and an organic radical (Eqn. 32),

$$R{-}Co(Chelate) \rightarrow [R\cdot] + Co^{(II)}(Chelate) \tag{32}$$

and mode III cleavage to produce a cobalt(III) complex and a carbanion (or incipient carbanion) (Eqn. 33) [72].

$$R{-}Co(Chelate) \rightarrow [R^-] + Co^{(III)}(Chelate) \quad (33)$$

As there are reasonably well documented cases of each of these three modes of bond cleavage we can deal with each independently.

(a) Mode I cleavage of carbon–cobalt bonds

Perhaps the best studied mode I cleavages are the base-catalyzed β-elimination reactions of 2-substituted alkylcobalt complexes with electron-withdrawing substituents (Eqn. 34).

$$XCH_2CH_2Co(Chelate)L \xrightarrow{OH^-} Co^{(I)}(Chelate)L^- + XCH{=}CH_2 \quad (34)$$

Such reactions are formally the reverse of the alkylation of cobalt(I) nucleophiles by suitably activated olefins (Eqn. 8). Indeed, Schrauzer et al. [53] have presented spectroscopic as well as other evidence that for cobaloximes where X = —CN or —$COOCH_2CH_3$ these reversible reactions proceed via intermediate formation of a cobaloxime(I)–olefin π complex, i.e. the microscopic reverse of Eqns. 10 and 11. However, Barnett et al. [73] have studied the kinetics of the analogous base-catalyzed elimination of 2-cyanoethylcobalamin to produce cob(I)alamin and acrylonitrile. These authors found no general base catalysis and a rate law which was first order in organocobalamin and first order in hydroxide ion and determined a second-order rate constant of 230/M/min. As these authors pointed out, this rate constant is several orders of magnitude greater than the second-order rate constant for ionization of acetonitrile so that the mechanism must either by a concerted E2 elimination (or possibly direct elimination of hydridocobalamin) or, if stepwise, the rate of β-proton dissociation must be substantially enhanced by the cobalt-containing substituent.

In what appears to be a similar process, adenosylcobalamin decomposes in base to give cob(I)alamin and β-elimination products of the 5′-deoxyadenosyl moiety (Eqn. 35) [74].

OH OH / O, Ad / CH_2 — Co(Cobalamin) $\xrightarrow{OH^-}$ $Co^{(I)}$(Cobalamin) + OH OH / O, Ad / $=CH_2$ (35)

However, in an interesting contrast, alkoxyethylcobaloximes do not, apparently, undergo β elimination in base, producing ethylene in an apparent mode III process (Eqn. 36) instead of the expected vinyl ethers [75].

$$ROCH_2CH_2Co(D_2H_2)OH_2 \xrightarrow{OH^-} Co^{(III)}(D_2H_2)(OH)_2^- + CH_2{=}CH_2 + ROH \quad (36)$$

The mechanisms of these reactions are still under study.

Via an apparently different mechanism, 2-hydroxyalkylcobalt complexes have been shown to decompose to aldehydes or ketones and cobalt(I) chelates in base (Eqn. 37) [74].

$$\mathrm{RCHOHCH(R')Co(Chelate)L} \xrightarrow{\mathrm{OH^-}} \mathrm{R{-}\overset{O}{\overset{\|}{C}}{-}CH_2R' + Co^{(I)}(Chelate)L^-} \quad (37)$$

Schrauzer and Sibert [74] have proposed a 1,2-hydride shift mechanism for this reaction (Eqn. 38 for hydroxyethylcobalt)

$$\begin{array}{l} \mathrm{O^-} \\ | \\ \mathrm{HC{-}H} \\ | \\ \mathrm{CH_2} \\ | \\ \mathrm{Co(Chelate)L} \end{array} \longrightarrow \mathrm{H_3CCHO + Co^{(I)}(Chelate)L^-} \quad (38)$$

via the ionized 2-hydroxyalkyl species. As experiments with the appropriately deuterated compounds have not been performed to prove this mechanism, the possibility of concerted β elimination of hydridocobalt species remains. However, the failure of alkoxyethylcobalt complexes to undergo base-catalyzed elimination of cobalt(I) (Eqn. 36) would seem to make it unlikely that 2-hydroxyalkylcobalt reagents decompose in this fashion. The possibility of an intermolecular (rather than intramolecular) hydride shift mechanism has also not been tested.

There is at least one confirmed case of concerted *cis* elimination of hydridocobalt species from the decomposition of the relatively unstable 1-phenyl-2-hydroxyethyl(pyridine)cobaloxime in neutral solution to give phenylacetaldehyde and a cobaloxime(I) species as studied by Naumberg et al. [57]. These workers used the known, rapid *cis* addition of hydrido(pyridine)cobaloxime to phenylacetylene (Eqn. 14) to form 1-phenylvinyl(pyridine)cobaloxime to confirm the formation of hydridocobaloxime during this reaction. When 2-dideuterio-1-phenyl-2-hydroxy(pyridine)cobaloxime was decomposed in the presence of phenylacetylene, 2-deuterio-1-phenylvinyl(pyridine)cobaloxime was obtained (Eqn. 39), thus confirming the *cis* elimination of hydrido- (or deuterido-) cobaloxime.

$$\begin{array}{l} \mathrm{HOCD_2CH(\phi)Co(D_2H_2)py + \phi\text{-}C{\equiv}CH} \\ \quad \rightarrow \mathrm{HDC{=}C(\phi)Co(D_2H_2)py + \phi CH_2CHO} \end{array} \quad (39)$$

Furthermore, decomposition of the unlabeled cobaloxime in CH_3OD provided mainly 1-deuterio-phenylacetaldehyde, indicating the involvement of solvent in formation of the aldehyde product. Thus the mechanism must involve a *cis* elimination of hydridocobaloxime to form an enol (Eqn. 40)

$$\begin{array}{l} \phi \quad \mathrm{CH_2OH} \\ \quad \mathrm{CH} \\ \quad | \\ \mathrm{Co(D_2H_2)py} \end{array} \longrightarrow \mathrm{HCo(D_2H_2)py + \phi{-}CH{=}CHOH} \quad (40)$$

followed by solvent-catalyzed tautomerization of the enol product to the aldehyde (Eqn. 41).

$$\phi\text{—CH=CHOH} \rightarrow \phi\text{—CH}_2\text{—CHO} \tag{41}$$

Simple alkylcobalt complexes (i.e. those without heteroatomic substituents on the alkyl ligand) are also known to undergo eliminations of cobalt(I) species. Thus Grate and Schrauzer [42] have studied the decomposition of unstable, sterically strained secondary alkyl- and cycloalkyl-cobalamins to form olefins and hydridocobalamin in neutral and acidic solution. In this case the formation of hydridocobalamin was inferred from the observation of monodeuteriohydrogen gas when undeuterated alkylcobalamins were decomposed in DCl/D_2O, presumably via Eqns. 42 and 43.

$$(CH_3)_2CHCo(\text{cobalamin}) \rightarrow HCo(\text{cobalamin}) + CH_3CH{=}CH_2 \tag{42}$$

$$HCo(\text{cobalamin}) \xrightarrow{D^+} HD + Co^{(II)}(\text{cobalamin}) \tag{43}$$

Brown and Hessley [76] have studied the base-catalyzed decomposition of ethyl(aquo)cobaloxime in aqueous base at 50°C. This extremely slow reaction ($T_{1/2}$ ca. 25 h in 1.0 N base) leads to mixtures of ethane and ethylene. The formation of the olefin product in D_2O was accompanied by substantial inverse solvent deuterium isotope effects ($k_2^{H_2O}/k_2^{D_2O}$ 0.24 ± 0.02 for ethyl(aquo)cobaloxime and $k_2^{H_2O}/k_2^{D_2O}$ 0.33 ± 0.02 for ethyl(hydroxo)cobaloxime). Such inverse solvent deuterium isotope effects are common, if a bit extreme, for base-catalyzed E2 eliminations of poor leaving groups. In addition, exchange of deuterium from D_2O into the alkyl ligand during decomposition was found to be at least 2 orders of magnitude slower than the rate of ethylene formation leading to the conclusion that the mechanism is probably a concerted E2 elimination.

Finally it should be pointed out that a number of studies have been published purporting to show the nucleophilic displacement of cobalt(I) chelates from alkylcobalt complexes by thiolate anions in base (Eqn. 44) [77,78].

$$RS^- + R'Co(\text{Chelate})L \rightarrow RSR' + Co^{(I)}(\text{Chelate})L^- \tag{44}$$

This mechanism is not intellectually satisfying given the extremely high nucleophilicity of cobalt(I) reagents and the lack of any evidence that they are facile leaving groups. Several other investigators have been unable to duplicate these results, and although some workers have demonstrated cobalt-to-sulfur methyl group transfer upon reaction of methylcobalamin with thiols [79,80], these reactions occur only in the presence of air, suggesting either a radical mechanism or possible involvement of disulfides (from air oxidation of thiolate anions) in an electrophilic cleavage reaction (mode III).

(b) Mode II cleavages of carbon–cobalt bonds

The prototypical mode II cleavage of carbon–cobalt bonds is the photolytic decomposition of methylcobalt species. Under strictly anaerobic conditions this reaction, although very slow, proceeds cleanly to give a cobalt(II) complex and methyl radicals (Eqn. 45).

$$CH_3\text{—}Co(Chelate)L \overset{h\nu}{\rightleftharpoons} \cdot CH_3 + Co^{(II)}(Chelate)L \quad (45)$$

Subsequent reactions of the methyl radicals produce the stable organic products ethane and methane. Schrauzer et al. [81] have shown that the ratio of ethane to methane is sensitive to the nature of both the equatorial and axial ligands. In addition, studies of methylcobalt photolyses in D_2O showed that methane arises via two competing processes: direct abstraction of a hydrogen atom from the equatorial ligand system (Eqn. 46, in which $Co^{(II)}(Chelate)L\cdot$ represents the cobalt complex minus one hydrogen atom)

$$\cdot CH_3 + Co^{(II)}(Chelate)L \rightarrow CH_4 + Co^{(II)}(Chelate)L\cdot \quad (46)$$

and reduction of the methyl radical by the cobalt center followed by protonation of the resulting methyl carbanion by solvent (Eqns. 47 and 48).

$$\cdot CH_3 + Co^{(II)}(Chelate)L \rightarrow Co^{(III)}(Chelate)\overset{+}{L} + CH_3^- \quad (47)$$

$$CH_3^- + D_2O \rightarrow CDH_3 + OD^- \quad (48)$$

The rate of recombination of methyl radicals and cobaloxime(II) (i.e. the reverse of Eqn. 45) has been measured by flash photolysis and found to be very high (k_2 ca. 5–8 $\times 10^7$/M/sec at 25°C) [82,83]. This presumably explains both the slow rate of photolytic decomposition under strictly anaerobic conditions as well as the marked acceleration of the photolytic decomposition rate in the presence of air due to trapping of the radicals by oxygen. In the latter case the final products are cobalt(III) (i.e. aquocobalamin from methylcobalamin) and formaldehyde, with traces of methanol, methane and formic acid.

A second example of photolytic mode II cleavage is the photolysis of 5′-deoxyadenosylcobalamin which leads cleanly to cob(II)amin and a 5′-deoxyadenosyl radical, which rapidly cyclizes to give 8,5′-cyclic adenosine [84]. In this case the rate of photolytic decomposition is not substantially affected by the presence of oxygen (presumably due to a much lower rate of recombination of the radical product with the cobalt(II) species) but the products are, of course, altered to aquocobalamin and a mixture of 8,5′-cyclic adenosine and adenosine-5′-carboxaldehyde [85], the ratio of the latter two being dependent on the oxygen concentration.

It should be pointed out that while all organocobalt complexes are photolabile it is not at all clear that all such reactions proceed via mode II cleavages. Duong and coworkers [54] have argued that alkylcobaloximes containing at least one β-hydrogen undergo concerted elimination of hydridocobaloxime when photolyzed (or thermolyzed) to form olefinic products. Such a mechanism is not only consistent with the products observed by these workers and the great similarity of the photolytic and thermolytic decomposition reactions, but also explains the apparent paradox that the quite thermally labile secondary alkylcobaloximes studied should require a high energy homolytic cleavage photolysis mechanism. These results are in accord with those of Brown and Ingraham [52] who found that the anaerobic photolysis of 2-hydroxy-*n*-propylcobaloximes forms acetone (Eqn. 49) (presumably via its enol from elimination of hydridocobaloxime)

$$CH_3CHOHCH_2Co(D_2H_2)L \xrightarrow{h\nu} HCo(D_2H_2)L + CH_3{-}\overset{\overset{\displaystyle O}{\|}}{C}{-}CH_3 \qquad (49)$$

while 2-hydroxyisopropylcobaloximes, which have two proton-bearing β-carbons, produce mixtures of allyl alcohol and propionaldehyde (Eqn. 50)

$$HOCH_2(CH_3)CHCo(D_2H_2)L \xrightarrow{h\nu} HCo(D_2H_2)L + CH_2{=}CHCH_2OH + CH_3CH_2CHO \qquad (50)$$

(again, presumably via its enol). Traces of isopropanol formed in the former case were believed to be due to subsequent reduction of acetone by hydridocobaloxime in as much as the amount of isopropanol obtained was found to increase with increasing photolysis time. The observation of substantial dependence of the ratio of allyl alcohol to propionaldehyde on the nature of the axial ligand, L, in the latter case (Eqn. 50) is consistent with a concerted elimination mechanism.

Chemaly and Pratt [86] have recently described the oxygen promoted decomposition of neopentylcobalamin which was ascribed to the occurrence of an equilibrium mode II cleavage to cob(II)alamin and neopentyl radicals (Eqn. 51).

$$(CH_3)_3CCH_2Co(Cobalamin) \rightleftharpoons (CH_3)_3CCH_2\cdot + Co^{II}(Cobalamin) \qquad (51)$$

Apparently the rate of radical recombination is high so that under anaerobic conditions the compound is stable, but in air, radical trapping by oxygen promotes net decomposition. The unusual lability of neopentylcobalamin towards spontaneous mode II cleavage was attributed to steric compression around the coordinated carbon in 6 coordinate complexes as evidenced by the fact that the compound is stable in acid (even in the presence of O_2) where the pendant ligand is dissociated and protonated (see Section 5.3).

The previously mentioned study of the base-catalyzed decomposition of ethylcobaloxime by Brown and Hessley [76] represents an interesting example of a base-catalyzed mode II cleavage. While the ethylene product of this reaction is undoubtedly formed via a mode I β elimination, the ethane product was originally thought to be formed by a mode III cleavage in analogy with the base-catalyzed formation of methane from methylcobaloxime (see Section 5.2 (c)). However, subsequent studies of the time dependence of the formation of monodeuterioethane from decomposition in alkaline D_2O and from the decomposition of ethylcobaloxime extensively deuterated in the equatorial ligand in both D_2O and H_2O showed conclusively that a homolytic, mode II process was responsible for the ethane product [87]. Thus, in analogy with Schrauzer's results for the photolysis of methylcobaloximes (Eqns. 46–48), ethyl radicals generated by a primary mode II cleavage either abstract hydrogen from the equatorial ligand or are reduced by the cobalt center and subsequently protonated by solvent.

Finally, M.D. Johnson and coworkers have recently demonstrated a number of interesting homolytic displacement reactions of organocobaloximes in which a polyhalomethyl radical (Eqn. 52) [88],

$$CCl_3C\cdot + \phi CH_2Co(D_2H_2)L \rightarrow \phi CH_2CCl_3 + Co^{(II)}(D_2H_2)L \tag{52}$$

or an arene sulphonyl radical (Eqn. 53) [89]

$$\begin{aligned} &ArSO_2\cdot + (CH_3)_2C{=}CHCH_2Co(D_2H_2)L \\ &\quad \rightarrow ArSO_2C(CH_3)_2CH{=}CH_2 + Co^{(II)}(D_2H_2)L \end{aligned} \tag{53}$$

attacks the γ-carbon of an allyl or propadienylcobaloxime or the α-carbon of a benzylcobaloxime with either concerted or subsequent loss of cobaloxime(II). These intriguing chain reactions, which are still being developed by these workers, have considerable utility in the regiospecific and stereospecific syntheses of organic reagents.

(c) Mode III cleavages of carbon–cobalt bonds

5′-Deoxyadenosylcobalamin has long been known to be acid-labile, decomposing in 0.1 N HCl at 100°C with elimination of aquocobalamin (Eqn. 54) [90].

OH OH (ribose ring, O, Ad; CH_2–Co(Cobalamin)) $\xrightarrow{H^+}$ H_2O(Cobalamin) + $H_2C{=}CHCHOHCHOHCHO$ + adenine (54)

This is, in fact, a particular example of a general class of reactions in which organocobalt complexes with β-oxy substituents (including 2-hydroxy-, 2-alkoxy- and 2-alkyloxyalkyl-cobalts) [52,91–93] are decomposed in acid via elimination of aquocobalt complexes (Eqn. 55).

$$ROCH(R')CH_2Co(Chelate)L \xrightarrow{H^+} ROH + R'CH{=}CH_2 + H_2OCo(Chelate)L \tag{55}$$

It has long been postulated that these facile reactions occur via formation of a cationic intermediate (upon acid-catalyzed loss of ROH) which can be formulated either as a σ-bonded alkylcobalt carbonium ion or a cobalt(III)–olefin π complex. Recently, firm kinetic evidence has been obtained for the occurrence of an intermediate in the acid-catalyzed decomposition of 2-hydroxy- and 2-alkoxyethylcobaloximes [94]. Thus, while 2-hydroxyethylcobaloxime decomposes with strictly first-order kinetics in mildly acidic H_2SO_4/H_2O mixtures, the alkoxy derivatives show a substantial lag followed by a first-order decay which is slower than that for the hydroxyethyl complex. In strongly acidic mixtures ($H_0 < -5$) all compounds show a rapid burst of absorbance change, followed by a slower first-order decay which is identical for all compounds whether measured spectrophotometrically or manometrically. These observations support the mechanism shown in Eqn. 56.

$$ROCH_2CH_2Co(D_2H_2)OH_2 \xrightarrow[-ROH]{+H^+} \underset{Co(D_2H_2)OH_2}{\overset{+}{H_2C{=}CH_2}} \longrightarrow H_2C{=}CH_2 + Co(D_2H_2)(OH_2)_2^+ \quad (56)$$

$$HOCH_2CH_2Co(D_2H_2)OH_2 \underset{+HOH}{\overset{+H^+}{\rightleftharpoons}} \underset{Co(D_2H_2)OH_2}{\overset{+}{H_2C{=}CH_2}}$$

Evidence that the intermediate formed is a π complex rather than an alkylcobaloxime carbonium ion was obtained by direct observation of the intermediate (in D_2SO_4/D_2O) by ^{1}H-nmr spectroscopy.

5′-Deoxyadenosylcobalamin also undergoes a mode III cleavage when treated with alkaline cyanide (Eqn. 57) [95].

$$\text{(ribose: OH, OH, Ad; } CH_2\text{–Co(Cobalamin))} \xrightarrow{CN^-} (CN)_2Co(\text{Cobalamin}) + \text{adenine} + H_2C{=}CH{-}(CHOH)_3{-}CN \quad (57)$$

This reaction proceeds via displacement of the axial pendant ligand by cyanide, followed by either an E2 elimination of the organic group or its nucleophilic displacement by CN^- to form adenine and D-*erythro*-2,3-dihydroxy-Δ4-pentenal which reacts with excess CN^- to form the two epimeric cyanohydrins. Some simple alkyl cobalamins are similarly cleaved in cyanide while many simply undergo axial ligand substitution [91]. Hogenkamp et al. [91] found that only alkylcobalamins with suitably electron-withdrawing substituents (such as the methyl esters of carboxymethyl- and carboxyethyl-cobalamin and cyanoethylcobalamin) undergo cyanolysis, presumably indicating that substantial polarization of the carbon–cobalt bond toward the carbon atom is required to allow elimination of the organic group as a carbanion.

Brown [72] has studied the decomposition of methyl(aquo)cobaloxime in aqueous base at 50°C, an interesting example of a base-catalyzed model III cleavage. Methane is the organic product, formed by solvent protonation of the methyl carbanion cleavage product, but its yield is always less than stoichiometric due to the base-catalyzed formation of a base-stable methylcobalt side product in which water has added across one of the carbon–nitrogen double bonds of the equatorial ligand. Studies of the hydroxide ion dependence of the methane-forming reaction showed that only methyl(hydroxy)cobaloxime, and not the aquo complex, was active in forming methane. Interestingly, addition of various *N*- and *S*-liganding axial ligands (which displace axial aquo or hydroxo ligands) depressed the rate of methane production by at least 4 orders of magnitude. It is not yet at all clear why only the hydroxo complex is active in this reaction, nor is there an explanation for the fact that base-catalyzed methane formation from methylcobaloximes proceeds via a different mechanism than base-catalyzed ethane formation from ethylcobaloximes (see Section 5.2 (b)).

Organocobalt complexes are also known to undergo electrophilic cleavage by a variety of inorganic electrophiles including mercury(II) [96,97] (e.g. Eqn. 58 for methylcobalamin),

$$CH_3Co(Cobalamin) + HgX_2 \rightarrow H_2OCo(Cobalamin) + CH_3HgX \quad (58)$$

thallium(III) [97,98] platinum(IV) and gold(III) [98], palladium(II) [99], other cobalt(III) chelates [100], and bromine [101,102]. Among the cobalamins, methylcobalamin is attacked much more readily than other alkyl derivatives, presumably due to steric inhibition of the approach of bulky electrophilic reagents to the carbon–cobalt bond. These electrophilic mode III cleavages are generally thought to occur via bimolecular electrophilic substitution (S_E2) and although such electrophilic substitutions are generally expected to occur with retention of configuration, both Jensen et al. [102] and Tada and Ogawa [101] found the electrophilic cleavage of substituted cyclohexylcobaloximes by bromine to occur with inversion at the α-carbon. Similarly the electrophilic cleavage of both secondary [101] and primary [96] organocobaloximes by mercury(II) has been shown to proceed with inversion. Most investigators feel that inversion of configuration in these reactions is mandated by the bulky nature of the chelated cobalt-leaving group.

5.3. Axial ligand substitution reactions

Thusius has studied the kinetics of cobalamin-ligand substitution reactions *trans* to the pendant 5,6-dimethylbenzimidazole ligand (Eqn. 59) for a variety of inorganic ligands including thiocyanate, azide, iodide and bromide ions [103].

$$\left[\begin{array}{c} OH \\ | \\ Co(Corrin) \\ \uparrow \\ Bz \end{array}\right. + L \underset{k_r}{\overset{k_f}{\rightleftharpoons}} \left[\begin{array}{c} L \\ | \\ Co(Corrin) \\ \uparrow \\ Bz \end{array}\right. + H_2O \quad (59)$$

These reactions are, for the most part, extremely rapid, requiring stopped flow techniques, in contrast to the ligand-substitution reactions of simpler (inorganic) cobalt(III) complexes which are substitution-inert. Thusius' principal finding was that while the forward-rate constants for 7 incoming ligands varied by only slightly more than a factor of 10, the reverse-rate constants varied by more than 7 orders of magnitude. These results are consistent with a transition state in which both incoming and leaving ligands are loosely bound to cobalt, i.e. a dissociative interchange, or I_d, mechanism. Reenstra and Jencks [104] recently reached a similar conclusion about the more complicated substitution reactions of aquocobalamin by cyanide ion which can displace both axial ligands. In addition, Thusius found that the substantially increased lability of aquocobalamin towards axial ligand substitution compared to simpler cobalt(III) chelates was due primarily to differences in activation entropies.

In contrast Brown and coworkers [105,106], who have intensively studied ligand substitution *trans* to the organic group in organo(aquo)cobaloximes by various nitrogen and sulfur donors (Eqn. 60)

$$RCo(D_2H_2)OH_2 + L \underset{k_r}{\overset{k_f}{\rightleftharpoons}} RCo(D_2H_2)L + H_2O \tag{60}$$

have concluded that this reaction is a purely dissociative (D) or S_N1 mechanism (Eqns. 61 and 62) in which a 5-coordinate intermediate is produced. In this case, both the forward- and reverse-rate constants decrease markedly with increasing electron-withdrawing ability of the organic group, R. In addition, non-linear structure–reactivity correlations as well as spectroscopic evidence suggest that the 6-coordinate–5-coordinate equilibrium of Eqn. 61 may be substantially displaced toward the 5-coordinate alkylcobaloximes for sufficiently electron-donating organic ligands. Structure–reactivity correlations in this work also permitted the calculation of the percentage of Co—N bond formation in the transition state of the ligand-addition step (Eqn. 62) which was found to be an "early" transition state with only about 20% bond formation.

$$RCo(D_2H_2)OH_2 \rightleftharpoons RCo(D_2H_2) + H_2O \tag{61}$$

$$RCo(D_2H_2) + L \rightleftharpoons RCo(D_2H_2)L \tag{62}$$

By far the most well studied ligand-substitution reaction *trans* to the organic ligand in organocobalamins is the simple displacement of the pendant axial 5,6-dimethylbenzimidazole in acid, the so-called "base-on"–"base-off" transition of organocobalamins (Eqn. 63).

R — Co (Corrin) ← Bz $\underset{}{\overset{+H_3O^+}{\rightleftharpoons}}$ R — Co (Corrin) — OH_2 … BzH^+ (63)

This ligand exchange causes a marked change in the electronic spectrum from the red base-on species to the yellow base-off and has been referred to as the "red–yellow transition". Apparent pK_a's for this process have been measured for a number of alkylcobalamins [91,107] and found to range from about 3.9 for ethylcobalamin to as low as 0.1 for cyanocobalamin. Interestingly, Hogenkamp and coworkers [91] found that correlation of these apparent pK_a's with the Hammet σ_m substituent parameter for the organic ligand (as a measure of the inductive effect) produced two lines with the same slope (ca. -1.5) but different intercepts, one for substituted methylcobalamins and one for substituted ethylcobalamins, the latter correlation having the higher intercept. Although this phenomenon has never been explained, it seems possible that the higher apparent pK_a's of the substituted ethylcobalamins reflect steric compression of the corrin moiety due to the presence of a tetrahedral β-carbon (and its substituents, when present) leading to increased lability of the pendant ligand's N—Co bond. This seems reasonable in light of recent observations that sterically crowded secondary alkylcobalamins are partially or completely base-off even in neutral and basic solution (i.e. where the pendant benzimidazole is unprotonated) [42,108].

Very little work has been done on the kinetics of axial ligand substitution *trans* to the organic ligand in organocobalt corrins primarily due to the excessive speed of these reactions. However, two noteworthy attempts have been made to study the kinetics of benzimidazole displacement in methylcobalamin (Eqn. 63). Milton and T.L. Brown [109] studied the temperature dependence of the ^{1}H-nmr resonance of the cobalt-bound methyl group and the ^{13}C-nmr resonance of 90% ^{13}C-enriched methylcobalamin. From analyses of the line shapes, the activation parameters for benzimidazole dissociation were calculated to be $\Delta G^{\neq} = 12.7 \pm 0.1$ kcal/mole, $\Delta H^{\neq} = 11.1 \pm 0.6$ kcal/mole, and $\Delta S^{\neq} = -5.9 \pm 2.4$ eu, leading to a calculated rate constant of 2060/sec at 25°C. In contrast, K.L. Brown and coworkers [110] studied the pH dependence of this reaction (Eqn. 63) at 5°C by temperature-jump spectroscopy and observed two relaxations, the faster of which ($T_{1/2}$ ca. 4.4 μsec) was pH-independent over the range pH 0.7–7.5 while the slower relaxation showed a slight pH dependence with a half-time of about 50 μsec at pH 5.5–7.5 increasing to about 75 μsec at pH 1.6 and below. These authors assigned the faster relaxation to the reversible loss of water from the axial coordination position and the slower, pH-dependent relaxation to the reversible dissociation of benzimidazole, i.e. an S_N1, or D mechanism (Eqn. 64).

$$\begin{bmatrix}\mathrm{CH_3} \\ | \\ \mathrm{Co(Corrin)} \\ | \\ \mathrm{OH_2} \\ \mathrm{BzH^+}\end{bmatrix} \underset{+\mathrm{H^+}}{\overset{\text{fast}}{\rightleftharpoons}} \begin{bmatrix}\mathrm{CH_3} \\ | \\ \mathrm{Co(Corrin)} \\ | \\ \mathrm{OH_2} \\ \mathrm{Bz}\end{bmatrix} \rightleftharpoons \begin{bmatrix}\mathrm{CH_3} \\ | \\ \mathrm{Co(Corrin)} \\ \mathrm{Bz}\end{bmatrix} \rightleftharpoons \begin{bmatrix}\mathrm{CH_3} \\ | \\ \mathrm{Co(Corrin)} \\ \uparrow \\ \mathrm{Bz}\end{bmatrix} \qquad (64)$$

This analysis leads to a value of $9.07 \pm 0.70 \times 10^3$/sec for the benzimidazole dissociation-rate constant at 5°C, a value substantially higher than that calculated by Milton and T.L. Brown [109] at 25°C. Furthermore, the data permit a calculation of the apparent formation constant of base-on methylcobalamin from the base-off, but unprotonated species. This equilibrium constant has the startling value of 0.43, indicating that "base-on" methylcobalamin (i.e. above pH 5) is only about 30% coordinated to benzimidazole. The discrepancies between these two kinetic studies have yet to be resolved.

5.4. Reactions of cobalt-bound organic ligands

Although many examples of reactions of cobalt-bound organic ligands exist, most of these have been carried out for synthetic purposes rather than as studies of mechanism or of the effect of the covalently bound cobalt moiety. Although many such transformations are quite simple, they frequently lead to organocobalt complexes which are difficult, or impossible, to obtain by direct alkylation. For example, Schrauzer and Windgassen [51] prepared carboxymethyl- and carboxyethyl-(pyridine)cobaloxime by hydrolysis of the corresponding methyl esters in concentrated sulfuric acid. Similarly, Brown and coworkers [58] prepared *p*- and *m*-carboxyphenyl(aquo)cobaloximes from their methyl esters by hydrolysis in 0.5 N KOH in aqueous methanol.

Acetal hydrolysis has also been successfully used to affect such organic ligand modifications. Thus Silverman and Dolphin [111] obtained both formylmethylcobaloxime and formylmethylcobalamin by hydrolysis of both the 2,2-diethoxyethyl- and 1,3-dioxa-2-cyclopentylmethyl-cobalt complexes (Eqn. 65).

$$(CH_3CH_2O)_2CHCH_2Co(Chelate)L \xrightarrow{pH\,9} OHCCH_2Co(Chelate)L \quad (65)$$
$$\text{1,3-dioxa-2-cyclopentyl-}CHCH_2Co(Chelate)L \xrightarrow{pH\,9} OHCCH_2Co(Chelate)L$$

Formylmethylcobalamin has also been obtained by periodate cleavage of 2,3-dehydroxy-*n*-propylcobalamin in aqueous ammonia (Eqn. 66) [111].

$$HOCH_2CH(OH)CH_2Co(Cobalamin) \xrightarrow{NaIO_4} OHCCH_2Co(Cobalamin) \quad (66)$$

Golding and coworkers [112] have also obtained a series of dihydroxyalkylcobaloximes by hydrolysis of the appropriate cyclic acetals in aqueous hydrochloric acid (Eqn. 67).

$$(CH_3)_2C(-O-CH_2-O-)CH(CH_2)_nCH_2Co(D_2H_2)L \xrightarrow{HCl/H_2O} HOCH_2CHOH(CH_2)_nCH_2Co(D_2H_2)L \quad (67)$$

As a final example, Golding and coworkers [92] obtained 2-acetoxyethyl- and 2-acetoxy-*n*-propyl-(pyridine)cobaloxime by acylation of 2-hydroxyethyl- and 2-hydroxy-*n*-propyl-(pyridine)cobaloximes with acetic anhydride in pyridine.

A few more detailed kinetic and thermodynamic studies of the reactions of cobalt-bound organic ligands have been carried out. For instance, Golding et al. [92]

studied the kinetics of the hydrolysis and alcoholysis of the aforementioned 2-acetoxyalkyl(pyridine)cobaloximes and found them to be extremely labile acetates. These workers reported a rate constant of 4.37×10^{-6}/sec for the ethanolysis of 2-acetoxyethyl(pyridine)cobaloxime, an energy of activation of 19.9 kcal, and an enthalpy of activation of −18.2 eu. The latter finding was taken to suggest formation of an intermediate (which can be formulated as a cobaloxime(III) ethylene π complex, Eqn. 56) in which carbon–carbon bond rotation is restricted. Subsequent studies of the methanolysis of 2-^{13}C-enriched 2-acetoxyethyl(pyridine)cobaloxime [113] showed that the two methylene carbons for this compound become equivalent during this reaction (Eqn. 68), again suggesting formation of a symmetrical intermediate.

$$H_3C{-}\overset{O}{\overset{\|}{C}}{-}O\,{}^{13}CH_2CH_2Co(D_2H_2)py \xrightarrow{-OAc^-} \underset{\overset{+}{Co}(D_2H_2)py}{{}^{13}CH_2{\cdots}CH_2} \xrightarrow{+MeOH}$$

$$\underset{(50\%)}{H_3CO^{13}CH_2CH_2Co(D_2H_2)py} + \underset{(50\%)}{H_3COCH_2{}^{13}CH_2Co(D_2H_2)py} \qquad (68)$$

Brown and coworkers [58,114] have studied the rate of base-catalyzed hydrolysis of the methyl esters of *m*- and *p*-carboxyphenyl(ligand)cobaloximes for a number of inorganic and primary amine axial ligands. Correlation of these rate constants with those of other substituted methylbenzoates via the Taft dual substituent parameter equation showed that the cobaloxime-chelated cobalt centers behave as extremely electron-donating substituents in the inductive sense with values of σ_I as negative as −0.53 for thiocyanate liganded cobaloxime. For the aquo and hydroxo complexes the values of the resonance-substituent parameters were negligibly close to zero, but for cyanide, nitrite, thiocyanate and azide ligands the cobaloxime substituents were found to be significantly resonance-electron donating. Unfortunately, anomalous results for a series of primary amine ligands in which the values of $-\sigma_I$ were found to vary inversely with the basicity of the amine ligand cloud the interpretation of these results and suggest the possibility that the mechanism of hydrolysis of some of the cobaloxime-substituted methylbenzoates may not be the same as that of the bases set compounds.

These workers [58,115] have also studied the acidities of carboxyethyl- and carboxymethyl-cobaloximes, the former behaving like normal 2-substituted propionic acids, with pK_as varying from 4.71 to 5.11 for carboxyethyl(ligand)cobaloximes with 15 different axial ligands. However, carboxymethyl(ligand)cobaloximes (p$K_a = 6.30$ for $L = OH_2$, 6.54 for $L =$ py) as well as carboxymethylcobalamin (p$K_a = 7.2$) [115] are substantially weaker acids than would be expected even considering the extreme electron-donating ability of such cobalt centers. This is an example of the organometallic β effect and can be ascribed to substantial hyperconjugation (or $\sigma \rightarrow \pi$ conjugation) in these systems. This effect evidently also occurs in benzylcobaloximes [116] and is currently under study.

6. *Concluding remarks*

Remarkable progress has been made in almost all aspects of organocobalt chemistry since the initial discovery of naturally occurring organocobalt B_{12} derivatives. As pointed out frequently during this chapter, much work remains to be done particularly regarding mechanistic aspects of organocobalt synthesis reactions, carbon–cobalt cleavage reactions and the chemistry of cobalt-bound organic ligands. Although an ultimate understanding of the enzymic reactions which utilize B_{12} coenzymes can only come with increased study of the enzymic reactions themselves, an increasingly broad and detailed understanding of organocobalt chemistry in general will surely be needed to insure the success of such endeavors.

References

1 Rickes, E.L., Brink, N.G., Koniuszy, F.R., Wood, T.R. and Folkers, K. (1948) Science 107, 396–397.
2 Smith, E.L. and Parker, L.F. (1948) Biochem. J. 43, viii–ix.
3 Hogenkamp, H.P.C. (1975) in: B.M. Babior (Ed.), Cobalamin, Wiley, New York, pp. 21–73.
4 Pratt, J.M. (1972) Inorganic Chemistry of Vitamin B_{12}, Academic Press, London.
5 *Abbreviations:* $RCo(D_2H_2)L$ = organo(ligand)bis(dimethylglyoximato)cobalt(III) = organo(ligand) cobaloxime; $HCo(D_2H_2)L$ = hydrido(ligand)bis(dimethylglyoximato)cobalt(III) = hydrido(ligand) cobaloxime; B_{12r} = cob(II)alamin(cobalt(II)cobalamin); B_{12s} = cob(I)alamin(cobalt(I)cobalamin); py = pyridine.
6 IUPAC–IUB Commission on Biochemical Nomenclature (1974) Biochemistry 13, 1555–1560.
7 Hodgkin, D.C. (1958) Fortschr. Chem. Org. Naturstoffe 15, 167–220.
8 Hodgkin, D.C. (1965) Proc. Roy. Soc. A 288, 294–305.
9 Friedmann, H.C. (1975) in: B.M. Babior (Ed.), Cobalamin, Wiley, New York, pp. 75–109.
10 Lenhert, P.G. (1968) Proc. Roy. Soc. A 303, 45–84.
11 Imfeld, M., Townsent, C.A. and Arigoni, D. (1976) J. Chem. Soc. Chem. Commun. 541–542.
12 Battersby, A., Hollenstein, R., McDonald, E. and Williams, D.C. (1976) J. Chem. Soc. Chem. Commun. 543–544.
13 Scott, A.I., Kajiwara, M., Takahashi, T., Armatage, I.M., Demou, P. and Petrocine, D. (1976) J. Chem. Soc. Chem. Commun. 544–546.
14 Dolphin, D. (1971) Methods Enzymol. 18, 34–52.
15 Lexa, D., Saveant, J.M. and Zickler, J. (1977) J. Am. Chem. Soc. 99, 2786–2790.
16 Blaser, H.-U. and Halpern, J. (1980) J. Am. Chem. Soc. 102, 1684–1689.
17 Schrauzer, G.N. and Hashimoto, M. (1979) J. Am. Chem. Soc. 101, 4593–4601.
18 Beavin, G.H. and Johnson, E.A. (1955) Nature (London) 176, 1264–1265.
19 Espenson, J.H. and Martin, A.H. (1977) J. Am. Chem. Soc. 99, 5953–5957.
20 Schrauzer, G.N. and Lee, L.P. (1968) J. Am. Chem. Soc. 90, 6541–6543.
21 Dunne, C.P. (1971) Ph.D. Dissertation, Brandeis University, Waltham, MA.
22 Ogoshi, H., Watanabe, E. Koketsu, N. and Yoshida, Z. (1976) Bull. Chem. Soc. Jpn. 49, 2529–2536.
23 Schrauzer, G.N. and Holland, R.J. (1971) J. Am. Chem. Soc. 93, 4060–4062.
24 Bernhauer, K. and Müller, O. (1961) Biochem. Z. 335, 44–50.
25 Johnson, A.W., Mervyn, L., Shaw, N. and Smith, E.L. (1963) J. Chem. Soc., 4146–4156.
26 Simándi, L.I., Budó-Záhonyi, E. and Szeverenyi, Z. (1976) Inorg. Nucl. Chem. Lett. 12, 237–241.
27 Chao, T.-H. and Espenson, J.H. (1978) J. Am. Chem. Soc. 100, 129–133.
28 Lexa, D. and Savéant, J.M. (1975) Chem. Commun. 872–874.
29 Schrauzer, G.N. and Windgassen, R.J. (1966) Chem. Ber. 99, 602–610.
30 Yamada, R., Shimizu, S. and Fukui, S. (1968) Biochemistry 7, 1713–1719.

31 Kaufman, E.J. and Espenson, J.H. (1977) J. Am. Chem. Soc. 99, 7051–7054.
32 Simándi, L.I., Németh, S. and Budó-Záhonyi, E. (1980) Inorg. Chim. Acta 45, L143–L145.
33 Schrauzer, G.N. and Holland, R.J. (1971) J. Am. Chem. Soc. 93, 1505–1506.
34 Venerable, G.D. and Halpern, J. (1971) J. Am. Chem. Soc. 93, 2176–2179.
35 Stadtmann, T.C. (1971) Science 171, 859–867.
36 Babior, B.M. (1975) Accts. Chem. Res. 8, 376–384.
37 Babior, B.M. (1975) in: B.M. Babior (Ed.), Cobalamin, Wiley, New York, pp. 141–212.
38 Barker, H.A. (1972) Annu. Rev. Biochem. 41, 55–90.
39 Taylor, C.D. and Wolfe, R.S. (1974) J. Biol. Chem. 249, 4879–4885.
40 Dodd, D. and Johnson, M.D. (1973) J. Organometal. Chem. 52, 1–232.
41 Brown, K.L. (1983) in: D. Dolphin (Ed.), B_{12}, Wiley, New York, pp. 245–294.
42 Grate, J.H. and Schrauzer, G.N. (1979) J. Am. Chem. Soc. 101, 4601–4611.
43 Jensen, F.R. and Buchanan, D.H. (1973) Chem. Commun. 153–154.
44 Eckert, H., Lenoir, D. and Ugi, I. (1977) J. Organometal. Chem. 141, C23–C27.
45 Schrauzer, G.N. and Deutsch, E. (1969) J. Am. Chem. Soc. 91, 3341–3350.
46 Jensen, F.R., Maden, V. and Buchanan, D.H. (1970) J. Am. Chem. Soc. 92, 1414–1416.
47 Bock, D.L. and Whitesides, G.M. (1974) J. Am. Chem. Soc. 96, 2826–2829.
48 Fritz, H.L., Espenson, J.H., Williams, D.A. and Molander, G.A. (1974) J. Am. Chem. Soc. 96, 2378–2381.
49 Schäffler, J. and Rétey, J. (1978) Angew. Chem. Int. Ed. Engl. 17, 845–846.
50 Breslow, R. and Khanna, P.L. (1976) J. Am. Chem. Soc. 98, 1297–1299.
51 Schrauzer, G.N. and Windgassen, R.J. (1967) J. Am. Chem. Soc. 89, 1999–2007.
52 Brown, K.L. and Ingraham, L.L. (1974) J. Am. Chem. Soc. 96, 7681–7686.
53 Schrauzer, G.N., Weber, J.H. and Beckham, T.M. (1970) J. Am. Chem. Soc. 92, 7078–7086.
54 Duong, K.N.V., Ahond, A., Merienne, C. and Gaudemer, A. (1973) J. Organometal. Chem. 55, 375–382.
55 Halpern, J. and Wong, L.Y. (1968) J. Am. Chem. Soc. 90, 6665–6669.
56 Dodd, D., Johnson, M.D., Meeks, B.S., Titchmarsh, D.M., Duong, K.N.V. and Gaudemer, A. (1976) J. Chem. Soc. Perkin II, 1261–1267.
57 Naumberg, M., Duong, K.N.V. and Gaudemer, A. (1970) J. Organometal. Chem. 25, 231–242.
58 Brown, K.L., Awtrey, A.W. and LeGates, R. (1978) J. Am. Chem. Soc. 100, 823–828.
59 Brown, K.L. and LeGates, R. (1982) J. Organometal. Chem. 233, 259–265.
60 Halpern, J. and Phelan, P.F. (1972) J. Am. Chem. Soc. 94, 1881–1886.
61 Marzilli, L.G., Marzilli, P.A. and Halpern, J. (1971) J. Am. Chem. Soc. 93, 1374–1378.
62 Roussi, P.F. and Widdowson, D.A. (1979) J. Chem. Soc. Chem. Commun. 810–812.
63 Roche, J.S. and Endicott, J.F. (1974) Inorg. Chem. 13, 1575–1580.
64 Goedken, V.L., Peng, S.M. and Park, Y. (1974) J. Am. Chem. Soc. 96, 284–285.
65 Clarke, D.A., Dolphin, D., Grigg, R., Johnson, A.W. and Pinnock, H.A. (1968) J. Chem. Soc. C. 881–885.
66 Costa, G., Mestroni, G. and Stefani, L. (1967) J. Organometal. Chem. 7, 493–501.
67 Cummins, D., Higson, B.M. and McKenzie, E.D. (1973) J. Chem. Soc. Dalton Trans. 414–419.
68 Cummins, D. and McKenzie, E.D. (1976) Inorg. Nucl. Chem. Lett. 12, 521–525.
69 Cummins, D., McKenzie, E.D. and Segnitz, A. (1975) J. Organometal. Chem. 87, C19–C21.
70 Callot, H.J. and Schaeffer, E. (1978) J. Organometal. Chem. 145, 91–99.
71 Silverman, R.B. and Dolphin, D. (1976) J. Am. Chem. Soc. 98, 4626–4633.
72 Brown, K.L. (1979) J. Am. Chem. Soc. 101, 6600–6606.
73 Barnett, R., Hogenkamp, H.P.C. and Abeles, R.H. (1965) J. Biol. Chem. 241, 1483–1486.
74 Schrauzer, G.N. and Sibert, J.W. (1970) J. Am. Chem. Soc. 92, 1022–1030.
75 Brown, K.L. and Szeverenyi, Z., unpublished.
76 Brown, K.L. and Hessley, R.K. (1980) Inorg. Chem. 19, 2410–2414.
77 Schrauzer, G.N. and Windgassen, R.J. (1967) J. Am. Chem. Soc. 89, 3607–3608.
78 Schrauzer, G.N. and Stadlbauer, E.A. (1974) Bioinorg. Chem. 3, 353–366.

79 Agnes, G., Hill, H.A.O., Pratt, J.M., Ridsdale, S.C., Kennedy, F.S. and Williams, R.J.P. (1971) Biochim. Biophys. Acta 252, 207–211.
80 Frick, T., Francia, M.D. and Wood, J.M. (1976) Biochim. Biophys. Acta 428, 808–817.
81 Schrauzer, G.N., Sibert, J.W. and Windgassen, R.J. (1968) J. Am. Chem. Soc. 90, 6681–6688.
82 Endicott, J.F. and Ferraudi, G.J. (1977) J. Am. Chem. Soc. 99, 243–245.
83 Lerner, D.A., Bonneau, R. and Giannotti, C. (1979) J. Photochem. 11, 73–77.
84 Hogenkamp, H.P.C. (1963) J. Biol. Chem. 238, 477–480.
85 Hogenkamp, H.P.C., Ladd, J.N. and Barker, H.A. (1962) J. Biol. Chem. 237, 1950–1952.
86 Chemaly, S.M. and Pratt, J.M. (1980) J. Chem. Soc. Dalton, 2274–2281.
87 Brown, K.L. (1981) J. Chem. Soc. Chem. Commun. 598–599.
88 Crease, A.E., Gupta, B.D., Johnson, M.D. and Moorhouse, S. (1978) J. Chem. Soc. Dalton Trans., 1821–1825.
89 Cooksey, C.J., Crease, A.E., Gupta, B.D., Johnson, M.D., Bialkowska, E., Duong, K.N.V. and Gaudemer, A. (1979) J. Chem. Soc. Perkin I, 2611–2616.
90 Hogenkamp, H.P.C. and Barker, H.A. (1961) J. Biol. Chem. 236, 3097–3101.
91 Hogenkamp, H.P.C., Rush, J.E. and Swenson, C.A. (1965) J. Biol. Chem. 240, 3641–3644.
92 Golding, B.T., Holland, H.L., Horn, U. and Sakriker, S. (1970) Angew. Chem. Int. Ed. Engl. 9, 959–960.
93 Espenson, J.H. and Wang, D.M. (1979) Inorg. Chem. 18, 2853–2859.
94 Brown, K.L. and Ramamurthy, S. (1981) Organometallics, 1, 413–415.
95 Barker, H.A., Smyth, R.D., Weissbach, J., Toohey, J.I., Ladd, J.N. and Volcani, B.E. (1960) J. Biol. Chem. 235, 480–488.
96 Fritz, H.C., Espenson, J.H., Williams, D.A. and Molander, G.A. (1974) J. Am. Chem. Soc. 96, 2378–2381.
97 Abley, P., Dockal, E.R. and Halpern, J. (1973) J. Am. Chem. Soc. 95, 3166–3170.
98 Agnes, G., Bendle, S., Hill, H.A.O., Williams, F.R. and Williams, R.J.P. (1971) Chem. Commun. 850–851.
99 Scovell, W.M. (1974) J. Am. Chem. Soc. 96, 3451–3456.
100 Costa, G., Mestroni, G. and Cocerar, C. (1971) Tetrahedron Lett., 1869–1870.
101 Tada, M. and Ogawa, H. (1973) Tetrahedron Lett., 2639–2642.
102 Jensen, F.R., Madan, V. and Buchanan, D.H. (1971) J. Am. Chem. Soc. 93, 5283–5284.
103 Thusius, D. (1971) J. Am. Chem. Soc. 93, 2629–2635.
104 Reenstra, W.W. and Jencks, W.P. (1979) J. Am. Chem. Soc. 101, 5780–5791.
105 Brown, K.L., Lyles, D., Pencovici, M. and Kallan, R.G. (1975) J. Am. Chem. Soc. 97, 7338–7346.
106 Brown, K.L. and Awtrey, A.W. (1978) Inorg. Chem. 17, 111–119.
107 Hayward, G.C., Hill, H.A.O., Pratt, J.M., Vanston, N.J. and Williams, R.J.P. (1965) J. Chem. Soc. 6485–6493.
108 Chemaly, S.M. and Pratt, J.M. (1980) J. Chem. Soc. Dalton 2259–2266.
109 Milton, P.A. and Brown, T.L. (1977) J. Am. Chem. Soc. 99, 1390–1396.
110 Brown, K.L., Awtrey, A.W., Chock, P.B. and Rhee, S.G. (1979) in: Vit. B_{12} Proc. 3rd Eur. Symp. Vit. B_{12} Intr. Fac., pp. 199–202.
111 Silverman, R.B. and Dolphin, D. (1976) J. Am. Chem. Soc. 98, 4633–4639.
112 Golding, B.T., Kemp, T.J., Sell, C.S., Sellars, P.J. and Watson, W.P. (1978) J. Chem. Soc. Perkin Trans. II 839–848.
113 Silverman, R.B., Dolphin, D. and Babior, B.M. (1972) J. Am. Chem. Soc. 94, 4028–4030.
114 Brown, K.L. and Awtrey, A.W. (1980) J. Organometal. Chem. 195, 113–122.
115 Walker, T.E., Hogenkamp, H.P.C., Needham, T.E. and Matwiyoff, N.A. (1974) J. Chem. Soc. Chem. Commun. 85–86.
116 Brown, K.L. and Lu, L.-Y. (1981) Inorg. Chem. 20, 4178–4183.

CHAPTER 13

Reactions in micelles and similar self-organized aggregates

CLIFFORD A. BUNTON

Department of Chemistry, University of California, Santa Barbara, CA 93106, U.S.A.

1. Introduction

Micelles are self-organized aggregates of amphiphiles, i.e., of species which contain both apolar, hydrophobic (lipophilic), and polar, (hydrophilic), groups. The solutions are transparent, but contain aggregates of the amphiphile which scatter light. Some amphiphiles are naturally occurring, e.g., the bile salts, others are synthetic, and the hydrophilic group may be anionic, e.g., carboxylate or sulfate ion, cationic, e.g., an ammonium ion, non-ionic, e.g., hydroxyl or amine oxide, or zwitterionic *. A very useful overview of the role of micelles is given in ref. 10, and the historical background is discussed in ref. 11.

Micellar effects upon chemical equilibria in aqueous solution were recognized many years ago, and Hartley [12] explained them in terms of the ability of ionic micelles to attract counterions and repel coions. This general explanation was subsequently applied to micellar effects upon chemical reactivity in aqueous solution [13]. A very important monograph outlined the state of knowledge up to 1974, and also noted other associated species which could influence the rates of thermal, photochemical and radiation induced reactions [1]. The initial studies of micellar effects were made in water, but subsequently micelle-like aggregates were observed in non-aqueous solution. These aggregates can also influence chemical reactivity. In some respects the effects of micelles on reactivity are similar to those of cyclodextrins or synthetic polyelectrolytes.

Micellar effects have been studied so extensively during recent years that it is impossible to cover the subject in an extensive monograph, let alone a chapter. Therefore this discussion will be selective, and will consider largely thermal reactions in aqueous solutions of micelles and the factors which govern chemical reactivity in these systems. For these reasons the bibliography is limited, and recent, rather than older, publications are cited because this makes it easier to follow the literature.

---

* A comprehensive review of micellar structure and micellar effects on reactivity up to 1974 is given in ref. 1. More recent discussions are in refs. 2–9.

Michael I. Page (Ed.), The Chemistry of Enzyme Action

2. *Formation of normal micelles*

Micelles in water are formed by the self-association of surfactants, also called surface-active agents or detergents. These amphiphiles contain both hydrophilic and hydrophobic (lipophilic) moieties, so that one part of them tends to associate with water and the other part is repelled by water, but readily interacts with apolar molecules [11,12].

Some synthetic surfactants are shown in Scheme 1. Many of the surfactants are articles of commerce made on an industrial scale [14], although often the industrial materials are impure and contain isomers which are not readily separable.

Surfactant chemistry is replete with trivial names and acronyms, whose use has been sanctified by time, if not logic, and some examples are given in Scheme 1,

	synthetic surfactants	cmc (M)
$C_{11}H_{23}CO_2Na$	Sodium dodecanoate	2.6×10^{-3}
$C_{12}H_{25}OSO_3Na$	Sodium dodecylsulfate, SDS Sodium laurylsulfate, NaLS	8×10^{-3}
$C_{14}H_{29}OSO_3Na$	Sodium tetradecylsulfate	2×10^{-3}
$C_{14}H_{29}NMe_3Br$	Tetradecyltrimethylammonium bromide Myristyryltrimethylammonium bromide, MTAB	3.5×10^{-3}
$C_{16}H_{33}NMe_3Br$	Hexadecyltrimethylammonium bromide, HTAB Cetyltrimethylammonium bromide, CTAB	9×10^{-4}
 $C_{12}H_{25}\overset{+}{N}C_5H_5\ Cl^-$	Dodecylpyridinium chloride	1.5×10^{-2}
$C_{12}H_{25}NMe_2O$	Dodecyldimethyl amine oxide	2×10^{-3}
$RO(CH_2CH_2O)_nCH_2CH_2OH$	Brij, Igepal, Triton	

Scheme 1.

which also contains approximate values of the critical micelle concentration (cmc), which is the concentration at which micelles appear in water [15].

Some general structures of naturally occurring amphiphiles are shown in Scheme 2 [14].

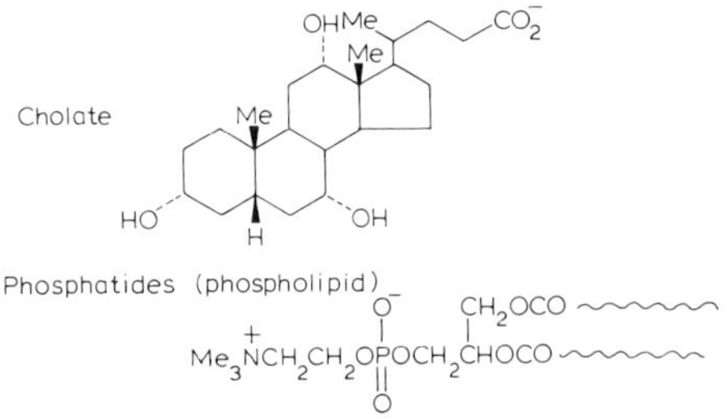

Cholate

Phosphatides (phospholipid)

$$Me_3\overset{+}{N}CH_2CH_2OP(O)(O^-)OCH_2CH(OCO\sim\sim)CH_2OCO\sim\sim$$

Scheme 2. Naturally occuring amphiphiles.

The surfactants shown in Schemes 1 and 2 are chemically inert, but active groups can be introduced giving functional surfactants, whose micelles are chemically reactive typically as nucleophiles or bases (Scheme 3). In many cases deprotonation gives the active reagent.

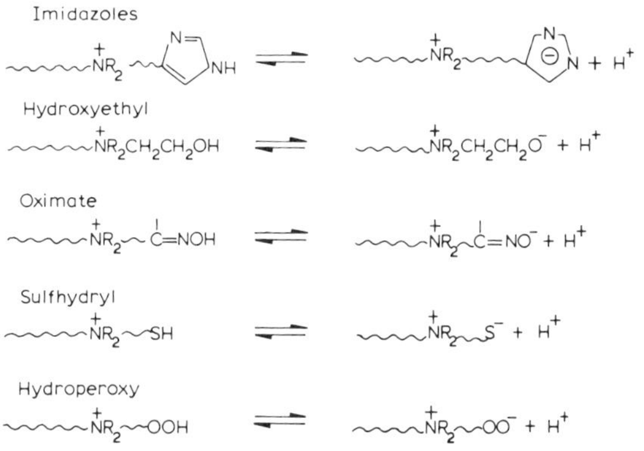

Imidazoles

$\sim\!\sim\!\sim\overset{+}{N}R_2\sim\text{(imidazole, N=, NH)} \rightleftharpoons \sim\!\sim\!\sim\overset{+}{N}R_2\sim\text{(imidazolate, }\ominus\text{)} + H^+$

Hydroxyethyl

$\sim\!\sim\!\sim\overset{+}{N}R_2CH_2CH_2OH \rightleftharpoons \sim\!\sim\!\sim\overset{+}{N}R_2CH_2CH_2O^- + H^+$

Oximate

$\sim\!\sim\!\sim\overset{+}{N}R_2\sim\overset{|}{C}{=}NOH \rightleftharpoons \sim\!\sim\!\sim\overset{+}{N}R_2\sim\overset{|}{C}{=}NO^- + H^+$

Sulfhydryl

$\sim\!\sim\!\sim\overset{+}{N}R_2\sim SH \rightleftharpoons \sim\!\sim\!\sim\overset{+}{N}R_2\sim S^- + H^+$

Hydroperoxy

$\sim\!\sim\!\sim\overset{+}{N}R_2\sim OOH \rightleftharpoons \sim\!\sim\!\sim\overset{+}{N}R_2\sim OO^- + H^+$

Scheme 3. Functional surfactants.

Many amphiphiles exist naturally and play a major role in biological processes. Phospholipids (Scheme 2) are of especial importance because their twin lipophilic groups allow them to form bilayers, so that they are the building blocks of membranes and they, and similar amphiphiles, will form vesicles in water [14,16,17].

Ionic surfactants are strong electrolytes in dilute aqueous solution, and non-ionic surfactants are monomers, but above the so-called critical micelle concentration (cmc) they spontaneously self-associate to form micelles [15]. Micellization in water is an example of the hydrophobic effect at work [18]. The phenomenon is more properly called the solvophobic effect, because it is important in associated solvents which have three-dimensional structure, and normal micelles form in 1,2-diols, or formamide [19] and micelles with a carbocationic head group form in 100% sulfuric acid [20], for example. However, we live in an aqueous world, and most normal micellar systems are studied in water, so we can reasonably retain the term "hydrophobic" with the hydrophobic "bond" dictated by water association.

When surfactants are dissolved in water the air–water interface becomes saturated with surfactant with its polar or ionic groups in the water, so that the surface tension is sharply reduced. Then, as more surfactant goes into the body of the solvent, the disruption of water structure is reduced by the hydrophobic groups coming together to form a micelle, and the hydrophobic "bonding" overcomes the repulsions of the polar or ionic head groups. At the same time counterions are attracted from the body of the solvent and will partially neutralize the charge of an ionic micelle, and reduce the head group repulsions [18,21,22].

The onset of micellization can be detected by changes in physical properties. For example there are breaks in the concentration dependence of such properties as surface tension or conductivity (of ionic surfactants), and of dye solubilization, at the cmc [1,15]. That these breaks correspond to the formation of micellar aggregates was demonstrated by a sharp increase in light scattering.

These observations suggest that the critical micelle concentration is the maximum concentration of monomeric surfactant in solution, and that additional surfactant simply generates micelles. In other words we treat the micelles as if they are a separate phase, although because the solutions are transparent, with no visible phase separation, and the micelles may carry an electrical charge, it is better to regard them as a pseudophase.

This two-state model, which assumes that surfactant exists either as monomers or micelles, is almost certainly an oversimplification. The mass action model assumes an equilibrium between monomers, *n*-mers and micelles, with the proviso that the bulk of the surfactant is present as monomers or micelles. In other words micellization is considered to occur over a narrow range of surfactant concentration, at least for aqueous micelles [1,2,23]. The thermodynamics of micellization have been discussed in terms of the hydrophobic interactions and the electrostatic interactions of the head groups, and, for ionic micelles, of the counterions with the ionic head groups [18,22,24].

3. Micellar structure in water

A widely used model of a normal micelle, in water, assumes that the aggregate is approximately spherical, with a radius similar to that of the length of the extended surfactant, a hydrocarbon-like core and the polar or ionic head groups at the surface. In an ionic micelle 70–80% of the head groups will be neutralized by counterions in the so-called Stern layer [1,2,22,25,26]. These counterions will be part of the kinetic micelle and will migrate with it, and the remainder of the counterions will be in the body of the solvent in the Gouy–Chapman layer. These counterions interact coulombically, at a relatively long range, with the micelle and with other ions.

However micelles are very mobile structures. Monomeric surfactant, counterions and solutes exchange rapidly between micelles and solvent [27–29], and it appears that monomers and incorporated solutes enter the micelle at a diffusion-controlled rate.

This very simplified model of micellization is illustrated in scheme 4 for a cationic surfactant. At concentrations below the cmc only monomeric surfactant is present, but at higher concentration the solution contains micelle, free surfactant and counterions which escape from the micelle. It is assumed that submicellar aggregates are relatively unimportant for normal micelles in water, although, as we shall see, this assumption fails in some systems. However it is probably reasonable for relatively dilute surfactant, although at high surfactant concentration, and especially in the presence of added salt, the micelle may grow, and eventually, new organized assemblies form, for example, liquid crystals are often detected in relatively concentrated surfactant [1]. However, this discussion will focus on the relatively dilute surfactant solutions in which normal micelles are present.

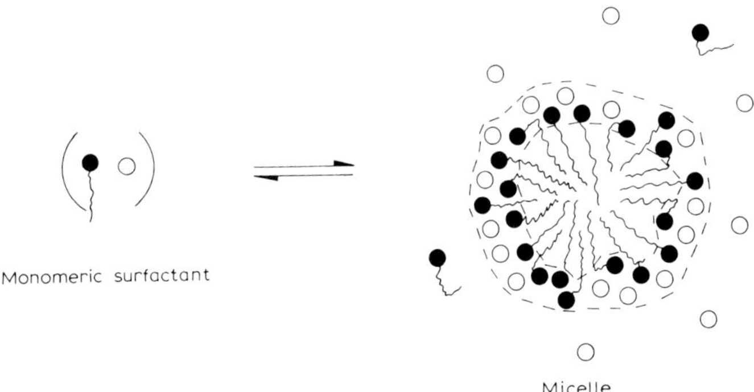

Scheme 4. The solid circles represent surfactant head groups and the open circles counterions. The Stern layer encompasses the head groups and associated counterions.

In this micellar model the head groups and associated counterions are extensively hydrated, so that the Stern layer could be regarded as akin to a concentrated electrolyte solution [25]. However, water penetration into the micelle has been assumed to be limited to only the first few methylene groups of the hydrophobic moiety [12]. There is considerable evidence that water molecules at the micellar surface are not much less reactive than those in water [30–33].

Although this model is consistent with many properties of micelles it has major shortcomings. For example, many workers have argued that geometrical constraints favor a spheroidal rather than a strictly spherical micelle, and extensive micellar growth is inconsistent with a spherical geometry [34].

An additional question is that of water penetration. Menger [35] attempted to construct a "micelle" using space-filling models, and found that if the chains are extended there will be extensive voids into which water molecules, or solutes, can penetrate. However, these voids will be reduced in size if some eclipsing is introduced into the hydrophobic alkyl groups. This "porous-cluster" model is supported by physical and chemical evidence of extensive water penetration into the micelle.

A key feature of Menger's model is that large portions of the alkyl groups in a micelle would, on the average, be exposed to water. Dill and Flory [36] have shown how an approximately spherical micelle can be formed, provided that some of the alkyl groups of the surfactant are eclipsed, and in this model parts of these alkyl groups will be exposed to water. Another model has been developed by Fromherz [37], based on the packing of surfactant, and it too involves some eclipsing in the alkyl groups, and their exposure to water. However, there is considerable controversy on this question of water penetration into normal micelles, and some question as to the validity of the evidence cited by the various protagonists. For example ^{19}F NMR relaxations have suggested that water does not penetrate deeply into perfluorinated anionic micelles [38].

Most of these models suggest that the micellar surface will be rough, with voids which will accept water, or guest solutes. The micellar structure is a very mobile one, so that individual head groups may easily move in and out of the aqueous region [39], and guest solutes may enter voids at the micellar surface, or penetrate this surface and create them [35]. This looseness of the micellar structure arises from the fact that there is a balance of opposing forces [34]. The hydrophobic effect is opposed by repulsions between the polar or ionic head groups, and interactions with counterions decrease coulombic repulsions between ionic head groups. The balancing of these opposing interactions means that there should be a distribution of micellar size, about some mean aggregation number [18,34], and size and shape will be affected by added ions, or non-ionic solutes which may interact with the micelle and stabilize it. As a result it is not easy to use added solutes as probes of micellar structure, because these solutes may themselves perturb the micelles.

The variations of the cmc with temperature have been used to estimate the thermodynamic parameters for micellization, and more rigorous direct measurements from heats of solution have also been made [1,2,7–9,34]. Tanford [18] estimates a contribution of ca. 700 cal/mole per methylene group to the free energy of micellization, which is consistent with hydrophobic interactions of hydrocarbons in water.

The importance of the hydrophobic effect in the self-association of amphiphilic solutes means that any factor which disrupts three-dimensional solvent structure also disrupts normal micelles. For example, addition of relatively large amounts of such organic solvents as ethanol or acetone to water makes the micelles smaller, and eventually they disappear and the surfactant behaves as a simple solute. On the other hand apolar solutes can enter the micelle and stabilize it and so lower the cmc, and added electrolytes also lower the cmc [15].

Although it is reasonable to believe that micelles in water are close to spherical at low surfactant concentrations, e.g., within an order of magnitude of the critical micelle concentration, there is evidence for micellar growth with increasing surfactant or electrolyte concentration. Such growth cannot easily be accommodated to a strictly spherical structure for the micelle, but requires it to become spheroidal. Solutions of some surfactants become very viscous on addition of salts. For example, addition of iodide, arenesulfonate or aromatic carboxylate ion to cationic surfactants sharply increases solution viscosity, suggesting that long, rod-like micelles are formed.

Quantitative evidence on micellar shape and size has in the past usually been obtained by light scattering [40], or by thermodynamic methods [41], although recently Turro and his coworkers developed an ingenious method, based on fluorescence quenching, for estimating micellar size [42,43].

Classical light scattering as a method for determining particle size depends on measurement of the intensity and radial dependence of scattered light. Recently quasi-elastic light scattering, also known as photocorrelation spectroscopy, has been used to determine micellar shape and size [44,45]. In this method the short term fluctuations in the intensity of scattered light, caused by Brownian movement of particles, are used to calculate the rates of diffusion of the particles and hence their hydrodynamic radius, using the Stokes–Einstein equation. When this method was used with micelles of sodium lauryl sulfate ($C_{12}H_{25}OSO_3Na$) it appeared that the micelles grew very sharply and became rod-like when electrolyte or surfactant concentrations were increased or the temperature decreased [44,45]. These results were in marked disagreement with results of other methods [41,42], and it is probable that the observed changes in rates of diffusion are not caused wholly by changes in micellar size, but are also affected by intermicellar interactions [46]. For example ionic micelles repel each other coulombically, and this repulsion increases their mobility, and therefore appears to decrease their size. Addition of electrolyte screens the micelles electrostatically, so that they diffuse more slowly, even though their size may not increase. At high electrolyte concentration van der Waals interactions between the micelles become important and micellar motion becomes correlated, which also appears to make the hydrodynamic radius of the micelles larger than their true radius.

A number of workers are now taking these interactions into account in their treatments [46–48], and it appears that micellar growth is less than had originally been supposed [45], and the evidence of quasi-elastic light scattering leads to estimates of micellar size consistent with those from other methods [46–48].

A major advantage of quasi-elastic light scattering is that analysis of the data allows one to calculate the radius of the micelles and to examine the nature of the electrostatic and van der Waals interactions between the micelles [46–48]. It is important to remember that intermicellar interactions must also be taken into account when methods such as classical light scattering or sedimentation are used for determination of micellar size [41].

At the present time there does not seem to be sufficient firm data regarding the effects of surfactant and electrolyte concentration to allow us to generalize on micellar shape and size. However, it seems that these factors cause at least modest micellar growth, and the situation will become clearer as more data are treated rigorously. Qualitatively it appears that the growth depends very much upon the nature of the counterion, with the less hydrophilic ions having the largest effect, and upon the alkyl group. For example growth seems to be more important with C_{16} than with C_{12} surfactants [46–48].

A key property of normal ionic micelles in water is their charge, which attracts counterions to the micellar surface. For a given ionic micelle one can assume that a fraction, β, of the head groups will be neutralized by counterions in the Stern layer [22,25,26]. Alternatively one can write α as the fraction of counterions which will be lost from a (hypothetical) neutral micelle so that:

$$\alpha = 1 - \beta \tag{1}$$

and $N\alpha$ will be the micellar charge, with N being the (average) number of monomers in the micelle.

Micellar charge can be estimated by electrophoresis [22], electrochemically [49–51], or by analysis of the Brownian movement by quasi-elastic light scattering [46–48].

The use of ion-selective electrodes depends on the observation that the electrode responds to ions in the bulk solution, and not to those bound in the micellar Stern layer [49–51]. The ions in bulk solution will be equal to the concentration of monomeric surfactant, plus those generated by ionization from the micelle. (The concentration of monomeric surfactant is generally assumed to be given by the cmc.)

Values of α are generally in the range 0.1–0.3, and more importantly seem to be relatively independent of the presence of added electrolyte in the solution, and there is theoretical support for this supposition [22,26]. However, different experimental methods, and sometimes the same method in different hands, often give different values of α, and it may be that the value of this parameter does indeed depend on the methods used in its determination [25,26].

Non-ionic solutes which can enter the micelle increase α, because they reduce the density of charged head groups at the micellar surface, and therefore its ability to attract counterions.

The sum of the various observations on micellar structure in water is that micelles behave as if they have a lipid-like interior with a rough surface with the head groups in contact with water and counterions. The extent of water penetration is a matter of controversy, but in view of the rapid disruption and reforming of micelles one would expect that, on the average, alkyl groups of the surfactant will be exposed to water. Because of the cooperativity of the large number of relatively weak interactions which hold the micelle together we are dealing with highly mobile structures of rapidly changing conformation. Polar solutes appear to locate, on the average, close to the micellar surface [21,35], and micellar-bound solute may dictate its local environment. Surfactant and added solutes rapidly enter and leave the micelle, so that it is in equilibrium with monomeric surfactant, and solutes in the micelle are in equilibrium with those in bulk solvent [27,52]. This equilibration is rapid on the NMR time scale, so that NMR spectroscopy, for example, only gives us evidence on the average conformation of micelle or location of solutes [7,35]. However, this equilibration is also much faster than most thermal reactions, so that in dealing with micellar effects on these reactions we can assume that equilibrium is maintained between reactants in the micelle and in bulk solvent. This assumption is, however, not necessarily valid for photochemical reactions where the individual chemical steps may be very rapid.

4. *Kinetic and thermodynamic effects*

4.1. Micellar effects upon reaction rates and equilibria

Hartley and his coworkers systematically investigated micellar effects upon equilibria in aqueous media and rationalized them in a set of rules [12]. Anionic micelles should attract hydrogen ions from bulk solution, and cationic micelles should repel them. Consistently anionic micelles increase the extent of protonation of weak bases, and cationic micelles have the opposite effect.

It was evident that such coulombic interactions were very important in acid–base equilibria and, as evidence accumulated on micellar effects upon reaction rates, it became evident that Hartley's rules were qualitatively applicable here also [1–6,13].

For example, reactions between anions and neutral molecules are typically speeded by cationic and inhibited by anionic, micelles, whereas reactions between cations and neutral molecules are typically speeded by anionic and inhibited by cationic, micelles. However charge effects are not all important, because reactions between non-ionic reactants are often speeded by micelles, and rates of some ionic reactions are affected by non-ionic micelles. Some examples of these rate effects are in Table 1 [1–6].

Many of the early studies of rate effects of aqueous micelles were on reactions in which OH^- acted as a nucleophile, e.g., in ester saponification, but cationic micelles also speeded bimolecular eliminations in which OH^- is a base [53–55], and ester hydrolyses by the E1cB mechanism in which the first step of reaction is an equilibrium deprotonation [56].

An important feature of Hartley's approach is that it requires the micelle to incorporate solutes. The bonding of apolar solutes to aqueous micelles can easily be demonstrated [1–9,21]. For example increased solubility in aqueous surfactant solutions can be ascribed to micelles incorporating the solutes. Chromatography shows that these bound solutes move with the micelles, and allows estimation of the amount of bound solute [57,58]. Incorporation of a solute in a micelle often gives

TABLE 1

Qualitative rate effects of normal micelles [a]

	Cationic	Anionic	Non-ionic
Nucleophilic attack			
$R'CO \cdot OR + OH^-$	+	−	0
$2,4\text{-}(O_2N)_2C_6H_4Cl + OH^-$	+	−	0
$2,4\text{-}(O_2N)_2C_6H_4Cl + PhNH_2$	+	+	
$(PhO)_2PO \cdot OC_6H_4NO_2 + OPh^-$	+		
$(PhO)_2PO \cdot OC_6H_4NO_2 + F^-$	+	−	−
Water catalyzed reactions			
$(O_2NC_6H_4O)_2CO + H_2O$	−	−	
R–C=N–N(–CO–Ph)–C=N (ring) + H_2O	−	−	
Acid catalysis			
$R'CH(OR)_2 + H_3O^+$	−	+	
Unimolecular reactions			
$(O_2N)_2C_6H_3OPO_3^{2-}$	+	0	0
$2,4\text{-}(O_2N)_2C_6H_3OSO_3^-$	+	0	+
$RX(S_N1)$	−	−	

[a] +, − and 0 denote rate enhancement, inhibition and no effect, respectively.

spectral shifts, which can be treated quantitatively in terms of a distribution between water and micelles [59].

The importance of the hydrophobicity of the reactants was soon recognized, and it was generally accepted that the more hydrophobic the reactant the more readily it bound to normal micelles, and qualitatively all these observations were consistent with micelles behaving as a reaction medium separate from the bulk solvent. Thus micelles could bring reactants together and speed reaction, or could incorporate one reactant, and repel the other, and thereby inhibit reaction.

This conclusion, for ionic reactants, was strengthened by the observation that inert electrolytes generally reduced micellar rate enhancements [60]. The effect depended on the charge density of the inert counterion (Table 2), and qualitatively could be understood in terms of a competition between reactive and unreactive ions for sites on the micellar surface [1–6].

The importance of proximity effects upon bimolecular reactions is evident from the example of micellar rate effects shown in Table 1. The volume of micelles in dilute aqueous surfactant solution is much less than that of the aqueous component, so that concentration of both reactants into the small volume of the micelles should increase the rate of a bimolecular reaction [1–6,25,61].

However, micelles, acting as a pseudophase, may have solvent properties different from those of water, for example they are less polar than water [21,62,63], and therefore rate constants could be different in micelles and in the aqueous part of the solvent. One cannot a priori separate concentration and medium effects for a bimolecular reaction, but there is no problem in doing this for a spontaneous, unimolecular, reaction, whose first-order rate constants are independent of concentration.

There are a number of unimolecular reactions whose rate constants are different in micelles and in water, and some examples are shown in Table 1.

These micellar effects are consistent with micelles being less polar than water. For example, anionic decarboxylations [64,65] and decompositions of aryl sulfate anions [66] and aryl phosphate dianions [67,68] are faster in most organic solvents than in water, whereas the opposite is true for S_N1 reactions [32,54,55,69].

Therefore a key problem is the separation of concentration and medium effects of micelles for bimolecular reactions, because if micelles affect rate constants of unimolecular reactions they could also affect those of bimolecular reactions. The

TABLE 2
Salt effects upon reactions in normal micelles

Reaction	Surfactant	Salt inhibition
$R'CO_2R + OH^-$	Cationic	$NO_3^- > Br^- > Cl^- > F^- >$ no salt
$(PhO)_2PO{\cdot}OAr + \{OH^-, F^-\}$	Cationic	$OTos^- > NO_3^- > Br^- > Cl^- \sim MeSO_3^- >$ no salt
$PhC(OMe)_3 + H_3O^+$	Anionic	$R_4N^+ > Cs^+ > Rb^+ > K^+ > Na^+ > Li^+ >$ no salt

solution to this problem requires consideration of the distribution of both reactants between the aqueous and micellar pseudophases and estimation of second-order rate constants in each pseudophase.

4.2. Quantitative treatments of micellar rate effects in aqueous solution

The development of quantitative models of micellar effects upon reaction rates and equilibria was based on the concept that normal micelles in aqueous, or similar associated, solvents behave as a separate medium from the body of the solvent.

Therefore, provided that there is an equilibrium between the micellar and aqueous pseudophases, the problem resolves itself into estimation of the distribution of reactants between aqueous and micellar pseudophase and calculation of the rate constants of reaction in each pseudophase. Menger and Portnoy [70] developed an equation which successfully accounted for micellar inhibited saponification of 4-nitrophenyl alkanoates. This model was also applied to spontaneous, unimolecular, hydrolyses of dinitrophenyl sulfate monoanions [66] and phosphate dianions [68] which are speeded by cationic micelles in water.

The treatment given here (Scheme 5) generally follows that originally developed [70], although there are differences in details. Provided that the properties of the micelle are not materially perturbed by incorporation of the substrate, S, the equilibrium distribution of S between the micellar and aqueous pseudophases can be written in terms of Eqn.. 2 and Scheme 5.

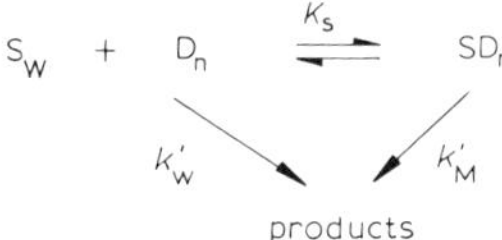

Scheme 5

where SD_n is the substrate–micelle complex, and K_s the binding constant written in terms of the concentration of micellized surfactant, $[D] - cmc$, where D is the surfactant (detergent):

$$K_s = [SD_n]/[S_w]([D] - cmc) \tag{2}$$

The equilibrium constant can alternatively be written in terms of the concentration of micelles. The equilibrium constant is then NK_s, where N is the average aggregation number of the micelle, and equilibrium constants are often written in these terms [70].

The first order rate constants for reaction in the aqueous and micellar pseudophases are k'_w and k'_M, so that the overall first-order rate constant, k_ψ, is given by:

$$k_\psi = \frac{k'_w + k'_M K_s([D] - cmc)}{1 + K_s([D] - cmc)} \tag{3}$$

Eqn. 3 adequately fits data for unimolecular micellar-catalyzed reactions [66,68], and for micellar-inhibited reactions, where for bimolecular reactions, k'_M is usually small so that the second term in the numerator of Eqn. 3 can be neglected [70]. In some cases, for example with very hydrophobic substrates, micellar rate effects are observed at [D] < *cmc*, so that in these cases we cannot equate the concentration of monomeric surfactant with the cmc, probably because the substrate promotes micellization or interacts with submicellar aggregates, and modified forms of Eqn. 3 have been used [71].

However, the general validity of Eqn. 3, for the reactions in which it is valid, is an impressive test of the pseudophase hypothesis [32,66,68,70,71], and kinetic estimates of the binding constant, K_s, are in reasonable agreement with binding constants measured directly. It appears that substrates do not materially affect micellar structure, except perhaps with very hydrophobic substrates in very dilute surfactant. Generally speaking values of K_s are ca. 50/M for such substrates as 4-nitrophenyl acetate or 2,4-dinitrochlorobenzene, and increase sharply as substrate hydrophobicity is increased.

Eqn. 3 has the same general form as the Michaelis–Menten equation of enzyme kinetics, and it is interesting to see that binding constants, per micelle, are often as large as those found for substrate binding to enzymes [70]. However the analogy should not be taken too far, because generally rate constants in micelles are much lower than in enzymes, and micellar systems generally show only limited substrate specificity.

4.3. Quantitative treatment of bimolecular reactions

Although Eqn. 3 is successful in treating unimolecular and micellar inhibited bimolecular reactions it cannot be applied directly to micellar catalyzed bimolecular reactions, because it considers only the distribution of one reactant between aqueous and micellar pseudophases. The rate of a bimolecular reaction will depend on the distribution of *both* reactants between the two pseudophases [4–6,25,61].

A problem in estimating second-order rate constants of solution reactions is that they depend upon the units of concentration, which are chosen arbitrarily. Typically molarity is the chosen unit, because it is a very convenient measure of concentration. However, even in non-micellar systems, assessment of solvent effects upon second-order rate constants depends directly upon our choice of the concentration unit. Relative second-order rate constants in a variety of solvents will, in general, be different if the concentration unit is molarity instead of mole fraction, for example.

We have exactly the same problem in micellar systems. The important parameter is concentration in the micellar pseudophase, which can be calculated in various ways. One approach is to calculate molarity in terms of moles of reagent per l of micelles, and the micellar volume can be estimated from micellar density, which is close to unity. But the micelle is probably not uniform in composition, and if reaction occurs in the Stern layer, at the micelle–water interface, it might be more reasonable to use the volume of that layer, rather than that of the whole micelle. In any event the volumes of the micelle and the Stern layer probably are within a factor of 2 [25], so the arbitrary choice of the volume element of reaction does not materially affect the overall conclusions.

Another approach, which will be followed here, is to evade the question by defining concentrations in terms of the mole ratio of reagent to micellar head group. This concentration unit is unambiguous, but rate constants so determined cannot be compared directly with those in the bulk solvent which have the units of reciprocal molarity. However, one set of units can be converted into the other using the molar volume of the micelle, or its Stern layer.

For a bimolecular reaction of a substrate, S, with a reagent, N, for example a nucleophile, Scheme 5 is modified to Scheme 6:

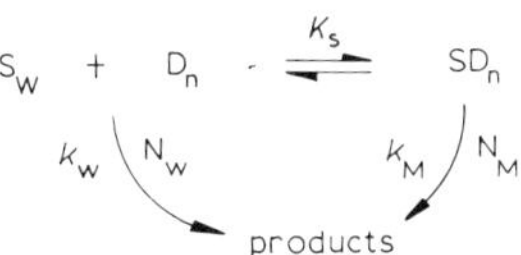

Scheme 6

where W and M denote reactant in the micellar and aqueous pseudophases, respectively. The first order rate constants (Scheme 6) are given by:

$$k'_{w} = k_{w}[N_{w}] \tag{4}$$

and

$$k'_{M} = k_{M} m_{N}^{s} = k_{M}[N_{M}]/([D] - cmc) \tag{5}$$

In Eqns. 4 and 5 k_w has the dimensions M^{-1}/sec and k_M, sec^{-1}, because the concentration of N in the micelles, m_N^s, is written as a mole ratio.

Combination of Eqns. 3–5 gives:

$$k_{\psi} = \frac{k_{w}[N_{w}] + k_{M} K_{s} m_{N}^{s}([D] - cmc)}{1 + K_{s}([D] - cmc)} \tag{6}$$

$$= \frac{k_{w}[N_{w}] + k_{M} K_{s}[N_{M}]}{1 + K_{s}([D] - cmc)} \tag{7}$$

In Eqns. 6 and 7 the quantities in squared brackets are molarities written in terms of total solution volume.

The derivation can alternatively be given using a second-order rate constant in the micellar pseudophase based on the volume of that pseudophase [25,61,72]. The rate equation then includes the micellar molar volume, and this formulation is often used.

The variations of k_{ψ} with surfactant concentration can be fitted to Eqn. 6 or 7, or similar equations, provided that the distribution of the reagent, N, between micellar and aqueous pseudophases is known.

The distribution can often be directly measured. For example, by standard methods for organic solutes, assuming that incorporation of one reactant in the micelle does not affect binding of the other, or the micellar structure [1–6,61]. For reactions of hydrophilic ions the distribution can sometimes be determined electrochemically [49–51], but the problem is more complex for ions such as OH^-, whose concentration in the micelle cannot be determined directly.

A major advance in the treatment of reactions of such ions was made by Romsted [25], who suggested that ions competed for the micellar surface, and that this competition was governed by an ion-exchange equilibrium:

$$N_M^- + S_w^- \rightleftharpoons N_w^- + X_M^-$$

where N^- is a reactive, and X^- an inert, ion. A similar treatment can be applied to the distribution of cations.

$$K_X^N = [N_w^-][X_M^-]/([N_M^-][X_w^-]) \tag{8}$$

Romsted assumed that β, the extent of charge neutralization of the micellar head groups, is independent of the nature and concentration of counterions, i.e., the micelle is saturated with counterions, which allows estimation of the amounts of micellar bound ions for a given exchange constant, K_X^N, because:

$$\beta = ([N_M^-] + [X_M^-])/([D] - cmc) \tag{9}$$

Eqns. 8 and 9 were combined with the mass–balance relationship to predict the relation between the concentration of micellar bound N^- and concentrations of surfactant or added reactive or inert ions for given values of K_X^N. This allowed a prediction of the effect of these quantities upon the rate constant, k_ψ, provided that the distribution of substrate between aqueous and micellar pseudophases was known, or could be estimated.

In practice computer simulation has generally been used to predict the variation of k_ψ with concentration of reactant, surfactant, or added electrolyte in terms of various values of the parameters, k_M, K_X^N and β. This simulation procedure has been used as an indirect method for the determination of the ion exchange constant K_X^N, and, for example, for the competition between various counterions for micelles, there is reasonable agreement between the values obtained kinetically and by other methods [25,72–79].

Although this ion exchange model fits the data one has to be cautious regarding its generality. The fitting of the kinetic data involves several parameters, e.g., k_M, K_X^N, β, which are assumed to be constant, and whose values are uncertain, and often the data can be fitted equally well using various combinations of values of the parameters [75]. It is also not obvious why β should be independent of the nature and concentration of added counterions, when these counterions have different affinities for the micelle, i.e., $K_X^N \neq 1$. For example, there is considerable evidence that the affinities of anions for a cationic micelle follow the sequence $NO_3^- \sim Br^- >$

TABLE 3
Relative second order rate constants for reactions of ionic reagents [a]

Reaction	Surfactant	k_{rel} [b]	Ref
CN^- + 3-carbamoylpyridinium (N^+—R, $CONH_2$)	CTABr CTACN	2	90
$OH^- + (4\text{-}O_2NC_6H_4O)_2CO$	CTABr CTAOH	0.7	96
$OH^- + (PhCO)_2O$	CTABr CTAOH	0.06	96
$OH^- + C_{12}H_{25}\overset{+}{N}Me_2CH_2CO_2Me$	CTACl	ca. 0.3	75
$H_3\overset{+}{O} + 4\text{-}O_2NC_6H_4CH(OEt)_2$	$C_{12}H_{25}OSO_3Na$ RSO_3H	ca. 0.05	77, 78, 87

[a] Based on a molar volume of the Stern layer of 0.14/M l.
[b] Relative to the second order rate constant in water.

$Cl^- > OH^-$. Some of these questions are considered further in this chapter.

However, despite the assumptions involved in the treatment, the ion exchange model accommodates a large number of experimental results in a consistent way, and some examples of reactions to which it has been applied are in Table 3, other examples are given in refs. 72, 73 and 76.

4.4. Second-order rate constants in the micellar pseudophase

As noted earlier comparison of second-order rate constants requires that one assume an arbitrary measure of concentration. Eqn. 6 was derived using concentration in mole ratios, which leads to a convenient form of the equation, but the rate constants, k_M, sec^{-1}, cannot be compared directly with values of k_w, M^{-1}/sec, in water, for example. Comparison can be made, provided that one estimates the volume element of reaction, i.e., the molar volume of the micelle or its Stern layer. Generally the value for the micelle has been taken as 0.3–0.37 l [25,61,72,76], and that of the Stern layer has been assumed to be ca. 0.14 l [79]. These estimates are no more than crude approximations. The micellar surface is almost certainly very rough and non-uniform, so the thickness of the Stern layer will be variable and probably perturbed by added solutes. The composition of a micelle is also non-uniform, and an apolar reactant should on the average be located more deeply in the micelle than a polar reactant [21,80,81], so that one probably cannot define an effective micellar volume appropriate for all reactions in aqueous micellar systems.

If one assumes that reaction occurs in the Stern layer, whose molar volume is 0.14

l, a second order rate constant, k_2^{m}, M^{-1}/sec is given by:

$$k_2^{\mathrm{m}} \approx 0.14\, k_{\mathrm{M}} \tag{10}$$

Values of k_2^{m} would be approximately doubled if a molar volume of ca. 0.3 l was used.

The second order rate constants, k_2^{m} and k_{w}, for reactions in the micellar and aqueous pseudophases have the same dimensions, and can now be compared directly, and within all the uncertainties of the treatment it seems that k_2^{m} and k_{w} are of similar magnitudes for most reactions, and in some systems $k_{\mathrm{w}} > k_2^{\mathrm{m}}$. This generalization is strongly supported by evidence for reactions of relatively hydrophobic nucleophilic anions such as oximate, imidazolide, thiolate and aryloxide, typically with carboxylate or phosphate esters [61,82–85]. These similarities of second-order rate constants in the aqueous and micellar pseudophases are consistent with both reactants being located near the micellar surface in a water-rich region. Therefore the micellar rate enhancements of bimolecular reactions are due largely to concentration of the reactants in the small volume of the micelles. Some examples are in Table 3 for reactions of H_3O^+ or hydrophilic nucleophilic anions and in Table 4 for reactions of more hydrophobic nucleophiles.

Examples of this general treatment are shown in Figs. 1 and 2 for saponification

TABLE 4
Relative second order rate constants in micellar and aqueous pseudophase

Reaction	Surfactant	k_{rel}
$ArS^- + CH_3CO \cdot OC_6H_4NO_2(4)$	$C_{16}H_{33}NMe_3Br$	0.6–1.1 [a,c]
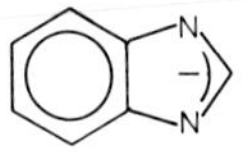 $+ C_6H_{13}CO \cdot OC_6H_4NO_2(4)$	$C_{16}H_{33}NMe_3Br$	ca. 10 [a,d]
$ArO^- + (PhO)_2PO \cdot OC_6H_4NO_2(4)$	$C_{16}H_{33}NMe_3Br$	ca. 0.3 [b,e]
$ArO^- + 2,4\text{-}(NO_2)_2C_6H_3Cl$	$C_{16}H_{33}NMe_3Br$	1.5 [b,f]
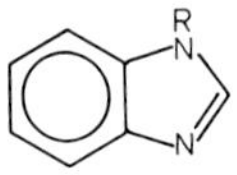 $+ C_6H_{13}CO \cdot OC_6H_4(NO_2)_2$	$C_{16}H_{33}NMe_3Br$	ca. 10^{-2} [a,d]
$PhNH_2 + 2,4\text{-}(NO_2)_2C_6H_3Cl$	$C_{12}H_{25}OSO_3Na$	0.1 [b,f]

[a] Based on micellar molar volume of 0.37/M l.
[b] Based on a Stern layer volume of 0.14/M l.
[c] Ref. 83.
[d] Ref. 61.
[e] Ref. 84.
[f] Ref. 82.

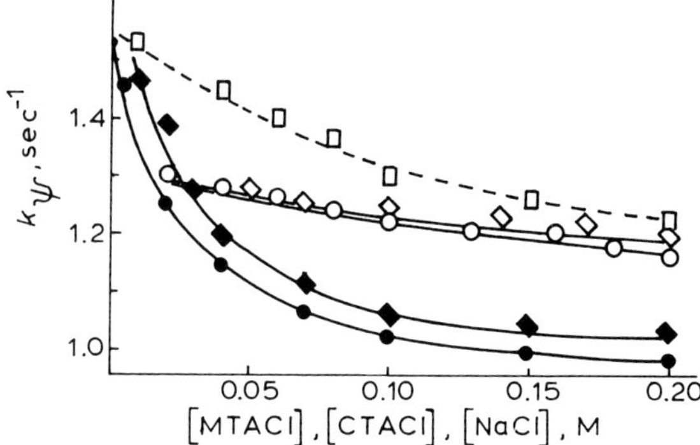

Fig. 1. Inhibition by cationic micelles of the saponification of $Me_3\overset{+}{N}CH_2CO_2Me$. The solid lines are predicted: ◆ in MTACl; ◇ in MTACl + 0.1 M NaCl; ● in CTACl; ○ in CTACl + 0.1 M NACl; □ in NaCl. (ref. 75. Reprinted with permission of the American Chemical Society.)

of betaine esters in cationic micelles of CTACl orMTACl [75].

$$R\overset{+}{N}Me_2CH_2CO_2Me + OH^- \rightarrow R\overset{+}{N}Me_2CH_2CO_2^- + MeOH$$

(1)

When the ester is very hydrophilic, as with (1), R = Me, it resides largely in the aqueous pseudophase and the cationic micelles inhibit reaction because they deplete the aqueous pseudophase in OH^- [72,75]. There is also a negative salt effect upon the reaction, and when both these effects are taken into account the predicted solid

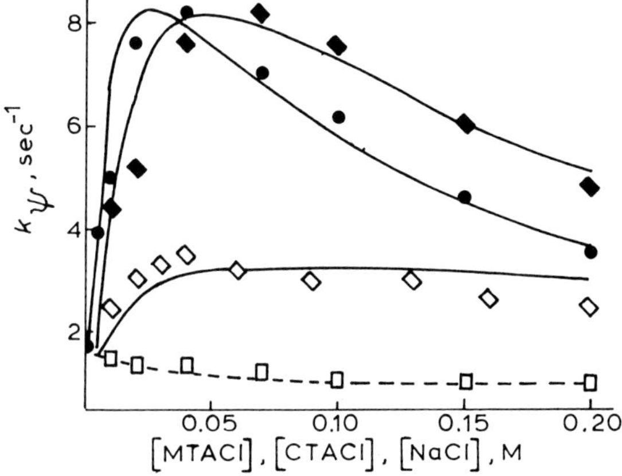

Fig. 2. Rate enhancement by cationic micelles of the saponification of $n\text{-}C_{12}H_{25}\overset{+}{N}Me_2CH_2CO_2Me$. The solid lines are predicted, symbols as in Fig. 1. (Ref. 75. Reprinted with permission of the American Chemical Society.)

lines in Fig. 1 are obtained. Rate enhancements are obtained with a more hydrophobic ester (1), $R = n\text{-}C_{12}H_{25}$, which is distributed between water and the micelles, and the solid lines are predicted by the ion exchange model (Fig. 2).

For both systems the data can be fitted with an ion exchange constant $K_{Cl}^{OH} \approx 4$, and the second order rate constant in the micellar pseudophase is about one-third of this in water, probably because the high ionic content of the Stern layer exerts a negative salt effect upon the reaction.

Very similar results have been obtained by a number of workers for reactions involving hydrophilic anions in cationic micelles [72,73,76]. An important point is that the ion-exchange constants determined from rates or equilibria of hydroxide ion reactions in cationic micelles agree reasonably well with independent estimates from physical measurements on the relative affinities of various anions for cationic micelles [74,86].

For bimolecular reactions of non-ionic reagents (Table 4) second-order rate constants are smaller than those in water [3–6,61,82]. This difference is readily understandable because the micelle is a less polar medium than water [21,63], and typically reactions between non-ionic reactants are inhibited by a decrease in the polarity, or water content, of the solvent [69].

Reactions of the hydrogen ion appear to be an exception to these generalizations (Table 3) because the second-order rate constants in anionic micelles are smaller than in water, even with relatively hydrophilic substrates [77,78,87]. These inhibiting effects suggest either, that micellized alkyl sulfuric and alkane and arene sulfonic acids are not strong, or, that there is strong hydrogen bonding between the sulfate or sulfonate head groups and the hydrogen ion which reduces its effective acidity.

The evidence cited thus far shows the key role of incorporation of both reactants in the small volume of the micelles, and the *overall* rate enhancements give little useful mechanistic information. In fact the overall rate enhancements for bimolecular reactions depend markedly upon the concentration of the reagent.

A detailed examination of relative second-order rate constants in micellar and aqueous pseudophases is outside the scope of this discussion, but for anionic reagents it seems that k_{rel} (Tables 3 and 4) deviates from unity when the substrate is very hydrophobic and the anion more hydrophilic, suggesting that, on the average, they are not located in the same region of the micelle. However, more evidence will be needed on micelle–solute interactions for this question to be answered.

So far as I know there was no reaction in the literature for which the second-order rate constants in the micellar pseudophase are much larger than those in water. However aromatic substitution by azide ion upon 2,4-dinitrochlorobenzene and naphthalene in cationic micelles is such a reaction. The second-order rate constants, k_2^m, in the micellar pseudophase are much larger than those in water [88]. However this high reactivity is not observed in deacylation or an S_N2 displacement and it is not obvious why aromatic substitution by azide ion should be an exception to generalizations about rate constants in aqueous and micellar pseudophases.

These micellar rate enhancements are generally discussed in terms of increased concentration of both reactants in the micellar pseudophase, but one can instead ascribe them to a more positive entropy of activation [89]. For example entropies of activation, calculated in terms of *total* reactant concentration, are typically less negative for bimolecular reactions in micellar solutions, as compared with those in water, and micellar enhancement of the oxidation of diethyl sulfide by iodine was explained in these terms [89]. In other words preincorporation of reactants in the micelle reduces the entropy loss when two reactants generate a transition state.

There is no real difference between those descriptions, one is merely choosing different standard states for the system, but I feel that it is most convenient to use the description based on concentration of reactants in the small volume of the micellar pseudophase.

5. Reactive counterion micelles

The previous section discussed reactions in which a reagent, e.g., a nucleophilic anion, is added to a cationic micelle of a surfactant whose counteranion is chemically inert, so that the two anions compete for the micelle. However one has a conceptually simpler situation when the only micellar counterion is also the reactant. There is then no interionic competition, and if the micelle is saturated with reactive counterions, i.e., if β is constant, the observed first order rate constants should increase as the substrate binds to the micelle, and will be constant once the substrate is fully micellar bound.

The predicted behavior is found in a number of reactions, for example, when the ionic reagent is H_3O^+, CN^- or Br^- [87,90–92].

Fig. 3 illustrates the effect of CTACN upon the first order rate constants for cyanide ion addition to pyridinium ions [90].

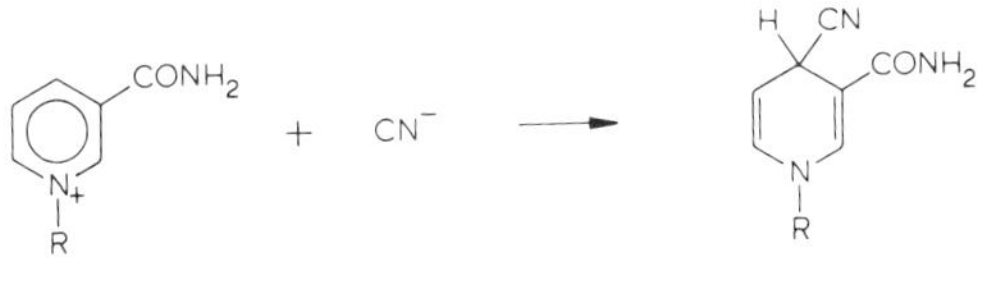

$R = n\text{-}C_2H_{25}$; $n\text{-}C_{14}H_{29}$; $n\text{-}C_{16}H_{33}$

This reaction is reversible [93], but the high concentration of CN^- at the micellar surface allows one to neglect the reverse reaction in CTACN. The rate constants for addition increase smoothly with increasing [CTACN] to values which are almost independent of the hydrophobicity of the substrate. The second-order rate constants in the micelle are very similar to that for reaction of the *n*-propyl derivative in water [93], consistent with the generalization that rate enhancements of bimolecular reactions are due to concentration of the reactants in the micellar pseudophase.

Unexpectedly the pattern of behavior is completely different when the reagent is a hydrophilic anion, e.g., OH^- or F^-, reacting with a triaryl phosphate or in aromatic nucleophilic substitution [87,92,94]. In these systems the first-order rate constants do not become constant when all the substrate is bound to the micelle, but continue to

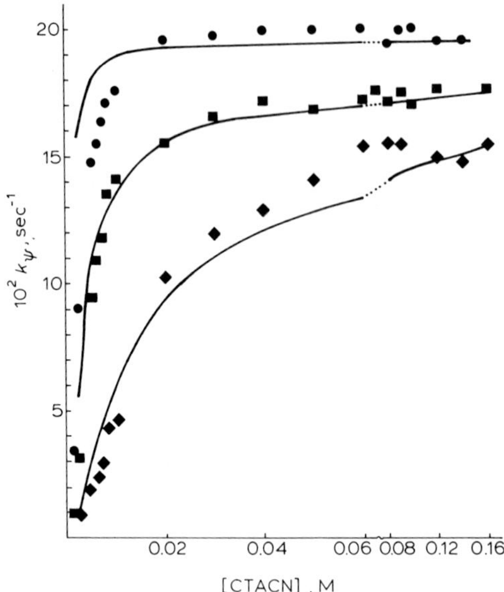

Fig. 3. Rate enhancements by CTACN of attack of cyanide ion upon *N*-alkyl-3-carbamoyl pyridinium ion: ◆, ■, ●, alkyl = *n*-$C_{12}H_{25}$, *n*-$C_{14}H_{29}$, *n*-$C_{16}H_{33}$ respectively. (Ref. 90. Reprinted with permission of the American Chemical Society.)

increase towards constant, limiting, values, and moreover at surfactant concentrations below those corresponding to the limit are increased by added nucleophilic anion. Some examples of this apparently anomalous behavior are shown in Fig. 4.

In the initial report of these observations a tentative explanation was given

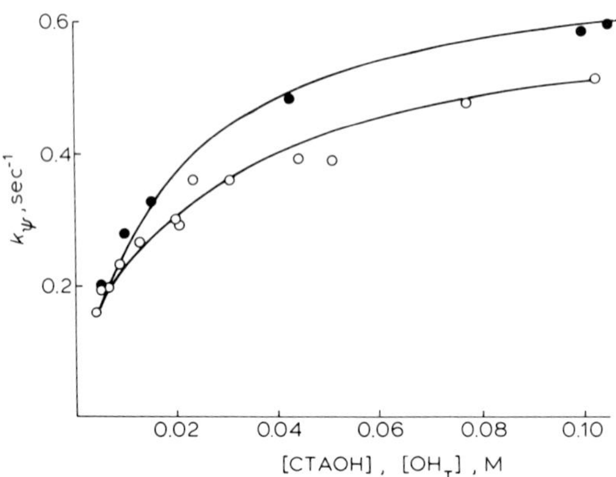

Fig. 4. Reaction of *p*-nitrophenyl diphenyl phosphate in CTAOH: ○, in absence of NaOH; ●, in 5.52×10^{-3} M CTAOH with added NaOH. [OH_T] denotes total concentration. (Ref. 94. Reprinted with permission of the American Chemical Society.)

assuming that there was reaction across the shear surface of the micelle, so that an anion in the water could react with micellar-bound substrate [87]. This was an explanation based on the clutching of straws, and little else, and it does not explain the subsequent observation of "normal" behavior with less hydrophilic anions, e.g., CN^- or Br^- [90–92]. However recent observations on the reaction of hydroxide ion with DDT and its derivatives in CTABr micelles are consistent with this suggestion. The results fit the ion-exchange model, but only with dilute hydroxide ion, and it was suggested that more concentrated hydroxide ion in water attacked substrate in the micelle [95].

An alternative explanation assumes that β is not constant when the anion is very hydrophilic, but that the distribution of a reactive nucleophilic ion, N^-, between water and micelles follows Eqn. 11 [94]:

$$K'_N = [N_M^-]/\{[N_w^-]([D] - cmc - [N_M^-])\} \tag{11}$$

This equation is formally similar to Eqn. 2 for micellar binding of non-ionic solutes. In this treatment β, i.e., $[N_M^-]/([D] - cmc)$, has the formal limits of 0 and 1, but the lower limit will never be reached because counterions are always present in solution, even when [surfactant] < *cmc*, and the upper limit would be reached only at [surfactant] well above the solubility limit.

This treatment accounts for a large amount of data for aromatic nucleophilic substitution, dephosphorylation and deacylation in reactive counterion micelles involving OH^-, F^- and RCO_2^- [94,96] (Table 5). Examples are shown in Table 5 together with values of K'_N. It is important to note that if $K'_N > 500/M$, as it is for Br^- or CN^-, Eqn. 11 predicts that β will vary little over a wide range of

TABLE 5
Reactions involving hydrophilic anions [a]

Reaction	K'_N (M^{-1})	k_{rel} [b]	Ref.
$OH^- + (PhO)_2PO \cdot OC_6H_4NO_2(4)$	55	0.2	94
$F^- + (PhO)_2PO \cdot OC_6H_4NO_2(4)$	40	0.6	94
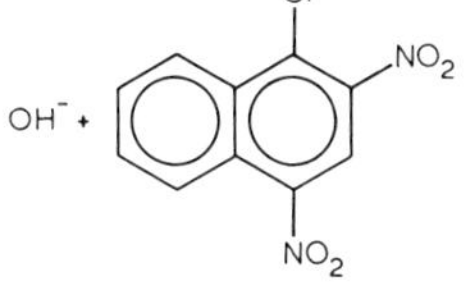 OH^- + Cl, NO_2, NO_2	55	2.2	94
$HCO_2^- + (PhCO)_2O$	80	0.2	96
$OH^- + (PhCO)_2O$	55	0.06	96

[a] In cetyltrimethylammonium micelles with a reactive counterion.
[b] Relative to the second order rate constant in water.

concentration [94], and under these conditions the predictions of this model are very similar to those of the ion-exchange model.

Two points should be emphasized regarding this treatment. First, the second-order rate constants for reactions of anions in the micellar pseudophase estimated using reactive counterion surfactants are not very different from those obtained with mixtures of reactive and inert counterions [87,90,94,96].

Second, we have to be cautious regarding the physical significance of Eqn. 11. The parameter, β, is an average property, and it may be that with surfactants such as CTAOH, or CTAF, in water, we have a mixture of very small micelles which only weakly bind counterions, and larger, "normal", micelles which bind them strongly, and with increasing surfactant or counterion concentration the population shifts from small to normal micelles and β increases. When the counterion is less hydrophilic, e.g., Br^-, the "normal" micelles would predominate.

It seems at first sight that the ion-exchange models are mutually incompatible, because the ion-exchange model assumes constancy of β (Eqn. 9) and the mass action model assumes that β can vary (Eqn. 11). But it may be that each model is satisfactory within certain limits. For example, most tests of the ion-exchange model have been based on kinetic data from experiments in water in which reaction of a hydrophilic ion, e.g., OH^-, in low concentration, was mediated by ionic micelles in which the counterion, e.g., Br^-, binds much more strongly to the micelle than does the reactive ion. In this situation the properties of the Stern layer, and the value of β, will be controlled by the more hydrophobic anion, e.g., by Br^- rather than OH^-, so that β may vary little over the concentrations used in the experiments which have been made using relatively low, and often fixed, concentration of reactive ion, e.g., OH^-, and the concentration of the surfactant, e.g., CTABr, has been varied widely. It is probable that the relatively simple ion-exchange models discussed here are applicable only for relatively low concentrations of added electrolyte, because high concentrations of electrolyte affect micellar structure and intermicellar interactions.

6. Reactions in functional micelles

In all the bimolecular reactions considered thus far the surfactant has been chemically inert, but a functionalized surfactant will generate a micelle in which reactant is covalently bonded (Scheme 3). The functional groups are basic or nucleophilic, and include amino, imidazole, oximate, hydroxamate, thiolate or hydroxyl [3–6,97–108]. In some cases comicelles of a functional and an inert surfactant have also been used. The reactions studied include deacylation, dephosphorylation, nucleophilic aromatic substitution, and nucleophilic addition to preformed carbocations, and some examples are shown in Scheme 7.

Most of the reactive groups are effective nucleophiles and in many cases the initial step of the reaction involves formation of a covalent intermediate, which subsequently breaks down, regenerating reactive surfactant in the micelle. Several techniques have been used to separate the initial nucleophilic attack and subsequent

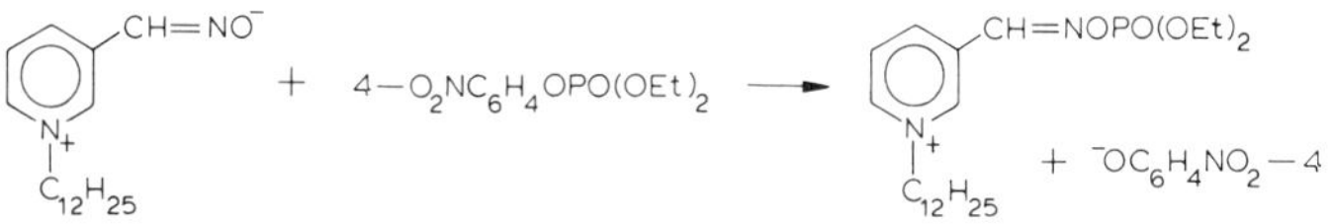

$$\text{(pyridinium-3-CH=NO}^-\text{, N-C}_{12}\text{H}_{25}) + 4\text{-}O_2NC_6H_4OPO(OEt)_2 \rightarrow \text{(pyridinium-3-CH=NOPO(OEt)}_2\text{, N-C}_{12}\text{H}_{25}) + {}^-OC_6H_4NO_2\text{-}4$$

Epstein, J., Kaminski, J.J., Bodor, N., Enever, R., Sowa, J. and Higuchi, T. (1978) J. Org. Chem. 43, 2816.

$$n\text{-}C_{16}H_{33}\overset{+}{N}Me_2CH_2CH_2\bar{S} + RCO\cdot OC_6H_4NO_2\text{-}4 \rightarrow n\text{-}C_{16}H_{33}\overset{+}{N}Me_2CH_2CH_2SCO\cdot R + \bar{O}C_6H_4NO_2\text{-}4$$

Moss, R.A., Bizzigotti, G.O. and Huang, C.-W. (1980) J. Am. Chem. Soc. 102, 754.

$$n\text{-}C_{16}H_{33}\overset{+}{N}Me_2CH_2CH_2\bar{O}_2 + CH_3CO\cdot OC_6H_4NO_2\text{-}4 \rightarrow n\text{-}C_{16}H_{33}\overset{+}{N}Me_2CH_2CH_2OOCO\cdot CH_3 + \bar{O}C_6H_4NO_2\text{-}4$$

Moss, R.A. and Alwis, K.W. (1980) Tetrahedron Lett., 1303.

$$n\text{-}C_{16}H_{33}\overset{+}{N}Me_2CH_2CH_2OH + (4\text{-}Me_2NC_6H_4)_2\overset{+}{C}Ph \xrightarrow{OH^-} n\text{-}C_{16}H_{33}\overset{+}{N}Me_2CH_2CH_2OCPh(C_6H_4NMe_2\text{-}4)_2$$

Ref. 5.

$$R'R_2\overset{+}{N}(CH_2)_nOH + R'\overset{+}{N}Me_2(CH_2)_mOCO\cdot Ar \overset{OH^-}{\rightleftharpoons} R'R_2\overset{+}{N}(CH_2)_nOCO\cdot Ar + R'\overset{+}{N}Me_2(CH_2)_mOH$$

Pillersdorf, A. and Katzhendler, J. (1979) Israel J. Chem. 18, 330.

$$n\text{-}C_{16}H_{33}\overset{+}{N}Me_2CH_2CH_2OH + 4\text{-}O_2NC_6H_4CO\cdot PO_4^{2-} \xrightarrow{OH^-} n\text{-}C_{16}H_{33}\overset{+}{N}Me_2CH_2CH_2OCO\cdot C_6H_4NO_2\text{-}4 + PO_4^{3-}$$

Ref. 5.

$$RCO\cdot NHOH + R'CO\cdot OC_6H_4NO_2 \xrightarrow{CTABr} RCO\cdot NHOCO\cdot R' + O^-\ C_6H_4NO_2\text{-}4$$

Ueoka, R. and Ohkubo, K. (1978) Tetrahedron Lett., 4131.

$$\text{(imidazole)} + H\text{—}O\text{—}H + (PhO)_2PO\cdot C_6H_4NO_2\text{-}4 \rightarrow \text{(imidazolium)} + (PhO)_2PO\cdot OH + {}^-OC_6H_4NO_2\text{-}4$$

Ref. 106.

Scheme 7. Functional micelles and comicelles.

regeneration of the reactive nucleophile, and they generally follow the methods used in studying enzymic reactions.

In some cases the intermediate can be detected spectrophotometrically. A very interesting example comes from the work of Moss, and Tonellato and their co-workers [97,98,101,102]. They used mixed micelles of a hydroxyethyl and an imidazole surfactant, or micelles of a surfactant which contained both functional groups, as deacylating agents. An *O*-acylated surfactant was formed and two reaction paths were considered (Scheme 8).

Reaction could involve acylation of the imidazole moiety, followed by $N \rightarrow O$-acyl transfer, alternatively the imidazole moiety could act as a general base and assist *O*-acylation in a single step reaction with no acylimidazole intermediate.

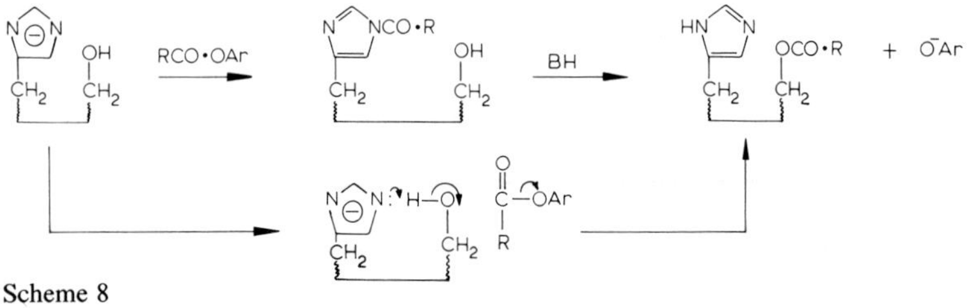

Scheme 8

The acylimidazole intermediate was detected spectrophotometrically, showing that reaction was stepwise rather than concerted, and both steps of the reaction were separated kinetically.

Micellized hydroxyethyl surfactants at high pH are effective reagents for the decomposition of 2,4-dinitrohalobenzenes and an intermediate ether can be detected spectrophotometrically [103]. The two steps of the reaction can be separated by following both the initial nucleophilic attack at the isosbestic point between ether and 2,4-dinitrophenoxide ion, and the formation of 2,4-dinitrophenoxide ion (Scheme 9).

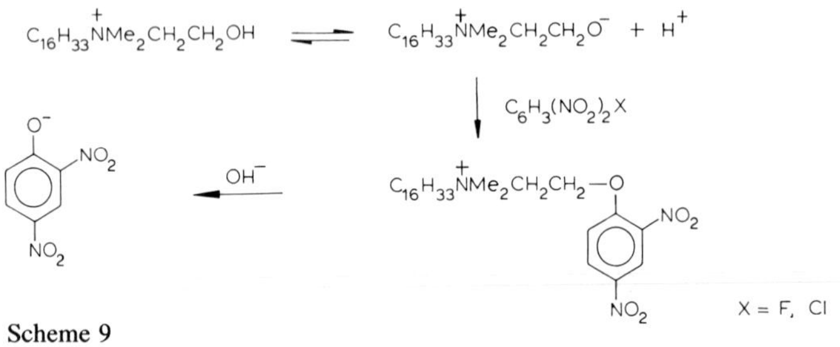

$$C_{16}H_{33}\overset{+}{N}Me_2CH_2CH_2OH \rightleftharpoons C_{16}H_{33}\overset{+}{N}Me_2CH_2CH_2O^- + H^+$$

$$\downarrow C_6H_3(NO_2)_2X$$

$$C_{16}H_{33}\overset{+}{N}Me_2CH_2CH_2{-}O{-}C_6H_3(NO_2)_2 \xrightarrow{OH^-} {}^-O{-}C_6H_3(NO_2)_2$$

X = F, Cl

Scheme 9

Another useful method for detecting formation of a covalent intermediate is that of "burst kinetics" [101,102,104]. In one example of this approach the formation of *p*-nitrophenoxide ion from 4-nitrophenyl acetate was followed in micelles of the hydroxyethyl surfactant (Scheme 10, **2**). The ester was in large excess over the surfactant and there was a rapid initial acylation of (2a), which quantitatively released *p*-nitrophenoxide ion (Scheme 10). This step was followed by a much slower turnover step which regenerated the reactive hydroxyethyl group [104].

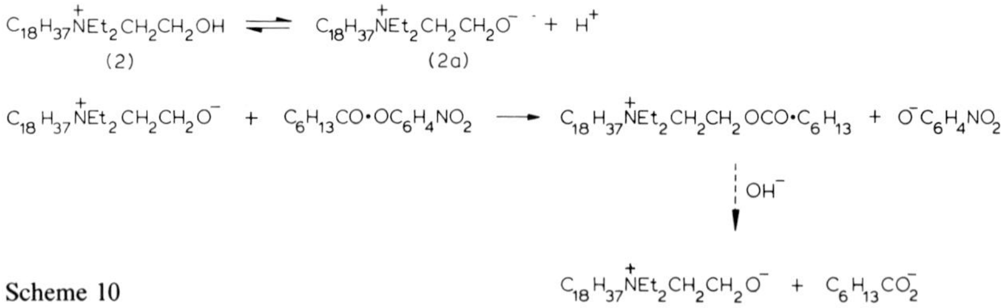

$$\underset{(2)}{C_{18}H_{37}\overset{+}{N}Et_2CH_2CH_2OH} \rightleftharpoons \underset{(2a)}{C_{18}H_{37}\overset{+}{N}Et_2CH_2CH_2O^-} + H^+$$

$$C_{18}H_{37}\overset{+}{N}Et_2CH_2CH_2O^- + C_6H_{13}CO{\cdot}OC_6H_4NO_2 \longrightarrow C_{18}H_{37}\overset{+}{N}Et_2CH_2CH_2OCO{\cdot}C_6H_{13} + {}^-OC_6H_4NO_2$$

$$\downarrow OH^-$$

$$C_{18}H_{37}\overset{+}{N}Et_2CH_2CH_2O^- + C_6H_{13}CO_2^-$$

Scheme 10

The nucleophilicity of the hydroxyethyl group in deacylation at high pH has also been demonstrated in transesterification [105].

Although most functional micelles act as nucleophiles it appears that imidazole derived micelles can also act as general bases and activate water towards a phosphoryl group, although at high pH, where the imidazolide moiety is the reagent, it probably acts as a nucleophile [106].

Very large rate enhancements by functional micelles have been observed, and sometimes, as in some deacylations, they are comparable in magnitude to those observed in enzyme catalyzed reactions, and it is important to understand the source of these rate enhancements.

A functional micelle can increase the rate of reaction over that in water in several ways.

(i) The functional group, which is usually a nucleophile, or a potential nucleophile, e.g., an oxime, hydroxamic or imidazole moiety, introduces a new reaction path.

(ii) Deprotonation of the functional group generates an effective nucleophile, and, assuming that one is working at $pH < pK_a$ of the functional group, incorporation of it into a cationic micelle will increase the extent of its deprotonation and therefore its reactivity.

(iii) Incorporation of the nucleophilic (or basic) group in a micelle may give a rate enhancement by increasing the effective concentration of the reagent or its inherent reactivity.

Only (iii) should be regarded as "micellar catalysis". (Strictly speaking the definition of a catalyst requires that we include the turnover step in any discussion of micellar rate enhancement. However the term "catalysis" is often loosely used to denote an increase of the rate of disappearance of reactant, and not necessarily of appearance of product.) We should note that relative rates in micellar systems, relative to those in water, are determined by a variety of factors, including pH. For example, for a reaction in water, in which the reagent is OH^-, the observed rate constant increases with increasing pH, but if under these conditions a micellar bound nucleophile does not change its state of ionization, then the overall micellar rate enhancement will *increase* with *decreasing* pH, although this change will tell us nothing about the sources of the micellar rate enhancements.

It is evident that for bimolecular reactions in non-functional micelles in water the key factor in rate enhancement is the increased concentration of the two reactants in the micellar pseudophase (Table 3 and 4) and the same effect should be at work for reactions in functional micelles. The problem is simply that of estimating the concentration of functional groups in the micellar pseudophase. For the simplest case, that of a functional micelle, not involving deprotonation equilibria, with a nucleophilic head group denoted as N, there will be one reactive group per micellar head group, and if the substrate is fully micellar bound we can apply Eqn. 12, derived for reaction in non-functional surfactants, where:

$$k_\psi = m_N^s k_M \tag{12}$$

because in this special case the mole ratio $m_N^s = 1$ [99].

In the more general case, in which a functional group is only partially deprotonated to the reactive nucleophile, a correction can easily be made for partial deprotonation. If a comicelle of functional and non-functional surfactant is used it is necessary to correct Eqn. 12 for the mole fraction of functional groups in the micellar pseudophase [107,108].

However, as for reactions in non-functional micelles, k_M, sec^{-1}, cannot be compared directly with second-order rate constants in water, whose dimensions are, conventionally, M^{-1}/sec. But this comparison can be made provided that one specifies the volume element of reaction, which can be taken to be the molar volume of the micelles, or the assumed molar volume of the micellar Stern layer. This choice is an arbitrary one, but the volumes differ by factors of ca. 2 [107,108], so it does not materially affect the conclusions. The rate constants in the micelle for dephosphorylation, deacylation and nucleophilic substitution by a functional hydroxyethyl surfactant are similar to those in water [99], and similar results have been observed for dephosphorylation using functional surfactants with imidazole and oximate [106,107]. Similar results have been obtained by Fornasier and Tonalleto [108] for deacylation of carboxylic esters by a variety of functional comicelles.

These results are illustrated in Table 6 where the second-order rate constants in the micelle of comicelle are given relative to those for reaction of a chemically similar monomeric nucleophile in water. (Different volume elements of reaction have been used by the various workers in this area, as footnoted in Table 6.)

TABLE 6

Relative rate constants of reactions in functional micelles and comicelles [a]

Functional Surfactant	Substrate	k_{rel}
$C_{16}H_{33}\overset{+}{N}Me_2CH_2CH_2O^-$	2,4-$(NO_2)_2C_6H_4F$	1 [b]
$C_{16}H_{33}\overset{+}{N}Me_2CH_2CH_2O^-$	$(PhO)_2PO\cdot OC_6H_4NO_2(4)$	0.3 [b]
$C_{16}H_{33}\overset{+}{N}Me_2CH_2CH_2O^-$	$CH_3CO\cdot OC_6H_4NO_2(4)$	0.4 [c]
$C_{16}H_{33}\overset{+}{N}Me_2CH_2$–(imidazol-4-yl, NH)	$C_5H_{11}CO\cdot OC_6H_4NO_2(4)$	1.9 [c]
$C_{16}H_{33}\overset{+}{N}Me_2CH_2$–(imidazol-4-yl, NH)	$(PhO)_2PO\cdot OC_6H_4NO_2(4)$	0.5 [d]
$C_{12}H_{25}\overset{+}{N}Me_2CH_2CPh=NO^-$	$(PhO)_2PO\cdot OC_6H_4NO_2(4)$	0.2–0.5 [e]

[a] Based on a molar volume of the Stern layer of 0.14/M l unless specified.
[b] Ref. 5.
[c] Ref. 108, based on a micellar molar volume of 0.[illegible]
[d] Ref. 106.
[e] Ref. 107.

Micelles have been prepared from surfactants which contain a complexed metal ion, which can catalyze hydrolysis of organic phosphates and therefore can be regarded as akin to functional micelles [109]. These micelles are effective catalysts for the hydrolysis of acetyl phosphate and one can draw an analogy between them and metal-dependent enzymes involved in phosphoryl transfer.

7. Stereochemical recognition

The striking efficiency of some enzymes in distinguishing between enantiomers encouraged investigators to attempt to develop micelles with enantiomeric selectivity. To date non-reactive chiral surfactants have provided only slightly different reaction rates for pairs of enantiomeric reactants [110], but chiral recognition has been obtained using chiral micelles or comicelles. The histidine group is very convenient for this purpose, because its imidazole moiety is an effective nucleophile in deacylation, and stereoselective catalysis has been observed with several micelles or comicelles having the histidine functional group [111–114]. However in all cases stereoselectivity is small.

Moss and his coworkers [115] have however successfully developed micelles of diastereomeric peptide-like surfactants and used them in deacylation, with, in some cases, considerable selectivity.

There is an interesting difference between the evidence for enantiomeric and diastereomeric recognition. The stereoselective deacylation of enantiomers with a chiral histidine surfactant seems to depend wholly upon transition-state interactions [111], but for diastereomeric recognition interactions between substrate and micelle also seem to play a part [115].

Micellized enantiomers and diastereomers can give a degree of stereochemical recognition, but in addition the stereochemical course of reaction can be influenced by carrying it out in a micellar pseudophase.

Deamination of optically active primary amines in water generally proceeds with extensive racemization, but some inversion of configuration. However, if the amine is sufficiently hydrophobic for its salt to be micellized net retention of configuration is observed, because micellization encourages water molecules to attack the carbocationic intermediates preferentially from the front rather than from the rear [116].

In much the same way the stereochemistry of S_N1 reactions of chiral alkyl arenesulfonates can be controlled by carrying out solvolyses in the presence of micelles, or by using self-micellizing substrates [117,118], and the general features of these reactions are as outlined by Moss et al. [116] for deamination of micellized amines.

8. Submicellar and non-micellar aggregates

The simple model of micellar effects upon chemical reactivity in water assumes no effect due to monomeric surfactant [70]. This assumption is reasonable, because for most of the surfactants used in kinetic studies the concentration of monomers, as given by the cmc, is very small, e.g., 10^{-3}–10^{-2} M, and too low for there to be any significant kinetic salt effect. For reactions of relatively hydrophilic substrates, at low concentration of ionic reactants, rate effects below the cmc are ususally very small, as predicted. But, with hydrophobic reactants, rate effects are often observed at surfactant concentrations below the cmc in water, which could be caused by reactant-induced micellization or to reaction in submicellar aggregates [71,79,84,105].

Both explanations are reasonable. Critical micelle concentrations are decreased by addition of both electrolytes and hydrophobic non-ionic solutes to water [15]. But submicellar aggregates could coexist in solution with monomeric and micellized surfactant, although their concentration is probably low [2,23]. They could interact with, and be stabilized by, hydrophobic substrates.

Piskiewicz [119] has developed a kinetic model of micellar catalysis, based on the Hill equation of enzyme kinetics, which assumes a cooperative interaction between reactants and surfactant to form reactive substrate–micelle complexes. This model is probably not applicable to systems in which the surfactant is in large excess over substrate, as in most micellar mediated reactions, but it gives a very reasonable explanation of the rate effects of very dilute surfactants.

There is a great deal of qualitative evidence for rate enhancements by small hydrophobic aggregates or monomers in water, although it is often difficult to distinguish between this situation and that of induced micellization.

One early example, that of enhanced rate of aminolysis of *p*-nitrophenyl-derived carboxylic esters, was originally ascribed to 1 : 1 hydrophobic binding of the reactants [120], but was subsequently shown to be due to self-micellization of the ester [121]. Guthrie and Ueda [121] have very carefully analyzed the problems of establishing rate enhancements due to formation of 1 : 1 complexes as a result of hydrophobic interactions.

Acylation of peroxide anions by *p*-nitrophenyl phenylacetate (3) is speeded by CTACl. With the hydroperoxide ion, (4), R = H, there is little rate enhancement

$$\underset{(3)}{PhCH_2CO \cdot OC_6H_4NO_2} + \underset{(4)}{O_2^-R} \rightarrow PhCH_2CO \cdot O_2R + O_2NC_6H_4O^-$$

below the cmc of CTACl, but with the more hydrophobic α-peroxycumyl ion (4) (R = $PhCMe_2$), there are large rate enhancements at [CTACl] well below the cmc [122]. In addition a substrate such as (3) reacts very rapidly with (4), R = $PhCMe_2$, even at very low concentration where micellization is improbable. In this reaction association between two hydrophobic ions could promote reaction.

Another example of catalysis involving a non-micellizing cation comes from deacylation studies by Kunitake and coworkers who used tri-*n*-octylmethylammonium chloride, (5) [123]. This ion in low concentration speeded deacylation of

$$\underset{(5)}{(n\text{-}C_8H_{17})_3N^+RX^-} \qquad R = Me;\ X = Cl$$

p-nitrophenyl acetate by functional surfactants containing the hydroxamate group.

The rate enhancements by (5) were larger than those found with the micelle forming surfactant, CTACl [123], and it was suggested that the quaternary ammonium ion generated "hydrophobic ion pairs" which were very good nucleophiles.

Dephosphorylation by benzimidazolide or naphthimidazolide ions is catalyzed by both cationic micelles of CTABr and the non-micellizing quaternary ammonium ions (6a, b) (Scheme 11) [85,124]. The initial step of the reaction involves phosphorylation of the imidazole moiety, which is followed by a turnover step which generates final products. However, (6a, b) were used under conditions in which only the first step has to be considered.

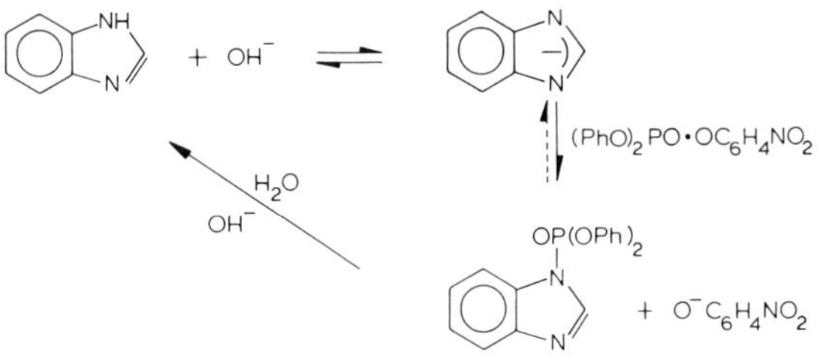

$+ OH^-$ ⇌

$(PhO)_2PO \cdot OC_6H_4NO_2$

H_2O

OH^-

$OP(OPh)_2$

$+ \; ^-OC_6H_4NO_2$

$(C_8H_{17})_3\overset{+}{N}Et\;Br^-$ (6a)

$(C_8H_{17})_3\overset{+}{N}Et\;MeSO_3^-$ (6b)

Scheme 11

The unexpected observation was that for a fixed concentration of areneimidazole, at constant pH, the first-order rate constants, k_ψ, went through maxima with increasing concentration of (6b). This observation suggests that there is a cooperative interaction between substrate, nucleophile and ammonium ion, and measurements of the substrate solubility and acid dissociation of benzimidazole allowed calculation of the amounts of the two reactants which were bound to each ammonium ion. Therefore the rate constant, k_M, for reaction in hydrophobic aggregate, could be calculated, using the treatment which was applied to micellar catalysis (Eqn. 6).

For reaction of benzimidazolide ion with p-nitrophenyl diphenyl phosphate in CTABr, $k_M = 7/\text{sec}$, and for reaction in (6b) k_M was between 11 and 31/sec, depending upon concentration. These observations suggest that catalysis by non-micellizing cations depends, as does micellar catalysis, on the bringing together of the two reactants, and in this particular case micellized and non-micellized ammonium ions seem to have similar catalytic efficiencies. However, the variations in k_M for reaction in the hydrophobic ammonium ion suggest that aggregates are forming with various compositions, rather than as 1 : 1 : 1 complexes and I believe that it is better to regard these reactions as occurring in hydrophobic clusters than with hydrophobic ion pairs.

Although in this system we see qualitative and quantitative similarities between micellizing and non-micellizing systems there are marked differences. In particular the catalytic effectiveness of non-micellizing hydrophobic cations is restricted to reaction of hydrophobic reagents for both deacylation [123] and dephosphorylation [124]. For example, cationic micelles effectively speed reactions involving nucleophilic anions, e.g., OH^-, but cations such as (5) and (6) are completely ineffective in this regard. Thus the high charge density of ionic micelles allows them to bind hydrophilic counterions, whereas single hydrophobic cations, or small clusters of them, do not have this ability.

It is easy to understand why reactions are very rapid in these hydrophobic clusters. The clusters are small, so that reactants which are bound to them are in very close proximity, even more so than micellar bound reactants, and this high concentration, of itself, enormously speeds reaction.

New types of amphiphiles have recently been investigated which contain several ionic residues and also hydrophobic groups, and which may behave like an ionic micelle, but be monomeric, or aggregates of a relatively few monomers. One such material is hexapus (7) [125].

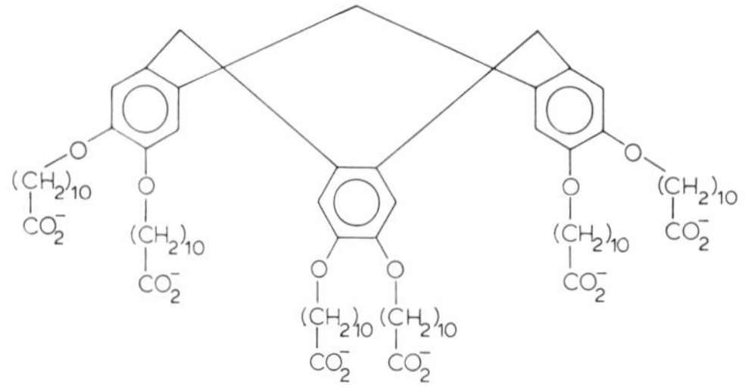

(7)

This material is not very surface-active, but it is very efficient at solubilizing organic solutes. The electronic spectra of absorbed dyes is shifted and hexapus will bind *p*-nitrophenyl butyrate and inhibit its reaction with OH^-. Hexapus apparently forms a hydrophobic cavity which can incorporate solutes.

9. *Micelles in non-aqueous systems*

Most investigators of micellar and related phenomena have used water as a solvent. It is abundant, cheap, and easily purified, and because biological reactions occur in aqueous media we are naturally interested in reactions in water which model biological reactions. However, micelles, or micelle-like aggregates, can form in non-aqueous solvents, and it is useful to distinguish between the normal micelles which form in solvents which have three-dimensional structure [19,20], and the so-called reverse micelles which form in apolar solvents [1,126,127].

9.1. Normal micelles in non-aqueous media

Micellization of both ionic and non-ionic surfactants occurs in solvents such as 1,2-diols, triols, and formamide which have sufficient hydrogen bonding centers that they form three-dimensionally associated structures [19,20]. Chemical reactivity has been little studied in these systems, although CTABr in H_2O: ethane-1,2-diol mixtures will speed nucleophilic attack by anions upon *p*-nitrophenyl diphenyl phosphate [128]. The rate-surfactant profiles are very similar to those observed in water, although with added OH^- the products are derived from reaction of both OH^- and $CH_2OH \cdot CH_2O^-$.

Carbocation or oxonium ion derived micelles form in concentrated sulfuric acid, a medium which is unsuitable for most organic solutes [20]. They are generated by dissolving the ether or alkene in sulfuric acid:

$$CH_3(CH_2)_{11}OMe \xrightarrow{H_2SO_4} CH_3(CH_2)_{11}O^+HMe$$

$$CH_3(CH_2)_{11}O-C_6H_4-C(=CH_2)Me \xrightarrow{H_2SO_4} CH_3(CH_2)_{11}O-C_6H_4-C^+Me_2$$

The effects of a variety of added organic solvents upon the thermodynamics of micellization have been investigated by Ionescu and Fung [129].

9.2. *Reverse micelles in aprotic solvents*

Aggregation of surfactants in apolar solvents, e.g., aliphatic or aromatic hydrocarbons, occurs provided that small amounts of water are present [1,126,127]. These aggregates are often called reverse micelles, although the solutions do not always appear to have a critical micelle concentration, and surfactant association is often governed by a multiple equilibrium, mass action, model with a large spread of aggregate sizes [130,131]. It has recently been suggested that the existence of a monomer $\rightleftharpoons$ *n*-mer equilibrium should be used as a criterion of micellization, and that this term should not be applied to self-associated systems which involve multiple equilibria [132].

Much of the work has been done using alkylammonium carboxylates or Aerosol OT (bis-2-ethylhexylsulfosuccinate). These types of surfactants readily form aggregates in apolar solvents.

The interior of the aggregate is aqueous, and the polar or ionic head groups are in the interior, which can be regarded as a "water pool" [133], and the hydrophobic residues of the surfactant are in contact with the apolar solvent. These aggregates therefore incorporate hydrophilic solutes into their aqueous interior, and, if the surfactants are ionic, the solutes will be in a concentrated ionic medium, which can also incorporate and concentrate hydrogen or hydroxide ions.

One of the first reactions to be examined in a reverse micelle was the mutarotation of 2,3,4,6-tetramethyl-α-D-glucose in solutions of dodecylammonium salts of carboxylic acids in benzene containing small amounts of water [134]. The substrate is relatively hydrophilic, and should be partitioned into the aqueous interiors of the micelles where it is exposed to the ammonium and carboxylate groups of the surfactant which could act as general acid–base catalysts. (Mutarotation in water is subject to general acid–base catalysis.)

Following these initial investigations many other kinetic studies have given very large rate effects on both organic and inorganic reactions. The requirements for rate enhancement seem to be that the reactants should be sufficiently hydrophilic to be concentrated in the small water pools in the interior of the micelles, so that the factors that control rate enhancements of bimolecular reactions in normal micelles seem to apply to reverse micelles [127,135]. Many organic and inorganic reactions have been studied mechanistically in reverse micelles, and in some cases very large rate enhancements were observed [135–138].

However, there are some special features of reactivity in reverse micelles which are very important.

The rates of proton transfer to excited state pyrene-1-carboxylate ion are extremely fast, which suggests that the solvated proton and the base are in very close proximity in the water pool [135].

Generally reactions in normal micelles in water occur in one micelle, and intermicellar reactions do not seem to be important. However, such intermicellar reactions have been observed in reverse micelles. They have been observed in imidazole-catalyzed hydrolysis of carboxylic esters [133], and in reactions involving energetic species generated photochemically [139] or involving electron transfer [140]. In these systems there is rapid transfer of reactants between water pools, so that the so-called reverse micelles may best be regarded as submicroscopic water droplets stabilized by surfactant.

These reverse micelles, or water pools, also provide excellent media for enzyme-catalyzed reactions [141,142], and recently the details of some enzymic reactions have been examined at low temperatures using supercooled reverse micelles [143]. Thus reverse micelles in apolar solvents have great potential for the study of enzymic reactions.

The formation of normal micelles in water, or in diols or triols or sulfuric acid is well established, as is the formation of aggregates in apolar solvents containing traces of water. But there are examples of aggregation in dipolar aprotic solvents or in aqueous–organic solvents of relatively low water content. In some cases reaction rates have been studied in these media, but generally only physical studies have been reported.

For a cationic (CTABr) and an anionic (NaLS) surfactant in a variety of dipolar aprotic solvents, e.g., dimethyl sulfoxide, dimethyl formamide, the variation of conductance with surfactant concentration is characteristic of micellization, and "critical micelle concentrations" were reported, which tend to be higher than those in water, especially for CTABr [144].

Variations of surface tension with surfactant concentration is another test of micellization, and there are sharp breaks in plots of surface tension against logarithm of the surfactant concentration at the cmc. Based on this criterion the cmc of CTABr increases steadily as dimethyl sulfoxide is added to water, but no break was observed above 0.275 mole fraction dimethyl sulfoxide [145].

Although there is only a limited amount of kinetic work in reactions mediated by surfactants in dipolar aprotic solvents E.J. Fendler and her coworkers [146] have observed that bile salts, e.g., sodium cholate in slightly aqueous DMSO, will speed a number of ionic reactions including the formation of *p*-nitrophenoxide ion from a phosphonate ester. The first-order rate constants increase smoothly with increasing sodium cholate concentration towards limiting values, and reaction is inhibited by water.

This reaction in DMSO is much faster than that in water, and reaction of an anionic nucleophile in water should be inhibited by an anionic surfactant, e.g., sodium cholate. The evidence is consistent with formation of a complex between substrate and steroid, and it was suggested that the carboxylate group of the cholate ion could act as a nucleophile or general base.

The nature of the aggregates formed from amphiphiles in dipolar aprotic solvents is not well established, probably because these systems have been less thoroughly studied than the normal micelles in water and the reverse micelles in apolar solvents. Bile salts (Scheme 2) have one face apolar, (hydrophobic), and the other face is hydrophilic because of the hydroxyl groups, therefore substrates could associate with the hydrophilic face by hydrogen bonding which could influence reactivity.

10. Related systems

10.1. Reactions in microemulsions

Microemulsions are transparent dispersions of oil in water (o/w), or water in oil (w/o), generally stabilized by both a surfactant and a cosurfactant, typically a medium chain length alcohol or amine [147]. However "surfactantless" microemulsions have been reported over limited ranges of solute concentration [148].

Although microemulsions are of great industrial importance, it is only in recent years that their role as media for organic and inorganic reactions has attracted attention [148–156].

Microemulsion droplets can be considered as akin to micelles, in that there is a lipid-like region with polar or ionic groups in contact with water. There seems to be no real difference between w/o microemulsions and reverse micelles in apolar solvents. In both systems the interior is water surrounded by surfactant and organic solvent, and the aggregates behave as aqueous dispersions stabilized by surfactant and cosurfactant.

In these systems, as in normal micelles, it is useful to distinguish between an aqueous and a non-aqueous pseudophase, except that the alcohol or amine in a microemulsion droplet may act as a nucleophile. Thus, as in micellar solutions, the aggregates can bring reactants together or keep them apart. For example, the complexing of metal ions to amino acid derivatives can be controlled by microemulsions [152], and the interior of a w/o microemulsion in these systems has been compared with the metal binding site of a protein.

Mackay and coworkers have used microemulsions extensively as reaction media, and have shown that in alkaline solution a hydroxyl function of an alcohol or a non-ionic surfactant can act as a nucleophile. For example in a microemulsion of

CTABr/*n*-butanol/hexadecane in alkaline solution diphenyl chlorophosphate gave

$$(PhO)_2PO \cdot Cl + OBu^- \rightarrow (PhO)_2PO \cdot OBu + Cl^- \quad (8)$$

and a similar reaction was probably given by *p*-nitrophenyl diphenyl phosphate, but the overall reaction was not very much faster than that in water [149,150,153].

A major problem in interpreting rate effects in microemulsions is that a hydrophobic reagent may bind in the non-polar part of the microemulsion droplet whereas a hydrophilic ion, e.g., OH^- or F^-, will be largely in the aqueous part of the medium. Therefore one can ask whether treatments similar to those applied to reactions in aqueous micelles are applicable to those in microemulsions. In addition it would be useful to know whether the interface between water and a microemulsion droplet has properties similar to those of a normal micelle in water.

Only limited evidence is available on these questions. For example Mackay and coworkers have estimated an effective dielectric constant of ca. 20 for a variety of microemulsion droplets, based on p*K* measurements [154], i.e., they appear to be somewhat less polar than normal micelles in water. However fluorescence shifts in microemulsions are similar to those in micelles, suggesting that the polarities of o/w microemulsions and micelles are similar [155]. We do not know whether these differences stem from the different probes used to estimate polarity, or whether there are marked differences between the surfaces of the various microemulsion aggregates.

There is some kinetic evidence that the surface of o/w microemulsion droplets based on CTABr are similar to those of normal cationic micelles. Decarboxylation of 6-nitrobenzisoxazole carboxylate ion (9) is a sensitive probe of the polarity of a solvent or of a submicroscopic interface [64,65], and rates of decarboxylation of (9) in cationic o/w microemulsions are very similar to those in normal micelles of CTABr [155].

(9)

The spontaneous water-catalyzed hydrolysis of bis-(4-nitrophenyl) carbonate, (10), is only weakly inhibited by cationic micelles of CTABr, suggesting that their surface is highly aqueous [32], and cationic microemulsions of octane, CTABr and *t*-amylalcohol behave in exactly the same way [155].

$$(O_2N{-}C_6H_4{-}O)_2CO + H_2O \longrightarrow 2\ O_2N{-}C_6H_4{-}O^- + CO_2$$

(10)

At first sight this evidence appears to be inconsistent with the demonstration of the high nucleophilicity of alkoxide ions in microemulsions [153], and the number of components in microemulsions makes it difficult to draw simple analogies between these systems and micelles.

There is limited evidence for the applicability of the pseudophase kinetic model to microemulsions. Amines can act as cosurfactants and stabilize o/w microemulsions, and these amines are effective nucleophiles towards 2,4-dinitrochlorobenzene, and the rate constants in the microemulsion droplets are not very different from those in water [156]. Reactions in which bromide ion acts as a nucleophile also have similar second-order rate constants in a microemulsion aggregate as in a micelle, suggesting that it should be possible to apply an ion exchange model to microemulsions [157]. However, Mackay and coworkers have treated ion binding in microemulsions in terms of the surface potential of the aggregates [153,154], following an approach which has been applied to ionic micelles in water [158].

At the present time it seems that, in considering chemical reactivity, one can regard o/w microemulsions as similar to normal micelles, with the proviso that the cosurfactant may be a reactant. However bimolecular ionic reactions will generally be slower in microemulsions than in micellar solutions, not necessarily because second-order rate constants are lower in the microemulsion droplet than in the micelle, but because reactant concentrations will generally be lower, at least for hydrophilic ions. Two factors will be at work, (i) the presence of cosurfactant in a microemulsion droplet will decrease its charge density, and therefore its ability to bind counterionic reactants, as compared with an ionic micelle, and (ii) the overall concentrations of the solute components of microemulsions tend to be high, as compared with the situation for micelles, so that bound reactants are distributed over a large volume, and their effective concentration in the droplet is low. On the other hand microemulsions can solubilize large amounts of solutes, and are much more effective than micelles in this regard, so that they have considerable potential as useful reaction media.

10.2. Reactions in vesicles

The liposomes generated from phospholipids can be regarded as models of membranes. Bilayered vesicles have been formed from a large number of synthetic amphiphiles, most of which contain two hydrophobic residues per head group, whereas the micelle-forming surfactants generally have one hydrophobic residue per head group [159]. However vesicles can be formed from single-chain surfactants, especially when the hydrophobic groups contains rigid sections [160]. A large number of different polar or ionic head groups have been used, as with micelle-forming surfactants.

Vesicles form when the hydrophobic interactions are increased relative to the ionic or dipolar repulsions, as happens when the second hydrophobic residue is present.

A discussion of the structure of vesicles is beyond the scope of this article, but vesicles can have large effects upon both thermal and photochemical reactions [159].

The reaction of OH^- with 5,5′-dithiobis(2-nitrobenzoic acid) is effectively catalyzed by vesicles of dioctadecyldimethylammonium chloride, as well as by micelles of CTABr, and the variation of rate constants with concentration of amphiphile can

$$RSSR + OH^- \longrightarrow RS^- + RSOH$$

$$2\,RSOH \xrightarrow{fast} RS^- + RSO^- + 2\,H^+$$

$$R \equiv \text{–}C_6H_3(CO_2^-)\text{–}NO_2$$

be fitted to a kinetic scheme, similar to that applied to micelle-catalyzed bimolecular reactions, in which the distribution of both reactants between the vesicle and water is taken into account [161].

Vesicles differ considerably from micelles in several ways. Micelles are very dynamic species, so that both monomeric surfactant and solubilized materials are constantly exchanging between the aqueous and micellar pseudophases. However, vesicles are relatively long-lived species, and can be separated from other materials by techniques such as gel-permeation chromatography, and they can be isolated and the structures determined by electron microscopy [159, 160]. In addition the bilayer has an inside and an outside surface, so that the transport of solutes across the bilayer is of considerable interest.

A recent important development has been the preparation of polymerized vesicles using surfactants which contain diacetylene moieties which can be cross-linked using free-radical activators or UV irradiation [162,163]. These polymerized vesicles maintain their integrity even in aqueous–organic media. Disruption of the water structure by addition of organic solvents generally leads to disruption of normal micelles and vesicles, but the integrity of the polymerized aggregates is such that they survive this treatment.

Although surfactants having rigidity in their hydrophobic group readily form bilayered vesicles Kunitake and his coworkers [164] have shown that relatively small changes in structure or environment can have a very great influence on the morphology of the aggregate, and they discuss the relation of their observations to the organization of biological structures.

11. Photochemical reactions

One of the motivations for the study of the effects of micelles and similar submicroscopic aggregates upon photochemical processes is the possibility of practical solar-energy conversion, for example by splitting water into hydrogen and oxygen [165]. It is relatively easy to use photochemical energy to split a molecule into highly energetic, and reactive, fragments, but it is often very difficult to prevent this energy from being dissipated, for example by recombination of the fragments [166]. Micelles, or micelle-like aggregates, e.g., vesicles, can help to solve this problem by bringing species together, or keeping them apart, just as in thermal reactions.

The ability of micelles, or similar aggregates, to increase the lifetime of a potentially useful energetic intermediate is illustrated by micellar effects upon the tris-bipyridyl ruthenium ion sensitized reduction of an amphiphilic methyl viologen (11) [167],

$$Ru(bpy)_3^{2+} \xrightarrow{h\nu} Ru(bpy)_3^{2+}*$$

$$RMV^{2+} \underset{Ru(bpy)_3^{+}}{\overset{Ru(bpy)_3^{2+}*}{\rightleftharpoons}} RMV^{+\bullet}$$

(11) $R = n\text{-}C_{14}H_{29}$

The hydrophobic methyl viologen (11) binds strongly to micelles of CTACl and conversion of $RMV^{+\bullet}$ back to the methyl viologen is inhibited.

Vesicles of dihexadecyl phosphate speed electron transfer from photoactivated $Ru(bpy)_3^{2+}*$ to methyl viologen, and it was possible to distinguish between reaction on the inner and outer surface of the vesicle. However although the forward step is very rapid so is the back reaction. A modification of this system in which $Ru(bpy)_3^{2+}$ is on the inner surface of a dihexadecyl phosphate vesicle will generate hydrogen, provided that the vesicle contains PtO which catalyzes the decomposition of water [168].

$$Ru(bpy)_3^{2+}* + MV^{2+} \rightarrow Ru(bpy)_3^{3+} + MV^{+\bullet}$$

$$2\,MV^{+\bullet} + H_2O \rightarrow H + OH^- + 2\,MV^{2+}$$

Photochemistry in micelles, vesicles and microemulsions is currently being studied by many groups [165–172]. The rigidity of the bilayered vesicles, as compared with a micelle, should make it easier to control the fate of high-energy species generated photochemically, and these systems seem to have considerable potential for the utilization of solar energy [165]. In addition, modification of the charge and hydrophobicity of the reactants makes it possible to control their location in solutions containing micelles or vesicles.

12. Isotopic enrichment

Turro and Kraeutler [173] have used micelles in an ingenious method for separating isotopes on the basis of their nuclear spins.

Photolysis of dibenzyl ketone gives 1,2-diphenyl ethane and CO (scheme 12).

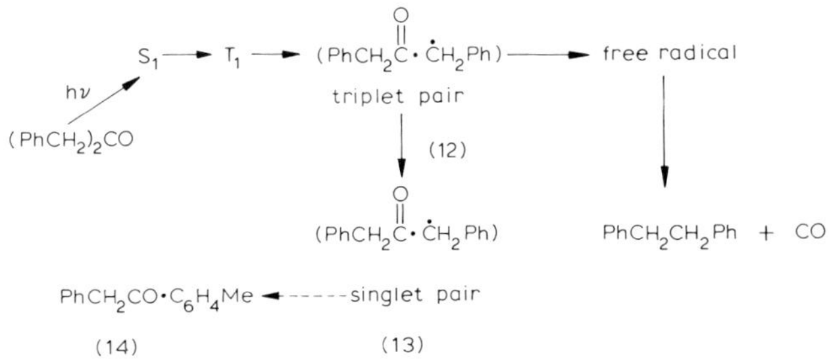

$$(PhCH_2)_2CO \xrightarrow{h\nu} S_1 \rightarrow T_1 \rightarrow (PhCH_2\overset{O}{\overset{\|}{C}}\cdot \dot{C}H_2Ph) \text{ triplet pair} \rightarrow \text{free radical} \rightarrow PhCH_2CH_2Ph + CO$$

$$(PhCH_2\overset{O}{\overset{\|}{C}}\cdot \dot{C}H_2Ph) \text{ triplet pair} \xrightarrow{(12)} (PhCH_2\overset{O}{\overset{\|}{C}}\cdot \dot{C}H_2Ph) \text{ singlet pair (13)} \dashrightarrow PhCH_2CO\cdot C_6H_4Me \text{ (14)}$$

Scheme 12

The first formed singlet goes via a triplet to the triplet pair (12). If (12) contains ^{13}C nuclei intersystem crossing is favored by nuclear hyperfine coupling so that the singlet pair (13) is generated and goes preferentially back to starting material plus some ketone (14). The triplet pairs which contain non-magnetic ^{12}C nuclei go preferentially to free radicals and finally to 1,2-diphenyl ethane depleted in ^{13}C. Thus ^{13}C is concentrated in starting material.

The isotopic enrichment is small in the absence of micelles, but increases sharply in CTABr because the micelles help the radical pairs to stay together, so that the magnetic effect of the ^{13}C nuclei controls the course of reaction.

13. Preparative and practical aspects

The ability of micellized surfactants to catalyze, or inhibit, reactions and to control stereochemistry and product composition, suggests that these agents could have a useful role in organic synthesis. A micelle can speed a desired reaction and inhibit an undesired one, and, for example, cationic micelles can control the ratio of unimolecular, S_N1, substitution to bimolecular, E2, elimination [54,55]. Micellization is of great importance in emulsion polymerization, but little use has been made of aqueous micelles in synthesis.

Cationic surfactants are often used as phase-transfer catalysts, but they are no more effective than other hydrophobic cations in this regard, suggesting that their ability to form micelles is not important in these systems [174]. Menger and his coworkers [175] showed that aqueous surfactants could improve yields in such reactions as alkaline hydrolyses of α,α,α-trichlorotoluene and permanganate oxidations, but it was not clear whether they acted by promoting better reactant mixing or by virtue of generating micelles [175]. The yields were generally worse when the surfactant was replaced by an inert organic solvent, which suggested that interfacial phenomena were important.

There are two obvious limitations to the use of surfactants in preparative chemistry. The first is that the high molecular weight of surfactants makes it impracticable to use them in large excess over reactants, and second, surfactants complicate product isolation. The isolation problem can often be solved by precipitating the surfactant, e.g., alkyl sulfates can often be precipitated from water as their potassium salts, or alkylammonium ions as their perchlorates, but both limitations are neatly solved using the two-phase system of phase-transfer catalysis.

One way to use surfactants in synthesis is to immobilize them. For example surfactant moieties can be bound covalently to a solid polymeric support [176]. Alternatively a surfactant or a hydrophobic cation can be supported on a solid support such as an ion-exchange resin [106] or silica gel [177].

In all these systems the aim is to design a reusable catalyst which can easily be separated from the products.

A different system, but akin to micellar catalysis, is the technique of triphase catalysis, in which reaction is carried out in a mixture of water and an apolar organic solvent and a polymer having cationic head groups which can assist reaction of the substrate with a nucleophilic anion [178]. Mass transfer across phase boundaries is important in these reactions, and some of the variables, such as stirring rates, have been explored in detail. A major advantage of this method for preparative chemistry is the ease of separation of the products and reuse of the "catalyst", and this technique has wide applications in synthesis.

A potentially important application of micelles or similar species is the extent to which they can control product composition. The ability of cationic micelles to speed E2 eliminations and inhibit S_N1 reactions has been cited [53–55]. But in addition micelles can control isomer distribution to some extent, as in the halogenation of phenolic ethers [179]. In another example Sukenik and coworkers have used micelles to mediate the isomeric composition of the products of oxymercuration of alkenes [180].

The limited concentrations used in micellar systems are not a problem in analytical chemistry, and micelles have been used very effectively to discriminate between desired and undesired analytical reactions, and this subject has been critically discussed by Hinze [181].

Armstrong [58] has recently pioneered the use of micellar solutions as the mobile phase in liquid chromatography. This technique appears to be very powerful and, as applied to high-pressure liquid chromatography, allows the use of cheap aqueous solvents instead of expensive, and often hazardous, organic solvents.

Specific ion electrodes have been used extensively to investigate counterion binding to micelles [49–51], but recently electrodes have been developed which are sensitive to monomeric surfactant, and can be used to study the aggregation of surfactants near the cmc [182].

Many drug formulations involve the use of surfactants, which not only solubilize the drug, but also, in some cases, appear to improve its therapeutic efficiency [14].

Attempts have been made to improve drug delivery by incorporating the drug in a liposome or surfactant vesicle, with the hope of directing the drug to a specific site in the organism. Another approach is to covalently bind the drug to a hydrophobic group. This approach has been used to develop site-specific or sustained delivery drugs which can penetrate the blood-brain barrier [183].

Surfactants, generally petroleum sulfonates, have been tested as a method of increasing tertiary oil recovery [184]. They act by reducing oil–water interfacial tensions and solubilizing the hydrocarbons. In some systems surfactant and an alcohol are used, so that microemulsions are present.

Note added in proof

Criticisms have recently been made of the quantitative treatment of micellar rate effects (section 4.3) and alternative formulations have been discussed. [Reddy, I.A.K. and Katiyar, S.S. (1982) ref. 92, p. 1017; Gensmantel, N.G. and Page, M.I. (1981) J. Chem. Soc. Perkin Trans. 2, 147, 155.]

Acknowledgements

My own work in the area has depended largely on the efforts of my coworkers, most of whom are noted in the references. This work has been supported by the National Science Foundation (Chemical Dynamics Program and Latin-American Program) and by the U.S. Army Office of Research.

References

1 Fendler, J.H. and Fendler, E.J. (1975) Catalysis in Micellar and Macromolecular Systems, Academic Press, New York.
2 Fisher, L.R. and Oakenfull, D.G. (1977) Chem. Soc. Rev. 6, 25.
3 Menger, F.M. (1977) in:E.E. van Tamelen (Ed.), Bioorganic Chemistry, III. Macro- and Multicomponent Systems, Academic Press, new York, p. 137.
4 Brown, J.M. (1979) in: D.H. Everett (Senior Reporter), Colloid Science, A Specialist Periodical Report, Vol. 3, Chemical Society, London, p. 253.
5 Bunton, C.A. (1979) Catal Rev. Sci.-Eng. 20, 1.
6 Sudhölter, E.J.R., van der Langkruis, G.B. and Engberts, J.B.F.N. (1979) Recl. Trav. Chim. Pays-Bas Belg. 99, 73.
7 Lindman, B. and Wennerström, H. (1980) Top. Curr. Chem. 87, 1.
8 (1977) K.L. Mittal (Ed.), Micellization, Solubilization and Microemulsions, Plenum, New York.
9 (1979) K.L. Mittal (Ed.), Solution Chemistry of Surfactants, Plenum, New York.
10 Mittal, K.L. and Mukerjee, P. (1977) in ref. 8, p. 1.
11 Hartley, G.S. (1977) in ref. 8, p. 23.
12 Hartley, G.S. (1948) Quart. Rev. 2, 152.
13 Duynstee, E.F.J. and Grunwald, E. (1959) J. Am. Chem. Soc. 81, 4540, 4542.
14 Elworthy, P.H., Florence, A.T. and MacFarlane, C.B. (1968) Solubilization by Surface Active Agents and its Application in Chemistry and the Biological Sciences, Chapman and Hall, London.
15 Mukerjee, P. and Mysels, K.J. (1970) Critical Micelle Concentrations of Aqueous Surfactant Systems, National Bureau of Standards, Washington, DC.
16 Bangham, A.D. (1968) Progr. Biophys. Mol. Biol. 18, 29.
17 Gregoriadis, G. and Allison, A.C. (1980) Liposomes in Biological Systems, John Wiley, New York.
18 Tanford, C. (1980) The Hydrophobic Effect, 2nd Edn., Wiley-Interscience, New York.
19 Ray, A. (1971) Nature (London) 231, 313; Ray, A. and Nemethy, G. (1971) J. Phys. Chem. 75, 809.
20 Menger, F.M. and Jerkunica, J.M. (1979) J. Am. Chem. Soc. 101, 1896.
21 Mukerjee, P., Cardinal, J.R. and Desai, N.R. (1977) in ref. 8, p. 241.
22 Stigter, D. (1975) J. Phys. Chem. 79, 1008.
23 Corrin, M.L. (1948) J. Colloid Sci. 3, 333; Mukerjee, P., Mysels, K.J. and Kapauuan, P. (1967) J. Phys. Chem. 71, 4166.
24 Jonsson, B. and Wennerström, H. (1981) J. Colloid Interface Sci. 80, 482.
25 Romsted, L.S. (1977) ref. 8, p. 509.
26 Gunnarsson, G., Jonsson, B. and Wennerström, H. (1980) J. Phys. Chem. 84, 3114.
27 Muller, N. (1979) in ref. 9, p. 267.
28 Robinson, B.H., White, N.C. and Mateo, C. (1975) Adv. Mol. Relaxation Processes 7, 321.
29 Almgren, M., Grieser, F. and Thomas, J.K. (1979) J. Am. Chem. Soc. 101, 279.
30 Bunton, C.A. and Huang, S.K. (1972) J. Org. Chem. 37, 1790.
31 de Albrizzio, J.P. and Cordes, E.H. (1979) J. Colloid Interface Sci. 68, 292.
32 Menger, F.M., Yoshinaga, H., Venkatasubban, K.S. and Das, A.R. (1981) J. Org. Chem. 46, 415.
33 Fadnavis, N. and Engberts, J.B.F.N. (1982) J. Org. Chem. 47, 152.
34 Tanford, C. (1978) Science 200, 1012, cf., Israelachvili, J.N., Mitchell, D.J. and Ninham, B.W. (1976) J. Chem. Soc. Faraday Trans. 2, 1525.

35 Menger, F.M. (1979) Accts. Chem. Res. 12, 111.
36 Dill, K.A. and Flory, P.J. (1980) Proc. Natl. Acad. Sci. (U.S.A.) 77, 3115.
37 Fromherz, P. (1980) Chem. Phys. Lett. 77, 460.
38 Ulmlus, J. and Lindman, B. (1981) J. Phys. Chem. 85, 4131.
39 Aniansson, G.E.A. (1978) J. Phys. Chem. 82, 2805.
40 Debye, P. and Anacker, E.W. (1951) J. Phys. Colloid Chem. 55, 644.
41 Kratohvil, J.P. (1979) Chem. Phys. Lett. 60, 238; (1980) J. Colloid Interface Sci. 75, 271.
42 Turro, N.J. and Yekta, A. (1978) J. Am. Chem. Soc. 100, 5951.
43 Infelta, P.P. (1979) Chem. Phys. Lett. 66, 88.
44 Mazer, N.A., Carey, M.C. and Benedek, G.B. (1977) ref. 8, p. 359.
45 Missel, P.J., Mazer, N.A., Benedek, G.B. and Young, C.Y. (1980) J. Phys. Chem. 84, 1044.
46 Corti, M. and Degiorgio, V. (1981) J. Phys. Chem. 85, 711.
47 Rohde, A. and Sackmann, E. (1979) J. Colloid Interface Sci. 70, 494; (1980) J. Phys. Chem. 84, 1598.
48 Dorshow, R., Briggs, J., Bunton, C.A. and Nicoli, D.F. (1982) J. Phys. Chem. 86, 2388.
49 Larsen, J.W. and Tepley, L.B. (1974) J. Colloid Interface Sci. 49, 113; Larsen, J.W. and Magid, L.J. (1974) J. Am. Chem. Soc. 96, 5664; (1975) J. Am. Chem. Soc. 97, 1988.
50 Zana, R. (1980) J. Colloid Interface Sci. 78, 330.
51 Bunton, C.A., Ohmenzetter, K. and Sepulveda, L. (1977) J. Phys. Chem. 81, 2000.
52 Bolt, J.D. and Turro, N.J. (1981) J. Phys. Chem. 85, 4029.
53 Minch, M.J., Giaccio, M. and Wolff, R. (1975) J. Am. Chem. Soc. 97, 3766.
54 Bunton, C.A., Kamego, A.A. and Ng, P. (1974) J. Org. Chem. 39, 3469.
55 Lapinte, C. and Viout, P. (1979) Tetrahedron 35, 1931.
56 Tagaki, W., Kobayashi, S., Kurihara, K., Kurashima, A., Yoshida, Y. and Yano, Y. (1976) J. Chem. Soc. Chem. Commun. 843; Al-Lohedan, H. and Bunton, C.A. (1981) J. Org. Chem. 46, 3929.
57 Herries, D.G., Bishop, W. and Richards, F.M. (1964) J. Phys. Chem. 68, 1842.
58 Armstrong, D.W. (1981) Am. Lab. August, 14; (1981) Anal. Chem. 53, 11.
59 Sepulveda, L. (1974) J. Colloid Interface Sci. 46, 372.
60 Bunton, C.A. (1973) in: E.H. Cordes (Ed.), Reaction Kinetics in Micelles, Plenum, New York, p. 73.
61 Martinek, K., Yatsimirski, A.K., Levashov, A.V. and Berezin, I.V. (1977) in ref. 8, p. 489.
62 Sudhölter, E.J.R. and Engberts, J.B.F.N. (1979) J. Phys. Chem. 83, 1854.
63 Cordes, E.H. and Gitler, C. (1973) Progr. Bioorg. Chem. 2, 1.
64 Thomson, A. (1970) J. Chem. Soc. B, 1198; Kemp, D.S. and Paul, K. (1970) J. Am. Chem. Soc. 92, 2555.
65 Bunton, C.A., Minch, M.J., Hidalgo, J. and Sepulveda, L. (1973) J. Am. Chem. Soc. 95, 3262.
66 Fendler, E.J., Liechti, R.R. and Fendler, J.H. (1970) J. Org. Chem. 35, 1658.
67 Kirby, A.J. and Varvoglis, A.G. (1967) J. Am. Chem. Soc. 89, 415; Bunton, C.A., Fendler, E.J. and Fendler, J.H. (1967) J. Am. Chem. Soc. 89, 1221.
68 Bunton, C.A., Fendler, E.J., Sepulveda, L. and Yang, K.-U. (1968) J. Am. Chem. Soc. 90, 5512.
69 Ingold, C.K. (1969) Structure and Mechanism in Organic Chemistry, 2nd Edn., Cornell University Press, Ithaca, Ch. 7.
70 Menger, F.M. and Portnoy, C.E. (1967) J. Am. Chem. Soc. 89, 4698.
71 Bunton, C.A. and Robinson, L. (1969) J. Org. Chem. 34, 773.
72 Quina, F.H. and Chaimovich, H. (1979) J. Phys. Chem. 83, 1844.
73 Chaimovich, H., Bonilha, J.B.S., Politi, M.J. and Quina, F.H. (1979) J. Phys. Chem. 83, 1851.
74 Bunton, C.A., Romsted, L.S. and Sepulveda, L. (1980) J. Phys. Chem. 84, 2611.
75 Al-Lohedan, H., Bunton, C.A. and Romsted, L.S. (1981) J. Phys. Chem. 85, 2123.
76 Funasaki, N. and Murata, A. (1980) Chem. Pharm. Bull. 28, 805.
77 Bunton, C.A. and Wolfe, B. (1973) J. Am. Chem. Soc. 95, 3742.
78 Bunton, C.A., Ramirez, F. and Sepulveda, L. (1978) J. Org. Chem. 43, 1166; Bunton, C.A., Romsted, L.S. and Smith, H.J. (1978) J. Org. Chem. 43, 4299.
79 Bunton, C.A., Carrasco, N., Huang, S.K., Paik, C. and Romsted, L.S. (1978) J. Am. Chem. Soc. 100, 5420.

80 Eriksson, J.C. and Gillberg, G. (1966) Acta Chem. Scand, 20, 2019.
81 Mukerjee, P. (1979) in ref. 9, p. 153.
82 Bunton, C.A., Cerichelli, G., Ihara, Y. and Sepulveda, L. (1979) J. Am. Chem. Soc. 101, 2429.
83 Cuccovia, I.M., Schroter, E.H., Monteiro, P.M. and Chaimovich, H. (1978) J. Org. Chem. 43, 2248.
84 Bunton, C.A., Ihara, Y. and Sepulveda, L. (1979) Israel J. Chem. 18, 298.
85 Bunton, C.A., Hong, Y.S., Romsted, L.S. and Quan, C. (1981) J. Am. Chem. Soc. 103, 5784.
86 Bartet, D., Gamboa, C. and Sepulveda, L. (1980) J. Phys. Chem. 84, 272; Gamboa, C., Sepulveda, L. and Soto, R. (1981) J. Phys. Chem. 85, 1429.
87 Bunton, C.A., Romsted, L.S. and Savelli, G. (1979) J. Am. Chem. Soc. 101, 1253.
88 Bunton, C.A., Moffatt, J.R. and Rodenas, E. (1982) J. Am. Chem. Soc. 104, 2653.
89 Young, P.R. and Hou, K.C. (1979) J. Org. Chem. 44, 947, and refs. cited.
90 Bunton, C.A., Romsted, L.S. and Thamavit, C. (1980) J. Am. Chem. Soc. 102, 3900.
91 Cowell, C. (1982) unpublished results.
92 Bunton, C.A. and Romsted, L.S. (1982) in: E.J. Fendler and K.L. Mittal (Eds.), Solution Behavior of Surfactants. Theoretical and Applied Aspects, Plenum, New York, p. 975.
93 Baumrucker, J., Calzadilla, M., Centeno, M., Lehman, G., Urdaneta , M., Lindquist, P., Dunham, D., Price, M., Sears, B. and Cordes, E.H. (1972) J. Am. Chem. Soc. 94, 8164.
94 Bunton, C.A., Gan, L.-H., Moffatt, J.R., Romsted, L.S. and Savelli, G. (1981) J. Phys. Chem. 85, 4118.
95 Nome, F., Rubira, A.F., Franco, C. and Ionescu, L.G. (1982) J. Phys. Chem,. 86, 1881.
96 Al-Lohedan, H. and Bunton, C.A. (1982) J. Org. Chem. 47, 1160.
97 Moss, R.A., Nahas, R.C. and Ramaswami, S. (1977) ref. 8, p. 603.
98 Tonellato, U. (1979) ref. 9, p. 541.
99 Bunton, C.A. (1979) ref. 9, p. 519.
100 Tagaki, W., Amada, T., Yamashita, Y. and Yano, Y. (1972) J. Chem. Soc. Chem. Commun., 1131.
101 Tonellato, U. (1977) J. Chem. Soc. Perkin Trans. 2, 822.
102 Moss, R.A., Nahas, R.C. and Ramaswami, S. (1977) J. Am. Chem. Soc. 99, 627.
103 Bunton, C.A. and Diaz, S. (1976) J. Am. Chem. Soc. 98, 5663.
104 Martinek, K., Levashov, A.V. and Berezin, I.V. (1975) Tetrahedron Lett. 1275.
105 Schiffman, R., Chevion, M., Katzhendler, J., Rav-Acha, Ch. and Sarel, S. (1977) J. Org. Chem. 42, 856; Schiffman, R., Rav-Acha, Ch., Chevion, M., Katzhendler, J. and Sarel, S. (1977) J. Org. Chem. 42, 3279.
106 Brown, J.M., Bunton, C.A., Diaz, S. and Ihara, Y. (1980) J. Org. Chem. 45, 4169.
107 Bunton, C.A., Hamed, F. and Romsted, L.S. (1980) Tetrahedron Lett. 21, 1217.
108 Fornasier, R. and Tonalleto, U. (1980) J. Chem. Soc. Faraday Trans 1 76, 1301.
109 Melhado, L.L. and Gutsche, C.D. (1978) J. Am. Chem. Soc. 100, 1850; Lau, H.-P. and Gutsche, C.D. (1978) J. Am. Chem. Soc. 100, 1857.
110 Moss, R.A. and Sunshine, W.L. (1974) J. Org. Chem. 39, 1083.
111 Brown, J.M. and Bunton, C.A. (1974) J. Chem. Soc. Chem. Commun. 969.
112 Ihara, Y. (1980) J. Chem. Soc. Perkin Trans. 2, 1483.
113 Ihara, Y., Nango, M. and Kuroki, N. (1980) J. Org. Chem. 45, 5009.
114 Ihara, H., Ono, S., Shosenji, H. and Yamada, K. (1980) J. Org. Chem. 45, 1623.
115 Moss, R.A., Lee, Y.-S. and Alwis, K.W. (1980) Tetrahedron Lett. 22, 282; J. Am. Chem. Soc. 102, 6646.
116 Moss, R.A., Talkowski, C.J., Reger, D.W. and Powell, C.E. (1973) J. Am. Chem. Soc. 95, 5215.
117 Okamoto, K., Kinoshita, T. and Yoneda, H. (1975) J. Chem. Soc. Chem. Commun. 922.
118 Sukenik, Ch.N. and Bergman, R.G. (1976) J. Am. Chem. Soc. 98, 6613.
119 Piskiewicz, D. (1977) J. Am. Chem. Soc. 99, 3067.
120 Blyth, C.A. and Knowles, J.R. (1971) J. Am. Chem. Soc. 93, 3017, 3021.
121 Guthrie, J.P. and Ueda, Y. (1976) Can. J. Chem. 54, 2745.
122 Brown, J.M. and Darwent, J.R. (1979) J. Chem. Soc. Chem. Commun. 169, 171.
123 Okahata, Y., Ando, R. and Kunitake, T. (1977) J. Am. Chem. Soc. 99, 3067.

124 Bunton, C.A., Hong, Y.S., Romsted, L.S. and Quan, C. (1981) J. Am. Chem. Soc. 103, 5788.
125 Menger, F.M., Takeshita, M. and Chow, J.F. (1981) J. Am. Chem. Soc. 103, 5938.
126 Kertes, A.S. and Gutmann, H. (1976) in: E. Matijevic (Ed.), Surface and Colloid Science, Vol. 8, John Wiley, New York, p. 194; Magid, L. (1979) ref. 9, p. 427.
127 Fendler, J.H. (1976) Accts. Chem. Res. 9, 153.
128 Bunton, C.A., Gan, L.-H., Hamed, F.H. and Moffatt, J.R. (1983) J. Phys. Chem., 87, 336.
129 Ionescu, L.G. and Fung, D.S. (1981) J. Chem. Soc. Faraday Trans. 1 77, 2907.
130 Muller, N. (1975) J. Phys. Chem. 79, 287; (1978) J. Colloid Interface Sci. 63, 383.
131 Lo, F.H.-Y., Escott, B.M., Fendler, E.J., Adams, E.T., Larsen, R.D. and Smith, P.W. (1975) J. Phys. Chem. 79, 2609.
132 Eicke, H.F. and Deuss, A. (1978) J. Colloid Interface Sci. 64, 386.
133 Menger, F.M., Donohue, J.A. and Williams, R.F. (1973) J. Am. Chem. Soc. 95, 286.
134 Fendler, J.H., Fendler, E.J., Medary, R.T. and Woods, V.A. (1972) J. Am. Chem. Soc. 94, 7288.
135 Escabi-Perez, J.R. and Fendler, J.H. (1978) J. Am. Chem. Soc. 100, 2234.
136 O'Connor, C.J., Lomax, T.D. and Ramage, R.E. (1982) in: E.J. Fendler and K.L. Mittal (Ed.), Solution Behavior of Surfactants. Theoretical and Applied Aspects, Plenum, New York, p. 803.
137 O'Connor, C.J. and Ramage, R.E. (1980) Austral. J. Chem. 33, 695.
138 El Seoud, O.A. and da Silva, M.J. (1980) J. Chem. Soc. Perkin Trans 2 127.
139 Eicke, H.F., Shepherd, J.C.W. and Steinemann, A. (1976) J. Colloid Interface Sci. 56, 168.
140 Robinson, B.H., Steytler, D.C. and Tack, R.D. (1979) J. Chem. Soc. Faraday Trans. 1 75, 481.
141 Menger, F.M. and Yamada, K. (1979) J. Am. Chem. Soc. 101, 6731.
142 Martinek, K., Levashov, A.V., Klyacho, N.L., Pautin, V.I. and Berezin, I.V. (1981) Biochim. Biophys. Acta 657, 277.
143 Douzou, P., Keh, E. and Balny, C. (1979) Proc. Natl. Acad. Sci. (U.S.A.) 76, 681.
144 Singh, H.N., Saleem, S.M., Singh, R.P. and Birdi, K.S. (1981) J. Phys. Chem. 84, 2191.
145 Ionescu, L.G., Tokuhiro, T., Czerniawski and Smith, E.S. (1979) ref. 9, pp. 487, 497.
146 Fendler, E.J., Koranek, D.J. and Rosenthal, S.N. (1979) in ref. 9, p. 575.
147 (1977) L.M. Prince (Ed.), Microemulsions: Theory and Practice, Academic Press, New York; Danielsson, I. and Lindman, B. (1981) Colloids Surf. 3, 391.
148 Barden, R.E. and Holt, S.L. (1979) in ref. 9, p. 707.
149 Mackay, R.A., Letts, K. and Jones, C. (1977) in ref. 8, p. 801.
150 Hermansky, C. and MacKay, R.A. (1979) in ref. 9, p. 723.
151 Robinson, B.H. and Steytler, D.C. (1978) Ber. Bunsenges, Phys. Chem., 82, 1012.
152 Smith, G.D., Barden, R.E. and Holt, S.L. (1978) J. Coord. Chem. 8, 157.
153 Mackay, R.A. and Hermansky, C. (1981) J. Phys. Chem. 85, 739.
154 Mackay, R.A., Jacobson, K. and Tourian, J. (1980) J. Colloid Interface Sci. 76, 515.
155 Bunton, C.A. and de Buzzaccarini, F. (1981) J. Phys. Chem. 85, 3139.
156 Bunton, C.A. and de Buzzaccarini, F. (1981) J. Phys. Chem. 85, 3142.
157 Bunton, C.A. and de Buzzaccarini, F. (1982) J. Phys. Chem. 86, 5010.
158 Fernandez, M.S. and Fromherz, P. (1977) J. Phys. Chem. 81, 1755; cf. Funasaki, N. (1979) J. Phys. Chem. 83, 1999.
159 Fendler, J.H. (1980) Accts. Chem. Res. 13, 7.
160 Kunitake, T. and Okahata, Y. (1980) J. Am. Chem. Soc. 102, 549; cf. Hargreaves, W.R. and Deamer, D.W. (1978) Biochemistry 17, 3759.
161 Fendler, J.H. and Hinze, W.L. (1981) J. Am. Chem. Soc. 103, 5439.
162 Regen, S.L., Czech, B. and Singh, A. (1980) J. Am. Chem. Soc. 102, 6638.
163 Bader, H., Ringsdorf, H. and Skura, J. (1981) Angew. Chem. Int. Ed. Eng. 20, 91.
164 Kunitake, T., Okahata, Y., Shimomura, M., Yasunami, S-i. and Takarabe, K. (1981) J. Am. Chem. Soc. 103, 5401.
165 Grätzel, M. (1981) Accts. Chem. Res. 14, 376.
166 Porter, G. (1978) Pure Appl. Chem. 50, 263.
167 Brugger, P.-A. and Grätzel, M. (1980) J. Am. Chem. Soc. 102, 2461.

168 Tunuli, M.S. and Fendler, J.H. (1981) J. Am. Chem. Soc. 103, 2507.
169 Almgren, M., Grieser, F. and Thomas, J.K. (1981) J. Am. Chem. Soc. 102, 3188.
170 Jones, C.E., Jones, C.A. and Mackay, R.A. (1979) J. Phys. Chem. 83, 805.
171 Laane, C., Ford, W.E., Otvos, J.W. and Calvin, M. (1981) Proc. Natl. Acad. Sci. (U.S.A.) 78, 2017.
172 Schmehl, R.H. and Whitten, D.G. (1980) J. Am. Chem. Soc. 102, 1938.
173 Turro, N.J. and Kraeutler, B. (1978) J. Am. Chem. Soc. 100, 7432; (1980) Accts. Chem. Res. 13, 369.
174 Weber, W.P. and Gokel, G.W. (1977) Phase Transfer Catalysis in Organic Synthesis, Springer, Heidelberg.
175 Menger, F.M., Rhee, J.V. and Rhee, H.K. (1975) J. Org. Chem. 40, 3803.
176 Brown, J.M. and Jenkins, J.A. (1976) J. Chem. Soc. Chem. Commun. 458.
177 Tundo, P. and Venturello, P. (1979) J. Am. Chem. Soc. 101, 6606.
178 Regen, S.L. (1979) Angew. Chem. Int. Ed. Eng. 18, 421.
179 Jaeger, D.A. and Robertson, R.E. (1977) J. Org. Chem. 42, 3298.
180 Link, C.M., Jansen, D.K. and Sukenik, Ch.N. (1980) J. Am. Chem. Soc. 102, 7798.
181 Hinze, W.L. (1979) ref. 9, p.79.
182 Kale, K.M., Cussler, E.J. and Evans, D.F. (1980) J. Phys. Chem. 84, 593.
183 Bodor, N., Farag, H.H. and Brewster, M.E. (1981) Science 214, 1370.
184 Bansal, V.K. and Shah, D.O. (1977) ref. 8, p. 87.

CHAPTER 14

Cyclodextrins as enzyme models

MAKOTO KOMIYAMA [a] and MYRON L. BENDER [b]

[a] *Department of Industrial Chemistry, Faculty of Engineering, University of Tokyo, Hongo Bunkyo-ku, Tokyo, Japan and* [b] *Department of Chemistry, College of Arts and Sciences, Northwestern University. Evanston IL 60201, U.S.A.*

1. Introduction

Cyclodextrins, sometimes called Schardinger dextrins, cycloamyloses, or cycloglucans, are cyclic oligosaccharides consisting of α-1,4-linked D-glucopyranose rings. α-, β-, and γ-cyclodextrins contain, respectively, 6, 7, and 8 glucose residues in a molecule. X-Ray crystallography shows that cyclodextrins have a doughnut shape, with the glucose units in the C1 conformation. The inner diameters of the cavities in α-, β- and γ-cyclodextrins (see Fig. 1) are approximately 4.5, 7.0 and 8.5 Å, respectively. The primary hydroxyl groups at the C-6 atom of a glucose unit are arranged in one of the open ends of the cavity, while the secondary hydroxyl groups at the C-2 and C-3 atoms of a glucose unit are arranged in the other open end (Fig. 2). The interior of the cavity contains a ring of C—H groups, a ring of glycosidic oxygens, and another ring of C—H groups. Therefore, the interior of cyclodextrin torus has hydrophobic (apolar) character, as do the binding sites of most enzymes.

Cyclodextrins (hosts) form inclusion complexes with many kinds of molecules and ions (guests), either in the solid phase or in solution. The stoichiometry of guest compounds to host compounds in inclusion complexes is usually 1 : 1 in aqueous solution.

One of the principal reasons for the utilization of cyclodextrins as models of enzymes is the formation of inclusion complexes between the catalyst and the substrate preceding the catalysis, which is comparable to the formation of a Michaelis–Menten complex in enzymatic reactions.

In addition to complex formation, the analogy between cyclodextrin-catalyzed reactions and enzymatic reactions takes many other forms: (1) Both reactions are carried out usually in aqueous media. (2) Both reactions show specificity with respect to both the substrate and product. (3) Both reactions show D,L specificity.

Recent studies have shown that cyclodextrins are still better models of enzymes than proposed previously. This review describes recent progress in cyclodextrin chemistry pointed toward enzyme models.

Michael I. Page (Ed.), The Chemistry of Enzyme Action

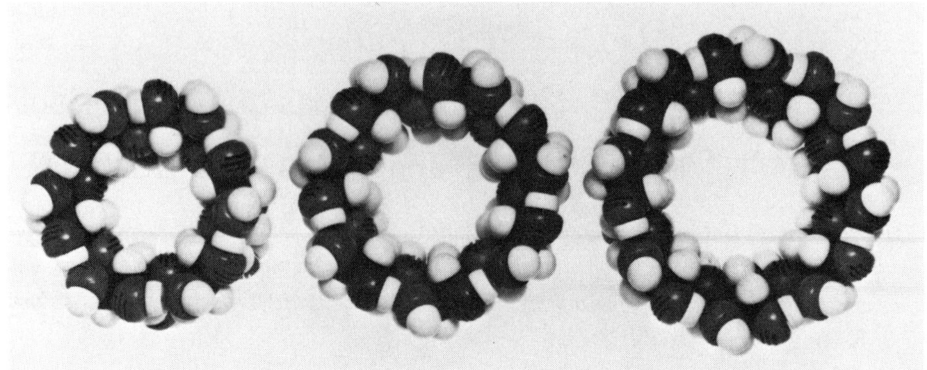

Fig. 1. From left to right, Corey–Pauling–Koltun molecular models of α-, β- and γ-cyclodextrins viewed from the secondary hydroxyl side of the torus. From Bender, M.L. and Komiyama, M. (1978) Cyclodextrin Chemistry, Springer, Heidelberg.

2. *Formation of an inclusion complex*

As described above, the most important reason for the increasing attention to cyclodextrins as enzyme models is the formation of inclusion complexes (the substrates are included in the cavities of cyclodextrins).

Inclusion complexes are formed in aqueous solution. In most cases, substrates with poor solubility usually form stronger complexes with cyclodextrins, indicating that the transfer of the substrate from aqueous medium to the apolar cyclodextrin cavity is an important driving force for complex formation [3].

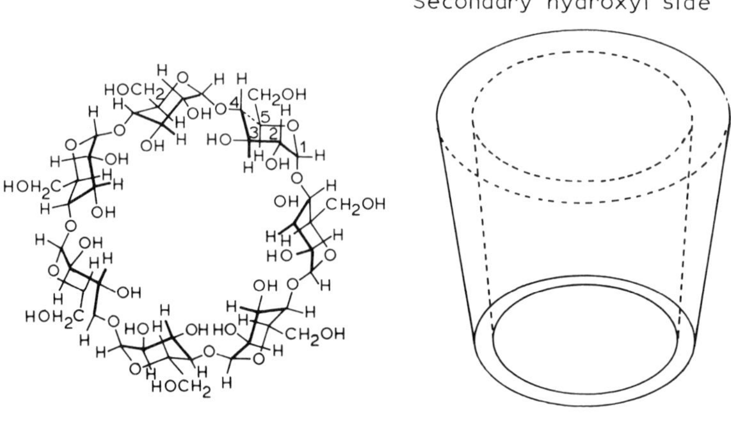

Fig. 2. The structure of a cyclodextrin.

Inclusion complexes are also formed in dimethylformamide and dimethyl sulfoxide, although they are less stable than those formed in aqueous solution [4]. However, no inclusion complexes are formed in other organic solvents such as alcohol, ether, dioxane or pyridine.

Capping of one side of the torus with an apolar moiety enhances the binding ability of a cyclodextrin. This effect is attributable to the increase in the apolar nature of the cavity due to the capping by the apolar groups. For example, capped β-cyclodextrin (1a) binds sodium 1-anilino-8-naphthalenesulfonate 24 times better than parent β-cyclodextrin [5]. The binding ability of another capped β-cyclodextrin (1b) changes on photoirradiation, since the conformation of β-cyclodextrin is varied due to the *cis–trans* photoisomerization of the azobenzene moiety [6].

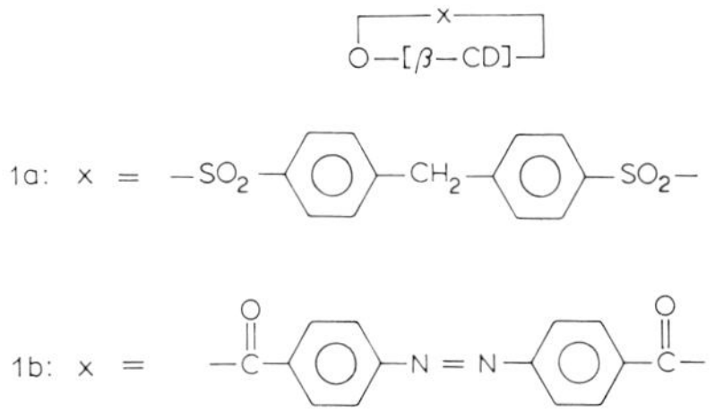

O—[β—CD]— (capped by X)

1a: X = $-SO_2-C_6H_4-CH_2-C_6H_4-SO_2-$

1b: X = $-C(=O)-C_6H_4-N=N-C_6H_4-C(=O)-$

One of the most important pieces of progress in recent years is the establishment of a method for the determination of the conformation of the inclusion complex in solution. The method, which takes advantage of the change of the ^{1}H-chemical shifts of cyclodextrin on complex formation with the aromatic guest compound (mostly due to the anisotropic shielding effect of the aromatic ring of the guest), is briefly described as follows [7–9].

(1) The changes of the chemical shifts of the protons of cyclodextrin on its complex formation with substrates are experimentally determined.

(2) The magnitudes of the anisotropic shielding effects by the aromatic rings of the guest compounds on the protons of the cyclodextrin are evaluated by use of the results of the calculation by Johnson and Bovey [10].

(3) The optimal position of the aromatic ring is determined by shifting the center of the aromatic ring along the longitudinal axis of the cavity of the cyclodextrin. At the optimal position, the agreement between the observed values of the chemical shift changes and the calculated values of the anisotropic shielding effects for all the protons of the cyclodextrin are maximal. The optimal position is taken as the time-averaged position of the aromatic ring of the guest compound.

Fig. 3 depicts the time-averaged position of phenyl acetates in the cavity of α-cyclodextrin, determined by the above method. In these time-averaged conformations, the centers of the aromatic rings of *p*-nitrophenyl acetate, phenyl acetate, and *m*-nitrophenyl acetate, respectively, are at the heights of 2.2, 1.9 and 1.7 Å with respect to the plane comprised of the 6 H-3 atoms of α-cyclodextrin. As shown in Table 1, the calculated values of the anisotropic shielding effects of the aromatic rings of the phenyl acetates on both the H-3 and H-5 atoms agree fairly well with the observed values.

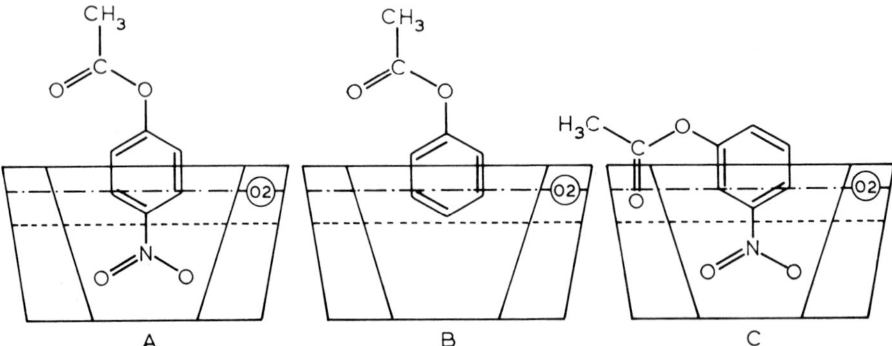

Fig. 3. Time-averaged conformations of the inclusion complexes of α-cyclodextrin with *p*-nitrophenyl acetate (A), phenyl acetate (B), and *m*-nitrophenyl acetate (C) in a 1 : 1 (v/v) mixture of 1 N deuterium chloride and dimethyl sulfoxide; - - - - - - and · — · — ·, respectively, refer to the planes comprised of the 6 H-3 and O-2 atoms of α-cyclodextrin; (O2) shows the position of 1 of the 6 O-2 atoms. From Komiyama, M. and Hirai, H. (1980) Chem. Lett., 1471.

Furthermore, the conformation of the complex between cyclodextrin and aromatic guest compounds by [^{13}C]NMR spectroscopy has also been accomplished [11]. This method is based on the correlation between the penetration depth of the carbon atoms of the aromatic guest compounds and the changes of the ^{13}C-chemical shifts as shown in Fig. 4. This correlation was found in complex formation between α-cyclodextrin and 4 guest compounds (see the legend for Fig. 4), where the depth of penetration was determined by the use of the [^{1}H]NMR method described above. Thus, 7 points except for those of the C-4 and C-9 atoms showed linearity. As the carbon atom was included more deeply in the cavity, its chemical shift moved from the values in the absence of α-cyclodextrin toward lower magnetic field to a greater extent. The C-4 and C-9 atoms, both of which have Z values of around -0.8 Å, showed almost identical Δ_C values (+0.1 ppm), although the corresponding points

TABLE 1

Observed and calculated values of the [^{1}H]NMR chemical shift changes on the complex formation of α-cyclodextrin with phenyl acetates [a]

Substrate	Change of the [^{1}H]NMR chemical shift (ppm) [b]			
	H-3		H-5	
	obs	calc	obs	calc
p-Nitrophenyl acetate	0.00	0.00	−0.06	−0.07
Phenyl acetate	+0.04	+0.04	−0.04	−0.07
m-Nitrophenyl acetate	+0.06	+0.06	−0.06	−0.08

[a] In a 1 : 1 (v/v) mixtures of 1 N deuterium chloride and dimethyl sulfoxide-d_6.

[b] A positive sign refers to an increase in % shielding.

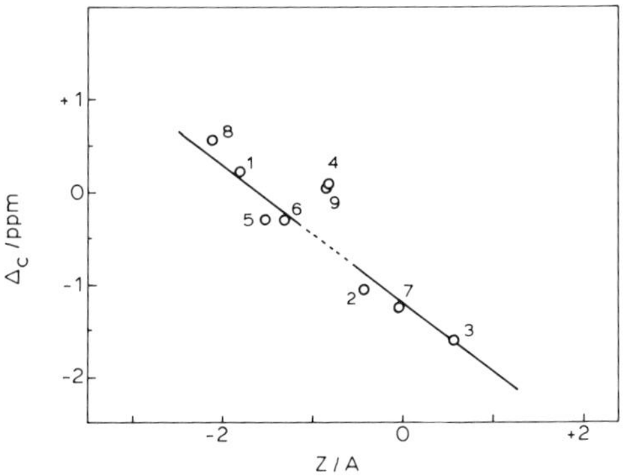

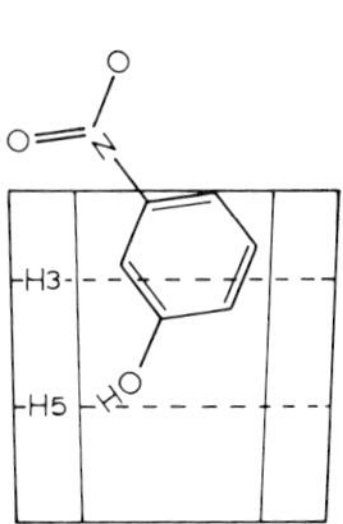

Fig. 4. Plot of the changes of the ^{13}C-chemical shifts (Δ_C in ppm) of the guest compounds on inclusion complex formation with α-cyclodextrin vs. the penetration depth (Z in Å); the positive sign in Δ_C shows an increase in shielding, and a positive sign in Z shows deeper penetration with respect to the plane ($Z = 0$) comprised of the 6 H-3 atoms of α-cyclodextrin; the numbering systems of the carbon atoms of the guest compounds are as follows:

From Komiyama, M. and Hirai, H. (1981) Bull. Chem. Soc. Jpn. 54, 828.

Fig. 5. The time-averaged conformation of the α-cyclodextrin-*m*-nitrophenol complex; - - -H3- - - and - - -H5- - - show the planes comprised of the 6 corresponding atoms of α-cyclodextrin. From Komiyama, M. and Hirai, H. (1981) Bull. Chem. Soc. Jpn. 54, 828.

deviated considerably from the straight line for the other 5 points.

The most important conclusion from Fig. 4 is that Δ_C is virtually governed only by the penetration depth in the cavity. The positions of the carbon atoms with respect to the substituents in the guest compounds are less important. Thus, the relationship in Fig. 4 is applicable to the determination of the conformation of the complexes of other guest compounds, especially to those of substituted benzenes.

For example, the time-averaged conformation of the inclusion complex between α-cyclodextrin and *m*-nitrophenol was determined by this method as shown in Fig. 5.

An advantage of the [^{13}C]NMR method over the [^{1}H]NMR method is that it can provide direct information on the positions of all the carbon atoms of the guest compounds and thus on the degree of penetration of the guest compounds in the cavity. However, only the position of the aromatic ring of the guest compound in the cavity can be determined by the use of the [^{1}H]NMR method.

TABLE 2
Reactions accelerated by cyclodextrins

Reaction	Substrate	Acceleration factor
Cleavage of esters	Phenyl esters	300
	Mandelic acid esters	1.38
Cleavage of amides	Penicillins	89
	N-Acylimidazoles	50
	Acetanilides	16
Cleavage of organophosphates	Pyrophosphates	> 200
	Methyl phosphonates	66.1
Cleavage of carbonates	Aryl carbonates	7.45
Cleavage of sulfates	Aryl sulfates	18.7
Intramolecular acyl migration	2-Hydroxymethyl-4-nitrophenyl trimethylacetate	6
Decarboxylation	Cyanoacetate anions	44.2
	α-Ketoacetate anions	3.95
Oxidation	α-Hydroxyketones	3.3

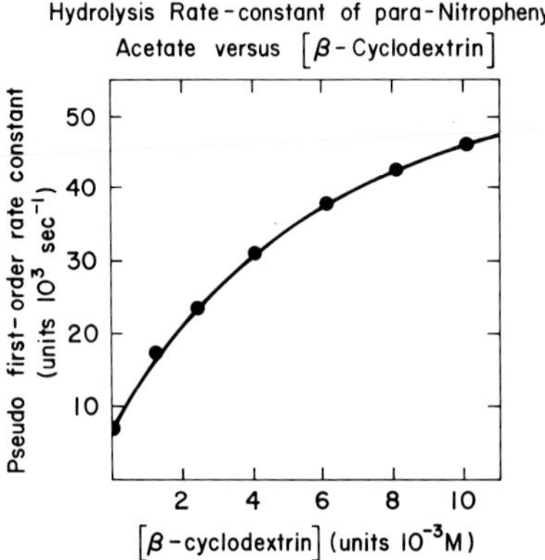

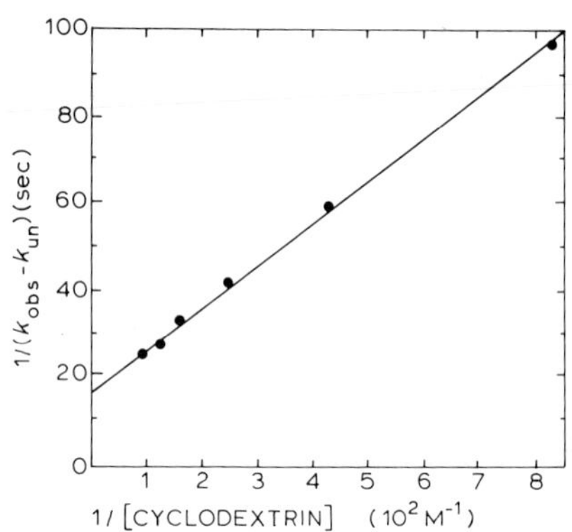

Fig. 6. The pseudo-first-order rate constant for release of phenol from *p*-nitrophenyl acetate at pH 10.6 plotted as a function of added β-cyclodextrin; 0.5% (v/v) acetonitrile–water, 25°, [*p*-nitrophenyl acetate] $\simeq 1 \times 10^{-4}$ M. From VanEtten, R.L., Sebastian, J.F., Clowes, G.A. and Bender, M.L. (1967) J. Am. Chem. Soc. 89, 3242, reprinted by permission.

Fig. 7. $1/(k_{obs} - k_{un})$ for *p*-nitrophenyl acetate decomposition is plotted vs. the reciprocal of the β-cyclodextrin concentration (data from Fig. 6). From VanEtten, R.L., Sebastian, J.F., Clowes, G.A. and Bender, M.L. (1967) J. Am. Chem. Soc. 89, 3242, reprinted by permission.

3. *Catalysis by cyclodextrins*

As shown in Table 2, cyclodextrins exhibit catalysis in many organic reactions. A typical rate vs. concentration plot for the catalysis by cyclodextrin is shown in Fig. 6, which is reminiscent of enzymatic saturation kinetics. A double reciprocal plot (of the same data) shows a straight line, just as an enzymatic reaction does, as shown in Fig. 7. This double reciprocal plot is a direct analog of a Lineweaver–Burk plot in enzymatic kinetics.

Catalysis by cyclodextrins may be divided into two categories: (1) Catalysis by the hydroxyl groups, in which the hydroxyl groups of the cyclodextrin function as intracomplex catalysts toward the substrates included in the cavity of the cyclodextrin. (2) Effect of the reaction field, in which the cavity of the cyclodextrin serves as an apolar and sterically restricted reaction field. Both of these catalyses are important in enzymatic reactions.

3.1. Catalysis by the hydroxyl groups

Three kinds of catalysis by the hydroxyl groups of cyclodextrin are known: (a) nucleophilic catalysis; (b) general base catalysis; and (c) general acid catalysis.

In nucleophilic catalysis, an anion of a *secondary* hydroxyl group of the cyclodextrin (CD-OH) attacks at the electrophilic center of the ester substrate included in the cavity of the cyclodextrin, resulting in the formation of acyl-cyclodextrin (2) together with the release of the leaving group (see Scheme 1 for ester hydrolysis). The catalysis is completed by the regeneration of the cyclodextrin through the hydrolysis of 2.

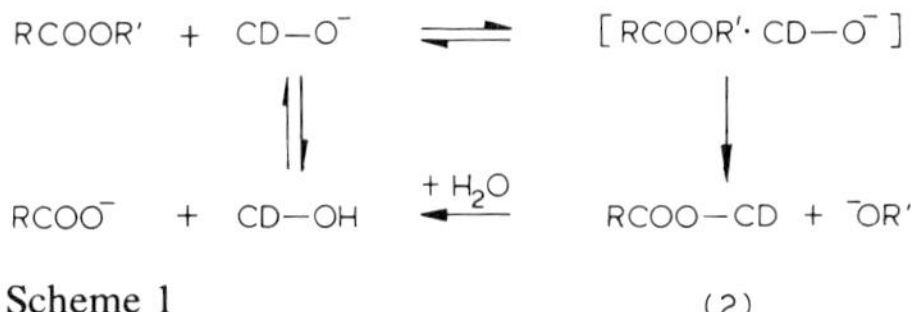

Scheme 1

Recent work indicates that the anion of the 3-hydroxyl group rather than the anion of the 2-hydroxyl group is the nucleophile. NMR studies of phenol-inclusion complexes [46] and the determination that the tosylation reaction of cyclodextrin takes place under basic conditions [47] lead to the conclusion that substitution occurs exclusively at the 3-position.

However, Breslow et al. [21] have found an acyl group both on the 2- and 3-positions; his reactions were carried out in basic solution where migration would be expected. Also, model studies can not differentiate between acylation at the 2- and 3-positions.

On nucleophilic attack by the secondary alkoxide ion, the tetrahedral intermediate (3) is formed in the rate-determining step. This is shown by the fact that the rate constants of α- or β-cyclodextrin-accelerated cleavage of (4a) (1.5×10^{-2} or 4.8×10^{-2}/sec at pH 10.5, 25°C) were almost identical with the values of (4b)

(1.6×10^{-2} or 4.2×10^{-2}/sec) [12].

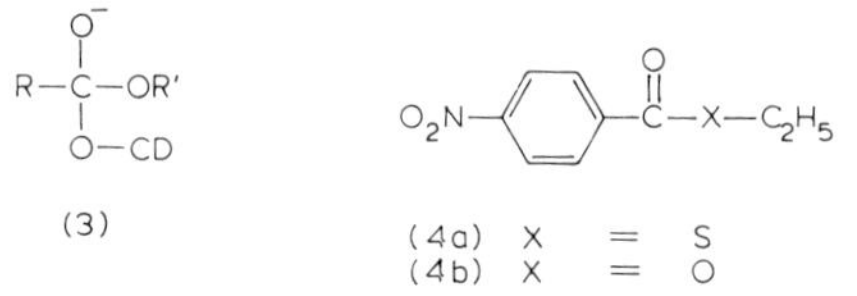

(3)

(4a) X = S
(4b) X = O

General base catalysis involves enhancement of the nucleophilicity of the water molecule by the abstraction of a proton. In the cyclodextrin case, general base catalysis was found for the first time in the hydrolysis of trifluoroethyl ester of *p*-nitrobenzoate [13], since no expected covalent intermediate is formed in the course of the reaction and since there is a 1.7-fold D_2O effect.

There have been no examples of reactions proceeding via general acid catalysis alone by cyclodextrin. In the hydrolysis of trifluoroacetanilide, however, general acid catalysis enhances the cleavage of the tetrahedral intermediate (5) formed by nucleophilic attack by a secondary alkoxide ion. General acid catalysis serves to convert the leaving group from an extremely unstable anion of aniline to a stable neutral aniline molecule (Scheme 2) [14].

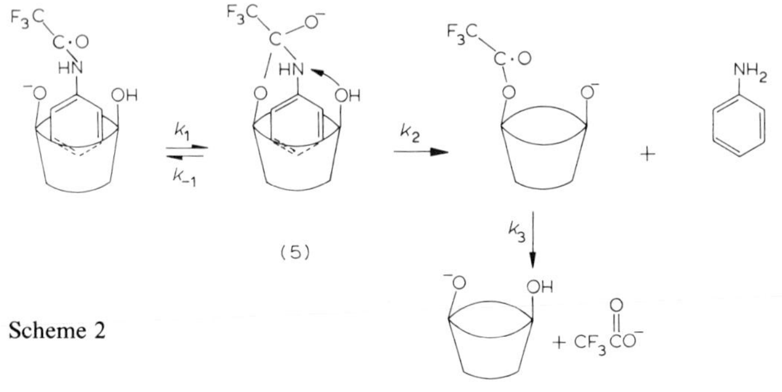

(5)

Scheme 2

This is the only example of the cyclodextrin-catalyzed hydrolysis of an amide. The rate-determining step in this process is acylation whereas the rate-determining step in the cyclodextrin-catalyzed hydrolysis of phenyl esters is deacylation. This dichotomy completely parallels the situation in chymotrypsin-catalyzed hydrolysis as shown in Table 3.

TABLE 3
Rate-determining steps in cyclodextrin- and chymotrypsin-catalyzed reactions

Catalyst	Reactant	Rate-determining step
Enzyme	Amide	Acylation
Enzyme	Ester	Deacylation
Cyclodextrin	Amide	Acylation
Cyclodextrin	Ester	Deacylation

3.2. Effect of reaction field

Catalysis by cyclodextrins does not always involve the catalytic functions of the hydroxyl groups. Sometimes, cyclodextrins simply provide the cavities as a reaction field. This effect is attributable to (a) a "microdielectric catalysis" due to the apolar character of the cavity or (b) a "conformational catalysis" due to the geometric requirements of inclusion.

A typical example of the microdielectric catalysis by cyclodextrins is the decarboxylation of anions of activated acids such as α-cyano and β-keto acids [15]. These reactions proceed unimolecularly via rate-determining heterolytic cleavage of the carbon–carbon bond adjacent to the carboxylate group (Scheme 3).

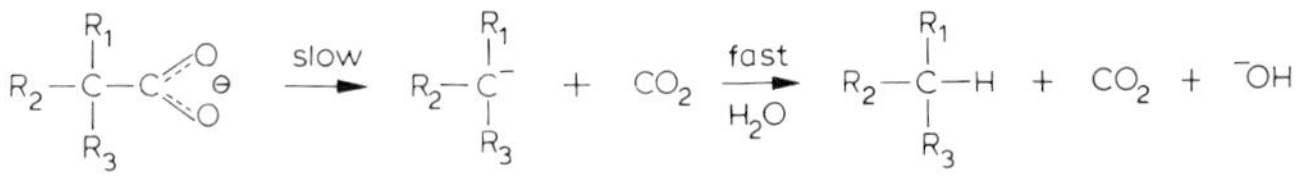

$$R_2C(R_1)(R_3)-CO_2^{\ominus} \xrightarrow{\text{slow}} R_2C(R_1)(R_3)^{-} + CO_2 \xrightarrow[H_2O]{\text{fast}} R_2C(R_1)(R_3)-H + CO_2 + {}^{-}OH$$

Scheme 3

Cyclodextrin can accelerate the first step through microdielectric catalysis, since the interior of the cavity has an apolar or ether-like atmosphere. In fact, the activation parameters for cyclodextrin-catalyzed reactions are almost identical to reactions in a 2-propanol–H_2O mixture.

An example of conformational catalysis is shown in the intramolecular acyl transfer of 2-hydroxymethyl-4-nitrophenyl trimethylacetate (6) (Scheme 4). The

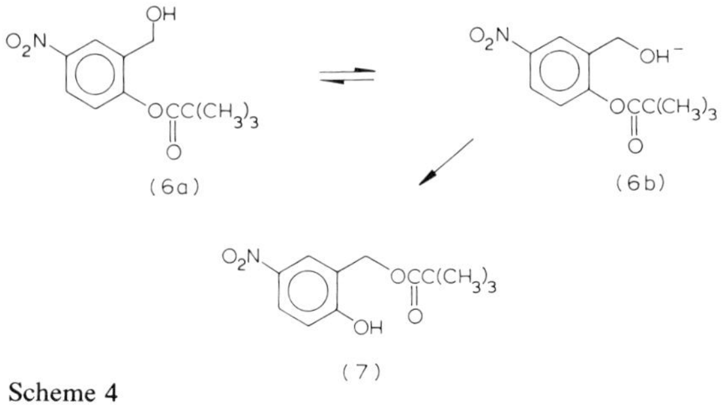

(6a) (6b) (7)

Scheme 4

inclusion of (6) within the cavity of α-cyclodextrin produces a 5–6-fold acceleration in the rate of conversion of (6) to (7). α-Cyclodextrin, by virtue of its ability to include (6) within a rigid binding site, perturbs the equilibrium between conformers (6a) and (6b) and therefore forces the reacting groups of (6) to assume a favorable conformation with respect to the reaction. Thus the binding forces between α-cyclodextrin and (6) are utilized to effect these conformational restrictions. The

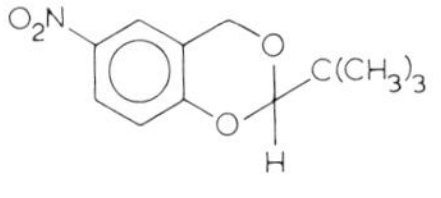

(8)

cyclodextrin-induced rate acceleration is entirely entropic in origin, with $\Delta S = 4.3$ gibbs. The inclusion complex of α-cyclodextrin with the "transition state analog" of the reaction (8) ($K_d = 1.22 \times 10^{-2}$ M) is more stable than the α-cyclodextrin–(6) inclusion complex by a factor which agrees exactly with the observed rate acceleration. Hence, the driving force, previously shown to be entropic in origin, can be attributed entirely to the affinity of the cyclodextrin for the activated complex [16], and thus Scheme 4 is the best explanation for conformational catalysis.

4. *Specificity in cyclodextrin catalysis*

Catalysis by cyclodextrins often shows specificity which is characteristic of enzymatic catalysis. Here, the specificities are divided into 3 categories: (1) substrate specificity, in which subtle changes in the structure of the substrates have large effects on the catalysis; (2) product specificity, in which the products of the catalyzed reactions are highly selective; and (3) D,L specificity, in which enantiomeric recognition is made by α-cyclodextrin.

4.1. Substrate specificity

The most striking specificity by cyclodextrins with respect to the substrate is found in the hydrolyses of phenyl acetates. As shown in Table 4, the magnitudes of the acceleration by α-cyclodextrin (k_{cat}/K_m) for *meta*-substituted compounds are 29-, 236-, 88-, and 13-fold larger than those for *para*-substituted compounds, for methyl, *tert*-butyl, nitro, and carboxyl substitution, respectively. Similar results are obtained also for β-cyclodextrin.

The larger magnitude of the acceleration of the cleavage of *meta*-substituted

TABLE 4
Catalytic rate constants and acceleration in the α-cyclodextrin-catalyzed hydrolyses of phenyl acetates [a,b]

Acetate	k_{cat} (10^{-2}/sec)	k_{cat}/k_{un}	K_d (10^{-2} M)
Phenyl	2.19	27	2.2
m-Tolyl	6.58	95	1.7
p-Tolyl	0.22	3.3	1.1
m-*tert*-Butylphenyl	12.9	260	0.2
p-*tert*-Butylphenyl	0.067	1.1	0.65
m-Nitrophenyl	42.5	300	1.9
p-Nitrophenyl	2.43	3.4	1.2
m-Carboxyphenyl	5.55	68	10.5
p-Carboxyphenyl	0.67	5.3	15.0

[a] Reprinted with permission from VanEtten, R.L., Sebastian, J.F., Clowes, G.A. and Bender, M.L. (1967) J. Am. Chem. Soc. 89, 3242. Copyright by the American Chemical Society.

[b] pH 10.6, carbonate buffer, $I = 0.2$ M: 1.5% (v/v) acetonitrile–water.

phenyl acetates (than the value for the cleavage of the corresponding *para*-compound) by cyclodextrins is due to a smaller activation enthalpy [17]. As shown in Fig. 8, the magnitude of acceleration increases with a decrease of the activation enthalpy, whereas the activation entropy term partly compensates this effect. (Comparison of $\Delta\Delta H^{\ddagger}$ and $\Delta\Delta S^{\ddagger}$ instead of $\Delta H^{\ddagger}$ and $\Delta S^{\ddagger}$ was made to normalize the values of different leaving groups.)

The origin of the above *meta–para* specificity is precisely investigated by determining the conformations of the inclusion complexes by [^{1}H]NMR spectroscopy as described in the previous section [9].

Table 5 shows the distances between the carbonyl carbon atoms, the electrophilic centers, and the ring comprised of the 6 O-2 atoms of α-cyclodextrin, the nucleophilic centers, estimated from the time-averaged conformations in Fig. 3, as well as the magnitude of the acceleration by α-cyclodextrin. Both values are determined in a 1 : 1 mixture of water and dimethylsulfoxide.

As clearly seen in Table 5, the magnitude of the acceleration by α-cyclodextrin is governed by the distance between the nucleophile (3-OH group) and the electrophile (carbonyl carbon atom). The order of the increase in the acceleration (*m*-nitrophenyl acetate ≫ phenyl acetate > *p*-nitrophenyl acetate) is identical with that of the decrease of the distance between the carbonyl carbon atom of the substrate and the O-3 atom of the cyclodextrin.

The strong dependence of the acceleration on the distance is associated with a

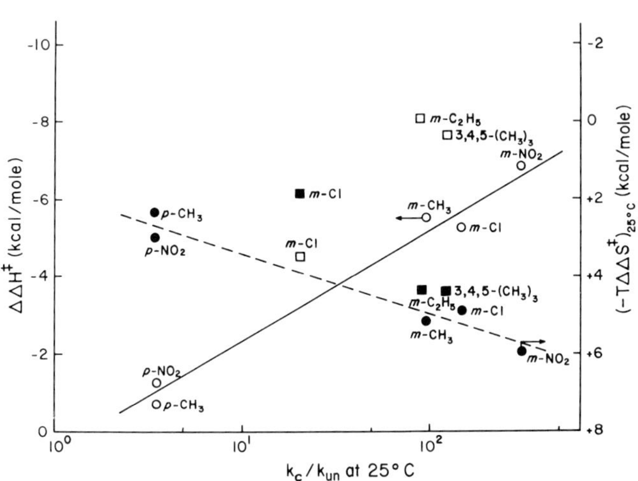

Fig. 8. Relations between the acceleration by cyclodextrins and the activation terms in the cyclodextrin-accelerated cleavages of phenyl acetates at 25°C: $\Delta\Delta H^{\ddagger}$ and $\Delta\Delta S^{\ddagger}$ are the differences between the values for the cyclodextrin reactions and those for the corresponding alkaline hydrolyses; the round points refer to α-cyclodextrin and the square points refer to β-cyclodextrin. From Komiyama, M. and Bender, M.L. (1978) J. Am. Chem. Soc. 100, 4576. Reproduced by permission of the American Chemical Society, copyright owners.

TABLE 5 [a]
Distances between the carbonyl carbon atoms of the phenyl acetates and the O-2 atoms of α-cyclodextrin and the magnitudes of the acceleration of the cleavages of the phenyl acetates by α-cyclodextrin

Substrate	Distance (Å) [b]	Magnitude of acceleration [c]
p – Nitrophenyl acetate	6.0	4.4
Phenyl acetate	5.8	14
m-Nitrophenyl acetate	3.4	220

[a] From Komiyama, M. and Hirai, H. (1980) Chem. Lett., 1471.

[b] The distances between the carbonyl carbon atoms of the substrates and the ring comprised of the 6 O-2 atoms of α-cyclodextrin are determined from the time-averaged conformations (Fig. 3) in the 1 : 1 (v/v) mixture of 1 N deuterium chloride and dimethyl sulfoxide-d_6.

[c] The ratios of the rate constants of the cleavages of the substrates incorporated in the α-cyclodextrin inclusion complexes to those in the absence of α-cyclodextrin are determined kinetically in the 1 : 1 (v/v) mixture of pH 8.5 buffer and dimethyl sulfoxide.

change in conformation of the inclusion complex in going from the initial state to the transition state. A conformational change makes access of the carbonyl carbon atom to the O-3 atom possible so that a nucleophilic reaction can result in the formation of the tetrahedral intermediate. The conformational change should accompany an increase in enthalpy, since the driving force of the formation of the inclusion complex is equivalent to a decrease of enthalpy. Thus, the conformational change partially compensates for a decrease in activation entropy coming from loss of the translational and rotational entropy due to complex formation between α-cyclodextrin and the substrate prior to chemical transformation. When no conformational change takes place during reaction, the loss of entropy by complex formation of the substrate with the catalyst shows up as a decrease of activation enthalpy, because of structural changes of the water molecules around the substrate and the catalyst. An inclusion complex with a smaller distance, which requires a smaller conformational change, shows a larger acceleration because of the large magnitude of the decrease of the activation enthalpy. This argument is supported by the thermodynamic study shown in Fig. 8.

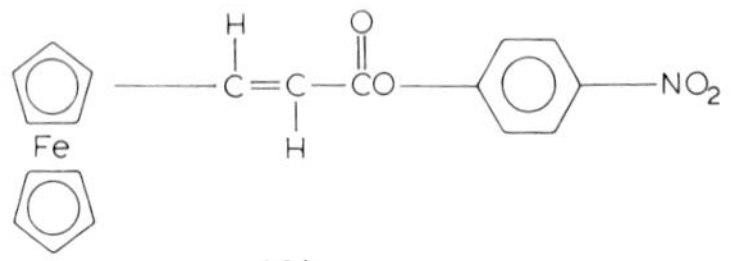

H O
C=C—CO— —NO_2
Fe H

(9)

This argument is further supported by the very fast acylation of β-cyclodextrin by bound *p*-nitrophenyl ferrocene acrylate (9) in dimethylsulfoxide–water. (60/40 v/v) [18]. In the β-cyclodextrin–(9) complex the acyl group of the ester rather than the leaving group is bound to the cyclodextrin. The acylation of β-cyclodextrin by this ester is accelerated by 750000-fold and the rate achieved is comparable with that for acylation of the enzyme chymotrypsin by *p*-nitrophenyl acetate. The much larger acceleration of the cleavage of (9) by β-cyclodextrin than that of *m*–*t*-butylphenyl acetate by β-cyclodextrin (260-fold) is explained in terms of structural changes of the complex in going from the initial state to the transition state. A molecular model study suggests that the *p*-nitrophenyl ferrocene acrylate–cyclodextrin complex can go to the tetrahedral intermediate for the acylation of the cyclodextrin with full retention of the optimum binding geometry in the cyclodextrin cavity, whereas this is not the case for the reaction of *m*–*t*-butylphenyl acetate (the aromatic ring has to be pulled somewhat out of the cavity to achieve the tetrahedral-like transition state). The importance of the structure of the substrates (and thus of the structure of the inclusion complex) in catalyses by cyclodextrin is vividly shown in hydrolyses of *p*-nitrophenyl esters involving an adamantyl group in the acyl portion of the ester. Thus, hydrolysis of (10a) is retarded 28-fold by β-cyclodextrin [19], in contrast with which a 3-fold acceleration by β-cyclodextrin in hydrolysis of (10b) is seen [20]. Furthermore, hydrolysis of (10c) is accelerated 2150-fold by β-cyclodextrin [21]. In

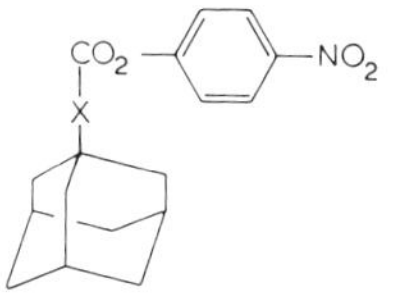

CO_2 — C_6H_4 — NO_2; X; adamantyl

(10a) X = none; (10b) X = —CH_2—; (10c) X = —C≡C—

the inclusion complexes of these substrates, the adamantyl portion is located inside the cavity because of its apolar character. Thus, the positions of the electrophilic centers (the carbonyl carbon atoms of the substrates) in the cavity are governed by X in (10), resulting in considerable specificity.

4.2. Product specificity

Cyclodextrins exhibit remarkable *ortho–para* selectivity in the chlorination of aromatic compounds by hypochlorous acid (HOCl) [22] (Scheme 5). Chlorination takes place via formation of a covalent intermediate, a hypochlorite ester of cyclodextrin. In the chlorination of anisole by hypochlorous acid, *para*-chlorination occurs almost exclusively in the presence of sufficient cyclodextrin, although in control experiments maltose had no effect on the product ratio. For example, selectivity for *para*-chlorination in the presence of 9.4×10^{-3} M α-cyclodextrin is 96%, which is much larger than that in the absence of α-cyclodextrin (60%). In the proposed mechanism, one of the secondary hydroxyl groups reacts with HOCl to form a hypochlorite ester, which attacks the sterically favorable *para* position of the anisole molecule included in the cyclodextrin cavity in an intracomplex reaction. The participation of one of the secondary hydroxyl groups at the C-3 position in the catalysis was shown by the fact that dodecamethyl-α-cyclodextrin, in which all the primary hydroxyl groups and all the secondary hydroxyl groups at the C-2 positions are methylated, exhibited equal or larger *ortho–para* specificity than native α-cyclodextrin [23].

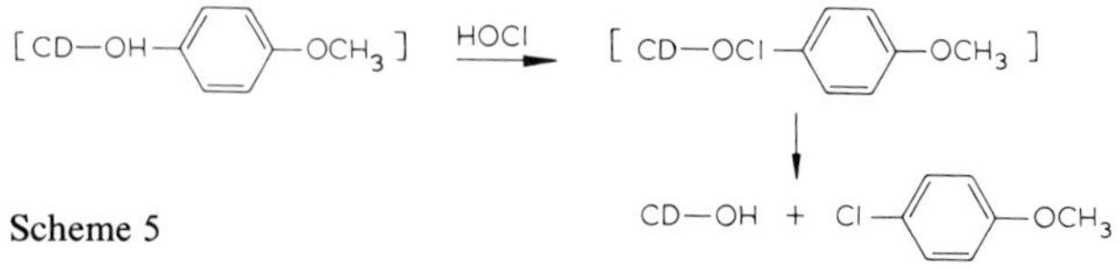

Scheme 5

β-Cyclodextrin catalyzes the electrophilic allylation of 2-methylhydronaphthoquinone with allyl bromide in aqueous medium, giving the vitamin K_1 or K_2 analog (11) in excellent yield (Scheme 6) [24]. The yield of (11) in the presence of 0.01 M β-cyclodextrin is 43%. The allylation reaction in the absence of β-cyclodextrin, however, gave the K_1 or K_2 analog in poor yield contaminated with considerable amounts of undesired by-products. The catalytic effect of the cyclodextrin is attributed both to the increase in nucleophilicity of the carbon atom on the naphthohydroquinone monoanion in the cyclodextrin cavity and to the protection against oxidative cleavage of included naphthoquinone derivatives.

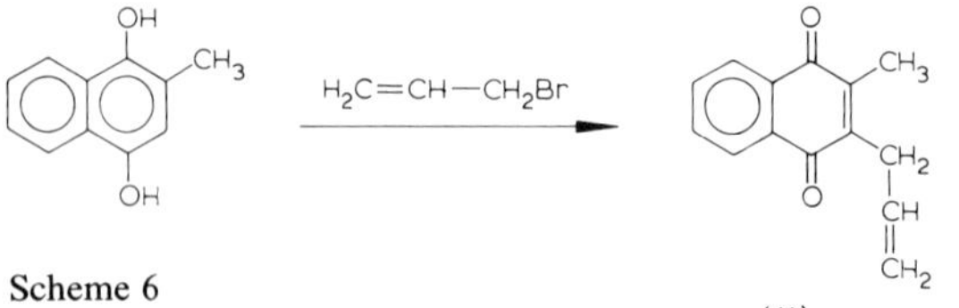

Scheme 6

Furthermore, β-cyclodextrin shows superb selectivity in the syntheses of 2.5-cyclohexadienones (12), which are important starting materials for the syntheses of physiologically active compounds, from *p*-substituted phenols, chloroform, and sodium hydroxide (Scheme 7) [25]. The selectivity of the production of 12 in the presence of β-cyclodextrin is virtually 100%, in contrast to the formation of large amounts (about 4–8 times as large as 12) of *ortho*-formulated compound 13 in its absence. The remarkable selectivity in the present reaction is probably due to the formation of dichlorocarbene from chloroform and hydroxide ion in the cavity of β-cyclodextrin. The *para*-substituted phenol should approach the cavity (and thus the carbene) from the side involving the *para*-carbon, resulting in a selective reaction. The penetration of this apolar side in the apolar cavity should be more favorable than the penetration of the polar side by the phenoxide atom.

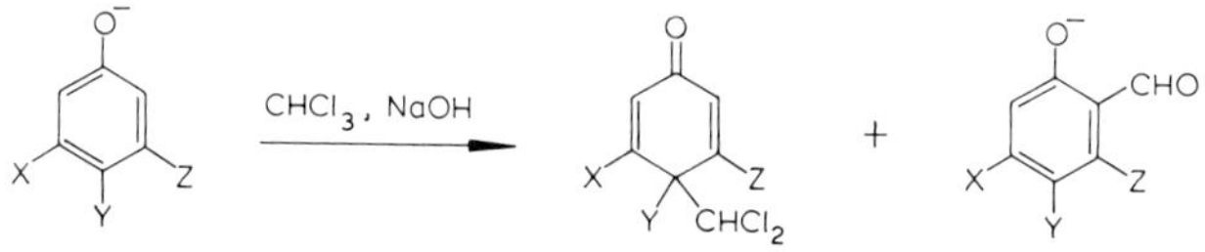

(12a) $Y=CH_3$, $X=Z=H$ (12b) $Y=C_6H_5$, $X=Z=H$ (12c) $Y=CH_3$, $X=Z=H$

Scheme 7

4.3. D,L *Selectivity*

The first observation of asymmetric specificity of cyclodextrins in catalysis was in the hydrolyses of ethyl mandelates; however, the asymmetric effect was quite small [26]. A much larger asymmetric specificity was observed in the cyclodextrin-accelerated cleavage of 3-carboxy-2,2,5,5-tetramethylpyrrolidin-1-oxy-*m*-nitrophenyl ester (14a), a substrate having an asymmetric carbon atom (like the mandelates) adjacent to the carbonyl group of the hydrolytically labile ester [27]. The hydrolysis of (14a) in the presence of cyclodextrin proceeds in the same way as that described for phenyl esters (Scheme 1); i.e. binding of the substrate with cyclodextrin using the phenyl portion of the substrate, acylation of the cyclodextrin, and deacylation of the acyl-cyclodextrin. The cleavage of racemic (14a) in the presence of α-cyclodextrin is biphasic with a fast first step followed by a slower second step. Because of the different rates at which the two complexed enantiomers are cleaved, the first and second phases, respectively, correspond to the cleavages of the (+) and (−) enantiomers. This assignment is based on the fact that both the catalytic rate constant, k_2, and the dissociation constant of the complex, K_d, determined from the dependence of the rates of the first phase on the cyclodextrin concentration, are equal to those determined by using the optically pure (+) enantiomer. The k_2 for the (+) enantiomer is 6.9 times larger than that for the (−) enantiomer. However, the K_d values are almost equal for the two enantiomers. Interestingly, in contrast to α-cyclodextrin, β-cyclodextrin exhibited no appreciable enantiomeric specificity. The loss of D,L specificity upon increasing the size of the cyclodextrin cavity indicates that D,L specificity is a function of the tightness of binding. The magnitude of enantiomeric specificity of k_2 shown by α-cyclodextrin is close to that shown by α-chymotrypsin (9.1 times) in the acylation step of the hydrolysis of the closely related ester, 3-carboxy-2,2,5,5-tetramethylpyrrolidinyl-1-oxy-*p*-nitrophenyl ester (14b).

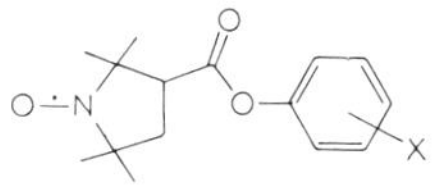

(14a) X = *m*-NO_2
(14b) X = *p*-NO_2

Cyclodextrins, however, show no enantiomeric specificity in the deacylation step. The rate of hydrolysis of acyl-α-cyclodextrin derived from the (+) enantiomer of (14a) is equal to that derived from the (−) enantiomer. This can be associated with the fact that the nitroxide function is not included in the cyclodextrin cavity, as shown by electron spin resonance. In the α-chymotrypsin-catalyzed hydrolysis of (14b), however, the acyl enzyme derived from the (+) enantiomer hydrolyzes 21 times faster than the acyl enzyme derived from the (−) enantiomer.

A much larger D,L specificity is exhibited by cyclodextrin in the cleavage of chiral organophosphates such as isopropyl methylphosphonofluoridate (Sarin) (15) [28] and isopropyl *p*-nitrophenyl methylphosphonate (16) [29]. The α-cyclodextrin-accelerated cleavages of these organophosphates proceed through nucleophilic attack of a secondary hydroxyl group of the cyclodextrin on the phosphorus atom, resulting in a phosphonylated α-cyclodextrin and hydrogen fluoride or *p*-nitrophenol.

$H_3C-P(=O)(F)-OCH(CH_3)_2$ (15)

$H_3C-P(=O)(OCH(CH_3)_2)-O-C_6H_4-NO_2$ (16)

In the α-cyclodextrin accelerated cleavage of (15) the catalytic rate constant for the (*R*) (−) enantiomer is 35.6 times larger than that for the (*S*) (+) enantiomer. This difference arises from the stereospecificity of the inclusion complexes, since the (*S*) (+) enantiomer, which is less accelerated, binds to α-cyclodextrin more strongly than the (*R*) (−) enantiomer, which is more accelerated. It was proposed that the stereospecific interaction(s) of the included enantiomers with the hydroxyl groups at the asymmetric C-2 atom and/or the asymmetric C-3 atom of the cyclodextrin in the inclusion complexes govern this asymmetric catalysis.

5. *Catalysis by modified cyclodextrins*

As shown previously, cyclodextrins in their native forms show many features characteristic of enzymes: (1) specificity; (2) the formation of catalyst–substrate complexes prior to chemical transformation; and (3) large accelerations.

However, cyclodextrins as enzyme models suffer from a shortcoming, namely their only catalytic group is the hydroxyl group, which restricts the scope of their applicability. Thus, many attempts have been made to introduce other catalytic functional groups into cyclodextrins. In these modified cyclodextrins, the introduced groups function as catalytic sites and the cavities of cyclodextrins serve as the binding sites for the substrates.

Table 6 lists the functional groups that have been introduced and their positions in modified cyclodextrins as well as their catalytic functions.

5.1. Models of hydrolytic enzymes

The first attempt by Dietrich and Cramer [43] at improvement of cyclodextrin catalysis of ester hydrolysis by the introduction of an imidazolyl group to cyclodextrin resulted in only a slight rate enhancement over hydrolysis by a combination of cyclodextrin and imidazole. The small effect of their modification can be attributed to the fact that the imidazolyl groups preferentially substituted the primary hydroxyl groups at the C-6 atoms of cyclodextrin rather than the secondary hydroxyl groups at the C-2 or C-3 atoms, which have been shown to be effective in catalysis. In this structure, the added imidazolyl groups cannot exhibit cooperativity with the secondary hydroxyl groups of cyclodextrin. However, a modified α-cyclodextrin (17 in Table 6), which has an histamine group on the secondary hydroxyl side, accelerates the hydrolysis of *p*-nitrophenyl acetate 80 times more than α-cyclodextrin itself and 6.3 times more than a mixture of α-cyclodextrin and histamine [30]. This modified α-cyclodextrin, which has a catalytic site (containing both the

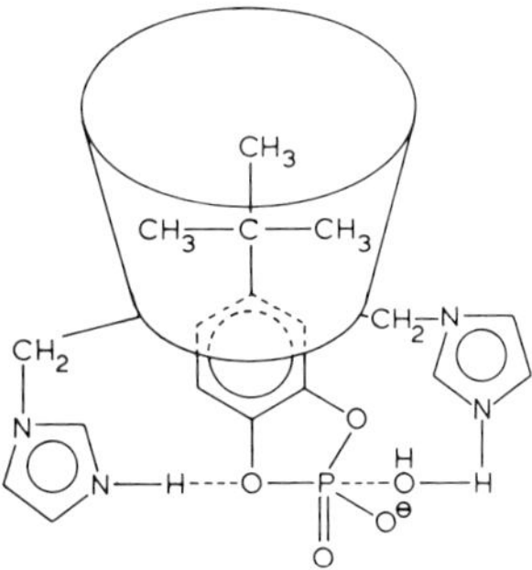

Fig. 9. Proposed mechanism of the (18)-catalyzed hydrolysis of a cyclic phosphate.

imidazolyl and hydroxyl groups) and a binding site (the hydrophobic cavity of cyclodextrin) as in serine proteases, is a better enzyme model than α-cyclodextrin itself, since it shows rate acceleration around neutrality.

Introduction of two imidazolyl groups on the primary side (18 in Table 6) makes cyclodextrin a good model of the enzyme, ribonuclease, since the two imidazolyl groups show cooperativity both as a general acid catalyst (in its acidic form) and as a general base catalyst (in its neutral form) [37]. The pH-rate constant profile of the hydrolysis of a cyclic phosphate ester has a bell-shape, showing that the catalytically most active form is monoprotonated, as expected for a mechanism with bifunctional acid–base catalysis. Similar profiles have been observed with the enzyme, ribonuclease, with pH-rate maxima near pH 7. The catalytic rate constant by (18) is much larger than that by cyclodextrin containing only one imidazolyl group. Fig. 9 shows the possible cooperativity of two imidazolyl groups in the (18)-catalyzed hydrolysis of a cyclic phosphate.

An interesting cooperativity is found in the hydrolysis of a phenyl ester by the α-cyclodextrin-2-benzimidazoleacetic acid (19) system [44]. This non-covalently modified system was used as a probe of the "charge-relay" system exhibited by the enzymes α-chymotrypsin and subtilisin. This system, like these enzymes, contains imidazolyl and carboxyl groups of 19 in addition to the hydroxyl groups of α-cyclodextrin. (19), which has both the imidazolyl and carboxyl groups in the same molecule, accelerates ester cleavage in the presence of α-cyclodextrin. On the other hand, neither benzimidazole (which has only an imidazolyl group) nor 2-naphthaleneacetic acid (which has only a carboxyl group) exhibits measurable acceleration. (19) does not form a complex with α-cyclodextrin. The pH-rate constant profile shows that catalysis by the α-cyclodextrin–(19) system involves the combination of a carboxylate ion, a neutral imidazolyl group, and the alkoxide ion. Probably,

O
HO C
N NH

(19)

TABLE 6
Modified cyclodextrins and their catalytic functions

Modification on the secondary side			Modification on the primary side		
Group R_1	Catalytic reaction	Ref.	Group R_2	Catalytic reaction	Ref.
$-NH(CH_2)_2-$(imidazol-4-yl) (17)	b	30	$(-CH_2-N$(imidazolyl)$NH)_2$ (18)	b	37
$-O-C(=O)-$(pyridine-2-carboxylato)···Ni···(pyridine-2-aldoxime, $CH=N-O^-$) (24)	b	31	$(-CH_2-N$(imidazolyl)$NH)_2$ (22)	Hydration of carbon dioxide	38
	b and keto-enol rearrangement	32	$(-NH(CH_2)_2-$(imidazol-4-yl)$)_2$ (23)		
$-O-P(=O)(O^-)-O^-$	Iodine oxidation	33	$-CH_2SCH_2-$(pyridoxamine: CH_2NH_2, OH, CH_3) (25)	Transamination	39

$-N$, H, $CONH_2$ (28)	Reduction	34			
			$(-CH_2-N(CH_3)-CHO)_6$ (20)	b	40
$-OCH_2C(=O)-N(OH)-(CH_2)_2N(CH_3)_2$	b	35			
			$(-CH_2-N(C_2H_5)-CHO)_6$ (21)		40
$-OCH_2C(=O)-N(OH)-CH_2-$(imidazol-4-yl, N, NH)	b	35			
			$-N(CH_3)_3$	b	41
			$-CH_2-SCH_3$	b	42
$-OCH_2C(=O)-N(OH)-CH_3$	b	36			
			$-CH_2-S(CH_3)_3$	b	42

a, R_1 R_2 b, ester hydrolyses.

nucleophilic attack by the imidazolyl group on the included phenyl ester is assisted by the carboxylate and alkoxide ions. Thus the mechanism is apparently different from that shown by the "charge-relay" system in serine proteases. Serine proteases exhibit two proton transfers involving the carboxylate ion, the imidazolyl group, and the alkoxide ion. Serine proteases show nucleophilic attack by alkoxide ion whereas the cyclodextrin process shows nucleophilic attack by imidazole.

In addition to modification by the introduction of catalytic groups, the introduction of certain (non-catalytic) groups on the cyclodextrin can make it a better enzyme model through improvement in the binding process rather than in the catalytic process. Compounds (20) and (21), which have 7 *N*-formyl groups at the bottom of the β-cyclodextrin torus (the primary alcohol group side), exhibit 10–20 times larger catalytic rate constants (k_{cat}) in cleavages of phenyl esters than native β-cyclodextrin does [40]. However, the modified cyclodextrins have equal or smaller binding constants with phenyl esters than native cyclodextrin. Molecular models show that the alkyl groups in (20) and (21) cluster close to the bottom of the cavity, forming an apolar floor on the cavity, and thus make the original 7 Å depth of the hydrophobic cavity of β-cyclodextrin shallower than the value (3.7 Å) for (20) and that (2.5 Å) for (21) (see Fig. 10). The larger catalytic rate constants of (20) and (21) are probably due to better binding of the substrates in the cavity for the nucleophilic attack of the secondary alkoxide ion at the carbonyl carbon atom of the substrate. The weaker binding of modified cyclodextrins can be attributed to too shallow a cavity for the substrates to be sufficiently included.

5.2. Model of carbonic anhydrase

Another enzyme that has been successfully mimicked with a cyclodextrin is carbonic anhydrase. To do this a compound that would bind CO_2, complex a zinc ion, and in addition, function as a general base was needed. To accomplish all of this bis(*N*-histamino) β-cyclodextrin (22) and bis(*N*-imidazolyl) β-cyclodextrin (23) were devised.

Complexes of these substituted cyclodextrins with zinc ion showed larger rates of hydration of carbon dioxide than the compounds without cyclodextrin although the absolute catalytic rate constants are lower than their enzymatic counterparts by many orders of magnitude, especially the second compound [38].

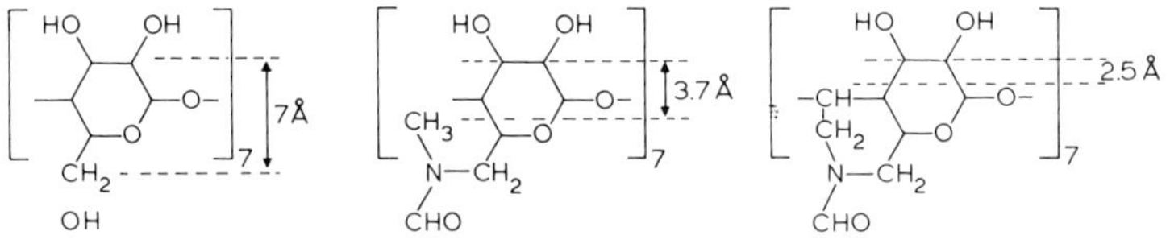

native β-cyclodextrin　　20　　21

Fig. 10. Depths of the cavities of modified β-cyclodextrins (20) and (21) as well as the native one, estimated by molecular model studies. From Breslow, R. et al. (1980) J. Am. Chem. Soc. 102, 762, reprinted by permission.

5.3. Model of metalloenzymes

Although metal ions have large catalytic effects, such effects are limited to substrates which can bind metal ions. Thus, it was expected that the combination of catalytic ability of a metal ion and the binding ability of a cyclodextrin would produce an excellent catalyst which mimics a metalloenzyme.

Using this hypothesis, modified α-cyclodextrin (24 in Table 6) was synthesized [31]. The observed rate constant for the hydrolysis of *p*-nitrophenyl acetate refering to the acetylation of the pyridine-carboxaldoxime (PCA) moiety in (24), is 4 times larger than the value for an equivalent concentration of the system without the cyclodextrin (PCA-Ni^{2+}), corresponding to a rate acceleration of greater than 10^3 over the uncatalyzed rate. The larger catalytic effect of (24) than the PCA-Ni^{2+} system is attributable to the binding of the substrate by the cyclodextrin moiety of (24). This is supported by the fact that 8-acetoxy-5-quinolinesulfonate, which does not fit in the cyclodextrin cavity, is only 57% as reactive toward (24) as toward PCA-Ni^{2+}. The smaller difference than expected is associated with the freezing of several degrees of rotational freedom on approach of the PCA oxygen atom (the catalytic site) to the carbonyl carbon atom of the bound substrate in the inclusion complex.

5.4. Introduction of a coenzyme moiety

Some of the analogs of coenzymes are introduced to cyclodextrin, where cyclodextrin mimics an apoenzyme.

A modified cyclodextrin (25), which contains a pyridoxamine moiety, catalyzes the transamination as shown in Scheme 8 [39]. The reaction between (25) and indolepyruvic acid (26) is 200 times faster than the reaction between pyridoxamine and (26), since the first reaction is an intracomplex one and the second is an intermolecular one. Transamination is completed by the reaction of (27), formed by the reaction between (25) and (26), with another amino acid, regenerating (25). Thus (25) which consists of a cyclodextrin and a pyridoxamine is a model of the enzyme transaminase.

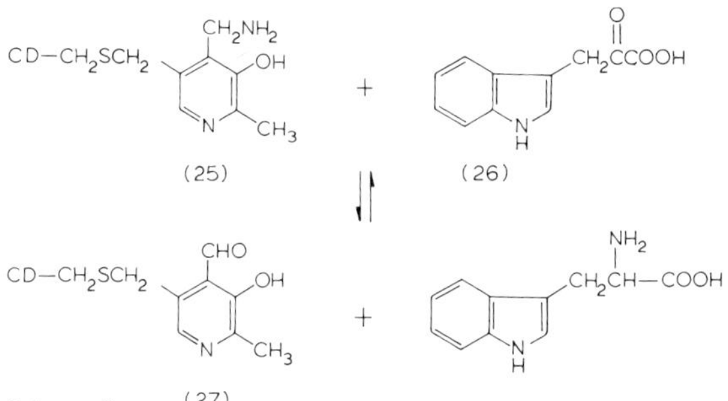

CD–CH_2SCH_2 (pyridine ring with CH_2NH_2, OH, CH_3) (25) + (indole with CH_2CCOOH, C=O) (26)

⇌

CD–CH_2SCH_2 (pyridine ring with CHO, OH, CH_3) (27) + (indole with $CH_2CH(NH_2)$–COOH)

Scheme 8

Reduction of ninhydrin by (28) (Table 6) is much faster than that by dihydronicotinamide. This difference is also attributable to inclusion complex formation between ninhydrin and the cyclodextrin moiety of (28) [34].

A visual pigment, rhodopsin, is mimicked by (29) [45]. (29) has an absorption maximum at 375 nm in its neutral form. However, the maximum shows a red shift of 100 ~ 120 nm on protonation of both the nitrogen atoms at low pH. The absorption spectrum of the diprotonated (29) is almost identical with that of native rhodopsin. Probably the retinal moiety of (29) is included in the cavity of cyclodextrin at low pH, resulting in the red shift due to a combination of an electrostatic effect and a microsolvent effect.

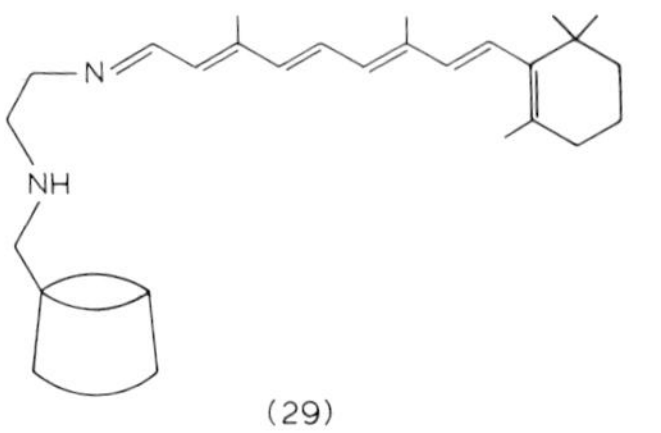

N

NH

(29)

6. Conclusion

Recently marked progress was made in cyclodextrin chemistry. In this review we dealt with only the materials relevant to the use of cyclodextrins as enzyme models since it is beyond the scope of this review to cover all the papers in this field. During this decade, many X-ray analyses of cyclodextrin inclusion complexes were made, giving much important information. Readers who are interested in them are recommended to refer to the book on cyclodextrin chemistry [1].

Acknowledgements

We thank Prof. Hidefumi Hirai at the University of Tokyo for his valuable comments in the preparation of this manuscript. M.L. Bender also wishes to thank the National Science Foundation (U.S.A.) for their continued support of this research.

References

1 Bender, M.L. and Komiyama, M. (1978) Cyclodextrin Chemistry, Springer, Heidelberg, 1978.
2 James, W.J., French, D. and Rundle, D.E. (1959) Acta Cryst. 12, 385.
3 Komiyama, M. and Bender, M.L. (1978) J. Am. Chem. Soc. 100, 2259.
4 Siegel, B. and Breslow, R. (1975) J. Am. Chem. Soc. 97, 6869.
5 Tabushi, I., Shimokawa, K., Shimizu, N., Shirakata, H. and Fujita, K. (1976) J. Am. Chem. Soc. 98, 7855.
6 Ueno, A., Yoshimura, H., Saka, R. and Osa, T. (1979) J. Am. Chem. Soc. 101, 2779.
7 Komiyama, M. and Hirai, H. (1980) Chem. Lett., 1467.

8 Komiyama, M. and Hirai, H. (1981) Polym. J. 13, 171.
9 Komiyama, M. and Hirai, H. (1980) Chem. Lett., 1471.
10 Johnson, C.E. and Bovey, F.A. (1958) J. Chem. Phys. 29, 1012.
11 Komiyama, M. and Hirai, H. (1981) Bull. Chem. Soc. Jpn. 54, 828.
12 Komiyama, M. and Bender, M.L. (1980) Bull. Chem. Soc. Jpn. 53, 1073.
13 Komiyama, M. and Inoue, S. (1979) Chem. Lett., 1101; (1980) Bull. Chem. Soc. Jpn. 53, 3334.
14 Komiyama, M. and Bender, M.L. (1977) J. Am. Chem. Soc. 99, 8021.
15 Straub, T.S. and Bender, M.L. (1972) J. Am. Chem. Soc. 94, 8875.
16 Griffiths, D.W. and Bender, M.L. (1973) J. Am. Chem. Soc. 95, 1679.
17 Komiyama, M. and Bender, M.L. (1978) J. Am. Chem. Soc. 100, 4576.
18 Breslow, R., Doherty, J.B., Guillot, G. and Lipsey , C. (1978) J. Am. Chem. Soc. 100, 3227.
19 Komiyama, M. and Inoue, S. (1980) Bull. Chem. Soc. Jpn. 53, 2330.
20 Komiyama, M. and Inoue, S. (1980) Bull. Chem. Soc. Jpn. 53, 3266.
21 Breslow, R., Czarñiecki, M.F., Emert, J. and Hamaguchi, H. (1980) J. Am. Chem. Soc. 102, 762.
22 Breslow, R. and Campbell, P. (1969) J. Am. Chem. Soc. 91, 3085.
23 Breslow, R., Kohn, H. and Siegel, B. (1976) Tetrahedron Lett., 1645.
24 Tabushi, I., Yamamura, K., Fujita, K. and Kawakubo, H. (1979) J. Am. Chem. Soc. 101, 1019.
25 Komiyama, M. and Hirai, H. (1981) Makromol. Chem., Rapid Commun. 2, 177.
26 Cramer, F. and Dietsche, W. (1959) Chem. Ber. 92, 1739.
27 Flohr, K., Paton, R.M. and Kaiser, E.T. (1975) J. Am. Chem. Soc. 97, 1209.
28 Van Hooidonk, C. and Braebaart-Hansen, J.C.A.E. (1970) Rec. Trav. Chim. Pays-Bas 89, 289.
29 Van Hooidonk, C. and Groos, C.C. (1970) Rec. Trav. Chim. Pays-Bas 89, 845.
30 Iwakura, I., Uno, K., Toda, F., Onozuka, S., Hattori, K. and Bender, M.L. (1975) J. Am. Chem. Soc. 97, 4432.
31 Breslow, R., Fairweather, R. and Keana, J. (1967) J. Am. Chem. Soc. 89, 2135.
32 Siegel, B., Pinter, A. and Breslow, R. (1977) J. Am. Chem. Soc. 99, 2309.
33 Eiki, T. and Tagaki, W. (1980) Chem. Lett., 1063.
34 Kojima, M., Toda, F. and Hattori, K. (1980) Tetrahedron Lett., 2721.
35 Kitaura, Y. and Bender, M.L. (1975) Bioorg. Chem. 4, 237.
36 Van Hooidonk, C., De Korte, D.C. and Reuland-Meereboer, M.A.C. (1977) Rec. Trav. Chim. Pays-Bas 96, 25.
37 Breslow, R., Doherty, J.B., Guillot, G. and Lipsey, C. (1978) J. Am. Chem. Soc. 100, 3227.
38 Tabushi, I., Kuroda, Y. and Mochizuki, A. (1980) J. Am. Chem. Soc. 102, 1152.
39 Breslow, R., Hammond, M. and Lauer, M. (1980) J. Am. Chem. Soc. 102, 421.
40 Emert, J. and Breslow, R. (1975) J. Am. Chem. Soc. 97, 670.
41 Matsui, Y. and Okimoto, A. (1978) Bull. Chem. Soc. Jpn. 51, 3030.
42 Fujita, K., Shinoda, A. and Imoto, T. (1980) J. Am. Chem. Soc. 102, 1161.
43 Dietrich, H.V. and Cramer, F. (1954) Chem. Ber. 87, 806.
44 Komiyama, M., Breaux, E.J. and Bender, M.L. (1977) Bioorg. Chem. 6, 127.
45 Tabushi, I., Kuroda, Y. and Shimokawa, K. (1979) J. Am. Chem. Soc. 101, 4759.
46 Bergeron, R.J. and Burton, P.S. (1982) J. Am. Chem. Soc. 104, 3664.
47 Inozuka, K.S., Kojima, M., Hattori, K. and Toda, F. (1980) Bull. Chem. Soc. Jpn. 53, 3221.

CHAPTER 15

Crown ethers as enzyme models

J. FRASER STODDART

Department of Chemistry, The University, Sheffield S3 7HF, Great Britain

1. Introduction

Enzymes can display formidable catalytic activities alongside daunting reaction stereoselectivities on substrates they choose to bind with varying structural specificities [1]. These are attributes which provide [2–6] an exciting challenge to the contemporary chemist to try and imitate with relatively low molecular weight synthetic molecular receptors. Aside from the quest for a better understanding of enzyme reaction mechanisms, the properties of enzymes which could be readily compromised or even forsaken [7] by the chemist include their high molecular weights, their constitutional complexities, their many functional groups, their numerous chiral centres, their limited solubilities in solvents other than water, and their lack of stability to reaction conditions outside of rather narrow pH, ionic strength and temperature ranges. The successful design and realisation of synthetic molecular receptors not only provides supramolecular systems [3] capable of functioning as co-receptors and co-carriers on account of their substrate binding, recognition, and transport properties but also permits the elaboration of highly efficient molecular co-catalysts capable of mimicing the actions of enzymes or fulfilling roles as new abiotic chemical reagents [3–6]. It is convenient to refer to the former co-catalysts as *enzyme mimics* and to the latter co-catalysts as *enzyme analogues*. Whereas with an enzyme mimic, the chemist attempts to imitate the known catalytic properties of a particular enzyme, with an enzyme analogue he is expressing his belief and ability to create abstractly a catalyst to catalyse a chosen chemical reaction. Although these goals are being advocated [2–8] and pursued [2–6] by a few, not all subscribe [9] to their relevance or indeed to their merits. Whatever the outcome of this somewhat futile and largely emotional debate, there is little doubt that synthetic molecular receptors — which have also been referred [3] to as supermolecules or mesomolecules — in the molecular weight range of 500–5000 have fascinating properties and deserve considerable attention. Already, research on them has given rise to much new chemistry beyond the molecule where the nature of the intermolecular non-covalent bond assumes an importance in solution which has hitherto been recognised mainly for biologically important macromolecules in the solution and solid states and just occasionally for smaller molecules, particularly in the solid state. There is little doubt that the steric and electronic aspects of the *intermolecular conformational analysis* of synthetic molecular receptors and their substrates is going to develop

Michael I. Page (Ed.), The Chemistry of Enzyme Action

dramatically as a discipline in chemistry during the next few years. Already, space-filling molecular models have played an important part in the early development of this new science and, as time passes and knowledge acrues, the modelling will inevitably become much more sophisticated [10,11]. Also, as the art of conventional organic synthesis is developed and perfected, the design of synthetic molecular receptors will become increasingly more ambitious. It would hardly be an understatement to claim that a relatively new area of chemistry which straddles the sub-disciplines of the subject in an immensely challenging manner for the active research investigator, is emerging.

This article sets out to highlight briefly some of the contributions that have been made by synthetic molecular receptors of the crown ether type to the development of enzyme mimics and analogues. However, before this aspect involving essentially binding at the transition state is reviewed, it is necessary to survey the ground-state phenomenon of complexation and decomplexation which characterises the first and last steps in an enzyme-catalysed reaction.

2. *Ground-state binding and recognition*

Clearly, a complementary steric and electronic relationship between a substrate species and a synthetic molecular receptor is desirable [3–6,10] if strong and highly ordered complex formation is to occur. Since the potential binding sites in the smaller substrate are necessarily going to be *divergent* [4,10], the larger molecular receptor has to be designed with matching *convergent* [4,10] binding sites. Although this kind of relationship is reminiscent [12] of the *lock* and *key* model which was proposed many years ago [13] to describe the interaction between an enzyme and its substrate, a more recent nomenclature system advocated by Cram and co-workers [4,10], that employs the terms *host* and *guest* respectively to refer to the crown ether molecular receptor and the substrate, has now gained popular acceptance.

2.1. *Binding forces*

In the context of crown ether hosts, non-covalent bonds of pole–pole, pole–dipole, and dipole–dipole types can all be employed [3–6] in the formation of host–guest complexes. Where the guest species is an alkali metal (i.e. Li^+, Na^+, K^+, Rb^+, Cs^+), alkaline earth metal (i.e. Mg^{2+}, Ca^{2+}, Sr^{2+}, Ba^{2+}), or harder transition or post-transition metal (e.g. Ag^+, Tl^+, Hg^{2+}, Pb^{2+}, La^{3+}, Ce^{3+}) cation [3–6,14], an electrostatic ($M^{n+} \cdots O$) pole–dipole interaction binds the guest to the host whilst the ($M^{n+} \cdots X^-$) pole–pole interaction with the counterion (X^-) is often retained. The features are exemplified by the X-ray crystal structure [15] shown in Fig. 1a for the 1:2 complex (1)·$(NaPF_6)_2$ formed between dibenzo-36-crown-12 (1) and $NaPF_6$. Molecular complexes involving metal cations have considerable strengths even in aqueous solution and a template effect involving the metal cation is often observed during the synthesis of crown ether derivatives.

In the binding of H_3O^+, NH_4^+, substituted ammonium (e.g. $MeNH_3^+$, $Me_3CNH_3^+$,

$PhCH_2NH_3^+$, $PhCHCO_2MeNH_3^+$, $PhCOCH_2NH_3^+$, $Me_2NH_2^+$, etc.), guanidinium, and imidazolium cations by crown ether hosts, hydrogen bonding of a (N^+–H ··· O) or (N^+–H ··· N) pole–dipole type to the heteroatoms of the ligands is sometimes

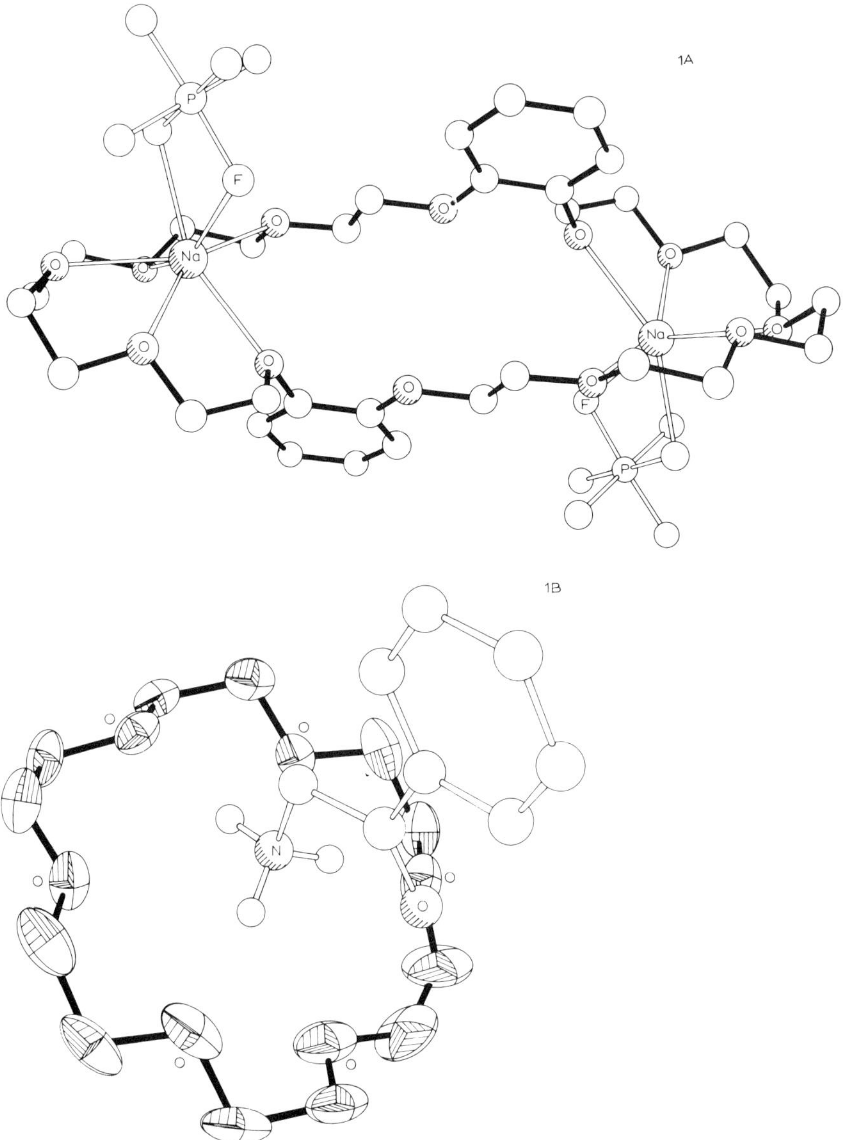

Fig. 1. Some X-ray crystal structures of complexes involving crown-ether receptor molecules. (1A), $[(1)\cdot(NaPF_6)_2]$. (1B), $[(2)\cdot(PhCOCH_2NH_3PF_6)]$. (1C), $[(3)\cdot(PhCH_2NH_3SCN)]$. (1D), $[(4)\cdot((NH_2)_2C{=}NHClO_4)]$. (1E), $[(5)\cdot(H_2O)]$. (1F), $[(2)\cdot(Me_2SO_2)_2]$. (1G), $[(6)\cdot((S){-}PhCHMeNH_3ClO_4)\cdot(H_2O)]$. (1H), $[(7)\cdot(Pt(bipy)(NH_3)_2(PF_6)_2\cdot(0.6H_2O))]$.

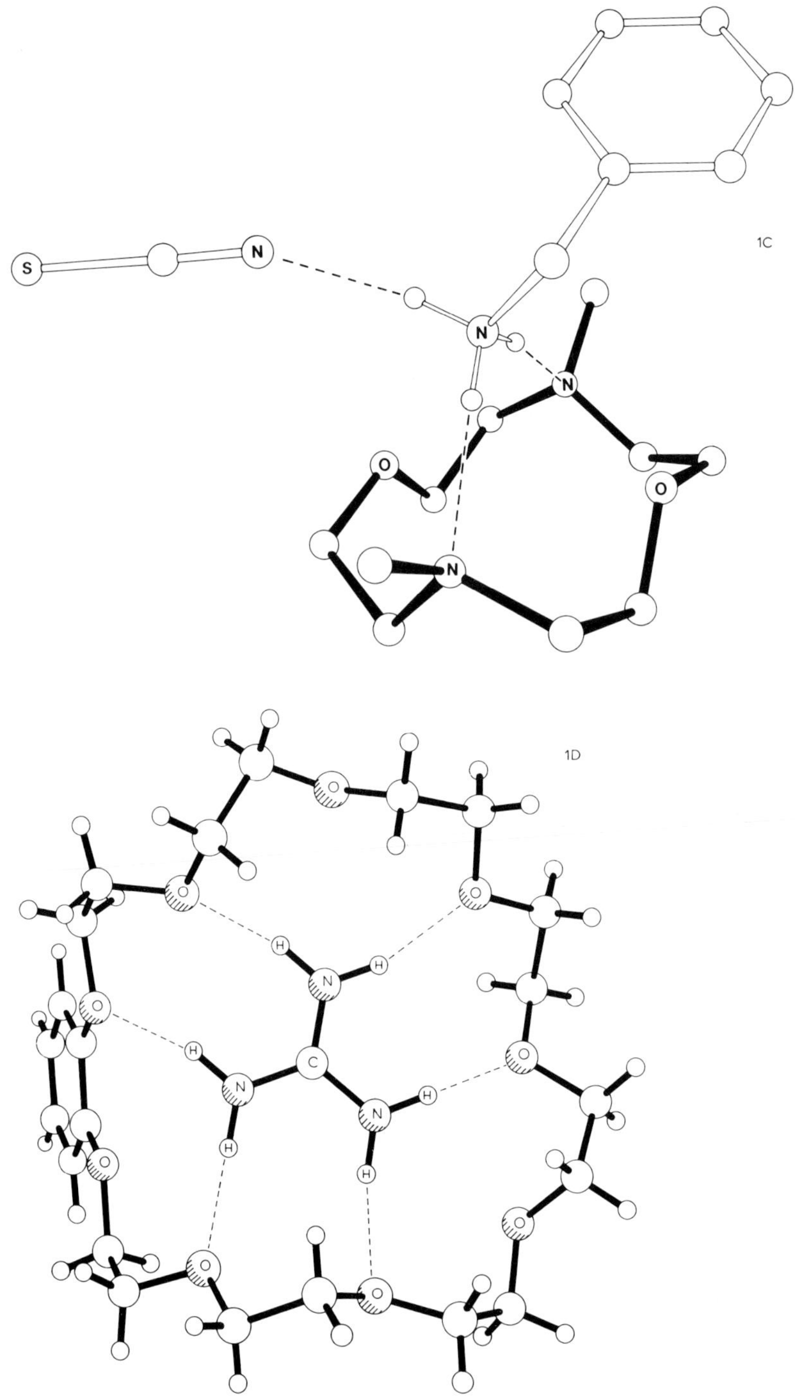
1C
S
N
N
N
N
O
O
1D
O
O
O
O
O
O
O
O
H
H
H
H
H
H
N
N
N
C

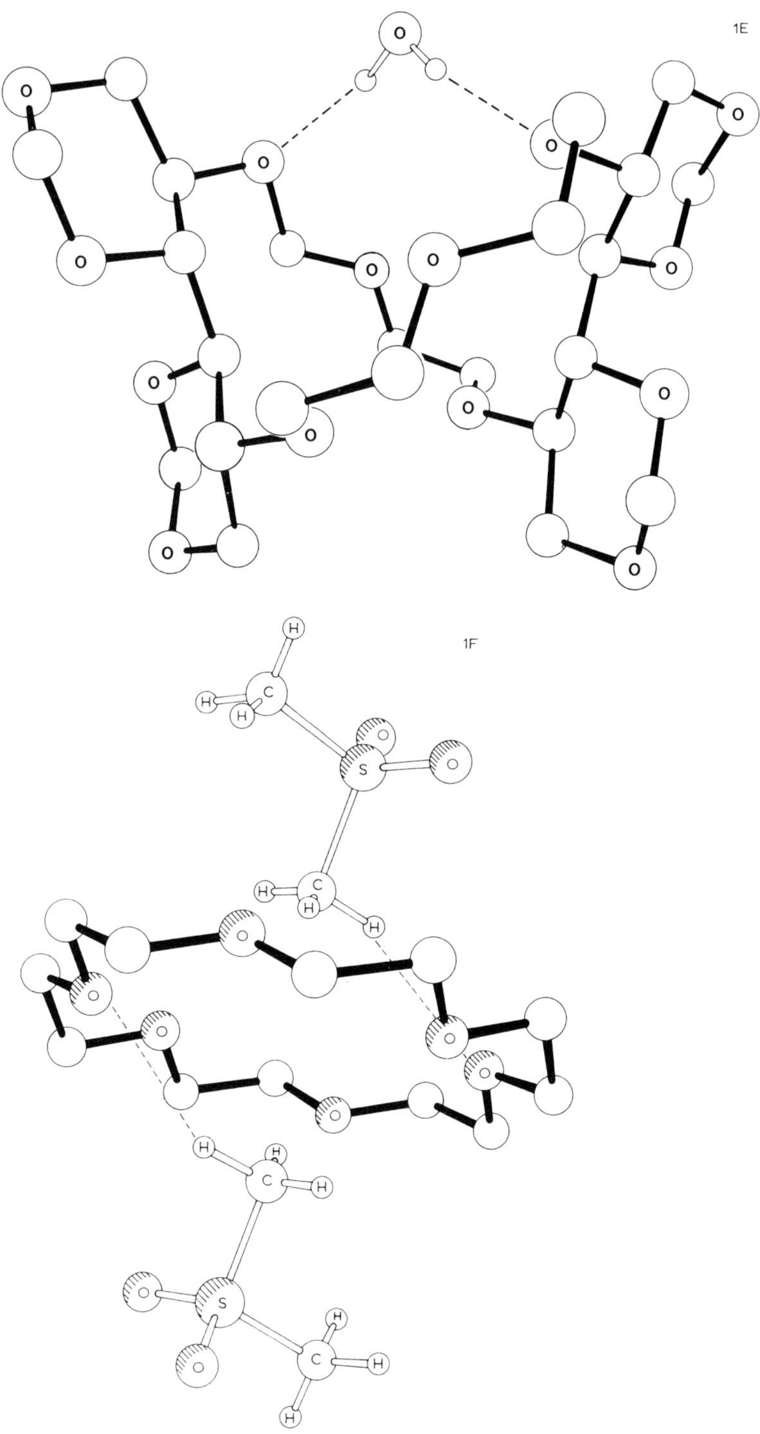
1E
1F

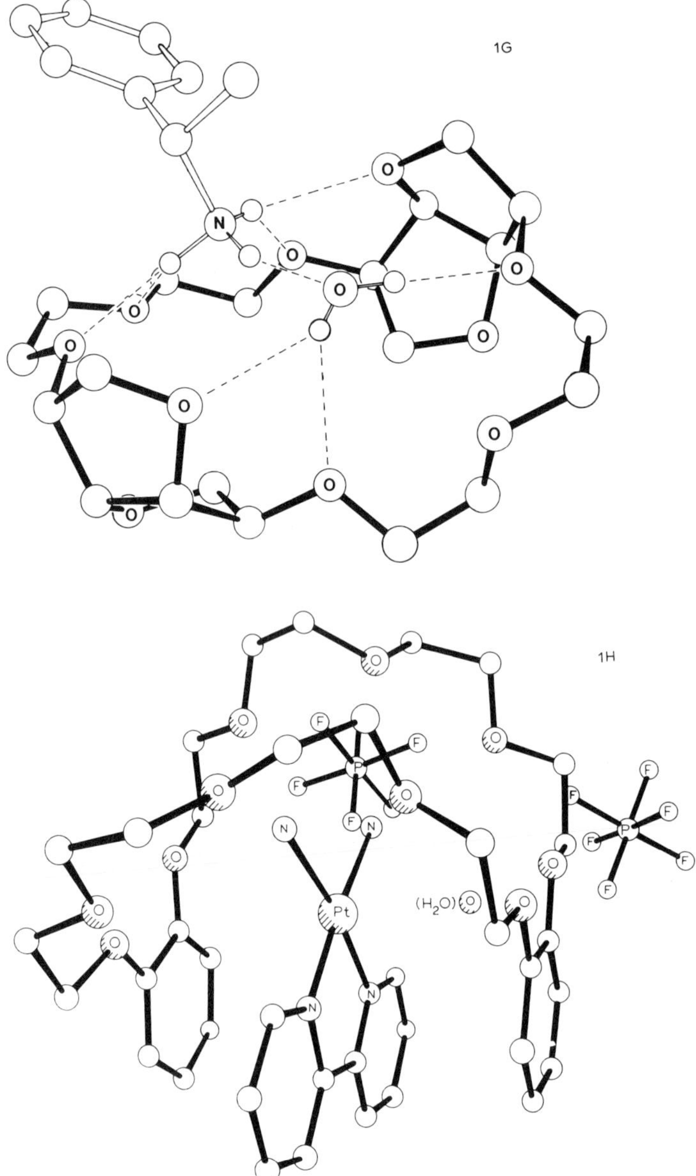

accompanied [3–6] by ($N^+–H \cdots X^-$) pole–pole hydrogen bonds to the counterion (X^-). The X-ray crystal structure of the 1 : 1 complex (2)·($PhCOCH_2NH_3PF_6$) formed between 18-crown-6 (2) and the $PhCOCH_2NH_3PF_6$ salt shown in Fig. 1b is illustrative of the typical [3–6] 3-point binding pattern characteristic of this class of host–guest complex where the counterion is not involved in ion pairing with the cationic complex, at least in the crystalline state. It should be noted that in this face-to-face type of complex, the 18-crown-6 host adopts the ubiquitous 'all-gauche' conformation [5] where the sequence of torsional angles (O–C, C–C, C–O) in the symmetrically independent portions of the ring read $ag^{+}aag^{-}a$. In this conformation with approximate D_{3d} symmetry, two assemblies of 3 oxygen atoms are oriented with their electron pairs directed towards opposite sides of the mean plane of the 18-membered ring. Replacement of oxygen atoms in crown ether hosts by either sp^2 or sp^3 hybridised nitrogen atoms, provides [4,5,17,18] even stronger (N^{+}–H ··· N) pole–dipole hydrogen bonds. Fig. 1c records an example of this stabilising interaction together with an (N^{+}H··· $\bar{N}$=C=S) ion pairing between the counterions in the X-ray crystal structure [19] of the 1 : 1 complex (3)·($PhCH_2NH_3SCN$) formed between *N*, *N*-dimethyl-1,7-diaza-4,10-dioxacyclododecane (3) and $PhCH_2NH_3SCN$. In the crystalline complexes shown in Fig. 1b and c, it is also possible that some electrostatic pole–dipole stabilisation is provided by the oxygen atoms in the host molecules not involved in hydrogen bonding. Moreover, it should be noted that although 12-crown-4 derivatives form stable complexes with both RNH_3^+ and $R_2NH_2^+$ cations, 18-crown-6 derivatives are highly selective in their binding towards RNH_3^+ cations. This observation has been exploited [20] rather elegantly (i) to effect acylation of secondary amines in the presence of complexed primary amines and (ii) to effect chemoselective acylation of secondary amino functions in diamines of the general type $RNH(CH_2)_nNH_2$ which also contain a primary amino group. Inspection of Corey–Pauling–Koltoun (CPK) space-filling molecular models has also led to the prediction [4,10] that arenediazonium (ArN_2^+) cations should be complexed by 18-crown-6 (2) and its derivatives by means of pole–dipole interactions between the $N\equiv N^{+}$ groups of the guest and the ethereal oxygen atoms of the host. Although spectroscopic evidence is available to support the existence of a 1 : 1 complex of this type in solution, no X-ray crystal structures are yet available. The situation is better, however, regarding the proposal [4,10], based again on CPK model examination, that 27-crown-9 and its derivatives should complex with the guanidinium ion utilising 6 ($N^{\delta+}$–H ··· O) hydrogen bonds and 3 ($N^{\delta+}$ ··· O) contacts. The original solution-spectroscopic evidence [10] for a 1 : 1 complex between benzo-27-crown-9 (4) and the guanidinium cation now has the reassuring support [21] of the X-ray crystal structure (see Fig. 1d) for (4)·($(NH_2)_2C{=}NHClO_4$). The imidazolium cation is also appreciably complexed with a 27-crown-9 hexacarboxylate [22] incorporating 3 L-tartaric acid residues. Two (N^{+}–H ··· O) hydrogen bonds involving oxygen atoms with a 1,4 or 1,5 relationship in the macrocyclic polyether ring of this host have been suggested [22] as a rationalisation for the binding of this biologically important cation.

Neutral molecules form very much weaker complexes with crown ethers usually as a result of (O–H ··· O), (N–H ··· O), and (C–H ··· O) hydrogen bonding of a dipole–dipole nature involving respectively OH-, NH-, and CH-acidic guest species [23]. Examples are provided in Fig. 1e and f respectively by the X-ray crystal structures [24,25] of (i) a 1 : 1 complex (5)·(H_2O) formed between a bisdi-*O*-methylene-D-mannitolo-22-crown-6 derivative (5) and water and (ii) a 1 : 2 complex (2)·[$(Me_2SO_2)_2$] formed between 18-crown-6 (2) and dimethyl sulphone. The former has

2 (O–H ··· O) and the latter 6 (C–H ··· O) hydrogen bonds. X-Ray crystallography has also revealed [26] that a water molecule is complexed (Fig. 1g) alongside an (*S*)-PhCHMeNH_3^+ cation in a 1 : 1 : 1 complex (6)·((*S*)-PhCHMeNH_3ClO_4)·(H_2O) and a bisdianhydro-D-mannitolo-30-crown-10 derivative (6). In all, 7 of the 10 oxygen atoms of the host are involved in hydrogen-bond formation. One of the NH_3^+ hydrogen atoms of the (*S*)-PhCHMeNH_3^+ cation is bonded to the oxygen atom of the H_2O molecule with both species residing side by side in the hydrophilic cavity of the host. The other two NH_3^+ hydrogen atoms and one of the H_2O hydrogen atoms are involved in bifurcated hydrogen bonds to 3 pairs of oxygen atoms leaving the other H_2O hydrogen atom to participate in hydrogen bonding to the 7th oxygen atom. It is believed that amines can also be bound alongside primary alkyl ammonium cations in this chiral host. Clearly, in the design of enzyme analogues, the capability to be able to bind two organic guests side by side in the same cavity of a chiral host is not unattractive [26].

Neutral and charged transition-metal complexes containing aqua, ammine, ethylenediamine, acetonitrile, etc. ligands readily form [27] 1 : 2 host–guest adducts with 18-crown-6 (2) and when the appropriate stereochemistry characterises the first sphere ligands, second sphere coordination can lead to face-to-face hydrogen-bonded chain copolymers with 1 : 1 stoichiometry. In the case of neutral guest molecules or cationic guests in which the positive charge is well dispersed, binding of 2 guests simultaneously to the opposite and homotopic faces of 18-crown-6 (2) can occur (cf. (2)·$(Me_2SO_2)_2$ in Fig. 1f), often with all 6 oxygen atoms in the host involved in hydrogen bonding. With RNH_3^+ cations, electrostatic repulsion between 2 guests associated with opposite faces of (2) renders this arrangement unstable [28]. The X-ray crystal structure [29] of $[Pt(bipy)(NH_3)_2 \cdot$ dibenzo-30-crown-$10]^{2+}[PF_6]_2^- \cdot$ $0.6H_2O$ [(7)·(Pt(bipy)$(NH_3)_2(PF_6)_2$·($0.6H_2O$))] portrayed in Fig. 1h reveals that, in addition to the expected (N–H ··· O) hydrogen-bond formation involving 3 of the 6 hydrogen atoms on the 2 *cis*-NH_3 ligands on square planar Pt(II) and 3 of the 10 oxygen atoms, there are π–π stabilising interactions between the 2 benzene rings in the host molecule (7) and 1 of the 2 pyridine rings of the 2,2′-bipyridyl ligand in the guest complex. This charge transfer interaction between a π-acidic and a π-basic component is an example of *secondary binding* [5] as distinct from *primary binding* which is associated directly with crown ether receptor sites. Other examples [3–5,10,30–32] of this kind of secondary binding are known. Examples [33–39] of other kinds of secondary interactions are also known. They vary considerably in strength. Electrostatic interactions of the pole–pole type between anionic groups (e.g. CO_2^-) in the side chains of hosts and cationic guests lead [3–6,22,31,33–35] to complexes with very high stabilities and well-defined structures. Much weaker interactions of a dipole-induced dipole type between RNH_3^+ cations containing polarisable phenyl groups with appropriate constitutional dispositions and dipolar acetal fragments (e.g. 1,3-dioxane rings in *O*-methylene [36] and *O*-benzylidene [37,38] derivatives and β-anomeric centres [39]) in host carbohydrate molecules have been detected [5] in CD_2Cl_2 solution at temperatures below 0°C by NMR spectroscopy. For example, the 1 : 1 complex [D-(8)·($PhCH_2NH_3ClO_4$)] between the di-*O*-

methylene-D-mannitolo-20-crown-6 derivative D-(8) and the $PhCH_2NH_3^+$ cation is stabilised [36] to the extent of ca. 1 kcal/mole by this dipole-induced dipole

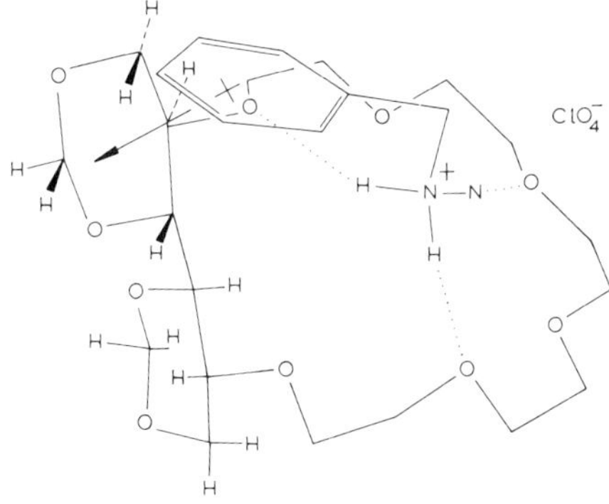

D-(8)·($PhCH_2NH_3ClO_4$)

interaction. Although dipole–dipole interactions, such as those between a carbonyl group and an electron pair on an amine, provide [4,5] structuring possibilities, no evidence for any appreciable binding between neutral amines and either oxo-12-crown-3 (9) or oxo-18-crown-6 (10) has been found [40]. However, these hosts do

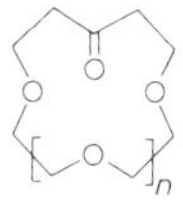

(9) $n = 1$
(10) $n = 3$

complex with RNH_3^+ cations. The future design of host molecules of the crown-ether type will most certainly revolve around the introduction of the appropriate secondary binding sites alongside suitably constituted primary binding sites, e.g. *N*,*N*′-dimethyl-1,7-diaza-12-crown-4 [19] (3) and *N*,*N*′,*N*″-trimethyl-1,7,13-triaza-

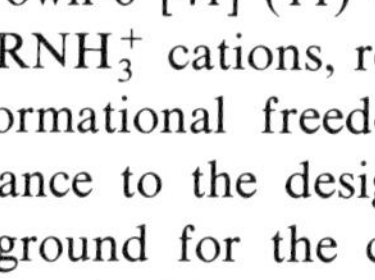

(11)

18-crown-6 [41] (11) derivatives for the highly selective complexation of $R_2NH_2^+$ and RNH_3^+ cations, respectively. The role of secondary interactions in reducing the conformational freedom available to a bound substrate molecule has obvious relevance to the design of enzyme models. In acquiring the necessary structural background for the construction of highly structured host–guest complexes, the happy coincidence between many solid- and solution-state structures is welcome. It is in this knowledge that complexation strengths and structures of complexes in solution can now be discussed and correlated.

2.2. Complexation

The organic cations which have been most widely investigated in relation to the strengths of their complexes in solution with crown ethers are the RNH_3^+ cations, particularly $MeNH_3^+$, $Me_3CNH_3^+$, $PhCHMeNH_3^+$ and $PhCH(CO_2Me)NH_3^+$. In one of the more convenient semi-quantitative methods of assessment, their salts (X^- = SCN^-, ClO_4^-, *Pic*rate$^-$, etc.) are distributed [42,43] at 25°C between H_2O and $CHCl_3$ in the absence and presence of the host in the chloroform layer. From the various distribution constants, which can be obtained spectroscopically, values of the association constants (K_a/M^{-1}) for the equilibrium,

$$RNH_3X + \text{Crown} \underset{CHCl_3}{\overset{K_a}{\rightleftharpoons}} RNH_3\,\text{Crown}\,X$$

can be calculated and hence the free energies of complexation (ΔG^0/kcal/mole) are obtained from the expression,

$$\Delta G^0(R, X^-) = -RT \ln K_a$$

The ΔG^0 values which are recorded under hosts (12)–(17) reveal that there is a

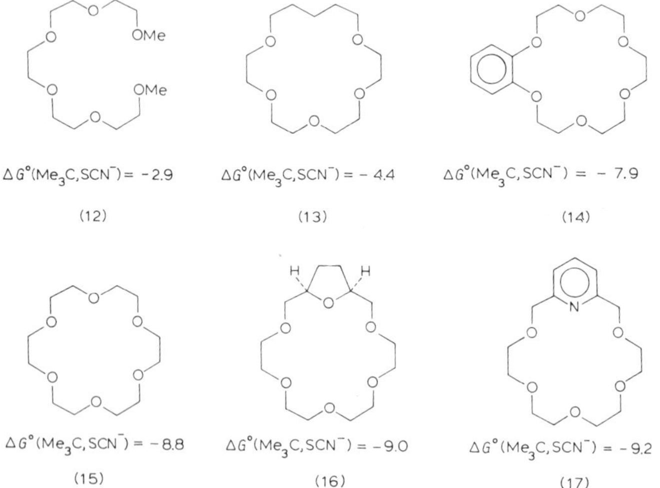

substantial dependence of complexation strengths upon the constitution of the host. Analysis of a range — including (13), (15), (16) and (17) — of 18-crown-6 derivatives containing pentamethylene, *m*-xylyl, *p*-phenylene, furan-2,5-dimethylyl, tetrahydrofuran-2,5-dimethylyl, and 2,6-pyridine-dimethylyl units in terms of identifiable host–guest parameters has revealed [42,43] the additive nature of the contributions to the ΔG^0 values from the different constitutional fragments in these hosts. The ΔG^0 (Me_3C, *Pic*$^-$) values (i) for the 5 possible diastereoisomers (18)–(22) of dicyclohexano-18-crown-6 and (ii) for the 6 diastereoisomeric D-glycosido-18-crown-6 derivatives α-D-(23)–α-D-(26) with the α- and *β-gluco*, α- and *β-galacto*, *α-manno* and

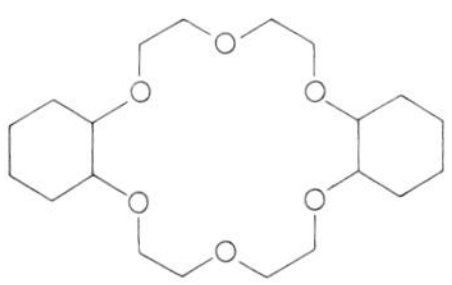

cis – cisoid – cis (18) $\Delta G^0(\mathrm{Me_3C}, Pic^-) = -7.6$
cis – transoid – cis (19) $\Delta G^0(\mathrm{Me_3C}, Pic^-) = -6.8$
trans – cisoid – trans (20) $\Delta G^0(\mathrm{Me_3C}, Pic^-) = -4.9$
trans – transoid – trans (21) $\Delta G^0(\mathrm{Me_3C}, Pic^-) = -4.3$
trans – cis (22) $\Delta G^0(\mathrm{Me_3C}, Pic^-) = -5.2$

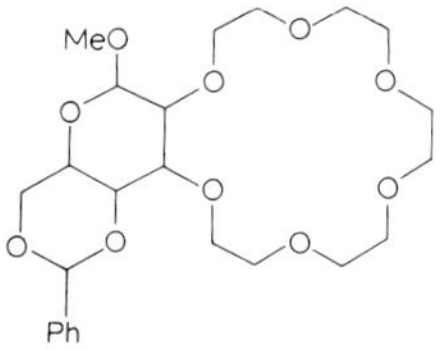

MeO

Ph

α-D-gluco α-D-(23) $\Delta G^0(\mathrm{Me_3C}, Pic^-) = -4.7$
β-D-gluco β-D-(23) $\Delta G^0(\mathrm{Me_3C}, Pic^-) = -4.1$
α-D-galacto α-D-(24) $\Delta G^0(\mathrm{Me_3C}, Pic^-) = -6.2$
β-D-galacto β-D-(24) $\Delta G^0(\mathrm{Me_3C}, Pic^-) = -5.9$
α-D-manno α-D-(25) $\Delta G^0(\mathrm{Me_3C}, Pic^-) = -5.9$
α-D-altro α-D-(26) $\Delta G^0(\mathrm{Me_3C}, Pic^-) = -3.1$

α-altro configurations also demonstrate [5,44] that configurational differences between hosts are highly significant in relation to their complex-forming potentials. The relative strengths of the 1 : 1 complexes are believed [45] to be associated with the relative ease with which these two groups of diastereoisomeric 18-crown-6 derivatives can attain the desirable 'all-*gauche*' conformation with averaged D_{3d} symmetry which characterises crystalline complexes (e.g. $[(2) \cdot (\mathrm{PhCOCH_2NH_3PF_6})]$ in Fig. 1b) between $\mathrm{RNH_3^+}$ cations and 18-crown-6 (2) and is also probably the preferred structure in solution [46]. In particular, the fact that *trans–transoid–trans*-dicyclohexano-18-crown-6 (21) is prevented, for configurational reasons, from adopting this conformation correlates with this host forming the weakest 1 : 1 complexes amongst the diastereoisomers (18)–(22). Other factors influence [3–6] complexation strengths besides the structure of the host. These include the temperature of the solution, the nature of the solvent, and the identities of the guest cation and counterion. Complexation strengths usually increase [42,43] with decreasing temperature. On the other hand, they decrease as solvent polarity is increased or as the ability of the counterion to compete with the host for hydrogen bonding to the guest cation is increased, i.e. complexation strengths decrease in the order $Pic^- > \mathrm{PF_6^-} > \mathrm{ClO_4^-} > \mathrm{SCN^-} > \mathrm{Cl^-}$ for $\mathrm{X^-}$. Differences ($\Delta\Delta G^0$/kcal/mole) in ΔG^0 values for the binding of $\mathrm{MeNH_3}Pic$ and $\mathrm{Me_3CNH_3}Pic$ provide [47,48] a useful probe of the steric inhibition by the host towards complexation of an $\mathrm{RNH_3^+}$ cation. The essentially planar naphtho-18-crown-6 derivative [48] (27) has a $\Delta\Delta G^0$ ($\mathrm{MeNH_3^+}$, $\mathrm{Me_3CNH_3^+}$) value of -0.6 kcal/mole whereas the sterically bulky (*RR*)-bisbinaphthyl-22-crown-6 hosts [30,35,48–51] (*RR*)-(28) and (*RR*)-(29) give values of -1.1 and -1.7 kcal/mole respectively for this parameter [48]. The fact that these two chiral host molecules show high structural recognition suggests that they could exhibit significant enantiomeric differentiation in binding guest cations of the type $\mathrm{LMSC^*NH_3^+}$

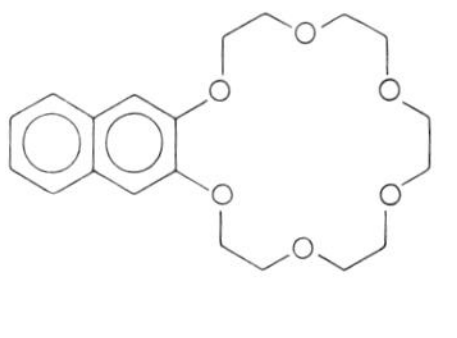

$\Delta G°(\mathrm{Me}, Pic^-) = -7.5$ $\Delta G°(\mathrm{Me_3C}, Pic^-) = -6.9$ $\Delta\Delta G°(\mathrm{MeNH_3^+}, \mathrm{Me_3CNH_3^+}) = -0.6$

(27)

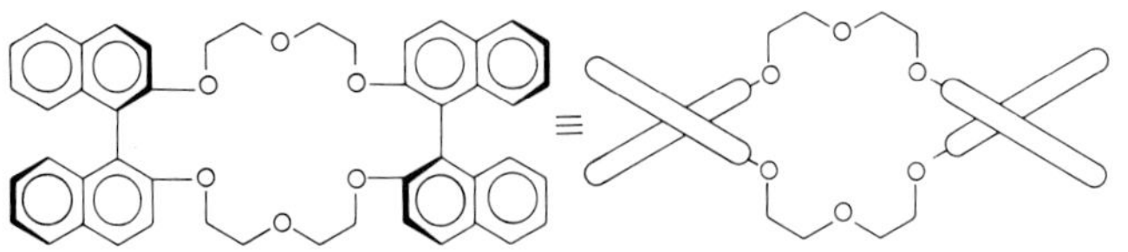

$\Delta G°(Me, Pic^-) = -3.8$ $\Delta G°(Me_3C, Pic^-) = -2.7$ $\Delta\Delta G°(MeNH_3^+, Me_3CNH_3^+) = -1.1$

(*RR*)-(28)

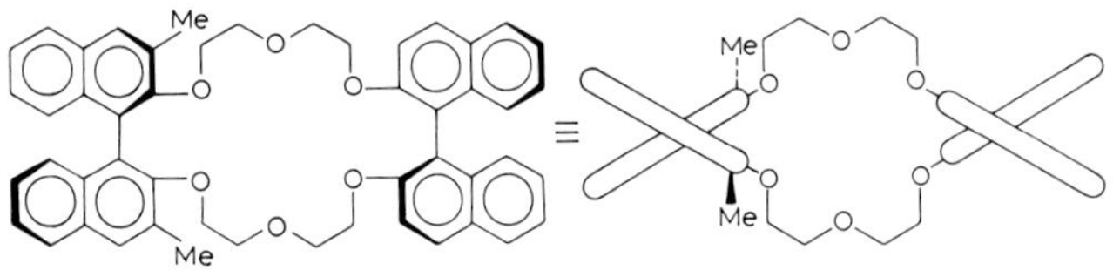

$\Delta G°(Me, Pic^-) = -4.4$ $\Delta G°(Me_3C, Pic^-) = -2.7$ $\Delta\Delta G°(MeNH_3^+, Me_3CNH_3^+) = -1.7$

(*RR*)-(29)

where L, M, and S are large, medium-sized, and small substituents respectively attached to a chiral centre (C*).

2.3. Enantiomeric differentiation

Equilibration experiments have been carried out to assess the degree of chiral recognition exhibited by (*RR*)-(28) and (*RR*)-(29) towards racemic amino acid ester salts and other substituted ammonium salts [4,30,51,52]. The optically pure hosts dissolved in $CHCl_3$ were equilibrated with H_2O solutions of the racemic salts at 0°C and the amounts of each enantiomer extracted into the $CHCl_3$ layers were estimated polarimetrically after isolation. An Enantiomeric Distribution Constant (EDC), defined as the ratio, D_A/D_B, where D_A and D_B are, respectively, the distribution constants between the two layers for the more and less complexed enantiomers, can then be calculated. The difference in the free energies of the diastereoisomeric complexes in the $CHCl_3$ layer is given by

$$\Delta(\Delta G^0) = -RT\ln(\mathrm{EDC})$$

Moreover, for chiral recognition to occur the counterion X^- in $LMSC^*NH_3X$ must be unable to hydrogen bond strongly with the NH_3^+ centre. This condition is fulfilled by ClO_4^- and PF_6^- ions and the host (*RR*)-(29) with the extended chiral barriers shows, as expected, somewhat better chiral recognition than does the host (*RR*)-(28). In the case of the guests $RCH(CO_2Me)NH_3PF_6$, the more complexed enantiomer has the D-configuration [51] and chiral recognition decreases as the R groups are changed in the order Ph $(-1.9) >$ p-HOC_6H_4 $(-1.4) >$ $Me_2CH \sim PhCH_2$ $(-0.87) >$ $MeSCH_2CH_2$ (-0.42) as indicated by the $\Delta(\Delta G^0)$ values given in parentheses. The proposed structure for the more stable (*RR*)-(29)-D-$RCH(CO_2Me)NH_3PF_6$ complex is shown in Fig. 2. Three-point binding of the NH_3^+ centre by hydrogen bond formation to alternately placed oxygen atoms on the macrocycle is accompanied by orientation at the chiral carbon atom of the R group (L) into the largest cavity of (*RR*)-(29) leaving the hydrogen atom (S) to rest against

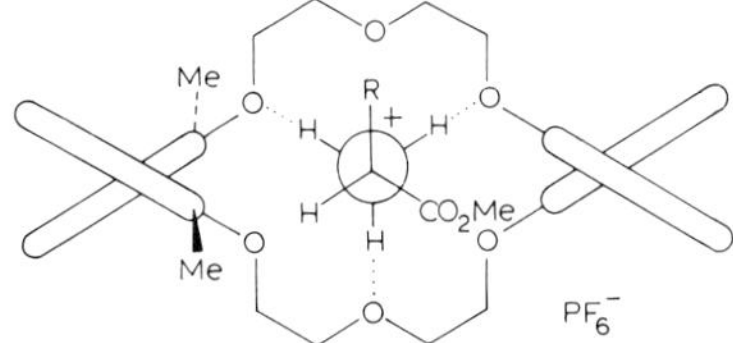

Fig. 2. The more stable (*RR*)-(29)-D-RCH(CO_2Me)NH_3PF_6 complex.

an extended chiral barrier and the CO_2Me group (M) to lie alongside the naphthalene wall such that a charge transfer interaction can occur. There is evidence for this secondary interaction in the *less stable* diastereoisomeric complex formed between (*SS*)-(28) and D-PhCH(CO_2Me)NH_3PF_6 which fortunately crystallised [30]. The X-ray structure [53] shown in Fig. 3 reveals the presence of a π-acid to π-base interaction between the CO_2Me group and a naphthalene ring as well as a secondary (C–H $\cdots$ O) hydrogen bond. Clearly, the electron withdrawing NH_3^+ and CO_2Me groups render the hydrogen atom bonded to the chiral centre sufficiently acidic to enter into hydrogen-bond formation. The chiral recognition exhibited by the optically pure hosts (28) and (29) has led to the design of an amino acid-resolving machine [54] and chromatographic systems in which (*RR*)-(28) and (*RR*)-(29) have been covalently attached to silica gel [55] and macroreticular polystyrene resin [56],

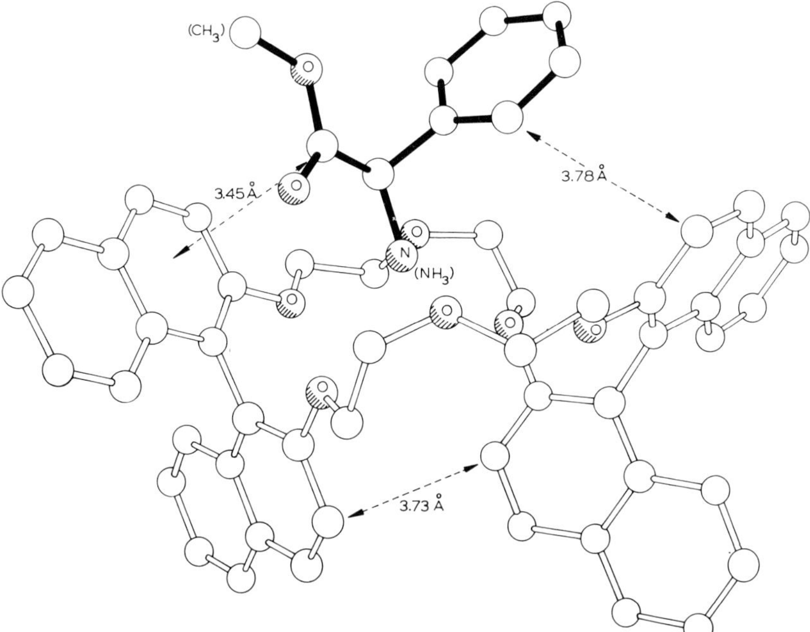

Fig. 3. The X-ray structure [53] of the (*SS*)-(28)-D-PhCH(CO_2Me)NH_3PF_6 complex.

respectively. Both chromatographic media have been employed to resolve racemic amino acid and ester salts analytically and in some cases preparatively. One disadvantage of the chiral host molecules (28) and (29) is their relatively weak binding towards both enantiomeric guest species: this dictates an upper limit to their ability to discriminate between enantiomers. Another disadvantage is the need to obtain host molecules (28) and (29) chiral starting from achiral precursors. Progress has been made in the direction of overcoming both of these problems. Initial optimism surrounding an apparent dramatic improvement [57] in chiral recognition by adding CH_3CN to aid the extraction of the racemic salts into the $CHCl_3$ layer was not supported by later experiments and chiral recognition was found [52] to decrease with increasing solvent polarity as expected. However, it has proved possible [52] to resolve (*RR*)(*SS*)-(29) into its enantiomers by crystallisation of the more stable diastereoisomeric complexes between firstly (*RR*)-(29) and D-$PhCH(CO_2H)NH_3ClO_4$ and then (*SS*)-(29) and L-$PhCH(CO_2H)NH_3ClO_4$. A similar chiral breeding cycle has also been devised [48] to effect the resolution of the substituted binaphthyl-20-crown-6 derivative (*RS*)-(30). The optically pure (*R*)-(30)

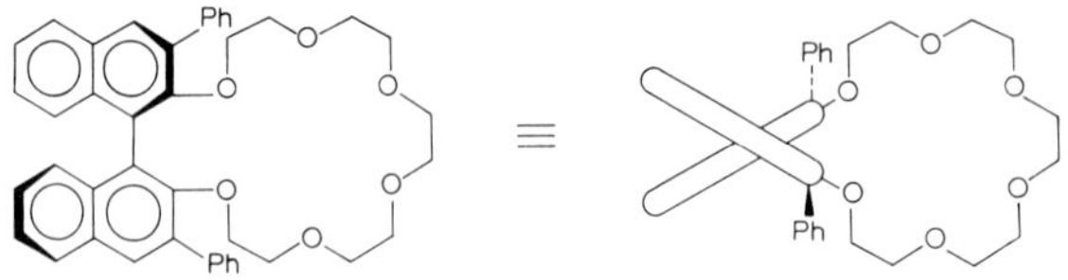

$\Delta G°(Me, Pic^-) = -6.2$ $\quad$ $\Delta G°(Me_3C, Pic^-) = -4.6$ $\quad$ $\Delta\Delta G°(MeNH_3^+, Me_3CNH_3^+) = -1.6$

(*R*)-(30)

and (*S*)-(30) hosts have shown [48] the highest chiral recognition yet towards D,L-$RCH(CO_2R^1)NH_3ClO_4$ ($R^1 = Me$ or H). The $\Delta(\Delta G^0)$ values for 6 amino acid salts and 6 amino acid ester salts ranged from -1.6 kcal/mole with R = Ph and $R^1 = H$ or Me to a low of -0.7 kcal/mole with $R = R^1 = Me$. The crystalline (*R*)-(30)-D-$PhCH(CO_2H)NH_3ClO_4$ and (*S*)-(30)-L-$PhCH(CO_2H)NH_3ClO_4$ complexes are also the more stable complexes in solution. The enhanced enantiomer-differentiating properties of optically pure (30) reflects in part the stronger complexing ability of this host compared with that of (29).

Modest chiral recognition in ground-state complexation has also been achieved [5,12,58] with crown ethers incorporating carbohydrate residues as their source of chirality. However, no attempt has been made to optimise their enantiomer-differentiating ability.

2.4. *Substrate recognition*

The ability of monocyclic crown-ether derivatives with steric barriers to discriminate [47,48] between $MeNH_3^+$ and $Me_3CNH_3^+$ cations has already been discussed. Recently the cylindrical doubly bridged macrotricycles [59,60,61] (31–40) and the triply bridged macrotetracycle [62] (41), based upon the *N*,*N*′-diaza-12-crown-4 and -15-crown-5, and the *N*,*N*′-diaza- and *N*,*N*′,*N*″-triaza-18-crown-6 units shown in Fig. 4, have been found to exhibit remarkable chain-length selectivity

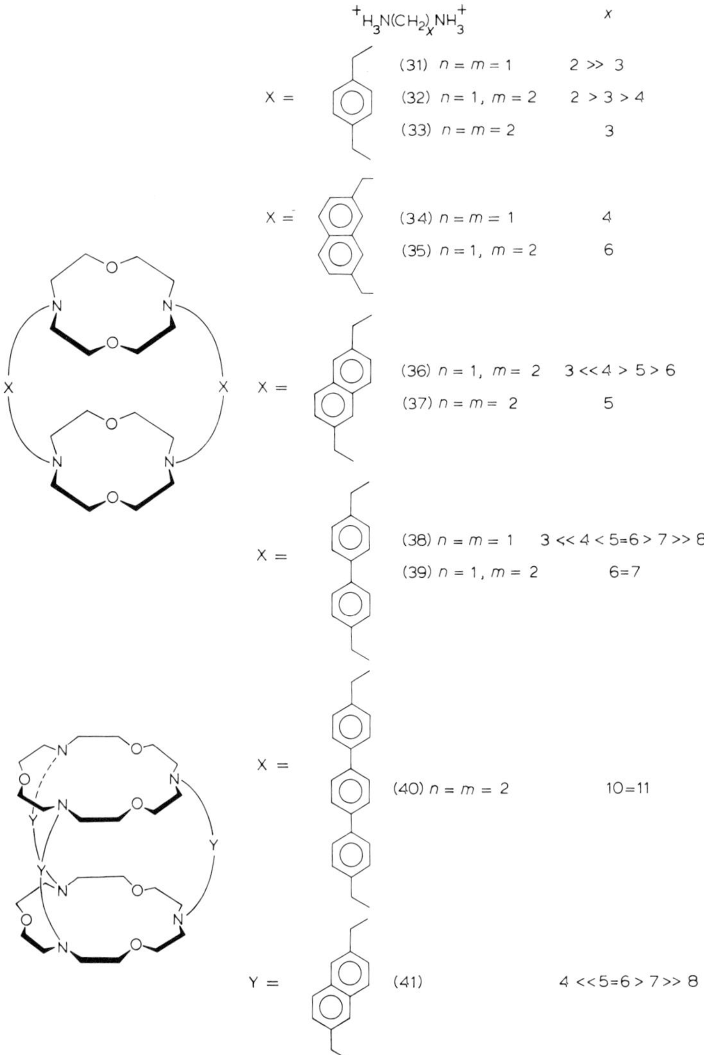

Fig. 4. The substrate recognition of the macropolycycles (31)–(41) for bis primary alkylammonium cations.

towards bis primary alkylammonium salts, $H_3N(CH_2)_xNH_32X$. Various NMR spectroscopic experiments involving (i) the mapping [59–62] of the ^1H and ^{13}C chemical shifts associated with the methylene groups in the guest cations, (ii) the determination [62] of ^{13}C relaxation times for 1 : 1 complexes and employing these to deduce correlation times and dynamic couplings between hosts and guests, and (iii) competition studies [59–62] on different guest dications for hosts at temperatures in the slow NMR site-exchange limits have been used to indicate when complementary relationships are being optimised. The results are summarised in Fig. 4. The success of this investigation is not only a consequence of good steric fits between bifunctional guests and hosts with two geometrically well-defined primary binding sites but also depends [17,59] on the highly stereoselective directing influence which characterises the 12- and 15-membered rings of (31), (32), (34), (35), (36) and (38) and the 18-membered rings of (41) and directs the guest dications *inside* the cavities of these hosts. Only a *syn* relationship between a guest cation and the *N* substituents attached to these macrocycles is tolerated [17,59]. Host asymmetry, as found in (36), can also be reflected [59] in different environments of ligands in a complexed guest which are identical in the free guest. This phenomenon has obvious analogies with the distinction that is often drawn by enzymes between enantiotopic ligands and faces on substrates.

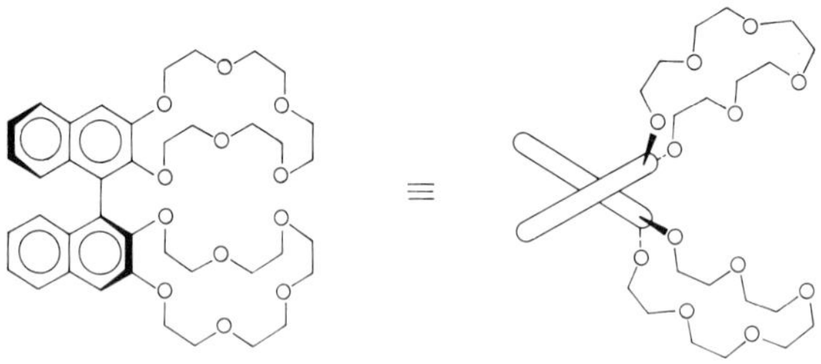

(*S*)- (42)

A number of other crown ether hosts with more than one primary binding site have been synthesised [17,63]. CPK molecular models of the chiral bismacrocycle (*S*)-(42) reveal that this host possesses a jaws-like receptor site in which the binaphthyl unit can serve as a hinge. The X-ray crystal structure [64] of the 1 : 1 complex ((*S*)-(42)-$H_3N(CH_2)_4NH_32PF_6$) shown in Fig. 5 possesses the expected structure. The host (*S*)-(42) has been used [5] to bind other bifunctional guests.

2.5. Allosteric effects

The regulation of enzyme action and function can be accomplished by allosteric effects. The binding of an effector at an allosteric site can induce conformational changes at an active site and alter the reactivity of an enzyme towards its substrate. This feature of enzymic catalysis should be capable of imitation in simpler synthetic systems.

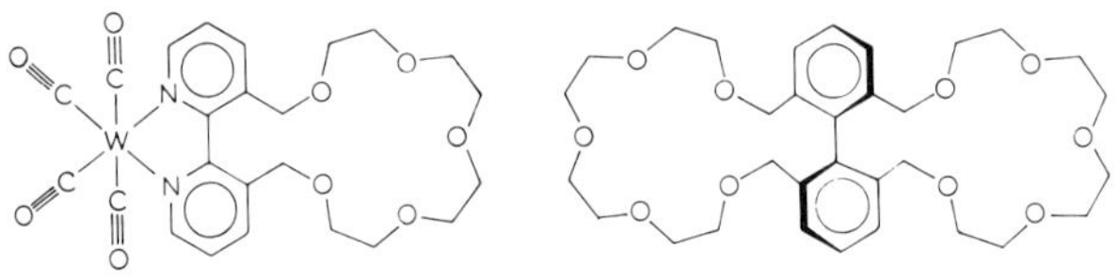

(43) (44)

Since the binding of appropriate cationic guests is sensitive to changes in conformation [45], it should be possible to model allosteric behaviour. This may

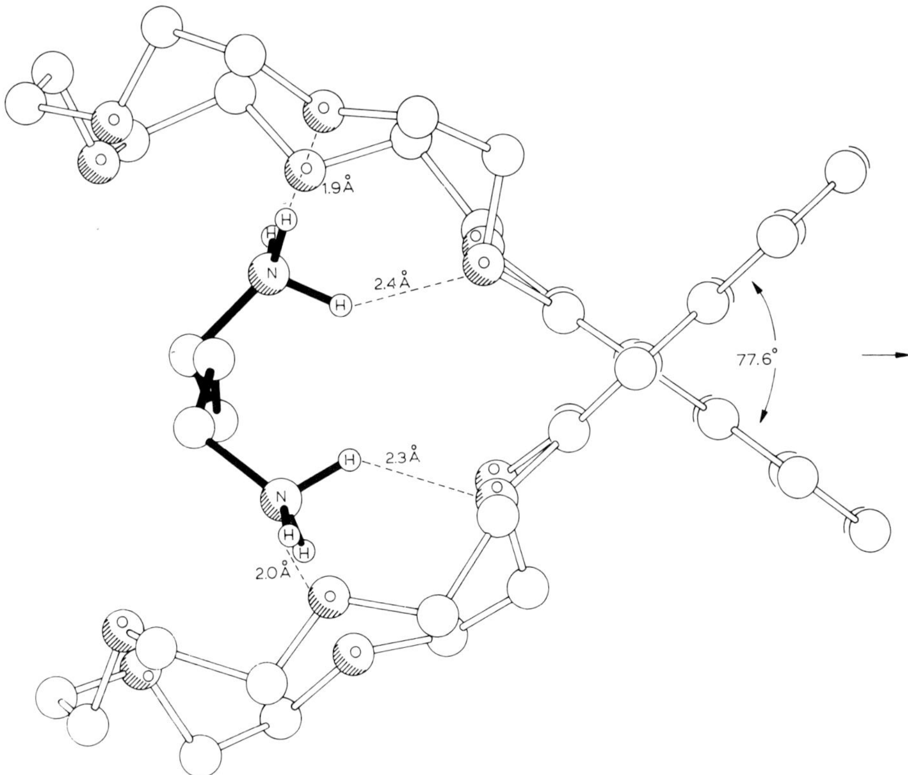

Fig. 5. The X-ray structure [64] of the (S)-(42)-($H_3N(CH_2)_4NH_32PF_6$) complex.

have been achieved [65] with the host molecule (43) in which chelation at the bipyridyl nitrogen atoms by $W(CO)_4$ induces planarity upon the bipyridyl unit; the consequent misalignment of the electron pairs on the benzylic oxygen atoms leads, it is suggested, to the observed small decrease in binding of the host towards $NaBPh_4$. This system, which exhibits *negative* cooperativity, allows [65] remote control of Li^+, Na^+, and K^+ ion transport to be effected. A search for *positive* cooperativity between the two primary binding sites of the bis-crown (44) in complexing $Hg(CF_3)_2$ has demonstrated [66] that the two sites act independently of each other. However, the receptivity of the host (44) towards a second $Hg(CN)_2$ guest is enhanced [67] 10-fold by the binding of a first $Hg(CN)_2$ guest. This small but perceptible effect suggests that even greater cooperativity should be available to systems composed of less flexible subunits.

3. *Binding and recognition at the transition state*

Suitably designed synthetic molecular receptors permit the investigation of reactions known to be catalysed by enzymes. In their roles as *mimics*, the synthetic counterparts offer the opportunity of gaining a better understanding of the mechanism of action of enzymes. Other synthetic molecular receptors, which perform a catalytic function not necessarily related to that of any particular enzyme, may simply be regarded as new chemical reagents. As *analogues* of enzymes, they rely upon the imagination and skill of the chemist for their conception and realisation. Obviously, both mimics and analogues have potential application in synthesis and both may be tailored by structural modification to meet a given set of requirements. A successful enzyme mimic or analogue has to fulfil at least some of the following criteria: (i) specific substrate binding, (ii) fast and selective reaction with the bound substrate, (iii) fast association and dissociation of the substrate and product, (iv) regeneration of the reactive site after the reaction, and (v) a high turnover reflecting the high stability of the catalyst to the reaction conditions. In addition to binding the substrate, the enzyme mimic or analogue can be designed to complex with the reagent, i.e. both binary and ternary catalytic sites can be envisaged. Many enzymes display high stereoselectivities in reactions carried out on bound substrates. Stereodifferentiating reactions which lead to essentially pure stereoisomers—enantiomers or diastereoisomers—can be effected by isomer discrimination of substrates (e.g. kinetic resolutions) or discrimination between stereoheterotopic ligands or faces in prochiral substrates. In this section, enantiomeric and enantiotopic differentiation at the transition state by chiral crown ethers will be discussed along with their properties as catalysts, paying particular attention to three classifications of reactions.

3.1. Enzyme mimics: hydrogen-transfer reactions

The hydride-donating properties of 1,4-dihydropyridine derivatives have obvious analogies with the well-known coenzymes, NADH and NADPH. The incentive to incorporate 1,4-dihydropyridine residues into crown ethers comes [68] from the knowledge that positive ions (e.g. H^+, Zn^{2+}, Mg^{2+}) can catalyse hydride transfers from 1,4-dihydropyridines to pyridinium rings and to readily reducible carbonyl groups. An X-ray structural investigation [69] has confirmed that the macrocyclic 1,4-dihydropyridine-bislactone (45) forms a 1 : 1 complex with $NaClO_4$. In the crystalline state, a molecule of acetone and the ClO_4^- counterion interact with the Na^+ ion from opposite faces of the complex. The possibility also exists of forming reactive binary complexes with organic cationic substrates which contain carbonyl groups. This objective has been realised [70]. The crown-ether derivative (45) reacts stoichiometrically (Fig. 6) with substituted phenacylsulphonium salts, e.g. $PhCOCH_2SPhMeBF_4$, to give acetophenone, methyl phenyl sulphide, the macrocyclic *N*-methylpyridinium-bislactone salt (47), and the rearranged isomer (48). The molecular receptor (45) reacts much faster than the reference compound, 3,5-bis(ethoxycarbonyl)-1,2,6-trimethyl-1,4-dihydropyridine, does with sulphonium salts.

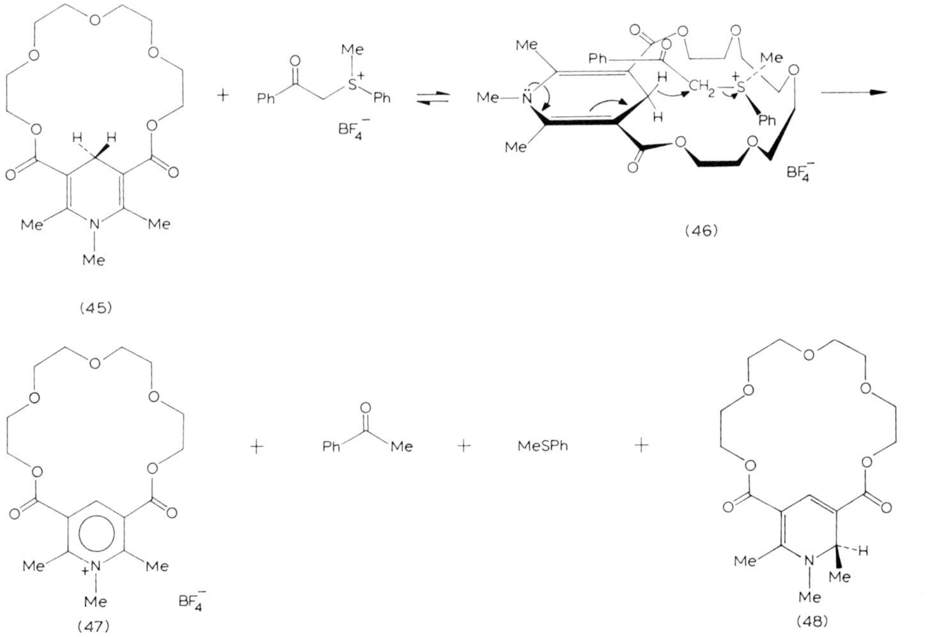

Fig. 6. The reaction between the macrocyclic 1,4-dihydropyridine-bislactone (45) and $PhCOCH_2SPhMeBF_4$.

Activation parameters for hydride transfers in the crown-assisted reaction ($\Delta G^{\neq}_{25°} = 23.0$, $\Delta H^{\neq}_{25°} = 31.6$, and $T\Delta S^{\neq}_{25°} = +8.6$ kcal/mole) and in the reference reaction ($\Delta G^{\neq}_{25°} = 26.0$, $\Delta H^{\neq} = 17.7$, and $T\Delta S^{\neq}_{25°} = -8.3$ kcal/mole) indicate that the rate enhancement is largely a result of a favourable increase in the entropy of activation in the case of the former reaction. Extrapolation of the kinetic data indicates that hydride transfer is 2700 times faster for the crown-assisted reaction. The observations indicate that (45) and $PhCOCH_2\overset{+}{S}MePh$ cations associate to form a binary complex (46) in a non-rate-determining step and that reduction occurs in this complex (Fig. 6). This proposal is supported by the fact that $NaClO_4$, which is a better complexing agent for (45), acts as a competitive and reversible inhibitor. In order to complete an imperfect catalytic cycle, the *N*-methylpyridinium bislactone salt (47) can be reduced back to the 1,4-dihydropyridine-bislactone (45) with aqueous dithionite [69]. An analogous catalytic cycle is shown in Fig. 7 for the reduction [71] of a prochiral carbonyl compound with a 1,4-dihydropyridine (1,4-DHP Crown) residue fused to a crown ether capable of complexing with a metal ion, M^+. The encapsulated M^+ ion forms a ternary complex with the carbonyl compound (RCOR′) in which the carbonyl group is activated towards hydride acceptance by its coordination to M^+. Reduction of the pyridinium salt (Pyr^+) formed during the hydride transfer with aqueous dithionite completes the cycle. The chiral macrocycle L,L-(49), incorporating a 1,4-dihydropyridine unit and two L-valine

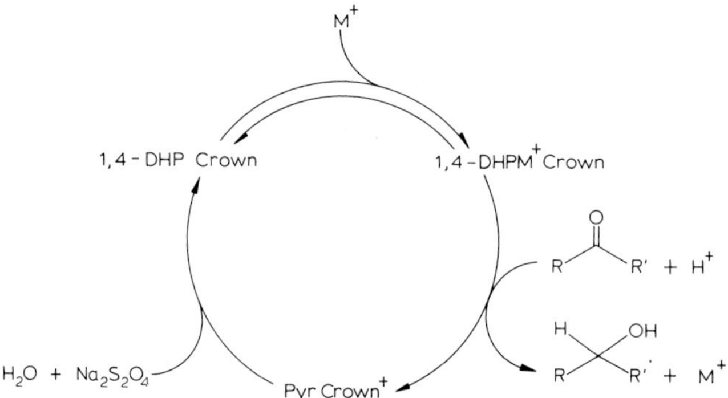

Fig. 7. A catalytic cycle.

residues, has been employed [71] with a molar equivalent of $Mg(ClO_4)_2 \cdot 1.5\ H_2O$ in CH_3CN to effect enantioselective reductions on prochiral carbonyl groups in stoichiometric amounts of the substrates, $PhCOCF_3$, $PhCOCO_2Et$, $PhCOCONH_2$ and PhCOCONHEt. The isolated products (*S*)-$PhCHOHCF_3$, (*S*)-$PhCHOHCO_2Et$, (*S*)-$PhCHOHCONH_2$ and (*S*)-PhCHOHCONHEt were obtained with enantiomeric excesses of 68, 86, 64 and 78%, respectively. The Mg^{2+} ion, which is essential for reaction, is believed to be complexed close to the diethylene glycol bridge in L,L-(49) and to provide a binding site for the carbonyl group in the substrate. Assuming that the carbonyl carbon atom is oriented towards the 4-position of the 1,4-dihydropyridine unit and that the phenyl group occupies the substituent-free side of the macrocycle, the observed (*S*) configuration of the chiral centre in the products can be predicted from examination of CPK molecular models.

L,L-(49)

Complexes between *N*-(ω-aminoalkyl)pyridinium salts (50) and (51) and the 18-crown-6 derivative L,L-(52), synthesised from L-tartaric acid and carrying four 1,4-dihydropyridine side chains, have been shown [32] to display enhanced rates of internal hydrogen transfer from the host to the guest. Evidence for face-to-face stacking of the pyridinium ring in the guests (50) and (51) with an indole ring in the host L,L-(53) comes [31] from the observation of charge-transfer bands in the UV spectrum of the 1 : 1 complexes. When the pyridinium salts (50) and (51) are added to CH_3CN solutions of L,L-(52) at 25°C, reduction to the corresponding 1,4-dihydropyridine derivatives occurs. Kinetic analysis shows that the reactions are first order:

for (50) $k_1 = 1.2 \times 10^{-4}$/sec and for (51), $k_1 = 2.5 \times 10^{-4}$/sec. This implies that the hydrogen transfer is *intramolecular* and takes place in a preformed binary complex.

COMe N+ $NH_3^+ 2BF_4^-$

(50)

COMe N+ $NH_3^+ 2BF_4^-$

(51)

L,L-(52) R = $CONH(CH_2)_2$–N CONHBu^n

L,L-(53) R = $CONH(CH_2)_2$ (3-Indole)

Inhibition is observed when excess of KBF_4 is added: the hydrogen transfer becomes *intermolecular* and follows second-order kinetics with identical rate constants (k_2 = 1.2×10^{-2}/M/sec) for hydrogen transfers between L,L-(52) and (50) and L,L-(52) and (51). This results from the exclusion of the substrates from the crown ether-receptor sites by the more strongly complexed K^+ ion. When inhibition is caused by a

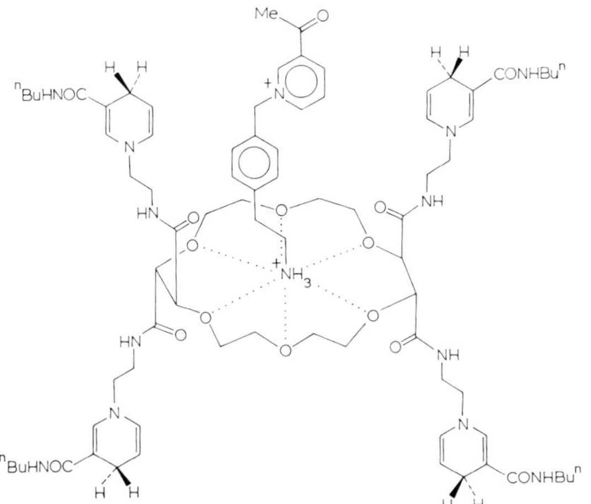

Fig. 8. A schematic representation of the 1 : 1 complex L,L-(52·51) between the molecular receptor L,L-(52) and the substrate (51).

presence of an excess of $H_3N(CH_2)_4NH_3\ 2BF_4$, the reaction between L,L-(52) and (51) becomes very slow ($k_2 < 10^{-4}$/M/sec) probably because of combined displacement and electrostatic effects (i) disfavouring the formation of a positively charged group attached to the crown ether and (ii) hindering the approach of the charged substrate. A schematic representation of the complex L,L-(52 · 51) in which hydrogen transfer takes place between the receptor L,L-(52) and the substrate (51) is shown in Fig. 8.

3.2. Enzyme mimics: acyl transfer reactions

In part at least, enzymes are thought to act as catalysts by stabilising rate-determining transition states through complexation. The transacylases, e.g. trypsin and papain, have been some of the most extensively studied enzymes. In the first stage of the enzyme-catalysed reactions, esters or amides are complexed; in the second stage, the acyl groups in the substrates are transferred to either serine hydroxyl groups or cysteine sulphydryl groups; in the third stage, the acyl groups in the covalently bound enzyme–substrate complexes are transferred to water.

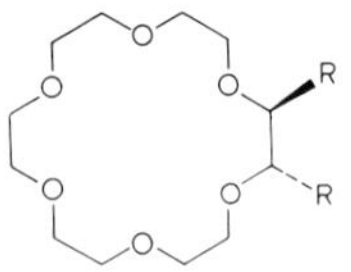

L-(54) R = CH_2SH
L-(55) R = $CH_2CH_2CH_2SH$
L-(56) R = $CH_2OCH_2CH_2SH$

Several crown ether receptors have been synthesised [3,4,7,72–75] with side chains which terminate in sulphydryl groups. In their enhancements of the rates of transesterification of amino acid *p*-nitrophenyl esters, chiral crown ether catalysts have not only been shown to exhibit substrate specificity but also enantiomeric differentiation. Substrate specificities have been observed [72] in the transacylations between the crown ether derivatives L-(54), L-(55), and L-(56), prepared from L-tartaric acid and a number of amino acid *p*-nitrophenyl ester hydrobromides. The results are summarised in Table 1. Formation of binary complexes as a prelude to fast reactions is supported by the following observations: (i) The relative instability of complexes between 18-crown-6 derivatives and secondary dialkylammonium salts compared with those between 18-crown-6 derivatives and primary alkyl ammonium salts is reflected in the substantially higher rate of release of *p*-nitrophenol from the glycine ester than from the *N*-methyl glycine ester. (ii) The relative lengths of the side arms in the 18-crown-6 derivatives L-(54) and L-(55) and of the amino acid ester chains influence the rates of the transacylations in a predictable manner. (iii) The fact that L-(56) reacts faster with the glycine ester than it does with L-(54) suggests that the oxygen atoms in the side arms of L-(56) become involved (Fig. 9) in pole–dipole interactions with the bound NH_3^+ centre. It is concluded that a more highly structured binary complex, in which the side arms of L-(56) converge upon the substrate molecule in the desired manner, is formed as a result of these additional interactions.

Chiral differentiation between enantiomeric amino acid *p*-nitrophenyl ester hydrobromines was observed [9,73] in transacylations catalysed by the chiral 3,3-

TABLE 1

Pseudo-first-order rate constants ($10^5\ k$/sec) for *p*-nitrophenol release [72] from amino acid *p*-nitrophenyl ester hydrobromides, $RCO_2C_6H_4$-*p*-$NO_2 \cdot Br^-$ [a]

Guest (R)	Host					
	None	18-Crown-6 (2)	18-Crown-6 (2)+BuSH [b]	L-(54) [c]	L-(55) [c]	L-(56) [c]
H_3N^+-CH_2	3	0.9	1	1170	50	2500
MeH_2N^+-CH_2	5	5	4	6	4	37
H_3N^+-$(CH_2)_2$	< 0.1	< 0.05	< 0.05	0.4	7	2
H_3N^+-$(CH_2)_3$	310	1	0.9	6	42	41
H_3N^+-$(CH_2)_5$	< 0.05	< 0.05	< 0.05	< 0.05	< 0.05	< 0.05

[a] 20% EtOH—CH_2Cl_2 buffered with 0.01 M AcOH and 0.005 M pyridine (pH 4.60 in H_2O) at 25°C, 10^{-4} M in substrates, 5×10^{-3} M in host.

[b] 10^{-2} M.

[c] Corrected for buffer solvolysis in the presence of 18-crown-6 (2).

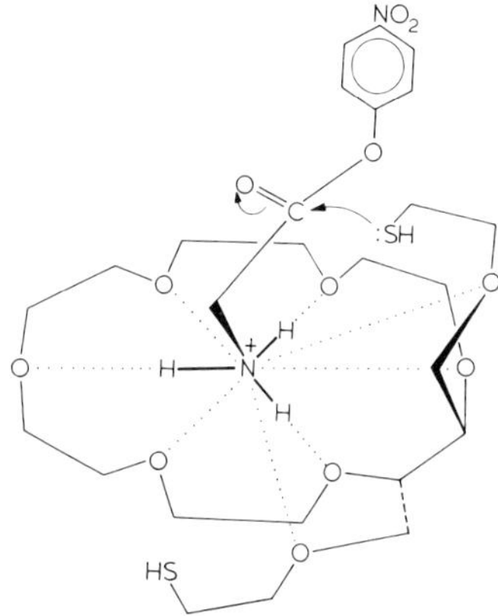

Fig. 9. A schematic representation of the 1 : 1 complex between the molecular receptor L-(56) and the glycine *p*-nitrophenyl ester cation.

bis(mercaptomethyl)binaphthyl-20-crown-6 derivative (*S*)-(57). The extent of crown ether catalyses has been measured with reference to the non-cyclic compound

(S)-(57) (S)-(58)

(*S*)-(58) as catalyst. The pseudo-first-order kinetics were investigated for acylation of large excesses of (*S*)-(57) and (*S*)-(58) by the nitrophenyl ester salts of glycine,

L-alanine, L-leucine, L-valine, L-proline, and D- and L-phenylalanine in a range of buffered solutions. The rate of the thiolysis reaction is independent of the buffer concentration — indicating the absence of general acid or base catalysis — and RS^- is believed to be the active nucleophile. The subsequent solvolysis of the intermediate thiol ester was shown to be slower by a factor of 10^4 compared with the original thiolysis. The overall reaction clearly pursues the following simplified course:

$$\begin{array}{ccccc} R-\overset{*}{C}H(NH_3^+)-C(=O)OAr & & R-\overset{*}{C}H(NH_3^+)-C(=O)SR' & & R-\overset{*}{C}H(NH_3^+)-C(=O)OEt \\ + & \xrightarrow[\text{Fast}]{k} & + & \xrightarrow[\text{Slow}]{\text{EtOH}} & + \\ R'SH\ [(S)\text{-}(57)] & & ArOH & & R'SH\ [(S)\text{-}(57)] \end{array}$$

The rate enhancements for the slow thiolysis step by (*S*)-(57) with reference to (*S*)-(58) can be attributed to transition state stabilisation by complexation. Selected data are presented in Table 2. The following specific comments can be made concerning the results. (i) They demonstrate that, with the exception of the substrate derived from L-proline, which, of course, bonds very weakly with (*S*)-(57), the reaction takes place in a binary complex at rates about 10^2–10^3 faster than with (*S*)-(58). These results indicate that the free energy of the rate-limiting transition state for the reaction is lowered by complexation that is dependent on the host (*S*)-(57) being cyclic and the guests each possessing 3 hydrogen-bonding protons. (ii) In the more hydrophilic medium (40% H_2O-CH_3CN), the rate factors decrease to 10. This trend indicates that the host–guest complex is less highly structured in this medium because of the hydrogen bond formation between the NH_3^+ centre and the solvent molecules. (iii) in high concentrations, K^+ ions act as a competitive inhibitor eliminating the rate accelerations caused by highly structured complexation. (iv) The rate factors in the (*R*)-host–L-guest series are 290, 46, 17 and 1 and in the (*S*)-host–L-guest series are 32, 32, 15 and 1 as the side chains of the amino acid ester salts are changed from Me to Me_2CHCH_2 to $PhCH_2$ to Me_2CH. Thus, the larger the steric requirements of the amino acid side chains in the vicinity of its complexing and reacting groups, the slower is the reaction rate. The maximum rate factors of 290 and 32 for the L-alanine ester salt provide respectively $-\Delta(\Delta G^{\neq})$ values of 3.4 and 2.1 kcal/mole for structural recognition. The 1.3 kcal/mole difference in structural recognition between these two diastereoisomeric complexes must reflect the maximum chiral recognition capacity of (*R*)- and (*S*)-(57) for these amino acid ester salts. (v) In the lipophilic medium, the host (*S*)-(57) reacts faster with the L-amino acid ester salts than does the host (*R*)-(57). In 20% EtOH-CH_2Cl_2, as the side chains of the amino acid ester salts are changed from Me to Me_2CHCH_2 to $PhCH_2$ to Me_2CH, the rate factors emerge as 1, 6.4, 8.3 and 9.2. Thus, chiral differentiation increases with the growth of the steric requirements of the side chains in the immediate vicinity to the chiral centre. These observations can be explained (Fig. 10) in terms of the formation of the diastereoisomeric tetrahedral intermediates (*S*)-

TABLE 2
Pseudo-first-order rate constants, rate-acceleration factors, and activation free-energy differences for thiolysis [7,73] of $RCO_2C_6H_4$-*p*-$NO_2 \cdot Br^-$ (10^{-4} M) in the presence of 5.0×10^{-3} M solutions of (*S*)-(57) and (*S*)-(58) at 25°C

Medium [a]	Host	Guest R	Guest Config.	10^3k/sec	$k_{cycle}/k_{noncycle}$	$-\Delta(\Delta G^{\neq})$ [b] (kcal/mole)
A	(*S*)-(57)	$(CH_2)_3CHNH_2^+$	L	16	0.8	10.1
A	(*S*)-(58)		L	21		
A	(*S*)-(57)	$CH_2NH_3^+$		> 700	> 130	> 2.9
A	(*S*)-(58)			5.4		
A	(*S*)-(57)	$CH_3CHNH_3^+$	L	> 700	> 130	> 2.9
A	(*S*)-(58)		L	5.4		
A	(*S*)-(57)	$Me_2CHCHNH_3^+$	L	13	160	3.0
A	(*S*)-(58)		L	0.08		
A	(*S*)-(57)	$PhCH_2CHNH_3^+$	L	200	490	3.7
A	(*S*)-(58)		L	0.41		
A	(*S*)-(57)	$PhCH_2CHNH_3^+$	D	25	69	2.5
A	(*S*)-(58)		D	0.36		
A	(*S*)-(57)	$Me_2CHCH_2CHNH_3^+$	L	> 700	> 1170	> 4.2
A	(*S*)-(58)		L	0.6		
B	(*S*)-(57)	$CH_2NH_3^+$		110	20	1.8
B	(*S*)-(58)			5.4		
B	(*S*)-(57)	$PhCH_2CHNH_3^+$	L	18	8	1.2
B	(*S*)-(58)		L	2.2		

[a] A, 20% EtOH in CH_2Cl_2 (v) buffered with 0.2 M AcOH and 0.1 M Me_4NOAc (pH = 4.8 in H_2O). B, 40% H_2O in CH_3CN (v) buffered with 0.2 M AcOH and 0.2 M NaOAc (pH = 4.8 in H_2O).
[b] $\Delta(\Delta G^{\ddagger}) = \Delta G^{\ddagger}_{cycle} - \Delta G^{\ddagger}_{non\text{-}cycle}$.

(57)–L-guest and (*S*)-(57)–D-guest. The chiral bias observed in the thiolysis is consistent with the (*S*)-L-configurational relationship being more stable than the

Fig. 10. The diastereoisomeric tetrahedral intermediates in the thiolysis of L- and D-guests by (*S*)-(57). (a) (*S*)-(57) L-guest (more stable); (b) (*S*)-(57) D-guest (less stable).

(*S*)-D one, as predicted from an inspection of CPK molecular models. The maximum chiral recognition rate factor of 9.2 for valine corresponds to 1.3 kcal/mole. The question now arises as to whether this enantiomeric differentiation is associated with the pre-equilibrium complexation step as well as with the rate-limiting transition state. The ground state chiral recognition of (*S*)-(59) in $CHCl_3$ at 0°C for the

(*S*)- (59)

extraction of D,L-valine from H_2O amounts to 0.6 kcal/mole free-energy difference. Therefore, it is concluded that more of the 1.3 kcal/mole chiral recognition is coming from the transition state than from the ground state. This is a reasonable conclusion. The transition state with its partial covalent bonding in addition to the hydrogen bonding should be more highly structured than the complex which is organised primarily by hydrogen bonding.

L,L – L,L,L,L-(60) R = H

L,L – L,L,L,L (61) R = CH_2Ph

Enantiomeric differentiation during the thiolysis of α-amino acid ester salts by two thiol-bearing 18-crown-6 derivatives prepared from (1*R*, 2*R*, 3*S*, 4*S*)- and (1*R*, 2*S*, 3*R*, 4*S*)-camphane-2,3-diols has also been demonstrated [74]. Discrimination by factors of 1.7–1.9 in the rates of *p*-nitrophenol release from the enantiomers of alanine-*p*-nitrophenyl ester salts has been observed. By contrast, the tetrakis-L-cysteinyl methyl ester receptor molecule L,L-L,L,L,L-(60) exhibits [75] extremely high chiral recognition in its reaction with the enantiomeric glycyl-D- and -L-phenylalanine-p-nitrophenyl ester hydrobromide. Depending on the medium, the L-antipode reacts 50–90 times faster. Thus, starting from the racemic dipeptide ester hydrobromide, kinetic resolution to afford the D-antipode with high optical purity in good chemical yield can be achieved. It has not been established if this high chiral discrimination is occurring in the complexation or reactivity step of the process and it is recognised that, in common with other crown-ether receptor molecules bearing thiol groups, L,L-L,L,L,L-(60) will not be a true catalyst until rapid deacylation of the acyl-crown intermediate can be achieved. Nonetheless, it exhibits enhanced rates (Table 3) of intramolecular thiolysis for p-nitrophenyl esters of (i) amino acid hydrobromides derived from glycine, β-alanine, and D- and L-phenylalanine and (ii) dipeptide hydrobromides derived from glycylglycine and glycyl-β-alanine as well as from glycyl-D- and -L-phenylalanine. The L-prolylglycine-p-nitrophenyl ester salt, a secondary dialkylammonium substrate, is complex much more weakly than the

TABLE 3

Pseudo-first-order rate constants and relative rates for the release [75] of p-nitrophenol from amino acid and dipeptide ester hydrobromide substrates in the presence of L,L-L,L,L,L-(60) and L,L-L,L,L,L-(60) plus excess of KBr at 20°C

Medium [a]	Substrate	$10^5\,k$/sec			Relative rates [b]
		Solv	L,L-L,L,L,L-(60)	L,L-L,L,L,L-(60)+K^+	
A	Gly	4.8	400	27	15
A	D- or L-Phe	2.9	19	34	0.6
A	β-Ala	7.8	18	27	0.7
A	Gly-Gly	4.8	230	2.9	80
A	Gly-D-Phe	2.4	23	0.7	35
A	Gly-L-Phe	2.4	150	1.5	100
B	Gly-Gly	390	235 000	1700	140
B	Gly-β-Ala	4.3	120	38	3
B	L-Pro-Gly	115	125	590	0.2
B	CbO-Gly	5.5	14.5	220	0.07
C	Gly-Gly	59	2600	–	–
C	Gly-D-Phe	74	30	–	–
C	Gly-L-Phe	74	1600	–	–
C	L-Pro-Gly	77	0.17	–	–
D	Gly-D-Phe	15	1500	–	–
D	Gly-L-Phe	15	16.5	–	–

[a] A≡MeOH—CH_2Cl_2—H_2O (78.5 : 20 : 1.5), Py—PyHBr buffer 0.05 M (pH = 6.1 in H_2O). B≡MeOH—DMF—H_2O (78.5 : 20 : 1.5), AcOH—$AcONMe_4$ buffer 0.02 M (pH = 4.8 in H_2O). C≡CH_2Cl_2—MeOH—H_2O (97.9 : 2 : 0.1), CF_3CO_2H—N-ethylmorpholine buffer 0.3 M (pH = 7.0 in H_2O). D≡CH_2Cl_2—EtOH (95 : 5), CF_3CO_2H—N-ethylmorpholine buffer 0.03 M (pH = 7.0 in H_2O). In media A and B, substrate 10^{-4} M, L,L-L,L,L,L-(60) 3.5×10^{-3} M, KBr 1.4×10^{-2} M. In media C and D, substrate 5×10^{-5} M L,L-L,L,L,L-(60) 3.5×10^{-4} M.

[b] Relative rates refer to the ratio of the rate constants for L,L-L,L,L,L-(60) and L,L-L,L,L,L-(60)+K^+.

primary alkylammonium substrates and so reacts very much more slowly. No rate enhancement is observed when the tetra-*S*-benzyl derivative L,L-L,L,L,L-(61) is used instead of L,L-L,L,L,L-(60). This indicates that the SH groups are the reactive centres as in the cysteine enzyme, papain. The reaction of L,L-L,L,L,L-(60) with *N*-carbobenzoxyglycine-*p*-nitrophenyl ester hydrobromide, which cannot form a complex, is accelerated by a factor of ca. 15 when KBr is added. Thus, complexation of K^+ ions renders the catalyst more reactive, possibly by lowering the p*K* of the SH groups. If the complexation of the NH_3^+ centres has a similar effect, then activation of the catalyst by intermolecular interactions must be occurring. It is possible this is a general phenomenon which may be important in enzymic catalysis. The reaction of the glycylglycine-*p*-nitrophenyl ester salt with L,L-L,L,L,L-(60) exhibits pseudo-first-order kinetics changing to second-order kinetics on addition of excess of KBr. These data indicate conclusively that the reaction proceeds intramolecularly from a binary complex and only becomes intermolecular when the substrate is displaced by K^+ ions. The process is illustrated in Fig. 11 for the reaction of the *p*-nitrophenyl ester hydrobromide of glycylglycine with L,L-L,L,L,L-(60).

3.3. Enzyme analogues: Michael addition reactions

Very recently, it has been demonstrated [76] that the hosts (*RR*)-(29) and (*R*)-(59), complexed to $KOCMe_3$ or KNH_2, catalyse the Michael additions of methyl vinyl ketone and methyl acrylate to the phenyl acetic esters (61) and (62), and the β-ketoester (63) with high catalytic turnover numbers (CTN ≡ mmoles of product

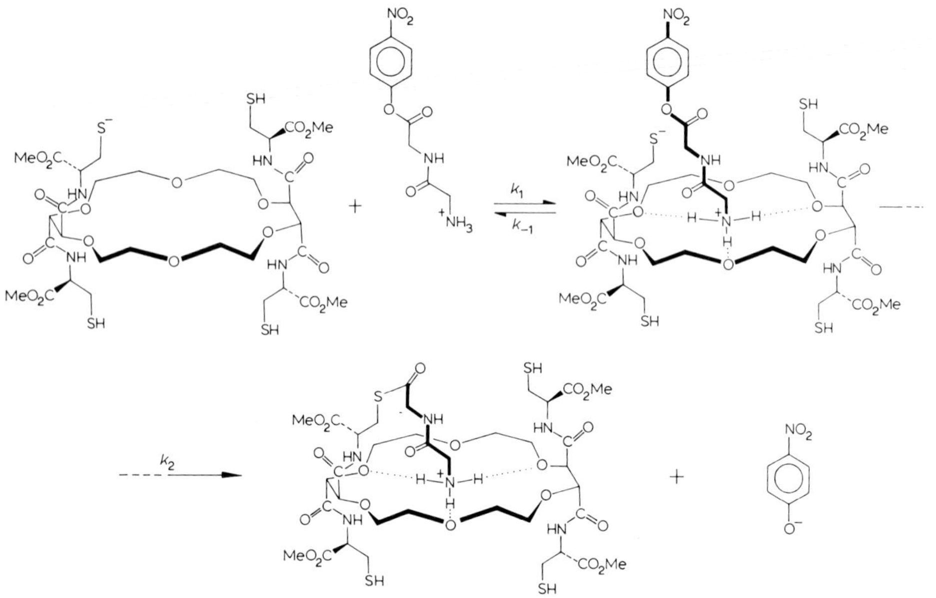

Fig. 11. The reaction of the p-nitrophenyl ester hydrobromide of glycylglycine with L,L-L,L,L,L-(60).

formed per mmole of catalyst complex employed). The reactions, some of which are summarised in Scheme 1, show that adducts have been obtained with optical yields

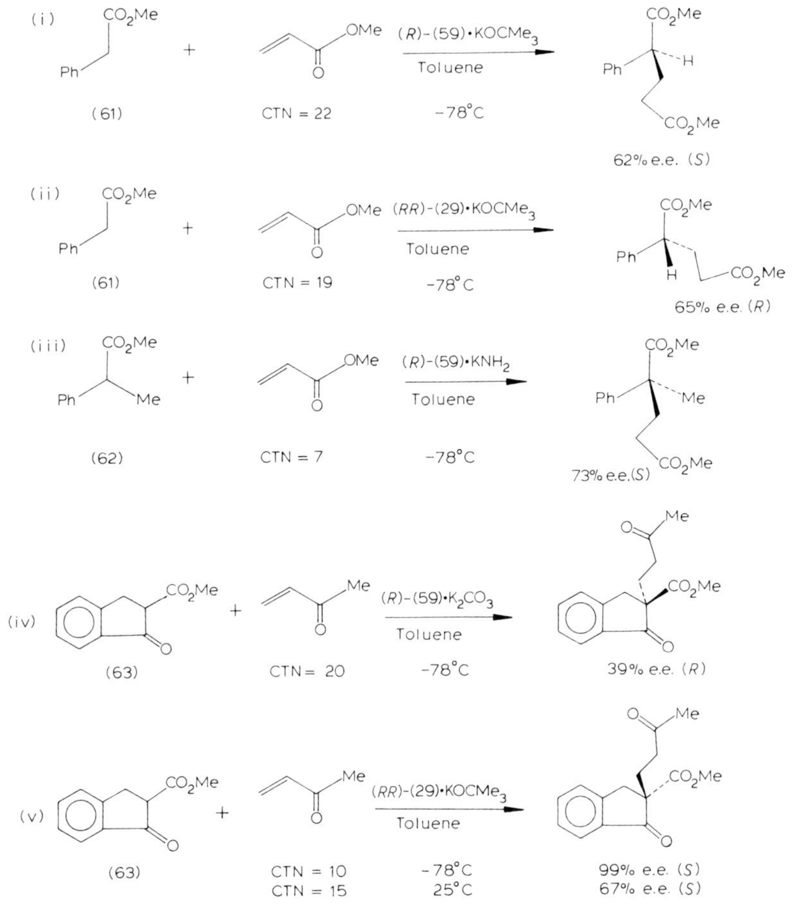

Scheme 1.

as high as 99% e.e. at low temperatures (−78°C) in toluene. The catalyses can be explained in terms of the chain reaction mechanism presented in Scheme 2 whilst the enantioselectivities can be rationalised on the basis of steric differences in the diastereoisomeric models for the complexes involved. In Fig. 12, the reactions between methyl acrylate and the potassium salts of the carbanions derived from (61) and (62) are analysed in the knowledge that (*R*)-(59) leads to an (*S*)-adduct and (*RR*)-(29) to an (*R*)-adduct. The ion pairs which may be complexed through the intermediacy of the K^+ ions to the best planes of the oxygen atoms in the hosts, can be symbolised as a rectangular 4-membered ring. A perpendicular approach of the ion pairs to the receptor sites on the hosts (*R*)-(59) and (*RR*)-(29) leads to the prediction that they should afford (*S*)- and (*R*)-adducts, respectively. This prediction is in accordance with the experimental results.

Initiation Host• K^+ B^- H—R → Host•K^+ R^- + BH

Addition Host• K^+ R^- → (Configuration determining step) R^* … O^- K^+• Host

Chain transfer R—H R^* … O^- K^+• Host → Host•K^+ R^- + R^* …O

Scheme 2.

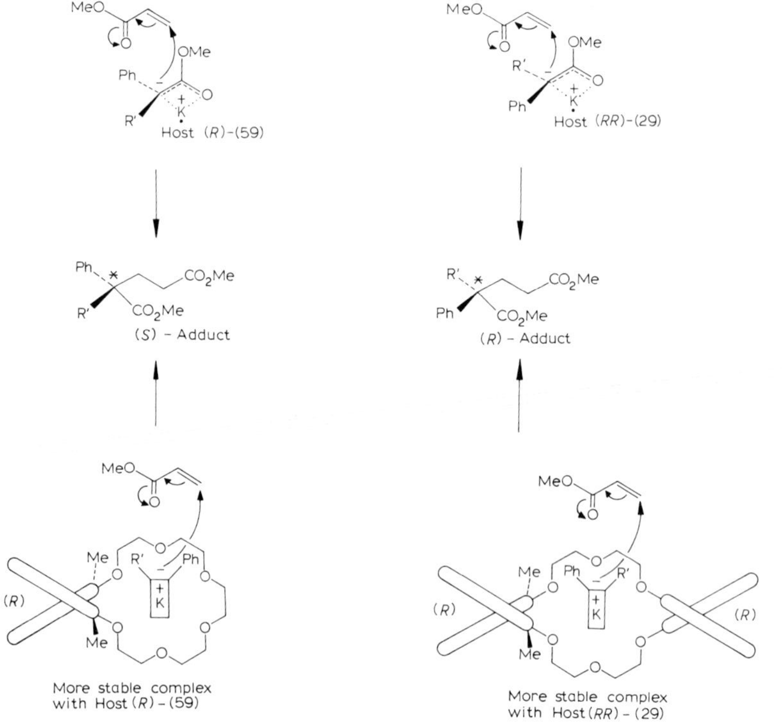

Fig. 12. An analysis of the stereochemical course of the Michael addition reactions.

4. Conclusion

The design and synthesis of more rigid synthetic molecular receptors of the crown-ether type should lead to better catalyses and improved stereoselectivities. Reaching this objective would appear to entail increasing the numbers and types of interactive sites in host molecules for the substrate guest species. Ultimately, the challenge will become not so much one in crown-ether chemistry as more one in synthetic molecular chemistry.

References

1 Kirby, A.J. (1979) in: Sir Derek Barton and W.D. Ollis (Eds.), Comprehensive Organic Chemistry, Vol. 5 (E. Haslam (Ed.)), Pergamon, Oxford, pp. 389–460; Fersht, A.R. and Kirby, A.J. (1980) Chem. Br. 136–142 and 156.

2 Bender, M.L. and Komiyama, M. (1978) Reactivity and Structure. Concepts in Organic Chemistry, Vol. 6, Cyclodextrin Chemistry, Springer Verlag, Berlin, pp. 1–96.

3 Lehn, J.M. (1973) Structure Bonding 16, 1–69; (1977) Pure Appl. Chem. 49, 857–870; (1978) Accts. Chem. Res. 11, 49–57; (1978) Pure Appl. Chem. 50, 871–892; (1979) Pure Appl. Chem. 51, 979–997; (1980) Pure Appl. Chem. 52, 2303–2319 and 2441–2459; (1980) in: A. Braibanti (Ed.), Bioenergetics and Thermodynamics: Model Systems, Reidel, Holland, pp. 455–461.

4 Cram, D.J. and Cram, J.M. (1974) Science 183, 803–809; Cram, D.J., Helgeson, R.C., Sousa, L.R., Timko, J.M., Newcomb, M., Moreau, P., de Jong, F., Gokel, G.W., Hoffman, D.H., Domeier, L.A., Peacock, S.C., Madan, K. and Kaplan, L. (1975) Pure Appl. Chem. 43, 327–349; Cram, D.J. (1976) in: J.B. Jones, C.J. Sih and D. Perlman (Eds.), Applications of Biomedical Systems in Chemistry, Vol. V, Wiley-Interscience, New York, pp. 815–873; Cram, D.J. and Cram, J.M. (1978) Accts. Chem. Res. 11, 8–14; Cram, D.J. (1978) in: P. Ahlberg and L.O. Sundeloef (Eds.), Struct. Dyn. Chem., Proc. Symp., pp. 41–56; Cram, D.J. and Trueblood, K.N. (1981) in: F. Vögtle (Ed.), Topics in Current Chemistry, Host–Guest Complex Chemistry 1, Springer Verlag, Berlin, pp. 43–106.

5 Stoddart, J.F. (1979) Chem. Soc. Rev. 8, 85–142; (1980) in: P. Dunnill, A. Wiseman and N. Blakebrough (Eds.), Enzymic and Non-Enzymic Catalysis, Ellis-Horwood, Chichester, pp. 84–110; (1980) in: A. Braibanti (Ed.), Bioenergetics and Thermodynamics: Model Systems, Reidel, Holland; (1980) in R.N. Castle and S.W. Schneller (Eds.), Lectures in Heterocyclic Chemistry, Vol. 5, pp. S47–S60; (1981) in: R.M. Izatt and J.J. Christensen (Eds.), Progress in Macrocyclic Chemistry, Vol. 2, Wiley-Interscience, New York, pp. 173–250; Crawshaw, T.H., Laidler, D.A., Metcalfe, J.C., Pettman, R.B., Stoddart, J.F. and Wolstenholme, J.B. (1982) in: B.S. Green, Y. Ashani and D. Chipman (Eds.), Chemical Approaches to Understanding Enzyme Catalysis, Elsevier, Amsterdam, pp. 49–65.

6 Reinhoudt, D.N. and de Jong, F. (1979) in: R.M. Izatt and J.J. Christensen (Eds.), Progress in Macrocyclic Chemistry, Vol. 1, Wiley-Interscience, New York, pp. 157–217; de Jong, F. and Reinhoudt, D.N. (1980) Adv. Phys. Org. Chem. 17, 279–433.

7 Chao, Y., Weisman, G.R., Sogah, G.D.Y. and Cram, D.J. (1979) J. Am. Chem. Soc. 101, 4948–4958.

8 Todd, Lord (1976) in: ICI at 50 Suppl. in Chem. Ind., pp. 31; (1981) Chem. Ind., 317–320; Suckling, C.J. and Wood, H.C.S. (1979) Chem. Br., 243–248.

9 Sinnott, M.L. (1979) Chem. Br., 293–297; Luisi, P.L. (1979) Naturwissenschaften 66, 498–504.

10 Kyba, E.P., Helgeson, R.C., Madan, K., Gokel, G.W., Tarnowski, T.L., Moore, S.S. and Cram, D.J. (1977) J. Am. Chem. Soc. 99, 2564–2571.

11 Bovill, M.J., Chadwick, D.J., Sutherland, I.O. and Watkin, D. (1980) J. Chem. Soc., Perkin Trans. 2, 1529–1543.

12 Curtis, D.W., Laidler, D.A., Stoddart, J.F. and Jones, G.H. (1977) J. Chem. Soc., Perkin Trans. 1, 1756–1769.

13 Fischer, E. (1894) Chem. Br. 27, 2985–2993.

14 Pedersen, C.J. and Frensdorff, H.K. (1972) Angew. Chem. Int. Ed. 11, 16–25.

15 Maud, J.M., Stoddart, J.F. and Williams, D.J. (1981) unpublished results.

16 Laidler, D.A. and Stoddart, J.F. (1980) in: S. Patai (Ed.), The Chemistry of Functional Groups, Suppl. E, The Chemistry of Ethers, Crown Ethers, Hydroxyl Groups, and their Sulphur Analogues. Part 1, Wiley-Interscience, Chichester, pp. 1–57.

17 Johnson, M.R., Sutherland, I.O. and Newton, R.F. (1979) J. Chem. Soc., Perkin Trans. 1, 357–371; Leigh, S.J. and Sutherland, I.O. (1979) J. Chem. Soc., Perkin Trans. 1, 1089–1103; Hodgkinson, L.C. and Sutherland, I.O. (1979) J. Chem. Soc., Perkin Trans. 1, 1908–1914; Hodgkinson, L.C., Johnson, M.R., Leigh, S.J., Spencer, N., Sutherland, I.O. and Newton, R.F. (1979) J. Chem. Soc., Perkin Trans. 1, 2193–2202; Pearson, D.P.J., Leigh, S.J. and Sutherland, I.O. (1979) J. Chem. Soc., Perkin Trans. 1, 3113–3126; Johnson, M.R., Sutherland, I.O. and Newton, R.F. (1980) J. Chem. Soc., Perkin Trans. 1, 586–600.

18 Lehn, J.M. and Vierling, P. (1980) Tetrahedron Lett., 1323–1326.

19 Metcalfe, J.C., Stoddart, J.F., Jones, G., Hull, W.E., Atkinson, A., Kerr, I.S. and Williams, D.J. (1980) J. Chem. Soc. Chem. Commun., 540–543.

20 Barrett, A.G.M. and Lana, J.C.A. (1978) J. Chem. Soc. Chem. Commun., 471–472; Barrett, A.G.M., Lana, J.C.A. and Tograie, S. (1980) J. Chem. Soc. Chem. Commun., 300–301.

21 Uiterwijk, J.W.H.M., Harkema, S., Geevers, J. and Reinhoudt, D.N. (1982) J. Chem. Soc. Chem. Commun., 200–201.

22 Lehn, J.M., Vierling, P. and Hayward, R.C. (1979) J. Chem. Soc. Chem. Commun., 296–298.

23 Vögtle, F. and Weber, E. (1980) in: S. Patai (Ed.), The Chemistry of Functional Groups, Suppl. E, The Chemistry of Ethers, Crown Ethers, Hydroxyl Groups, and their Sulphur Analogues, Part 1, Wiley-Interscience, Chichester, pp. 59–156; Vögtle, F., Sieger, H. and Müller, W.M. (1981) in: F. Vögtle (Ed.), Topics in Current Chemistry, Host–Guest Complex Chemistry 1, Springer Verlag, Berlin, pp. 107–161.

24 Fuller, S.E., Stoddart, J.F. and Williams, D.J. (1982) Tetrahedron Lett., 1835–1836.

25 Bandy, J.A., Truter, M.R. and Vögtle, F. (1981) Acta Crystallogr., Sect. B 37, 1568–1571.

26 Metcalfe, J.C., Stoddart, J.F., Jones, G. Crawshaw, T.H., Gavuzzo, E. and Williams, D.J. (1981) J. Chem. Soc. Chem. Commun., 432–434.

27 Colquhoun, H.M. and Stoddart, J.F. (1981) J. Chem. Soc. Chem. Commun., 612–613; Colquhoun, H.M., Stoddart, J.F. and Williams, D.J. (1981) J. Chem. Soc. Chem. Commun., 847–849, 849–850 and 851–852; (1982) J. Am. Chem. Soc. 104, 1426–1428.

28 Trueblood K.N., Knobler, C.B., Lawrence, D.S. and Stevens, R.V. (1982) J. Am. Chem. Soc. 104, 1355–1362.

29 Colquhoun, H.M., Stoddart, J.F., Williams, D.J., Wolstenholme, J.B. and Zarzycki, R. (1981) Angew. Chem. Int. Ed. 20, 1051–1053.

30 Kyba, E.P., Timko, J.M., Kaplan, L.J., de Jong, F., Gokel, G.W. and Cram, D.J. (1978) J. Am. Chem. Soc. 100, 4555–4568.

31 Behr, J.P., Lehn, J.M. and Vierling, P. (1976) J. Chem. Soc. Chem. Commun., 621–623.

32 Behr, J.P. and Lehn, J.M. (1978) J. Chem. Soc. Chem. Commun., 143–146.

33 Newcomb, M., Moore, S.S. and Cram, D.J. (1977) J. Am. Chem. Soc. 99, 6405–6410.

34 Cram, D.J., Helgeson, R.C., Koga, K., Kyba, E.P., Madan, K., Sousa, L.R. Siegel, M.G., Moreau, P., Gokel, G.W., Timko, J.M. and Sogah, D.G.Y. (1978) J. Org. Chem. 43, 2758–2771.

35 Timko, J.M., Helgeson, R.C. and Cram, D.J. (1978) J. Am. Chem. Soc. 100, 2828–2834.

36 Laidler, D.A. and Stoddart, J.F. (1979) Tetrahedron Lett., 453–456.

37 Laidler, D.A. and Stoddart, J.F. (1977) J. Chem. Soc. Chem. Commun., 481–483.

38 Pettman, R.B. and Stoddart, J.F. (1979) Tetrahedron Lett., 461–464.

39 Pettman, R.B. and Stoddart, J.F. (1979) Tetrahedron Lett., 457–460.

40 Beresford, G.D. and Stoddart, J.F. (1980) Tetrahedron Lett., 867–870.

41 Lehn, J.M. and Vierling, P. (1980) Tetrahedron Lett., 1323–1326.

42 Timko, J.M., Moore, S.S., Walba, D.M., Hiberty, P.C. and Cram, D.J. (1977) J. Am. Chem. Soc. 99, 4207–4219.

43 Newcomb, M., Timko, J.M., Walba, D.M. and Cram, D.J. (1977), J. Am. Chem. Soc. 99, 6392–6398.

44 Stoddart, J.F. and Wolstenholme, J.B. (1982) unpublished results.

45 Coxon, A.C., Laidler, D.A., Pettman, R.B. and Stoddart, J.F. (1978) J. Am. Chem. Soc. 100, 8260–8262.

46 Live, D. and Chan, S.I. (1976) J. Am. Chem. Soc. 98, 3769–3778.

47 Helgeson, R.C., Weisman, G.R., Toner, J.L., Tarnowski, T.L., Chao,, Y., Mayer, J.M. and Cram, D.J. (1979) J. Am. Chem. Soc. 101, 4928–4941.
48 Lingenfelter, D.S., Helgeson, R.C. and Cram, D.J. (1981) J. Org. Chem. 46,, 393–406.
49 Kyba, E.P., Gokel, G.W., de Jong, F., Koga, K., Sousa, L.R., Siegel, M.G., Kaplan, L., Sogah, G.D.Y. and Cram, D.J. (1977) J. Org. Chem. 42, 4173–4184.
50 Cram, D.J., Helgeson, R.C., Peacock, S.C., Kaplan, L.J., Domeier, L.A., Moreau, P., Koga, K., Mayer, J.M., Chao, Y., Siegel, M.G., Hoffman, D.H. and Sogah, G.D.Y. (1978) J. Org. Chem. 43, 1930–1946.
51 Peacock, S.C., Domeier, L.A., Gaeta, F.C.A., Helgeson, R.C., Timko, J.M. and Cram, D.J. (1978) J. Am. Chem. Soc. 100, 8190–8202.
52 Peacock, S.C., Walba, D.M., Gaeta, F.C.A., Helgeson, R.C. and Cram, D.J. (1980) J. Am. Chem. Soc. 102, 2043–2052.
53 Goldberg, I. (1977) J. Am. Chem. Soc. 99, 6049–6057.
54 Newcomb, M., Toner, J.L., Helgeson, R.C. and Cram, D.J. (1979) J. Am. Chem. Soc. 101, 4941–4947.
55 Sousa, L.R., Sogah, G.D.Y., Hoffman, D.H. and Cram, D.J. (1978) J. Am. Chem. Soc. 100, 4569–4576.
56 Sogah, G.D.Y. and Cram, D.J. (1979) J. Am. Chem. Soc. 101, 3035–3042.
57 Peacock, S.C. and Cram, D.J. (1976) J. Chem. Soc. Chem. Commun., 282–284.
58 Curtis, W.D., King, R.M., Stoddart, J.F. and Jones, G.H. (1976) J. Chem. Soc. Chem. Commun., 284–285.
59 Johnson, M.R., Sutherland, I.O. and Newton, R.F. (1979) J. Chem. Soc. Chem. Commun., 306–308 and 309–311; Mageswaran, R., Mageswaran, S. and Sutherland, I.O. (1979) J. Chem. Soc. Chem. Commun., 722–724; Jones, N.F., Kumar, A. and Sutherland, I.O. (1981) J. Chem. Soc. Chem. Commun., 990–992.
60 Kotzyba-Hibert, F., Lehn, J.M. and Vierling, P. (1980) Tetrahedron Lett., 941–944.
61 Kintzinger, J.P., Kotzyba-Hibert, F. Lehn, J.M., Pagelot, A. and Saigo, K. (1981) J. Chem. Soc. Chem. Commun., 833–836; see also Pascard, C., Riche, C., Cesario, M. Kotzyba-Hibert, F. and Lehn, J.M. (1982) J. Chem. Soc. Chem. Commun., 557–560.
62 Kotzyba-Hibert, F., Lehn, J.M. and Saigo, K. (1981) J. Am. Chem. Soc. 103, 4266–4268.
63 Helgeson, R.C., Tarnowski, T.L. and Cram, D.J. (1979) J. Org. Chem. 44, 2538–2550.
64 Goldberg, I. (1977) Acta Crystallogr. Sect. B 33, 472–479.
65 Rebek Jr., J., Trend, J.E., Wattley, R.V. and Chakravorti, S. (1979) J. Am. Chem. Soc. 101, 4333–4337; Rebek Jr., J. and Wattley, R.V. (1980) J. Am. Chem. Soc. 102 4853–4854.
66 Rebek Jr., J. Wattley, R.V., Costello, T., Gadwood, R. and Marshall, L. (1980) J. Am. Chem. Soc. 102, 7400–7402.
67 Rebek Jr. J., Wattley, R.V., Costello, T., Gadwood, R. and Marshall, L. (1981) Angew. Chem. Int. Ed. 20, 605–606.
68 van Bergen, T.J. and Kellogg, R.M. (1976) J. Chem. Soc. Chem. Commun., 964–966; Piepers, O. and Kellogg, R.M. (1978) J. Chem. Soc. Chem. Commun., 383–384.
69 van der Veen, R.H., Kellogg, R.M., Vos, A. and van Bergen, T.J. (1978) J. Chem. Soc. Chem. Commun. 923–924.
70 van Bergen, T.J. and Kellogg, R.M. (1977) J. Am. Chem. Soc. 99, 3882–3884.
71 de Vries, J.G. and Kellogg, R.M. (1979) J. Am. Chem. Soc. 102, 2759–2761; see also Jouin, P., Troostwijk, C.B. and Kellogg, R.M. (1981) J. Am. Chem. Soc. 103, 2091–2093.
72 Matsui, T. and Koga, K. (1978) Tetrahedron Lett., 1115–1118; (1979) Chem. Pharm. Bull. 27, 2295–2303.
73 Chao, Y. and Cram, D.J. (1976) J. Am. Chem. Soc. 98, 1015–1017.
74 Saski, S. and Koga, K. (1979) Heterocycles 12, 1305–1310.
75 Lehn, J.M. and Sirlin (1978) J. Chem. Soc. Chem. Commun., 949–951.
76 Cram, D.J. and Sogah, G.D.Y. (1981) J. Chem. Soc. Chem. Commun., 625–628.

Subject index

Abortive complexes 99
Acetal hydrolysis 404, 408–420
Acetate kinase 253
Acetoacetate decarboxylase 273, 288–291
Acid catalysis 213–214, 230–240, 398–426
Acids 229
Activation energy, decreasing 22–27
Active-site directed agents 273
Active-site titration 122
Acylation 511–517
Acyl–enzyme mechanism 121, 188
Adenosine triphosphate 252–253
S-Adenosyl-L-methionine decarboxylase 311
Adrenaline 308
Aggregates, non-micellar 487–490
  self-organised 461–499
  submicellar 487–490
Ainsworth notation 94
Alcohol dehydrogenase 257–258
Aldolases 246–249, 271–276, 279–286
Allosteric control 84
  effects 84–86, 544–545
Amino acid decarboxylases 306
Amino acid, oxidation 260
Aminoacyl tRNA synthetases 63
γ-Aminobutyrate 315
Aminolevulinate synthetase 327–331
Aminotransferases 304, 314–331
Ammonium cations, binding 538
Amphiphiles 461
Analogue to digital conversion 116
Anomeric effect 391–395
Anti-Hammond behaviour 181–183
Approximation 16–22
Ascorbate oxidase 268
Aspartate aminotransferase 112
Aspartate transcarbamylase 83
ATP 252–253

Baeyer–Villiger reaction 262
Bases 229
Binding, geometry of 517, 530–550
  non-sequential 91–93
  primary 536
  random 92–93
Binding, (*continued*)
  reactions 119–120
  secondary 536
  sequential 91–93
Binding energy 22, 33, 530–537
  estimation of 50–53
  utilisation of 27–34
Biopterin 373–385
Bond-angle bending 35
Bond stretching 35
Braunstein–Snell hypothesis 303–306
Briggs–Haldane kinetics 7, 73
Brönsted correlation 128–142, 234, 278
  anomalies 141–142
  curvature in 137–141
  extended 132
  interpretation 129–130, 133–137
Burst kinetics 121–123

Cahn–Ingold–Prelog nomenclature 306
Cannizzaro reaction 256
Carbanions 142, 242, 246–252, 260, 276
Carbinolamine 273
Carbohydrate metabolism 286
Carbon–carbon bond fission 246–252, 279–286
Carbon–carbon bond formation 246–252, 279–286
Carbon–cobalt bonds 439–452
Carbonic anhydrase model 524
Carbonyl groups, activation 241–251, 276–279
Carboxylic acids 260
Carboxypeptidase 245
Catalase 266
Catalysis 229–268
    see also individual types
  conformational 513
  covalent 32, 240–242
  hydrogen bonding 236
  microdielectric 513
  preassociation 235–236
  trapping 231–235
Changes in mechanism 189–194
Charge effects 27
Charge neutralisation 30
Charge transfer 38
Charge-relay system 521
Charton parameters 159–161, 165

Chemical catalysis 12–14
Chemical flux 10
Chemical mechanisms 229–268
Chymotrypsin 74, 166, 169, 188, 217, 512
Claisen condensation 255
Cleland notation 94
Cobalt complexes 433–456
Coenzyme A 248, 253–255
Coenzyme catalysis 246–255
Coenzyme Q 256
Complementarity 30, 49, 487, 530
Complexation 538–539
Concerted catalysis 236–237
Conformational catalysis 513
Continuous flow apparatus 107
Cooperativity 114, 545
Copper-proteins 267–269
Cornish–Bowden plot 82
Correlation coefficients 130–131
Corrin macrocycles 435
Counterions 479
Coupled reactions 124
Covalent catalysis 32, 213–214, 240–242
Critical micelle concentration 466
Cross-correlation 179–183
Crown acyl transfer model 550–556
Crown ethers 529–559
Crown hydride transfer model 546–550
Crown Michael addition model 556–558
Cyanocobalamin 433–456
Cyclodextrins 505–527
  catalysis 509–514
  modified 520–526
  specificity 514–520
Cystathionine-γ-synthase 344
Cytochrome C 256
Cytochrome P450 265–267, 385

Data analysis 118
  collection 118
Dead-end inhibitors 93, 104
Decarboxylases 246–252, 271, 293, 306–314
Decarboxylation 246–252, 276, 289, 292
Dehydrogenases 257
  pyridoxal 304
Dehydroquinase 295
Deoxymyoglobin 123
Deoxynucleoside hydrolysis 423–424
Desolvation 31
Diamine oxidase 310
Diastereotopic hydrogens 258
Diffusion 137–141, 231–235
Dihydroflavins 259–262
Dihydrofolate reductase 69, 375–376
Dihydroxyacetone phosphate 276, 279
Dihydroxyacetone sulphate 276, 284
Dipole–dipole interactions 56, 537
Dispersion forces 45–46, 65
Distortion 27, 34–37, 49
Distribution functions 5
Disulphide bonds 4, 36–37
L-Dopa 308
Dopamine 268, 308

E parameter 175–177
Eadie–Hofstee plot 79
Edwards relationship 177–179
Effective charges 133–137
  additivity of 135
Effective concentrations 16–22, 237–240
Eigen plots 137–141, 195–197
Eisenthal plot 82
Elastase 217
Electron donors 229
Electron transfer 256–268
Electroneutrality principle 229
Electronic effects 144–166, 390–397
Electrophiles 229
Electrophilic catalysis 243–252, 398–420, 421–426
Electrostatic catalysis 407–408, 419
  effects 30, 38, 40–44, 49, 56, 62–64, 392
  molecular potentials 43–44, 63–66
  repulsion 536
  stabilisation 407–408, 419
Elimination 248–249, 295
Elimination–deamination reaction 336–342
Elucidation of mechanism 186–197
Enamine formation 242, 247
Enantiomeric differentiation 519, 539
Enediols 257
Energy diagrams 22–27
Energy surface 182
Enolisation 241–243, 276
Entropic strain 32
Entropy 16–22, 27, 30, 56
  of activation 16–22, 409
  of rotation 18–21, 516
  of translation 18–21, 516
  of vibration 18–21
Enzymes, active conformation 14
  allosteric 85
  cavities in 52
  extracellular 7–9
  flexibility 4, 27, 48
  fluctuations in 5
  inhibitors of 86–91

Enzymes, (*continued*)
intracellular 7–9
kinetics 73–109
mobility 3
non-reacting part 13
packing density 3, 52
position of equilibrium and 10
rigidity 2–6, 27, 31
size 4, 14
structure 2–6, 66
unfolding 2
Ethyl dichloroacetone hydrolysis 91
Exchange repulsion 38, 56
Experimental determination of rate constants 91–107

FAD 259–262
Ferricytochrome C 120
Flavin coenzymes 256, 259–262
Flavomonooxygenases 262
FMN 259–262
Folate 373–385
Folic acid cofactors 373–385
Force fields 34–37, 38–46
Forces, attractive 37, 57
dispersion 45–46, 57
electrostatic 30, 38, 40–44, 49, 56, 62–64
intermolecular 37–46
intramolecular 34–37
non-covalent 55–71
repulsive 37
Formyl transfer 379–381
Fractionation factors 214–218
Franck–Condon principle 262
Free energy of activation 11, 27
Free-energy correlations 127–201
diagrams 182
Frontier orbitals 394
Fructose diphosphate 279
Functional surfactants 463

GABA 315
Galactose oxidase 267
Galactosidase 401
General acid base catalysis 213–214, 230–240, 413–416, 512
intramolecular 237–240, 417–420
Geometry, distortion of 27–28, 29–30, 34–37, 49
of ground states 28–30
of transition states 28
Glutamate dehydrogenase 326
Glyceraldehyde phosphate 280
Glycine, decarboxylation 311
Glycoside hydrolysis 404, 408–420
synthesis 402–403
Glycosylamine hydrolysis 422
Glycosyl fluoride hydrolysis 426
Glycosyl transfer 389–427
Glyoxylase 257
Gouy–Chapman layer 464
Ground-state, destabilisation 22, 28
geometry of 28, 34–37
recognition 530
stabilisation 22, 30
Grunwald–Winstein relationship 173–174
Guest–host complexes 530

Haemoproteins 266
Haldane relationship 100
Halogenation of acetone 129
Hammett equation 143–161
sigma values 145–163
Hanes plot 79
Hansch parameters 166–173
Hemithioacetal hydrolysis 425–426
High spin complexes 265
Hooke's Law 47
Horseradish peroxidase 266
Host–guest complexes 530
Hydration 51
Hydride transfer 256–259, 267, 546
Hydrogen bonding catalysis 236
Hydrogen bonds 3, 39–40, 55, 236, 530–544
Hydrogen peroxide 261
Hydrolytic enzyme models 520–524
Hydrophilic groups 461
Hydrophobic effects 44–45, 66–69, 166–173, 463
non-linear 170
parameters 166–173
pockets 166
Hydroxy acids, oxidation 260
Hydroxylation 381
Hydroxyl group catalysis 21, 511
Hyperacid 243

Imine formation 241–242, 246–249, 271–298
Inclusion complex 506
Induced fit 14–16
Inductive effects 148
Inhibition 86–91, 123, 282
competitive 86–87
mixed 90–91
non-competitive 89–90, 294
product 93, 105
uncompetitive 87–89

Interfacing instruments 116
Intermolecular conformational analysis 529
Intramolecular force fields 34–37
Intramolecular general acid base catalysis 237–240, 417–420
Intramolecular nucleophilic catalysis 16–22, 255, 406–413
Intramolecular reactions 16–22
Intrinsic barrier 139–141
Ionisation constants 183–186
Iron-containing proteins 265
Isoalloxazine 260
Isomerisation 246–252
Isoracemisation 223
Isotope effects 203–226, 290
  equilibrium 207–209
  fraction factors 214–218
  heavy atom 218–219
  measurement 205–207
  primary 209–213
  secondary 219–220
  solvent 213–218
  theory of 203–205
Isotope exchange 106, 284
Isotope labelling techniques 220–226
Isotopic enrichment 226, 497

Jencks diagrams 182, 415

Ketal hydrolysis 408–420
α-Ketal transfer 251
Kinetic ambiguity 195
Kinetics 73–109, 111–125
  burst 121–123
  errors in 116–119
  measurements 114–125
  multi-substrate 91–107
  pre-steady state 111–125
  rapid 107–109
  relaxation 123–125
  steady state 76–107
  temperature jump 123–125
  whole time-course 106–107
King–Altman procedure 97
Koshland–Némethy–Filmer model 85
Kyneurinase 342

Labelling techniques 220–226
Lactate dehydrogenase 93, 259
Leffler index 136–137
Ligand binding 119
Linear free-energy relationships 127–201
Lineweaver–Burk plot 79, 97, 509
Lipophilic groups 461
Lock and key model 33
Lone pairs 390–391
Low spin complexes 265
Lysozyme 74, 407

Macropolycycles 543
Marcus model 139–141
Mechanism of reactions 127–201
  change in 189–194
  elucidation 186–197
Mechanistic identity 186–187
Menger–Portnoy model 471
Metal complexes 263, 530
Metal-ion catalysis 243–246
Metalloenzymes 244
Metalloenzyme models 525
Metalloproteins 264
Metaphosphate 253
Methyl transfer 381
Methylene transfer 376–379
Micellar structure 464–468
Micelles 461–499
  effect upon reaction rate 468–479
  functional 482–487
  non-aqueous 490–493
  reactive counterion 479–482
  reverse 491–493
  stereochemical recognition 487
Michaelis–Menten kinetics 6, 76–78, 111, 472, 505
  adherence to 82–86
Michaelis–Menten parameters 11, 78, 112
Microcomputers 115
Microdielectric catalysis 513
Microemulsions 493–495
Molar refractivity 170
Molecular receptors 505–527, 529–559
Monoamine oxidase 310
Monod–Wyman–Changeux model 84–86
Monooxygenases 261, 267–268, 385
More O'Ferrall diagrams 182, 415
Morse curves 205
Multi-substrate kinetics 91–107, 111

NADH 107, 256–259, 375
Negative cooperativity 114, 545
Nicotinamide adenine dinucleotide 256–259
NMR spectroscopy 507
No-bond resonance 394
Non-bonded interactions 37
Non-covalent forces 55–71

Non-productive binding 14
Non-sequential binding 91–93
Nucleophiles 229
Nucleophilic catalysis 213–214, 240–242, 406–407, 413
Nucleoside hydrolysis 423–424

Organo cobalt complexes, reactivity 438–456
  structure 433–438
  synthesis 439–444
Outer sphere mechanism 263
Oxidases 264–269
Oxidation 256–268
Oxocarbonium ions 394–396, 398–399
Oxygen exchange 281
Oxygen, reductive activation 261–262
Oxygenases 261
Oxymyoglobin 123

Pantetheine 253
Papain 68
Partition coefficient 166–172
Penicillin 238
Peroxide ion 264, 383
Phenylalanine hydroxylase 373, 382
Phosphoglucose isomerase 336
Phosphotransacetylase 253
Photochemical reactions 494–497
Photo-oxygenation 283
Ping-pong mechanism 91–93, 100, 108
Polarisation 38
Porphyrins 265–267
Positive cooperativity 545
Preassociation mechanism 235, 399–402
Pre-steady state kinetics 111–125
Primary binding 536
Primary isotope effects 209–213
Prochirality 258
Product inhibitors 93, 105
Product specificity 517
Proteins, fluorescence 107
  folded state 2, 66
  globular 3, 66
  mobility 3
  packing density 3, 52
  structure 2–6
  unfolding 2
Proton transfer 128–129, 137–141, 195–197, 203–226, 222
  concerted 236–237
  diffusion controlled 137–141, 231, 235
  hydrogen bonding 236
Proton transfer, (*continued*)
  intramolecular 237–240, 340
  linear 211–213
  preassociation 235, 399–403
  stepwise 231–235
  transition state structure 137–141, 195–197, 211–213, 230–237, 413–416
Pyridoxal dehydrogenases 304
Pyridoxal phosphate 246–249, 303–385
  aminotransferases 314–331
  decarboxylation 306–314
  elimination–deamination 336–342
  racemases 319–320
  replacements 331–336
  structure 349–367
Pyridoxamine 304
Pyridoxamine phosphate 303
Pyridoxine 304
Pyrimidine biosynthesis 377
Pyruvate-containing enzymes 291–294

Racemases 319–320
Radical trapping 260
Random-order binding 92
Rapid reaction kinetics 107–109
Rate enhancements 12–14, 510
Rate-limiting step, change in 191–194
Reaction fields 513
Reactivity–selectivity 183
Recognition 30, 49, 487
  ground-state 530–542
  substrate 542–544
  transition state 546–558
Redox metalloproteins 264
Redox reactions linked 124
Reduction 256–268
Regulation, and thermodynamics 9
  and control 84
Relaxation methods 123–125
Replacement reactions 331–336
Reporter groups 175–177
Resonance effects 148
Retention mechanism 283
Retroaldol reaction 276
Riboflavin 256, 259–262
Ribonuclease 74
Romsted model 472

Saturation kinetics 6–9, 74–76
Schiff bases 241–249, 246–249, 303–385
Secondary binding 536
Secondary isotope effects 219–220
Selectivity–reactivity 179–183

Selectivity with cyclodextrins 517
Self-organised aggregates 461–499
Sequential binding 91–93
Serine hydroxymethylase 373
Serine hydroxymethyltransferase 320–327
Serine proteases 217
SHMT 320–327
Solvation 27, 30, 31, 51, 59
Solvent effects 173–174
Solvent isotope effects 213–218, 408
Specific acid–base catalysis 230
Specificity 11–12, 12–14, 14–16, 28, 63, 514, 542
Standard states 22
Statistical treatments 130–132
Steady-state assumption 77
Steady-state kinetics 76–107
Stereochemical recognition 30, 49, 487–530
Stereochemistry 307, 346
Stereoelectronic control 248, 305, 376, 396–397
Steric effects 142–166
Stern layer 464
Stopped-flow apparatus 107, 119–123
Strain 27–28, 34, 47–49
Strain energy 21, 27
Stress 47–49
Structural rearrangements 123
Substituent effects 142–166
  additivity 171–172
  separation of 156
Substrate, accessible surface area 44–45
  concentration in vivo 9
  distortion of 27–30, 34–37, 49
  effect of concentration on rate 6, 82–86
  non-reacting part 13
  recognition 542
  specificity 514
  weak binding of 26
Succinyl-CoA–acetoacetate transferase 13
Surfactants 462
  functional 463
Swain–Scott relationship 177–179

Taft's steric parameter 161–166
Temperature jump 123–125
Ternary complexes 91–97, 355, 366
Tetrahydrofolate 311, 373–385
Theorell–Chance mechanism 91–93, 102
Thiamine pyrophosphate 249–251
Thioesters 254
Thioglycoside hydrolysis 425–426
Thiols, oxidation 261
Threonine deaminase 83
Threonine synthetase 344
Thymidylate synthetase 373
Thyroxine–prealbumin complex 66
Torsional strain 35
Transaldolase 286–288
Transamination 246–249, 314–331
Transient intermediates 112
Transition state 12
  analogues 49
  effective charges in 133–137
  index 136
  loose 20, 33, 238
  product-like 128
  reactant-like 128
  recognition 546–558
  stabilisation 22–27, 30
  structures 181–183, 211–213, 413–416
  tight 20, 34, 238
Trapping mechanism 231–235
Trypsin 63, 217
Tryptophan synthase 337–338
Tryptophanase 338–339
Tyrosine decarboxylase 309

Ubiquinone 256

Van der Waals forces 37–38, 56
Vesicles 495–496
Vitamins 256
Vitamin B_2 259–262
Vitamin B_6 246, 303
Vitamin B_{12} 433–456

Y parameter 173–177
Yukawa–Tsuno equation 155–156

Z parameter 175–177
Zero point energy 205, 210, 219